Applied Space Systems Engineering

SPACE TECHNOLOGY SERIES

*This book is published as part of the Space Technology Series,
a cooperative activity of the United States Department of Defense
and the National Aeronautics and Space Administration.*

Wiley J. Larson
Managing Editor

From Kluwer and Microcosm Publishers:

Space Mission Analysis and Design Second Edition by Larson and Wertz.

Spacecraft Structures and Mechanisms: From Concept to Launch by Sarafin.

Reducing Space Mission Cost by Wertz and Larson.

From McGraw-Hill:

Understanding Space: An Introduction to Astronautics by Sellers.

Space Propulsion Analysis and Design by Humble, Henry, and Larson.

Cost-Effective Space Mission Operations by Boden and Larson.

Fundamentals of Astrodynamics and Applications by Vallado.

Applied Modeling and Simulation: An Integrated Approach to Development and Operation by Cloud and Rainey.

Modeling and Simulation for Space Systems by Rainey.

Human Spaceflight: Mission Analysis and Design by Larson and Pranke.

Applied Project Management for Space Systems by Chesley, Larson, McQuade, and Menrad.

APPLIED SPACE SYSTEMS ENGINEERING

Space Technology Series

Edited by:

Dr. Wiley J. Larson
Managing Editor
Space Technology Series,
USAF Academy

Dr. Doug Kirkpatrick
Technical Editor
Space Technology Series,
USAF Academy

Dr. Jerry Jon Sellers
Senior Space Systems Engineer
Teaching Science & Technology, Inc.

Dr. L. Dale Thomas
Constellation Program Deputy Manager
NASA Marshall Space Flight Center

Dr. Dinesh Verma
Dean, School of Systems and Enterprises
Stevens Institute of Technology
Hoboken, NJ

McGraw Hill **Learning Solutions**

Boston   Burr Ridge, IL   Dubuque, IA   New York   San Francisco   St. Louis
Bangkok   Bogotá   Caracas   Lisbon   London   Madrid
Mexico City   Milan   New Delhi   Seoul   Singapore   Sydney   Taipei   Toronto

Applied Space Systems Engineering
Space Technology Series

2 3 4 5 6 7 8 9 0 DOC DOC 10 9 8 7 6 5 4 3 2 1

ISBN-13: 978-0-07-340886-6
ISBN-10: 0-07-340886-7

Learning Solutions Manager: Anne Kaiser
Production Editor: Carrie Braun
Cover Design: Mary Tostanoski
Cover Illustrations: © 2009 Teaching Science & Technology Inc.
Printer/Binder: RR Donnelley

Table of Contents

List of Authors and Editors

William H. Arceneaux. Constellation Program Test and Evaluation Director, NASA Johnson Space Center, Houston, Texas. B.S. (Computer Engineering) Florida Institute of Technology; B.S. (Space Sciences) Florida Institute of Technology. Chapter 11—*Verification and Validation.*

Bruce G. Barker. Associate Industry Professor, Stevens Institute of Technology, Hoboken, New Jersey. M.S. (Systems Engineering) Stevens Institute of technology; B.A. (Computer Sciences) University of Texas at Austin; B.A. (Economics) University of Virginia. Chapter 4—*Engineering and Managing System Requirements;* Chapter 5—*System Functional and Physical Partitioning.*

Jaya Bajpayee. Program Executive, Astrophysics Division, NASA Headquarters, Science Mission Directorate, Washington, DC. M.S. (Space Systems Engineering) Delft University of Technology, Delft, Netherlands; B.S. (Electrical Engineering) University of Pittsburgh, Pennsylvania. Chapter 12—*Product Transition.*

Mary Jane Cary. President, Cyber Guides Business Solutions, Inc., Naples, Florida. M.E.A. (Engineering Administration) George Washington University; B.S. (Nuclear Engineering) Kansas State University. Chapter 16—*Manage Configuration.*

Bruce C. Chesley. Director, Strategic Architectures, The Boeing Company, Colorado Springs, Colorado. Ph.D. (Aerospace Engineering) University of Colorado, Boulder; M.S. (Aerospace Engineering) University of Texas at Austin; B.S. (Aerospace Engineering) University of Notre Dame.

Paul J. Componation. Professor, Industrial and Management Systems Engineering, The University of Alabama in Huntsville, Huntsville, Alabama. Ph.D. (Industrial and Management Systems Engineering) West Virginia University; M.S. (Management) Troy State University. Chapter 6—*Decision Making.*

Steven H. Dam. President, Systems and Proposal Engineering Company (SPEC), Marshall, Virginia. Ph.D. (Physics) University of South Carolina; B.S. (Physics) George Mason University. Chapter 3—*Concept of Operations and System Operational Architecture.*

Michael S. Dobson. Principal, Dobson Solutions, Bethesda, Maryland. B.A. (English) University of North Carolina at Charlotte. Chapter 6—*Decision Making.*

Riley Michael Duren. Principal Systems Engineer, Jet Propulsion Laboratory, Pasadena, California. B.S. (Electrical Engineering) Auburn University. Chapter 11—*Verification and Validation.*

Katherine Erlick. Director of Mission Assurance, Integrated Defense Systems (IDS), The Boeing Company, Seal Beach, California. M.S. (Engineering Management) West Coast University; B.S. (Engineering, Mechanical) Cal Poly Pomona. Chapter 12—*Product Transition.*

Jody Fluhr. Chief Engineer, Fluhr Engineering, LLC, Crestwood, Kentucky. M.Eng. and B.S. (Engineering Math and Computer Science) University of Louisville. Chapter 19—*FireSAT End-to-End Case Study*.

Edward B. Gamble Jr. Jet Propulsion Laboratory/California Institute of Technology, Pasadena, California. Ph.D. (Electrical Engineering and Computer Science) Massachusetts Institute of Technology. Chapter 11—*Verification and Validation*.

Eberhard K.A. Gill. Professor Dr., Delft University of Technology, Faculty of Aerospace Engineering, Chair Space Systems Engineering, Delft, Netherlands. Ph.D. (Theoretical Astrophysics) Eberhard-Karls-University Tübingen, Germany; M.A. (Space Systems Engineering) Delft University of Technology/SpaceTech post-graduate programme. Diploma (Physics, Astrophysics) Eberhard-Karls-University Tübingen, Germany; Chapter 10—*System Integration*.

Vernon Wayne Grapes. Director, Systems Engineering, National Reconnaissance Office, Chantilly, Virginia. M.A. (Engineering Management) George Washington University; B.A. (Electrical Engineering) Johns Hopkins University.

Joseph W. Hamaker. Senior Cost Analyst, SAIC, Rockeldge, Florida. Ph.D. (Systems Engineering) University of Alabama Huntsville. Chapter 7—*Lifecycle Cost Analysis*.

Eric C. Honour. President, Honourcode, Inc., Pensacola, Florida. Ph.D. candidate (Systems Engineering) University of South Australia; M.S.E.E. (Control Systems Theory) Naval Postgraduate School; B.S.S.E. (Systems Engineering) United States Naval Academy. Chapter 14—*Technical Direction and Management: The Systems Engineering Management Plan (SEMP) in Action*.

Ivy Hooks. CEO, Compliance Automation, Inc., Boeme, Texas. M.S. and B.S. (Mathematics) University of Houston. Chapter 4—*Engineering and Managing System Requirements*.

Scott P. Hutchins. Aerospace Flight Systems Engineer, NASA Marshall Space Flight Center, Huntsville, Alabama. M.S. (Systems Engineering) University of Alabama in Huntsville; M.S. (Physical Science) University of Houston, Clear Lake; M.B.A. (Business Administration) New Mexico Highlands University; B.S. (Mechanical Engineering) University of Arkansas; B.A. (Mathematics) Ouachita Baptist University; B.A. (Business) Ouachita Baptist University. Chapter 6—*Decision Making*.

Peter C. Kent. Manager, Software Development, United Space Alliance, LLC, Cape Canaveral, Florida. M.B.A. (Business Administration) University of South Florida; B.S. (Applied Mathematics, Engineering and Physics) University of Wisconsin. Chapter 17—*Manage Technical Data*.

Douglas H. Kirkpatrick. Technical Editor, Ensign Group, Inc., Colorado Springs, Colorado. Ph.D. (Aerospace Engineering) University of Texas at Austin. *Technical Editor*.

David Y. Kusnierkiewicz. Chief Engineer, Space Department, Johns Hopkins University, Applied Physics Laboratory, Laurel, Maryland. M.S. and B.S. (Electrical Engineering) University of Michigan. Chapter 9—*Product Implementation.*

Wiley J. Larson. Space Technology Series Editor for the U.S. Air Force Academy's Space Mission Analysis and Design Project, President, CEI, Colorado Springs, Colorado. D.E. (Spacecraft Design) Texas A&M University; M.S. and B.S. (Electrical Engineering), University of Michigan. Chapter 1—*Space Systems Engineering.*

Robert E. Lee. REL Consulting, Severna Park, Maryland. M.S.E.E. (Electromagnetic Fields) Purdue University; B.S.E.E. (Electrical Engineering) Purdue University. Chapter 4—*Engineering and Managing System Requirements.*

Ted A. Leemann. President, The Center for Systems Management, Vienna, Virginia. M.S. (Logistics Management) Florida Institute of Technology; B.A. (Business and Economics) Westminster College Fulton, Missouri. Chapter 6—*Decision Making.*

Randy Liefer. President and Partner, Teaching Science & Technology, Inc., Colorado Springs, Colorado. Ph.D. (Aerospace Engineering) University of Kansas; M.S. (Aeronautics and Astronautics) Massachusetts Institute of Technology; B.S. (Aeronautical Engineering) United States Air Force Academy. Chapter 12—*Product Transition.*

James E. Long. CEO and Chief Methodologist, Vitech Corporation, Vienna, Virginia. M.S. (Astronautics) Purdue; B.S. (Mechanical Engineering) General Motors Institute. Chapter 5—*System Functional and Physical Partitioning.*

Perry D. Luckett. Director and Chief Consultant, Executive Writing Associates, Colorado Springs, Colorado. Ph.D. (American Studies) University of North Carolina at Chapel Hill; M.A. and B.A. (English) Florida State University. *Technical Editor.*

Marilyn McQuade. Self-employed, Colorado Springs, Colorado. M.S. and B.S. (Nuclear Engineering) Massachusetts Institute of Technology. *Technical Editor.*

Wade Douglas Morris. Aerospace Engineer, NASA Langley Research Center, Hampton, Virginia. M.S. (Oceanography) Old Dominion University; B.S. (Aerospace Engineering) North Carolina State University. Chapter 7—*Lifecycle Cost Analysis.*

Gerrit Muller. Professor, Buskerud University College, Kongsberg, Norway. Embedded Systems Institute, Eindhoven, Netherland. Ph.D. (Technology Management) University of Delft; M.A. (Physics) University of Amsterdam. Chapter 10—*System Integration.*

Michael B. Nix. Lead Operability Engineer, Qualis Corporation, Huntsville, Alabama. Ph.D. (Industrial and Systems Engineering) University of Alabama in Huntsville; M.A. (Civil Engineering) Mississippi State University. Chapter 7—*Lifecycle Cost Analysis.*

Michael C. Pennotti. Associate Dean for Academy Programs, Stevens Institute of Technology, Hoboken, New Jersey. Ph.D. (Electrical Engineering) Polytechnic Institute of New York; M.S.E.E. (Electrical Engineering) Polytechnic Institute of Brooklyn; B.E.E. (Electrical Engineering) Manhattan College. Chapter 18— *Technical Assessment and Reviews.*

Robert Samuel Ryan. Engineering Consultant, Madison, Alabama. M.S. (Engineering Mechanics) University of Alabama; M.A. (Secondary Education George Peabody College; B.S. (Secondary Education) University of North Alabama. Chapter 15—*Manage Interfaces.*

Michael G. Ryschkewitsch. Chief Engineer, NASA Headquarters, Washington DC. Ph.D. (Physics) Duke University; B.S. (Physics) University of Florida. Chapter 1—*Space Systems Engineering.*

Mark P. Saunders. Director, Independent Program Assessment Office, NASA Headquarters, Hampton, Virginia. B.S. (Industrial Engineering) Georgia Institute of Technology. Chapter 18—*Technical Assessment and Reviews.*

Christopher J. Scolese. Associate Administrator/Acting Administrator, NASA, Headquarters, Hampton, Virginia. M.S.E.E. (Electrical Engineering) George Washington University; Maxwell National Security Management Course, Syracuse University; Naval Nuclear Program, Bettis Reactor Engineering School; B.S.E.E. (Electrical Engineering) State University of New York at Buffalo. Chapter 1—*Space Systems Engineering.*

Dawn Schaible. Manager, System Engineering Office, NASA Langley Research Center, Hampton, Virginia. M.S. (Engineering and Management) Massachusetts Institute of Technology; B.S. (Mechanical Engineering) Bradley University. Chapter 1—*Space Systems Engineering.*

Jerry Jon Sellers. Senior Space Systems Engineer, Teaching Science & Technology, Inc., Manitou Springs, Colorado. D.Phil. (Satellite Engineering) University of Surrey, UK; M.Sc. (Aero/Astro Engineering) Stanford University; M.Sc. (Physical Science) University of Houston, Clear Lake; B.S. (Human Factors Engineering) United States Air Force Academy. Chapter 11—*Verification and Validation;* Chapter 19—*FireSAT End-to-End Case Study.*

Joey Dewayne Shelton. Chief Technical Officer, TriVector Services, Huntsville, Alabama. Ph.D. and M.S.E. (Mechanical Engineering) University of Alabama in Huntsville; B.S. (Aerospace Engineering) Auburn University. Chapter 15—*Manage Interfaces.*

Robert Shishko. Principal Systems Engineer/Economist, Jet Propulsion Laboratory, California Institute of Technology, Pasadena, California. Ph.D. and M.Phil. (Economics) Yale University; S.B. (Economics/Political Science) Massachusetts Institute of Technology. Chapter 8—*Technical Risk Management.*

Peter A. Swan. Vice President and Partner, Teaching Science and Technology, Inc., Paradise Valley, Arizona. Ph.D. (Space Systems Design) University of California at Los Angeles; M.S.S.M. (Systems Management) University of Southern California. Chapter 14—*Technical Direction and Management: The Systems Engineering Management Plan (SEMP) in Action.*

Lawrence Dale Thomas. Deputy Manager, Constellation Program, NASA Marshall Space Flight Center, Huntsville, Alabama. Ph.D. (Systems Engineering) University of Alabama in Huntsville; M.S. (Industrial Engineering) North Carolina State University; B.S.E. (Industrial and Systems Engineering) University of Alabama in Huntsville. Chapter 13—*Plan and Manage the Technical Effort.*

Peter M. VanWirt. Senior Space Systems Engineer, Teaching Science & Technology, Inc., Monument, Colorado. Ph.D. (Electrical Engineering) Utah State University. Chapter 19—*FireSAT End-to-End Case Study.*

Dinesh Verma. Professor of Systems Engineering, Stevens Institute of Technology, Castle Point on Hudson, Hoboken, New Jersey. Ph.D. (Industrial and Systems Engineering) Virginia Tech. Chapter 1—*Space Systems Engineering;* Chapter 2—*Stakeholder Expectations and Requirements Definition;* Chapter 3—*Concept of Operations and System Operational Architecture.*

Preface

Today's challenges in designing, developing, and implementing complex aerospace systems are staggering. The Department of Defense, the National Aeronautics and Space Administration, National Reconnaissance Office (NRO), and the Federal Aviation Administration alone invest up to $50 billion (in FY2008 dollars) annually in space-related products and services. Globally, governments invest about $74 billion every year. And commercial spending on communications spacecraft, navigation systems, imaging systems, and even human spaceflight equals or exceeds government spending in some years. The complexities of technology, multiple organizations, geographical distribution, various political entities, assorted budget cycles, and intercultural concerns challenge even the most experienced systems engineers and program managers.

The purpose of *Applied Space Systems Engineering* (ASSE) is to provide inspiration, processes, approaches, tools, and information for systems engineers that are leading the way in complex aerospace-system design, development, and operation. An extensive author and editor team created this book based on a complete and rigorous set of systems engineer competencies rooted in the experiences and philosophies of seasoned space systems engineers from across the community.

This book presents the "how-to" necessary to "systems engineer" complex aerospace-related projects, along with information to help the aspiring or current systems engineer achieve a higher level of understanding and performance. It's geared to practitioners as they work through projects, but may also serve as a primary text or reference for graduate-level courses and development programs. Many aerospace-related case studies, examples, and lessons learned are spread throughout ASSE to provide historical insights and practical applications. The aerospace-related example, FireSAT, appears frequently, to provide systems details at each technical baseline, and to highlight what's expected from the systems engineer.

For the editors, ASSE is a dream come true—and not just because we're finished! After decades of working in the aerospace community, we were able to persuade the "best of the best" performing system engineers to contribute their wealth of experience, successful tools and approaches, and lessons learned to this project. This is good for us, but REALLY good for you, because you'll benefit from the 350-plus years (that's over three and a half centuries) of collective systems engineering experience!

Applied Space Systems Engineering is the 17th book produced by the US Air Force Academy's Space Technology Series team. The Department of Astronautics at the Academy has continued to provide the leadership for this project. Our deepest gratitude goes to Brigadier Generals Bob Giffen and Mike DeLorenzo and Colonel Marty France for their vision and persistence in leading this community effort. Lieutenant Colonels Mark Charlton and Timothy Lawrence supplied the hands-on leadership to make ASSE a reality. Thank you!

The Space Technology Series is sponsored and funded by the country's space community—and has been for more than 20 years. Twenty-three national organizations have championed this effort since it began in 1988—many of them on an annual basis. Leadership, funding, and support from Air Force Space and Missile Systems Center (SMC), numerous program offices and AF Wings, NRO, NASA Headquarters (including several mission directorates), NASA/Goddard Space Flight Center (GSFC), the Federal Aviation Administration, Naval Research Laboratory, and a host of others, have made this work possible. Thank you for your continued support!

Several sponsors that provided pivotal inspiration, guidance, and motivation for this particular book include: Mike Ryschkewitsch, NASA's Chief Engineer; Doug Loverro, (Chief Technical Officer); Colonel Jim Horejsi, Chief Engineer of SMC; and Vernon Grapes, Chief Engineer of NRO.

Here we must recognize the forbearance and long suffering of the 68 authors of this text. For four years they outlined, drafted, reorganized, reviewed, and re-crafted their material to suit the editors' demands. This book would not be possible without their expertise, persistence, and patience. As editors, we have strived to create a useful, professional book of which the authors can be proud.

The true heroes of this effort are Anita Shute and Marilyn McQuade. As a team, they edited every word, developed and finalized all graphics, and did everything necessary to create a camera-ready copy that truly represents the quality of the author and editor team. Mary Tostanoski created many complicated graphics, as well as the cover, and Perry Luckett provided early editing. Anita Shute personally entered and formatted every word and sentence in this book, as she has 13 others—over 11,000 pages of text! We could not have done this book with without you, Anita!

A team of 17 seasoned Government and industry systems engineers, guided by Mark Goldman of GSFC, laid the foundation for this book. Working with NASA's Academy of Program/Project and Engineering Leadership, his team painstakingly documented the capabilities and competencies that systems engineers must possess to be successful in complex systems design, development, and operation. The resulting capabilities were vetted with and reviewed by representatives of the International Council on Systems Engineering (INCOSE), Defense Acquisition University, and NASA's Systems Engineering Working Group.

Multiple teams (over 100 individuals) critiqued drafts of this book to ensure credibility—Constellation's EVA systems team, and NASA Ames's and Johnson Space Center's (JSC's) participants in the Graduate Certificate in Space Systems Engineering. Maria So led a highly effective review team at GSFC to conduct a detailed review of the book's contents, providing a complete set of integrated comments and improvements. We sincerely thank all of these folks for their conscientious contributions to the overall quality of the book!

Finally, we want to express our heartfelt appreciation to our spouses, our children, and our friends for their unswerving support and encouragement, while sharing our dream with us!

The authors and editors of ASSE, and of its companion, *Applied Project Management for Space Systems*, hope that our colleagues—old hands and neophytes alike—will find useful and insightful information and approaches here to help them meet the challenges in engineering complex aerospace systems for national defense, civil application, and commercial enterprise.

Dr. Wiley J. Larson
Managing Editor
Space Technology Series
USAF Academy, 2009
719-333-4110
wileylarson@comcast.net

Dr. Doug Kirkpatrick, USAF Academy
Dr. Jerry Jon Sellers, Teaching Science & Technology
Dr. Dale Thomas, NASA JSC
Dr. Dinesh Verma, Stevens Institute

Chapter 1

Space Systems Engineering

Dawn Schaible, Chris Scolese, and Michael Ryschkewitsch,
National Aeronautics and Space Administation
Dinesh Verma, *Stevens Institute of Technology*
Wiley J. Larson, *United States Air Force Academy*

This work culminates years of experience in systems engineering (SE) and focused discussions among systems engineering leadership, systems engineers, and systems engineering trainers within the aerospace community [Schaible, 2009]. One consistent theme in these experiences and discussions is that the space community uses many definitions and descriptions of SE. We use the terms and job titles of chief engineer, mission systems engineer, systems engineering and integration manager, system architect, vehicle integration, and so on for various pieces of the complete SE function. We need to agree on a common understanding of systems engineering. And no matter how we divide the roles and responsibilities, we must make sure that they're clear and executed as a functional whole. Our objectives are to provide a clear definition of SE, describe the highly effective behavioral characteristics of our best systems engineers, delineate fundamental systems engineering processes and approaches, and make explicit the expectations of systems engineers within the space community.

Systems engineering is both an art and a science. We can compare it to an orchestra. Most people understand what music is, but not everyone can play an instrument. Each instrument requires different expertise and skill. Some musicians spend their entire careers mastering a single instrument, but sophisticated music involves many different instruments played in unison. Depending on how well they come together, they may produce beautiful music or an ugly cacophony.

We can think of a symphony as a system. The musicians apply the science of music: they translate notes on a page to play their instruments. But an orchestra conductor, a maestro, must lead them to connect the process of playing to the art of creating great music. Maestros do a lot more than just keep time! They:

- Know and understand music—such matters as pitch, rhythm, dynamics, and sonic qualities—as well as the capabilities of various instruments and musicians
- Are necessary once the orchestra reaches a certain size and complexity
- Have typically mastered one or more musical instruments
- May be composers
- Select and shape the music that an orchestra plays
- Interpret a composer's music in light of the audience
- Strive to maintain the integrity of the composer's intentions
- Organize and lead the musicians
- Are responsible for the success of the performance

The systems engineer is like the maestro, who knows what the music should sound like (the look and function of a design) and has the skills to lead a team in achieving the desired sound (meeting the system requirements). Systems engineers:

- Understand the fundamentals of mathematics, physics, and other pertinent sciences, as well as the capabilities of various people and disciplines
- Are necessary once a project reaches a certain complexity
- Have mastered a technical discipline and are familiar with multiple disciplines
- Must understand the end game and overall objectives of the endeavor
- Create a vision and approach for attaining the objectives
- May be architects or designers
- Select and shape the technical issues to be addressed by multidisciplinary teams
- Must often interpret and communicate objectives, requirements, system architecture, and design
- Are responsible for the design's technical integrity
- Organize and lead multidisciplinary teams
- Are responsible for the successful delivery of a complex product or service

The similarities between maestros and systems engineers are useful in describing the latter's desired behavioral characteristics and capabilities.

1.1   The Art and Science of Systems Engineering

Systems engineering is the art and science of developing an operable system that meets requirements within imposed constraints. This definition is independent of scale, but our discussion here concentrates on developing complex systems, such as aircraft, spacecraft, power plants, and computer networks. Systems engineering is holistic and integrative. It incorporates and balances the contributions of structural, mechanical, electrical, software, systems safety, and power engineers, plus many others, to produce a coherent whole. It's about tradeoffs and compromises, about generalists rather than specialists [Schaible, 2009].

Systems engineering is not only about the details of requirements and interfaces among subsystems. Such details are important, of course, in the same way that accurate accounting is important to an organization's chief financial officer. But accurate accounting doesn't distinguish between a good financial plan and a bad one, nor help to make a bad one better. Similarly, accurate control of interfaces and requirements is necessary to good SE, but no amount of care in such matters can make a poor design concept better. Systems engineering is first and foremost about getting the right design—and then about maintaining and enhancing its technical integrity, as well as managing complexity with good processes to get the design right. Neither the world's greatest design, poorly implemented—nor a poor design, brilliantly implemented—is worth having.

The principles of SE apply at all levels. Similar activities are essential to the architecture, design, and development of elements and subsystems across the broad spectrum of space system and related developments. But for the remainder of this discussion, we use the term systems engineering in the context of complex, multidisciplinary system definition, development, and operation.

In his 2007 presentation, "Systems Engineering and the 'Two Cultures' of Engineering," Mike Griffin described how the complexities of today's aerospace systems and the ways they fail have led to branching within the industry [Griffin, 2007]. For our purpose, we divide SE into technical leadership and its ally, systems management.

- *Technical leadership* focuses on a system's technical design and technical integrity throughout its lifecycle
- *Systems management* focuses on managing the complexity associated with having many technical disciplines, multiple organizations, and hundreds or thousands of people engaged in a highly technical activity

Technical leadership, the art of systems engineering, balances broad technical domain knowledge, engineering instinct, problem solving, creativity, leadership, and communication to develop new missions and systems. The system's complexity and the severity of its constraints—not just its size—drive the need for SE.

Today's space systems are large and complex, requiring systems engineers to work in teams and with technical and other professional experts to maintain and enhance the system's technical integrity. The creativity and knowledge of all the people involved must be brought to bear to achieve success. Thus leadership and communications skills are as important as technical acumen and creativity. This part of SE is about doing the job right.

For large complex systems, there are literally millions of ways to fail to meet objectives, even after we have defined the "right system." It's crucial to work all the details completely and consistently and ensure that the designs and technical activities of all the people and organizations remain coordinated—art is not enough.

Systems management is the science of systems engineering. Its aim is to rigorously and efficiently manage the development and operation of complex systems. Effective systems management applies a systematic, disciplined approach that's quantifiable, recursive, repeatable, and demonstrable. The emphasis here is on organizational skills, processes, and persistence. Process definition and control are essential to effective, efficient, and consistent implementation. They demand a clear understanding and communication of the objectives, and vigilance in making sure that all tasks directly support the objectives.

Systems management applies to developing, operating, and maintaining integrated systems throughout a project's or program's lifecycle, which may extend for decades. Since the lifecycle may exceed the memory of the individuals involved in the development, it's critical to document the essential information.

To succeed, we must blend technical leadership and systems management into complete systems engineering. Anything less results in systems not worth having or that fail to function or perform.

1.1.1   The Scope of Systems Engineering

Since the late 1980s, many aerospace-related government and industry organizations have moved from a hard-core, technical leadership culture (the art) to one of systems management (the science). But history shows that many projects dominated by only one of these cultures suffer significant ill consequences. Organizations biased toward systems management often create products that fail to meet stakeholder objectives or are not cost-effective. The process often becomes an end in itself, and we experience "process paralysis." Organizations biased toward technical issues often create products or services that are inoperable, or suffer from lack of coordination and become too expensive or belated to be useful.

To achieve mission success, we must identify and develop systems engineers that are highly competent in both technical leadership and systems management. That's why we focus on the complete systems engineer, who embodies the art and science of SE across all phases of aerospace missions—a type reflected in Figure 1-1. The scope of systems engineering and the associated roles and responsibilities of a systems engineer are often negotiated by the project manager and the systems engineer. They should be understood and documented early in the project.

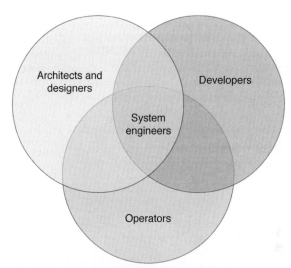

FIGURE 1-1.   **The Scope of Systems Engineering.** Systems engineers often concentrate on one lifecycle phase like architecture and design versus development or operations, but good systems engineers have knowledge of and experience in all phases.

1.1.2   The Characteristics Needed in a Systems Engineer

Here we describe the characteristics, some innate and others that can be developed, that enable select people to "systems engineer" complex aerospace missions and systems—to design, develop, and operate them. We then discuss how to further develop systems engineers to help them deal better with the complexities of sophisticated missions and systems. Figure 1-2 depicts the personal behavioral characteristics of effective systems engineers [Adapted, Lee, 2007].

Intellectual curiosity. Perhaps the most important personal characteristic of successful systems engineers is intellectual curiosity. People who prefer boundaries around their work, know what they know, and enjoy a focused domain may want to consider another occupation. Systems engineers continually try to understand the what, why, and how of their jobs, as well as other disciplines and situations that other people face. They are always encountering new technologies, ideas, and challenges, so they must feel comfortable with perpetual learning.

Ability to see the big picture. Good systems engineers maintain a big-picture perspective. They understand that their role changes throughout a project's lifecycle. At any point they must be fully cognizant of what's been done, what's necessary, and what remains to be done. Each phase has a different emphasis:

- Concept—mission and systems architecture, design, concept of operations, and trade studies

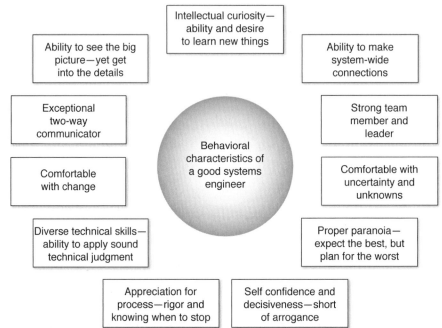

FIGURE 1-2.   Characteristics of a Good Systems Engineer. The characteristics are shown in decreasing priority from top to bottom. Some of them are innate, whereas others can be learned and honed [Adapted, Lee, 2007].

- Development—maintaining technical integrity throughout all lifecycle phases: preliminary design review, critical design review, verification, validation, and launch
- Operations—making sure that the project meets mission requirements and maintains technical integrity

Systems engineers pay particular attention to verification and validation. Verification answers the question: "Did we build our system right?" Verification is successful if our product meets the requirements. We emphasize the hard-earned lesson, "Test like you fly, fly like you test." Validation, on the other hand, answers the question: "Did we build the right system?" It's successful if the system does what it's supposed to do, which often goes well beyond just meeting requirements!

Good systems engineers are able to "translate" for scientists, developers, operators, and other stakeholders. For example, "Discover and understand the relationship between newborn stars and cores of molecular clouds," is meaningful to a scientist. But developers and operators would better understand and use this version: "Observe 1,000 stars over two years, with a repeat cycle of once every five months, using each of the four payload instruments." The systems engineer that knows the project's objectives, helps determine how to meet them, and maintains the system's technical integrity throughout its lifecycle has a good chance of succeeding.

Ability to make system-wide connections. First-rate systems engineers understand the connections among all elements of a mission or system. They must often help individuals on the team to see how their systems and related decisions connect to the bigger picture and affect mission success. For example, the star tracker's designer must understand that the star tracker is part of an attitude control system—specifically, of an attitude estimator used to take precisely pointed observations—and that the star tracker's output determines whether or not the proper images are obtained. If the designer doesn't understand this, the project is in trouble. Good systems engineers can anticipate the impact of any change in the system or project, and describe the nature and magnitude of the impact throughout their system.

Exceptional two-way communicator. Communications skills are the great enabler. Systems engineers need to be able to get out of their offices and communicate well—listen, talk, and write. George Bernard Shaw once stated that England and America are "two countries separated by a common language," but engineers are separated by their distinct languages—even more so since the advent of electronic communications. Systems engineering helps bridge the communication gaps among engineers and managers with consistent terms, processes, and procedures. A key to success is the ability to communicate the big picture, and to help others develop a big-picture view.

Strong team member and leader. Systems engineers have to be skilled leaders as well as managers. They must meet the special challenge of commanding diverse technical knowledge, plus managing and leading effectively!

Comfortable with change. Systems engineers should be comfortable with change. They understand that change is inevitable. They anticipate change, are able to understand how it affects their systems, and deal with those effects properly, ideally without losing sleep at night.

Comfortable with uncertainty. A companion characteristic is being comfortable with uncertainty—indeed, embracing uncertainty. We usually don't know when we will finish a task, or even a mission. Requirements aren't complete, so we have to interpret them. This is the simple side of uncertainty. But uncertainty has a more complex side, so a strong background in probability and statistics is important. A good systems engineer understands and encourages quantification of uncertainty. For example, if the mission objective is to land a probe on a comet, the location and severity of jets or debris may be unknown or the comet's albedo may be uncertain. The systems engineer must be able to work with a team to design a system that accommodates the uncertainties.

Proper paranoia. This means expecting the best, but thinking about and planning for the worst. The systems engineer is constantly checking and crosschecking selected details across the system to be sure that technical integrity is intact.

Self confidence and decisiveness. Systems engineers must be self-confident. They know what they know, are aware of what they don't know, and aren't afraid to own both. It doesn't mean never making mistakes!

Appreciate the value of process. Good systems engineers appreciate process. That doesn't mean SE is just one process, plus another, plus another—like recipes in a cookbook. Processes are tools that provide them with a common frame of reference, help them manage design, cost, schedule and risk, and allow the team to work together to produce the right system. But processes alone don't guarantee a great product.

Diverse technical skills. A systems engineer must be able to apply sound principles across diverse technical disciplines. Good systems engineers know the theory and practice of many technical disciplines, respect expert input, and can credibly interact with most discipline experts. They also have enough engineering maturity to delve into and learn new technical areas.

Herein lies the art—how well does the maestro lead the people and use the tools provided? Maestros know how to bring out the best in their musicians; they know how to vary the tempo and the right moment to cue the horn section to draw in the listeners. The same is true for systems engineers. It's what we DO with the processes and talents of the team that matters. Ideally, as systems engineers gain experience, they are able to deal with more complex systems through:

- Breadth of technical knowledge and expertise, combined with excellent execution

- Passion for the mission and challenges, combined with force of personality and leadership ability

- Creativity and engineering instinct—ability to sense the right way to attack a problem while appreciating inherent risks and implications

- Ability to teach and influence others

A successful systems engineer knows how to balance the art of technical leadership with the science of systems management. The behavioral characteristics described above are necessary to meet the many challenges facing space systems engineers today and in the future. Later in this chapter we describe the capabilities, i.e., the skills, that systems engineers need to do their jobs.

1.1.3   Realities of Complex System Design

To this point, we've defined systems engineering as a combination of technical leadership and systems management. We've established that highly effective systems engineers share certain behavioral characteristics. These elements feed into successful mission and system design: the ability to get a system's design right initially and to maintain its technical integrity throughout its lifecycle. We use the following definitions:

- *Architecture* encompasses the fundamental organization of a system, embodied in its components, their relationships to each other and the environment, and the principles governing its design and evolution

- *Design* is creating a product or system, as well as a plan to develop and use it

For our purpose, architects provide the rules; designers create solutions using those rules. Systems engineers do both; they help create the design and maintain its integrity throughout its lifecycle. Designing new aerospace missions and systems is a very creative, technical activity. Most engineers use a variation of a fundamental thought process—1) define the problem, 2) establish selection criteria, 3) synthesize alternatives, 4) analyze alternatives, 5) compare alternatives, 6) make a decision, and 7) implement (and iterate, for that matter). Though not usually mandated, we use this process, or one like it, because it produces good, useful results—it works!

The first credible design for a space mission and its associated systems is usually the product of a few individuals or a small design team. They:

- Determine stakeholders' needs and success criteria
- Identify critical top-level requirements (normally 3 to 7) and understand the acceptance criteria
- Create a mission concept as well as physical and functional architectures
- Develop a concept of operations and integrate it with the mission concept, architecture, and top-level requirements
- Design critical interfaces among the architecture's elements
- Develop clear and unambiguous requirements that derive from the mission concept, architecture, concept of operations, and defined interfaces

The result of this intense, highly iterative, and creative activity is a first credible design that's consistent with basic physics and engineering principles and meets top-level requirements. It's a baseline from which we apply systems management processes to do tradeoffs and more detailed quantitative analyses that focus on enhancing the design detail. We also continue to identify and mitigate technical, cost, and schedule risks.

Defining the interfaces is key. We have to keep the number of interfaces to an acceptable minimum; less is usually more as long as the appropriate level of isolation of interaction is maintained. We should also keep them as simple as possible and, when confronted with a particularly difficult interface, try changing its characteristics. And of course, we have to watch out for Murphy's Law! Designers and systems engineers engaged in this early activity, and indeed, throughout the lifecycle, follow several hard-learned principles:

- Apply equal sweat
- Maintain healthy tension
- Manage margin
- Look for gaps and overlaps
- Produce a robust design
- Study unintended consequences
- Know when to stop

Apply equal sweat. The concept of equal sweat is to apportion the required performance or functional requirements in such a way that no single subsystem has an insurmountable problem. Figure 1-3 provides an example of the allocation and flow down of the top-level requirement for mapping error. If the mapping error is misallocated, it can easily drive the cost and complexity of one element up significantly, while allowing another element to launch with excess margin. Good engineering judgment and communication are required to allocate a requirement across subsystem boundaries in such a way that each element expends equal sweat in meeting the requirement. We must be prepared to reallocate when a problem becomes unexpectedly difficult in one area. To achieve this, team leaders must maintain open communications and the expectation that the members can and should raise issues. The concept of equal sweat applies to many aspects of space systems. Table 1-1 lists a few.

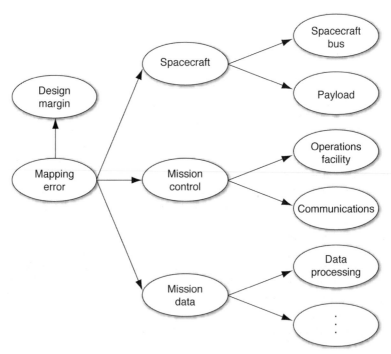

FIGURE 1-3.   **An Example of Equal Sweat.** Here we see a potential mapping error allocation for a space system. Mapping error represents how well the system is expected to pinpoint the location of an image created by the system. (Zero mapping error is perfection.) The goal, after setting sufficient design margin aside, is to allocate the mapping error to elements of the system in such a way that no element has an insurmountable challenge and each element expends roughly equal effort (sweat) in meeting the requirement.

TABLE 1-1.   **Application of Equal Sweat.** This is a partial list of space system aspects for which we must apportion requirements more or less evenly.

System	Aspects
Spacecraft	• Pointing stability • Pointing knowledge
Payload	• Line-of-sight stability • Optical stability • Thermal stability • Structural stability
Operations	• Command complexity • Recorder management
Communications	• Total data volume • Data latency
Data processing	• Artifact removal • Data integrity verification • Throughput • Reprocessing
Dissemination	• Metadata management • Archive management

Maintain healthy tension. A project must simultaneously meet its cost, schedule, and technical objectives. This often creates conflict. How much should we spend to make the system better and how good is good enough? How much time and money must we spend in chasing down a problem? What risk will we take if we eliminate a test and how well do we understand that risk? NASA's system of checks and balances is designed to ensure balancing of these objectives. If engineering overemphasizes creating the perfect system, project management must push on cost and schedule. If project management goes too far in minimizing testing to reduce schedule, then engineering or safety and mission assurance must push on the technical integrity. Discussions sometimes become extremely complex and passionate but we need to keep the common goal of mission success at the forefront.

Constructive dialogue among respectful peers along with the appropriate mention of impasses is critical. We have to allow sufficient time for productive discussion while making timely decisions to move forward. The interactions may be individually stressful and even seem wasteful, but the space community has a long history of mission successes when we maintained healthy tensions among all of the parties, and conversely, a number of major failures when we didn't.

Similar healthy tension occurs in many areas—across organizations, across subsystems or elements, and between mission phase requirements. This tension plays out during the design phase, when the operators try to be sure that the system will be operable and maintainable while the designers work to balance significant near term constraints such as cost, schedule, or mass. It also plays out during development and operations, when teams balance design changes and workarounds with ensuring safe, successful systems. Throughout the lifecycle continual tension helps maintain the proper requirements, constraints, and testing. For example, we must strike a balance between too little and too much testing. Not enough testing adds risk to a program, whereas testing too much can be very costly and may add unnecessary run-time to the equipment. These healthy tensions are a key to creating and maintaining the environment that will produce the best-balanced system, and the systems engineer must embrace and foster them.

Manage margin. Good systems engineers maintain a running score of the product's resources: power, mass, delta-V, and many others. But more importantly, they know the margins. What exactly does margin mean? Margin is the difference between requirements and capability. If a spacecraft must do something (have some capability), we allocate requirements. If we meet requirements, test effectively, and do the job correctly, we create a capability. One way to add margin is to make the requirements a little tougher than absolutely necessary to meet the mission's level-one requirements, which some people call contingency.

In Figure 1-4, the outer shape defines the capability, the inner shape represents the requirement, and the space between the two represents margin. The requirement comes very close to the capability on the right side of the diagram, so there we have a minimum margin. (The figure also applies to characteristics.)

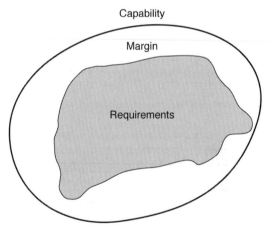

FIGURE 1-4.   **Requirements, Capability, and Margin.** Where the requirements come close to the capability (as on the right side of the figure), we have little margin [Adapted, Lee, 2007].

Look for gaps and overlaps. Once we begin feeling comfortable and confident about our design, looking for gaps and overlaps will help us recover from our comfort and confidence. What have we forgotten? Which requirements are incomplete? Where are the disconnects among our project's top-level requirements, architecture, design, and concept of operations? We must also carefully consider all system interfaces and look on both sides of these interfaces to identify what could interfere with our system. When we do this we often find that our system of interest or the design's scope is not sufficiently defined.

Create a robust design. Robust design is a proven development philosophy for improving the reliability of systems, products, and services. Terms that characterize it include resilient, stable, flexible, and fault tolerant. Robust design is vital to successful missions and systems. It must be an early and integral part of development. Our objective is to make our systems resistant to factors that could harm performance and mission success. A robust design performs consistently as intended throughout its lifecycle, under a wide range of conditions and outside influences, and it withstands unforeseen events. In other words, a robust design provides stability in the presence of ignorance!

Study unintended consequences. A key to success in spaceflight is to rigorously analyze failure modes and effects to determine how the system will perform when individual elements, subsystems, or components fail. Good systems engineers study failures of complex systems to gain insights into their root causes, ripple effects, and contributing factors. Hardware, software, interfaces, organizations, and people introduce complexity, so we study failures to avoid them. Henry Petroski, a professor at Duke University and author of *Success Through Failure,* points out that studying failures helps us better assess our design's unintended consequences [Petroski, 2006]. In *Apollo: The Race to the Moon,* Murray and Cox offer a stirring account of the Apollo 13 oxygen tank's explosion—a significant anomaly that resulted in mission failure. It shows how flight and ground crews creatively worked together to save the lives of the astronauts [Murray and Cox, 1989]. The systems engineer should study as many failures as possible to develop good engineering judgment.

Know when to stop. At some point in the project, we'll have discussed the philosophy of mission and system design, reviewed hard-earned wisdom about design, and even applied what we have learned from previous failures and created our "first credible" design. But we may hesitate to show the design to others until we've enhanced it a little more, a little more, and even a little more. It's hard to stop tweaking the design to make it better. Eventually, though, because of such realities as lack of money or time, we have to say, "Better is the enemy of good enough." The principle of knowing when to stop applies during the whole life of the project.

In universities, engineers learn to optimize designs, especially in the traditional electrical and mechanical disciplines. But in a large, complex system design, competing requirements and constraints often make it inappropriate to optimize subsystems. We need a balanced design that meets stakeholder needs as well as top-level critical requirements and constraints. However, system constraints such as mass often require the overall system to be optimized.

1.1.4   Processes for Systems Engineers

Design involves managing constraints, organizing system complexity, developing effective interfaces, managing resources and margin, and injecting leading-edge technology when and where appropriate. Creating an architecture and design for a complex system often requires the use of proven processes to manage complexity.

Getting the design right is a key first step in creating, developing, and operating aerospace systems. It represents the 10-percent inspiration associated with an acceptable solution. Maintaining technical integrity and managing complexity, using good solid processes, represents the 90-percent perspiration necessary to deliver the needed products and services. No matter how brilliant the design, we must still understand and properly apply rigorous processes and procedures throughout the project's lifecycle. Otherwise, what seemed like the right design will probably fail to meet its intended mission, within cost and schedule. Systems engineers must be able to deal with broad technical issues and apply rigorous processes and procedures, especially as projects get larger and more complex. NASA has documented its systems engineering policy in NPR 7123.1a, NASA *Systems Engineering Processes and Requirements*.

Experience shows that technical teams tend to ignore policy documents ("cookbooks") that dictate what processes they must follow unless the documents are tailored to project circumstances and supported by complementary elements. Examples of these elements are "how-to" resources (such as NASA SP-6105— *Systems Engineering Handbook* and this book), education and training materials, on-the-job learning activities, and appropriate tools.

We need to preserve lessons learned and tried-and-true processes, as well as enhance communication and consistently apply processes and procedures throughout the space community. Solid processes enable good systems engineering of complex systems. Systems engineers should own the processes and tools, and know when and how to use them, but not be owned by them. A lack of process rigor can easily lead to disaster, but too much can lead to rigor mortis. So our challenge is to develop systems engineers with good engineering judgment who know how to take a balanced approach. The goal of a process is to provide the needed product or service!

1.1.5   Meeting the Systems Engineering Challenge

For our purposes, a space systems engineer, **the** person responsible for the technical integrity of the entire system, needs to fully understand the technical red line (how technologies relate) through the system: the hardware and software as well as the basic physics and engineering associated with the system; the potential ripple effect of changes throughout the system; and key parameters and their sensitivities, to name a few. Many books capture at least a portion of this technical red line— *Space Mission Analysis and Design*, *Space Vehicle Design*, *Human Spaceflight*, *Space Launch and Transportation Systems*, and *Understanding Space*.

With the increasing complexity of space missions and systems, the challenge of engineering systems to meet the cost, schedule, and performance requirements within acceptable levels of risk requires revitalizing SE. Functional and physical interfaces are becoming more numerous and complicated. Software and embedded hardware must be integrated with platforms of varying intricacy. Pre-planned project development and the extension of system applications drive higher levels of integration. Another driver of increasing system complexity is the significant reduction of operations staff to reduce lifecycle cost and the incorporation of their workload into the system. In addition, systems are becoming more autonomous with stored knowledge, data gathering, intra- and inter-system communications, and decision-making capabilities.

While rising to the greater challenge, we must also address concerns over past failures. The need for more rigorous guidance and approaches is driven both by past experience and evolving project and program requirements. Drawing on the result of reports and findings of space-related experience since the mid-1990s, DOD, NASA, and industry have revamped their approach to SE to provide for future missions. Achieving our goals in space requires systems-level thinking on the part of all participants.

This book is intended to provide the processes, tools, and information that a space systems engineer can use to design, develop, and operate space systems. To this end we focus on "how to" as we proceed through the topics listed in Table 1-2.

The flow of activities from top to bottom looks linear, but nothing could be further from the truth. Each activity has many iterations, which have impacts on other activities, which are constantly changing.

1.2   Apply Systems Engineering Across the Lifecycle

This section sets the framework for this book by reviewing the NASA, DoD, and industry approaches to project phase-based lifecycles. From this we define a common frame of reference to examine a broad array of SE topics. Systems engineering is a discipline, so the systems engineer should always seek additional resources and training aids to increase or maintain proficiency. Therefore, we also provide guidance on where to obtain additional information to help succeed in this challenging occupation.

1.2.1   Introduction to Project Lifecycles

Many of the systems engineer's activities are driven by the organization's governance and the project manager. Our description of project lifecycles is taken largely from *Applied Project Management for Space Systems* [Chesley et al., 2008], Chapter 1. Several approaches are available for managing the development of a space-based system. In technology development efforts "rapid prototyping" is the usual method. We can describe this as "build a little, test a little, repeat until complete." While it's possible to use this model for other types of projects, it's inherently high-risk. It only affords management and stakeholders insight into critical problems once they're extremely expensive to resolve or effectively

Table 1-2. The Applied Space Systems Engineering Flow. This table provides an overview of the topics in this book. The early emphasis is on getting the right design, followed by realizing the system and managing the development and implementation of the system.

Description	Activities	Where Discussed
DESIGN the system	Define needs and stakeholder expectations Generate concept of operations and operational architecture Develop system architecture—functional and physical Determine technical requirements, constraints and assumptions Make decisions and conduct trade-off analyses Estimate lifecycle cost Assess technical risk	Chap. 2 Chap. 3 Chap. 4 Chap. 5 Chap. 6 Chap. 7 Chap. 8
REALIZE the system	Integrate the system Implement the system Verify and validate the system Transition the system into use	Chap. 9 Chap. 10 Chap. 11 Chap. 12
MANAGE creation, development, and implementation of the system	Plan and manage the technical effort Develop and implement a systems engineering management plan Control interfaces Maintain configuration Manage technical data Review and assess technical effort	Chap. 13 Chap. 14 Chap. 15 Chap. 16 Chap. 17 Chap. 18
Document and iterate		Chap. 1

insurmountable. This is unacceptable for a mission to Mars constrained by a launch window every two years, or a military system intended to support the war fighter waiting in the field.

Therefore, organizations and stakeholders mostly rely upon the "phased" lifecycle approach for the development of implementation-based projects. Also known as the "waterfall" model, this approach allows us to manage risk throughout the development, not just near the end. Problems surface early, allowing the project manager, systems engineer, organization, and stakeholders to agree on remedies. And since implementation-based projects are primarily the application of technology—as compared to the exclusive development of new technologies—their structures and elements are repeatable. This has the corollary benefit of allowing managers to accumulate experiences over many projects. They can then define reasonable expectations for what resources a project will need to complete a given set of activities, within a prescribed period of time, for a specified degree of complexity.

These are powerful reasons for using the phase-based approach for development of space-based flight systems. Systems engineers and project managers must be prepared to do more than simply manage projects under this approach; we need to be able to master it to ensure that our team succeeds.

If the phase-based approach were monolithic, it would be easy. We would simply review experiences accrued at lower levels in past projects and apply them to our own project. But the approach isn't monolithic and the communities of civilian space, military space, and private industry have tailored the process to meet the specific nuances associated with their communities. New systems engineers and project managers must master the processes their own communities use to execute the phase-based model. Before we go further, it's useful to define some items common to most lifecycle models.

Definitions

- *Decision authority*—The highest-ranking individual assigned by the organization with the authority to approve a project's formal transition to the next lifecycle phase
- *Gates*—Events, usually mandatory, that a project must participate in during which it will be reviewed by the stakeholders, user community, or decision authority
- *Key decision point (KDP)*—A milestone at which the decision authority determines the readiness of a project to formally proceed into the next phase of the lifecycle. The decision is based on the completed work of the project team as independently verified by the organization.
- *Lifecycle*—All the phases through which an end-item deliverable (i.e., system) passes from the time it's initially conceived and developed until the time it's disposed of
- *Milestone*—An important event within a project, usually the achievement of a key project deliverable or set of deliverables, or the demonstration of a group of functionalities. Milestones can be "internal"—self imposed by the project team or "external"—imposed by the stakeholder, user community, or decision authority.
- *Phases*—In projects, a specific stage of the lifecycle during which the majority of the project's resources are involved in the same primary activity for a common goal or deliverable, such as requirements definition, design, or operations
- *System*—The combination of elements—flight and ground—that function together in an integrated fashion to produce the capability required to meet an operational need. Each element comprises all the hardware, software, firmware, equipment, facilities, personnel, processes, and procedures needed for the element's contribution to the overall purpose.
- *Technical authority*—The individual, typically the project's systems engineer, that the decision authority assigns to maintain technical responsibility over establishment of, changes to, and waivers of requirements in the system

The NASA Model

Figure 1-5 illustrates the NASA-specified phase-based lifecycle approach that governs all the agency's flight projects. Here the project is segregated into two major pieces or agency-level phases, Formulation and Implementation. The agency places heavy emphasis on both phases; the new systems engineer or project manager should not think that formulation activities are any less rigorous than implementation activities.

The agency-level phases are in turn subdivided into project-level phases. These phases are where the project team carries out its work. They're also where processes, procedures, and products are subjected to external reviews (see Chapter 18) and key decision points before the project may officially progress into the next phase.

Each project-level phase has a set amount of work that must be completed to reach the level of maturity necessary to continue. So this accounting serves as the primary input into the integrated master schedule (see Chapter 3, *Applied Project Management for Space Systems*). Since this information is subject to change as the approach evolves, we don't repeat it here. Rather, the systems engineer and project manager should obtain the latest version of NPR 7120.5d, *NASA Space Flight Program and Project Management Requirements* for the current requirements. *NASA Systems Engineering Processes and Requirements*, NPR 7123.1a, is the guiding document for systems engineering associated with NASA projects and programs.

The DoD Model

Figure 1-6 illustrates the DoD-specified phase-based lifecycle approach governing its projects and programs. Here the project is segregated into three DoD-level phases: Pre-Systems Acquisition, Acquisition, and Sustainment. DoD emphasizes all phases, requiring diligence by the systems engineer and project manager to demonstrate that the project team has treated each with equal rigor. The DoD-level phases are subdivided into project-level phases. These phases are where the project team carries out its work. Here the processes, procedures, and products are subjected to reviews and key decision points before the project progresses into the next phase. Each project-level phase has a set amount of work that must be completed to reach the maturity necessary to continue. This serves as the primary input into the DoD project's integrated master schedule (see Chapter 3, *Applied Project Management for Space Systems*).

The NASA and DoD lifecycle models exhibit a great deal of similarity in how they encapsulate specific work-types into named phases. The differences between the models are less about the actual work being done and more about a community's expectations and product list dictating what to complete within a given phase. That is to say, from a purely practitioner point of view, an activity such as "requirements definition" or "production" is essentially the same in either community.

Therefore, we can set aside the community-specific expectations for any one phase and concentrate on the significant similarities that any systems engineer will

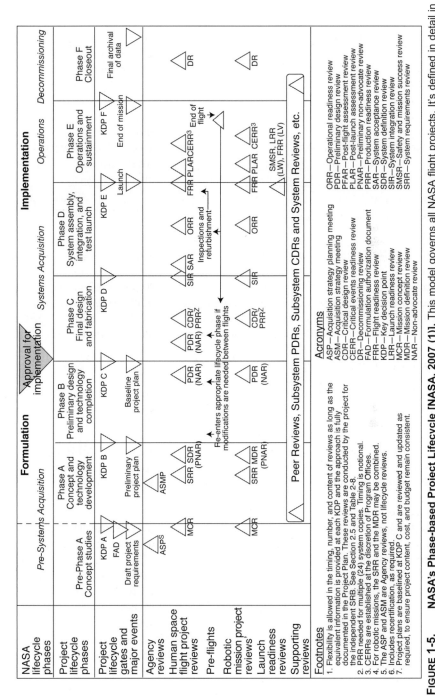

FIGURE 1-5.   **NASA's Phase-based Project Lifecycle [NASA, 2007 (1)].** This model governs all NASA flight projects. It's defined in detail in the agency's policy document 7120.5D, *NASA Space Flight Program and Project Management Requirements* [NASA, 2007 (2)].

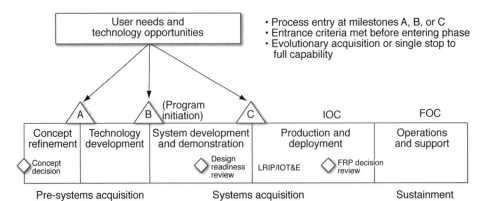

FIGURE 1-6.   **The Defense Acquisition Management Framework [OSD, 2008].** DoD projects progress through five project-level phases during their evolution. (IOC is initial operational capability; FOC is full operational capability; LRIP is low rate initial production; IOT&E is initial operational test and evaluation; FRP is full rate production.)

encounter in a given activity. Where practical, the subject matter experts have focused on the elements for success in terms of systems engineering as a discipline, less on the specific nuances of any one community. Where information is specific to one community, the author makes that distinction clear. Members of other communities would be wise to study these instances to gain a richer understanding of the trade space. Otherwise, this book directs the reader to the core SE competencies that are fundamental ingredients for success. We leave it to the systems engineer to take this knowledge and apply it within his or her organization's environment.

1.3   Create and Maintain the Technical Baseline

Space system design evolves from the abstract notion of a mission concept to an operational system. As described in Section 13.2.1, technical work encompasses a broad array of artifacts, including various documents (plans, specifications, drawings, and so on) and the system's own end or enabling products. Clearly some of these artifacts must precede others; for instance, engineering drawings must come before fabrication. So a technical artifact must be complete or mature enough to become the basis for succeeding artifacts and thus be part of the technical baseline. Once we baseline a technical artifact, it can change only through formal change control procedures (Chapter 16) because it influences technical work products under development.

At any point in the system's lifecycle, a *baseline* divides the work done from the work in progress. The baselining of each technical artifact contributes to the project's technical baseline. Thus, tens or even hundreds of technical baselines will exist throughout the development cycle as we baseline technical artifacts one by one. Still, a space project's technical baseline typically corresponds to the system's development phases and therefore consists of seven discrete parts:

- **Mission baseline**—Fully articulated and stakeholder-affirmed mission needs and objectives, as well as the concept for meeting those objectives
- **System baseline**—Functional and performance requirements characterizing systems that perform the mission concept
- **Functional baseline**—The system design, with functional and performance requirements allocated to its elements
- **Design-to baseline**—The fully specified system design, with functional and performance requirements and design or construction standards fully allocated and flowed down to the configuration item (hardware and software) level
- **Build-to or code-to baseline**—The complete system design as described by engineering drawings or CAD models, software design documents, and plans for fabrication, assembly, integration, and test
- **As-built or as-coded baseline**—The realized system as verified, validated, and certified, and ready for deployment
- **As-deployed baseline**—The operational system as launched or otherwise deployed, activated, and checked out—including calibration—and ready to proceed with full, routine operations

Figure 1-7 provides representative artifacts and documents associated with each of the baselines described above. It's imperative that at each milestone we ensure the integrity of the technical aspects of the project or program. While many of the artifacts shown are documents—products that must be developed and delivered—we must focus on the technical design, both hardware and software, associated with our system baseline.

This book focuses on these seven baselines, each one building on the previous one. The status of each baseline is represented by a set of artifacts (some of them documents). We usually establish these baselines at certain milestones in the lifecycle, as shown in Table 1-3. It's important to decide which baselines we will use to control the technical integrity of the project. Chapter 18 provides more details on baselines. Figure 1-8 highlights the differences between an early baseline and a baseline from mid-project.

Project managers often talk about the integrated baseline. This includes the work breakdown structure (WBS), WBS dictionary, integrated master schedule, lifecycle cost and workforce estimates, the project's cost analysis data requirements (CADRes), and the technical performance baseline or mission content. Details of the WBS and integrated master schedule are provided in Chapter 3 of *Applied Project Management for Space Systems* [Chesley et al., 2008].

Baseline	Representative Technical Artifacts	Established at
Mission	Need, goals, and objectives; Mission requirements	MCR
System	Concept of operations; System requirements; SEMP; IMP; Natural environment definition; Safety and mission assurance plan	SRR
Functional	Baseline configuration; Interface requirements; Allocated requirements; Test and evaluation master plan; Software management plan; Technology development plan	SDR
Design-to	Component specifications; Software specifications; Ground support equipment specifications; Interface control documents; Verification plans; Manufacturing plan	PDR
Build-to	Engineering drawings; Operational limits and constraints; Manufacturing process requirements; Acceptance plans and criteria; Verification procedures; Integration and assembly plan; Test plans; Training plan	CDR
As-built	End products; Enabling products; Acceptance data package; Operations procedures; In-flight checkout plan	SAR
As-deployed	Integration and test anomaly resolutions; Mission support training and simulation results; Scientific instrument calibration results; System activation results	ORR FRR PLAR

FIGURE 1-7.   Typical Technical Baselines for Space Systems and Missions. Our discussion centers on seven baselines in the project lifecycle. Each has a set of artifacts or documents that represent its status. These baselines are established at certain milestone reviews. (MCR is mission concept review; SRR is system requirements review; SDR is system design review; PDR is preliminary design review; CDR is critical design review; SAR is system acceptance review; ORR is operational readiness review; FRR is flight readiness review; PLAR is post-launch assessment review.)

1.4   Establish a Systems Engineering Framework

The rigors of the space domain, the unprecedented nature of space missions, and the lack of tolerance for system risk call for the application of a systematic, disciplined engineering approach. The approach is supported by a robust framework that allows the iterative activities of synthesis (development of alternatives and trades, supported by domain knowledge and creativity), analysis (assessment of alternatives, supported by an understanding of constraints—technical, legacy, cost, and schedule—and design drivers), and evaluation (decision making, supported by an understanding of the underlying uncertainties and risks).

Figure 1-9 depicts the interrelation among these activities. The approach must be quantifiable, recursive, iterative, and repeatable for the development, operation, and maintenance of systems integrated into a whole throughout the lifecycle of a project or program. Systems engineering emphasizes safety in achieving stakeholders' functional, physical, and operational performance

TABLE 1-3.   **Typical Baseline Items from NASA's NPR 7123.1a.** This table shows a more exhaustive list of system-related products and artifacts and tells when they're initially developed and baselined. (KDP is key decision point; FAD is formulation authorization document; ASM is Acquisition Strategy Meeting; NEPA is National Environmental Policy Act; EIS is environmental impact statement; CADRe is cost analysis data requirement; SRB is standing review board; MD is mission director; CMC is Center Management Council; PMC is program management committee.)

Products	Pre-Phase A KDP A	Phase A KDP B	Phase B KDP C	Phase C KDP D	Phase D KDP E	Phase E KDP F
Headquarters and Program Products						
1. FAD	Approved Draft					
2. Program requirements on the project (from the program plan)		Baseline	Update			
3. ASM minutes		Baseline				
4. NEPA compliance documentation			EIS			
5. Interagency and international agreements			Baseline			
Project Technical Products						
1. Mission concept report	Preliminary	Baseline				
2. System level requirements		Preliminary	Baseline			
3. Preliminary design report			Baseline			
4. Mission operations concept		Preliminary	Baseline			
5. Technology readiness assessment report			Baseline			
6. Missile system pre-launch safety package			Preliminary	Baseline	Update	
7. Detailed design report				Baseline		
8. As-built hardware and software documentation					Baseline	
9. Verification and validation report					Baseline	
10. Operations handbook				Preliminary	Baseline	
11. Orbital debris assessment		Initial	Preliminary	Baseline		
12. Mission report						Final

TABLE 1-3. **Typical Baseline Items from NASA's NPR 7123.1a. (Continued)** This table shows a more exhaustive list of system-related products and artifacts and tells when they're initially developed and baselined. (KDP is key decision point; FAD is formulation authorization document; ASM is Acquisition Strategy Meeting; NEPA is National Environmental Policy Act; EIS is environmental impact statement; CADRe is cost analysis data requirement; SRB is standing review board; MD is mission director; CMC is Center Management Council; PMC is program management committee.)

Products	Pre-Phase A KDP A	Phase A KDP B	Phase B KDP C	Phase C KDP D	Phase D KDP E	Phase E KDP F
Project Planning, Cost, and Schedule Products						
1. Work agreement for next phase	Draft	Baseline	Baseline	Baseline	Baseline	Baseline
2. Integrated baseline		Preliminary	Baseline			
3. Project plan		Preliminary	Baseline	Update		Update
4. CADRe		Preliminary	Baseline			
5. Planetary protection plan		Planetary protection certification	Baseline			
6. Nuclear safety launch approval plan		Baseline				
7. Business case analysis for infrastructure		Preliminary	Baseline			
8. Range safety risk management plan			Preliminary	Baseline		
9. System decommissioning and disposal plan			Preliminary	Preliminary		Baseline
KDP Readiness Products						
1. SRB report	Final	Final	Final	Final	Final	Final
2. Project manager recommendation (includes response to SRB report)	Final	Final	Final	Final	Final	Final
3. CMC recommendation	Final	Final	Final	Final	Final	Final
4. Program manager recommendation	Final	Final	Final	Final	Final	Final
5. MD-PMC recommendation (Category I projects)	Final	Final	Final	Final	Final	Final
6. Governing PMC recommendation	Final	Final	Final	Final	Final	Final

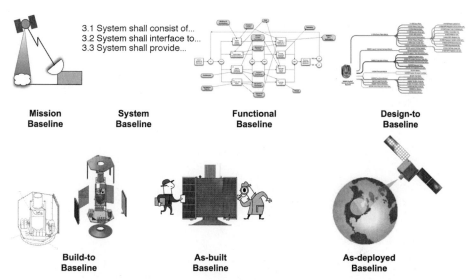

3.1 System shall consist of...
3.2 System shall interface to...
3.3 System shall provide...

**Mission   System   Functional   Design-to
Baseline   Baseline   Baseline   Baseline**

**Build-to   As-built   As-deployed
Baseline   Baseline   Baseline**

FIGURE 1-8.   Another Perspective on Technical Baselines. This figure captures the evolution of the system design through the various technical baselines. Space systems engineering takes the relatively crude concepts in the Mission Baseline and uses the design, manage, and realize processes to deliver the capability in the as-deployed baseline.

requirements in the intended environments over the system's planned life within cost and schedule constraints. To make this happen, all project participants need to have a systems perspective.

Synthesis

Analysis

Evaluation

FIGURE 1-9.   Systems Engineering Activities. The systems engineering framework (discussed below) supports iterative and recursive synthesis, analysis, and evaluation activities during complex systems development. Domain knowledge, systems thinking, and engineering creativity are key prerequisites.

The SE framework itself consists of the three elements shown in Figure 1-10. The integrated implementation of common technical processes, workforce, and tools and methods provides greater efficiency and effectiveness in the engineering of space systems. This section describes each element. Systems engineering processes are one element in a larger context, including workforce (that's the team), and tools and methods to produce quality products and achieve mission success. Together, these elements constitute the SE capability of an organization. Furthermore, the SE processes themselves represent a framework for the coordinated evolution and increasing maturity of the mission, the required operational scenarios and performance requirements, and the constraints on the one hand; and the conceived solution and its architecture, design, and configuration on the other.

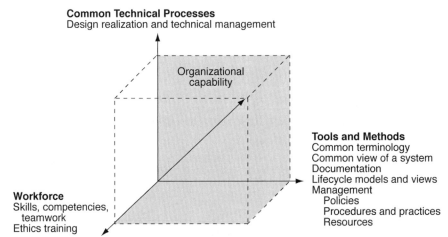

FIGURE 1-10.   Systems Engineering Framework. Our goal is to provide the products and services required by our stakeholders. This requires people, processes, tools, and methods [NASA, 2007 (2)].

Technical and Technical Management Processes. Common technical processes help us treat hardware, software, and human elements of a system as one integrated whole. They also provide the concepts and terminology that are basic to consistent application and communication of systems engineering practices. NPR 7123.1a provides an example of common technical processes in the space arena.

Tools and Methods. The right tools and methods enable the efficient and effective application of systems engineering. Most organizations have identified and documented the tools and methods that work best for them.

Workforce—People and Their Capabilities. Well-trained and knowledgeable systems engineers, working with the technical workforce and the key stakeholders, and steeped in the necessary domain knowledge, are essential for successful application of SE. We hope that this book will provide the necessary knowledge, processes, tools and methods to make systems engineers even more valuable to their organizations.

1.4.1   Systems Engineering Technical and Technical Management Processes

This section establishes a core set of common technical and technical management processes for space projects in engineering system products and services during applicable product-line lifecycle phases to meet phase exit criteria and project objectives. A study of the literature suggests considerable agreement on the main systems engineering technical and technical management processes. Some of the standards and handbooks are:

- ISO/IEC 15288, *Systems and Software Engineering—System Life Cycle Processes*
- ANSI/EIA 632, *Processes for Engineering a System*
- IEEE 1220, *Application and Management of the Systems Engineering Process*
- EIA 731, *Systems Engineering Capability Model*
- CMMI SWE/SE/IPPD/SS, *Capability Maturity Model-Integration for Software Engineering, Systems Engineering, Integrated Product and Process Development and Supplier Sourcing*
- NASA 7123.1a, *NASA Systems Engineering Processes and Requirements*
- DoDI 5000.02, *Defense Acquisition Guidebook (DAG)*

The Defense Acquisition Guidebook and NPR 7123.1a describe 16 and 17 systems engineering processes, respectively. The two guides exhibit significant consistency and convergence, as shown in Table 1-4. Figure 1-11 depicts the 17 processes in NPR 7123.1a in the context of an SE engine. In this figure, the technical processes have been divided into system design precesses and product realization processes. They're iterative within any phase of the development and recursive at lower and lower levels of the system structure.

Systems engineering processes allow an orderly progression from one level of development to the next more detailed level through the controlled baselines. We use them for the system, subsystems, and components as well as for the supporting or enabling systems used for the production, operation, training, support, and disposal of the system. During technical management processes and activities, such as trade studies or risk management activities, we may identify some requirements, interfaces, or design solutions as non-optimal and change them to increase system-wide performance, achieve cost savings, or meet deadlines. The value of these processes is not only in translating stakeholder expectations into detailed specifications and further into test plans, but also as an integrated framework within which the requirements can be defined, analyzed, decomposed, traded, managed, allocated, designed, integrated, tested, fielded, and sustained [OSD, 2008].

The application of these processes, including methods and tools in the context of a space-based system, is the emphasis of this book. Our intent is not only to convey the "what" of systems engineering, but also to illustrate the "how to" aspects of each process.

TABLE 1-4.   **Systems Engineering Processes.** There is considerable agreement on the definition of the core set of SE processes, classified into technical management processes and technical processes.

Systems Engineering Processes per OSD Defense Acquisition Guidebook	Systems Engineering Processes per NPR 7123.1a—SE Processes and Requirements
Technical Management Processes	Technical Management Processes
• Decision analysis	• Decision analysis
• Technical planning	• Technical planning
• Technical assessment	• Technical assessment
• Requirements management	• Requirements management
• Risk management	• Technical risk management
• Configuration management	• Configuration management
• Technical data management	• Technical data management
• Interface management	• Interface management
Technical Processes	Technical Processes
• Requirements development	• Stakeholder expectation definition
• Logical analysis	• Technical requirements definition
• Design solution	• Logical decomposition
• Implementation	• Physical solution
• Integration	• Product implementation
• Verification	• Product integration
• Validation	• Product verification
• Transition	• Product validation
	• Product transition

1.5   Identify Capabilities of the Systems Engineer

Let's switch gears and answer the question, "So, what are space systems engineers supposed to be able to do?" In this section, we discuss the capabilities that they must have and their associated performance levels. Because SE is both an art and a science, many of the skills and abilities needed to be highly effective in complex systems are not learned in school; they are gained through experience. Processes and tools are important, but they can't substitute for capable people. Following processes and using tool sets doesn't automatically result in a good systems engineer or system design. Entering requirements into a database does not make them the right requirements. Having the spacecraft design in a computer-aided design (CAD) system does not make it the right design. Capable and well-prepared people make the difference between success and failure. So systems engineers face a broad range of expectations, including:

- Prepare the documents necessary for key project milestones
- Act as the technical conscience of a project
- Design, realize, and manage the creation, development, and implementation of the system

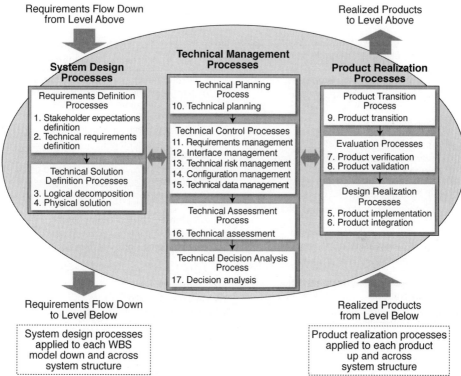

FIGURE 1-11.   The Systems Engineering Engine, Depicting the Common Systems Engineering Sub-Processes, Per NASA NPR 7123.1a. These processes apply at any level of the development effort. The three groups of processes shown here are used to develop, produce, and deliver products and services.

- Know the technical "red line" through the system; be able to anticipate the impact of changes throughout the system and lifecycle
- Maintain the big picture perspective, integrating all disciplines
- Establish and maintain the technical baseline of a system
- Maintain the technical integrity of the project, whatever it takes

Besides this broad range of expectations, we need a better defined set of capabilities and performance descriptions. Systems engineers in any field have similar capabilities and use similar processes; they just focus on different types of system. A space systems engineer, of course, needs to be steeped in the technical aspects of space.

The leaders and managers of the organization should describe each item in Table 1-5 in some detail to help develop systems engineers within the organization. The aspiring systems engineer should ask, "What should I be able to do for each item in the table?" We can discuss each item and the integrated whole in terms of performance levels. Table 1-6 lists four levels of performance.

TABLE 1-5.   Systems Engineer Capabilities. In the life of a project, a systems engineer should be able to deal with the items shown here.

1.0 Concepts and architecture	**5.0 Project Management and Control**
1.1 Mission need statement	5.1 Acquisition strategies and procurement
1.2 System environments	5.2 Resource management
1.3 Trade studies	5.3 Contract management
1.4 System architecture	5.4 Systems engineering management
1.5 Concept of operations	
1.6 Technical baseline development and management	**6.0 Internal and external environments**
	6.1 Organization structure, mission, and goals
	6.2 PM and SE procedures and guidelines
2.0 System design	6.3 External relationships
2.1 Stakeholder expectation development and management	
2.2 Technical requirements definition	**7.0 Human capital management**
2.3 Logical decomposition	7.1 Technical staffing and performance
2.4 Design solution definition	7.2 Team dynamics and management
	8.0 Security, safety, and mission assurance
3.0 Production, system transition, operations	8.1 Security
3.1 System implementation	8.2 Safety and mission assurance
3.2 System integration	
3.3 System verification	**9.0 Professional and leadership development**
3.4 System validation	9.1 Mentoring and coaching
3.5 Operations	9.2 Communications
	9.3 Leadership
4.0 Technical management	
4.1 Technical planning	**10.0 Knowledge management**
4.2 Requirements management	10.1 Knowledge capture and transfer
4.3 Interface management	10.2 International Traffic in Arms Regulations
4.4 Technical risk management	
4.5 Configuration management	
4.6 Technical data management	
4.7 Technical assessment	
4.8 Technical decision analysis	

As an individual progresses through performance levels 1 to 4, the focus of activities, as shown in Table 1-5, changes. While systems engineers operating at level 3 must have capabilities that span the list, they must often focus more on technical planning, managing teams, staffing, and external relationships than on the day-to-day details of requirements management. So, depending on performance level, some capabilities may figure more prominently than others.

Another point is that not all organizations require all these capabilities. Many have their own tried and true approaches and processes. It's important to understand one's organization's approach to SE to learn and apply the most valuable capabilities. In general systems engineers should:

TABLE 1-6.   **Systems Engineer Performance Levels.** The focus of this book is to help develop systems engineers that can operate at performance levels 1–3. Those who enhance their capabilities and gain experience may increase their performance level and be candidates for more significant responsibilities.

Performance Level	Generic Title	Description	Example Job Titles
1. Know	Project team member	An individual performing as part of a project team that's developing or operating a portion of a system	Electrical engineer, thermal engineer, software engineer, technical resource manager
2. Apply	Subsystem engineer	An individual responsible for the development, integration, or operation of a complete portion of a spacecraft or space mission	Guidance, navigation and control engineer, thermal control subsystem engineer, power subsystem engineer
3. Manage	Program or project systems engineer	An individual responsible for the systems engineering activity for an entire program or project, or for a major element of a large mission or system	Systems engineer for Hubble Space Telescope, Systems engineer for the Space Shuttle Orbiter, Mars Pathfinder Systems engineer
4. Guide	Chief engineer	An individual responsible for the systems engineering effort within a large organization. This person develops guidance and oversees all SE activities and systems engineers.	Chief engineer of the National Reconnaissance Office, Chief engineer of NASA

- Understand their technical domain better than anyone else, and continually learn about other technical domains and technologies
- Learn to deal effectively with complexity
- Maintain a "big picture" perspective
- Strive to improve in written and oral communications
- Strive to improve interpersonal skills and team member and team leadership abilities

1.6   Introduction to the End-to-End Case Study— FireSAT

The difference between systems engineering as an academic subject and as an integral part of everything we do in the space business is its application. It's not enough to "talk the talk"; we have to "walk the walk." One of our primary goals in this book is to focus on the "how" of SE, not just the "what." To that end, we have included numerous real world space examples from our own experiences. While useful and necessary, these examples are inherently cherry-picked from dozens of different missions. But SE processes take a project from cradle to grave. And seemingly innocent decisions made during early development may come back to haunt us during integration or operations. To better convey the living nature of space systems engineering in every phase of the lifecycle, this book presents the FireSAT mission as a integrated case study.

FireSAT may well be the most famous space mission that has never flown. The concept was first developed by editors of *Space Mission Analysis and Design* (SMAD) [Larson and Wertz, 1999]. SMAD had a similar need to provide a consistent example across widely different analysis and design methods and techniques. FireSAT was called to duty again in *Understanding Space* [Sellers, 2004] as a running example illustrating basic principles of astronautics. As this book was first taking shape, the editors looked for a single example that could tie together the 17 systems engineering processes to give readers a better sense of how to pull SE out of the book and into their day-to-day professional lives. FireSAT again gave us the right mix of reality, relevance, and rigor. Any end-to-end example must have enough complexity to illustrate key issues without being so complicated that the reader gets lost in the details. FireSAT provides that balance.

What is FireSAT? The FireSAT Project was conceived as a system to detect and monitor potentially dangerous wildfires throughout the US. Every year, wildfires claim hundreds of lives, threaten thousands more, and lay waste to millions of acres, causing losses in the billions of dollars. FireSAT was envisioned to provide space-based "fire scouts" that would survey the US daily to give the Forest Service a means for earlier detection to increase the probability of containment and to save lives and property.

The FireSAT example lets us address one of the first questions any systems engineer faces: What "system" are we talking about? The FireSAT mission architecture is shown pictorially in Figure 1-12, and in Figure 1-13 in a hierarchy diagram.

This book uses the term *system of interest* to refer to the current focus of discussion. At any given time the FireSAT system of interest could include the entire architecture, or *system of systems* as this level is sometimes referred to, or to any element within that architecture, or to any sub-element, such as a subsystem.

Along with the system of interest, the systems engineer is typically most concerned with interfaces one level above and below the system of interest. The meaning of "system" in the space domain is not helped by our confusing habit of naming projects after their spacecraft (or vice versa). The Space Based Infrared System (SBIRS), for example, is both the name of a major Air Force Program and the name of the satellites that are part of that program. Similarly, when people speak of the "Space Shuttle System" they could be talking about the most visible part that launches into orbit, or the entire Space Shuttle mission architecture that includes numerous facilities and equipment spread across all the NASA centers, as well as integral contractor support.

This approach is understandable given the high profile of the space element of the mission architecture, as well as the nearly complete reliance on legacy systems for the launch and ground elements. It can, however, lead to confusion about what system of interest is currently under discussion. Throughout the book we shift our focus through different levels of the FireSAT Project. Initial discussions use

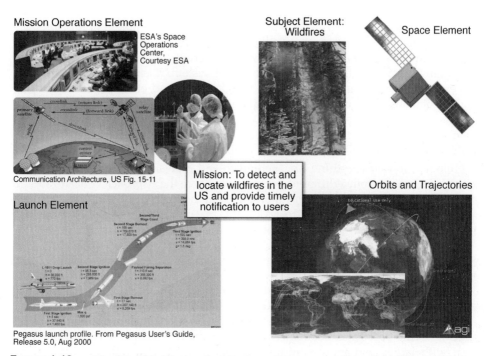

FIGURE 1-12.   The FireSAT Mission Architecture. Here we see the various elements of the complete mission architecture.

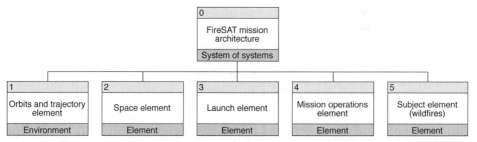

FIGURE 1-13.   FireSAT Mission Architecture (aka Physical Architecture) Hierarchy Diagram. This is another way to depict information similar to that in Figure 1-12.

"FireSAT" to mean the project or mission level. However, for the most part our primary emphasis is on the space element—the FireSAT spacecraft and its subsystems. On occasion we even dip down to the part level to illustrate a given point, but the objective is always to reinforce how the approaches and application of SE processes apply at any level.

Chapter 2 begins the FireSAT saga, starting with the initial needs of the stakeholder and leading us through the development of stakeholder or mission-level requirements. Subsequent chapters in the "Design" unit of the book continue through architecture development, requirements engineering, logical decomposition and eventually the physical solution of the FireSAT system. Then in the "Manage" unit we explore how decision analysis, risk management, and other processes support both design and implementation throughout the lifecycle. In the "Implement" unit, we first examine the issues associated with FireSAT buy, build, or reuse decisions, then move on to integration, verification, and validation. The transition chapter includes examples of how FireSAT goes from delivery to operations.

Finally, to tie everything up, the end-to-end case study chapter looks at the FireSAT example from a different perspective. There the approach concentrates on each of the seven primary technical baselines in the lifecycle, to illustrate how the 17 SE processes come to bear to tackle the real world challenges of moving a program from concept to launch to operations, one major milestone at a time.

Summary of Key Points

Space systems engineering efforts are usually more successful in a systems engineer that knows the technical domain—hardware and software, interactions, and sensitivities. We described above the highly effective behavioral characteristics of some of the best systems engineers in the country. Good processes enable us to deal with larger and more complicated systems. We also pointed out that SE provides a framework for problem solving—the more complicated the system or problem, the more useful the processes are in helping systems engineers do their job.

We discussed the capabilities that practicing space systems engineers should possess and described four performance levels—know, apply, manage and guide—in the hope that they will help in bettering career planning and development. As we practice systems engineering, we realize that it's not enough to know our domain and all the processes; we must leverage strong teamwork and leadership skills as well as exceptional communications skills to be successful.

References

Chesley, Julie, Wiley J. Larson, Marilyn McQuade, and Robert J. Menrad. 2008. *Applied Project Management for Space Systems*. New York, NY: McGraw-Hill Companies.

Griffin, Michael D. March 28, 2007. *Systems Engineering and the "Two Cultures" of Engineering*, NASA, The Boeing Lecture.

Larson, Wiley J. and James R. Wertz. 1999. *Space Mission Analysis and Design*. 3rd Ed. Dordrecht, Netherlands: Kluwer Academic Publishers.

Lee, Gentry. 2007. *So You Want to Be a Systems Engineer*. DVD, JPL, 2007.

Murray, Charles and Catherine Bly Cox. 1989. *Apollo: Race to the Moon*. New York, NY: Simon and Schuster.

NASA. 2007 (1). NPR 7123.1a—NASA Systems Engineering Processes and Requirements. Washington, DC: NASA.

NASA. 2007 (2). NPR 7120.5D—Space Flight Program and Project Management Requirements. Washington, DC: NASA.

Office of the Secretary of Defense – Acquisition, Technology and Logistics (OSD – AT&L). December 8, 2008. DoDI 5000.02, Defense Acquisition Guidebook. Ft. Belvoir, VA: Defense Acquisition University.

Personal Interviews, Presentations, Emails and Discussions:

Michael Bay, Goddard Space Flight Center

Harold Bell, NASA Headquarters

Bill Gerstenmaier, NASA Headquarters

Chris Hardcastle, Johnson Space Center

Jack Knight, Johnson Space Center

Ken Ledbetter, NASA Headquarters

Gentry Lee, Jet Propulsion Laboratory

Michael Menzel, Goddard Space Flight Center

Brian Muirhead, Jet Propulsion Laboratory

Bob Ryan, Marshall Space Flight Center

Petroski, Henry. 2006. *Success through Failure: The Paradox of Design.* Princeton, NJ: Princeton University Press.

Schaible, Dawn, Michael Ryschkewitsch, and Wiley Larson. *The Art and Science of Systems Engineering.* Unpublished work, 2009.

Additional Information

Collins, Michael. *Carrying the Fire: An Astronaut's Journeys*, New York, NY: Cooper Square Press, June 25, 2001.

Defense Acquisition University Systems Engineering Fundamentals. Ft. Belvoir, Virginia: Defense Acquisition University Press, December 2000.

Derro, Mary Ellen and P.A. Jansma. *Coaching Valuable Systems Engineering Behaviors.* IEEEAC Paper #1535, Version 5, December 17, 2007.

Ferguson, Eugene S. 1992. *Engineering and the Mind's Eye.* Cambridge, MA: MIT Press.

Gladwell, Malcolm. *Blink: The Power of Thinking Without Thinking.* New York, NY: Back Bay Books, April 3, 2007.

Gleick, James. *Genius: The Life and Science of Richard Feynman.* New York, NY: Vintage Publications, November 2, 1993.

Griffin, Michael D. and James R. French. 2004. *Space Vehicle Design.* 2nd Ed., Reston, VA: AIAA Education Series.

Johnson, Stephen B. 2006. *The Secret of Apollo: Systems Management in American and European Space Program (New Series in NASA History).* Baltimore, MD: Johns Hopkins University Press.

Kidder, Tracy. 2000. *The Soul Of A New Machine.* New York, NY: Back Bay Books.

Larson, Wiley J., Robert S. Ryan, Vernon J. Weyers, and Douglas H. Kirkpatrick. 2005. *Space Launch and Transportation Systems*. Government Printing Office, Washington, D.C.

Larson, Wiley J. and Linda Pranke. 2000. *Human Spaceflight: Design and Operations*. New York, NY: McGraw-Hill Publishers.

Logsdon, Thomas. 1993. *Breaking Through: Creative Problem Solving Using Six Successful Strategies*. Reading, MA: Addison-Wesley.

McCullough, David. 1999. *The Path Between the Seas: The Creation of the Panama Canal 1870–1914*. New York, NY: Simon and Schuster.

Menrad, Robert J. and Wiley J. Larson. *Development of a NASA Integrated Technical Workforce Career Development Model*. International Astronautical Federation (IAC) Paper IAC-08-D1.3.7, September 2008.

Perrow, Charles. 1999. *Normal Accidents: Living with High-Risk Technologies*. Princeton, NJ: Princeton University Press.

Rechtin, Eberhard. 1991. *System Architecting*. Prentice Hall, Inc.

Ryan, Robert. 2006. *Lessons Learned—Apollo through Space Transportation System*. CD, MSFC.

Sellers, Jerry Jon. 2004. *Understanding Space*. 3rd Ed. New York, NY: McGraw Hill.

Squibb, Gael, Wiley J. Larson, and Daryl Boden. 2004. *Cost-effective Space Mission Operations*. 2nd Ed. New York, NY: McGraw-Hill.

USGOV. Columbia Accident Investigation Board, Report V. 5, October 2003 (3 Books), PPBK, 2003, ISBN-10: 0160679044, ISBN-13: 978-0160679049.

Williams, Christine and Mary-Ellen Derro. 2008. NASA Systems Engineering Behavior Study, October, 2008. Publication TBD.

Chapter 2

Stakeholder Expectations and Requirements Definition

Dinesh Verma, *Stevens Institute of Technology*

The top priority of any new program or project is to define and scope the customer's expectations. These expectations lay out the mission, the required capability, or the market opportunity. They drive all of design, development, integration, and deployment. Customer expectations, established by the mission need statement or capability definition, represent the problem space for systems engineering. We use them to identify relevant stakeholders and the technical requirements that describe what the mission is to achieve. Failure to adequately capture these initial expectations results in a mismatch between what is needed and what is delivered. As Yogi Berra famously said, "If you don't know where you're going, you'll probably end up some place else."

Evolving initial customer expectations into a set of stakeholder requirements, and translating these into technical requirements and system architecture is highly iterative, as Figure 2-1 shows. In fact, developing solution concepts and architectures often helps better define the need or capability itself.

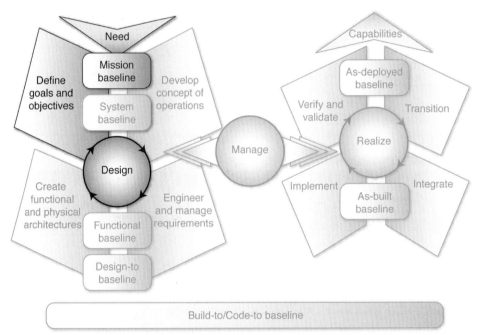

FIGURE 2-1.   **Iterative Evolution of the Problem and Solution Spaces.** The early elements of systems engineering are iterative and interdependent, rather than procedural and linear.

Figure 2-2 provides a context for the discussions in Chapters 2 through 5, and emphasizes their interdependence. A "process oriented" view of systems engineering suggests a procedural and linear implementation. In reality, these processes are iterative and interdependent. While we mustn't confuse the problem with the solution, better understanding of one leads to a better definition of the other. An architecture often helps clarify the scope of the problem and the stakeholder expectations, while the concept of operations helps us understand the requirements that an architecture must satisfy.

Stakeholder requirements, their expectations, and major concepts drive the development of the concept of operations which helps us assess system concepts in an operational context. It also validates stakeholder expectations and mission requirements. Table 2-1 reflects these activities and also outlines this chapter. We start by seeing how to define the customer's initial expectations, then step through the other activities. From there we can begin to define concepts and operational architectures, and eventually technical requirements, as described in subsequent chapters. Figure 2-2 diagrams this process.

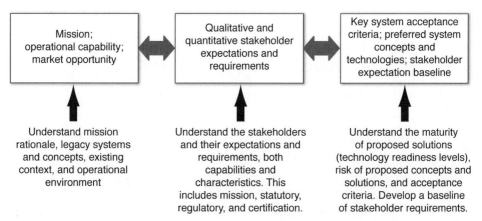

FIGURE 2-2.   An Abstraction of the Framework Presented in Table 2-1. While the systems engineering process is iterative, our understanding of the problem and solution matures progressively.

TABLE 2-1.   A Framework for Developing Stakeholder Expectations and Requirements. A general process for developing stakeholder expectations and requirements is shown here along with the sections that discuss them.

Step	Description	Where Discussed
2.1	Define initial customer expectations	Section 2.1
2.2	Identify the stakeholders	Section 2.2
2.3	Solicit and synthesize stakeholder expectations and requirements	Section 2.3
2.4	Articulate stakeholder expectations and requirements	Section 2.4
2.5	Rank order mission requirements and identify critical acceptance criteria	Section 2.5
2.6	Synthesize essential mission requirements	Section 2.6
2.7	Validate and baseline requirements. Translate into technical requirements.	Section 2.7

2.1   Define Initial Customer Expectations

A common understanding of customer expectations is essential to the project's success. We must first understand a customer's underlying motivation, such as national security concerns, science, social or political issues, profit, or technology development. Our understanding depends on knowing the environment in which the capability will be deployed, operated, and sustained. It also provides the systems engineer with insights into the rationale for the system. Clear understanding helps us assign priorities to the more detailed stakeholder expectations and requirements, identify critical mission requirements or acceptance criteria, and conduct the inevitable trade studies.

2.1.1   Sources of Customer Expectations

The initial customer expectations[*] may be in response to:

- A current functional deficiency
- An existing operational deficiency
- A desire to leverage new technology breakthroughs to enhance mission capability or market positioning
- An evolving threat or competition
- A need to improve the capability based on behavior of current systems and their operators or maintainers

A *functional deficiency* is a lack of a certain capability in the current operational or business environment. The capability is required but not provided by any existing systems. Earth orbiting satellites, for example, often fill functional deficiencies for persistent coverage of large parts of the Earth for communications or weather observations. Airborne or other terrestrial assets can't practically do this. The FireSAT mission, developed as a case study in this book, is envisioned to fill a functional deficiency in the current wildfire detection system. A science mission such as a comet rendezvous would also fill a functional deficiency.

An *operational deficiency* describes an operational problem or insufficiency with a current system. The current system provides the necessary functionality, but not as well as desired. This may be due to increased user demands. For example, new satellite communications missions provide enhanced capabilities to address demand for increased bandwidth to feed our digital world. New satellite communication (SATCOM) missions increase data rates or expand geographical coverage, or both. No new functionality is needed (we're still providing geostationary communication links)—the need is to do the same thing better.

Often a technology breakthrough and its potential exploitation for the mission or a competitive market advantage becomes the basis of a development program. For example, as the reliability of ion and plasma rockets has increased, geostationary SATCOM customers have begun using them in place of less efficient chemical rockets, greatly increasing the on-orbit lifetime.

An intimate understanding of the threat environment or the competitive landscape may also drive the definition of necessary missions and capabilities. A detailed threat assessment or competition analysis must provide the rationale. An increasing number of international players in the commercial remote sensing market, for example, may spur companies to develop new missions.

---

[*] For our purposes, customer expectations encompass the mission or capability desired, or the market opportunity being targeted.

Finally, a study of how operators and maintainers use a system may suggest ways to make this interaction more efficient and the overall system more effective. In a complex system, the weakest link is often the human operator or maintainer. A study of this link may be the basis for system improvements—in functionality, reliability, safety, and security. Many of the efforts in NASA's Constellation Program are aimed at reducing the cost and complexity of Ares 1 ground processing. The automobile industry has designed many improvements based on behavior analysis. These include automatic tilt-down of passenger side view mirror (to aid in parallel parking), auto-dimming rear view mirrors, and cut-out trunks for easier storage, to name a few.

2.1.2   Articulate Customer Expectations

The sponsors (including political leaders), customers, operators, and mission specialists often lead the definition of the initial mission or capability. The influence of initial customer expectations on the development program suggests some guidelines:

- The mission or opportunity exists in the domain of the customer, the operator, and the mission specialist, not that of the developer or engineer. The mission statement describes what the customer or sponsor wants and so represents the problem space or the opportunity space. We may develop a system or solution to satisfy the need. The system itself is not the need; it's a response to the mission or market opportunity. In all cases, a customer is ultimately buying a capability, not a satellite. The satellite is simply the means to that end.

- The capability should support a current or future business objective or mission. This is the rationale for the system need and helps document the business case. The capability definition should also document what is lacking in current business or mission processes. This is particularly important when we have scarce resources and competing priorities.

- A mission need exists independent of any solution to that need—the customer has a need whether or not it can be met. It's also independent of whether the solution is performed in hardware or in software or with changes in doctrine and processes. In fact, in many cases, the need can be met with a change in doctrine, organization, training, leadership, policy, personnel, or facilities, rather than a physical solution (which is usually the most expensive).

- The capability and mission need should be expressed in the language of the customer, not that of the designers. A military need should be expressed in terms of the military mission. An astronomer's need should be expressed in the astronomer's language (not the language of the solution provider). At this early stage, we must focus on describing the expectation in terms of the required capability, rather than the system that will respond to it.

- We can use a common format to write a good mission need statement. "[A customer] has a need to [do something]." This is a succinct statement of the operational need. We can then describe the rationale or business case. This form expresses a need to "have" or "do" something, not a need "for" something. This way, we focus on the need. If we state a need "for" something, the focus becomes the solution. For example, we could write the statement as "The US Forest Service needs to more effectively **detect and monitor** potentially dangerous wildfires." This is better (less solution-dependent) than writing it as "The US Forest Service has a need **for** a fire detection satellite system".

Some examples of solution-independent capability statements include:

- Individual households need to insulate themselves from intermittent power outages. The energy shortage is only going to get worse, so here is an opportunity to develop and sell a capability to supplement the energy from the public utility with an independent energy source.

- NASA scientists have a need to conduct *in-situ* measurements of Mars soil morphology.

- Every high-rise office building in the US needs to provide security to its workers. People who live and work in such buildings are afraid they won't be able to escape in the event of a terrorist attack or other disaster. Since 2001, many metropolitan workers have either: 1) changed jobs or careers to avoid working in high-rise office buildings, or 2) experienced anxiety and lost productivity after deciding to stay. Moreover, any building above seven stories makes firefighters' jobs very difficult. It is in the interest of both building and business owners to provide security to workers and visitors in their buildings.

With this in mind, let's review the initial customer expectations for the FireSAT project, as defined by the customers (US Forest Service (USFS), NASA, and the National Oceanic and Atmospheric Administration (NOAA)), and stated as goals with associated objectives. Goals are generally qualitative and descriptive. Objectives are specific and quantitative expansions on goals.

For a systems engineer, Table 2-2 is the beginning of understanding stakeholder expectations. We have to clarify, validate, and further detail the list in the table by eliciting the expectations from all the key stakeholders. We must also understand the constraints that the legacy environment and the regulations impose upon any solution. For example, what does "potentially dangerous wildfires" mean? Will our system just detect and monitor certain wildfires, or will it also be able to assess the potential "danger" associated with them? The difference will have a significant impact on the system we conceive. Furthermore, some of the objectives as stated may actually be compound stakeholder requirements, or multiple requirements included in a single stakeholder expectation. We have to resolve this matter as well.

TABLE 2-2.   **FireSAT Need, Goals, and Objectives.** This is the beginning of the process of understanding stakeholder expectations.

Mission Need: *The US Forest Service (USFS) needs a more effective means to detect and monitor potentially dangerous wildfires*	
Goals	**Objectives**
1. Provide timely detection and notification of potentially dangerous wildfires	1.1. Detect a potentially dangerous wildfire in less than 1 day (threshold), 12 hours (objective)
	1.2. Provide notification to USFS within 1 hour of detection (threshold), 30 minutes (objective)
2. Provide continuous monitoring of dangerous and potentially dangerous wildfires	2.1. Provide 24/7 monitoring of high priority dangerous and potentially dangerous wildfires
3. Reduce the economic impact of wildfires	
4. Reduce the risk to firefighting personal	4.1. Reduce the average size of fire at first contact by firefighters by 20% from 2006 average baseline
	4.2. Develop a wildfire notification system with greater than 90% user satisfaction rating
5. Collect statistical data on the outbreak, spread, speed, and duration of wildfires	
6. Detect and monitor wildfires in other countries	
7. Collect other forest management data	
8. Demonstrate to the public that positive action is underway to contain wildfires	

2.1.3   The Legacy Context

To better understand the required capability represented by these expectations, we have to understand how they fit into the current operational environment and the existing systems landscape. This perspective brings clarity to the required capability and its scope, exemplified through its logical and physical boundaries.

A completely unprecedented mission or operational capability depends less on legacy concepts and systems. But a new mission or operational capability usually involves one or more of the following:

- An upgrade to an existing system
- An upgrade to an existing business process or operational concept
- An interface to a legacy system
- An interface to an existing business process or operational concept

In these cases, the systems engineer must understand the legacy systems and concepts while defining the new mission, the new operational capability, and its contextual setting. This drives the definition of the boundary of the new system. Undocumented or implicit interfaces and interactions pose significant risk and can lead to unanticipated test and integration challenges. So we must understand and make explicit the as-is contextual setting, business processes, and operational environment. To fully understand the legacy system context, we should gather three types of information:

- The legacy processes, including the logical interfaces
- The legacy resources, including the physical interfaces
- Limitations within the legacy context

The *legacy processes* describe how the legacy system is used. They specify the business or mission flow of work as the users perform their duties. The processes may be defined with a graphical representation (e.g., use cases, interaction diagrams, etc.) or in text. Processes show how the business uses the legacy systems, and the logical interfaces between these systems.

The *legacy resources* describe the physical assets and resources used to perform the legacy processes and the interactions between these resources. Physical resources include hardware, software, and people, as well as the interfaces between these resources.

The *legacy context limitations* describe any defects, limitations, or potential risks within the legacy processes. This information allows the systems engineer to better understand the strengths and weaknesses of current processes and provide a better solution to meet the operational need.

The legacy context described above may not apply to a completely new mission or operational capability, but we still have to understand how the operational need fits within the mission or business processes. The systems engineer must understand the environment that the current need lives within to better understand the need and, subsequently, the more detailed operational requirements. We must also know who will interface with the system, to determine possible stakeholders.

Consider a military SATCOM with a need to increase geostationary communications bandwidth and coverage. A lot of capability already exists within the legacy context. Any new systems built to support the mission must fit within this context. The processes for spacecraft command and control have plenty of heritage so the new satellite system has to use the same operational processes (with minor differences to account for expanded capabilities). Similarly, we would face considerable pressure to leverage the existing military infrastructure that supports geostationary satellite communications command, control, and operations. Remote tracking and data relay sites, command and control centers, and operations centers represent massive investments and years of design and development and will only be modified or replaced reluctantly. Finally, compatibility with these existing processes and resources imposes a host of constraints on the new system. In many cases, we can't exploit the latest technologies or different concepts of operations because we have to work within the existing architecture. Few development projects are unconstrained by legacy architectures, and from a systems engineering perspective they enjoy greater freedom.

The FireSAT mission likewise has some legacy issues. The USFS has long-established procedures for disseminating wildfire notifications. Regional field offices throughout the US have extensive communication infrastructure to send out wildfire "911" calls and coordinate firefighting efforts. And since FireSAT is a joint effort between USFS, NASA, and NOAA, existing NOAA command and control stations in Fairbanks, AK and Wallops Island, VA are part of this system. The advantages or limitations of these processes and resources depend on their nature and whether development of the FireSAT system identifies any weaknesses in them that would limit exploitation of the space element. If so, the stakeholders must consider the advantages and disadvantages of any changes. Budgetary and schedule considerations often drive the trade studies.

Let's examine a likely FireSAT scenario. One customer expectation is that the satellite will notify users of a wildfire in less than 30 minutes. But if the messages must pass through existing NOAA ground sites, this may be impossible. A satellite must wait until one of the two sites is in range to send a message. We might ask the customer to consider having messages broadcast directly to regional field offices instead. But this conflicts with their expectation that any new systems comply with existing procedures.

Systems engineers have a number of tools to help synthesize and analyze the stakeholders' detailed expectations and requirements. *Voice of the Customer* (VOC), also called concept engineering, is a formal process to identify and define a more detailed set of stakeholder expectations in response to a defined mission, operational capability, or business opportunity. We can use it to either develop stakeholder requirements or to understand existing ones. The systems engineer interfaces with the customer and customer representatives to better understand their expectations and requirements, both stated and unstated, and to delineate the system acceptance criteria. Section 2.3 tells how to apply this technique.

2.2   Identify the Stakeholders

To translate the initial customer expectation, mission, or capability into a more complete set of qualitative and quantitative expectations and requirements, we must first know who the stakeholders are. A *stakeholder* is a person or organization that has an interest in, or can be affected by, the system or project. Since stakeholders have an interest in the need, the mission, or the operational capability, they help us translate the high-level mission statement into a more detailed set of requirements. We must understand who the sponsors are, and rank order their expectations and requirements appropriately.

There are two groups of stakeholders: active and passive. Particularly important are the sponsor, the acquirer, and the bill-paying customers (including venture capitalists). Sponsors can be active or passive depending on their role in the program.

NASA NPR 7123.1a characterizes the two sets of stakeholders as relevant stakeholders and other interested parties [NASA, 2007]. Relevant stakeholders are those designated in a plan for stakeholder involvement, or people that contribute to a specific task; other interested parties are those that may be affected by the resulting system.

2.2.1   Active Stakeholders

Active stakeholders are individuals and organizations that actively interact with the system of interest once it is operational and in use [SDOE, 2006]. They include system users, system maintainers, and logistics personnel, as well as other systems that will interact with the system of interest. These other systems (or the owners) provide operational requirements and constraints (e.g., interoperability, data formats, display formats) on our system. Active stakeholders are represented on a context diagram and should be included in the developed system concept of operations. They help define the logical and functional boundary of the system of interest. Examples of active stakeholders for the Space Shuttle include the astronauts, mission control personnel, and the International Space Station Program.

2.2.2   Passive Stakeholders

Passive stakeholders are other individuals and organizations that influence the success of the deployed system [SDOE, 2006]. They don't actively interface with the system of interest, but can provide many requirements and constraints to it. Passive stakeholders don't use the system, but they influence its definition by driving requirements into it. Passive stakeholders can also be influenced by the system's performance. They usually put some nonfunctional requirements on the system—constraints, regulations, etc. Examples of passive stakeholders are: local government authorities that issue building regulations (a building must meet them or it cannot be built), and organizations that enact air worthiness or certification requirements.

In the space world, the Federal Communications Commission (FCC) and the International Telecommunications Union (ITU) are passive stakeholders in that they impose constraints on available frequencies for radio frequency communications. Launch vehicle providers are another good example of passive stakeholders since they impose many requirements that shape the design and necessary test levels.

2.2.3   Sponsors

Sponsors control the program development and procurement funding or resources. They either buy the product or fund it with a specific mission in mind or a profitability objective. There may be more than one sponsor for a given need (Table 2-3). A system may be targeted for several industries or to multiple sponsoring stakeholders (e.g., Boeing's 787 has multiple stakeholders at many airlines, cargo air carriers, aircrew members, ground crew members, etc.).

As the holder of the purse strings, the sponsor is arguably the most important stakeholder. A customer may be either a passive or an active stakeholder. It depends on whether the customer interacts with the system when it is operational. An example of a sponsor who's an active stakeholder is a person who buys a computer for personal use. This person is a sponsor (purchases the product) and an active stakeholder (uses the product). An example of a passive customer is the Army Acquisition Office, which buys the soldiers' weapons. The Acquisition Office is the sponsor (buys the product), but does not use the product (the soldiers use it).

TABLE 2-3.   Stakeholders and Their Roles for the FireSAT Mission. Sponsors provide funding, and may be either active or passive.

Stakeholder	Type
Congress	Sponsor, Passive
US Forest Service	Active
National Oceanic and Atmospheric Administration	Active
National Aeronautics and Space Administration	Passive
Prime contractor	Passive
Taxpayers	Sponsor, Passive
People living near forests	Active
State governments	Passive

Identifying the stakeholders is crucial to synthesizing a complete set of stakeholder expectations. Given the iterative nature of systems engineering in problem definition and solution synthesis, the list of FireSAT stakeholders is likely to change during the lifecycle. For example, the question might arise whether the information gathered and analyzed by FireSAT should be made available to the public or researchers. If so, these groups would become stakeholders and any liability issues with this decision would then have to be considered.

2.3   Solicit and Synthesize Stakeholder Expectations and Requirements

After identifying the stakeholders, the systems engineering team must define their expectations and requirements. Stakeholder requirements are sometimes called "operational requirements", "customer requirements", "business requirements", or "mission requirements." These all mean the same thing—a set of statements defining what is required to fulfill the mission or business objective. *User requirements*, another common term, is a subset of the stakeholder requirements (those defined by the actual users of the system).

Stakeholders often have trouble articulating their requirements. They may know what they want or what's wrong with what they have, but expressing these ideas as succinct, actionable requirements isn't always easy. They sometimes have a specific solution in mind, and state the requirement in terms of the solution. Or they try to be vague in stating their requirements, so that, as their need evolves over time, their "implicit" requirements can change along with it. The systems engineering team must explicitly capture mission requirements in a way that allows system development to proceed.

2.3.1   Properties of Stakeholder Expectations and Requirements

We discuss characteristics of requirements in Chapter 4 in some detail, and highlight key points here. Stakeholder expectations and requirements:

- Should support the operational need, mission, or business objectives. The requirements are a detailed description from the stakeholders' perspective of what is required to meet the operational need.

- State **what** must be done in operational terms. Customers often express their expectations in terms of the existing solution or **how** to satisfy the need, not from the perspective of **what** is needed. It's the systems engineer's job to get to the real expectations and requirements.

- Are written using the language of the business or stakeholder

- Ought to be solution-independent

The mission requirements should state what the product should do, which current products can't do or don't do well enough. The requirement should state some capability that is missing or what current capability is deficient and the nature of this deficiency (it can't be done fast enough or cheap enough, it's not robust enough for its intended environment, or it's not available enough of the time). We should state the requirement in terms of the operational outcome rather than the system response. Operational requirements must address both the functional (capabilities) and nonfunctional (characteristics) needs of the stakeholders.

Mission requirements should apply for the entire lifecycle of the solution. Different stakeholders have requirements for different phases of the lifecycle. Users and other active stakeholders normally define requirements for the operational phase. Sponsors may define requirements for all phases (cost and schedule for the development phase, cost for the production phase, ease of upgrade for the operational phase, ease of disposal for the retirement phase, etc.). Passive stakeholders define constraining requirements for the development, manufacturing, and production phases. We must include all phases of the lifecycle when we capture stakeholder requirements. The word *capture* indicates that we must take control of and own the stakeholder requirements. No matter who writes the original requirements, the systems engineering team must own, manage, and control those requirements.

The system context (e.g., government vs. commercial, NASA or European Space Agency vs. DoD) determines how we document requirements. A sponsor may document operational requirements and provide them to the development team in the form of a request for proposal. A marketing team may help a customer develop stakeholder or market requirements and give them to the development team as a marketing or planning document, or an internal statement of work. The development team may develop a requirements document jointly with the stakeholders. In any case, we must take control of and understand the requirements. We must understand what each requirement means, the operational environment, and the application domain of the intended system.

As mentioned earlier, we have two types of mission requirements: capabilities (or functional needs) and characteristics (or nonfunctional needs) [SDOE, 2006]. We need both types to define a complete set.

Capabilities or functional requirements—Some stakeholder expectations and requirements involve the need for a *capability* or functionality or a system with a certain behavior. These are dynamic requirements, and we often categorize them as functional needs or mission capabilities. They define the functionality, the services, the tasks, and the activities that the stakeholders need. Because active stakeholders use or interact with the system when it is operational, they usually define the functional requirements. Questions that can elicit functional needs or capabilities from stakeholders include, "What would you like to be able to do that you can't do now?" or "What else do you wish you could do?" Some examples of mission capabilities can be derived from Table 2-2 as follows:

- Users need persistent coverage of remote areas
- Users need the system to detect wildfires
- Users need the system to notify the US Forest Service if it detects a wildfire

Characteristics or nonfunctional needs—While some mission requirements suggest a need for capabilities and solutions, others suggest the need for certain solution attributes or impose certain constraints—organizational, technological, statutory or regulatory, standards or protocols, and so on. *Nonfunctional needs* or *characteristics* are quality attributes expected or imposed by the stakeholders, including cost and schedule constraints. We must capture a comprehensive set of such requirements. Inadequate or incomplete nonfunctional requirements may put the entire program in jeopardy. We must take a very broad view when defining them, including the acquisition or development strategy, and the organizational business model. The physical design of the solution must address the complete set of nonfunctional requirements.

Both passive and active stakeholders are sources for nonfunctional requirements. Sponsors usually provide cost and schedule requirements. Normally, each functional requirement entails at least one nonfunctional requirement, so after a stakeholder has defined a functional requirement, we should look for the associated nonfunctional ones. These tend to state how well the function must perform. Questions such as the following can elicit nonfunctional requirements: How fast? How many? How big? How accurate? or When? Some examples of stakeholder expectations that are system characteristics include:

- If a wildfire is detected, the US Forest Service should be notified within an hour of detection. If possible, the notification time should be reduced to 30 minutes.
- Reduce the average size of a wildfire first addressed by firefighters by an average of 20%
- Once detected and classified as a wildfire, it should be monitored 24/7

2.3.2   The Voice of the Customer

The *Voice of the Customer* allows us to translate the high-level operational need or mission into a more complete and detailed set of expectations and requirements. There are many techniques to capture mission requirements. These include direct discussion or interviews, surveys, analysis, focus groups, customer feedback, observation, rapid prototyping, warranty data, and field reports. Indirect techniques tend to be easier and cheaper, but individual or group interviews tend to provide the most complete set of requirements. Table 2-4 summarizes some techniques for obtaining and rank ordering operational requirements.

Individual or group interviews usually allow us to capture detailed operational and mission requirements more easily. Focus groups are a common way to conduct a group interview with stakeholders. A *focus group* brings together a group of stakeholders to enable us to elicit and develop requirements. It allows stakeholders to discuss their needs and ideas, and even resolve conflicting expectations and priorities. Group interviews have several advantages over individual interviews. They are often more time and cost effective, since we can obtain the requirements for several stakeholders in parallel. Also, group interaction may lead to deeper discussion as one stakeholder enhances the opinions of others. Group interviews enable the group to drive the focus of the discussion. Individual interviews have advantages as well. Stakeholders can more freely state their requirements without fear of being overridden by groupthink. Group interviews tend to drive toward consensus instead of individual requirements or ideas.

Individual interviews allow the interviewer to more easily keep the discussion on topic. No matter which technique we use to capture the voice of the customer, we should adhere to the following guidelines:

- Use the customer's actual words, preferably collected during a face-to-face visit
- Employ structured processes and tools to help customers identify and express their true requirements
- Understand the mission requirements from multiple perspectives— operational, installation, maintenance and support, acquisition
- Keep the requirements solution-independent to allow innovative concepts and solutions

TABLE 2-4. Information Gathering Tools and Techniques. These tools and techniques are invaluable in obtaining information from active and passive stakeholders and sponsors.

Technique	Description or Purpose
Structured interview	Iterative working sessions using a protocol to guide verbal information gathering (individual or group)
Cost / benefit analysis	Helps prioritize business requirements by estimating cost of delivery against anticipated business benefits. Helps refine scope and architecture.
SWOT (strengths, weaknesses, opportunities, and threats) analysis	Identifies internal and external forces that drive an organization's competitive position in the market. Helps rank order system development and commercialization phases: time to market (market business opportunity) and probability of success.
Brainstorming or white board session	Method to develop creative solutions for a problem. We can also use it to define a set of problems.
Field data and analysis	Analyzes existing and analogous systems and missions to understand pros and cons. Analyzes field data, when available, to deepen insights into the strengths and weaknesses of existing and analogous solutions.
Surveys	Questions distributed to all stakeholders. Must be qualitative. Usually takes a lot of work to create unambiguous questions and to gather true requirements.
Customer feedback or comment cards	Provides information on current systems—potential improvements or problem areas.
Other techniques	Impact analysis or root cause analysis or rapid prototyping

During interviews, stakeholders provide their requirements, and also often paint verbal "pictures" or images of their current or desired environment. Both types of input are important. Mission requirements define capabilities not available today. They help us understand the "why" of the desired features or functions, and this in turn makes it easier to synthesize differentiated solutions. Table 2-5 shows some examples.

Stakeholders invoke images to describe the current or desired environment and operations. An image is a word picture. It's a verbal, impressionistic description of the customers' environment that tells what life is like for them or what motivates them. An image doesn't express the requirement; it supplies supporting information or background. Table 2-5 provides some examples.

Images and requirements are often expressed together, after the interviewee is asked a probing question. Images imply system capabilities and characteristics, but do not specify them directly. They hint at how to interpret the requirements, or what may be required in the sales process. Table 2-6 lists differences between requirement voices and images. Let's look at some requirement voice examples from FireSAT:

"We need to reduce the loss of life and property caused by forest fires each year"

"We need to be able to detect small fires before they get out of hand"

"Our rangers need daily updates on fire status so we know how to allocate our firefighting resources"

TABLE 2-5. **Examples of Requirements and Images.** Here are some classic examples of requirement statements and images that can help us understand the current or desired situation. (ISDN = Integrated Services Digital Network.)

Stakeholder Requirement	Stakeholder Environment—Images
"I need to have the same communication system at home as I have at work"	"I've got phone wires in every nook and cranny"
"I'd like to be able to go to the manual and find out how to add an ISDN line"	"A typical day for me is 24 hours long"
"I wish I could separate Caribbean, domestic, and Latin American calls"	"Sometimes, our receptionist, you hear her screaming, 'Oh my God!'"
	"I have to wear ear plugs in order to concentrate in this center"
	"My customer has a mobile phone that works anywhere in the world. It also sends and receives email."

TABLE 2-6. **Differences Between Requirements and Images.** Requirements are known, explicit needs. Images are unknown, latent needs.

Requirement Voices	Images
Describe a specific need or demand that will solve a current problem	Describe the context and condition of a solution's use
Usually give statements relating to a solution, i.e., features, functionality, benefits, or attributes	Describe customer's environment
Are usually not emotional	Are emotional and vivid; draw on customer's frustrations, anxieties, likes, and dislikes
	Conjure up a picture

FireSAT image voices might say:

> "Its frustrating to find out that a small fire got out of control and burned someone's house down because we had our firefighters in the wrong place"

> "The forest service should be out in front of the forest fire problem, focusing on containing the small fires while we can"

> "My home was destroyed by a forest fire; why isn't the USFS doing something?"

> "Forest fires are critical to the health of the forest, but too much of a good thing can be dangerous to people"

Both types of information help us better understand the true stakeholder expectations.

2.3.3 Stakeholder Interviews

A personal interview with a stakeholder allows us to gather information more easily. We should plan and design the interview process with this in mind. The steps include:

1. Prepare an interview guide
2. Conduct the interview
3. Debrief after the interview (and possibly do follow-up interviews)
4. Document the requirements

Below we describe each step in detail. Although the process allows for tailoring to the situation, following these steps increases the probability of a successful interview.

1. Prepare an Interview Guide—The guide should prepare the interviewers for the stakeholder interviews. It should define the interviewers' roles, list the people to be interviewed, and give the schedule and location of each interview. The guide should also provide a set of expectations for each interviewee—how long the interview will take and its purpose (to gather operational requirements), and what is expected of the interviewer (preparation or documentation). It may also supply a list of questions to guide the interview and ensure that we obtain the necessary information. Although the interview guide provides the roadmap of the interview process, the process should be flexible enough to accommodate unexpected glitches. We may need different guides that reflect questions appropriate to specific users. The interview guide should include such detailed topics as:

- Know whom to call later if an issue comes up
- Provide an opening statement describing the purpose, ground rules (e.g., audio taping), and outcomes; do introductions

- Start with a question about the interviewee's role with regard to the need
- Maintain a proper balance between specificity and generality (we can think of the guide as a discussion outline with prompts and reminders)
- Define the topic areas (6–9 is ideal) to be covered, summarized by overarching themes
- Document the interview in outline form with topics and low-level questions; subjects include past breakdowns and weaknesses, current needs, future enhancements
- Include a few probes (2–4 per topic area); probes should flow from conversation with the stakeholder; variations on "why" phrases
- Avoid closed-ended questions ("will," "could," "is," "do," "can" or other question leads that can be answered with a simple yes or no)
- Have questions in the guide elicit both images and requirements, not potential solutions
- Provide a closing statement and a plan for future contact or follow-up

2. Conduct the Interview—We must remember the goal of the interview: to elicit mission requirements. Interviewers should not use the interview to sell or develop solutions, and should prevent the interviewees from describing solutions. At this point, we want to define the problem; then we can design solutions. We have to cover the topics and discussion points defined in the interview guide, but also allow for open-ended discussion. The interview should include at least two people besides the interviewees—the interviewer (moderator) and the note taker. Role of the moderator:

- Manage and conduct the interview
- Lead the interview (if there is more than one interviewer)
- Build rapport with the customer
- Execute the interview guide
- Take very few notes
- Write observations
- Probe the interviewees for their true requirements

Role of the note taker:

- Take **verbatim** notes. Do not filter the gathered data. Get vivid, specific, and rich nuggets of information
- Rarely interrupt
- Capture the images within the notes (the emotions and verbal pictures)
- Write observations

The moderator needs to probe to gather the true requirements. Failing to do so is the principal shortcoming of interviewers. As Figure 2-3 shows, a primary factor for success in Voice of the Customer is having the persistence to ask "Why?" until the customer's true need is identified.

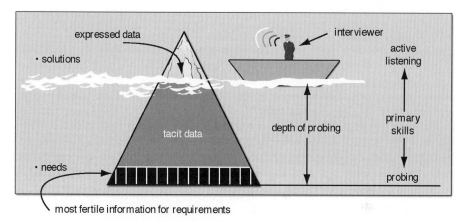

FIGURE 2-3.   Probing for Requirements. It's often difficult to determine a stakeholder's true needs—we must persist and probe until we're sure we've identified the needs. Repeatedly asking "Why?" increases the depth of probing to get past expressed and tacit data to the underlying need.

Unless we probe for answers, stakeholders will talk in terms of features they think they need (e.g., "I need a database to store our customer information"). The interviewer needs to find out why that feature is important and to determine what the stakeholder need truly is (e.g., "I need to be able to access our customer information more quickly"). While probing for requirements:

- The interviewer's favorite words and phrases should be:
    - Why?
    - Could you explain...?
    - Could you share...?
    - Describe
    - Give me an example
    - What else...?
- The interviewer should not be afraid to go for the emotional response. The emotional response will start the interviewee's discussion on the problem (the image) and often leads to the requirement voice. What did you hate most about X? What was the worst, best, most loved, etc.
- The interviewer should not allow the interviewee to stop with vague responses. We must probe to gather the real nuggets of information that describe a requirement. "Easy", "fast", and "good" aren't good enough.

- The interviewer should ask relevant and eye-opening questions, even if they're not in the interview guide. They may point to new insight or to hidden customer requirements (e.g., "Why are all those phones lined up in the center of the room?")

Another probing technique used in problem solving and root cause analysis is called the "Five Whys" technique. The notion is that if we ask "why" five times we will get to the root cause or the real requirement.

It's best to interview at the operational site, if possible. By interviewing where the system or product will be used, we get a chance to observe the operational environment. This allows us to gain more background information on the use of the system or product, which can further clarify the stakeholders' requirements. If the interview can't take place at the operational site, the team should try to schedule a formal observation period (plant tour or a tour of the user's work place). All interviewers should record their observations of the operational site.

3.   Debrief after the Interview (and Possibly Do Follow-up Interviews)—After the interview, the interview team should debrief as soon as possible, while the information is still fresh in their minds. They should review the notes (or tapes, if the interview was recorded), fill in any gaps, and emphasize the important items. They should determine if any information is unclear or incomplete, and schedule any needed follow-up interviews. Also, the interview guide and process can be reviewed and updated if necessary.

4.   Document the Requirements—Finally, we must summarize the interview notes, including requirements and images, and the interview observations, capturing them as mission requirements. To do this, we start with the stakeholders' requirement voices, since these state the actual need. We then supplement the requirement voices with the image voices and observations noted during the interviews. We should focus on words that describe the stakeholders' business or mission (like "customer account", "threat", "unfriendly target", "preliminary order", etc.). This helps ensure that the operational requirements are defined using the customer's language.

We have to consolidate the individual stakeholders' requirements, obtain missing information, and resolve conflicts in documenting the set of operational requirements. Similar requirements from different stakeholders may be combined into one. The systems engineer must ensure consistency between requirements. If two stakeholder requirements are inconsistent, we need a mediator among the stakeholders to resolve the issue. As the one who controls the money, the customer (or a designee) usually has the final say. We must work with the arbiter to determine the true requirements set. The result should be a set of operational requirements that completely defines the operational need.

2.4   Articulate Stakeholder Expectations and Requirements

Mission requirements should express the stakeholders' needs in a simple, concise, verifiable, and understandable format. We should state them in terms of operational and mission outcomes, rather than implementation and solution concepts.

Mission requirements should be in a specific format that the stakeholders agree on. One good format is "[Subject] requires [Functionality] [Modifier]." The subject is "the business", "the user", or "the mission", not "the system." By focusing away from the system, the operational requirements are more likely to be solution-independent. We could use "shall" or "needs" in place of "requires", but it's good to use just one of these for all requirements. "Functionality" expresses what the stakeholder needs. And finally, "Modifier" quantifies the desired performance or measurable result. For example, "The user requires notification of a potentially dangerous wildfire within 30 minutes of detection."

Many systems engineers and customers prefer the following format for operational requirements: "[Subject] shall [Active Verb] [Modifier]". Here the subject is "the system". An example of this format is the following: "The system shall provide notification of a potentially dangerous wildfire within 30 minutes of detection."

2.5   Rank Order Mission Requirements and Identify Critical Acceptance Criteria

As we consolidate stakeholder requirements, we should also determine their priority. A requirement's priority should reflect the stakeholders' perspective. Rank ordering supports and further clarifies the mission goals and objectives and the associated business case. From here we develop the priorities of the technical requirements and supporting trade studies. This ranking ensures that the highest value requirements are satisfied and weighted appropriately in the inevitable trade offs and conflict resolutions. Other benefits include:

- The stakeholders and development team better understand the need. The systems engineer and the development team receive insight into the customer value system. Thus, the development team can create a technical, cost, and schedule solution that provides the best value to the stakeholders.

- It helps resolve requirement conflicts. It's easier to resolve conflicts—even subtle differences between similar requirements—when we know their priorities.

- Understanding requirement priorities makes the change management process easier. Stakeholders often change or add to their requirements during development.

2.5.1   Priorities of Mission Requirements

We can rank order mission requirements relative to one another or relative to their value to the mission. In the first case, we rank each requirement from 1 to X, where X is the number of requirements. This technique gives the most information about the importance of each requirement to the stakeholder, but it's not always easy to be so precise. The second method is more common. It uses a more generic scale, usually with 3, 5, or 10 values, to indicate how important the requirement is to the success of the solution. IEEE STD 830 scales the priorities as follows [IEEE, 1998]:

- Essential—The system will not be acceptable unless these requirements are met
- Conditional—These requirements would enhance the system or product, but their absence would not make the product unacceptable
- Optional—These requirements may or may not be worthwhile

Let's look at a partial set of mission requirements for the FireSAT project and their priority:

- FireSAT shall detect wildfires greater than 150 m in any dimension with a confidence level of 95% (essential)
- FireSAT shall cover the entire United States, including Alaska and Hawaii (essential)
- FireSAT shall revisit the coverage area at least once every 12-hour period (essential)
- FireSAT shall send notification data on a wildfire to the USFS within 60 minutes of detection (essential)
- FireSAT shall provide geo-location information on detected wildfires to within 5000 meters (2 sigma) (essential)
- FireSAT shall have a 98% total availability, excluding weather, with a maximum 3 day outage (conditional)
- Average user satisfaction with the FireSAT notification system shall be greater than 90% (conditional)
- FireSAT shall operate through existing NOAA ground stations (optional)

The above requirements contain a source of confusion common on space missions, that is, using the same name for the project and the spacecraft. Here, "FireSAT" refers to all the elements that make up the project's mission architecture: spacecraft, launch systems, and ground systems. Derived requirements should clearly specify the element, e.g., "The FireSAT spacecraft shall..."

2.5.2   Acceptance Criteria

Once we've listed mission requirements in order of priority, we establish acceptance criteria. Each requirement should identify the criteria by which the stakeholders will agree that the solution meets the requirement or its sign-off condition. Acceptance criteria constitute an agreement between the stakeholders and the development team—it says, "If the solution meets these criteria, it meets the stakeholder requirement." Effective acceptance criteria reduce the risk to both the development team and the customer, since they point to the core capabilities and characteristics needed to satisfy the mission. Acceptance criteria afford several benefits to the systems engineer and the development team. They:

- Provide the basis for the acceptance or sign-off of the mission requirements. During solution validation and acceptance, the solution is deemed acceptable for a requirement if it meets the acceptance criteria associated with that requirement.

- Clarify or add detail to "fuzzy", vague, or poorly written mission requirements, and help the development team and the customer agree on their meaning. If a mission requirement contains words like "easy to use", "world class", "fast", "inexpensive" or other vague expressions, the acceptance criteria help clarify the meaning. For example, the business may want a "world class web site". What does this mean? The customer's idea of "world class" (much better than others) may not be that of the development team (our usual product). Or if the functionality has to be "easy", what does that mean? The acceptance criteria can clarify "easy" by saying that the functionality will be provided "in fewer than 3 steps" or "in less than 5 seconds." By specifying the acceptance criteria, the stakeholder and development team agree on the meaning of the vague requirement.

- Provide a basis for validation and user acceptance test plans. Acceptance criteria provide all stakeholders (including the test team) insight into the necessary number and complexity of systems tests.

The stakeholder requirements, their rationale, and related acceptance criteria can be captured in the form of a matrix, as shown in Table 2-7. This is subject to change given the iterative nature of the early systems engineering phases. The synthesis of a concept of operations is likely to highlight other stakeholder requirements, validate already identified stakeholder requirements, and result in the elimination of some of them. The table depicts a concise way of articulating the stakeholder expectations so that the engineering effort can be aligned and focused accordingly. Such a table should include the following:

- A stakeholder requirement number—a unique number for each stakeholder expectation or requirement

- The requirement statement—sometimes a table such as this may only include a summary description of the requirement along with a pointer to a different document or database

- The acceptance criteria for each stakeholder requirement

We can later expand this table into a verification matrix as we develop the derived system technical requirements from the stakeholder expectations, and it becomes a critical part of system verification and validation. This is discussed in more detail in Chapter 11.

TABLE 2-7.   **Stakeholder Requirements, Acceptance Criterion, and Rationale.** We can expand this table into a verification matrix. The matrix affords a compact summary of requirements and their associated acceptance criteria.

Requirement	Acceptance Criterion	Rationale
1. Detection	The FireSAT system shall detect potentially dangerous wildfires (defined to be greater than 150 m in any linear dimension) with a confidence interval of 95%	The USFS has determined that a 95% confidence interval is sufficient for the scope of FireSAT
2. Coverage	The FireSAT system shall cover the entire United States, including Alaska and Hawaii	As a US Government funded program, coverage of all 50 states is a political necessity
3. Persistence	The FireSAT system shall monitor the coverage area for potentially dangerous wildfires at least once per 12-hour period	The USFS has determined that this revisit frequency is sufficient to meet mission objectives for the available budget
4. Timeliness	The FireSAT system shall send fire notifications to users within 30 minutes of fire detection (objective), 1 hour (threshold)	The USFS has determined that a 1-hour to 30-minute notification time is sufficient to meet mission objectives for the available budget
5. Geo-location	The FireSAT system shall geo-locate potentially dangerous wildfires to within 5 km (objective), 500 m (threshold)	The USFS has determined that a 500-m to 5-km geo-location accuracy on detected wildfires will support the goal of reducing firefighting costs
6. Reliability	FireSAT space elements shall be single-fault tolerant ("fail-ops") for critical mission functions consistent with a Class B NASA mission as per NASA NPR 8705.4	FireSAT is considered class B based on its priority, national significance, cost, complexity, and lifetime
7. Design Life	The FireSAT system shall have an operational on-orbit lifetime of 5 years. The system should have an operational on-orbit lifetime of 7 years.	The USFS has determined that a minimum 5-year design life is technically feasible. 7 years is a design objective.
8. Initial/Full Operational Capability	The FireSAT system initial operational capability (IOC) shall be within 3 years of Authority to Proceed (ATP) with full operational capability within 5 years of ATP	The on-going cost of fighting wildfires demands a capability as soon as possible. A 3-year IOC is reasonable given the scope of the FireSAT system compared to other spacecraft of similar complexity.

TABLE 2-7.   **Stakeholder Requirements, Acceptance Criterion, and Rationale. (Continued)** We can expand this table into a verification matrix. The matrix affords a compact summary of requirements and their associated acceptance criteria.

Requirement	Acceptance Criterion	Rationale
9. End-of-Life Disposal	FireSAT space elements shall have sufficient end-of-life Delta-V margin to de-orbit to a mean altitude of <200 km (for low-Earth orbit missions) or >450 km above geostationary belt (for GEO missions)	End-of-life disposal of satellites is required by NASA policy.
10. Ground System Interface*	The FireSAT system shall use existing NOAA ground stations at Wallops Island, Virginia and Fairbanks, Alaska for all mission command and control. Detailed technical interface is defined in NOAA GS-ISD-XYX.	The NOAA ground stations represent a considerable investment in infrastructure. The FireSAT project must be able to leverage these existing assets, saving time, money, and effort.
11. Budget	The FireSAT system total mission lifecycle cost, including 5 years of on-orbit operations, shall not exceed $200M (in FY 2007 dollars)	This is the budget constraint levied on the project based on projected funding availability

* Item 10 is a constraint, not an acceptance criterion.

2.6   Synthesize Essential Mission Requirements

We identify selected essential mission requirements as key performance parameters (KPPs), key acceptance criteria, or measures of effectiveness (MOEs). Organizations typically use one or more of these terms for their essential mission requirements. This varied and inconsistent terminology sometimes poses problems in communication among organizations. There are normally three to seven of them, representing the "sacred" stakeholder expectations. For example, the key acceptance criteria for the Apollo Program were: a) Put a man on the moon; b) Return him safely to earth; c) By the end of the decade. Apollo had many important stakeholders and mission requirements, but these three "sacred" mission requirements provided the necessary focus, and drove the inevitable trade studies and tradeoffs.

All key stakeholders must agree to these KPPs or MOEs early in the life of a project, because these are the select few, critical, and non-negotiable criteria that the solution must satisfy to be acceptable. They represent the absolutely critical subset of measurable and observable capabilities and characteristics that the solution must meet. Developing KPPs, MOEs, or key acceptance criteria is a joint effort between the systems engineer and the key stakeholders. Since they inherently imply user acceptance, we must have a very good understanding of the business or mission to effectively synthesize them. This synthesis is difficult, but allows us to distinguish a need from a want, and focus the development effort and resources accordingly. The Department of Defense [OSD (AT&L), 2006] defines

KPPs in the initial capabilities document and validates them in the capabilities description document. Defining KPPs often takes the collaboration of multiple stakeholders, but it's critical to providing focus and emphasis on complex multi-year, multi-agency, multi-center (within NASA) development programs.

TABLE 2-8.   **Key Performance Parameters (KPPs) for the FireSAT Project.** These are the sacred requirements for the FireSAT Project from the stakeholder's perspective.

No.	KPP
1	Detection—The FireSAT system shall detect potentially dangerous wildfires (defined to be greater than 150 m in any linear dimension) with a confidence of 95%
2	Coverage—The FireSAT system shall cover the entire United States, including Alaska and Hawaii
3	Persistence—The FireSAT system shall monitor the coverage area at least once per 12-hour period
4	Timeliness—The FireSAT system shall send fire notifications to users within 30 minutes of fire detection (objective), 1 hour (threshold)
5	Geo-location Accuracy—The FireSAT system shall geo-locate potentially dangerous wildfires to within 500 m (objective), 5 km (threshold)

2.7   Validate and Baseline Requirements. Translate Into Technical Requirements.

Mission requirements are validated to ensure "correctness". One way to do this is with a concept of operations. The concept of operations reflects the "as-is" environment, and helps visualize the desired environment and the mission, together with the system concepts being proposed and assessed. It lets us visualize the operational environment as a function of the required operational scenarios, the context of the system of interest, and the proposed system concepts. This is the focus of Chapter 3. Chapters 4 and 11 discuss requirements verification and validation further.

We must also translate the stakeholder expectations and requirements into more specific, measurable, and observable technical requirements. We first select an implementation concept, as Chapter 3 discusses. After selecting the appropriate implementation concept and the operation architecture, we translate mission requirements into technical requirements, the focus of Chapter 4. We have to maintain traceability between the mission requirements and the technical requirements to ensure they're complete and aligned with the engineering effort and stakeholder expectations.

Some of the concepts discussed in this chapter are in a continual state of flux. On any complex development program, the set of stakeholders is evolving, and new ones are likely to emerge over the life of a program. Stakeholder priorities are also likely to evolve with the environment (changing political, social, national security, and environmental priorities). Selected concepts, such as the key acceptance criteria, enable us more easily to manage these evolving priorities and realities. When discussing systems engineering, given its recursive nature, it's important to clarify the system of interest. This ensures that we address an appropriate and consistent level of abstraction among the various stakeholders and the systems engineering team. The output from the activities discussed in this chapter includes the following:

- A clear articulation of the stakeholders' need, goals, and objectives
- A clear articulation of the critical stakeholder requirements or mission requirements, including a sense for the stakeholders' priorities
- A list of the key performance parameters (KPPs), the measures of effectiveness (MOEs), or the key acceptance criteria
- An articulation of the stakeholders' risk tolerance with regard to performance, safety, cost, and schedule

References

Institute of Electrical and Electronic Engineers (IEEE). 1998. Std 830–1998 — IEEE Recommended Practice for Software Requirements Specifications.

National Aeronautics and Space Administration (NASA). March 2007. NPR 7123.1a — "NASA Systems Engineering Processes and Requirements."

Office of the Under Secretary of Defense for Acquisition Technology, and Logistics (OSD (AT&L)). July 24, 2006. *Defense Acquisition Guidebook, Version 1.6.*

Stevens Institute of Technology (SDOE). July 2006. Course Notes: "SDOE 625— Fundamentals of Systems Engineering, System Design and Operational Effectiveness Program."

Chapter 3

Concept of Operations and System Operational Architecture

Steven H. Dam, *Systems and Proposal Engineering Company*
Dinesh Verma, Ph.D., *Stevens Institute of Technology*

One of the weakest links in applying systems engineering concepts and principles is translating stakeholder expectations into technical system requirements. We improve our translations by robustly articulating the system of interest within its reference context from an operational or user perspective. This context becomes clear as we define the system's physical and logical boundaries, which in turn helps us establish operational scenarios and capabilities for its operators and users. Scenarios, capabilities, and associated performance attributes are vital to the concept of operations and operational architecture.

We develop a concept of operations iteratively while establishing the stakeholder expectations, technical requirements, and system architecture, as shown in Figure 3-1. This concept reflects the "as-is" and the "to-be" (inclusive of the conceived solution) mission environments. A good concept of operations

65

verbally and graphically reflects stakeholders' expectations, so it becomes a platform for validating the system's architecture and technical requirements. According to ANSI G-043-1992 [ANSI/AIAA, 1993], its purpose is to

- Describe the system's operational characteristics
- Help users, customers, implementers, architects, testers, and managers understand system goals
- Form a basis for long-range operations planning
- Guide how system definition documents, such as system and interface specifications, develop
- Describe how the user's organization and mission relate to the system

The concept of operations also helps us create, analyze, and evaluate competing concepts for applying the system and its technologies. The system's operational architecture consists of its major elements, their positioning within the reference universe, and tracings of key operational scenarios between the system, its elements, and the context.

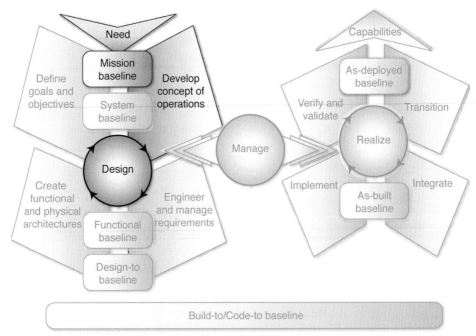

FIGURE 3-1.   The Concept of Operations. Developing a concept of operations is a major step in translating stakeholder expectations and requirements into system and technical requirements.

This chapter describes how to develop a concept of operations (Table 3-1). We start by defining the mission's scope and the system's boundaries. The next steps prepare us to define the system's technical requirements, as well as its logical and physical architectures.

TABLE 3-1. **Framework for Developing the Concept of Operations.** This table outlines a process for developing the mission's concept of operations and the system's operational architecture.

Step	Description	Where Discussed
1	Validate the mission scope and the system boundary	Section 3.1
2	Describe the system's operational environment, primary constraints, and drivers	Section 3.2
3	Develop operational scenarios and timelines	Section 3.3
4	Synthesize, analyze, and assess key implementing concepts for the system and its elements	Section 3.4
5	Document the concept of operations and the system's operational architecture; apply architectural frameworks	Section 3.5
6	Validate and baseline the system's operational architecture	Section 3.6

3.1   Validate the Mission Scope and the System Boundary

We start developing a concept of operations by engaging the relevant stakeholders to validate the scope and the conceived system boundary. Chapter 2 concludes having

- Defined the mission and associated objectives
- Consolidated stakeholders' expectations and requirements (capabilities and characteristics)
- Set the acceptance criteria
- Isolated key performance parameters (KPPs) or measures of effectiveness (MOEs)

These items are important to understanding the mission's scope. Clarity is critical, and the concept of operations helps validate the mission scope, the KPPs, and other stakeholder expectations. Defining the active stakeholders for the system helps us develop a system context diagram to understand the mission's scope and the system's boundary. Our definition of scope must be explicit because ambiguity leads to fluctuating requirements, miscommunication among stakeholder communities, and wasted engineering effort.

We identify active stakeholders to more quickly understand the mission scope and system boundary. For example, let's revisit the FireSAT system's need statement from Chapter 2: The US Forest Service needs a more effective means to detect and monitor potentially dangerous wildfires. This mission has the goals and objectives shown in Table 3-2.

TABLE 3-2.   **FireSAT Need, Goals, and Objectives.** This is the beginning of the process of understanding stakeholder expectations.

Mission Need: *The US Forest Service needs a more effective means to detect and monitor potentially dangerous wildfires*	
Goals	**Objectives**
1. Provide timely detection and notification of potentially dangerous wildfires	1.1. Detect a potentially dangerous wildfire in less than 1 day (threshold), 12 hours (objective)
	1.2. Provide notification to USFS within 1 hour of detection (threshold), 30 minutes (objective)
2. Provide continuous monitoring of dangerous and potentially dangerous wildfires	2.1. Provide 24/7 monitoring of high priority dangerous and potentially dangerous wildfires
3. Reduce the economic impact of wildfires	
4. Reduce the risk to firefighting personal	4.1. Reduce the average size of fire at first contact by firefighters by 20% from 2006 average baseline
	4.2. Develop a wildfire notification system with greater than 90% user satisfaction rating
5. Collect statistical data on the outbreak, spread, speed, and duration of wildfires	
6. Detect and monitor wildfires in other countries	
7. Collect other forest management data	
8. Demonstrate to the public that positive action is underway to contain wildfires	

Ideally, our preconceived notion of how to apply the system doesn't constrain the concept of operations. Instead, we start by describing the current operational environment: inability to detect fires quickly and accurately, to forecast a fire's spread tendencies, and so on. Then we describe the envisioned system to transition the current environment, with its drawbacks, to the envisioned environment, with its benefits.

If we understand FireSAT's scope and goals, as well as the active stakeholders, we can develop a simple context diagram for the system of interest and show how to apply systems engineering principles. Because this book emphasizes space-based systems, we take some shortcuts and make assumptions to arrive at an implementing concept that includes a space-based element. But a different concept is imaginable, such as more unpiloted airborne vehicles or fire-lookout towers.

Given the intent of the concept of operations, we begin with a simple black-box view of the system and likely active stakeholders that represent the system's context, as reflected in Figure 3-2. For complex systems, the development teams, users, and operators often have an implementation concept and system elements in mind, based on legacy systems and experience with similar systems. If a reference operational architecture is available, and the community accepts the system elements, we reflect this system-level perspective (Figure 3-3). Two other perspectives on the initial context diagram for FireSAT are in Figures 3-3 and 3-4. Diverse views are common and necessary to discussions with stakeholders and within the development team. After several iterations, we converge on the final context diagram, mission scope, and system boundary—based on trade studies of the implementing concepts, scenarios, and so on.

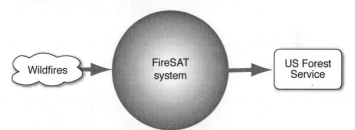

FIGURE 3-2.   **A Simple Context Diagram for the FireSAT System.** A context diagram reflects the boundary of the system of interest and the active stakeholders.

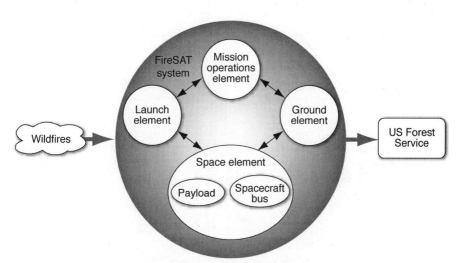

FIGURE 3-3.   **Context Diagram for the FireSAT System, Including Likely Reference System Elements.** If we have a reference architecture for the system of interest, the context diagram can reflect it.

Preconceived notions about implementation concepts have obvious drawbacks. They can impede innovative thinking and technology applications, inventive packaging ideas and architectures, and even ways of creating new system capabilities.

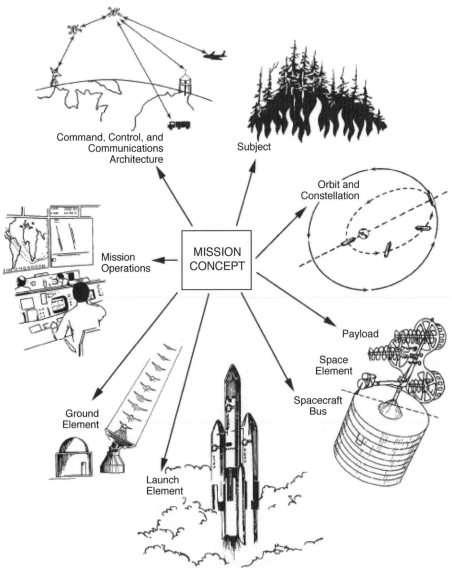

FIGURE 3-4.   **Another Perspective on the FireSAT System's Mission Scope and Boundary.**
During a complex development's early stages, people have different views of the mission scope and system boundary. (Adapted from Wertz and Larson [1999].)

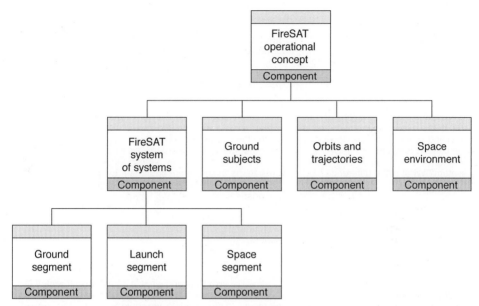

FIGURE 3-5.   Yet Another Perspective on the FireSAT System's Mission Scope and Boundary.
We use many graphical tools to depict the mission scope and system boundary.

Figures 3-2, 3-4, and 3-5 reflect typically diverse views that lead to the kinds of questions summarized in Table 3-3.

As discussions and decisions develop from these types of questions, stakeholders deepen their understanding of mission and system boundaries. This is iterative, and we may have to proceed to the next step before we answer some of these questions.

3.2   Describe the System's Operational Environment, Primary Constraints, and Drivers

A concept of operations must describe the "as-is" and "to-be" (intended) operational environment. A crucial aspect of this description is the current abilities available to the stakeholders for the primary mission: detecting forest fires, tracking them, analyzing their spread and trends, and so on. It should also include capabilities that current and legacy systems don't provide. Furthermore, it should isolate aspects of the current environment that constrain the conceived system and drive which implementing concepts we select. Thus, we consider the operational and support environment from a broad perspective in six main areas:

- The current operational doctrine, including relevant policies, regulations, and operational procedures

TABLE 3-3.   Questions that Stimulate Discussion. These questions raise key issues, which lead to decisions that move the program forward.

Questions	Implications of Decisions
Should we show the wildfires as being within the system boundary?	It's important to separate FireSAT's sensing function from the subject. We apply it using a number of technologies based in space, in the air, or on the ground. So the wildfires remain outside the system boundary.
What about the launch element? Is this truly a system element or just an enabling system? Are we developing the context diagram for FireSAT's program and mission or the FireSAT system? What is the difference?	This is a critical decision. If we're concerned about the entire FireSAT mission, we should include systems such as the launch element as parts of the overall system. On the other hand, if we're interested mainly in the FireSAT capability, we should view launch as an enabling system outside the main system boundary. Either way, we must clearly define the system's physical, logical, and temporal boundaries.
Are the mission-operations element and the ground element not part of the command, control, and communication architecture? Can we articulate it better?	We need to show clearly whether the mission operations and ground elements are part of the system's command, control, and communication element.
Do orbits and constellation truly represent a system element, a contextual element, or just a constraint on the space element as part of our selected concept for implementing the space element (such as low versus geosynchronous Earth orbit)?	Here we're also clarifying what is an actual element of the system versus simply the characteristics of one of these elements. In this example, we could argue that the FireSAT constellation represents the space element environment or context, rather than a specific system element.
Is the mission-operations center part of NOAA or separate? Does FireSAT have a dedicated ground element for the FireSAT mission?	We must be clear about the boundaries between the new system and the existing and legacy systems. We often need to think of the system boundary in logical and physical terms. NOAA may have a physical mission-operations center, but we still have to decide whether to embed new functions for FireSAT in this center. Would that be an advantage? Do we need to logically separate or combine them?

- Legal requirements or regulations for privacy, security, and safety
- Organizational or legal charters, as well as the roles and responsibilities of organizations that help realize the mission's goals and objectives
- Operational elements that we've changed or adapted to improve performance, such as latency of response; legacy elements that will constrain the system we're developing
- Legacy operational and support resources, as well as systems and infrastructure, such as the locations of regional Forest Service offices, communication infrastructure, and limits on bandwidth, latency, and reliability
- Skill levels and competencies of available operators and support people

Our description of the operational environment must also address known constraints on schedule, funding, and technology. For example, the system may have to coexist with legacy processing and communication standards to interoperate within a larger context. The sponsors may require certain hardware and software elements (such as displays or command and control software) to be used within the ground systems. Outputs and alerts may need to conform to a certain data and communication protocol, so we can integrate FireSAT into the larger enterprise. In any event, clearly defining these inputs and outputs (Chapter 10) makes system integration much easier.

Constraints on the system strongly affect how we implement it. Examples include timeliness of response and alerts to the firefighters, and resolving this requirement with the alert procedure and protocol between NOAA and the Forest Service's regional offices. Table 3-4 shows an example of dependency and correlation between performance and operational concepts. FireSAT has three other main constraints:

- FireSAT shall achieve initial operational capability within five years of authority to proceed (ATP) and final operational capability within six years of ATP

- FireSAT's total mission lifecycle cost, including five years of on-orbit operations, shall not exceed $200M (in FY 2007 dollars)

- FireSAT shall use existing NOAA ground stations at Wallops Island, Virginia, and Fairbanks, Alaska, for all mission command and control. NOAA GS-ISD-XYX defines the technical interface.

The funding and schedule constraints, plus necessary changes to the legacy doctrine, may require the concept of operations to reflect an evolutionary transition involving airborne reconnaissance between the current and intended states. NASA's Constellation Program and the Vision for Space Exploration are other examples of evolving concepts of operations. In this vision, the Shuttle is first to be replaced by the Orion crew module lifted by the new Ares 1 launch vehicle to service the Space Station in low-Earth orbit (LEO). LEO transportation would eventually be augmented by the commercial carriers under the Commercial Orbital Transportation Systems effort. Finally, a lunar mission concept of operations would evolve using the Orion and Ares 1 elements along with the new Ares 5 heavy lift vehicle to place lunar landers and other large payloads into LEO where they would dock with the Orion vehicles for departure to the Moon and beyond.

For the FireSAT mission, Figure 3-6 captures the current environment, and Figure 3-7 shows the envisioned environment. In summary, the concept of operations should capture the current and envisioned operational environment in six areas:

- Current and conceived systems and infrastructure

- Current and conceived organizations, roles, and responsibilities

TABLE 3-4.   **FireSAT's Key Performance Parameters.** Performance drivers and constraints affect the concept of operations. [Wertz and Larson, 1999]

Key Performance Parameters	First-order Algorithm (low-Earth orbit)	First-order Algorithm (Geosynchronous)	Performance Drivers
Persistence—Monitor the coverage area for potentially dangerous wildfires at least once every 12 hours	(Number of spacecraft)/ 12 hr	Scan frequency	Number of spacecraft for low orbit
Timeliness—Send fire notices to users within 30 minutes of fire detection (objective), 1 hour (threshold)	Onboard storage delay + processing time	Communications + processing time	Storage delay (if applicable)
Geo-location accuracy—Geo-locate potentially dangerous wildfires to within 5 km (threshold), 500 m (objective)	Distance x ((wavelength/ aperture) + control error)	Distance x ((wavelength/ aperture) + control error)	Altitude, aperture, control accuracy
Coverage—Cover the entire United States, including Alaska and Hawaii	Constellation design	N/A (continuous from geosynchronous orbit)	Altitude, inclination, number of satellites
Availability	Cloud cover interval	Cloud cover interval	None (weather dominated)

- Current and conceived doctrine, including regulations, policies and procedures, and standards
- Conceived system's performance, cost, and schedule drivers
- Legacy and constraining data formats and communication protocols and standards
- Current and conceived personnel system, including numbers, skills, and competencies

3.3   Develop Operational Scenarios and Timelines

The next element in developing a concept of operations is to better understand what operators and users want and how the conceived system and its elements will meet their needs. We build these operational scenarios—sometimes called functional threads, operational threads, or operational use-case scenarios—and trace them as much as possible to the stakeholders' expectations in Chapter 2. But the steps in Figure 3-1 are iterative, so in developing the concept of operations we're likely to discover new capabilities and characteristics that stakeholders will find necessary.

By overlaying operational scenarios on the system context diagram, we make sure they'll capture the capabilities that the users want. This technique also helps

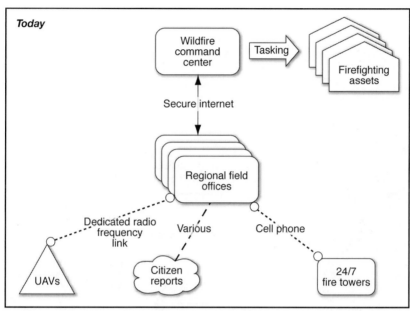

FIGURE 3-6. **The "As-is" Environment to Be Addressed by the FireSAT Mission.** It's important to understand the legacy (current) context (such as systems, policies, procedures, and doctrine) when developing a concept of operations. (UAV is unmanned aerial vehicle.)

- Identify interactions needed among the system of interest, its reference elements, and its active stakeholders
- Discover gaps in the envisioned concept of operations
- Validate the scenarios and capabilities

For example, FireSAT users mainly want to **detect and monitor forest fires**. So we analyze these capabilities to understand their performance characteristics, as shown in Table 3-5. These characteristics reflect how well the system must perform.

Performance characteristics combine to reflect complex variations on basic operational scenarios that support users, and some scenarios have overlapping characteristics and functions. We must define those that are "architecturally significant"—representing the conceived system's performance thresholds. We easily define the simple scenarios and build logic toward the most complex ones. For FireSAT, the simplest scenario is a single satellite, seeing a single wildfire, on a clear day (no weather issues) and in a benign environment. A more complex scenario might involve detecting and monitoring multiple wildfires, using several FireSAT satellites (a complete constellation), with each wildfire happening in a different unfavorable environment.

Once we've defined the operational scenarios, we trace and sequence them to understand how stakeholders must interact with system elements to apply the

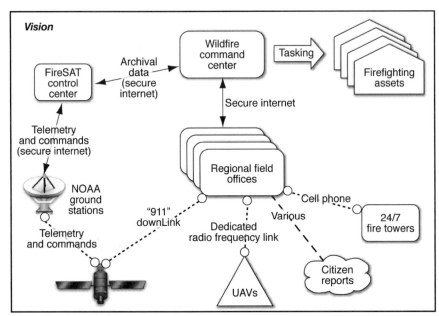

FIGURE 3-7.   The FireSAT Mission's Envisioned Environment. Although we show the as-is and envisioned environments in physical terms, we must also consider the logical perspective through operational scenarios. (UAV is unmanned aerial vehicle.)

TABLE 3-5.   Performance Characteristics for FireSAT Users' Main Need: To Detect and Monitor Forest Fires. These characteristics help us understand the boundary conditions for each capability and identify likely operational scenarios.

Characteristic	Description
Number of targets	How many targets (simultaneous wildfires) the system is tracking
Benign or hostile environment	A benign environment requires no special considerations. A hostile environment involves weather problems or solar flares and other potentially anomalous conditions.
Number of platforms	A platform is a single instance of the system being designed, in this case the FireSAT spacecraft
Number of air-ground stations	Number of stations, or other active stakeholder assets, that the satellites will need to communicate with
Nature of activity	The system must support various types of activities, such as target tracking, search and rescue, and environmental situations, that may require special behavior or functions from the system

scenarios successfully. Words and simple cartoons help everyone understand and agree on how the system might operate and how we might assign system elements to parts of the operation. These sequence diagrams are also useful for system-level testing and integration planning.

Next, we consider the stakeholder capability "detect wildfire," assuming a simple scenario: one satellite detecting a single wildfire. We also assume the wildfire starts at some time and grows while remaining under the detection threshold. When it reaches the detection threshold ($t = X$), a FireSAT satellite within range should detect it and alert the firefighters and NOAA. Figure 3-8 is a simple cartoon of this scenario, but we also show it in a block diagram (Figure 3-9).

| Fire starts | Fire becomes large enough to detect | Fire detected on next pass | "911" warning message to regional field offices | Warnings to wildfire command center for action | Detailed observation data to NOAA ground station | Data received by mission control for further action and archiving |

FIGURE 3-8.   Simple Cartoon of Timeline for "Detect Wildfire" Capability. Pictures quickly capture a simple scenario that shows how a fire starts, becomes larger, and is detected by the FireSAT system, which then alerts NOAA and the firefighters. (NOAA is National Oceanic and Atmospheric Administration.)

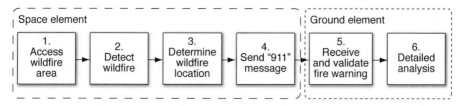

FIGURE 3-9.   Block Diagram for an Operational Scenario That Responds to the "Detect Wildfire" Capability. We use many informal and formal graphical methods to describe operational scenarios. In this context, a "911" message is an alarm.

Table 3-6 describes the sequence for our first operational scenario and several unresolved concerns, and Figure 3-10 is a simple illustration of this scenario in a context diagram. This figure visually communicates interactions between system elements and active stakeholders, but more formal schemes exist. For example, graphics like Figure 3-11 make the sequence of threads explicit and enable discussions that lead to agreed-on functional threads from end to end. Another option is use-case scenarios, as we describe in Chapter 4.

TABLE 3-6.   **Using Descriptive Language to Capture an Operational Scenario.** This approach is important because it gives users and systems engineers a shared understanding of how the system handles a capability. It also makes it easier to validate users' expectations and requirements.

Seq. #	Event and Description	Remarks and Questions
1	**Access wildfire area**—the satellite comes into view of a wildfire that's above the detection threshold	Timing of access to a wildfire by a given satellite will depend on the implementing concept defined by the orbit and constellation selected. Number of satellites? Coverage gap?
2	**Detect wildfire**—spacecraft detects the wildfire	A satellite's ability to detect a wildfire above a certain size depends on the payload resolution. Detection threshold value? Processing time?
3	**Determine wildfire location**—spacecraft determines the wildfire's size and geo-location.	A satellite's ability to determine a detected wildfire's location depends on its attitude, navigation accuracy, and other parameters. Geo-location accuracy? Use of GPS?
4	**Send "911" message**—spacecraft sends to the ground a wildfire-detection message containing time and location of the detected wildfire	A satellite's ability to send a warning message to the ground depends on its line-of-sight access to the ground stations and its communication subsystem. Frequencies? Location of ground stations? Satellite orbits?
5	**Receive and validate fire warning**—ground station receives wildfire "911" message and validates it independently to preclude "false positive" messages	A ground station's ability to receive the message depends on the line of sight from it to the satellite and its communication system.
6	**Analyze further**—ground-based element further analyzes the wildfire characteristics to validate its criticality. If it meets selected thresholds, it transmits information to mission operations. They also log the information into the archive of wildfire characteristics.	How long is the archived data maintained? How should we validate the wildfire's characteristics while satisfying latency requirements? Have we identified all the data format and communication protocols?
7	**Process information**—mission-operations element may further process this information. If it meets selected thresholds, it alerts NOAA's Command Center and the US Forest Service's regional office that has jurisdiction, based on the wildfire's location.	Are all the handoffs necessary? Can we come up with better packaging concepts? What information should go to NOAA and the US Forest Service's regional offices? How often can we update wildfire information? What level of automation should the alert system include, while satisfying the procedural and regulatory requirements?

The sequence in Table 3-6 and Figures 3-9 through 3-11 helps us create timelines for different aspects of this operational scenario, which we use to analyze a capability's latency requirements (how quickly the system must respond). Latency influences how we package and apply the capability, as we describe in Section 3.4. Capturing an operational scenario in words and graphics helps users and systems engineers agree and share an understanding of the capability. Figures 3-12 and 3-13 show how timeline analysis works. "Detection lag" depends on Objective 1.1—Detect a wildfire in less than 1 day (threshold), 12 hours (objective).

And the "notification lag" depends on Objective 1.2—Notify USFS within 1 hour of detection (threshold), 30 minutes (objective). This timeline provides a time budget for each of these periods, which later analyses will allocate to the functions that complete each aspect of the scenario.

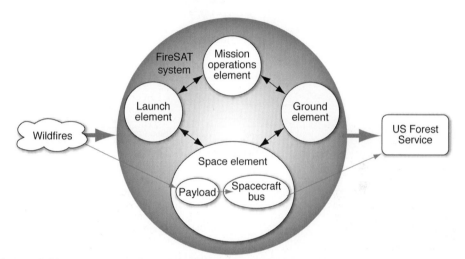

FIGURE 3-10.   Tracing an Operational Scenario on a Context Diagram. This type of diagram brings users and systems engineers together in understanding required capabilities and the system that will realize them.

Sequencing an operational scenario also helps us understand likely system anomalies. The candidate operational scenario described in Table 3-6 and Figures 3-8 through 3-10 assumes the system will succeed, with all elements available and operational. But we analyze it to understand likely failures, interrupts, aborts, and other system anomalies. Example anomalies for the operational scenario in Table 3-6 are loss of power to the ground-based element and failure of the command and data handling sub-element. We develop more scenarios to define how the system reacts to these anomalies and then iterate through all the system anomalies to make the system more robust.

These sequence diagrams and timelines, with our analysis of the performance drivers and constraints (Section 3.2), help us decide how we'll use system elements. For example, operational scenarios and implementing concepts may suggest that we need to supplement space-based sensors with air-based ones (Section 3.4).

By developing operational scenarios, we improve understanding among the stakeholders—users, operators, systems engineers, architects, testers, and developers—of the main logical handoffs between system elements during operations. These scenarios clarify the sequence of events necessary to satisfy users and operators. They also help identify the major timeline drivers, which enables us to develop targeted trade studies and select implementing concepts for the system elements.

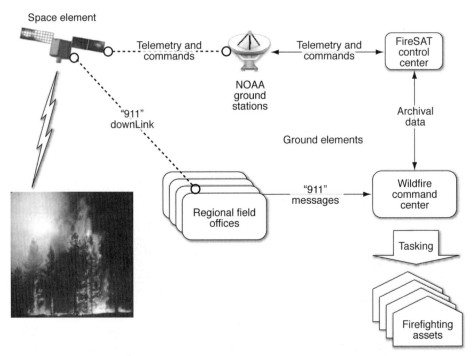

FIGURE 3-11.   Another Graphical Depiction of the Operational Scenario for "Detect Wildfire." Graphically depicting scenarios is critical to capturing and validating capabilities that stakeholders want.

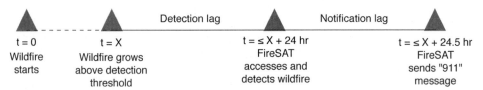

FIGURE 3-12.   Timeline Analysis for the Operational Scenario. Analyzing how long detection takes is the first step in allocating the time budget.

3.4   Synthesize, Analyze, and Assess Key Implementing Concepts for the System and Its Elements

Most complex developments start by upgrading, evolving, or leveraging part of an existing system. So we naturally think about similar systems when exploring how to build a new system of interest. Because implementing concepts suggest technical approaches to satisfy system-level requirements, we must define alternatives with consistent detail. For large, complex systems (including

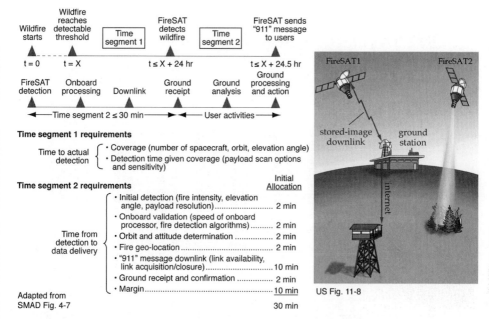

FIGURE 3-13.   More Detailed Timeline Analysis for the Operational Scenario. This input is critical to assessing and applying implementation concepts for the system elements. [Wertz and Larson, 1999; Sellers, 2004]

FireSAT), we may need to identify more detailed concepts for selected subsystems. Representative subsystems for FireSAT are its elements: ground-based, space-based, and mission analysis. Thus, we must develop and analyze alternatives for each element and then select a preferred concept.

For space missions, basic implementing concepts often develop around the orbit or constellation design. (For a complete discussion of different orbit types see *Understanding Space* [Sellers et al, 2005]). The same mission requires a different satellite for every orbit type, and each orbit has advantages and disadvantages, depending on the mission's goals and objectives. Table 3-7 summarizes alternatives for the space-based element.

Generating conceptual solutions is highly subjective, so experience and familiarity with similar systems often influence this process, which involves creative thought and innovation. McGrath [1984] notes that people working separately on concepts "generate many more, and more creative ideas than do groups, even when the redundancies among member ideas are deleted, and, of course, without the stimulation of hearing and piggy-backing on the ideas of others." On the other hand, groups are better for evaluating and selecting concepts. Design teams often use techniques such as brainstorming, analogy, and checklists to aid creative thinking and conceptual solutions:

TABLE 3-7. **Examples of Orbit-defined Implementing Concepts for Typical Earth Missions.** The orbit we choose strongly affects which concept to use for space missions. This table summarizes typical special and generic orbits with their advantages and disadvantages. We use such a table to trade off implementing concepts for the system elements. (ISS is International Space Station.)

Orbit Type	Definition	Advantages	Disadvantages	Typical Applications
Geosynchronous (GEO)	• Period ~24 hr • Inclination ~0 deg	• Maintains relative fixed position over a selected spot on the equator—simple to track • Covers nearly 1/3 of Earth	• High altitude is more expensive to reach • Fixed coverage area • Little or no coverage of high latitudes • High altitude means lower imaging resolution and higher space losses	• Communications • Weather
Sun-synchronous	• Period ~100 min • Inclination ~95 deg (depending on altitude)	• Provides nearly constant Sun angles for consistent Earth imagery	• High inclination is more expensive to reach	• Remote sensing
Molniya	• Period ~12 hr • Inclination = 63.4 deg	• Fixed apogee and perigee locations • Great high-latitude coverage	• High altitude is more expensive to reach • High altitude means lower imaging resolution and higher space losses	• Communications for high arctic regions
Low-Earth orbit (LEO); single satellite	• Period: 90–120 min • Inclination: 0–90 deg	• Low altitude is less expensive to reach • Low altitude means higher imaging resolution and lower space losses	• Limited coverage • More complex tracking	• Science missions • ISS • Space Shuttle
LEO constellation	• Period: 90–120 min • Inclination: 0–90 deg • Multiple planes, multiple satellites per plane	• Same as LEO single satellite • More satellites expand coverage	• Multiple satellites and possibly multiple launches means higher cost	• Communications (such as Iridium)

- Brainstorming is the most common technique. Alex Osborn developed it so a group could attempt "to find a solution for a specific problem by amassing all the ideas spontaneously contributed by its members" [Osborn, 1957].

- Analogy involves translating a conceptual solution across disciplines and adapting it to solve similar problems

- Checklists use thought-provoking questions to enhance innovation and creative thought

Once we have several conceptual solutions, we analyze each one to see if it's physically possible, matches the mission's functions, achieves market goals, and meets financial and budgetary constraints. Our objective is to screen out concepts that won't work, so only feasible ones enter the more rigorous evaluation and selection phase. But we must discard only the unlikely or impossible ones, in keeping with the "principle of least commitment" in system design and development [Asimow, 1964]. Committing to a concept too early may harm the project. We carefully evaluate workable concepts to select the best for further development.

We describe concept selection as though it moves sequentially through synthesis, analysis, and evaluation, but continual feedback and iteration are always part of the process. Pugh first introduced formality to evaluating conceptual designs by proposing a way of creating and evaluating concepts [Pugh, 1991]. In Chapter 6 we describe several methods for trade-offs and decision analysis.

We analyzed the five concepts for FireSAT's space-based element in Table 3-7 to identify two strong candidates, as shown in Figures 3-14 and 3-15. These figures also include some important selection criteria. After establishing evaluation criteria, we define both concepts in the same detail, to avoid bias or premature decisions.

Designers and operators must often defer choosing an implementation concept until they have better technology or more information, such as from feasibility studies and tests. Meanwhile, they develop several concepts. Although this approach may require more time and money, it often reduces risk to the overall development. No amount of rigor in the later engineering stages compensates for an inferior concept.

3.5 Document the Concept of Operations and the System's Operational Architecture; Apply Architectural Frameworks

Several approaches and standards have described the documented content of a concept of operations. For example, the ANSI/AIAA G-043-1992 standard [ANSI/AIAA, 1993] states: "...a good Operations Concept Document (OCD) should tell a story; that is, it should be a narrative, pictorial description of the system's intended use." This same description could apply to a good science fiction story—which many concept of operations appear to be! It goes on to suggest that the concept of operations should describe the "*What, Where, When, Who,* and *How* of

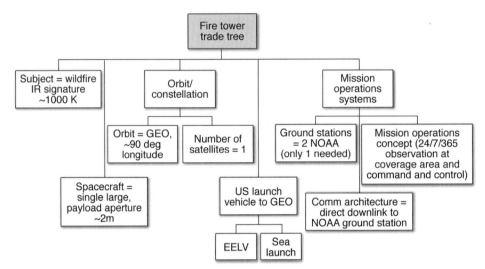

FIGURE 3-14. **First Concept for Implementing the Space-based Element.** This concept is one we synthesized for the FireSAT mission. (IR is infrared; GEO is geosynchronous orbit; Comm is communication; EELV is Evolved Expendable Launch Vehicle.)

system operations." A documented concept of operations should include an introduction, business need, system justification, system concepts, operational scenarios, business impacts, rationale, and conceptual model. Yet another guide, *Systems Engineering Fundamentals* [DSMC, 2001] says a concept of operations document should

- Define the operational need
- Analyze the system mission
- Show operational sequences
- Show operational environments
- Include conditions or events to which a system must respond
- State the operational constraints on the system
- Describe the mission's performance requirements
- Specify the user and maintainer roles, as well as the structure of organizations that will operate, support, and maintain the system
- Define the operational interfaces with other systems

Table 3-8 contains our recommended outline for a concept of operations.

Several architecture frameworks have been developed recently to consistently document efforts in systems engineering and architecture, thus improving communication among stakeholder communities. Many departments and agencies in the US federal government have developed or adopted a framework to

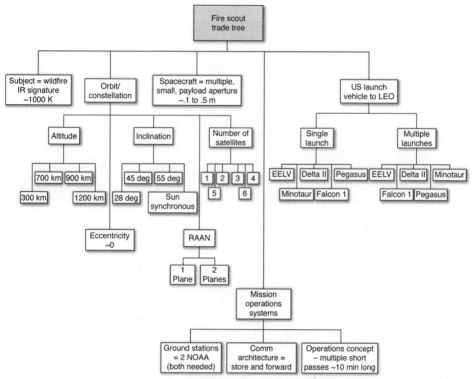

FIGURE 3-15.   Second Concept for Implementing the Space-based Element. This is another way of meeting FireSAT's mission. Now we must assess it against the concept in Figure 3-14 and select the better one for development. (IR is infrared; LEO is low-Earth orbit; EELV is Evolved Expendable Launch Vehicle; RAAN is right ascension of the ascending node; Comm is communication.)

control how contractors and other government organizations prepare and present architecture information.

One of the first frameworks was the Department of Defense's Architecture Framework, or (DoDAF [Dam, 2006]), which compares architectures from concept definition through system design. Several agencies involved in space use the DoDAF. The National Security Agency, National Geospatial Intelligence Agency, and National Reconnaissance Office are a few examples.

Table 3-9 summarizes the DoDAF 1.0 artifacts; it offers a special version of some of the products, such as SV-5, to focus on net-centricity. DoDAF 1.0's artifacts fall into four categories of views. First are the "all views" or common views, consisting of only two products: the AV-1, which is an executive summary of the architecture, and the AV-2, which captures all architecture definitions. This glossary isn't the only relevant one, but it also includes the metadata and schema for collecting the architecture information.

TABLE 3-8.   **Suggested Outline for a System Concept of Operations.** This table consolidates the major aspects of a concept of operations.

Section #	Section Name	Remarks
1	Executive summary	Briefly summarize the mission, organizations' roles and responsibilities, and key capabilities and performance characteristics the stakeholders want. Outline main findings and significant departures from legacy systems; include a brief rationale in each case.
2	Mission description	Describe the mission, its goals and objectives, and the underlying mission and business rationale. Identify relevant stakeholders and their main expectations.
3	System operational context and reference operational architecture	These elements clarify the system's boundary and establish the initial reference architecture based on what similar systems have used. Reference active stakeholders in the context diagram, together with the reference system's elements and sub-elements. Describe the "as-is" and "to-be" contexts to help clarify the conceived system's value.
4	System drivers and constraints	Describe performance drivers and constraints; constraints resulting from the existing systems and infrastructure; doctrine (policy, procedures, and processes); organizational roles and responsibilities; and regulatory requirements. Explicitly defining drivers helps us assess different concepts for the system, its elements, and its sub-elements.
5	Operational scenarios	Create the main operational scenarios to support capabilities the stakeholders expect, considering the system context and reference operational architecture, as well as the proposed operational architecture. Use language and graphics to ensure that systems engineers understand the stakeholders' expectations. Develop timelines for each operational scenario to understand latency thresholds and gain insights into other concepts for partitioning and implementation.
6	Implementation concepts selected and rationale	Synthesize different ways to partition and carry out stakeholders' intentions in system elements and sub-elements. Document why we selected the preferred application, addressing especially the important drivers and constraints, including funding and schedule.
7	Proposed system operational architecture	Document changes to partitioning of the system, its elements and sub-elements, in a proposed system operational architecture.
8	Organizational and business impact	Analyze and present the effects of a changed operational architecture on legacy doctrine to ensure appropriate decision making. Doctrine includes policy, procedures, and processes; organizational roles and responsibilities; necessary skills and competencies; and workload changes.
9	Risks and technology readiness assessment	Document risks for the proposed operational architecture. Include technology readiness levels for the principal implementing concepts. Also assess schedule and funding risks resulting from the proposed approach.

TABLE 3-9.   **Products of the Department of Defense's Architecture Framework (DoDAF).** The OV-1 represents the system's operational architecture.

Applicable View	Framework Product	Framework Product Name	General Description
All Views	AV-1	Overview and summary information	Scope, purpose, intended users, environment depicted, analytical findings
All Views	AV-2	Integrated dictionary	Architecture data repository with definitions of all terms used in all products
Operational	OV-1	High-level operational concept graphic	High-level graphical and textual description of operational concept
Operational	OV-2	Operational node connectivity description	Operational nodes, connectivity, and information exchange need lines between nodes
Operational	OV-3	Operational information exchange matrix	Information exchanged between nodes and the relevant attributes of that exchange
Operational	OV-4	Organizational relationships chart	Organizational roles or other relationships among organizations
Operational	OV-5	Operational activity model	Capabilities, operational activities, relationships among activities, inputs, and outputs; overlays can show cost, performing nodes, or other pertinent information
Operational	OV-6a	Operational rules model	One of three products that describe operational activity—identifies business rules that constrain operation
Operational	OV-6b	Operational state transition description	One of three products that describe operational activity—identifies business process responses to events
Operational	OV-6c	Operational event-trace description	One of three products that describe operational activity—traces actions in a scenario or sequence of events
Operational	OV-7	Logical data model	Documentation of the system data requirements and structural business process rules of the operational view
Systems	SV-1	Systems interface description	Identification of systems nodes, systems, and system items and their interconnections, within and between nodes
Systems	SV-2	Systems communications description	Systems nodes, systems, and system items, and their related communications lay-downs
Systems	SV-3	Systems-systems matrix	Relationships among systems in a given architecture, can be designed to show relationships of interest, e.g., system-type interfaces, planned vs. existing interfaces, etc.
Systems	SV-4	Systems functionality description	Functions performed by systems and the system data flows among system functions
Systems	SV-5	Operational activity to systems function traceability matrix	Mapping of systems back to capabilities or of system functions back to operational activities
Systems	SV-6	Systems data exchange matrix	Provides details of system data elements being exchanged between systems and the attributes of those exchanges
Systems	SV-7	Systems performance parameters matrix	Performance characteristics of systems view elements for the appropriate timeframes

TABLE 3-9.   Products of the Department of Defense's Architecture Framework (DoDAF).
(Continued) The OV-1 represents the system's operational architecture.

Applicable View	Framework Product	Framework Product Name	General Description
Systems	SV-8	Systems evolution description	Planned incremental steps toward migrating a suite of systems to a more efficient suite, or toward evolving a current system to a future implementation
Systems	SV-9	Systems technology forecast	Emerging technologies and software and hardware products that are expected to be available in a given set of time frames and that will affect future development of the architecture
Systems	SV-10a	Systems rules model	One of three products that describe system functionality—identifies constraints on systems functionality due to some aspect of systems design or implementation
Systems	SV-10b	Systems state transition description	One of three products that describe system functionality—identifies responses of a system to events
Systems	SV-10c	Systems event-trace description	One of three products that describe system functionality—identifies system-specific refinements of critical sequences of events described in the operational view
Systems	SV-11	Physical schema	Physical implementation of the logical data model entities, e.g., message formats, file structures, physical schema
Technical	TV-1	Technical standards profile	Listing of standards that apply to systems view elements in a given architecture
Technical	TV-2	Technical standards forecast	Description of emerging standards and potential impact on current systems view elements, within a set of time frames

Second are the operational views (OVs). These include a high-level concept diagram (OV-1), interface diagrams (OV-2 and OV-3), organizational chart (OV-4), functional analysis products (OV-5 and OV-6), and the logical data model (OV-7). We say the OV-1 is the "as is" of the contextual setting for FireSAT (Figure 3-3b).

Third are the system views (SVs). They include several versions of interface diagrams (SV-1, SV-2, SV-3, and SV-6); functional analysis products (SV-4, SV-5, and SV-10); a performance matrix (SV-7); transition planning diagrams (SV-8 and SV-9); and finally the physical data model (SV-11).

Last are the two standard technical views: one showing the current (or near-term projected) standards and the other forecasting standards that may appear over the life of the architecture.

The DoDAF is just one of the architectural frameworks used within industry and government. It was originally designed for Command, Control, Communications, Computers, Intelligence, Surveillance, and Reconnaissance (C4ISR) but now applies generally to DoD systems. The Zachman Framework (Figure 3-16) focuses on enterprise architectures. Many of the ideas and concepts from Zachman have become part of the Federal Enterprise Architecture Framework [CIO, 2001], that now includes reference models and has been applied to projects within the US federal government.

3.6   Validate and Baseline the System's Operational Architecture

Most concepts of operations include an initial notion of the system and elements that will satisfy active stakeholders. Figures 3-2 through 3-6 represent such a notion for the FireSAT system. Based on experience with similar systems in the past, and with existing systems, we assume the system consists of four elements: space-based, launch, ground-based, and mission operations. We also assume the space-based element consists of two sub-elements: a satellite bus and a special-purpose payload. This constitutes the reference architecture for FireSAT. We use it for early discussions about identified performance drivers and constraints, operational scenarios developed with users and operators, and the implementation concepts we're considering for the elements and sub-elements. The first proposed FireSAT architecture is just one of the system's interdependent aspects that we define, synthesize, and assess.

FireSAT's performance drivers and conceived capabilities may cause systems engineers to consider concepts very different from existing ones for similar systems. Examples of performance drivers that could push innovative solutions are latency and accuracy requirements for detecting wildfires. One conceived capability might be minimal human intervention in wildfire alerts and tracking. Technology advances in sensors and processors allow engineers to consider such novel ideas.

These ideas may prompt systems engineers to consider concepts for a FireSAT system with different elements and sub-elements. For example, the ground and mission-operations elements may combine into one and the space-based element may combine space- and air-based sub-elements. As we describe in Section 3.4, we have to synthesize, analyze, and assess these alternatives to produce the operational architecture.

Figures 3-2 through 3-6 represent various perspectives on the FireSAT system's initial reference architecture. But Figures 3-17 and 3-18, along with Table 3-10, represent its operational architecture based on analyzing the drivers and constraints, operational scenarios, and different ways of implementing its elements and sub-elements. This architecture becomes the basis for validating capabilities for users and operators, as well as the supporting operational scenarios.

The main deliverable from developing a concept of operations is an updated operational architecture, reflecting any changes to the initial reference architecture. These changes may be new concepts for implementing and partitioning the system to leverage new advances in technology, while satisfying key performance drivers and other constraints. This process includes

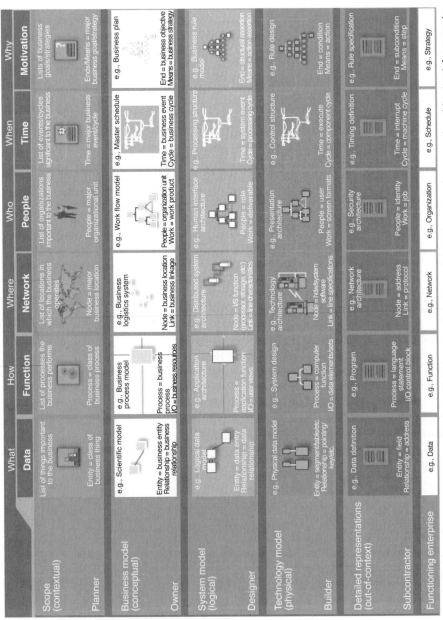

FIGURE 3-16.   **The Zachman Framework.** This framework is the point of departure for several architectural frameworks proposed since the late 1990s. [Zachman, 1987]

- A validated set of operational scenarios that support the capabilities that users and operators want
- A clear understanding of the system's contextual setting and boundary

A complete, coherent concept of operations is vitally important. It's one of the foundational concepts to ensure that the engineering effort aligns properly with stakeholders' expectations. The technical requirements we describe in Chapter 4 use this concept to validate system requirements, as well as for system-level acceptance tests. Because the concept of operations is so critical, consistency and clarity are vital to documenting and communicating it across diverse communities and stakeholders.

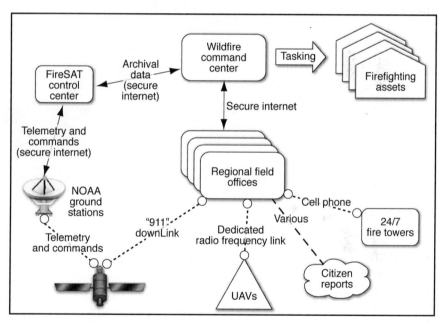

FIGURE 3-17.   High-level Context and Concept View of FireSAT. This view clarifies the envisioned system's "physical boundary" within the larger context. (UAV is unmanned aerial vehicle.)

FireSAT space element	Telemetry (packetized data, RF link)		"911" Messages (packetized data, RF link)		
Commands (packetized data, RF link)	**NOAA ground stations**	Telemetry (packetized data, secure internet link)			
	Commands (packetized data, secure internet link)	**FireSAT C2**		Archival data (storage media, e.g., DVD)	
			Regional field offices	Recommend-ations and requests (email)	
		Archival data requests (email)		**Wildfire command center**	Taskings (email)
					Firefighting assets

FIGURE 3-18.   **An N×N Diagram Showing the Connectivity among Envisioned System Elements.** This diagram further clarifies the interfaces among the system elements. (RF is radio frequency; C2 is command and control.)

TABLE **3-10.   The Operational Nodes within the FireSAT System.** This table lists the operational nodes and a loose connectivity mapping of the system's physical elements.

Operational Node	Activities Performed
FireSAT space element	• Detect wildfires • Generate and send "911" warning messages • Generate and send telemetry • Apply commands
NOAA ground stations	• Receive telemetry from space element • Send commands to space element
FireSAT control center	• Analyze telemetry • Generate commands • Compile and store archival data
Regional field offices	• Receive "911" warning messages • Validate messages • Collate warnings from multiple assets • Send recommendations and requests to command center
Wildfire command center	• Receive recommendations and requests from field offices • Rank order needs • Generate tasking orders to firefighting assets • Request and receive archival data
Firefighting assets	• Receive orders from command center • Fight wildfires

References

ANSI/AIAA G-043-1992. 1993. Guide for the Preparation of Operational Concept Documents. Reston, VA: American Institute of Aeronautics and Astronautics.

Asimow, Morris. 1964. *Introduction to Design.* Englewood Cliffs, NJ: Prentice Hall, Inc.

Chief Information Officers Council (CIO). February 2001. A Practical Guide to Federal Enterprise Architecture, version 1.1; available at the CIO website.

Dam, Steven H. 2006. *DoD Architecture Framework - A Guide to Applying System Engineering to Develop Integrated, Executable Architectures.* Marshall, VA: SPEC Publishing.

Defense Systems Management College (DSMC). 2001. Systems Engineering Fundamentals. Fort Belvoir, VA: Defense Acquisition University Press.

Larson, Wiley J. and Wertz, James R., eds. 1999. *Space Mission Analysis and Design,* 3rd Edition. Torrance, CA: Microcosm Press and Dordrecht, The Netherlands: Kluwer Academic Publishers.

McGrath, J. E. 1984. *Groups: Interaction and Performance.* Englewood Cliffs, NJ: Prentice Hall, Inc.

Osborn, A. 1957. *Applied Imagination – Principles and Practices of Creative Thinking.* New York, NY: Charles Scribners Sons.

Pugh, S. 1991. *Total Design: Integrated Methods for Successful Product Engineering*. New York, NY: Addison-Wesley, Inc.

Sellers, Jerry Jon. 2004. *Understanding Space*. 3rd Edition. New York, NY: McGraw-Hill Companies, Inc.

Zachman, John. A Framework for Information Systems Architecture. IBM Systems Journal, Vol. 26, No. 3, pp. 276–290, 1987.

Chapter 4

Engineering and Managing System Requirements

Robert E. Lee, *REL Consulting*
Ivy Hooks, *Compliance Automation, Inc.*
Bruce G. Barker, *Stevens Institute of Technology*

One key to effective system development is to reduce the effects of requirement changes by establishing defect-free (sometimes called bug-free) requirements early on. A study commissioned by the European Systems and Software Initiative states:

> We found that the vast majority of bugs (85%) were due to missing or changed requirements; and misunderstood requirements by developers and testers....We found that requirements-related bugs accounted for 51% of all the bugs [Vinter et al., 1999].

Requirements are the communication link between the customer, the developer, and the tester to ensure that we build the system right. We have a good chance of doing so if we start with the right requirements—and if the customer, developer, and tester all understand these requirements consistently. But if we define and manage the wrong requirements or misinterpret them, the chances of failure increase.

In this chapter we discuss a process to develop defect-free requirements, organize them, and manage them over the system lifecycle. Table 4-1 outlines best practices for developing and managing system requirements. We need to apply them whenever we're developing, defining, or refining (changing) requirements. The iterative application of these best practices cannot be overemphasized, as reflected in Figure 4-1. Articulating stakeholder expectations and the concept of operations facilitates alignment with system requirements, ensuring that they're complete and correct. The process of developing the requirements should lead us to revisit and clarify stakeholder expectations and the assumptions we made while building the concept of operations. It should also identify additional trade studies.

TABLE 4-1.   Best Practices for Developing and Managing Requirements. The steps below focus effort on the proper issues. (APMSS is *Applied Project Management for Space Systems* [Chesley et al., 2008].)

Step	Description	Where Discussed
Identify basic scope	Define need, goals, and objectives; identify stakeholders; capture stakeholder expectations; assign authority and responsibility	Section 4.1.1 and Chap. 2
Develop operational concepts	Develop plain-language scenarios for how the system might behave and be used, expanding them to cover all lifecycle phases	Section 4.1.2 and Chaps. 3–7
Identify drivers	Determine constraints, regulations, standards, and resources (schedules, budgets, and technology)	Section 4.1.3 and Chap. 3
Identify external interfaces	Determine the boundary between the system and the rest of the world, clarifying the system's inputs and outputs	Section 4.1.4, Chaps. 2, 3, 5, and 15
Develop system requirements to meet stakeholder expectations	Develop a set of system requirements derived from the stakeholders' expectations (capabilities and characteristics). These requirements define what the solution must do to meet the customer's needs.	Section 4.2, Chaps. 2 and 5
Write defect-free system requirements	Guide system design, in unambiguous terms, toward what the stakeholders expect. Expose assumptions and factual errors; capture corporate knowledge. Write all requirements at the correct level, being sure that we can trace them back to their origins. Avoid implementation. Identify each requirement's verification technique and associated facilities and equipment.	Section 4.2, Chaps. 2 and 11
Organize system requirements	Include the appropriate types of requirements, making sure that the development team can locate them within a document or database. Organize system requirements in a way that helps the reader to understand the system being specified.	Section 4.3
Baseline system requirements	Validate that requirements are correct, complete, consistent and meet project scope and stakeholders' expectations; don't add gold plating	Section 4.4, Chaps. 2 and 11, APMSS Chaps. 5, 6, and 8
Manage system requirements	Control foregoing activities; develop defect-free requirements to reduce unnecessary changes caused by poor requirements; capture management information; control necessary requirement changes	Section 4.5, Chaps. 2, 3, 15, and 16, APMSS Chaps. 7, 8 and 14

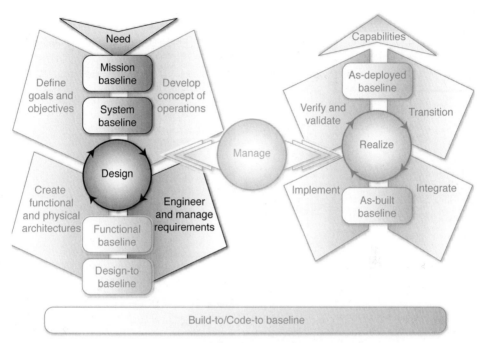

FIGURE 4-1. **Engineer and Management Requirements.** Iterative application of SE best practices leads to the development of a consistent, traced, and traceable, set of technical system requirements.

The activities in Table 4-1 appear organized as a linear step-by-step process. But Figure 4-2 shows concurrency and feedback between the scoping and requirements-development activities.

Writing good requirements is hard work: it takes time and resources. Writing defective requirements is hard work too, but we have to do it over and over again. The time and resources required to fix bad requirements, as well as the rework necessary to implement them in the system, all contribute to cost and schedule overruns and possibly project failure. A system built on defective requirements won't meet the needs of the customer and will cost more and take longer to develop than one that starts with a good set of validated requirements. Understanding the scope and stakeholder expectations, as discussed in Chapter 2, and then translating this understanding into a coherent concept of operations, as discussed in Chapter 3, goes a long way to providing the necessary focus and alignment when defining the technical requirements for a system.

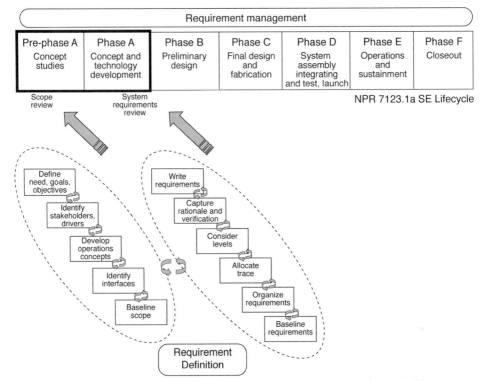

FIGURE 4-2.   Requirement Definition Related to NPR 7123.1a—Systems Engineering Lifecycle. This figure provides perspective that's useful in understanding the challenges associated with requirements.

4.1   Before Requirements

The first step to writing good requirements isn't writing them. We must begin with some or all of the following activities to ensure that we have the necessary information:

- Analysis
- Research
- Prototyping or modeling
- Trade studies

The information we gather from these activities creates the foundation for good requirements—the system scope. Much of the **project** scope, which focuses on what organizations or people will do to develop the system, translates directly to the **system** scope. The system scope emphasizes what the system must do to fulfill organizational goals and the project scope's objectives. We must document it and share it with the whole team so everyone has the same vision and viewpoint when generating and reviewing the requirements.

4.1.1   Understand Scope

The system scope identifies and defines several key aspects of a project:

- The need—why we need to develop this system
- Goals and objectives—what we want to accomplish
- Stakeholders (active and passive)—who will use the system, who the customers are, who can influence the system, and who has authority and responsibility for the scope and ensuing requirements
- Operational scenarios—the primary sequences of activities or steps that the system will engage in, in response to the identified stakeholder capabilities
- Operational concepts—how the system will be developed, tested, deployed, used, and disposed of
- Drivers—outside influences on the system, such as regulations, existing systems, cost, and schedule
- Interfaces—systems external to our own that we need to do our job and systems that need our system to work so they can do their jobs

A coherent description and discussion of these items leads to a sound concept of operations, as we describe in Chapter 3.

Why Define the Scope?

We must establish a common vision before writing the first "shall" statement, i.e., the first requirement. Writing requirements without first determining the scope leads to divergent views of the system and results in frustration and unnecessary battles over requirements during requirement reviews. At that point, we'll have to revisit the scope's elements to resolve conflicting requirements and begin to realign requirement reviewers (and authors) to what the system is and is not. So we can "pay now" and define the scope up front, or "pay later" by not defining scope and having to correct defective requirements throughout the system's lifecycle [Hooks and Farry, 2001].

The scope defines what system will be developed within a given schedule for a given budget. It also sets the system's limits—its boundaries and interfaces. Scope provides the basis for the system requirements. In this section we discuss the first two aspects of scope development: 1) identifying stakeholders and 2) establishing the need, goals, and objectives.

Who Defines the Scope?

The scope-definition team, comprising a small number of people with diverse experiences, defines the scope. The team could be as few as two people for small projects and up to twelve people for large ones. The team should

- Represent the viewpoints of multiple stakeholders
- Not be afraid to ask questions
- Avoid confronting stakeholders as they gather the scope elements
- Have strong skills in oral and written communication
- Be available to work together without interruption for relatively long periods

Once the team has gathered the scope elements from all stakeholders, they prepare the scope document and review it with stakeholders to get their buy-in. A representative from each stakeholder organization signs the scope document, which then goes under configuration management just like any other project document.

Why are Stakeholders Important?

A *stakeholder* is a group or individual that's affected by or is in some way accountable for an undertaking's outcome. "Relevant" stakeholders are an important subset. They're people or roles designated in a plan for stakeholder involvement [NASA, 2007]. The scope team must determine the relevant stakeholders and get them to help develop the scope. Typical stakeholders for most system developments are

designers	developers
end users	trainers
testers	maintenance personnel
reliability personnel	safety personnel
operations personnel	customers

We must identify relevant stakeholders by office code if possible and plan how we will manage their expectations and acquire the information we need from them. Chapter 2 suggests dividing stakeholders into active and passive stakeholders. *Active stakeholders* **use** the system and typically generate capabilities the system must have. *Passive stakeholders* **influence** the system and often generate characteristics that the system must have. Each stakeholder has a unique perspective and vision about the system, and each is a source of requirements. By gathering stakeholder needs and expectations during scope development, we reduce the likelihood of surprise later in the development cycle. Incorporating new expectations (requirements) into the system late in the lifecycle entails a lot of rework and resources—staffing, cost, and schedule.

Not all stakeholders are created equal. For example, the stakeholder holding the purse strings, the sponsor, probably has the biggest influence. The end users offer the most detail about how they work and hence what they expect the system to do. We should use our list of stakeholders as a starting point and tailor it for the system. Although we have to consult many people, missing just one during scope development could result in considerable pain later on. Finding missing requirements during development or testing—or worse yet, during operations— guarantees cost and schedule overruns.

How Do we Engage the Stakeholders?

The systems engineering team must collaborate with stakeholders to encourage them to share their information. We don't recommend gathering all stakeholders in a room and asking what their needs are. Instead, we work with them in small groups or individually. Problems, issues, and disagreements occur, but the team must work with stakeholders to resolve these problems and establish a common vision. Chapter 2 details several other approaches.

What Are the Project's Need, Goals, and Objectives (NGO)?

The need, goals, and objectives (NGO) for a system establish a clear vision to guide its development. All activities related to the system should support the NGO, which we use to keep stakeholders focused throughout the system's lifecycle.

The *need* is a single statement that drives everything else [Hooks and Farry, 2001]. It should relate to the problem that the system is supposed to solve, not be the solution. The need statement is singular. If we try to satisfy more than one need, we'll have to trade between the two, which could easily result in failing to meet at least one, and possibly several, stakeholder expectations. After establishing the need and getting agreement on it, we must put it in writing and make it available to all stakeholders. At this point, the need shouldn't change. If it does, we must re-examine everything else we've done and probably start over.

Goals elaborate on the need and constitute a specific set of expectations for the system. They further define what we hope to accomplish by addressing the critical issues identified during the problem assessment. Goals need not be in a quantitative or measurable form, but they must allow us to assess whether the system has achieved them.

Objectives are specific target levels of outputs the system must achieve. Each objective should relate to a particular goal. Generally, objectives should meet four criteria:

- Specific—Objectives should aim at results and reflect what the system must do, but they don't outline how to implement the solution. They must be specific enough to provide clear direction, so developers, customers, and testers will understand them.
- Measurable—Objectives must be quantifiable and verifiable. The project must monitor the system's success in achieving each objective.

- Aggressive, but attainable—Objectives must be challenging but reachable, and targets must be realistic. At first, we may include objectives "To Be Determined" (TBD)—until trade studies occur, operations concepts solidify, or technology matures. But we must be sure that objectives are feasible before we start writing requirements and designing systems.
- Results-oriented—Objectives must focus on desired outputs and outcomes, not on the methods used to achieve the target (what, not how).

Table 4-2 lists several situations that produce poorly written, inappropriate, or inadequate requirements.

TABLE 4-2.   Scenarios Leading to the Development of Defective Requirements. Such scenarios lead to significant design rework or the development of a system not aligned to the real mission need, goals, and objectives.

Scenarios	Remarks
Requirements are received from another organization or individual, without any rationale or "traceability" back to the mission need and goals	This decoupling often results in the development of requirements that are not focused or aligned with the mission need and goals. It may lead to misinformed trade-offs and priorities. Furthermore, in the event of ambiguities no good clarifications exist, resulting in incorrect assumptions.
Requirements are solicited from an unresponsive set of stakeholders and customers	This unhealthy scenario might reflect a lack of commitment on the part of the customers and stakeholders. They may respond with requirements that speak more to implementation than to the real need and goals.
Requirements take a back seat to design and implementation	In this approach, we simply assume that the engineers and developers inherently and consistently understand the real need, goals, and objectives. This highly risky approach is likely to develop a system that's irrelevant to the mission, or unresponsive to the market opportunity.
Requirements are developed by engineers and developers without engaging with stakeholders and customers	This scenario reflects unhealthy engineering arrogance. The engineers and developers assume knowledge of stakeholder and customer needs and expectations. This approach is also very risky.

Although the goals are in general terms, we can verify they've been met. Likewise, the objectives are in measurable terms, which the team can verify while assessing the as-built system. Another way to look at goals and objectives is that goals state the problem that a solution must solve (the **need**), and objectives state the acceptable or expected performance.

4.1.2   Synthesize the System Concept of Operations

An operational concept describes in plain language or graphical depiction a day in the life of our system [Hooks and Farry, 2001]. Chapter 3 provides more specific approaches for developing operational concepts.

What Are the Benefits of Synthesizing the System Concept of Operations?

Synthesizing and documenting the concept of operations, as discussed here and in Chapter 3, has a high return on investment. Each stakeholder's story offers a wealth of information for scope definition. As we document and integrate these stories, we identify conflicts to be resolved, which facilitates a common vision before writing requirements. Without this common vision, we may find ourselves deep into the system's lifecycle with multiple requirements for a function or capability that's incorrect or not even needed.

As the stakeholders tell their stories about how they'll use the system or what they expect it to do, and what their constraints are, we learn much of the information needed to write requirements:

- Interfaces—We see where the system interacts with other systems or processes, so we identify the interfaces early in requirement development

- Assumptions—We see what the stakeholders have assumed about the system or their processes, allowing us to discover whether the assumptions are true or conflict with those of other stakeholders

- "What if" situations—We ask the stakeholders to describe their expectations of the system during multiple situations in normal and abnormal conditions. Then we document how they expect the system to behave if something goes wrong. These "what-if" scenarios expose many issues to resolve during the scope phase. They are the source of requirements from which we produce a system that meets the stakeholders' expectations.

- Issues and risks—We identify issues among stakeholders as well as system risks, allowing us to address them early in requirement development

By developing a concept of operations, including alternative implementation concepts that encompass the system's lifecycle and that consider normal and abnormal conditions, we achieve six important goals:

- Promote a common vision—Instead of having different visions about the system from different stakeholders, the operational concept will be the single vision for how to develop, test, deploy, operate, maintain, and dispose of the system

- Avoid omitting requirements—As the stakeholders step through all scenarios for each of the system's lifecycle phases, we get a complete picture of what the system needs to do, thus avoiding requirement omissions

- Increase understanding among stakeholders—The stakeholders' stories expand knowledge and understanding for all

- Aid communication—The operational concept sets definitions for main system terms, leading to a common language for better understanding. This process begins a glossary for use throughout the requirement activities.

- Resolve conflicts early—We must identify issues and address them early in requirement development, instead of waiting until the requirement review, or a major design review, to uncover a problem

- Set stakeholder expectations—The concept of operations establishes the stakeholders' expectations and precludes surprises when we deliver the system

4.1.3   Identify Implementation Drivers and Constraints

Many limits on our system result from externally mandated rules and regulations. These drivers usually come from passive stakeholders, and are normally beyond the project's control. Because we can't control them, we must identify them early in requirement development to avoid unpleasant surprises later. Examples of requirement drivers include these seven:

- Higher-level requirements—Requirements imposed by a higher-level organization or system

- Standards and regulations—Standards and regulations within an industry, such as the Environmental Protection Agency, Federal Communication Commission, Occupational Health and Safety Administration, Food and Drug Administration, or the parent organization

- Cost—Project budgets that restrain us from developing requirements we can't afford to implement, such as producing an expensive launch vehicle when the budget demands less costly access to space

- Schedule—Schedule constraints that keep us from developing requirements we don't have time to implement

- Existing systems—Existing systems and processes outside our control often affect our system's features or functions. The FireSAT satellite has to use existing infrastructure, such as the communications architecture from NOAA's Command Data Acquisition system. If it's to be launched from the Kennedy Space Center, we need to follow processes required for range safety.

- Technology—The technology needed to develop the system may not be available today, but it must be mature enough to integrate into the system according to our schedule. We can't pick a technology that won't be ready on time or that presents too great a risk to the schedule.

- User expectations—Users of the FireSAT sensor may believe that integrating their sensor into the satellite will take only one week. We need to identify how their expectations drive the system's features or functions. If they don't fit the scope, we need to reset them.

Table 4-3 shows how these drivers affect FireSAT.

TABLE 4-3.   **FireSAT Key Performance Parameters and Illustrative Constraints.** Performance drivers and constraints impact the selection of implementation concepts. [Larson and Wertz, 1999] (IOC is initial operational capability; ATP is authority to proceed; FOC is final operational capability; NOAA is National Oceanic and Atmospheric Administration; GS-ISD-XYX is Ground System Interface System Document -XYX.)

Key Performance Parameters	Performance Drivers
Persistence—FireSAT shall monitor the coverage area for potentially dangerous wildfires at least once per 12-hour period	Number of spacecraft for low orbit
Timeliness—FireSAT shall send fire notifications to users within 30 minutes of fire detection (objective), 1 hour (threshold)	Storage delay (if applicable)
Geo-location Accuracy—FireSAT shall geo-locate potentially dangerous wildfires to within 5 km (objective), 500 m (threshold)	Altitude, aperture, control accuracy
Coverage—FireSAT shall cover the entire United States, including Alaska and Hawaii.	Altitude, inclination, number of satellites
Availability	None (weather dominated)

Illustrative Constraints for FireSAT include:
- FireSAT shall achieve IOC within 5 years of ATP, FOC within 6 years of ATP
- FireSAT total mission lifecycle cost, including 5 years of on-orbit operations, shall not exceed $200M (in FY 2007 dollars)
- FireSAT shall use existing NOAA ground stations at Wallops Island, Virginia and Fairbanks, Alaska for all mission command and control. Detailed technical interface is defined in NOAA GS-ISD-XYX.

Except for the full operational capability schedule, all drivers depend on assumptions about how the FireSAT spacecraft will be used. These assumptions may also be program or project mandates, but they remain assumptions until verified.

4.1.4   Identify External Interfaces

Most systems involve an interface to some external system, so one of the biggest requirement drivers is external interfaces. We must identify them during the scope phase to capture and control interface requirements as early as possible. Defining the active stakeholders helps with the early determination of external interfaces.

Our concept of operations helps define external interfaces by identifying external systems our system needs, along with other entities (ground systems, launch systems, users, and others) that our system will interact with. This depiction must show the complete set of active stakeholders. For example, FireSAT needs a launch vehicle to get to orbit, so the launch provider is an active stakeholder. We must also identify systems that need our system to do their job, such as the rangers' need for FireSAT's data at their locations. We must gather this information for all the lifecycle phases and identify which external systems exist and which ones don't. If any are to be developed, we have to work closely with the other developing systems to ensure compatibility at the interface. In addition to user interfaces, the external interfaces include such items as structural support, physical connections, communication buses, power supplies, command lines, data lines, operating systems, and data bases.

Context diagrams are a good first way to represent the system and its external interfaces. As we get into the details of requirement development, we must include details about each interface—what our system provides to other systems and what it receives from them. Figure 4-3 is a context diagram for the FireSAT spacecraft that shows interfaces for pre-launch ground support, launch, and on-orbit operations.

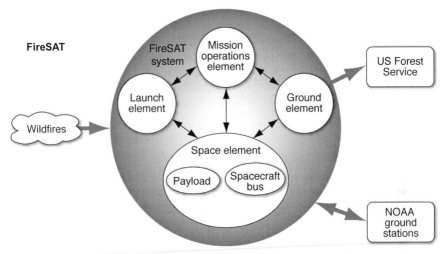

FIGURE 4-3.   Context Diagram for the FireSAT Spacecraft's External Interfaces. This diagram shows the major players in the FireSAT system, including the external interfaces to wildfires, the US Forest Service, and the NOAA ground stations.

4.2   Engineer the System Requirements

After we define the mission need and collect a set of stakeholder expectations, we must translate this information into technical system requirements (articulated from the perspective of the developers—hardware and software, the testers and integrators, and so on.) This translation is often the weakest link in the implementation of systems engineering principles. Developing the concept of operations facilitates this translation. Clarity in the definition of the system boundary, the scope, and the key operational drivers and constraints is critical.

The definition of a system depends on the stakeholders' point of view; we usually call it the *system of interest* (SOI). To the customer, the SOI may include all of the components in Figure 4-3. To the developer of the FireSAT spacecraft, the SOI is the 'Space element' component in Figure 4-3; all the other components are external systems. To the developer of the bus propulsion subsystem, that subsystem is the SOI.

While each of these systems of interest may be subsystems to the higher level system, their developers consider them to be systems and thus use the same processes to develop their requirements. As we show in Figures 4-1 and 4-4 (and in Chapters 2, 3, and 5), technical requirements development is an iterative process in conjunction with architecture development and synthesis of the concept of operations to define the solution in response to a customer or mission need.

Through these processes, the information becomes less abstract and more detailed as well as solution- and implementation-specific. A set of technical requirements at any level of abstraction is based on (and should be traceable to) the requirements and the architecture at the next higher level. The "how oriented" requirements at any level of abstraction become "what oriented" requirements at the next level down. Chapter 5 describes how to develop the solution architecture, and this section describes how to concurrently develop the solution (system) requirements.

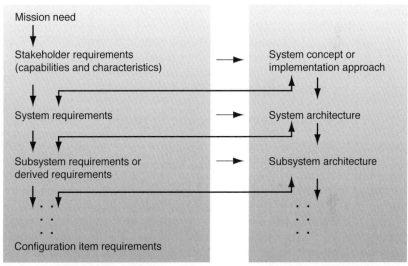

FIGURE 4-4.   **The Recursive Nature of the Requirements and Architecture Development Process.** We cannot develop the requirements in a vacuum. We must iterate among stakeholder requirements, system concepts, system requirements, and system architecture.

4.2.1   Capturing the Customer Need

Chapter 2 and Section 4.1 describe the significance of clearly articulating the mission need, goals, and objectives, which in turn triggers a solution response, and the application of systems engineering principles. The need drives the iterative articulation of stakeholder expectations, requirements, development of the concept of operations, and synthesis of alternative implementation concepts. The baselined and rank-ordered stakeholder expectations, expressed as capabilities and characteristics, represent *what* is required to achieve the business or mission need, goals, and objectives. The expectations should state what the business or mission needs, but currently cannot realize. The requirement should state some missing capability or what current capability can't be done well enough (it can't be performed fast enough, cheap enough, or it's not available enough of the time). The expectations should be stated in terms of the operational outcome rather than the system response. These stakeholder expectations must address the functional (capabilities required) and nonfunctional (characteristics required along with imposed constraints) needs of the stakeholders.

The capabilities define the functionality, the services, the tasks, and the activities that the stakeholders need. Because the active stakeholders use or interact with the system when it's operational, they usually specify the capabilities.

Nonfunctional expectations and constraints (cost, schedule, and legacy implementations) are the operational characteristics; they define the necessary quality attributes of the system. Nonfunctional expectations include performance, assumptions, dependencies, technology constraints, security and safety, human factors and other system "ilities", standards and regulations, and cost and schedule constraints. Most nonfunctional expectations are measurable. Some are stated in terms of constraints on the desired capabilities (e.g., constraints on the speed or efficiency of a given task). We often use capability-based expectations to develop one or more performance or quality characteristics.

4.2.2   Deriving and Developing the System Requirements

System requirements analysis is the process of analyzing stakeholder expectations and deriving the system requirements from them, based on insights developed from the concept of operations, and within the constraints and context of the selected implementation concepts. Engineering judgement, creativity, and domain knowledge, supported by analysis and tradeoffs, modeling and prototyping, simulation, and optimization, support this translation. Stakeholder expectations are written in the stakeholders' language and from the stakeholders' point of view. The system requirements are written from the system's point of view. The relationship between stakeholder requirements and system requirements may be one-to-one, one-to-many, or many-to-one, but we usually have more system requirements than customer expectations. Also, the relationships between stakeholder expectations and system requirements must be traceable. Table 4-4 shows how the two main kinds of expectations map to requirement types.

Figure 4-5 illustrates a two-prong approach to developing the system requirements from the stakeholder expectations. We translate the capabilities into a set of system requirements in the form of input requirements, output requirements, and functional requirements. We translate the characteristics into a set of system requirements also, such as performance requirements, quality and usability requirements, cost and schedule requirements, technology and standards related constraints, and other constraints that might be imposed by the legacy systems. Different methods and tools are appropriate for analyzing capabilities and characteristics. Section 4.4 has a more detailed discussion on the types of requirements.

TABLE 4-4.   **Stakeholder Expectations and Requirements.** Capabilities and characteristics translate to different types of requirements.

Stakeholder Expectations	System Requirements
Capabilities—these are translated into methods used, including use-case scenarios and interaction diagrams	• System input requirements • System output requirements • System functional requirements
Characteristics, including constraints—these are translated into methods that facilitate this translation, including prototyping, simulation, physical and mathematical modeling, market studies, benchmarking, etc.	• System performance requirements • System usability and quality requirements • Cost and schedule requirements • Technology constraints and requirements • Physical requirements (e.g., power, mass, volume, and so on) • System assurance-related requirements • System safety and security requirements • And so on

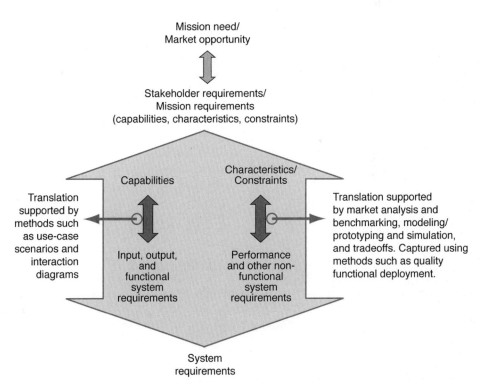

FIGURE 4-5.   **Translation of Stakeholder Expectations into System Requirements.** Different methods and tools support the translation of capabilities and characteristics into system requirements.

Translating Stakeholder Capabilities into Input, Output, and Functional System Requirements

As shown in Figure 4-5, the system input, output, and functional requirements typically map to the stakeholder expectations. These system requirements define what inputs the system needs and must react to, what functions, processing, and services the system must perform, and what outputs the system must produce. The context diagram and the use-case scenarios, developed as part of the concept of operations, are the primary tools we use to develop system requirements.

The context diagram supports the understanding of the scope of the system. In Chapter 3, we develop an illustrative scenario in response to the "Detect and Monitor Forest Fires" capability expected by the stakeholders. We must now translate the capability into an appropriate set of system requirements.

At the system level, use cases (use-case scenarios or operational scenarios) often represent system capabilities. We represent each stakeholder expectation by one or more use cases that describe a session (or a partial session) that a set of active stakeholders, such as the users, has with the system. The use-case scenarios are developed from the concept of operations—sometimes they're even defined within it—and they document how the system will be used. We further analyze each use case using an interaction diagram (also sometimes called a swim-lane diagram or sequence diagram), which documents the detailed interactions between the system of interest and the users, active stakeholders, or external systems. Section 5.2.5 discusses interaction diagrams in more detail, and Figure 4-6 gives examples for two FireSAT scenarios.

A use-case scenario depicts a user or other set of active stakeholders using the system as it exhibits a given desired capability. To completely articulate all the uses of a system, we must develop many related interaction diagrams. Use-case scenarios provide a lot more detail on the inputs, outputs, and functionality of the system than the context diagram. The latter shows all the external systems and users of a system within a single but static picture. Each use-case scenario shows only the external users and systems needed for its purpose. But if we document all the scenarios, the inputs and outputs (I/O) for every external system and user will be included in at least one of them.

The system input, output, and functional requirements are developed from these interaction diagrams. All arrows into the SOI describe input requirements. All arrows out of it describe output requirements. The functional requirements are what the system must do to transform the inputs into the outputs. We describe these functions by the arrows that begin and end at the system of interest in the use-case scenarios.

Developing the Nonfunctional Requirements

As shown in Figure 4-5, the nonfunctional requirements map to the system characteristics. These requirements tell how well the system must perform while also describing the physical, operational, and development characteristics of the

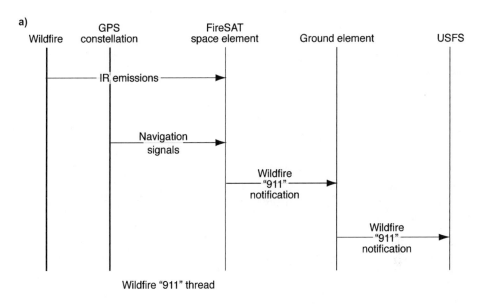

Wildfire "911" thread

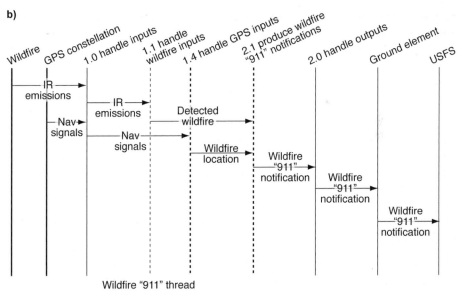

Wildfire "911" thread

FIGURE 4-6.   **The Wildfire "911" Scenario. a) At the System Level.** We show the inputs and outputs of the scenario for the system. **b) At the Subfunction Level.** The system level inputs and outputs from a) are now allocated to the subfunctions. Inputs and outputs between the subfunctions complete the scenario.

solution. They also reflect constraints on the system—performance, usability and human factors, availability, technology, security and safety, technical, size, mass, cost, and schedule. Input/Output matrices, requirement checklists, and quality function deployment (QFD) are examples of tools and practices to develop these system requirements.

4.2.3   Tools for Developing Requirements

Operational requirements specify not what the system does but how it should tolerate or respond to the operational environment throughout its lifecycle—from design and development through operations and disposal. We consider five key areas and representative questions for each area:

Transportation—How will we transport the system? What are the limits on size or weight? Should we be concerned with the transportation environment? Will the system need a shipping container?

> *Example:* The FireSAT spacecraft shall fit into a Type 2 Space Flight Equipment Shipping Container for transport from the manufacturing final assembly plant to Kennedy Space Center (KSC).

Facilities—Where will teams develop, test, assemble, integrate, and deploy the system? Do any facility restrictions affect it, such as size, weight, or cleanliness?

> *Example:* The FireSAT spacecraft shall pass fully assembled through the payload entry and exit doors of the Spacecraft Final Assembly Building #166 at KSC.

Training and personnel—Who will maintain, operate, or otherwise use the system? What are their skills? What must the system do to fit their skills or needs in terms of language, complexity, colors, or other attributes?

> *Example:* The FireSAT spacecraft shall be maintainable by a Class 2 Engineering Technician, as defined in KSC WP-3556, Technician Qualifications Manual.

Maintenance—What is the maintenance concept? What features or functions must the system have to meet it?

> *Example:* The FireSAT spacecraft shall accept on-orbit upgrades to its navigation and attitude control software.

Environment—What is the system's environment (induced and natural)? What features or functions does it need to survive and operate in its environment?

> *Example:* The FireSAT spacecraft shall meet its requirements after exposure to the launch environment defined in the Evolved Expendable Launch Vehicle's Standard Interface Specification, paragraphs 3.6 through 3.11.

4.2.4   "-ility" Requirements

The "-ility" (a subset of the nonfunctional) requirements are often the largest cost drivers and can't be taken lightly or just copied from other specifications without thoroughly reviewing how they affect the system. The following are common "-ility" terms.

Reliability involves attributes that enable the system to perform and maintain its functions under specified conditions for a given period, such as mean time between failure. Examples of these attributes are the probability that it will perform as specified, the failure rate, and the lifetime.

> *Example:* The FireSAT spacecraft shall have an on-orbit lifetime of
> at least five years.

Maintainability is associated with the mean time to repair (or replace), so this requirement covers the system's attributes that make it easy to maintain.

> *Example:* The FireSAT spacecraft shall require the removal (or
> opening) of no more than ten fasteners (panels) to
> replace any Line Replaceable Unit (LRU) listed in Table
> 3-2, Major Line Replaceable Units, during pre-launch
> ground operations.

Operability concerns the system's attributes that improve the ease of everyday operations.

> *Example:* The FireSAT spacecraft shall maintain payload pointing
> accuracy limits with no more than one ground update of
> spacecraft attitude in any consecutive 72-hour period.

Availability means the percentage of time a system needs to be operating or the probability that it will be operating when needed. It's sometimes stated in such terms as 24 hours a day, seven days a week, capacity during peak operations, or operational availability.

> *Example:* The FireSAT spacecraft shall have an operational
> availability of 98%, excluding outages due to weather,
> with a maximum continuous outage of no more than 72
> hours.

Supportability involves the system's attributes that enable technical support people to debug or analyze root causes to resolve a failure.

> *Example:* The FireSAT spacecraft shall downlink spacecraft and
> payload health and status data for the previous 60
> minutes at least every 45 minutes.

Manufacturability covers the system's attributes that support manufacturing goals, including those related to materials, welding versus screws, and restrictions on parts.

> *Example:* The FireSAT spacecraft shall contain no Class 1 ozone-depleting substances.

Security and safety relate to attributes that enable the system to comply with regulations and standards in these two areas.

> *Example:* The FireSAT spacecraft shall meet payload safety requirements in Chapter 3 of EWR 127-1, Eastern and Western Range Safety Requirements.

Interoperability specifies attributes that enable the system to exchange information with one or more systems and to use the exchanged information.

> *Example:* The FireSAT spacecraft shall use the Space Link Sub-network Data Transfer Services protocols defined in CCSDS-701.0-B-2 for ground-to-space communications.

Input/Output (I/O) Matrices

Input/output matrices document the details or performance necessary for every system-level input and output requirement. For each input and output requirement, the matrix determines the performance criteria for the actual input or output. These criteria define the details of the input or output such as accuracy, data rate, range of values, unexpected data, etc. The types of information depicted in the matrix are in Table 4-5.

TABLE 4-5.   Input/Output (I/O) Matrix Information. A well thought out I/O matrix reduces surprises later in the lifecycle.

	Intended or Expected	Unintended or Unexpected
Inputs	These inputs (in a particular form and format, with a particular latency and bandwidth, etc.) are expected by the system for it to produce the necessary outputs or system behavior. The constraints associated with these inputs should also be articulated here.	These unintended, and sometimes undesired, inputs, are also important. They're often beyond the system control, which defines the environment in terms of operating conditions, facilities, equipment and infrastructure, personnel availability, skill levels, etc.
Outputs	Descriptions of desired outputs. Upon development and deployment, the system should produce these outputs.	A newly developed system inevitably produces some undesirable outputs that, if anticipated in time, can be minimized. Examples include unintended electromagnetic and other emissions.

We show an example of an I/O matrix for different types of inputs and outputs in Table 4-6. In this example, the systems engineer writes I/O performance requirements for the inputs and outputs pertaining to each type of requirement. For example, if the SOI is to receive a signal from an external system, the systems engineer would write I/O characteristic requirements describing the pulse shapes of the input signal, its data rate, its signal-to-noise ratio, etc.

TABLE 4-6.   An Example of an Input/Output (I/O) Matrix. We replace the generic descriptions in the matrix cells with specific or quantitative requirements data as it becomes known.

| | Inputs | | Outputs | |
	Intended	Unintended	Desired	Undesired
Radio Frequency (RF) Signals	Ground commands	RF Interference	Telemetry signal	RF Interference
Electrical	Ground support equipment power	Surge voltages	None	Electrostatic discharge
Mechanical	Launch vehicle loads, shock, vibration, and acoustic noise	Transportation or handling loads, shock	Solar array articulations	Structural deformation
Environmental	Solar electromagnetic radiation, charged particles, vacuum, atmospheric drag	Micrometeroids or orbital debris	Heat, thruster exhaust	Outgassing, propellant leaks

Checklists

The other approaches to translate the desired system characteristics into system requirements depend on such factors as the maturity of an organization with developing similar systems and the newness of the implementation concepts. For example, organizations with experience in developing certain types of systems (e.g., Ford Motor Company and automobiles, IBM and IT systems and solutions, Nokia and cell phones) have the necessary domain knowledge and experience, and understand the correlations between desired characteristics and related system requirements. They have put together checklists delineating these correlations for use by new programs. After all, their development projects have much in common, and this way the development teams don't have to go through an extensive exploration and analysis to understand and derive the nonfunctional system requirements. Illustrative examples of such checklists are included in Figure 4-7. MIL-STD-490X–*Specification Practices* provides a set of guidelines and checklists to support the development of nonfunctional system requirements in the defense sector.

Unprecedented systems, or new technologies and new implementation concepts, call for more engineering exploration, benchmarking, analysis and modeling, prototyping, and tradeoffs. The engineering team has to address a larger number of unknown variables and correlations. Models, prototyping,

Design/System Parameters Impacting System Effectiveness:

• Performance
• Reliability
• Maintainability and maintenance
• Ergonomics
• Design life/shelf life
• Size and shape
• Transportability and installation
• Weight
• Packaging
• Potential to pollute, disposability
• Safety
• Skill level/training
• Requirements
• Distribution
• Energy consumption
• Aesthetics
• Design and development time

Design/System Parameters Impacting System Lifecycle Cost:

• Lifecycle cost
• Design and development cost
• Manufacturing cost
• Operating cost
• Maintenance and support cost
• Unit sale cost (acquisition cost)
• Phase-out and disposal cost

System Requirements Review (SRR) Template:

System Requirements Categories

• Functional requirements and functions to be performed
• Non-functional requirements
  – Performance requirements
  – Volume estimations
  – Performance and volume monitoring
  – Languages and location requirements
  – Growth and data migration requirements
  – Disaster recovery requirements
  – Availability and reliability, along with other specialty requirements
  – Security
  – Fault tolerance; probes: monitoring, measuring, and management
  – Cost and schedule
  – Testing, verification, and validation requirements
• External interface requirements
• Usability requirements
• Other
  – Technology constraints/customer desired
  – Standards and protocols
  – Dependencies, assumptions, and issues

FIGURE 4-7.   **Checklists as an Aid to Developing Nonfunctional Requirements.** Checklists contain distilled organizational wisdom, and help prevent "reinventing the wheel" when developing requirements.

benchmarking and trade-offs are vital. For example, one stakeholder expectation might be, "The cell phone should feel good to the person holding it." The engineering team needs to address this statement, and in the absence of domain knowledge and experience with such matters, they have to make certain engineering judgments. They build models and prototypes to test the correlations and then perhaps leverage benchmarking to validate these correlations. The engineering parameters they finally identify might relate to the physical shape of the phone, its weight and center of gravity, its texture and thermal conductivity, and so on. While each of these engineering parameters is important, equally important are the correlations between them.

Quality Function Deployment

Quality function deployment (QFD) helps translate desired system characteristics into nonfunctional system requirements. In this method, we try to be sure that the system requirements address stakeholder expectations appropriately. It provides a simple, direct correlation between them. Figure 4-8 depicts the QFD process. The output is a set of system nonfunctional requirements traced from the stakeholder characteristics. This practice is iterative at any level, and recursive across abstraction levels during complex system development. It also applies generally to the practice of systems engineering principles. Figure 4-9 shows a partial QFD matrix for the FireSAT system. This matrix is developed using a simple spreadsheet.

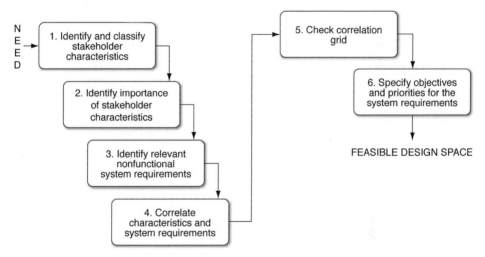

FIGURE 4-8.   Quality Function Deployment (QFD) Process Steps. We follow this process to move from the program need to the nonfunctional system requirements.

The I/O matrices help us develop the input and output performance or characteristic requirements, but not the other types of nonfunctional requirements. These requirements (e.g., reliability, supportability, maintainability, size, mass, usability, etc.) call for another technique. Requirement checklists apply to systems in environments where similar systems have been previously developed. And QFD is a technique for all types of nonfunctional requirement (I/O and others) in which a direct correlation exists between the stakeholder requirements and the system requirements. Table 4-7 lists the steps to develop a QFD matrix.

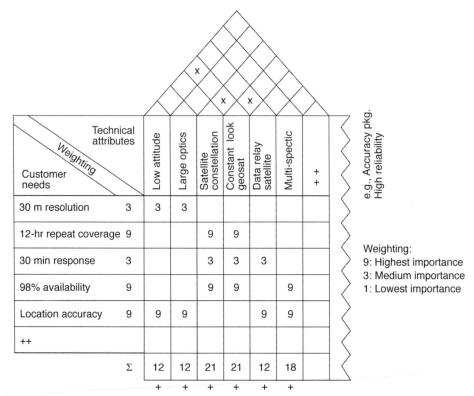

FIGURE 4-9.   **Simplified Quality Function Deployment (QFD) for FireSAT Mission.** This limited set of needs and corresponding technical attributes is expanded significantly in the complete design process. [Larson and Wertz, 1999]

Qualify the System Requirements

While we define the system requirements, we must think about how to qualify them, and for this we normally use a requirements verification matrix (RVM). Sometimes we use a requirements traceability and verification matrix (RTVM), where requirements traceability information is added. In its simplest form, the RVM is similar to Table 4-8. It shows each system requirement, the acceptance criteria for the requirement, and the test method to qualify or verify the requirement.

The acceptance criteria define how we determine if the requirements have been met. They're pass/fail. When defined at the same time as the requirement, they also help us understand the requirement statement. If the requirement is written badly (more on this in Section 4.3), the acceptance criteria provide additional information to ensure that all stakeholders have a similar understanding of the meaning of the requirement statement. The verification method describes how to determine whether the acceptance criterion is met.

TABLE **4-7.**   **Steps to Develop a Quality Function Deployment (QFD) Matrix.** The matrix reveals correlations between expectations and nonfunctional requirements.

Step	Description of the Step
1. Identify and classify characteristics	These entries are the row headers of the matrix. From the list of stakeholder requirements, we find the characteristics (nonfunctional requirements). We then classify or group them into a hierarchy. The grouping may be based on several criteria, depending on the types and conciseness of the requirements. Possible ways to sort are: by requirement details (abstract to more detailed), by types of requirements (cost, availability, performance, etc.), or by the stakeholder (user, customer, external systems, operations and maintenance, etc.). In all cases, the goal is a firm definition of WHAT we want from the system.
2. Identify importance of characteristics	As Section 4.5.1 points out, we should rank order all requirements. The priorities may be relative ("this requirement is more important than that one") or absolute ("this is the most important requirement, then this one, then..."). Either way, the priority represents the customer's value system. This effort lets the systems engineer know which requirements are more important than others.
3. Identify relevant design parameters or system objectives	These entries are the column headers. This step is critical because the design parameters describe HOW the system will address the stakeholder requirements. They become system requirements. While the stakeholder requirements define what the stakeholder needs (in their language), the design parameters reflect what the system will do to meet the stakeholder requirements. The design parameters are the system requirements that relate directly to the customer requirements and must be deployed selectively throughout the design, manufacturing, assembly, and service process to manifest themselves in the final system performance and customer acceptance. Thus, the design parameters must be verifiable (they are often quantifiable) and meet the requirement 'goodness' characteristics defined in Section 4.3.1.
4. Correlate characteristics and design parameters or system objectives	This step populates the cells in the QFD matrix. We analyze each design parameter for its influence on customer requirements. The correlation matrix has several levels of correlation. Depending upon how much resolution we need, we use three to five levels of correlation.
5. Check correlation grid	We have to see that the design parameters address and impact every characteristic, eliminating any that do not. We look at each cell for: • Blank rows that represent characteristics not addressed by the design parameters • Blank columns that represent possibly unnecessary design parameters • Rows and columns with very weak correlations to indicate possible missing correlations or design parameters • A matrix that shows too many correlations, which typically suggests that the stakeholder requirements need more detail
6. Specify design priorities	This step helps us understand whether the potential solution will meet the design parameters (system requirements). It also helps us understand the design parameter priorities (which one is most important in meeting the customer's need). This step may drive trade studies or design studies into potential solutions to meet the design parameters.

TABLE 4-8. **A Sample Requirements Verification Matrix.** Here we list sample system requirements and acceptance criteria for the FireSAT system.

System Requirement	Acceptance Criteria	Verification Method
The system shall provide current system status when the user logs on	When the user logs on, the system must display the current status of all fires	Demonstration
The system shall provide the current position of all fires within 5 km (threshold), 500 m (objective)	Using a simulator to inject a potential fire, the position of the leading edge of the fire must be within 5 km (threshold), 500 m (objective) of the injected position	Simulation and modeling
The system shall have an availability of at least 98%, excluding outages due to weather	The system must be operational for a 100-hour demonstration time period	Demonstration

Typical verification methods are: formal test (T), demonstration (D), inspection (I), analysis (A), and simulation or modeling (S/M). We may add information to the matrix during the system lifecycle as it becomes known, information such as:

- Traceability from stakeholder to system to component requirements
- Current development status of the requirement
- Development artifacts that the requirement affects (architecture, hardware or software design documents, the code modules, etc.)
- Testing information (test phases in which the requirement will be tested, test plans, test cases, test results, etc.)

Define, Review, and Baseline the System Requirements

Once we develop the system functional and nonfunctional requirements, we must document them (see Sections 4.3 and 4.4). Then the requirements, along with the system-level architecture, are reviewed and baselined by all stakeholders (Section 4.4). We have to rank order the system requirements, based on their traceability to the rank-ordered stakeholder requirements and the design priorities defined in the QFD matrix.

The system requirements review (SRR) is where the stakeholders review and baseline the system requirements and architecture, as well as the associated cost and schedule to design, implement, test, manufacture, deliver, and maintain the system. From the system requirements we conduct the system verification. The risk associated with the system requirements and the technical decisions made to develop the requirements also should be discussed at the SRR. System risks may be technical, cost, and schedule risks. For the technical risks, we should review the technical performance measures. These identify deficiencies in the solution in meeting some top-level requirement (stakeholder or system requirement). They provide an early warning of potential program problems. After finding these deficiencies, we monitor them throughout the system lifecycle. We should discuss tradeoff decisions so all stakeholders understand the system requirements. After the SRR, the system requirements must be managed (see Section 4.5).

4.2.5   Develop the Component Requirements

The system architecture and its associated requirements completely define the system of interest. As Figure 4-4 shows, developing them is an iterative and parallel process. (Chapter 5 deals with how to develop a system architecture.) The requirements should be traced from the higher levels of abstraction to the lower levels. At any level, the various artifacts—concept of operations, stakeholder expectations, implementation concepts, technical system requirements, and system architecture—are developed iteratively, and there's significant interdependency among them. Accordingly, depending on the scope and complexity of the program or project, we may need to develop one or more levels of component requirements.

Figure 4-4 alludes to the multiple levels of abstraction. Just like the requirements, the functional and physical architectures are also defined at multiple levels. Requirements are typically defined at each level of the architecture to ensure traceability through the development of the solution. As Section 5.3 discusses further, the configuration item level is the lowest level for developing an architecture. So it's also the lowest at which the requirements are defined, and sets what the development teams (hardware, software, and process development teams) need to design and develop. The components are the physical elements of the system, the pieces of the physical architecture. Requirements must be developed for each one.

As previously mentioned, a system of interest depends on the stakeholder's point of view. One stakeholder's system of interest may be just a subsystem of another stakeholder's system of interest. Since the requirements and architecture processes are iterative at any abstraction level, and recursive across levels, we should use the same (or similar) process for developing component requirements as for system requirements.

4.3   Document the Requirements

Requirements fall into two major categories: those for systems and those for statements of work. System requirements state what the system must do or what features and qualities it must have (the functional and nonfunctional requirements). Statement of work requirements state what people have to do. This chapter covers system requirements, but the following rules and characteristics of "good" requirements apply equally to those for a statement of work. Table 4-9 shows examples of system and statement-of-work requirements.

Projects require both types of requirements, but we keep them separate, either in different documents or within individual sections of a single document. We also want to avoid stating project needs in our system requirements.

TABLE 4-9.   **Examples of System and Statement-of-Work Requirements for FireSAT.** Here we list system requirements and requirements for the statement of work. A complete list for a small project could comprise dozens of items.

System Requirements	Statement of Work Requirements
The FireSAT sensor shall operate on 28 Vdc	The contractor shall submit cost and schedule status each month according to...
The FireSAT spacecraft shall have a lifetime of at least five years	The contractor shall do trade studies to determine...
The FireSAT spacecraft shall have a diameter of not more than one meter	The contractor shall provide technical status following...

4.3.1   What is a Good Requirement?

A good requirement clearly states a verifiable and attainable need. With technical and corporate knowledge from the scope and this working definition of a "good" requirement in hand, how then do we write good requirements?

Mandatory Characteristics

Just because a sentence contains the word "shall" doesn't mean it's an appropriate requirement. Every requirement must have three characteristics:

- Needed—We want to have a **necessary and sufficient** set of requirements in the system specification. We should ask, "What's the worst thing that could happen if I delete this requirement?" If we can't think of anything bad, we probably ought to delete it. This precaution guards against over-specifying the system.

- Verifiable—We must be able to verify that the system does what the requirement says or the statement isn't a requirement

- Attainable—If a requirement is technically impossible or can't be obtained within the current budget and schedule, we shouldn't include it

We must analyze all requirements to determine if they're needed, verifiable, and attainable before committing to a major design initiative. We have to eliminate or rewrite those that don't have these mandatory characteristics.

Other Characteristics

Requirements communicate what the project or the customer wants from the provider or developer. The key to good requirements is good communication, which depends on meeting five criteria:

- One thought—Limit each requirement to a single thought
- Concise—Keep it "short and sweet" by avoiding long, complex statements

- Simple—The same words should mean the same things. A gate is a gate, not a door, portal, or opening. Good requirements can be boring to read; but we shouldn't dive into a thesaurus to make them more interesting.
- Stated positively—Use positive statements. Verifying that a system **doesn't** do something is next to impossible. Some exceptions exist, but in general we should avoid negative requirements.
- Grammatically correct—Requirements are hard enough to understand without introducing poor grammar

Effective Syntax for Requirements

Good requirements state **who shall do what**. In the case of system requirements, the **who** is the system. For example:

- The transmitter shall operate at a power level of...
- The telemetry receiver shall acquire data from...
- The launch vehicle shall withstand loads of...

Following "shall" are active verbs stating "what" the system is to do, and what must be verified. When a requirement includes conditional clauses (if or when or under what circumstances), these clauses should follow the requirement statement—after identifying the **who**. All too often requirements with an opening conditional clause omit the **who**, leaving readers to determine who must complete the stated action.

Things to Avoid

To improve requirements, avoid **ambiguous terms** and don't include words that refer to implementation (**how** to do something) and operations. Ambiguous terms aren't verifiable, so statements containing them fail the "Is it verifiable?" test. Some examples [Hooks, 2007] of ambiguous terms are

adequate	easy	et cetera
high speed	including, but not limited to	maximize
minimize	quickly	robust
sufficient	ultra-low power	user-friendly

There are many others. We should create a list for our requirement specification, search all the requirements for these terms, and replace them with verifiable terms.

Including implementation in a requirement means saying how to provide something rather than what we need. Requirements that tell the developers **how** forces them to use a particular design or solution that may not be the best one. Although we may believe that stating the solution covers all real requirements or needs, we could be missing the true need. The developer may deliver what stakeholders ask for, but it may not be what they wanted.

For example, one of the requirements in the original specification for the DC-3 aircraft was: "The aircraft shall have three engines." This statement imposed a solution of three engines. The developer (Douglas Aircraft) didn't think it made sense and asked the airline: "Why did you specify three engines?" The answer was that if the aircraft lost one engine, it still needed to be able to land safely. So the real requirement was: "The aircraft shall meet its requirements with a single engine out." The initial requirement could have led to an aircraft with three engines that crashed or became unstable if one of the three engines failed—totally missing the stakeholder's intent.

Sometimes, we do have to state how to implement something, such as when a higher-level requirement directs a particular solution or imposes a specific regulation or standard. Still, best practices dictate that requirements typically state what customers want, not how to provide it, so developers can deliver the best solution.

Operations show up in requirements when they blur the line between operational concept and requirement. Operational concepts are an important part of understanding what the requirements are, or the need for a particular requirement, but they aren't requirements. In other words, we can't rewrite operational-concept statements with **shalls** and turn them into requirements. Table 4-10 shows how we must rewrite a requirement that includes an operational statement.

TABLE 4-10. **Operations Statement Versus True Requirement.** The requirement should state what the product is to do, not how it will fit into an operational scenario.

Incorrect (Contains Operations)	Correct (True SE Requirement)
The FireSAT sensor operator shall be able to receive a FireSAT sensor's data download no more than 45 [TBD] seconds after initiating an "out-of-cycle download" command.	The FireSAT spacecraft shall downlink sensor data no later than 30 [TBD] seconds after receipt of an "out-of-cycle download" command from the ground.
	The FireSAT system shall display updated sensor data no more than 45 [TBD] seconds after initiation of an "out-of-cycle download" command.

We must remember that requirements need to follow the **who shall do what** format, and that we write requirements for the system, not for people. The phrase **shall be** or **shall be able to** often indicates an operations requirement. The requirements should capture functions the system needs to do or features it must have to meet the operational concept.

Correctly representing information in an operations statement may take more than one requirement. The **who** for each of the new requirements in the example refers to two systems: the FireSAT spacecraft and the FireSAT system as a whole. For this reason, operations statements can cause requirement omissions throughout the system hierarchy, so we need to remove them from all system requirements.

4.3.2   How Can We Improve Requirements?

Besides writing defect-free requirements that communicate more effectively, other ways exist to promote understanding of the requirements and improve their management. One way is to document and associate with each requirement certain "requirement quality" attributes: rationale, verification method, allocation, and traceability. We should maintain this information manually in a spreadsheet or automatically using one of a host of automated tools. Having these attributes documented and available for reviewers benefits the process as we go through requirement review and approval. It also reduces questions or assumptions by those who develop, verify, and maintain the system.

Include the Rationale

While writing requirements, we often need to include explanations that help developers and verifiers understand what they mean. Rationale explains the need, assumptions, design efforts or trade studies that drove a requirement, the source of any numerical values it contains, and other information to help understand it. All change control processes require a reason for making changes to baselined requirements, so we can't afford to be any less conscientious when we're writing original requirements.

A rationale has many benefits. It helps shorten the review time for the requirements by increasing understanding and reducing the interpretation problems that could otherwise arise. Many developers and testers make assumptions about the requirements due to lack of information, and good rationale overcomes this problem and reduces development and test problems. After the system is developed and has been used for some time, the rationale aids maintenance and upgrades because we know the requirement's original basis. Also, as people move up or away, their documented knowledge stays with the requirements, thereby preserving corporate knowledge.

Authors often place this information in the requirement statement, which leads people to confuse the requirement with its rationale. So where does this information go then, if we're to keep each requirement to a single concise thought? Typically, we place a rationale either below a requirement in its own paragraph and in a different font or beside the requirement in a table. The first way of including a rationale looks like the following (notice the clearly different font for the rationale):

> The FireSAT spacecraft shall meet the payload safety requirements in Chapter 3 of EWR 127-1, Eastern and Western Range Safety Requirements.
>
> Rationale: Nothing can be launched from the Kennedy Space Center unless it complies with the ground safety standards in Chapter 3 of EWR 127-1.

Table 4-11 shows an example of how we could list requirements in table format, with the requirement statement in the left column and the rationale directly across from it in the right column.

TABLE 4-11. **Example FireSAT Requirement with Rationale.** Including the rationale beside the requirement connects every requirement to the reason it exists. This process reduces misunderstandings during the entire program lifecycle.

Requirement	Rationale
The FireSAT spacecraft shall meet the payload safety requirements in Chapter 3 of EWR 127-1, Eastern and Western Range Safety Requirements.	Nothing can be launched from the Kennedy Space Center unless it complies with the ground safety standards in Chapter 3 of EWR 127-1.

Make Sure That the Rationale Is Unique and Brief

To keep the rationale brief and unique, we develop and document it as we write the requirement. Adding a rationale "after the fact" is at best difficult and at worst impossible, if the author isn't available to provide it. Writing a rationale at the same time as the requirement easily improves requirement understanding. This approach also keeps us from making the rationale

- A rewrite of the requirement
- A place to hide other requirements
- A copy of the rationale from another requirement (each requirement is unique, so its rationale should be unique)
- Everything we know about the requirement (brevity is important)

Define the Verification Methods

Verification proves that the "as built" system does what the requirements said it would do. Requirements must be verifiable. That doesn't mean we must write the entire test plan for each requirement, but as we write the requirement, we should think about which of the five commonly accepted verification methods applies: test, demonstration, analysis, modeling and simulation, or inspection.

Going through this thought process improves the quality of the requirements by driving out implementation language and ambiguity. Consider the example requirement for three engines on the DC-3 aircraft. We could verify the aircraft has three engines using inspection—just count the engines—but we really want to verify that the aircraft will fly with a failed engine.

An added benefit to defining the verification method while writing requirements is that we start thinking about what facilities and equipment we need to verify them. Will what we have do the job, or do we need new or modified capabilities? Addressing facilities and equipment needs early helps in overall project planning. We reduce verification time and cost if we also consider the requirement's precision. For example, verifying a narrow tolerance requirement may be very expensive, but an assessment may indicate that a wider tolerance is acceptable. If we change our requirement to the wider tolerance, we may improve cost and schedule during verification. Chapter 11 provides details on developing verification requirements that address the "what," "when," and "how well" of determining that the system has met a requirement.

Allocate Requirements to System Levels

Before discussing allocation and multiple levels, we need to define the concept of system of interest (SOI). An SOI is closely integrated with the concept of system and system scope. Figure 4-10 demonstrates the idea of levels of SOIs. For each level of the FireSAT system's hierarchy, we define an SOI. For example, the FireSAT spacecraft SOI includes every subsystem that makes up the spacecraft. The space segment SOI includes the SOIs for the FireSAT spacecraft and the mission control center (MCC). These SOIs have different scopes, but requirements written for each one must begin: "The [SOI name] shall...."

Allocating requirements means assigning requirements defined at one level of the architecture to a portion of the architecture's next level. Thus, we allocate requirements for the space segment either to the FireSAT spacecraft, to the MCC, or to both. But allocations don't skip levels, so we allocate every requirement at one level of the architecture only to the next SOI level. Allocation can't occur until we define most of the requirements and develop an architecture for the next level. We also use the term *allocation* to refer to apportioning resources (mass, reliability, timing) among the lower-level SOIs.

Besides ensuring that all system requirements apply to at least some part of the architecture and guiding writers of requirements for the lower-level SOIs, allocation helps us

- Identify possible internal interfaces
- Find redundant or inconsistent requirements
- Ensure completeness of the requirements

Make Requirements Traceable

Traceability means identifying a requirement's source, at whatever level the requirement exists. As with rationale, we should identify traceability as we write the requirement. The requirement's source (parent) is often in the requirement document for the next higher-level SOI. But it may be in another document, such as the function to which the performance requirement responds. The parent may also be in other documents, including the operational concept document or in a standard or regulation mandated by the program or project plan. A requirement may have more than one parent, but this isn't usually the case.

Examining traceability quickly roots out orphan requirements and mismatches between traceability and allocation (e.g., requirements appearing in a lower-level system that trace to requirements that were not allocated to the next

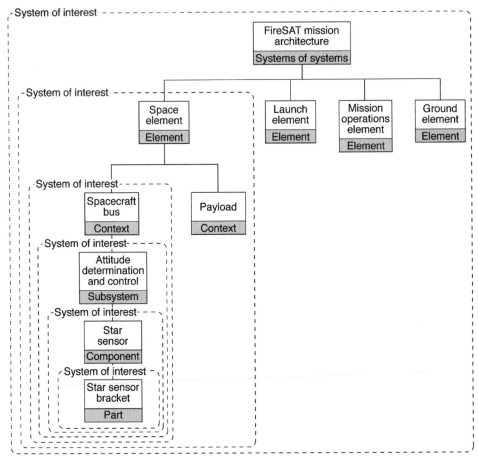

FIGURE 4-10. **Notional System-of-Interest Hierarchy for FireSAT.** This diagram shows the various system-of-interest levels for the FireSAT program. Each level has stakeholders and needs requirements prepared to ensure it integrates well with the next higher level and contributes to mission success.

higher-level system). It gives us an early opportunity to correct omissions or misunderstandings and to prevent gold-plating. It also helps us analyze the impact of changing a requirement once we've baselined the requirements and placed them under configuration control.

Proper traceability requires that we assign unique, unchanging identification (ID) numbers to each parent and child requirement. We also need to assign ID numbers to parents that aren't requirements, such as a statement in an operations concept or in any other document making up the approved scope. We use these ID numbers to determine unambiguously the child-to-parent relationship between a requirement and its source. Using document paragraph numbers for requirement IDs isn't appropriate because paragraph numbers are likely to change as the documents change. Small projects may be able to allocate and trace requirements in a spreadsheet, with precautions for keeping ID numbers unique, but larger projects need to use a requirement management tool to track everything properly.

4.4   Baseline the Requirements

The methods used to baseline requirements vary among organizations, depending on their culture and the types of systems they deliver. But for a method to succeed, it must:

- Validate scope—Document the scope, validate it with all stakeholders, and get sign-off at the mission concept review [NASA, 2007]
- Validate requirements—With the baselined scope as a foundation, write and document the requirements. Validate each requirement statement (including requirement attributes) as it's written. Chapter 11 discusses this activity in detail.
- Baseline requirements—When the requirements are complete, validate them as a complete set with all the stakeholders and baseline them at the system requirements review

Validating the scope and the resulting requirements is the key to developing quality, defect-free requirements. We want to remove requirement defects before, not after, we baseline.

4.4.1   Validate Scope

To develop a defect-free requirement baseline, we have to review the scope, but first we need to address questions in nine important areas:

- System boundary—Have we identified all the external interfaces? Are they developed and mature?
- Stakeholders—Have we talked to all stakeholders?
- System lifecycle—Have we addressed all the system's lifecycle phases?
- Disagreements or conflicts—Have we resolved disagreements or conflicts among the stakeholders?
- Technical issues—Have we resolved the technical issues or issues about technology availability?
- Schedule—Is the schedule too optimistic considering the technologies and complexity of the development?
- Budget—Is the budget too lean to develop the product within the given schedule?
- Uncertainties—Have we resolved concerns about the scope?
- Assumptions—Have we documented all assumptions? Has everyone reviewed the assumptions and do they understand them?

The answers to these questions quantify the scope's completeness. We use the above list to decide if we're ready to hold the scope review. If we answer no to any of these questions, we must assess the risk of holding a review. A scope review should confirm that we've answered these questions to the stakeholders' satisfaction and have agreement that we're ready to move on to writing requirements. Our last step in the scope review is to document agreement among the stakeholders, with the understanding that all scope documentation then goes under formal configuration control.

4.4.2   Validate Requirements to Reach a Baseline

Requirement validation gives us confidence that the requirements we write are needed, verifiable, and attainable; clear and concise; and correct, consistent, and complete. It is either continuous or discrete:

- Continuous validation—The day-to-day activities that control the quality of the requirements as they're written. It starts with the first requirement we write and continues throughout the system's lifecycle. Continuous validation holds everyone (writers, reviewers, developers, testers, and managers) responsible for the requirements' quality and is the best way to produce defect-free requirements.

- Discrete validation—The activities to prepare for and conduct formal requirement reviews. The first formal review of the requirements is the system requirements review (SRR). Other major reviews requiring updated requirements follow, such as those for preliminary design, critical design, and test readiness. The process defined to prepare requirements for the system requirements review applies equally to other reviews.

4.4.3   Make Validation Continuous

Continuous validation is crucial to ensuring continuous process improvement. We don't want to wait until we're preparing for a major milestone review, especially the SRR, to find out we have a **big bad** set of requirements. However, if our schedule is tight, we may baseline the requirements even when the SRR exposes many problems with them. But baselining defective requirements always wastes resources because we have to fix the requirements and redo any work completed before we discovered the defect.

Incremental work-system inspections are a key to continuous validation. They're more commonly called Fagan inspections—named for Dr. Michael Fagan, who originated the process at IBM in 1976. They identify and correct requirement defects before they become system defects, which are instances "in which a requirement is not satisfied" [Fagan, 1986].

Requirement inspections are highly structured, formal peer reviews of selected subsets of the total requirements by a small group of reviewers (no more than six) including a moderator. Moderators work with requirement authors to set up and manage the review, and they follow up to make sure that defects are addressed.

Criteria for the review are determined in advance, and reviewers measure requirements only against those criteria. Example criteria might be: "The requirements are free of ambiguous terms, have the proper **who shall do what** format, include rationales, and identify the verification method." So we inspect a reduced set of the requirements and focus on specific defects we wish to remove.

We use inspection results not only to remove the requirement defects but also to help determine their root causes, so we can improve how we define our requirements. For example, if we find that several authors repeatedly use certain ambiguous terms, we should update our ambiguous-terms checklist. Inspections are a proven method and integral to practices under the Capability Maturity Model® Integration (CMMI®).

Work-system inspections help us do more than goodness or quality checks on the requirements. We also focus them on the requirements' content. For example, are the requirements correct and consistent? If we're involved in model-based systems engineering (MBSE), do they reflect the systems engineering models, and do the models reflect what the requirements state? Validating the requirements is essential to validating the evolving MBSE design. Even if we use MBSE, however, we have to build the system and verify that it meets the requirements. The models aren't sufficient to build the system; they simply help us understand the requirements the same way, improve communications, and reduce misunderstandings.

In other words, we use requirements to generate the model, which in turn helps us more efficiently determine the completeness of our requirements by using the automated modeling tool capabilities to help us do consistency checks, generate N×N interface diagrams, simulate system behavior, and complete other automated cross-checking activities. But the model exists and is **valid** only insofar as it correctly reflects the requirements.

4.4.4   Follow a Process for Discrete Validation

Preparing for a milestone review, especially the SRR, is a major undertaking. It's often the first time many of the stakeholders view the requirements as a complete set. Unlike the incremental inspections, this review focuses on the content of all requirements, not on whether the requirements meet "requirement standards" criteria.

The SRR isn't just a point in time on a project's milestone chart. Most of the work for it doesn't occur at the meeting itself. The requirements must be ready for distribution to the reviewers several weeks before, with instructions on what we want them to review and how we're going to address and resolve their changes. Requirement changes that we resolve to the satisfaction of all stakeholders are accepted and placed in the document or requirement database. Ones that aren't resolved before the SRR become agenda items at the review. After the SRR, with final disposition of requirements resolved, we baseline them and place them under configuration control. In other words, all of the approved requirements become version 1 (or a)—identified collectively as "requirement baseline Ver. 1" or whatever numbering system we choose.

Review of the requirements for a milestone review takes a lot of time and money, so we want the requirements to be in the best condition possible. Table 4-12 depicts a process to reduce the time to conduct the review and correct the document, limit the effort of individual reviewers, and produce the best results [Hooks and Farry, 2001].

TABLE 4-12. 4-1/2-Step Process for a Requirement Review. Preparing for a requirements review takes a staff of specialists and most of the stakeholders to review the documents, make and review corrections, and publish the best documents possible for the official review.

Step	Review For	Who Does It	How Many
1	Editorial	Person with editorial skills	1–2
2	Goodness	People who know requirement rules and have some technical knowledge	2–3
3	Content	All relevant stakeholders	As many as needed
4	Risk	Person with technical and management knowledge	2–3
4 -1/2	Editorial	Person with editorial skills	1–2

The 4-1/2-step process establishes the initial requirements baseline, such as for a system requirements review. It's equally effective in preparing a scope review or any other set of formal documents for review. Before distributing the requirements in a specification or requirements database, someone must edit the working, and afterward technical specialists must check the contents for goodness. If we inspect all along, these two steps shouldn't require much time or resources. Then the stakeholders have a good set of requirements to review and won't waste time with misspellings, bad formatting, or unclear wording. With all changes resolved and approved, we need to assess risk using the requirement risk factors:

- **Unclear**—The requirement is ambiguous or unverifiable, or has poor grammar, the wrong format, or inaccurate or missing rationale
- **Incomplete**—The requirement is missing related requirements or requirement attributes
- **Subject to change**—The stakeholders are not in agreement, the requirement depends on an undefined interface, or the selected technology is not ready

If we do the 4-1/2-step process with proper management controls in place, we'll never again have to apply this exact process to this document. We'll do incremental changes and update versions, but we won't need a complete review.

Before we baseline the requirements, we can change them as needed with few resources, such as inspections. But afterwards, a configuration change control board will manage changes, which requires a lot more resources and more strongly affects the schedule.

4.5   Manage the Requirements

Managing requirements consists of having a management structure in place that ensures that each requirement:

- Is structured correctly and written at the correct level (in **who shall do what** format, with the **who** matching the system for which we're writing requirements)

- Meets all characteristics of a good requirement (one thought, concise, simple, positively stated, grammatically correct, and unambiguous)

- Has its **requirement quality** attributes (rationale, verification method, traceability, and allocation) documented and validated

- Has its **requirement management** attributes (owner, author, risk, priority, validation status, change status) documented and validated

- Is accessible with all of its attributes to all stakeholders for review and status, either through direct access in an automated tool for managing requirements or through periodic status reports

- Is properly assessed to determine how changes affect it

Previously, we discussed the first three bullets and how we use them to improve the quality of our requirements. But what are **requirement management** attributes and how do we use them to manage requirements?

4.5.1   Manage and Measure Requirement Attributes

Quality attributes (third bullet in the above list) give us the means to validate the requirements. But they're also valuable as metrics, along with the **requirement management** attributes (fourth bullet in list above), to show us the status of our requirement development activities. For example, if we're trying to decide if we're ready for the system requirements review (SRR), we could look at the total number of

- Requirements (counting the number of requirement IDs)

- Requirements allocated (counting the number of requirements with the allocation field filled in)

- Requirements validated

Assuming the allocations are correct (valid), we could safely assume we're nearly ready for the SRR, if we have allocated and validated nearly 100 percent of the requirements.

The following is a list of requirement attributes whose metrics provide us with management information. The list isn't all-inclusive, but does suggest the kind of information we should maintain with each requirement, so we have statistics that are invaluable in managing the system development.

- Owner—The owner of the requirement is the relevant stakeholder (as defined in Section 4.1.1) who has the authority and the responsibility to change the requirement

- Author—The requirement writer. This person may be the owner or someone else. This attribute is invaluable in assessing "continuous process improvement." It's also important for resource management. Someone should always be the designated author, so if the original author leaves, we must name a replacement and transition responsibility for maintaining information on requirement attributes.

- Risk—This risk is that the requirement will change, not risk to the project. It could be as simple as H (high), M (medium), and L (low) or 1, 2, and 3. The requirement risk is a subjective measure based on the **requirement risk** factors defined in Section 4.4.4.

- Priority—All requirements aren't equal. Rank ordering assigns a relative importance to requirements. We should keep it simple; otherwise, assigning and managing the priorities becomes too time-consuming. Rank ordering allows us to handle unforeseen events, such as a budget cut or a forced schedule change, without going into crisis mode. In combination with risk, priority focuses development effort or trade studies on requirements essential to the system's success.

- Validation status—Indicates whether the requirement has been validated

- Version number—Even before we baseline, we should start keeping version numbers of the requirement to see the amount of churn (instability). If many of our requirements have many versions, we need more churn management.

The quality and management attributes enable us to manage requirements during the development phase (before the system requirements review) and throughout the system's lifecycle. Let's look at just five ways we use requirement attributes to manage the requirements:

- Consistency—Cross check allocation with traceability to ensure that subsystem requirements that trace to a parent system requirement match with the subsystem allocations of the parent requirement.

- Missing requirements—Determine the number of allocated requirements without child requirements to assess completeness of requirement definition at the next level down. Generate a list of allocated requirements that don't have child requirements for management follow-up.

- Requirement completeness—Determine the number of requirements with validation status **complete**, which shows whether we're ready for the system requirements review. Use traceability to generate a report showing all child requirements of a parent requirement. Include the text and rationale for the parent requirement and the child requirements. Assess whether the child requirements are sufficient to implement the parent.

Cross check against system models for projects using model-based systems engineering.

- Requirement volatility—Determine the number of requirements with a change number entered in the change status field, and evaluate over time to determine if they need management attention. (This is most effective after baselining the requirements.)

- Risk assessment—Determine the number of requirements with high (H), medium (M), and low (L) risk; evaluate over time to determine how well we're meeting the plan to reduce risk.

  Note: In all of the above examples the team needs access to more than the requirements for their own SOI. To manage their SOI requirements effectively, they must also have access to the requirements (and related requirement attributes) in the first level above and below their level of the architecture.

4.5.2   Manage Change

By having a valid scope, we develop a good set of requirements. Using that scope for guidance, we continue managing both requirement changes and stakeholder expectations. The best way to minimize change is doing the best possible job the first time—by defining and baselining the scope before writing requirements and by not baselining a bad requirements document. Chapter 16 addresses the configuration control activity that is key to managing change.

No matter how well we develop defect-free requirements, however, we'll probably still need to change some of the requirements after they're baselined. Budget cuts, system upgrades because of new technology affecting long-term projects, unplanned changes in an interface, or changes in a regulation or a standard are just a few events that necessitate changing the requirements. To help us manage these necessary changes, five requirement attributes are most valuable:

- Requirement text—statement of who shall do what: the content of the existing requirement (if one exists) and the change proposed

- Rationale—the reason for the existing requirement (if one exists) and the reason for changing or deleting it or the reason for a new requirement

- Traceability—other requirements related to the requirement being changed or added

- Priority—how important the requirement is in meeting stakeholders' expectations

- Risk—a subjective measure of the expectation that the requirement will change

Of the five attributes, traceability (child-to-parent and parent-to-child) is the key to evaluating how a requirement change will affect the system. We must trace up to the parent to ensure we're still in compliance and down to the child requirements to obtain a complete list of the requirements affected by the change. Traceability, when combined with the rationale for each requirement, allows us to assess the impact of making or not making the change.

Impact assessment, however, includes more than just counting the number of requirements affected. We also need to know how important they are. The conventions used to assign priority and risk to the requirements for any system development are in the project's systems engineering management plan (SEMP). But in the discussion that follows we assume priority has one of three values:

Priority 1—not meeting it will cause us to miss meeting a stakeholder's expectation

Priority 2—identifies the trade space

Priority 3—has alternatives, such as meeting it using manual operation rather than automation or waiting until the system's next version

When we generate the trace to all the requirements that a change affects and find that many are **priority 1**, we may have a serious problem. This situation is particularly difficult to resolve if the change is driven by decisions outside our project's span of control. Examples include a change in an interface owned by another organization, a correction to a safety hazard, and violation of a regulation or standard.

Alternatively, if we respond to a budget cut or schedule reduction, we can search for requirements that have **priority 3**. Again, we look at the requirement traces to determine how many requirements are affected (some of which may be **priority 1**) and then assess which set of **priority 3** requirements will least affect the **priority 1** requirements.

Before we baseline the requirements, we manage requirement risk by focusing on developing a valid scope and inspecting requirements to reduce the impact of requirement change risk factors (defined in Section 4.4.4) for "understood one way", "incomplete", and "subject to change". As we describe in the 4-1/2-step process, we assign a value for requirement risk as we baseline the requirements. Before then, requirements are too volatile and management overhead too large for the measure to be useful. Requirements traceable to key performance parameters (KPPs) identified during scope development always carry a risk rating of high or medium because a change in them affects whether we meet stakeholders' expectations.

We also must manage requirements that continue to contain an outstanding To Be Determined (TBD) after they're baselined. Some organizations automatically rate such a requirement as high risk, with good reason. In fact, we can quickly check whether requirements are ready for baselining by determining the number of TBDs: more than one or two percent of the total requirements being TBD should be a red flag. Still, we manage these much as we do the requirements themselves. First, we generate a list (or matrix) of all To Be Determined requirements in any SOI. The list needs to contain the requirement ID, requirement text, rationale, and another requirement attribute called *TBD resolution date*. Rationale should have the source of any numbers in the requirement and refer to the document that defines its value. We then take them from the list and place them in a table or spreadsheet for tracking.

If we don't have a requirement-management tool, we just search for To Be Determined requirements in our document and manually build a table containing a TBD management matrix. For tracking, we could enter their numbers in a tracking action log, but the requirement ID should be sufficient to track and identify which one we're commenting on. Once the To Be Determined closes, we won't need the TBD number. Table 4-13 is an example of a TBD tracking table.

TABLE 4-13.   **Example Management Matrix for To Be Determined (TBD) Requirements.** Because their uncertainty poses a risk to the project, we need to track TBD requirements until we can resolve them.

ID Number	ID of Requirement Containing the TBD	Planned Date for Resolving TBD	Actual Date When TBD was Resolved	Rationale for Closing the TBD
TBD001	FSRD065	Feb 2007	Apr 2007	Completed trade study
TBD002	FSRD235	May 2008		Awaiting results of technology assessment

Based on Table 4-13, the requirement in FireSAT's system requirement document (SRD) looks like this:

> [FSRD235] The FireSAT system shall display updated sensor data no more than 45 [TBD002] seconds after initiation of an "out-of-cycle download" command.

We notice that the example TBD requirement also includes a value. We must have a "current estimate" and place it in the requirement. This process at least bounds the problem for the reader and gives a clue about the units, such as seconds or milliseconds, centimeters or meters, grams or kilograms. By combining the requirement statement (which identifies the requirement), the TBD related to it, and the To Be Determined management matrix, we have all the information needed to assess the status of our TBDs.

Any requirement for which we have an unresolved To Be Determined should also have a high or medium risk, depending on the requirement's priority, how soon we need the information to avoid affecting the schedule, and how the unknown value may affect meeting stakeholders' expectations. When we know the value, we enter it into the requirement and update its rationale to reflect the justification for selecting the final value. Whether we change the risk attribute's value depends on other risk factors.

To manage change we need to assess risk continually; we don't want to wait until a problem occurs. One way to provide insight into the risk is a risk-priority matrix, such as the one shown in Figure 4-11 for the FireSAT's system requirements document. The risk-management plan states what to enter in the blocks, but normally we identify requirements with high priority and high-to-medium risk by requirement number. For tracking, the matrix depicts only the number of requirements meeting other lower-risk criteria.

Priority \ Risk	Low	Med	High
1	(70)	FSRD011	FSRD045 FSRD134 FSRD056
2	(45)	(19)	FSRD155 FSRD003
3	(20)	(11)	(0)

FIGURE 4-11. **Notional Risk (Likelihood)—Priority (Impact) Matrix for FireSAT Requirements.** Such a matrix provides a snapshot of the project's current risk status.

The foregoing discussion focuses on developing and managing requirements, starting at the system level down through several levels of the architecture during the requirements definition phase (Pre-Phase A and Phase A). However, these management techniques and methods—in particular establishing traceability between requirement levels, and between system design and test requirements—are invaluable in managing change throughout design (Phase B), build (Phase C), and test (Phase D).

Traceability between system definition requirements and design requirements, and subsequently between design requirements and test requirements, is a powerful tool for assessing project status as well as the total impact of requirement changes. We use it to ensure that the design is implementing the requirements, thereby eliminating late-breaking changes during integration and test. It precludes disasters where one part is changed but a related part is unchanged. It helps us see changes needed in verification associated with a requirement change, which might dramatically drive up the cost of the change. It also ensures that the integration and test program addresses requirements verification.

We must maintain discipline while defining, writing, organizing, baselining, and managing requirements. Writing requirements is hard work, but if we write bad ones we have to go back and do it over again until we get it right. Before writing, we must develop the scope and make it available to everyone. If we don't, we will develop the scope while writing, but it's much easier to gather the knowledge first.

If we take enough time up front to define scope before writing requirements and stay consistent while defining product requirements, we increase our chances of delivering a quality product. Otherwise, we're guaranteed "big bad" requirement documents and a failed or very unsatisfactory product.

References

Chesley, Julie, Wiley J. Larson, Marilyn McQuade, and Robert J. Menrad. 2008. *Applied Project Management for Space Systems.* New York, NY: McGraw-Hill Companies.

Fagan, Michael E. July 1986. Advances in Software Inspections. *IEEE Transactions on Software Engineering*, Vol. SE-12, No. 7, pp. 744–751.

Hooks, Ivy F., and Kristin A. Farry. 2001. *Customer Centered Systems.* New York, NY: AMACOM.

Hooks, Ivy F. 2007. *Systems Requirements Seminar.* Boerne, TX: Compliance Automation Inc.

Larson, Wiley J. and James R. Wertz, eds. 1999. *Space Mission Analysis and Design*, 3rd edition. Torrance, CA: Microcosm Press and Dordrecht, The Netherlands: Kluwer Academic Publishers.

NASA, NPR 7123.1a. March 2007. NASA Systems Engineering Processes and Requirements.

Vinter, Otto, Soren Lauesen, and Jan Pries-Heje. 1999. *A Methodology for Preventing Requirements Issues from Becoming Defects.* ESSI Project 21167. Final Report, Brüel & Kjær Sound & Vibration Measurement A/S, DK-2850 Nærum, Denmark.

Chapter 5

System Functional and Physical Partitioning

Bruce G. Barker, *Stevens Institute of Technology*
James E. Long, *Vitech Corporation*

Partitioning a system into its functions and components is vital to defining a problem and solution. System requirements that drive partitioning depend on the concept of operations and the stakeholders' expectations. Partitioning is iterative, as depicted in Figure 5-1, and it enables a better understanding of the technical requirements. It also clarifies and validates the system's scope, and highlights any concerns about the stakeholders' needs. For unprecedented systems, functional and physical partitioning may have a minimal set of technology constraints, but most legacy developments and upgrade programs mandate reusing system elements, accommodating existing interfaces, or using specific commercial-off-the-shelf (COTS) technologies.

Functional and physical partitioning gives us a chance to see whether the system requirements are feasible, to identify critical trade studies, and to clarify the system's scope. It helps show whether the system can support required operational capabilities and scenarios, while also ensuring traceability for easier test and integration. We expect a partitioned system to satisfy requirements and expectations over the lifecycle, so we must be sure that the system is scalable,

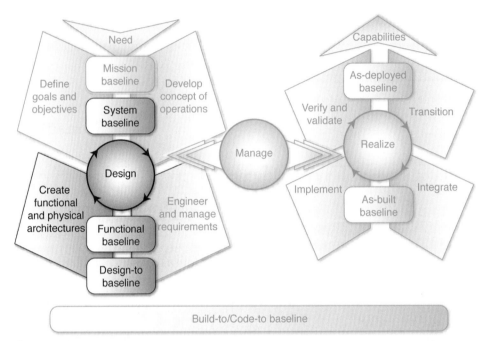

FIGURE 5-1.   Problems and Solutions Evolve Together. The early elements of systems engineering are highly iterative and interdependent, not sequential or linear.

testable, available, flexible, and able to meet requirements. This robustness is particularly important for systems that must operate well during changing environments, stakeholder expectations, and component technologies. The system architecture synthesis, definition, and documentation:

- Communicate to stakeholders. They ensure that the system will meet their expectations, and present options and opportunities for them to consider.
- Facilitate long-term corporate commitment
- Further the necessary financial commitment and investment, and the sponsors' decision to implement it
- Establish a concept and a framework as the basis for detailed designs and applications
- Incorporate all functional and other requirements and constraints

An architecture is essential to a system's success because it enables us to develop and maintain the system effectively. A system architecture requires several types of inputs:

- Rank-ordered stakeholder expectations (capabilities, characteristics, and legacy desired)

- The concept of operations
- The operational architecture, including operational scenarios and the system's inputs and outputs
- System and subsystem implementation concepts defined to best meet the operational scenarios
- Rank-ordered system requirements
- Known constraints, dependencies, assumptions, risks, issues, and measures of technical performance

Once developed, the functional and physical architecture comprise the following seven outputs as a minimum:

- Architectural models that describe the system's functions, its physical components, and the interfaces between functions and components. Depending on a system solution's complexity, these models have one or more levels, each level further detailing the solution.
- Derived requirements for the development teams (typically hardware and software), from which they begin designing and building their components
- Traceability of the system requirements to the hardware and software specifications
- Allocation of system requirements to the various architectural elements (such as functions, components, and interfaces)
- Traceability of the requirements to the high-level verification plan
- Documents capturing technical decisions and trade-offs
- Updated risks, issues, and technical performance measures

To create a useful architecture, we must have an effective process. This chapter discusses steps that designers follow to make sure that an architecture meets requirements. Table 5-1 lists these steps and the sections that discuss them.

A "black-box" view of a system includes the project's active stakeholders and the system's concept of operations. It helps delimit the problem's scope and system boundaries, plus inputs and expected outputs, given certain performance and physical constraints. But functional and physical partitioning represents the "white-box" view. This phase of systems engineering addresses how functions and components are arranged within the system boundary. Here we translate inputs into expected outputs, while respecting constraints on performance, cost, schedule, and other aspects of the project. This process repeats and may advance in layers—top-down, bottom-up, or some combination of both (Figure 5-2).

Figure 5-3 also shows systems engineering's recursive nature, as the process refines requirements and architectures in more and more detail. As described in Chapters 2 through 4, stakeholder requirements further refine the customer's need, so we use them to develop our first ideas about a system architecture in the system concept. We then use this to create a concept of operations and an

TABLE 5-1.   **A Framework for Developing Functional and Physical Partitioning of a System.** This table shows a general process to develop the functional and physical architecture for a system.

Step	Description	Where Discussed
1	Specify the system context	Section 5.1
2	Define the functional partitions	Section 5.2
3	Create the physical partitions and allocate functions to the physical components	Section 5.3
4	Assess the architecture	Section 5.4
5	Allocate requirements to the components and generate detailed specifications	Section 5.5
6	Trace requirements	Section 5.6

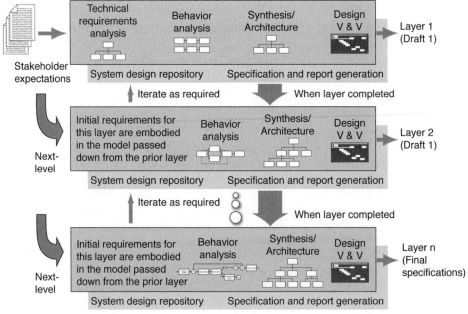

FIGURE 5-2.   **Recursive Partitioning of Functions and Physical Components.** This process repeats and may advance in layers—top-down, bottom-up, or some combination of both. (V&V is verification and validation.)

operational architecture that meet stakeholder's requirements. System requirements develop from the system concept, operational architecture, and stakeholder requirements.

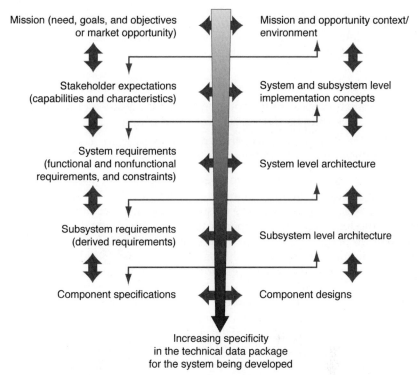

FIGURE 5-3. **Developing System Architectures and Requirements.** While the left side of the figure reflects the evolution in expectations, requirements, and specifications, the right side reflects the evolution in the concept of operations, the architectures, and the system design. These activities are iterative at any level, and recursive across levels.

A system's complexity determines whether we need one or several levels for the architecture and requirements. The architecture defines the subfunctions and subsystems for which we develop requirements. We then develop the functional and physical architecture for each subsystem by specifying its functions and components. This process continues until the architecture is complete.

The architecture and corresponding requirements are complete when we've detailed the solution to a level—usually in terms of "configuration items" (see definition in Section 5.3.1—that development teams use to begin designing, building, and testing their components. Partitioning incorporates seven distinct actions:

1. Analyze the system's technical requirements, as derived from the stakeholders' expectations and validated by the concept of operations
2. Develop functions to define how the system will process each class of inputs
3. Combine these functions to create an integrated system logic
4. Complete all details for functional requirements, which include input/output (I/O) relationships, processing steps, I/O attributes, and the functional control logic
5. Select and decompose the system's physical components
6. Allocate behaviors to the components and indicate all interfaces between them, including hardware, software, and people
7. Capture this information in a model to produce consistent, complete specifications for the detailed design activity

But listing the main actions for partitioning doesn't mean the process is sequential—in fact, the actions are often concurrent. Also, if we're doing reverse engineering, we may start with the existing physical description and derive the functional model. Planning and performing systems engineering activities are much safer in increments (or in layers, like peeling an onion). Each layer completely defines the system solution with increasing detail, but all layers contain requirements, behaviors, physical architecture, requirements definition, and validation.

The above discussion assumes we're developing a completely new (unprecedented) system, but that's rare. Normally we must use legacy components or commercial items off the shelf for new or upgraded systems, so we have to tailor our partitioning approach for real projects. Still, we must understand the activities that make up the approach to allow this tailoring. Designers and stakeholders must be part of functional and physical partitioning. Key stakeholders include:

- Architects and developers of hardware or software
- People who handle integration and test, reliability and maintainability, safety and security, and operations and logistics
- Users or customers
- Managers of projects or programs

The last two groups are especially important because they must understand the developing solution, as well as the cost and schedule. To reduce program risk we have to address important concerns early on, while partitioning is under way. We must also identify and perform trade studies that will affect our results.

Figure 5-4 shows the importance and relationships of the various architectures. The *operational architecture* establishes the solution from the users' point of view—how it will operate. The *functional architecture* shows the capabilities, functions, and services our solution provides to satisfy the operational architecture. The *physical architecture* specifies the solution's components—the people, hardware, and software that will perform functions to meet operational needs.

As a rule, we don't develop these three architectures serially. We develop them concurrently because together they lead to a complete solution. As one architecture develops, it offers insight into the others. Then updating the other architectures adds insight to the first. Thus, developing the architectures in parallel produces greater overall understanding and therefore a more integrated, complete solution.

Figure 5-4 also shows some of the many diagrams we use to describe the three architectures. Two major design methods are: 1) structured analysis and design (SAAD) (also called functional decomposition) and 2) object-oriented design (OOD). For the most part, systems engineering has used SAAD, and software design and engineering has employed OOD. But because many systems built in recent years include a lot of software development, systems engineers have started to use OOD more.

Both methods are effective in designing systems from stakeholder requirements or in developing complete system models. Structured analysis and design uses diagrams (shown in Figure 5-4) for integrated definition (IDEF), data flow, functional flow, and N2—as well as others—to describe the functional or logical architecture. Object-oriented design applies the class, use case, and activity diagrams—among others. For the physical architecture, SAAD may use the physical block diagrams, whereas OOD may use those for deployment and component structure. Both of these methods should be part of the systems engineer's or architect's toolbox.

In this book, we follow the structured analysis and design method. It's more traditional for systems engineering and common to many systems. For more information on object-oriented design, see Lykins et al. [2001] and SysML [ongoing].

For this chapter, rather than partitioning the entire FireSAT solution, we focus on one of its elements. Figure 5-5 depicts FireSAT and its individual elements. To describe the process for developing the functional and physical architectures, let's treat the space element as our system of interest.

5.1   Specify the System Context

The system *context diagram* enables reviewers to understand all system inputs and outputs, which in turn affords a basic understanding of the system's boundaries and sets a baseline for developing the internal architecture. By setting the system boundaries in this diagram, we

- Understand the system boundary and scope (by defining external interfaces)
- Portray external events and inputs to which the system must respond
- Delineate the data or information the system produces
- Understand the system context—the environment in which it has to operate

Operational architecture
- Shows how the operator will use the system
- Shows inputs and outputs to users and other systems
- Usually described by:
  – Operational concept
  – Context diagram
  – Use cases and scenarios
  – Sequence diagrams
  – High-level data model

Functional architecture
- Defines the capabilities, the services, or the functions provided by the system
- Shows the messages and data between functions
- Usually described by:
  – IDEFO diagrams
  – Data flow, functional flow, and N2 diagrams
  – Classes, methods, use cases, etc.

Physical architecture
- Allocates resources (hardware and software)
- Shows the interconnections between the resources
- Usually described by:
  – Physical block diagrams
  – Physical interface definitions

Instantiated physical block diagram (Fig. 5-29)

Enhanced functional flow block diagram (EFFBD) (Fig. 5-24)

Interaction diagram (Fig 5-21b)

Iteratively define all three views and allocate the requirements to the elements within these views to 'completely' define the architecture.

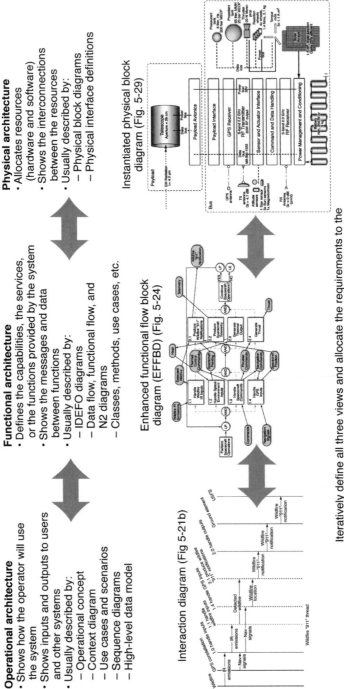

FIGURE 5-4.   Architecture Viewpoints and Work Products. By developing these three architectures at the same time, we gain insights and produce a more complete solution. (IDEFO is integrated definition for function modeling.)

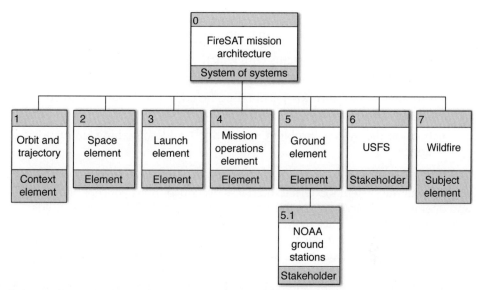

FIGURE 5-5. **Space Element as Part of the FireSAT System of Systems.** We focus on the space element of the FireSAT System to illustrate the concepts discussed in this chapter. (USFS is US Forest Service.)

A system's technical requirements, scope and context, and external interfaces are important inputs to the context diagram. They come from an operational view of the architecture and corresponding operational scenarios that we use with stakeholder requirements to develop the system requirements.

The initial system context diagram begins the systems engineering process and is an important opening step in functional and physical partitioning. Our first attempt at it may raise questions we must eventually answer concerning the system boundaries, external systems, and interfaces among the systems. We develop a complete system context diagram by

- Clarifying the system boundary, the external systems or active stakeholders, the external interfaces, and the context and environment
- Verifying the list of active stakeholders
- Determining what each active stakeholder will lend to the process (inputs) and expect from it (outputs)

Knowing the interfaces between our system and external systems or users identifies our system's inputs and outputs. And ascertaining the stakeholders' requirements gives us significant information that helps us shape the system context diagram.

5.1.1   Features of the System Context Diagram

The system context diagram, also referred to as the external systems diagram,

- Portrays all external systems with which our system must interface and the mechanisms for interfacing
- Provides a structure for partitioning behaviors and data assigned to the interfaces
- Bounds the system problem so we don't add something unintended or unneeded

The diagram encapsulates all the system's operational interfaces, and so compiles all the system's external interfaces described in the concept of operations and operational view. It shows the system at its center, with active stakeholders or other actors that interface with the system arranged around that center. The interfaces are either physical links between the components or functional connections, such as data, information, or signals transferred between the system and the actors.

Figure 5-6 shows an example of a context diagram for our system. It depicts the system, the external systems with which it interfaces, and the inputs and outputs between it and each of the external systems.

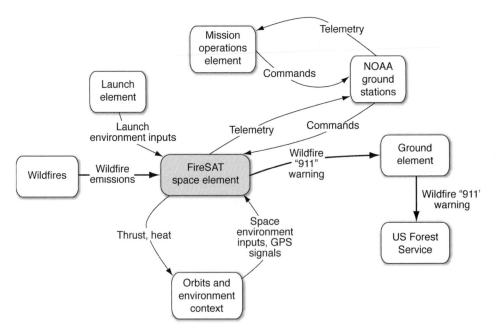

FIGURE 5-6.   **System Context Diagram for the FireSAT Space Element.** The system of interest is at the center, linked by inputs and outputs (arrows) to and from external systems.

5.2   Define the Functional Partitions

The functional architecture (partitioning a system's functions) is crucial to developing the solution architecture. Many system architects say they don't have a functional architecture, and some successful systems don't appear to have a diagram for it. But that doesn't mean it doesn't exist. This architecture describes what the software and hardware do, and no one can design, maintain, or upgrade a system without this information. So whenever system, software, and hardware architects describe what their components do, they're describing the functional architecture.

The question is whether the functional partitioning is on paper or in the architects' heads. Documenting the functional and physical architectures is a way of communicating the complete solution to stakeholders—a major goal. If the architecture remains only in the architects' heads, it may be as good as a documented one, but communication to the stakeholders is less effective. Documenting the architecture improves knowledge transfer.

5.2.1   Understand Terms and Elements

As a very high-level architecture, the system context diagram shows the system as a black box, with the inside unknown. But it

- Encapsulates operational scenarios by showing the interfaces to external systems and users, each of which exchanges information, data, signals, or other material with the system across some physical interface
- Shows system requirements that show what the system must do (the functional requirements)

What we **can't** tell from this diagram is how the system transforms the inputs it receives to the outputs it sends or what physical elements allow it to do so.

Developing the system's functional architecture is the first step in framing how a system will meet its requirements and operational scenarios. To do so, we must understand each *function* as a **process** that transforms inputs into outputs. A function describes an **action** from the system or one of its elements, so it should begin with a verb. We must not confuse the thing that performs the function with the function itself. For example, a database is not a function, but we can use one to perform the function of **Store and retrieve data**. Accordingly, the following are not functions:

- Computers or software applications—we assign them the functions of Process data or Compute current navigation state
- Cold-gas thrusters—they provide the function of **Generate thrust**
- Batteries—they have the function **Provide electrical power**

Thus, when we develop a functional architecture for a system we're partitioning based on functions, services, or tasks. The functional architecture describes how the system meets its input and output requirements, and more specifically "what the system does." So it fills two roles:

- A partition of the system into more detailed functions
- A model indicating how the system transforms its inputs into outputs and achieves its ability to realize required behaviors

Just as the system context diagram fully defines the system at a highly abstract level, the functional architecture must completely define the system's functions at increasing levels of detail. To ensure that it does, we follow two key principles:

- The sum of the lower-level functions is the same as the sum of the upper-level functions
- External inputs and outputs are conserved, meaning subfunctions must have the same number and type of inputs and outputs as the function

A data model usually captures these inputs and outputs. Thus, the functional architecture shows:

- Elements—functions or tasks; what the system does
- Relationships among elements—the data, information, material, or energy that functions exchange to carry out their tasks
- Relationships among functions—the chronological order in which functions execute and how one action depends on another. As described in Section 5.5, we use this relationship to develop an executable model of the architecture.

Developing this architecture requires three major steps:

1. Determine the functions
2. Determine which functions receive the external inputs and generate the external outputs
3. Determine the inputs and outputs between functions and verify that the resulting functional architecture is correct and complete

We use several different types of diagrams to describe the functional architecture, as summarized in Table 5-2.

Figures 5-7 and 5-8 show examples of functional architecture diagrams for FireSAT. They capture the system of interest, the external interfaces, and the inputs and outputs between external systems and our system.

System-level inputs consist of items that cross the system's boundary from any external system and "stimulate" it. That is, the system must accept and respond to them by either generating an output or changing its state. Each element handles a particular task. For example, the system of interest (space element) handles the high-level function **Perform space mission**. The launch element covers the function **Launch spacecraft**. And the mission operations element carries out the function

TABLE 5-2.   Characteristics of Diagrams for Functional Architectures. The architects should develop the diagrams that best communicate the functional architecture to the stakeholders. The diagrams may vary by the type of system or even by the preferences of the architect and stakeholders.

Functional Diagram	Characteristics of the Diagram
IDEF (integrated definition)	Shows • Functions as boxes on a diagonal • Inputs as arrows into the function box's left side • Outputs as arrows out of the box's right side • Controls (inputs or constraints that trigger a function) as arrows into the box's top side • Mechanisms (physical components that perform the functions) as arrows pointing up to the box's bottom edge
N×N (also called N2)	Functions are on a diagonal. The top row contains external inputs into each function, with external outputs in the far right column. Shows each input and output between functions in the function's row or column.
FFBD (functional flow block diagram)	Shows functions in the order in which they interact with one another throughout the functional flow.

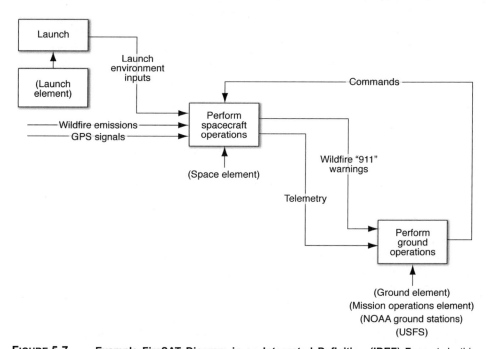

FIGURE 5-7.   Example FireSAT Diagram in an Integrated Definition (IDEF) Format. In this format, we decompose the FireSAT mission functions into the primary subfunctions shown here. We then represent the interactions between these functions with the arrows showing the inputs and outputs.

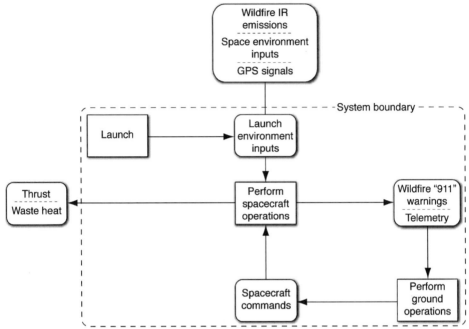

FIGURE 5-8.   FireSAT Mission Context Diagram in a Modified NxN Format. Mission inputs are at the top, and outputs external to the system are on the left. Mission functions are on the diagonal. The inputs and outputs between the functions are in the function's row and column.

Perform ground operations. The diagrams also show inputs and outputs between each external function and the system function.

Items that move across the system's boundary to any external system are system-level outputs. They may develop whenever the system processes a stimulus or generates and transmits a message or command to any external system.

5.2.2   Determine the Approach to Functional Partitioning

When determining the functions of a system, we must consider both the basic partition approach and how to define the functions. For the former, we have two choices: decomposition and composition. Systems engineers and architects have traditionally decomposed their systems; recently, they've also started using composition.

Decomposition—This top-down method starts with a single definition of the system function, such as the system context diagram, and partitions the system one level at a time. As Figure 5-9 shows, this approach is orderly and methodical.

A rule of thumb for decomposition is not to decompose a function into more than six subfunctions. If a function has more than six, we should decompose it at

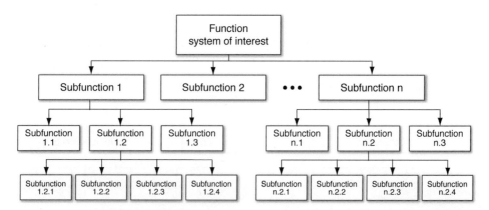

FIGURE 5-9.   **Functional Decomposition Approach.** This method starts with a single definition of the system function, and allocates functionality systematically through the lower levels.

more than one level. Figure 5-10 shows an example of this situation. Rather than breaking the one function into eight subfunctions, we decompose it into four and then further break two of the subfunctions into six more subfunctions. By limiting the number of subfunctions, we reduce complexity and enable reviewers of the architecture to comprehend it better. But a rule of thumb isn't a mandate. Sometimes, seven or more subfunctions come from a single function, if the decomposition isn't too complex.

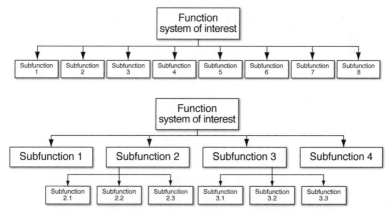

FIGURE 5-10.   **Functional Decomposition Approach to Multiple Levels.** Following a rule of thumb, rather than breaking one function into eight subfunctions, we decompose it into four and then further break two of the subfunctions into six more subfunctions.

Composition—This method is bottom-up. It starts by determining a complete set of bottom-level functions that the main function incorporates. Then the architect synthesizes these bottom-level functions into a functional hierarchy (Figure 5-11).

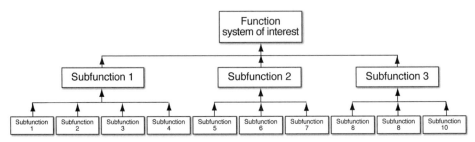

FIGURE 5-11.   Functional Composition Approach. This method starts by defining bottom-level functions and builds them into a functional hierarchy.

Picking an approach. Systems engineers and architects use decomposition to develop their systems' functional architecture whenever the system is very complex or incompletely defined. In this case, it's difficult to determine all the bottom-level functions. Partitioning the system one level at a time makes complex systems simpler and undefined systems easier to understand. Each level of partitioning creates a clearer understanding of the previous level, so we iterate and update that previous level with new information. Thus, we make the system architecture clearer at a higher level, rather than waiting until we get to the lowest level, as we must with composition.

The composition approach applies better when architects know all (or most) of the bottom-level functions. It also allows more reuse of functions. By combining subfunctions into a higher-level function, we look for functions that do the same or similar things. If we discover this commonality at one architecture level, we can reuse functions at the lower levels of design and development. We typically use a combination of approaches. We decompose the system to a certain level of subfunctions (where we know the functionality well). At that point, we aggregate the subfunctions using composition.

5.2.3   Determine the Functions

Architects use decomposition, composition, or a combination of both, but they must still determine the best set of subfunctions for any function. To do this, they

- Use operating modes
- Use inputs and outputs
- Use stimulus-response threads (cases or operational scenarios)
- Segment fundamental and custodial functions
- Apply the Hatley-Pirbhai template (see Figure 5-16)

- Use function states to determine the subfunctions
- Use processing rates
- Use organizational structure
- Use functional requirements
- Consider stakeholder priorities
- Match the physical architecture

Use operating modes. An *operating mode* is an operating capability of the system during which it may do all or some of its functions fully or partially. We partition a function into its subfunctions based on these operating modes. For example, the system function for a cell phone could partition into subfunctions for each of its modes, such as **Make and receive phone calls, Play games, Access the Internet,** or **Upgrade the system.** Figures 5-12 and 5-13 illustrate this method for the space element's function **Perform space mission.**

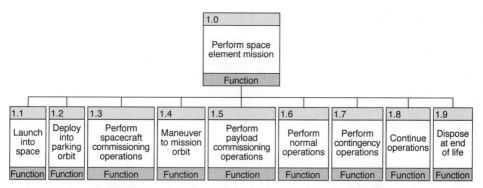

FIGURE 5-12.   **Functional Hierarchy Based on Operational Modes.** This diagram shows how we partition the space element's **Perform space mission** function into subfunctions based on operating modes. This hierarchy violates the 'no more than six subfunctions' rule of thumb, but having more than six subfunctions makes sense in this case.

Use inputs and outputs. We sometimes base a function's partitions on its inputs and outputs (I/O) or the **types** of inputs and outputs, thus capturing its external interfaces. For example, a computer's subfunctions might be: **Provide visual output; Provide audio output; Handle inputs for keypad, touch screen, and mouse;** and **Interface with peripherals** such as the printer, memory stick, fax machine, and so on. Because this method focuses on the external I/O, our decomposition must account for internal processing or control of the system in some function. We often add a separate processing and control subfunction (see the Hatley-Pirbhai template below). Figure 5-14 shows how to use I/O for the space example.

Use stimulus-response threads (use cases or operational scenarios). *Operational scenarios* or system *use cases* illustrate how the system will be used. We partition the system's functions to match these scenarios, so users and customers

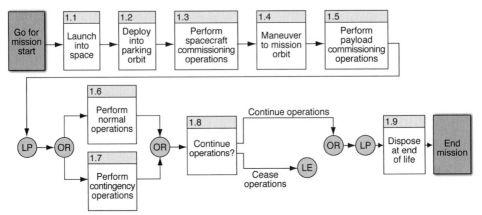

FIGURE 5-13.   Functional Flow Block Diagram Based on Operational Modes. This diagram helps us visualize the logical sequencing of system functions. (LP is loop; OR is an "or" decision point; LE is loop exit.)

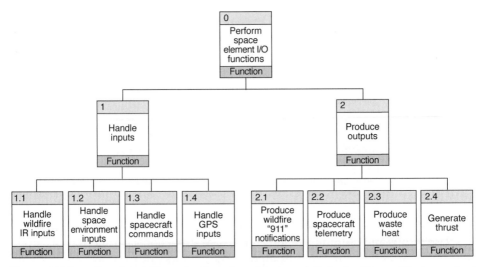

FIGURE 5-14.   Functional Hierarchy Based on Inputs and Outputs. This partitioning method emphasizes a system's external interfaces by focusing on how it handles inputs and produces outputs. (IR is infrared; GPS is Global Positioning System.)

better understand the architecture. If the scenarios don't interact much, the resulting architecture is fairly simple. But if they do similar things or have similar interfaces with users, decomposition by this method may hide possibilities for reuse. An example of using stimulus-response threads is decomposing an automatic teller machine system into functions such as **Login, Perform withdrawal, Perform deposit,** and **Perform balance inquiry.** Figure 5-15 shows the FireSAT example.

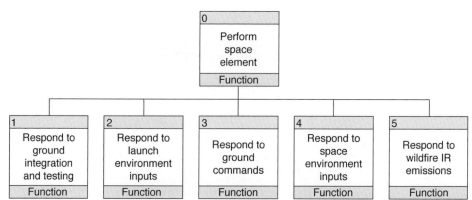

FIGURE 5-15.   **Functional Partitioning Based on Use Cases.** This example uses stimulus-response threads to partition FireSAT's **Perform FireSAT mission** function into its subfunctions.

Segment into fundamental and custodial functions. This simple method separates custodial functions from the normal operational functions. In this case, we segment subfunctions for maintenance, supportability, and operability from the basic functions for users. Fundamental and custodial functions are relatively independent of one another. In other words, they have few mutual interfaces in normal daily use of the system, because the interfaces are usually part of the status and update messages. Thus, partitioning them in the architecture makes sense.

Apply the Hatley-Pirbhai template. This template states that any function partitions into the subfunctions shown in Figure 5-16, though we may combine or eliminate some of them. For example, if the system interfaces only with users, we would drop the subfunctions related to non-user inputs or outputs. If the system interfaces with only one external interface, we might combine the subfunctions **Receive non-user inputs** and **Produce non-user outputs** into a single subfunction called **Interface with the external system**. Figure 5-17 shows how we should apply this method to FireSAT.

Use the function states to determine the subfunctions. A *function state* is a static snapshot of the metrics or variables needed to describe what that function is doing. Many functions progress through a series of states to complete their actions. We usually know the starting state based on some input and the ending state based on an output. For example, many common systems, such as automatic teller machines, debit cards, computers, and access to classified areas, use an authorization process. The subfunctions for any of these systems (using function states) may be: **Receive request for use**, **Ask for unique identification**, **Receive unique identification**, and **Authorize user**.

Use processing rates. Some functions occur periodically, based on a timer or a constant interrupt; others occur irregularly, based on input from the user or some other source. At some point, developers decouple these two types of processing because a single software program or piece of hardware rarely does both. Architects

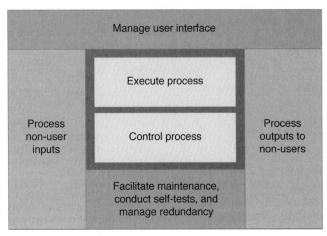

FIGURE 5-16.   **The Hatley-Pirbhai Template.** This template suggests that we can partition any function into the generic subfunctions shown here.

either partition the processing in the architecture or leave it to the developer, but the former better describes the triggers for each function and the throughput needed for the architecture's physical components (described in Section 5.4).

Use organizational structure. Eventually, a project's functional architecture gets divided into parts that development teams design and build or commercial products handle. Thus, at the lowest level, every function should be assigned to a single development team (or a purchased product). If we know the commercial products and development-team structure before or while developing the architecture, we should simply partition it to match the organizational structure.

We could also match the customer's organization—a common approach. For example, because Department of Defense contracts often call for developing systems of systems, several customers may fund a single solution. In this case, partitioning to match the functions these customers request may be helpful and effective. That way, customers can visualize the system from their own perspectives and better understand what they're getting for their money.

Use functional requirements. Because a system's functional requirements describe what it must do, we use them to determine its subfunctions. Requirements and architectures develop iteratively, so we use the same method for lower-level functions. Each subsystem's functional requirements help us define its subfunctions. Figure 5-18 shows an example for FireSAT.

Consider stakeholder priorities. For any project, the stakeholders' rank-ordered requirements trace to a set of rank-ordered system requirements. We use this information when creating the system's functional architecture. For example, we isolate functions with very high or low priority to customize services for our customers. By isolating high-priority functions and specifying their interfaces, we make sure that they're developed first. On the other hand, partitioning low-priority

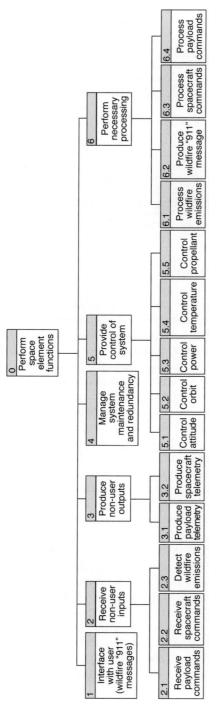

FIGURE 5-17. FireSAT Example Using the Hatley-Pirbhai Template. Here we show how the subfunctions for **Perform space element functions** come from the template, and those subfunctions divide into particular items that FireSAT must receive, produce, control, and process. (Functions 1 and 4 also have partitioning at this level; for simplicity, we don't show it in this figure.)

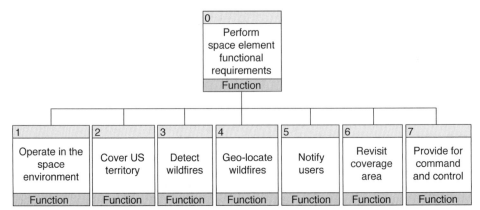

FIGURE 5-18.   Functional Hierarchy Based on Functional Requirements. We base this partition on the requirements that state what the system must do.

functions lets us reduce total risk. If cost or schedule issues arise, we can drop these functions or move them to a later delivery.

Match the physical architecture. If we know a system's physical architecture (Section 5.4), we can allocate functions to physical resources—hardware, software, and people. Several conditions could enable architects to know these physical components early in development. For example, we may be updating a legacy system whose hardware and software are established. To avoid affecting this system, we should partition the functions based on these components. Sometimes stakeholders require certain physical resources. For example, they may require a specific piece of hardware, software, or commercial-off-the-shelf product.

Architects may also want to match the physical architecture to limit costs for support and maintenance. Suppose we know before developing the functional architecture that a system will need to refresh many components while operating. In this case, we partition the architecture to create distinct interface boundaries between the physical components, so the refreshes will be easier and cheaper. Figure 5-19 shows a way to partition system functions based on the physical architecture, first dividing it into two main parts and then partitioning each part into lower-level functions.

Table 5-3 summarizes benefits and shortcomings of the 11 functional-partitioning methods discussed in this section. Architects may use several ways to partition a function because each method develops a different functional architecture. They may also combine techniques. For example, they may decompose a function into a set of subfunctions with the use-case method, but all of the use cases may have a common input or output function. The best method depends on the functions being partitioned and the external interfaces to users and outside systems; it may be the simplest, the most modular, or the most detailed. (Section 5.4 discusses techniques and metrics for assessing architectures.)

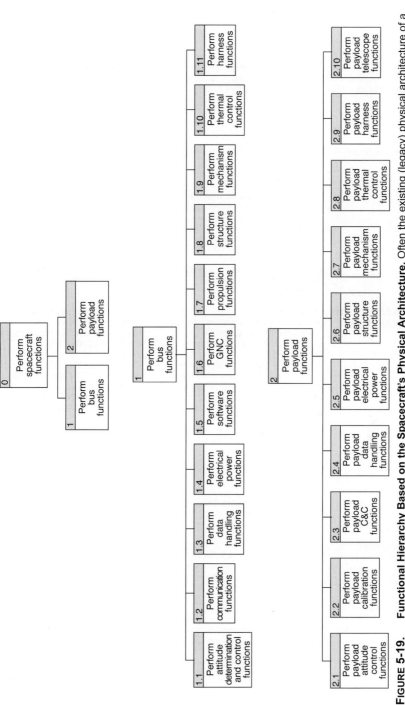

FIGURE 5-19. **Functional Hierarchy Based on the Spacecraft's Physical Architecture.** Often the existing (legacy) physical architecture of a system imposes constraints, so we structure the functional architecture to accommodate these constraints. (GNC is guidance, navigation, and control; C&C is command and control.)

TABLE 5-3. **Benefits and Shortcomings of Functional Partitioning Methods.** We partition functions in many ways, depending on the available information. Awareness of each method's strengths and weaknesses helps us choose the best one.

Method	Benefits of Method	Shortcomings of Method
Use operating modes	• Easy for operators and customers to understand • Addresses all major functions	• May hide opportunities to reuse functions across similar operating modes • Gets very complicated if operating modes interact a lot
Use inputs and outputs	• Addresses (conserves) all inputs and outputs (I/O) • Provides common functions for similar types of I/O • Good for functions that don't handle a lot of complicated processing or control	• Doesn't cover complicated control and processing, so we may need to combine this method with another for complex functions
Use stimulus-response threads (use cases or operational scenarios)	• Matches the operational architecture and thus the users' point of view • Describes all uses of the system • Simple to understand if scenarios don't interact a lot	• May hide opportunities to reuse functions across scenarios • Can get very complicated if many scenarios or similar scenarios exist
Segment into fundamental and custodial functions	• Is logical • Used in many systems because custodial functions differ from those for normal operations	• Partitions only part of the system's functions • Must use with other methods for complete partitioning
Apply the Hatley-Pirbhai template	• A simple method that applies to many different types of functions • Effectively partitions at a high level	• Doesn't offer detailed understanding of processing or control functions • May need another method at a lower level to understand processing
Use the function states	• Very good to describe functions with multiple steps • Similar to a state machine that is familiar to many engineers • Easy for most people to comprehend • Can help architects discover the details of functions not completely known	• Architects must understand the start and end states up front • Gets complicated for long processing strings or complex functions
Use processing rates	• Is simple • Matches how functions will apply, because hardware or software separates them by processing rates	• Partitions only part of the system's functions • Must use with other methods for complete partitioning
Use organizational structure	• Matches how functions will apply, because they eventually allocate to components that different organizations will handle • Used in organizations whose structure is set or difficult to change	• Because it forces functions to meet the organization's requirements, may inhibit creativity or trade-offs for a more efficient solution (see discussion of the Morphological Box in Section 5.3.2)

TABLE 5-3.   **Benefits and Shortcomings of Functional Partitioning Methods. (Continued)** We partition functions in many ways, depending on the available information. Awareness of each method's strengths and weaknesses helps us choose the best one.

Method	Benefits of Method	Shortcomings of Method
Use functional requirements	• Good when functional requirements are well-defined and complete	• May develop a partial solution, because functional requirements don't encompass nonfunctional characteristics
Consider stakeholder priorities	• Enables effective support and maintenance of systems • Supports giving customers the "best bang for the buck" by creating an architecture that meets their priorities	• Partitions only part of the system's functions • Must use with other methods for complete partitioning
Match the physical architecture	• Matches how functions will apply • Normally used (for all or part) when we must incorporate a legacy system, commercial-off-the-shelf items, or purchased products	• Because it forces one-to-one mapping of the functional and physical architectures, may inhibit creativity or trade-offs for a more efficient solution (see discussion of the Morphological Box in Section 5.3.2)

5.2.4   Determine Which Function Receives the External Inputs and Generates the External Outputs

Each level of functional partitioning should represent the system completely, so subfunctions must address their corresponding function. To verify this match, architects must first be sure that subfunctions have the same external inputs and outputs as the function. Starting at the system level, they use the system context diagram and operational scenarios to determine the system's inputs and outputs. Then they determine which system subfunctions receive particular inputs or produce certain outputs.

Following this process affords the architects better understanding of each subfunction. In many cases, they even uncover system-level inputs and outputs that they previously missed. They should then update the system context diagram and operational scenarios with their discoveries. Or they may determine that the developed subfunctions are incomplete or wrong. In this case, they update the architecture (using the same or a different partitioning method) by developing a set of subfunctions that better handle the system's inputs and outputs.

Figure 5-20 shows a simple functional decomposition of the space element. Each of the space element inputs is received by one of the subfunctions and each of the outputs is generated by one of the subfunctions. Thus, all the external I/O of the space element (as shown in the context diagram for the function **Perform spacecraft operations** in Figure 5-8) is captured by the set of these subfunctions.

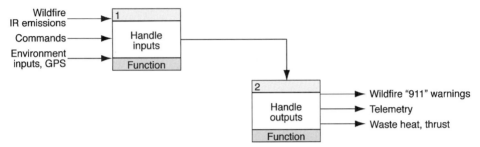

FIGURE 5-20.   Allocating External Inputs and Outputs to the Subfunctions. The external inputs and outputs are assigned to the appropriate subfunction. Here we simply partition the space element into two subfunctions. These are later partitioned into more descriptive subfunctions that better describe what the system does. (IR is infrared; GPS is Global Positioning System.)

5.2.5   Use Scenario Tracing to Verify Functional Interactions

In the previous step, we partitioned the function into its subfunctions and verified that these cover all external inputs and outputs. From an outside view, we've partitioned our function successfully into more detail. But the functional partitioning is incomplete at this level until we capture all the functional interactions. A method called *scenario tracing* helps us verify the functional architecture's completeness.

For each system use case in the concept of operations, a scenario trace through the functional partitioning shows whether the architecture correctly generates the system's required outputs based on its inputs. These scenario traces show

- Shortfalls or absences of functions in the partitioning
  - Inability to receive all inputs
  - Inability to produce required outputs
  - Insufficient feedback or control to produce required outputs
- Overlapping or redundant functions

Shortfalls usually occur because inputs and outputs between subfunctions are missing, so architects must add them to the functional partitions. Overlaps occur when several subfunctions receive the same inputs, do the same (or similar) tasks, or send the same outputs. To fix this problem, architects must change which subfunctions receive the external inputs or generate the external outputs. They may even have to swap the set of subfunctions with a different set. Scenario tracing helps them discover interaction problems and to partition completely to meet the system's operational needs. It has five phases: 1) trace the operational scenarios defined at the system level; 2) trace other functional scenarios we need to develop; 3) derive an integrated system behavior model; 4) resolve differences in function definitions; and 5) integrate the system's scenarios and generate its functional architecture.

1. **Trace the operational scenarios defined at the system level.** We start with system-level operational scenarios to verify whether each level of functional partitioning generates required outputs based on given inputs. These scenarios show the system's sequential interactions with operators and other external systems as the operators use it. By tracing them through the functional architecture, we show that the functional interactions meet their requirements.

Figures 5-21 and 5-22 show two operational scenarios. Each figure depicts the scenario as a sequence diagram at the system level and at the functional partition level. In both figures, part a) shows the scenario at the system level. The sequence diagram shows the system of interest (FireSAT space element) and the external systems (users) as vertical lines. System inputs are arrows that end at the FireSAT space element line. System outputs are arrows that begin with the FireSAT Space element line. Time passes downward in the vertical axis. Thus, in Figure 5-21, the system receives the input Navigation Signals before it produces the output Wildfire "911" Notification. The scenario shows the sequence of inputs and outputs to and from the system of interest.

Parts b) of Figures 5-21 and 5-22 illustrate the operational scenarios traced through the functional partitions of the space element as described in Figure 5-14. In these examples, the space element subfunctions appear as dashed vertical lines instead of the solid vertical lines that represent the space element and the external systems.

External inputs and outputs (I/O) come into or emerge from one of the subfunctions. The diagrams also show internal I/O between the subfunctions, as the functions execute during the scenario. For example, in Figure 5-21b, when the subfunction 1.1 **Handle Wildfire Inputs** receives the external input IR Emissions, it sends the internal message Detected Wildfire to the subfunction 2.1 **Produce Wildfire "911" Notifications**. For the system to produce the actual "911" notification, it must know the location of the wildfire. When the subfunction 1.4 **Handle GPS Inputs** receives the external input Nav Signals, it sends the internal message Wildfire Location to the subfunction 2.1 **Produce Wildfire "911" Notifications**. This function now has all of the information it needs to generate the external output Wildfire "911" Notification to the Ground Element external system.

Whenever the system receives an external input, we must show a connection between the subfunction that receives the input and the subfunction that generates the subsequent external output. The system level scenarios (part a of the figures) are fully realized at the subfunction level (part b of the figures).

2. **Trace other functional scenarios we may need to develop.** Sometimes the system-level operational scenarios don't cover the system's external inputs and outputs. In other cases, internal functions drive external outputs, such as internal error processing that leads to an error status being reported to the operator. Planners may not have developed all the operational scenarios because, in the interest of cost and schedule, they've

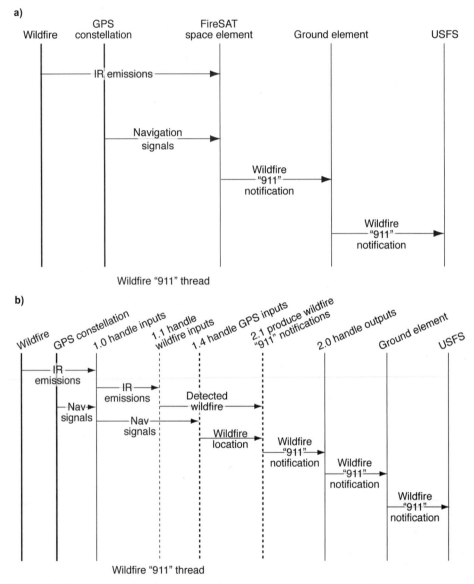

FIGURE 5-21.   The Wildfire "911" Scenario. a) At the System Level. We show the inputs and outputs of the scenario for the system. **b) At the Subfunction Level.** The system level inputs and outputs from a) are now allocated to the subfunctions. Inputs and outputs between the subfunctions complete the scenario. (GPS is Global Positioning System; IR is infrared; USFS is US Forest Service.)

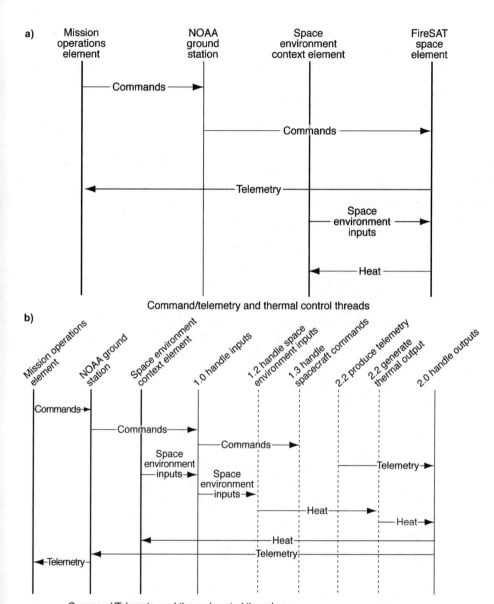

FIGURE 5-22.   The Control Scenario. a) At the System Level. Here we show the scenario inputs and outputs for the system. **b) At the Subfunction Level.** The system-level inputs and outputs from a) are now allocated to the subfunctions. Inputs and outputs between the subfunctions complete the scenario. (Space environment inputs and waste heat are produced continuously, although we represent them discretely here.) (NOAA is National Oceanic and Atmospheric Administration.)

settled for only the architecturally significant ones at the system level. But we must still address all external inputs and outputs in the functional architecture by defining functional scenarios and tracing them through the architecture. We then use these scenarios to determine if we need more functions or more I/O to develop a complete functional architecture. "Complete" means that the architecture covers all the system's external inputs, outputs, and required functions.

We also have to consider the internal interactions between scenarios. For instance, a security system typically has a user authorization or login scenario that's separate from the main system. Examples are a Web-based system to buy a plane ticket and the login process for an ATM machine. But these scenarios must interact: the user authorization scenario must pass the user's identification or authorization information to the main scenarios. The functional architecture must include these interactions between scenarios to verify that the subfunctions have the information they need to complete their tasks.

To generate a thread we start by selecting inputs or outputs that the operational scenarios didn't address. Then we create the sequence of system functions that must occur to completely respond to that input or generate that output.

3. **Derive an integrated system behavior model.** When we've derived all functional scenarios to augment the operational ones, we have a complete set of system scenarios and have defined all necessary I/O for the functional partitions (the subfunctions). That is, we have a complete partitioning of one level of the functional architecture to the next detailed level. We're ready to integrate the scenarios and functional architecture to verify its completeness. Figure 5-23 shows functional architectures updated from Figures 5-21 and 5-22, with the subfunctions' internal I/O added to satisfy the operational scenarios.

4. **Resolve differences in function definitions.** One source of integration problems is different groups of engineers generating different scenarios. Functions that seemed to be almost the same may have different names, boundaries, and inputs and outputs. We must resolve these differences to integrate the scenarios into a common single architecture.

Functions with differing inputs and outputs or exit conditions are different functions, so we must account for their differences when we identify the unique system functions while integrating threads. At the same time, we should limit the size and complexity of our final integration. We should also try not to duplicate functions and logic streams. Although integrating these scenarios would be easy if we kept them in their original form and linked them in parallel, that approach makes it difficult to understand how system functions interact in an integrated way. One reason to integrate the scenarios into a single structure or architecture is to make it understandable to the systems engineer. If we don't integrate well, design engineers won't understand the interactions as system scenarios run concurrently.

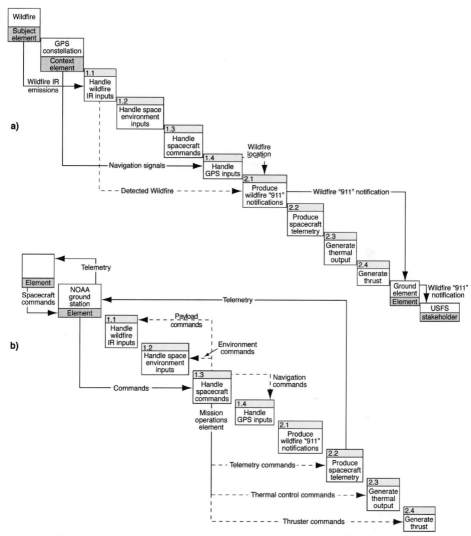

FIGURE 5-23.   **Functional Partitioning. a) The Wildfire "911" Scenario. b) The Control Scenario.** We combine the partitioning of all the scenarios to show a complete functional partitioning of the system (see Figure 5-24 for an example). (GPS is Global Positioning System; IR is infrared; NOAA is National Oceanic and Atmospheric Administration.)

System scenarios identify how the system responds to each input independent of all other inputs. But these individual scenarios don't account for functions that work together or for interactions and logical controls between scenarios. Integrating them is one way to address this apparent hole. This activity is challenging but essential to creating efficient, understandable system logic.

5. **Integrate the system's scenarios and generate its functional architecture.**
    Figure 5-24 shows an integrated functional architecture from the previous
    scenarios. We combined the subfunctions' inputs and outputs from each
    scenario to provide a complete set of inputs and outputs for each subfunction.
    Thus, this functional architecture can perform all of the scenarios.

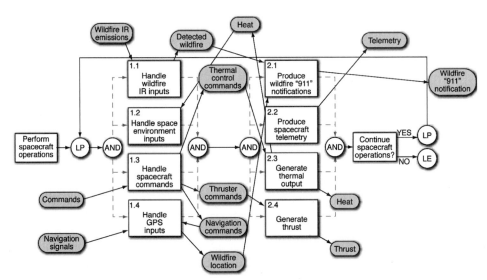

**FIGURE 5-24.   From Multiple Scenarios to an Integrated Architecture of Behaviors and
Functions.** Here we show a complete functional architecture that includes the
interactions from all the individual scenarios as well as the interactions between the
scenarios. The boxes are functions, the ovals are inputs, commands and outputs,
and the circles and lines represent the logical execution flow of the functions. The
"LP" flow circle represents an execution loop and "LE" represents the loop exit
(condition when the execution flow exits the loop). (IR is infrared; GPS is Global
Positioning System.)

5.2.6   Iteratively Apply the Same Process throughout the Functional Architecture

Now we need to go beyond one level of partitioning for the functional
architecture. The architecture may include several partition levels, so we should
follow the same process for each function at every level. Figure 5-24 shows the
system partitioned one level: decomposed into its first-level subfunctions. The
subfunctions receive or produce all external inputs and outputs at the system level,
and we complete the I/O between subfunctions by tracing the scenarios through
the subfunctions.

Following the same process, we partition all or some of the subfunctions to the
next level. For example, suppose we want to partition the subfunction **Handle
wildfire IR inputs** in Figure 5-24. We begin with three steps:

FIGURE 5-25.   System Context for the Subfunction Handle Wildfire IR Inputs. We must understand the context of a function before partitioning it to the next level of detail. (IR is infrared.)

1. Define the system context (Figure 5-25) for this subfunction, which is now the "function of interest"

2. Determine the subfunctions of the "function of interest" (Figure 5-26)

3. Determine which of the subfunctions receives or sends each of the system's external inputs and outputs

Architects complete the functional architecture when they've partitioned functions to enough detail for development teams to start designing, building, and testing. That means developing the physical architecture and allocating the functional architecture to the physical architecture. As described in Section 5.3.1, architects stop at the configuration item level of detail, which we describe further in Section 5.3.

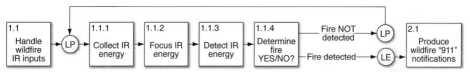

FIGURE 5-26.   The Subfunctions of the Function Handle Wildfire IR Inputs. We implement the same process described above to determine the subfunctions of every function that must be partitioned to a lower level of detail. (LP is loop; LE is loop exit.)

5.3   Create the Physical Partitions and Allocate Functions to the Physical Components

Because the physical architecture stipulates the components (people, hardware, or software) that are going to be part of the solution, we haven't finished designing the solution until we determine a physical architecture.

5.3.1   Understand How to Define a Physical Architecture

Thus far, we've described how to develop system subfunctions and their interactions to meet the system's operational uses. The next step is to decompose the system's physical structure and then use the results to map or allocate the functional architecture onto elements of the physical architecture. To define the physical architecture, we

- Determine the physical elements or partitions
- Allocate the functional architecture to these physical elements
- Establish the interfaces between the system's physical elements

As with the functional architecture, these steps aren't sequential. We typically do them in parallel. Completing part of one step gives us insight into the other steps, so we start working on one or more of the other steps, from which we gain further insight into the first step. The physical solution that best meets requirements unfolds as we work through the process.

Also, as with the functional architecture, architects generally define the physical architecture in multiple levels of detail. The process apportions the system architecture down to the configuration item level, at which point the development teams begin their designs. A *configuration item* is the lowest level of design for which we create a bill of materials—or the level to which we would manage the system configuration. After we decide on the configuration items, we must trace requirements to them and then pass the requirements to the development teams that will start producing hardware and software. To define the physical architecture at the configuration item level, we must understand what that level is:

- For configuration items to be developed, it's the level at which the hardware or software teams start to design
- For configuration items that don't have to be developed, it's the level at which the project must buy or otherwise obtain an already existing item such as a commercial-off-the-shelf item, a government furnished item, a government off-the-shelf item, or any other hardware or software item to reuse.

Thus, the configuration item level includes different amounts of detail, depending on the customer, contract, contractor's development environment, and even the solution to be built. To help visualize this configuration item level, we describe three examples below.

Example 1. Suppose we're building a system with a hardware chassis that includes several processors the contractor must develop. What is the configuration item level for this hardware? If the hardware development team starts with the chassis, we should partition the physical and functional architectures to the chassis level—this is the configuration item. But what if one development team produces the chassis and separate teams develop each chassis drawer? Then we should define the architecture down to the drawer level. In this example, the hardware development organization determines the configuration item level.

Example 2. Suppose we're building a system that has a lot of software in a single processor. If a single development team is building all the software, the configuration item is the "software in the processor," and we would match the architecture and requirements to this configuration item. If several subcontractors develop the software, we should partition the architecture to a low enough level to allocate requirements to each subcontractor. In this example, the development environment and teaming agreements determine the configuration item level.

Example 3. Suppose we're building a system that uses an existing, non-developed item—commercial-off-the-shelf or otherwise. Here, we should partition the architecture functionally and physically to match what the project will buy or obtain. The partitioning should stop after we define:

- This item as a configuration item, or physical component
- The interfaces between this component and others in the architecture
- The functions this component will perform
- Functional and other requirements the component must meet

Figure 5-27 shows another example of a context diagram describing the system's physical components and its physical links to external systems. This view doesn't depict the information, data, signals, or material being transferred across the physical links.

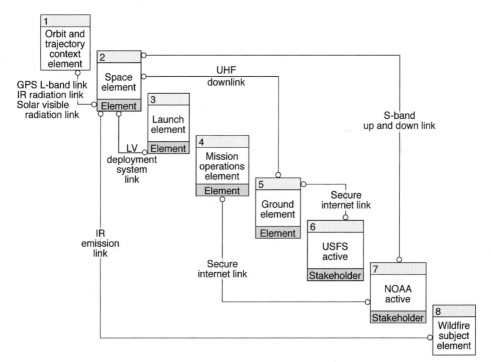

FIGURE 5-27.   A System Context Diagram Represented as a Physical Block Diagram. The diagram shows the space element, the external systems with which it interfaces, and the physical interfaces with those external systems. (GPS is Global Positioning System; IR is infrared; LV is launch vehicle; USFS is US Forest Service; NOAA is National Oceanic and Atmospheric Administration; UHF is ultra-high frequency.)

5.3.2   Create the Physical Partitions

The physical architecture partitions a system of interest into its subparts or components. The physical components are the system resources that perform its functions: hardware, software, facilities, people, or procedures. Whereas we represent functions with names that begin with a verb, the physical components have names represented by a noun phase: what or who is completing a function. Architects typically work with two types of physical architectures: generic and instantiated.

Generic physical architecture—This architecture partitions the physical elements for a system without specifying the performance characteristics of the physical resources that compose each element. It's sometimes known as a "reference" or "platform" architecture. As the examples in Figure 5-28 illustrate, we fix the types of physical components for the solution but not the actual components and requirements they must meet. For instance, the generic architecture states that we need a telescope, a pressure tank, a propellant tank, etc., but it doesn't give the details of these components. Many industries or contractors use generic physical architectures as templates or reference architectures for their final products. Cellular and computer companies have platform architectures that state their basic components. When they build a specific cell phone or computer, it's usually just a version of the generic architecture.

Instantiated physical architecture—This architecture adds to the generic physical architecture by specifying the resources' performance characteristics and requirements for the physical items. To create an instantiated physical architecture, we allocate functions to the generic architecture, determine the interfaces between its physical components, and derive requirements for these components from the system requirements. We now model this complete physical architecture—sometimes called the "product" or "solution" architecture—and begin developing it.

Figure 5-29 depicts a high-level example of an instantiated physical architecture. In this example, we define the physical components: hardware, software, and interfaces. The callout boxes give the ID numbers of the requirements that these components will meet.

In previous steps, we define the generic physical architecture, because we want to understand what type of physical components our solution will include. Next, we allocate functions to the physical components, establish the physical interfaces between them, and derive and allocate requirements to them. The result is an instantiated or product architecture—the one that captures stakeholder and system requirements.

5.3.3   Allocate Functions to the Physical Components

At this point, we have a functional architecture that describes what the system of interest will do. We've also created a generic physical architecture that describes the types of components we think we need and the expected interfaces between them. Now we need to relate these two architectures by allocating the functions to

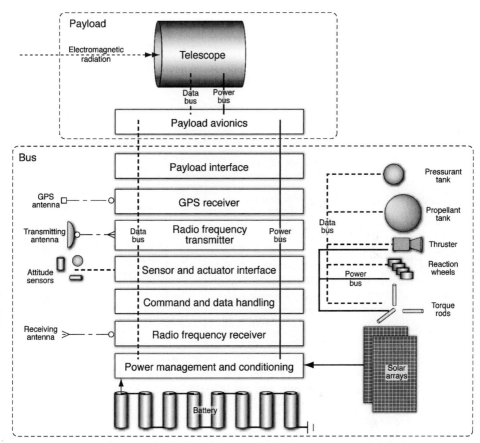

FIGURE 5-28.   Example of Generic Physical Architectures. We use generic architectures as templates (reference architecture) for many products. They show the types of components needed for the system. (GPS is Global Positioning System.)

the physical components. We decide what functions each of the hardware, software, people, process, or infrastructure components should do. This step is essential, because it defines the physical solution to build and the interfaces within physical components. (We describe the interfaces in Section 5.3.4.) The physical solution drives the solution cost, schedule, performance, and associated risks.

This step may also require a lot of iteration as we alter the functional architecture to more efficiently fit a physical solution. Trade-offs occur among solutions until we find the best one. That solution may not be the cheapest, the quickest to develop, or the "sexiest" technically. But it's the best one if it meets the system requirements and, therefore, the customer's needs.

All functions must be allocated to some physical component, if we are to understand how each function gets done. If a function is allocated to multiple

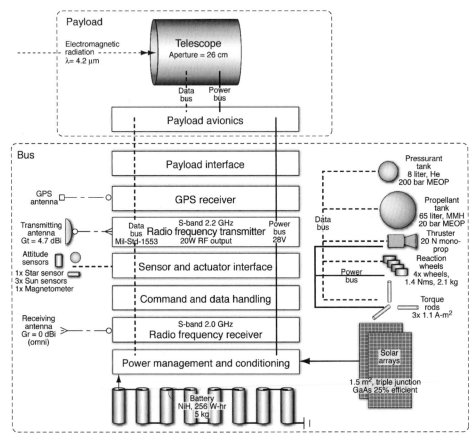

FIGURE 5-29.   **High-level Diagram of an Instantiated Physical Architecture.** Instantiated architectures describe specific products or systems. Normally they are developed from generic architectures and show the details of the physical components. (λ is wavelength; GPS is Global Positioning System; Gt is transmission gain; Gr is receiving gain; NiH is nickel hydride; GaAs is gallium arsenide; MEOP is maximum expected operating pressure; MMH is monomethyl hydrazine; Mil-Std is military standard.)

physical components, we should decide whether to split the function into several subfunctions, so we can allocate each subfunction to a single component. This step is important because the physical architecture drives requirements for the components and the interfaces between them. If we split a function across components, the requirements also split across their developers, so the interface between them may be nebulous. As described in Sections 5.3.4 and 5.4, an architecture is usually better when it's modular and the interfaces are well defined.

Sometimes splitting a function handled by more than one physical component doesn't make sense. For example, a system may have a function, **Provide information to the user**, that puts system responses on a display for viewing. Two or more physical components may cover this function—for example, a display monitor and a software application. It may not be cost-effective to split this function into two functions (one performed by the display monitor and one performed by the application software) because the development teams know the functionality of each of these components and the interfaces between them. Distributed systems are another case in which several physical components may do the same function. As long as we know the interfaces between the components, changing the functional and physical architectures to further document this solution may not add value.

Figure 5-30 shows how we allocate functions to the physical architecture. A single system function corresponds to the system of interest and each subfunction allocates to a single subsystem.

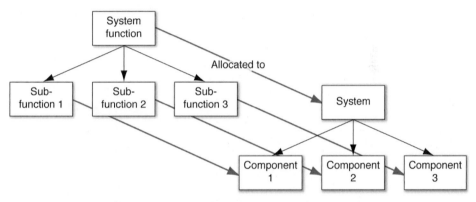

FIGURE 5-30.   **Allocating Functions to the Physical Components.** The functions in the functional architecture must be allocated to the components in the physical architecture. Every function must be performed by some component.

Although some people say each physical component should perform only one function, this is typically not the case in real life, nor should it be. Architects must understand functions in the functional architecture and allocate them in the best way to the physical architecture. This process seldom produces a one-to-one mapping of functions to components. Figure 5-31 shows an example of multiple functions allocated to physical components. But all physical components must normally perform at least one function. Reasons to group functions in a single physical component include:

- Functions are related or handle similar tasks
- Performance will improve if functions combine in a single component
- Existing components (commercial-off-the-shelf or otherwise) cover several functions
- Functions are easier to test
- Interfaces are less complex

- Technology risk is lower
- Allocation meets future performance requirements
- It matches potential future technology

Figure 5-31 also shows a generic example of mapping functions to components. Here component c_3 isn't doing anything (performing no functionality). It may be extraneous (not needed), a function may be missing from the functional architecture, or there may be other reasons for c_3 to be part of the system's physical architecture. This solution may be reusing a legacy system, a reference architecture, or some platform architecture that has been successful in previous solutions. Even though the current solution doesn't need this component, it's cheaper and quicker to keep it. We may also keep an extra component if some future function will require it and keeping it now is easier than adding it later. Still, whenever a component isn't doing something, we should verify whether the system needs it.

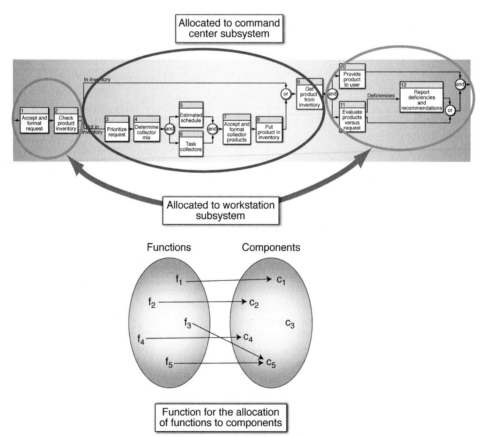

FIGURE 5-31.   Other Examples of Allocating Functions to Components. Here we show generic examples of allocating functions to physical components.

Allocating functions or generic components to the physical components creates fertile ground for trade studies, technical decision making, and iterative updates to the functions or generic components. One effective technique for allocating functions to components uses a *morphological box* to determine how functions or generic components may apply and instantiate the best physical components to match these applications. Table 5-4 shows a morphological box. The functionalities are the column headers, and the rows indicate possible physical solutions for each function. We then select the best way to implement the function based on technical, cost, and scheduling requirements for the system of interest. Figure 5-32 shows the physical element-to-function relationship for the FireSAT mission, including some of the FireSAT elements and the functions that they perform. We then decompose the space element into the ways that we can implement it by selecting:

- Which type of component or element will handle the function—hardware, software, infrastructure, or people and processes?
- What kind of hardware will we use—including commercial-off-the-shelf (COTS) versus developed, distributed versus centralized, and so on?
- What type of software will we use—including COTS versus developed, handled in the client versus the server, using a single processor versus distributed processors, and so on?

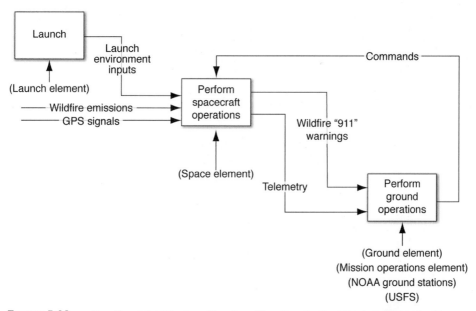

FIGURE 5-32.   **Functional Architecture Showing Allocation to the Physical Elements.** Here we show the physical components within the parentheses and the arrow pointing up to the function being performed by each component. (GPS is Global Positioning System; NOAA is National Oceanic and Atmospheric Administration; USFS is US Forest Service.)

TABLE 5-4.   Morphological Box for FireSAT's Spacecraft Systems Design. We use the morphological box to select the best physical implementation of the functional architecture. (UHF is ultra-high frequency; NiCad is nickel cadmium; GPS is Global Positioning System; Ni-H is nickel hydride; RTG is radioisotope thermoelectric generator; Li-Ion is lithium ion.)

Communication Uplink	Communication Downlink	Attitude Sensors	Attitude Actuators	Power Generation	Power Storage	Navigation Method	Propulsion Technology
None (total autonomy)	UHF	Sun sensor	None (free tumble)	None (primary battery only)	Primary battery	Ground-based	Cold-gas
UHF	S-band	Star sensor	Passive spinning	Photovoltaic	NiCAD	Space-based GPS	Mono-propellant
S-band	Ka-band	Magnetometer	Dual spin	Fuel cell	Ni-H	Space-based optical	Bi-propellant
Ka-band	K-band	Accelerometer	Gravity gradient boom	RTG	NiM hydride		Solid
K-band	X-band	Gyro	Permanent magnet	Nuclear reactor	Li-Ion		Nuclear thermal
X-band			Torque rods	Solar dynamic			Resistojet
			Reaction wheels				Arcjet
			Control moment gyros				Ion
							Hall effect thruster
							Pulse plasma thruster

Figure 5-33 shows a physical hierarchy of the architecture, with a block for each of the spacecraft's physical components. We haven't yet dealt with the interfaces between the components (Section 5.3.4).

5.3.4   Define the Physical Interfaces

With physical components selected and allocations complete, architects turn to the interfaces between the physical components. Interfaces connect physical components and transfer information, signals, data fluids, or energy between them (Chapter 15). We must define them carefully, because systems usually fail at the interfaces. As we better understand and lay out these interfaces, we may iterate back to change the functional architecture or reallocate functions to the physical components.

We need to establish interfaces whenever there's information, data, signals, or energy transfers between functions handled by different physical components or when two physical components need to be connected. To do so, we attend to the same details as in defining the external interfaces in Chapter 15. The following questions help architects determine the best way to apply an interface for a particular system (some information may not pertain to certain interfaces):

- What components are being connected?
- What is the physical approach: mechanical, electrical, data, and so on?
- What requirements, such as mean time between failures, must the interface and connected physical components meet?
- What are the technology limits of the potential implementations?
  - Throughput capacities
  - Delay times
  - Distance limits
- What are the interface's locations or environments?
- What protocols or standards are being used?
  - Transmission Control Protocol/Internet Protocol (TCP/IP) or Open Systems Interconnection (OSI) (7-layer) reference model
  - Data formats
  - Data bus standard, e.g., RS-232, Mil-Std-1553
- What is the information, signal, data, or energy being transferred across the interface (including type, amount, and frequency)? These details should be specified in the functional architecture, as part of the definition of information moving from one function (performed by one component) to another function (performed by another component)

Figure 5-34 depicts the interfaces between the elements of the physical architecture shown in Figure 5-33.

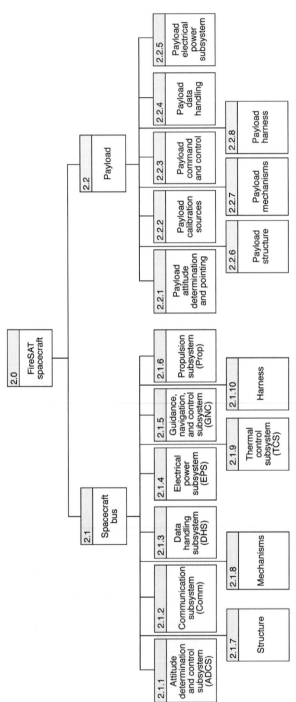

FIGURE 5-33.   Physical Hierarchy of the FireSAT Spacecraft. This figure shows the partitioning of the spacecraft to lower-level physical components.

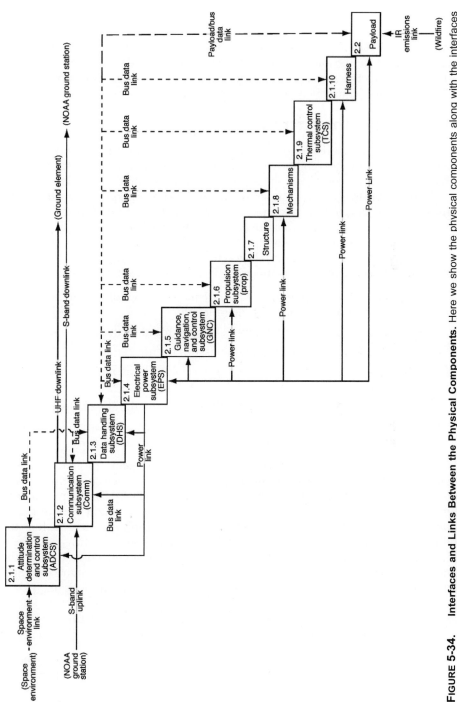

FIGURE 5-34.   Interfaces and Links Between the Physical Components. Here we show the physical components along with the interfaces between these elements. Now we understand the physical architecture. (IR is infrared.)

By establishing the physical components, allocating functions to these components, and defining the interfaces, we have completed the physical architecture (physically partitioned the system). As stated, the architecture should develop to the configuration item level, thus enabling development teams to design and build the physical components or to buy or otherwise obtain them (COTS or reusable components).

5.4   Assess the Architecture

The various methods for synthesizing a system's functional and physical elements suggest the possibility of developing multiple functional and physical architectural alternatives. Given the significant influence of an architecture on the organizational business strategy, operational mission governance and doctrine, and engineering and development complexity and resources, the design team should assess these alternatives. The objective is to synthesize and select the preferred system physical and functional constructs. The metrics that represent architectural preference or "overall architectural goodness" vary from project to project. Architectural assessments have many goals, but some basic ones include:

- Investigate design trade-offs: what architectural options were or weren't selected? And why? What was the rationale?
- Identify solution risks and potential ways to mitigate them
- Identify technical and organizational issues for further analysis
- Verify whether the solution reuses other solutions as much as possible or correctly integrates commercial-off-the-shelf products
- Communicate the solution and create buy-in from the stakeholders
- Will the architecture meet the system's performance requirements, satisfy the key constraints, and execute all the operational scenarios?

Architectural assessments also reveal potential problems to architects and other stakeholders, so they can address these problems before beginning the detailed design of configuration items, or the selection of COTS system elements. The best time to assess an architecture is before it's baselined and the solution is underway. We normally baseline the architecture in a formal review such as the preliminary design review (PDR), so we should conduct the assessment beforehand. If we don't, stakeholders can assess the architecture **during** the review. At this point, all stakeholders should agree that the architecture can meet system requirements and developers should start on the detailed design.

5.4.1   Establish Methods and Metrics for Architectural Assessments

Architectural assessment metrics typically consider selected and specific aspects of the system solution, such as performance or flexibility and scalability or testability and maintenance, and so on. The metrics may come from the system or stakeholder requirements or from characteristics that all (or most) systems meet. We generally either measure the system (or intended system) to determine whether its design meets the metrics or look at other system architectures to select an alternative that best meets them. In either case, basic methods include using the Software Engineering Institute's architecture trade-off analysis, applying quality attribute characteristics, and assessing an architecture based on metrics common to all or most systems.

Software Engineering Institute's Architecture Trade-off Analysis Method (ATAM). This method looks at business or mission drivers for an architecture. It describes each driver's measurable attributes and the scenarios in which it will be measured (or estimated) to determine whether the architecture meets the attribute. Figure 5-35 shows an example of this method.

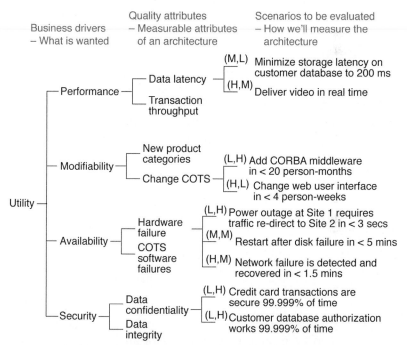

FIGURE 5-35.   The Software Engineering Institute's Architecture Trade-off Analysis Method (ATAM) for Assessment. Here we show an architecture assessment process that is used within industry. The pairs of letters (high, medium, low) in parentheses show, respectively, customer priority and expected difficulty in meeting the attribute. (CORBA is Common Object Requesting Broker Architecture.)

Quality attribute characteristics method. This method is similar to ATAM in that it looks at architecture attributes based on business or mission drivers (the system's quality attributes—see Figure 5-36). The method determines whether the architecture has or hasn't met the quality attribute by:

- Establishing the stimuli to which the architecture must respond
- Documenting which part of the architecture will meet the attribute
- Documenting how the architecture will be measured (or estimated)

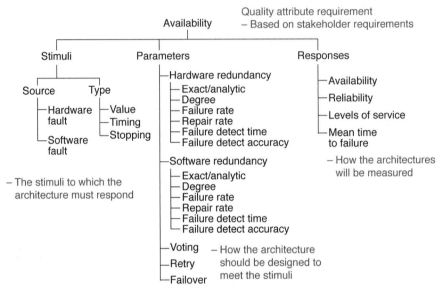

FIGURE 5-36.   Quality Attribute Characteristics Method. Here is another architecture assessment process used within industry.

Common metrics method. Another way to assess architectures uses metrics common to all or most systems. This methodology measures or estimates an architecture's "goodness" based on how well it meets these metrics. Although this method doesn't directly use mission requirements as an input, we vary it to incorporate them. Figure 5-37 lists some of the metrics under typical categories.

Figure 5-38 combines stakeholder priorities (or mission drivers) and common metrics to quantitatively measure or estimate an architecture. We use common metrics (step 1) to assess an architecture but rank order them based on the mission needs (step 2). In step 3, we quantitatively assess the architecture to determine how well it meets the metrics—typically using those shown in Figure 5-38. Finally, in step 4, we assign an assessment score. This score and the score for each metric enable us to determine the architecture's goodness and its risk of not meeting the customer's needs.

Commonality	Modularity	Standards Based	RMT

• Physical commonality (within the system)
  • Hardware commonality
    • Number of unique line replaceable units (LRUs)
    • Number of unique fasteners
    • Number of unique cables
    • Number of unique standards implemented
  • Software commonality
    • Number of unique software packages implemented
    • Number of languages
    • Number of compilers
    • Average number of software instantiations
    • Number of unique standards implemented
• Physical familiarity (from other systems)
  • % vendors known
  • % subcontractors known
  • % hardware technology known
  • % software technology known
• Operational commonality
  • % of operational functions automated
  • Number of unique skill codes required
  • Estimated operational training time—initial
  • Estimated operational training time—refresh from previous system
  • Estimated maintenance training time—initial
  • Estimated maintenance training time—refresh from previous system
	• Physical modularity
	  • Ease of system element upgrade
	    • Lines of modified code
	    • Number of labor hours for system rework
	  • Ease of operating system upgrade
	    • Lines of modified code
	    • Number of labor hours for system rework
	• Functional modularity
	  • Ease of adding new functionality
	    • Lines of modified code
	    • Number of labor hours for system rework
	  • Ease of upgrade existing functionality
	    • Lines of modified code
	    • Number of labor hours for system rework
	• Orthogonality
	  • Are functional requirements fragmented across multiple processing elements and interfaces?
	  • Are there throughput requirements across interfaces?
	  • Are common specifications identified?
	• Abstraction
	  • Does the system architecture provide options for information hiding?
	• Interfaces
	  • # of unique interfaces per system element
	  • # of different networking protocols
	  • Explicit versus implicit interfaces
	    • Does the architecture involve implicit interfaces?
	  • # of cables in the system
		• Open systems orientation
		  • Interface standards
		    • # of interface standards/# of interfaces
		    • Multiple vendors (more than 5) exist for products based on standards
		    • Multiple business domains apply/use standard (aerospace, medical, telecommunications)
		    • Standard maturity
		  • Hardware standards
		    • # of form factors/# of LRUs
		    • Multiple vendors (more than 5) exist for products based on standards
		    • Multiple business domains apply/Use standard (aerospace, medical, telecommunications)
		    • Standard maturity
		  • Software standards
		    • # of proprietary and unique operating systems
		    • # of non-standard databases
		    • # of proprietary middleware
		    • # of non-standard languages
		• Consistency orientation
		  • Common guidelines for implementing diagnostics and performance monitoring and fault localization
		  • Common guidelines for implementing OMI
			• Reliability
			  • Fault tolerance
			    • % of mission critical functions with single points of failure
			    • % of safety critical functions with single points of failure
			  • Critical points of delicateness (system loading)
			    • % processor loading
			    • % memory loading
			      • How critical is this?
			    • % network loading
			      • How critical is this?
			• Maintainability
			  • Expected mean time to repair
			  • Maximum fault group size
			  • Is system operational under maintenance?
			  • Accessibility
			    • Are there space restrictions?
			    • Are there special tool requirements?
			    • Are there special skills requirements?
			• Testability
			  • # of LRUs covered by BIT (BIT coverage)
			  • Reproducibility of errors
			    • Logging or recording capability
			    • Create system state at time of system failure?
			  • Online testing
			    • Is system operational during external testing?
			    • Ease of access to external test points?
			  • Automated input or stimulation insertion

FIGURE 5-37.   Commonly Used Metrics for Architectural Assessments. We use these metrics to assess architectures. These examples are from the Architecture Trade-off Analysis Method (ATAM) Business Drivers (Figure 5-35) or Quality Attribute Requirements (Figure 5-36). Middleware is a layer of software that enables open systems. (RMT is reliability, maintainability, and testability; OMI is operator machine interface; BIT is built-in test.)

Process steps	Architectural assessment for system XXXX			
① Enter the metrics to be used in the assessment ② Enter the relative priority of each metrics ③ Evaluate the architecture with respect to each metric ④ Determine the assessment score (multiply the evaluation times the normalized priority)	① Metrics to be measured	② Importance of metric—enter relative priority (higher number means higher priority)	③ Actual architecture evaluation—enter between 0 and 4 (4 being very good)	Total score
	–Responsiveness	10	2.75	
	–Scalability	30	3.25	
	–Modularity	50	2.25	
	–Availability	10	4.00	
	–Affordability	5	3.50	
	–Simplicity	50	4.00	
	–Functionality/performance	75	3.50	
	Total			82 ④

Total Score	Goodness Code
< 70	Red
between 70 and 85	Yellow
> 85	Green

FIGURE 5-38.   **A Quantitative Approach to Architectural Assessments.** The colors represent a qualitative indication of the "goodness" of the architectural alternative. A total score that is green represents a superior alternative, yellow represents an architecture that satisfies necessary objectives, but could be improved, while red represents an architecture that must be improved.

5.4.2   Model the System or Subsystem

As part of an architectural assessment, we analyze potential system resources and timing as a whole or just the system's risky aspects. To do so, we model or simulate the system or part of it, mainly to verify that the executable logic diagram meets the architecture requirements. This verification covers four main types of analysis:

- Dynamic analysis—Assesses the system behavior's dynamic consistency and ability to execute

- Timeline analysis—Establishes and analyzes integrated behavior timelines

- Resource analysis—Monitors the amounts and dynamics of system resources, such as people, computer million instructions per second, memory, supplies, power, radar pulses, and number of interceptors

- Flow analysis—Investigates how the system operation is affected based on the capacity of the link and the size of the items carried across this link

Models may also determine whether the system meets other technical requirements, such as availability, maintainability, size, or mass.

5.4.3   Update the System Architecture and Risk Plan Based on the Assessment

If we assess the architecture before baselining it, and discover that it won't meet the stakeholders' needs, we can update it before baselining it at PDR. But if we discover problems after PDR, we must determine whether cost, schedule, and technical risk compel us to update the architecture before implementing it. In all cases, project staff should integrate all risks and issues that the assessment captures into the risk and issue management for the project. They should also document and track technical performance measures through the development cycle to manage risk.

5.5   Allocate Requirements to the Components and Generate Detailed Specifications

We have to define and allocate derived requirements to the physical components, because these are the requirements we must provide to the development teams. They are the basis for software and hardware development or processes that system operators must complete. If we don't specify requirements, the development teams don't have a basis for designing, developing, and testing a solution.

5.5.1   Tracing Requirements to Physical Components

Developing an architecture results in requirement specifications for each development team. The developer (or acquisition agent for non-developed items) of every physical component must know what the component needs to do. For this reason, systems engineers and architects have to specify the functional and nonfunctional requirements for every system component. To illustrate, Figure 5-39 updates Figure 5-30 in Section 5.3.2 by tracing requirements to each component in the physical architecture.

As Figure 5-39 shows, architects trace many requirements to the functions allocated to the physical components. But why not trace requirements directly to elements in the physical architecture? This approach is common in the real world. Many very good architects often go from the requirements domain to the physical domain, and bypass the functional domain. They do so because they are vested in an application domain (e.g., sonar systems, avionics, bus architectures on a satellite), know the nuances of the application domain (key variables and constraints—operational, functional, and physical), know the component technologies, and know the main functions. They internalize the functional domain and do the mapping between the requirements, functions, and physical elements in an ad-hoc manner, without documenting this process. However, the result is an undocumented architecture with associated risks. We may trace requirements to the physical elements, without understanding the underlying rationale. The functions become largely implicit, and this becomes a constraint on

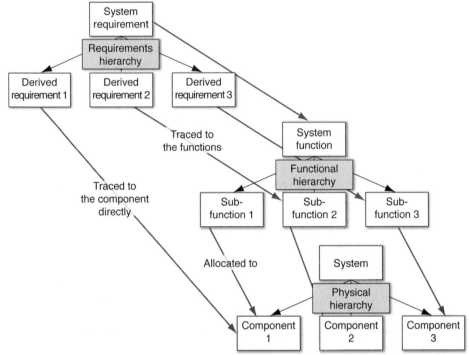

FIGURE 5-39.   Defining Requirements for the Physical Components. We generally trace functional and input/output (I/O) requirements to functions. We also trace nonfunctional requirements to functions or the physical components. Either way, the requirements for a component include the ones traced to the component's function and the ones traced directly to the component.

future evolution of the architectures, adding to the long-term risk with regard to integration and test, upgrade and refinement programs, and so on.

5.5.2   Handling Functional and Nonfunctional Requirements

Functional requirements include those for interfaces (input/output or technical) and processing. They are based on the tasks or the functionalities that the solution performs as described in the functional partitioning, the operational scenarios, and the tracing of the scenarios through the partitioning. We trace them to the functions. Nonfunctional requirements, including constraints, address such system aspects as:

- Usability (ease of use)
- Availability
- Maintainability, supportability, and reliability
- Security

- Size, shape, and mass
- Cost
- Schedule
- Interface protocols and standards

Some of these requirements—such as size, shape, and mass—depend solely on the physical components, while many derive from the components and their functions. For example, availability, maintainability, supportability, and reliability clearly relate to the physical components. Also, we normally define a different set of requirements for hardware components than for software components. For example, we may want to define a recurring cost and a nonrecurring cost for a hardware component since the component must be both designed and developed (non-recurring) as well as produced in quantity (a recurring cost). For software, only a development cost applies. Suppose some function isn't essential to the system's availability, meaning that even if it fails, the system doesn't. In this case, the availability requirement isn't the same as when the function is essential to the system's availability.

The derived functional requirements are based on the functional architecture and operational scenarios, but the derived nonfunctional requirements normally depend on nonfunctional requirements at the system level. Systems engineers and architects must determine which subfunctions or physical components a system-level requirement affects and generate derived requirements for the components using one of three main methods:

- Equivalence—the component requirement flows down from and is the same as the system requirement
- Allocation—spreads a system-level requirement, such as size or mass, among the system's components while maintaining the same units
- Synthesis—addresses situations in which the system-level requirement (such as availability or reliability) consists of complex contributions from the components, causing the component requirements to depend on some analytic model

This step results in specified requirements for each system component, including specifications for processing, interfaces, and nonfunctional requirements. They are the basis for teams to develop or acquire hardware, software, people, and processes.

5.6   Trace Requirements

This section discusses the tracing of system and derived requirements to the functional and physical architectures of the system and its subsystems. Tracing requirements gives us confidence that the solution meets the stakeholders' expectations. Traceability also begins the integration, verification, and validation planning (as detailed in Chapters 10 and 11).

5.6.1   Using Requirements Management Tools

Requirements management tools help us trace requirements. They may be as simple as general purpose software such as Microsoft Office or Apple iWork or dedicated tools such as DOORS™, Requisite Pro™, or CORE™. We use them to trace requirements through the system lifecycle and to document that the customer accepts these requirements. Figures 5-40 through 5-42 show output of a requirements management tool—tracing requirements from one level to lower levels, as well as to the design and to system risks and issues.

5.6.2   Developing a Requirements and Verification Traceability Matrix

Figure 5-43 shows a matrix that traces requirements through the system lifecycle. As discussed in Chapters 2, 3, and 4 the problem starts with a customer need that we express in a set of stakeholder requirements. We then follow clearly defined steps to complete the process:

1. Translate stakeholder requirements into system requirements

2. Validate and trace system requirements back to the stakeholder requirements

3. Determine acceptance (or pass/fail) criteria that the project team will use as a system measurement during test (Chapter 11)

4. Distribute and divide system requirements into component requirements down to the configuration item level while creating the functional and physical architectures

5. Verify and trace component requirements back to system-level requirements. During this time, determine acceptance criteria for the component specifications that the project team will use as a system measurement during test.

In Figure 5-42, stakeholder requirement C1 traces to system requirements S1.1, S1.2, and S1.3. System requirement S1.1 traces to component requirements CO.1.1.1, CO1.1.2, and CO1.1.3. The acceptance and pass-fail criteria for each system requirement would appear in the column marked "System acceptance criteria." The test method and test phase in which teams qualify the system requirement are in the columns under each test phase. System requirement S1.1 will be qualified by a formal test method (to verify that it meets its acceptance criteria) in phases for functional verification (FVT), application systems control and auditability (ASCA), system integration (SIT), and user acceptance (UAT). The acceptance criteria for component requirements CO1.1.1, CO1.1.2, and CO1.1.3 go into the column "Acceptance Criteria." These requirements will be qualified in the component verification test (CVT) phase using the "Test," "Simulation/Modeling," and "Demonstration" test methods, respectively.

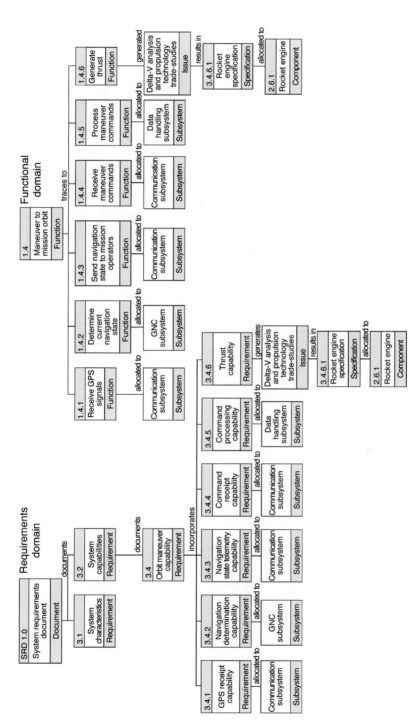

FIGURE 5-40.   Output of a Requirements Management Tool. Here we show the traceability of requirements 1) from one level to the next, 2) to the physical components, and 3) to issues associated with the requirement. (SRD is systems requirements document; GPS is Global Positioning System; GNC is guidance, navigation, and control.)

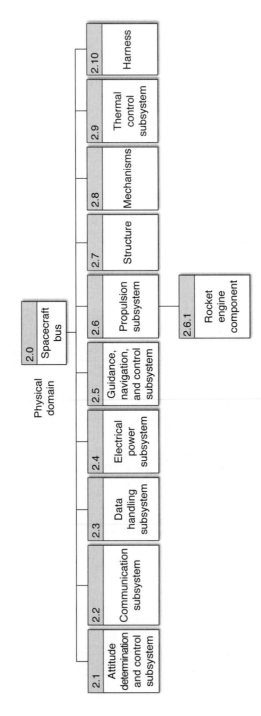

FIGURE 5-41.   Output of a Requirements Management Tool. This diagram shows the physical components of the space element to which we trace the requirements.

RTM

RVM

Test Type/Test Method

Stakeholder requirements	System requirements	System acceptance criteria	FVT	CVT	TVT	ASCA	SIT	PST	UAT	Pre-prod	Pilot
C1	SR 1.1		T			T	T		T		
	SR 1.2		I		I					I	
	SR 1.3			A			A				
C2	SR 2.1										

System requirements	Component requirements	Acceptance criteria	FVT	CVT	TVT	ASCA	SIT	PST	UAT	Pre-prod	Pilot
SR 1.1	CO 1.1.1		T								
	CO 1.1.2		S/M								
	CO 1.1.3		D								
SR 1.2	CO 1.2.1		D								

FVT = Functional verification test
CVT = Component verification test
TVT = Translation verification test
ASCA = Applications system control and auditability
SIT = System integration test
PST = Performance stress test
UAT = User acceptance test
Pre-prod = Pre-production test

Legend: Enter one or more of the following in each test method cell. A blank cell indicates that the test method is not required for that requirement.
A = Analysis
D = Demonstration
I = Inspection
S/M = Simulation and modeling
T = Test

FIGURE 5-42.   Example of a Requirements Traceability and Verification Matrix (RTVM). The RTVM shows a tracing of one level of requirements to the next level, the tracing of each requirement to its acceptance criteria (pass-fail criteria), and the tracing of requirements to the phase and method of verification. The requirements traceability matrix (RTM) documents the tracing through the various levels of the architecture. The requirements verification matrix (RVM) documents the tracing of each requirement to its verification information.

As this example illustrates, defining the derived requirements is also the start of establishing a qualification strategy. In addition, setting the acceptance or pass/fail criteria helps all stakeholders to better understand the requirements statements, which may mean different things to each stakeholder.

Summary

Developing the functional and physical architecture results in several outputs:

- Architectural models that describe what the system does, the physical components that perform these functions, and all interface definitions between the functions and components. Depending on the system's complexity, these models may be one level or many levels deep (with each level further detailing the solution)
- Derived requirements for the hardware and software development teams, from which they begin designing and building their components

- Traceability of the system requirements to the hardware and software specifications

- Traceability of the requirements to the architecture elements, such as functions, components, and interfaces

- Traceability of the requirements to the high-level verification plan

- Documentation of the technical decisions and trade-offs made during the process

- Updated risks, issues, and technical performance measures

Figure 5-43 shows how the development, test and integration, and systems management teams use these outputs to create their work products.

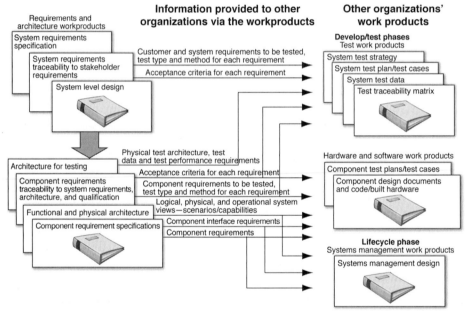

FIGURE 5-43.   Deliverables to Development and Test Organizations. This diagram shows the deliverables from the requirements and architecture development processes to the development and test processes. These deliverables provide a baseline from which the development and test processes begin.

Development organizations use the component requirements, architectures, and requirements traceability to test methods and phases to design, develop, and test their individual components. The integration and test teams use them to build the test architectures, as well as the detailed test plans and procedures. The systems management teams use the architectures and requirements to operate and maintain the system.

Not shown in the figure, but just as important, is how project or program managers use the architecture outputs. The architecture and component requirements give them the information to develop costs and schedules for the solution. Because these requirements specify the hardware and software components to build or acquire, the staff can develop the project or program plan from them. The same information enables managers to better understand the costs for developing, integrating, verifying, and validating the solution.

The process is iterative: each step uncovers more information about the solution and may require revisiting previous steps. Thus, the process produces important technical decisions and trade-off analyses that help project staff understand the current solution and support future decisions. As discussed in Section 5.4, all architectures create issues and risks. We assess architectures formally and informally to understand their risks and to verify that they meet the stakeholders' needs.

By developing the physical and functional architectures, as well as their corresponding requirements, we provide a technical baseline for the rest of the project or program's development. This baseline enables the project's team to produce a detailed cost and schedule, understand the solution's risk, and give the development, test, and systems management teams confidence that the solution addresses the stakeholders' needs.

References

Lykins, Howard, Sanford Friedenthal, and Abraham Meilich. 2001. "Adapting UML for an Object Oriented Systems Engineering Method (OOSEM)." International Council On Systems Engineering (INCOSE).

SysML.org. Ongoing. "System Modeling Language (SysML)-Open Source Specification Project." Available at the SysML website.

Chapter 6

Decision Making

Michael S. Dobson, *Dobson Solutions*
Paul Componation, Ph.D., *University of Alabama*
Ted Leemann, *Center for Systems Management*
Scott P. Hutchins, *NASA Marshall Space Flight Center*

"Nothing is more difficult, and therefore more precious, than being able to decide." – *Napoleon Bonaparte*

We make decisions every day, mostly using informal processes based on our intuition. So we might ask: "Why do we need a formal decision making process in space system design and management?" The reasons include [Clemen and Reilly, 2001]:

- The inherent complexity of problems we face in designing and managing space systems
- The uncertainty of situations under which we address the problems
- The need to satisfy multiple objectives
- The conflicting needs of multiple stakeholders

These decisions can clearly benefit from a formal, rigorous process:

1. Establish guidelines to determine which technical issues require formal analysis or evaluation

2. Define the criteria for evaluating solutions

3. Identify solutions to address decision issues

4. Select evaluation methods and tools

5. Evaluate solutions with the established criteria and selected methods

6. Select recommended solutions from alternatives based on the evaluation criteria

7. Report analysis results with recommendations, effects, and corrective actions

8. Capture work products from decision analyses

Decision making is too big a subject for a single chapter, or even for a single book. We can't prescribe one process for all environments and situations, so we encourage the reader to seek more resources on this vital topic. Organizations benefit from identifying common tools and approaches that improve the quality and consistency of mission-critical decisions. Table 6-1 describes the steps to consider in decision making. We discuss each step in this chapter, followed by pertinent questions to ask to be sure that decisions have followed the process steps.

TABLE 6-1.   Formal Process for Decision Making. A methodical decision process requires careful progression from defining the problem through testing and evaluating the solution. Formal and informal methods abound.

Decision Making Steps	Description	Where Discussed
1. Identify what we need to decide	The first step in getting the right answer is asking the right question. What is the decision? What's at stake? In what context will the decision take place? Who should decide?	Sections 6.1, 6.2.2, 6.7
2. Frame the decision	Decisions aren't made in a vacuum. Define the environmental, organizational, mission-related, and major uncertainty factors that influence the right choice.	Sections 6.2, 6.7
3. Select a method to evaluate alternatives	Establish criteria for evaluating the alternatives. Which tools are best for this decision? Formal processes take more time and effort to document the decision, help establish a consensus, and apply an objective method.	Sections 6.4, 6.7
4. Generate alternatives	Before settling on a single answer, look for several. Create backups in case the first-choice solution doesn't work.	Sections 6.3, 6.4, 6.7
5. Evaluate alternatives	Select evaluation methods and tools, and examine the solutions in light of established criteria. Look for cognitive and perceptual biases that may distort the process.	Sections 6.4, 6.5, 6.7
6. Choose the best solution	Select the best solution. Document the process, alternatives considered and rejected, and rationales. List action steps required, and then carry out the decision.	Section 6.5
7. Test and evaluate	Evaluate the applied decision's effects. Determine the likely outcome of rejected strategies. Prepare necessary documentation. Integrate lessons learned for future decisions.	Sections 6.6, 6.7

6.1   Identify What We Need to Decide

In Douglas Adams' novel, *The Hitchhiker's Guide to the Galaxy*, the computer known as "Deep Thought" spends 7.5 million years calculating the ultimate answer to life, the universe, and everything, only to come up with the answer "42." When challenged, Deep Thought replies, "[T]hat quite definitely is the answer. I think the problem, to be quite honest with you, is that you've never actually known what the question is." [Adams, 1979]

Establishing the correct question is the necessary first step in finding any answer. Too often, decision makers rush this step, assuming they already understand. For example, one might say: What's half of thirteen? Six and one half is the mathematical answer, but "thir" (or teen, or even 1 or 3) might be a better answer if we're designing the type for a brochure. The correct answer requires the correct question.

At the beginning of the process, we write down the decision issue. Is it clear and relevant to the problem at hand? Do the participants in the decision process understand it? What difference will a correct (or incorrect) decision make? Who must be part of the decision process? Who are the decision's owners, and who are contributors or reviewers? Is this decision part of a larger decision?

Without agreement on the basic premise, no decisions are meaningful. With agreement, we can scope the project. We can't assume others have taken this basic step; sometimes stakeholders skip the first steps in the rush to start engineering. Therefore, the first step in deciding is to identify the objectives, much as systems engineers define the boundary of the system and develop the system requirements. If we don't, we'll have what's known as a "type III error": the right answer to the wrong problem.

In a complex decision, identifying and defining objectives should be a team effort, using such resources as the concept of operations, mission statement, or requirements documents. We must ensure agreement on the objectives, their definition, and their relative priority. It's helpful to summarize results so decision participants share the same basis. Table 6-2 shows an example of possible objectives for an enterprise resource planning (ERP) system.

Objectives aren't created equal, nor are they necessarily compatible. We have to rank order them to make the right trade-offs and get the best results. This gets progressively more difficult as the list of objectives grows, the stakes get higher, and the number of stakeholders increases.

One way to deal with a large set of objectives is to have the design team rank them from most to least important, and then use a standardized rank ordering technique, such as rank sum or rank reciprocal (Table 6-3). Either technique allows us to quickly assign weights to objectives based only on their rank. The objectives' weights (priorities) sum to 100 percent with rank reciprocal or rank sum, but analysts use rank sum more commonly because it produces a relatively linear decrease in objective weights from most to least important.

At the beginning of the decision process, we can't always be sure we've defined the correct question. But once we reach a decision, it's good practice to go

TABLE 6-2.   **Objectives for an Enterprise Resource Planning (ERP) System.** Each objective is important when selecting this system.

Number	Title	Objectives for an Enterprise Resource Planning (ERP) System
1	Internal rate of return	The project's internal rate of return will meet or exceed the company's minimum attractive rate of return: 25 percent
2	ERP requirements	The computer-integrated manufacturing system will meet the requirements identified by the plant's departments, including production, logistics, scheduling, and accounts payable
3	Serviceability	The vendor will respond to and correct 95 percent of reported system problems within 24 hours of being notified
4	Engineering support	Engineering support, in labor hours, will meet the level required to support designing and installing the system
5	Risk	The project won't exceed this level of technical, cost, and schedule risk, as identified by the implementation team

TABLE 6-3.   **Using Rank Sum and Rank Reciprocal to Generate Priorities for Objectives.** These mathematical techniques enable decision makers to generate weights quickly based on the objectives' order of importance. Rank sum produces a relatively linear decrease in weights; rank reciprocal produces a non-linear weight distribution because it places greater importance on the first objective. (IRR is internal rate of return; ERP is enterprise resource planning.)

Objectives	Rank Sum			Rank Reciprocal		
	(A) Rank	(B) Inverted Rank	(C) Weight (B) / Σ (B)	(A) Rank	(B) 1/(A)	(C) Weight (B) / Σ (1/ (A))
IRR	1	5	5/15 = .33	1	1	1/2.28 = .44
ERP Requirements	2	4	4/15 = .27	2	.5	.5/2.28 = .22
Serviceability	3	3	3/15 = .20	3	.33	.33/2.28 = .14
Risk	4	2	2/15 = .13	4	.25	.25/2.28 = .11
Engineering Support	5	1 Σ (B) = 15	1/15 = .07	5	.20 Σ (B) = 2.28	.20/2.28 = .09

back to this step to be sure we chose appropriately. Defining an issue typically requires six questions:

- Who are the decision makers?
- What role does each play?
- What is the nature of the problem?
- Where and when must a decision be made?

- Why is this decision necessary?
- How will we measure its accuracy or applicability?

6.2   Frame the Decision

In *Smart Choices*, John Hammond observes: "A good solution to a well-framed problem is almost always smarter than an excellent solution to a poorly posed one." [Hammond et al., 1999] Throughout systems engineering, this theme reappears often. By frame, analysts mean the decision context, or how a problem is defined, which largely determines its solution process. For example, is time pressure a factor? Is the risk high or low? Does the solution meet a threat or opportunity? Is the issue complex? Do we have a lot of information available, or only bare bones? Is the data ambiguous or incomplete? Is the situation dynamic or fixed?

NASA's well known pre-launch telecon with Thiokol concerning *Challenger's* readiness reflects many of these decision frames. Time pressure was significant. Risk was high. The issue was complex, and the data was ambiguous and incomplete. The situation was dynamic, with continually changing stresses. Such factors hamper good decision making but are the stuff of life.

We need to recognize that decisions must reflect organizational goals under less than ideal circumstances. Sometimes, even a successful project may not benefit the organization, so the right decision could involve hampering or even destroying the project. Also, decision frames depend on the realities surrounding a project. If we have more money, different policies, or procedures that don't require this item, the frame changes—and likely, so does the decision. Technical professionals usually emphasize getting the job done correctly over meeting a schedule, but organizational circumstances may compel the latter.

The "five-whys" technique is a tool for improving decision framing. Each "why" question drills down closer to the real issue. Why are the Thiokol engineers reluctant to endorse the *Challenger* launch? The O-rings. Why are they concerned about the O-rings? The temperature has dipped below freezing. Why does the low temperature concern them? The O-rings may be less resistant to burn-through. Why do we care if the O-rings burn through? Hot gas will escape. Why do we care if hot gas escapes? It may cause a catastrophic explosion. At each stage, we amplify the "why" answers. What temperature-related effects have we seen? What evidence do we have that the O-rings could fail? What is the probability of escaping gas causing an explosion? What is the likely effect of an explosion in that system?

Another technique is to try different perspectives. Retired Milton-Bradley executive Mel Taft arranged for each member of the creative and marketing team— including himself—to work in a toy store for one week during each holiday season. He knew decision makers gain a complete perspective only by taking on other roles. Engineers need to see how the world looks to managers, and vice versa. The well-known conflict of perspectives between Thiokol engineers and managers before the *Challenger* accident kept each side from seeing the other's legitimate needs and, arguably, compromised the final decision [Dobson and Feickert, 2007].

A third technique is to list the candidate decisions. For *Challenger*, managers had several possible ones:

- Never launch with this configuration
- Launch tomorrow
- Don't launch tomorrow
  - Launch later when weather conditions are different
  - Launch after further analysis or study: How long? At what cost? To focus on what question?
  - Launch after taking a remedial action: What action? When? Why this action?

The completed decision frame should capture criteria for timeliness, precision, and cost; a list of desirable or mandatory goals; and accepted analysis methods. To make sure it's complete, we must keep in mind certain *framing questions*:

- Have we defined the problem correctly?
- Do we have consensus about this decision's priority?
- What organizational and external influences affect the decision?

6.2.1   Handling Risk and Uncertainty

Decision makers face a host of challenges. Critical information may be unavailable, circumstances may be uncertain, and consequences may be catastrophic. Stakeholders' interests may be incompatible. And always, the clock is ticking. As a result, people too often try to avoid difficult choices altogether, become trapped in "analysis paralysis," or pass the hot potato to someone else.

The twin challenges of risk and uncertainty complicate this process. In classic risk analysis, we know each outcome's probability and effects. Here, "risk" may refer to an undesirable event (threat) or a desirable one (opportunity). Risk management involves:

- Assessing the risk, including the probability of its occurrence and the consequence if it does occur
- Deciding how to act in response to that risk (for threats: avoid, mitigate, transfer, or accept; for opportunities: exploit, enhance, share, or accept)

In space projects, we don't necessarily know the range of possible outcomes and almost never have reliable probability information. Critical information for decision making is often subjective and based on values. An engineering evaluation might tell us an event has a 42-percent probability of happening and the consequence is losing $21.6 million and three lives. What such an evaluation doesn't tell us is whether the risk is **worth** running. Values—organizational, mission-related, or ethical—consider worth. Some are assumed and implicit. Some can be quantified, and others—usually political considerations—can't even be discussed on the record.

Decisions often require trade-offs because a perfect solution may not exist. Each potential choice has a downside, or includes risk or uncertainty. In some ways, choosing the least bad alternative from a set of poor ones takes greater skill and courage than making a conventionally "good" decision.

A decision's outcome, whether positive or negative, doesn't prove its quality, especially where probability is concerned. The odds may dramatically favor a positive outcome, but low-probability events occur. Conversely, a stupid decision that results in a good outcome is still stupid. A solid decision process improves our odds of achieving the desired outcome most of the time.

The decision making process must often be open and auditable. We have to know the decision and the process that led us to it. If the outcome is bad, someone else—a boss, a customer, a Congressional committee—uses "20-20 hindsight" to determine whether our decision was reasonable and appropriate. This second-guessing leads to the rational (but not always appropriate) strategy known as "CYA" (cover your assets), in which the decision focuses less on the mission perspective than on a method that ensures that blame and punishment fall elsewhere if an outcome is bad. As Fletcher Knebel points out, a "decision is what a man makes when he can't find anybody to serve on a committee." [Knebel, n.d.]

With enough time and resources, we can structure and study even the most complex problems in a way that leads to the best decision. Without them, however, we still must decide and remain accountable. That responsibility brings to mind the story of a person who called an attorney and received this advice: "Don't worry, they can't put you in jail for that." The person replied, "But counselor, I'm **calling** from jail!"

No formal process or method removes all risk from decisions. Many tools and techniques improve decision making, but ultimately it requires good judgment. Good judgment comes from experience combined with wisdom, but experience often comes from bad, unwise judgment. To improve on experience, we should focus on central questions about **risk and uncertainty**:

- How much do we know? How reliable is that knowledge?
- What do we know we don't know?
- How certain is a given outcome?
- Is there stakeholder or political conflict?
- What are the consequences of being wrong?
- How much time do we have to gather information?
- What values are important, and how much weight should each receive?
- Are there any unequivocally positive outcomes, or do all the options carry risk and uncertainty?

6.2.2   Understanding Organizational and Mission Context

Decision making contains two types of complexity. The first, and most obvious, is the issue's technical complexity and the trade-offs it may require. The second, though not always openly addressed, is organizational complexity: the number of people involved, departments or workgroups we must consult, relationships that shape communication among people and groups, organizational culture, and political pressure.

Decisions also vary in their importance, which we measure from their consequences and the constraints imposed on them. An incorrect decision isn't critical if the damage it causes is acceptably low, although still undesirable. Critical decisions—in which wrong choices have unacceptably serious consequences—fall into four categories:

- **Time-critical** decisions must occur within a narrow time window
- **Safety-critical** decisions may cause injury or death
- **Business-** or **financial-critical** decisions affect the organization's future or funding
- **Mission-critical** decisions affect the project's ability to achieve its objectives

Some critical decisions fall within the project's scope (they're ours to make), but others don't. Often, outside specialists must make certain critical decisions. For example, the counsel's office must decide on legal and regulatory matters. A configuration-control board may need to approve a configuration change. We may appeal their decision to higher management in some cases, but the burden of proof normally lies on our shoulders. We must determine whether the decision is properly ours and, if not, move it to the proper decision maker.

Decision making has attracted two contradictory slogans: "Don't just stand there, do something!" and "Don't just do something, stand there!" Either one can be worthwhile advice; the trick is determining which philosophy applies. The first question is whether a problem exists and, if so, its nature. The second question is whether we should act and, if so, how. These questions lead to four options, as shown in Figure 6-1.

At different times in decision making, we must consider the opportunities and negative consequences that can result from deciding to act or to wait. If the consequences of a missed opportunity are greater, the appropriate bias is toward action. If an inappropriate action could cause greater harm, the bias falls toward delay, so we can gather more information and reassess. To sift through these choices, we should ask key questions about **organizational and mission context**:

- Does the decision involve critical considerations, such as safety, significant business or financial consequences, legal liability, or major compromises to the mission?
- Does the decision depend on mandatory policy, procedural, regulatory, or legal constraints?
- Must we consider political or public information issues?
- How much organizational consensus do we need?
- Is the decision in the proper hands, or must it move to someone else?

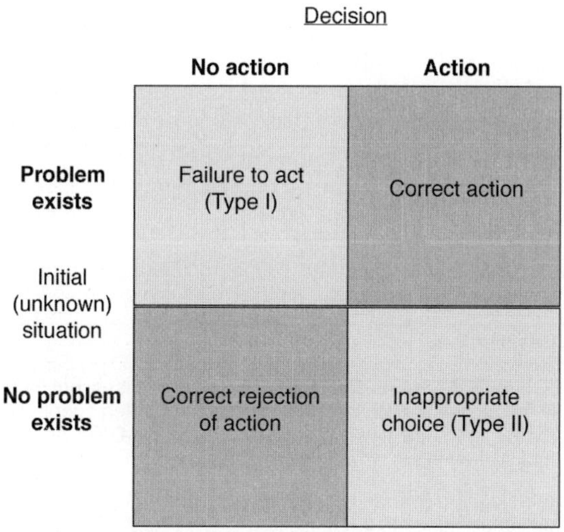

FIGURE 6-1.   **Options for Analyzing Decisions.** Decisions (including the decision not to act) fall into four categories—two appropriate and two inappropriate.

6.3   Select a Method to Evaluate Alternatives

The next step is to select a method to evaluate alternatives while understanding that potential outcomes are subject to residual and secondary risk. Residual risk is left over after we apply the solution. Secondary risk is new risk that our proposed solution generates. But we must not give up too quickly if secondary risk seems unacceptable, because we may be able to reduce it enough to move ahead with the solution.

Another way to think about risk is to divide it into "pure" and "business risk." Pure risk (also known as insurable risk) has only a downside. If the vehicle doesn't explode on launch, we aren't better off—we've simply avoided becoming worse off. If the risk doesn't occur, we get the status quo. Business risk can have positive outcomes. Usually, we want to eliminate pure risks if possible at an equitable cost, but sometimes we must add business risk in the hope of getting a positive outcome. Before trying to determine which method will best help us evaluate alternatives, we should ask some general questions:

- What trade-offs must we make for each solution?
- Does this alternative generate secondary risk? If so, can we reduce the risk enough to use it?
- How much time and resources will it cost to apply each alternative?
- Does this alternative carry a business risk (potential upside) or pure risk (only downside)?

Driving carpet tacks with a sledgehammer is usually a bad idea, so the method for a given decision must be in proportion. It can't take more time, energy, or resources than the decision's value allows, but it must be rigorous enough to support recommendations and choices. And if the outcome is bad, it must provide enough detail to evaluate the decision process. When it's a short-fuse decision, we may not have enough time to use the tools properly. Table 6-4 lists some ways to evaluate decision making, but it's not exhaustive. The case study in Section 6.7 uses a different method.

TABLE 6-4.   A Range of Ways to Evaluate Decision Making. Formal, informal, analytic, and intuitive evaluation tools all have a place in effective decision making.

Method	Description
Intuition	"Gut feel" decisions don't follow formal processes, but they may be the only tool available when time is very short. Intuition often supplements other, more formal, methods. If we've done a complete analysis, and our gut argues strongly that something is wrong, we're usually wise to listen.
Charts	A table of options, a "pro-con" list, or a force-field analysis (described below) all use charting techniques
Heuristics	Rules, often derived from practical experience, guide decisions in this method. For example, a traditional heuristic in problem solving is: "If understanding a problem is difficult, try drawing a picture."
Qualitative analytic models	These models are intended to completely describe the matter at hand
Quantitative analytic models	These models involve building complex statistical representations to explain the matter at hand
Simulations	Simulations imitate some real thing, state, or process by representing a selected system's main characteristics and behaviors
Real systems test	Ultimately, if we want to know whether the Saturn V engines work, we have to test the real thing. Properly developed heuristics, models, and simulations can be highly predictive but may omit variables that only a real test will reveal.

A number of factors influence our choice. What are the quality and completeness? What will it cost in time and resources to get missing data? We point out here that data isn't the same as knowledge. Numbers are not decisions; they're inputs. Subjective data reflects psychology or behavior, as in how the project's stakeholders feel about the amount of progress. We can make subjective data at least partly quantitative using surveys, interviews, and analytical hierarchies. But both types of input have value, depending on the nature of the question. Data comes in many forms and levels of certainty, so we must understand its nature and origin to use it properly. *Objective data* includes:

- Precise spot estimates (deterministic)
- Expected values (deterministic summary of probabilistic data)
- Range of estimates
- Probability distributions

Subjective data includes

- Personal (expert) opinions
- Relative rankings
- Estimates of ranges

Some examples of evaluation tools and techniques are force-field analysis, decision trees, decision criteria with Kepner-Tregoe analysis, rating and weighing, analytical hierarchies, and optimizing techniques. Most techniques work with quantitative and qualitative decisions. They differ in the amount of time and effort required and in the decision complexity they can handle. Decision trees and optimizing techniques work best with quantitative data.

6.3.1   Force-field Analysis

Listing pros and cons dates back at least to Benjamin Franklin, but force-field analysis adds insight to such a list. The current state of affairs is the balance point between forces that improve the state and those that retard it. We can alter the current state for the better by adding forces to the positive side, deleting them from the negative side, increasing the existing positive forces, or reducing the existing negative forces. Force-field analysis enables us to consider objective and subjective, as well as qualitative and quantitative, information. Table 6-5 shows a force-field analysis.

TABLE 6-5.   Force-field Analysis. This table shows forces that improve or worsen organizational decisions. To alter the situation, we must add or amplify forces on the positive side, or eliminate or mute them on the negative side.

FORCE FIELD ANALYSIS Quality of Organizational Decision Making	
Positive Forces $\Rightarrow$	$\Leftarrow$ **Negative Forces**
Formal decision making process	Lack of authority for decision makers
Technical knowledge and skill	Necessity for technical trade-offs
Available training courses	Too short-staffed to send people for training

Additive model. For the additive model [Philips, 1984], we add the products of the weights and scores to arrive at a single score for each design alternative. Figure 6-2 shows the general format for data collection.

Data collection to populate the model is usually one of the most time-consuming steps in decision analysis. Two items are particularly important. First, we must identify the performance range, threshold, and optimum for each objective. Second, we have to develop a standardized scale, so we can relate each objective's performance to other objectives. A five-point scale is common, and odd-number scales allow for a median. Table 6-6 provides an example.

		Alternatives		
Objective	Weight	I	II	III
1.				
2.				
3.				
4.				
Totals:				

FIGURE 6-2.   **Additive Model.** We customize this general form to collect data for widely varying decisions.

TABLE 6-6.   **Objective Evaluation Sheet.** This table illustrates how decision analysts might use a five-point standardized scale to relate each objective's performance to other objectives.

Grade	Score	Examples
Excellent	5	The technology is current in field operations undergoing type IV (stress) testing
Very Good	4	The technology has passed type III (prototype) testing
Good	3	The technology has been verified by type II (brassboard) testing to check for form, fit, and function
Fair	2	The technology has been verified by type I (breadboard) testing to check for functionality
Poor	1	The technology has been verified by analytical testing

6.3.2   Decision Trees

A *decision tree* is particularly valuable for financial decisions. It compares the expected monetary value (EMV) of different options. Figure 6-3 shows an example. The best EMV doesn't automatically equal the best decision. We can try altering one or more assumptions to see if the decision tree's result changes. Over a year, the numbers may be unattractive, but changing the horizon to two years may flip the decision. We must keep in mind that numbers are decision **inputs**, not decisions. The decision-tree method requires objective, quantitative data.

6.3.3   Decision Criteria

We should turn the "musts" and "wants" for the best outcome into decision criteria and then analyze each alternative to determine how well it meets those criteria. To do this, decision makers weigh each criterion based on its relative importance to major stakeholders. High-quality evaluation criteria must

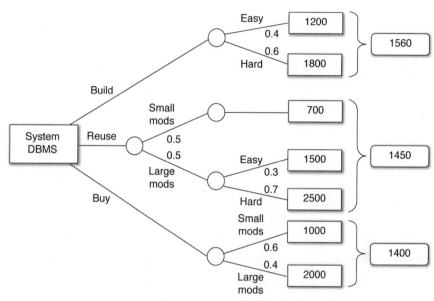

FIGURE 6-3.   **Decision Tree.** This decision tree compares the financial impact of three possible strategies to build a database management system (DBMS): build, reuse, or buy. Build, with an expected monetary value of 1560, appears to be the best financial option, but other information is usually necessary to arrive at the best decision.

discriminate meaningfully among alternatives, be independent of one another, and enable decision makers to define and apply them consistently. Musts and wants may be objective or subjective, qualitative or quantitative. The *Kepner-Tregoe Technique* provides a structure to display and analyze the data, as shown in Figure 6-4.

6.3.4   Rating and Weighting

After scoring each alternative on how well it meets each decision criterion, we then weight the decision criteria for impact. For example, in Table 6-7 we consider how well four design alternatives for the FireSAT satellite satisfy three criteria: cost, mass, and the quality of sensor data.

Alternative A has the highest cost and mass, and it scores third in quality, so its total weighted score is 21. Alternative D has the second-lowest cost and mass, as well as the best quality, so its total weighted score is 9. The lowest-scoring alternative offers the best balance among these weighted criteria. With a tie or very close scores, it's wise to look at the individual criteria. In this case, D, with two second-place results and a first place in the most important category, looks better than C that has a worse score in the most important category, as well as a third place in cost.

Decision Statement:													
Evaluation Criteria:		Candidate 1			Candidate 2			Candidate 3			Candidate 4		
Musts (Go/No-Go):													
Wants	Weight	Score			Score			Score			Score		
		Comments	Row (R)	R/W	Comments	Row (R)	R/W	Comments	Row (R)	R/W	Comments	Row (R)	R/W
Max Score (10xW):													
Total Score:													

FIGURE 6-4.   **Kepner-Tregoe Technique.** This tool allows managers to list, weight, and rank "musts," and "wants," weights, and scores as they apply to each decision alternative. Adapted from [Goodwin, 2005].

TABLE 6-7.   **Rating and Weighting Table.** This example identifies four design alternatives for the FireSAT satellite and how well they satisfy the three weighted criteria shown. In this example, quality is three times as important as cost in making the decision. The lower the score, the better.

	Alternative			
Criteria	A	B	C	D
Cost (Weight = 1)	4th	1st	3rd	2nd
Mass (Weight = 2)	4th	3rd	1st	2nd
Quality (Weight = 3)	3rd	4th	2nd	1st
Rank × Weight	21	19	11	9

6.3.5   Analytical Hierarchies

The previous example compared three variables with a weight given to each variable. Some decision matrices involve many more variables. The normal response to this complexity is to simplify, and cut back on the number of variables. But simplifying may omit critical information or important variables, so we recommend applying the pairwise comparison technique to build analytical hierarchies for decision making.

Pairwise comparison is a mathematical method that allows us to consider qualitative and quantitative aspects of decisions by reducing them to a series of one-to-one comparisons and then synthesizing the results. Organizations can buy any of several software packages that handle comparisons, especially for complex decisions with large numbers of values. Spreadsheet programs also allow pairwise comparison modeling. Figure 6-5 shows a multi-variable approach to selecting a launch vehicle. Table 6-8 summarizes the comparisons.

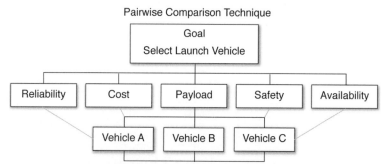

FIGURE 6-5.   **Applying the Pairwise Comparison Technique to Select a Launch Vehicle.** This diagram sets up the rating of three different launch vehicles against five quality criteria. Criteria can be quantitative and qualitative.

TABLE 6-8.   **Weights for Top-level Criteria Resulting from the Pairwise Comparison Technique.** This table summarizes a series of pairwise comparisons, showing relative weight for each criterion. (Weights shown total only to 99% due to rounding.)

	Payload	Reliability	Availability	Cost	Safety	Row Totals	Weight
Payload		0.20	0.33	1.00	2.00	3.53	8%
Reliability	5.00		2.00	4.00	6.00	17.00	41%
Availability	3.00	0.50		2.00	5.00	10.50	25%
Cost	1.00	0.25	0.50		8.00	9.75	23%
Safety	0.50	0.17	0.20	0.13		1.00	2%
						Grand Total	
Column Total	9.50	1.12	3.03	7.13	21.00	**41.78**	

The approach in Table 6-8 requires us to construct an L-shaped matrix and then judge the relative preference or importance of each criterion to every other criterion using a 1-10 scale. A score of "1" means the criteria are equal. A "5" shows a strong preference for the "row" criterion, and a "10" shows extremely strong preference. Then we

1. Add the rows
2. Sum the row totals
3. Divide each row total by the grand total of the columns
4. Convert the result to a percentage

Each percentage represents that criterion's relative weight, which we then factor into our launch vehicle selection, as depicted in Figure 6-6. Table 6-9 shows that we're now ready to compare alternatives numerically, with all the data rolled into a single total.

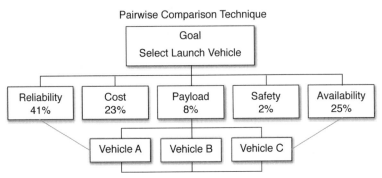

FIGURE 6-6.   **Relative Weighting of Criteria for Selecting a Launch Vehicle.** The criteria now have percentage weights, so their effect on the final answer will be proportional.

TABLE 6-9.   **Rating the Alternatives.** To rate alternatives, we calculate relative scores for each criterion and each launch vehicle. Then we multiply them by the weighting of each criterion to get totals for meaningful comparison. Vehicle C appears to be the best choice, even though it gets a 0% for payload capability.

	Reliability	Cost	Payload	Safety	Availability	TOTALS
Weight	41%	23%	8%	2%	25%	
Vehicle A	Good	Poor	Excellent	Marginal	Excellent	0.58
Vehicle B	Very Good	Excellent	Good	Very Good	Poor	0.62
Vehicle C	Excellent	Excellent	Poor	Excellent	Good	0.81

Excellent = 100%; Very good = 80%; Good = 60%; Marginal = 20%; Poor = 0%

6.3.6   Optimizing Techniques

When clear numeric values are available, various tools help analysts optimize a decision. Linear programming, integer programming, forming specialized networks, nonlinear programming, modeling, and calculus come to mind. These tools are beyond this chapter's scope, but many complex analyses use them for space system design decisions. With several kinds of analysis methods available, we should consider some pertinent questions when determining which one to apply for a given decision:

- What methods are appropriate to this decision?
- How much will each method cost in time and resources?

- How much time and resources can we afford to commit to this decision?
- What is this decision's value?
- What other decisions does the project require?
- What data is available?
- What is its quality and level of detail?
- Are the factors to consider objective, subjective, values-driven, or unknowns?
- Would small changes in assumptions result in big changes in outcomes or costs?
- Is the method we've selected giving us useful results?

Once we've answered these questions and analyzed alternatives using one of the methods described in this section, our next step is to select a solution.

6.4   Generate Alternatives

Miners must process a lot of ore to get a small amount of gold. Similarly, in problem solving and decision making, a large set of alternatives usually results in better-quality outcomes. Decision makers can hamper or kill choices by considering only a single alternative, business-as-usual approaches, or "perfect" solutions. We should always capitalize on knowledge within and outside our organizations. As Picasso observed: "Immature artists borrow. Great artists steal."

Brainstorming is one way to tap into this knowledge to generate better alternatives. It has many variations (Table 6-10), so if one type doesn't produce good results, we should try another. All variations have a common characteristic: no evaluating or criticizing ideas while brainstorming is under way. (Later, of course, we must evaluate.) To shape the process for identifying alternatives, we should ask these questions:

- What different decisions are possible?
- Can we build a matrix of possible choices?
- What are the bad or wrong choices? Why are they bad or wrong? Can we improve them?
- What new problems and risks may a solution create?

6.5   Choose the Best Solution

Because no evaluation tool can quantify everything, we should select a solution based on, but not directed by, the analysis from earlier steps. Human intuition, judgment, and wisdom play necessary roles. So for important decisions, especially in the safety-critical areas, we must document the analysis process we used, alternatives we considered, and the rationale for the final choice. Table 6-11 describes several ways to select a decision from alternatives; each has its time and place.

TABLE 6-10. **Brainstorming Techniques.** Several kinds of brainstorming techniques are available to decision makers as a way to create a larger set of alternatives, which typically results in better-quality decisions.

Brainstorming Type	Description
Conventional brainstorming	Conventional brainstorming is familiar to most professionals. With this technique, we • Go for quantity of ideas over quality • Keep the pace fast • Write down every suggestion or idea
Delphi brainstorming.	The Delphi technique has several variations to lessen "group think." In "silent" Delphi, for example, selected participants (usually subject-matter experts) write down ideas, which are then collected, reproduced, and distributed. Team members offer opinions on the ideas to the entire team.
Negative brainstorming	If the team is stuck in a negative frame of mind, we could consider asking the opposite question: "Why are we certain we won't succeed?" Brainstorm the obstacles, restrictions, barriers, and other handicaps to a decision, using the same "no criticism" rule as in conventional brainstorming. Afterward, during evaluation, we explore how we might overcome each obstacle. Some obstacles will prove intractable, but we may be able to overcome others once they're out in the open.
Backward brainstorming or planning scenarios	We imagine the project has succeeded beyond expectations and brainstorm all the reasons it did

TABLE 6-11. **Methods for Selecting a Decision.** The individual decision's nature and circumstances influence the choice of best method.

Type	Pros	Cons	When to Use
Authority rule	Fast	Reduced buy-in, possible cognitive bias	Crisis or emergency
Consultative	Efficient, less potential for bias	Subject to abuse	Daily operational decisions
Minority rule	Efficient	Reduced buy-in, potential group-think	Enlightened leadership coalition [When a minority should have veto power over high risk choices (e.g., safety)]
Majority rule	Efficient	Alienation of minority	Non-critical group decisions
Consensus	Group buy-in	Time-consuming	Buy-in essential

When teams decide, other considerations come into play. Conflict is not only inevitable but, if managed properly, desirable. The goal is to reach a consensus, which isn't necessarily 100-percent agreement but a decision that all team members can live with. Teams make effective decisions when they:

• Clearly define the problem or issue

- Agree on who has responsibility for deciding
- Are clear on their level of authority
- Are the right size for group decisions
- Communicate effectively
- Apply a decision process

6.5.1   Using Intuition Versus a Logical Process in Making Decisions

When choosing, we must recognize when we're using intuition or logic. *Intuition* is recognizing patterns without evident conscious thought. It's automatic thinking, as compared to controlled thinking or analysis. Both logical and intuitive strategies have value for decision making. In fact, the United States Marine Corps' Command and Control Doctrine [1996] recommends intuition as a decision tool in many situations:

> The intuitive approach is based on the belief that war being ultimately an art rather than a science, there is no absolutely right answer to any problem. Intuitive decision making works on the further belief that, due to the judgment gained by experience, training, and reflection, the commander will generate a workable first solution.... Intuitive decision making is generally much faster than analytical decision making.... The intuitive approach is more appropriate for the vast majority of typical tactical and operational decisions.

What makes intuition so valuable? We acquire much of our accumulated store of knowledge, wisdom, and insight without conscious effort. Daily exposure to successful and unsuccessful decisions, our colleagues' knowledge (and errors), organizational cultures, books, TV shows, and movies builds a substantial platform on which to base decisions.

To make intuition a stronger, more effective tool in our decision making toolbox, we consider and record an intuitive decision even when we will follow an analytical process in the end. Then, after we know the result, we assess it. By evaluating the intuitive part of thinking and comparing it to process, the quality of our intuitive decisions increases over time.

When time is short and decisions are crucial, a well-tempered intuition is especially valuable. For example, a Shuttle pilot crashes many times—in a simulator, where the experience is survivable and repeatable. Mental rehearsal, "lessons-learned" sessions, and similar techniques help build the foundation for better intuitive decision making.

We mustn't neglect either intuitive or formal methods. If our intuition is screaming that our formal process has led us to the wrong conclusion, we should look hard at the formal process. Conversely, if the formal process indicates that our intuition is dangerously wrong, we should look carefully at perceptual biases that may distort our judgment. To improve intuitive decisions, we ask certain shaping questions:

- Does my intuition recommend a particular decision?

- Does the formal process reach the same conclusion, or does it recommend another decision?

- What factors not in the formal process may I have considered by using intuition?

- Was the intuitive choice validated? Did it hamper my considering other, possibly better, alternatives?

- When I finally know the outcome, was the intuitive call more or less correct than the formal one?

6.5.2   Avoiding Perceptual Bias and Decision Traps

Other factors to consider when making choices are the traps and biases that silently influence our decisions. Our culture, language, cognitive biases, and stress play a part in every decision we make.

Culture. "Pardon him," said Julius Caesar in George Bernard Shaw's *Caesar and Cleopatra*, "He is a barbarian, and thinks that the customs of his tribe and island are the laws of nature." [Shaw, 1898] All culture—organizational and otherwise—consists of its members' values, beliefs, attitudes, behaviors, personalities, and priorities. Positive and negative cultural factors have complex and sometimes subtle effects on decisions. For example, NASA policy emphasizes minimizing risk to human life, but this standard isn't an automatic choice in every organization or culture, or during different times in any culture. It's a values decision. NASA arrived at it consciously, but other parts of organizational culture develop without conscious process.

A prime example of cultural influence is the fatal accident of Space Shuttle Columbia. The accident investigation board's report concluded that the NASA culture contributed to the accident because the "Apollo era created at NASA an exceptional 'can-do' culture marked by tenacity in the face of seemingly impossible challenges.... The culture...accepted risk and failure as inevitable aspects of operating in space." [CAIB, 2003] Although a "can-do" tenacity isn't wrong in itself, decisions based on factors no longer operative may be wrong. The Shuttle era differs from that of Apollo, but cultural norms are tenacious and strongly resist change.

Language. Language affects our perceptions. For example, prefixing an English word with "in" often turns it into its opposite: "ingratitude" is the opposite of "gratitude." But "flammable" and "inflammable" are synonyms. Linguist (and former insurance adjuster) Benjamin Whorf observed that this language-influenced misunderstanding was a leading cause of fires. We can't assume that others will always interpret our words as we intend.

Cognitive biases and decision traps. These impediments to good decision making come in many forms and often have profound implications. Table 6-12 lists some common examples. Simple awareness of these biases helps us overcome them. In addition, decision makers use several coping techniques:

- Collect more data
- Get reviews from others who don't advocate this decision
- Brainstorm a full list of alternatives before choosing a course of action
- Avoid having leaders state their positions as a discussion begins
- Appoint a team "devil's advocate" to introduce contrary views
- Refrain from exaggerating any switch from the status quo
- Increase team diversity to ensure different perspectives
- Define and examine assumptions used to forecast
- Look for direct statistical data
- Double check to include all evidence
- Don't ask leading questions that invite confirming evidence
- Deliberately view a problem from different perspectives
- Seek new data and differing opinions
- Elicit the views of people not involved in the decision
- Recognize that decisions affect only the future, not the past

Stress. People experience different levels and types of stress, and have different reactions to it. In general, the two characteristics that determine a person's perception of stress are lack of control and lack of closure.

Many factors contribute to stress on a space project. Some common ones are time pressure, personal consequences of the decision, amount of sleep and exercise, drugs such as caffeine, and issues in one's personal life. Stress is hard to avoid, so we need to understand that it affects decisions in various ways. Under stress, our emotions play a larger role because we tend to limit the information we take in and process it simply. We may give certain aspects of the problem too much attention, especially any negative information, and we're more subject to perception traps and biases.

We can't always eliminate stressors. In fact, some of the most demanding decisions occur under the worst stress conditions. But we may be able to take some extra time—even a few deep breaths have been known to help. And we should seek others' perspectives, although we should also check to see what level of stress they feel. To make our decisions less likely to suffer from perception traps and biases, we must ask ourselves several important questions:

- What values, beliefs, and culture-based perceptions do we (and others) bring to this decision?
- What kinds of stress are decision makers experiencing?

TABLE 6-12. Common Decision Traps and Biases. The way in which we look at the world in general is not always accurate. Various decision traps and biases often lead to serious error if not corrected.

Single- and multi-play decisions	If we launch a single satellite with some probability of failure, we sense that all our eggs are in one basket. Launching ten satellites with the same probability of failure per launch feels safer, but if the consequences of failure are undiminished, the decision situation hasn't improved.
Wishful thinking	People often overestimate probabilities of desirable events and underestimate the undesirable—*Challenger* is an example. This tendency results from optimism or "organizational cheerleading": the idea that imagining bad outcomes is disloyal.
Groupthink	Social psychologist Robert Cialdini [2006] describes this characteristic as "social proof": the idea that uncertain people decide what is correct by finding out what other people think is correct. Standing against the majority is difficult. Even determined nonconformists feel the pressure. If we're last in line to approve a launch, and everyone before us has said "yes," the pressure to agree is very strong.
Status quo bias	"If it ain't broke, don't fix it," says the proverb. People perceive change as inherently risky. With more choices available, the status quo bias amplifies. It also carries with it a type of immunity against wrong decisions: decision makers perceive it as carrying a lower risk of personal consequences. For example, at one time, managers of institutional technology said: "Nobody ever got fired for buying IBM."
Overconfidence	We tend to overestimate the organization's—and our own—abilities. Being in the driver's seat makes us comfortable, and criticizing the organization seems disloyal.
Availability bias	Decision makers often exaggerate more recent or more vivid information. For example, far more people die in automobile accidents each year than fall victim to terrorist attacks. Yet, the extreme vividness of such attacks makes them loom far larger for decision makers. Events that lead to losing a space vehicle, no matter how improbable, tend to have an exaggerated effect on the next launch.
Confirming evidence	People tend to seek information that supports their point of view and ignore or trivialize what contradicts it.
Anchoring bias	People tend to weigh disproportionately the first information received, especially when a large amount of data exists.
Sunk-cost bias	People tend to decide in a way that justifies past choices. Cialdini [2006] reports a study that bettors at a racetrack are much more confident their horse will win immediately after placing the bet than before. Once people have taken a stand, they believe they must stick to it.

- What steps have we taken to avoid common decision traps for ourselves and others?
- What decision does the process recommend?
- Is one alternative clearly superior, or are several choices almost equally good?

- Will applying a different method give us a different outcome?
- Is our intuitive judgment warning us that the proposed decision is incorrect?

6.6 Test and Evaluate

A good process doesn't necessarily result in a good consequence. Predicting the future is never an exact science, especially when probabilities are involved, so we must evaluate the decision process separately from the decision outcome. Hindsight is a useful tool, though teams must remember that what seems so clear in light of actual events looked different to the decision makers.

Still, comparing the actual to the intended helps us evaluate the process. If we see a discrepancy, the crucial question is "Why?" Knowing the results, would the team have done better with a different process? Should we change the process for future decisions? Do the outcomes—especially the bad ones—follow a trend? If so, processes may be faulty. Besides these basic questions, we recommend three others for test and evaluation:

- Was the actual outcome the same as the intended one?
- Was the actual outcome affected by random or unexpected events?
- Was the process for this decision correct?

6.7 Consider a Case Study

To apply the decision process to a real example, we've selected the International Space Station's (ISS) Node 3 Launch-to-Activity. It involves a contemplated change to the Space Station's operational constraints for extravehicular activity, which seriously threaten a major assembly. (It's the constraints that threaten Node 3; the contemplated change is meant to mitigate or eliminate the threat.) Let's examine how the decision process might play out.

ISS Node 3 (Figure 6-7) is a cylindrically shaped, pressurized module that gives the Space Station more docking ports. It also houses key avionics and life-support systems. The Space Shuttle delivers Node 3 to the Station.

The ISS program constrained high-priority extravehicular activity **after** teams completed the critical design review and started building the Node 3 hardware. These new constraints meant the construction sequence had to change. Instead of being moved from its temporary Node 1 location after twelve hours, Node 3 would have to stay as much as 24 more hours. Without power for its shell heaters, Node 3 would now risk violating its thermal constraints and "freezing" while temporarily docked to Node 1. Node 3 is one of the Space Station's last major elements, and its success or failure determines the Station's final operational capability. So the ISS program manager must make the final decision concerning any corrective action.

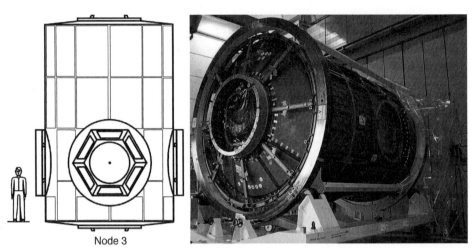

FIGURE 6-7.   **International Space Station's (ISS's) Node 3.** ISS Node 3 is a cylindrically shaped, pressurized module that gives the Space Station more docking ports. It also houses key avionics and life-support systems. New constraints on extravehicular activity for the assembly crew require a decision on how to protect the node from freezing while temporarily docked for many extra hours to Node 1. [HEM, ca. 2008; NASA, n.d.]

The ISS Program Office established a Node 3 "Lead Test Assembly (LTA) Team" (referred to here as the team) to assess the issue and offer alternative solutions. What would be the most effective way to keep Node 3 from freezing while temporarily docked to Node 1? The team consisted of representatives from the Node 3 Project Office and its hardware contractor, the EVA Office, and the ISS Program Office [Node 3 PMR, 2005].

6.7.1   Generate Alternatives

All the seriously considered alternatives involved designing, building, and installing a power cable that connects Node 1 power to Node 3's shell heaters. Figure 6-8 is a notional schematic of the LTA cable's routing.

- Alternative I: Install LTA cable on orbit—During EVA operation for the LTA, connect one end of a 12-meter LTA cable to the Node 3 shell heater's power connectors, route and secure it along the length of Node 3, and connect it to Node 1 power.

- Alternative II: Install LTA cable on the ground—Above Node 3's shields: Install the cable on the ground and secure it along the length of Node 3 above its protective shields. Then, connect each end of the cable during EVA.

- Alternative III: Install LTA cable on the ground—Below Node 3's shields: As in Alternative 2, but install cables below the protective shields.

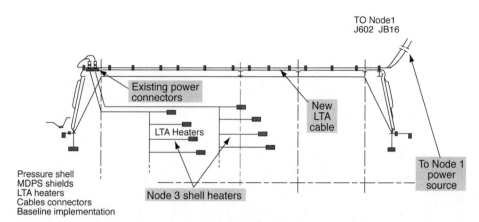

FIGURE 6-8.   Lead Test Assembly (LTA) Cable's Number on Its Routing to Node 3's Shell Heaters. The purpose of this modification is to keep Node 3 from freezing during its protracted docking to Node 1. [Alenia Spazio, 2004] (MDPS is meteoroid and debris protection system.)

6.7.2   Establishing Objectives and Framing the Decision

Reviewing information when time permits is always smart. In this case, the team reviewed 23 technical papers and peer-reviewed documents to help understand the main things that affect the EVA astronauts' work performance. They

- Verified the importance of the timeline and training as an EVA objective

- Selected four objectives and ranked them by consensus with input from technical experts

- Developed scoring criteria using a 5-point Liker scale (Table 6-13)

6.7.3   Select the Best Method

The team then determined the best approach for this decision. The goal was to get high-quality evaluation data at a relatively low cost. They selected three activities to evaluate the alternatives: simulate work-element tasks, interview experts, and use computer-based models to analyze worksites.

Simulate work-element tasks. The team ran a single medium-fidelity simulation as an add-on to already scheduled Node 3 EVA training in the Neutral Buoyancy Laboratory (NBL). (This is a large underwater training tank used to simulate EVA operations in the orbital free fall environment, with real extravehicular mobility units and mockups of ISS hardware.) The operational task elements simulated the three alternatives. Limits on the NBL's time and resources didn't allow simulating all alternatives, but a partial simulation could reveal time-study data for certain task elements and indications of task difficulty, potential fatigue factors, and future training requirements.

TABLE 6-13. Criteria for Evaluating Effects on Extravehicular Activity (EVA) Work Performance. This table illustrates how the Lead Test Assembly (LTA) Team used a 5-point Liker scale to determine the effects of decisions on four objectives for extravehicular activity to attach the Space Station's Node 3. (EMU is extravehicular mobility unit.)

Score	Effect on Timeline for EVA Operations (Ranked #1)	Potential for Significant Fatigue in EVA Astronauts (Ranked #2)	Task Complexity and EVA Crew Training (Ranked #3)	Potential for Issues with Work Access or Reach Envelope (Ranked #4)
Excellent (5)	Any effects on timeline for EVA stay within the planned 6.5-hour work day	Low concern that added EVA tasks will increase the significance of general body fatigue or local hand fatigue above current expected levels	New EVA tasks are simple—no new dedicated training of EVA crew likely	Initial worksite analysis indicates all anticipated gloved-hand access and EMU reach envelopes for LTA cable connections easily exceed EVA requirements
Very good (4)	Requires a planning timeline between 6.5 and 6.75 hours. Flagged as a concern, but acceptable with no EVA operational requirement waivers required.	Intermediate level	Intermediate level—expect one to two dedicated training sessions	Intermediate level
Good (3)	Requires planning timeline between 6.75 and 7 hours. Acceptable with approval of EVA operations management.	Moderate concern that the extra EVA tasks will increase the significance of general body or local hand fatigue above current expected levels	Moderate complexity—two to three dedicated training sessions expected for EVA crews	Anticipate several tight but acceptable working spaces for the EVA crew's gloved-hand and EMU reach access— pending a more detailed worksite evaluation
Fair (2)	Requires planning timeline between 7 and 7.25 hours. Requires formal waiver to EVA requirements approved by EVA-level management.	Intermediate level	Intermediate level—expect four to five dedicated training sessions	Intermediate level
Poor (1)	Requires planning timeline of more than 7.25 hours. Requires formal waiver to EVA operational requirements approved by senior-level ISS management.	High concern that the added EVA tasks will increase the significance of general body fatigue or local hand fatigue above current expected levels	New tasks are complex—six or more dedicated EVA crew training sessions are anticipated	Anticipate one or more tight working spaces for the EVA crew. Expect gloved-hand and reach-access requirements to be unacceptable after more detailed worksite evaluation. Likely to require developing unique tools or handling aids to help the EVA crew finish the task.

Interview experts. The team talked to the NBL's simulation EVA crew and training staff, covering results of the partial NBL simulation and experience with the LTA alternatives. They used this knowledge to extrapolate likely times for work elements that they couldn't measure directly, relative task complexity and training requirements, and fatigue-related issues.

Use computer-based models to analyze work sites. The team applied the worksite analysis to understand the extravehicular mobility unit's reach envelopes and reach-access conditions for each alternative.

6.7.4   Evaluate Alternatives

The LTA team then evaluated each alternative against the agreed-on criteria: effect on the timeline, potential for astronaut fatigue, task complexity, and reach or access. We've summarized the results below.

Effect on the timeline for EVA operations. The team performed a time study assessment to help determine the impact to the EVA timeline. For each alternative, they divided the anticipated task into major work elements.

Alternative I was expected to take longer than the other two, so the team defined the elements of this task first and then compared them to those for Alternatives II and III. They built a mock-up of the LTA cable hardware, and the training crew did the operational task for Alternative I at the end of a scheduled 5-hour training session for Node 3 EVA. This "piggyback" evaluation had the added benefit of replicating the anticipated sequence of on-orbit tasks. Installing and connecting a new power cable must occur at the end of the planned operational day, when crew fatigue is a real concern. So this simulation helped give the team reasonable data on task fatigue, timeline effects, and task complexity.

The team then determined a standard operational time for Alternative I using a time-study procedure described in *Work Design* [Freivalds and Niebel, 2003]. Limits on time and resources kept them to one cycle, in which they

- Recorded the observed time in minutes to complete the work task elements
- Adjusted the observed time measurements to account for rating factors
- Established time allowances using the time-study procedures and team's advice
- Determined that 57 minutes was the expected task time for EVA operations under Alternative I

Because the team couldn't run another simulation to measure Alternatives II and III, they estimated the times based on results for Alternative I and produced standard task times and criterion ratings for all three:

- Alt I: fair at 57 minutes
- Alt II: very good at 35 minutes
- Alt III: excellent at 19 minutes

Potential for significant astronaut fatigue. At the end of the simulation, the team interviewed the crew to discuss potential fatigue for each alternative. They told the crew that the task of connecting the LTA power cable must occur at the end of the current six-hour day, after completing other tasks in the sequence, just as in the five-hour training session.

The team used a variation of Borg's Subjective Ratings of Perceived Exertion with Verbal Anchors [Freivalds and Niebel, 2003] to help the crew evaluate the individual and cumulative effects of task elements for general body and local hand fatigue. Based on the simulation and past experience, the crew rated the elements for body and hand discomfort from fatigue on a scale of 0 (no discomfort) to 10 (extremely strong discomfort). The crew then reported an overall evaluation score for the second criterion: potential for significant EVA astronaut fatigue.

The team and crew compared results for Alternative I to those anticipated for Alternatives II and III. The crew estimated how much easier or harder the latter's element tasks would be compared to Alternative I and evaluated them in terms of general and local fatigue. Table 6-14 summarizes the fatigue assessment.

TABLE 6-14. Assessments of Astronaut Fatigue. Alternative III scored the best on this criterion. (LTA is lead test assembly.)

Alt I	Fair	A moderate to high concern about significant fatigue from the extra LTA tasks. A long work time with a lot of cable manipulation and translation. Estimate of 48 significant hand manipulations to secure and connect a loose cable.
Alt II	Good	Moderate concern about significant fatigue from the extra LTA tasks. Work time is more than half of Alt I. Requires inspection translations, as well as cable deployments at each end of Node 3's heaters. Only 20 hand manipulations to secure and connect the cable.
Alt III	Excellent	Low concern about significant fatigue from the extra LTA tasks. Short amount of added work time. Cable manipulations are the shortest; expect fewer than ten hand manipulations to secure and connect the cable.

Task complexity and crew training. The LTA team discussed with training experts the task complexity and estimated training required for each alternative. The group evaluated task complexity and potential for dedicated training based on the premise that the following cases may require unique training [Crocker et al., 2003]:

- The task is highly complex, requiring multiple steps and tools
- The task is safety-critical, requiring a very high probability for success
- The error margin is small with regard to avoiding hazards
- The sequence of events is time-critical for accomplishing tasks or because of the EMU's endurance limits

For all alternatives, time is critical to determining the need for more unique training. Installing the power cable before the life support systems reach their endurance limit is critical. Although none of the alternatives is terribly complex, all have at least some unique tasks that will require more training. Table 6-15 summarizes the results of the assessment for training effects.

TABLE 6-15. **Assessments of How the Alternatives Affect Training.** Alternative III scored the best on this criterion.

Alt I	Fair	The estimated one-hour time line and unique cable deployment and routing work elements will likely require four to five dedicated training sessions in the Neutral Buoyancy Laboratory (NBL) or other specialized training
Alt II	Good	Expect two or three dedicated training sessions to train for external cable inspections and unique tasks for routing end cone cables
Alt III	Very good	Expect only one or two training sessions in the NBL

Potential for issues with work access or reach envelope. The team assessed work access and reach envelope during the NBL run. They used the assessment for Alternative I to help indicate areas that might cause issues for a particular worksite. They were especially concerned about LTA task activities in the relatively confined work space where Node 3 joins Node 1.

The team asked a worksite analysis team to help identify potential reach and access concerns for the alternatives using computer modeling. The latter modeled expected crew positions and reach and access requirements for each work element in tight areas. Their analysis complemented the overall simulation, as summarized in Table 6-16.

TABLE 6-16. **Assessment of Potential Issues with Work Access and Reach Envelopes.** Alternative I scored the best on this criterion. (LTA is lead test assembly; EVA is extravehicular activity.)

Alt I	Very good	Using EVA to install the LTA cable results in very few close-access areas. No issues expected.
Alt II	Fair	The worksite analysis indicated a possible violation of access requirements for the EVA crew and extravehicular mobility unit. The crew must access the preinstalled LTA cable very close to Node 1 hardware, possibly keeping them from releasing the stowed cable. The project may require redesigning cable routes, developing new cable-handling tools, or using other operational workarounds to avoid the potential interference.
Alt III	Good	Expect reach envelopes and access clearances to be acceptable pending more detailed hardware information.

Summary evaluation. The LTA team used the rank reciprocal method [Composition, 2005] to determine the relative scoring for each alternative and recommend one. This method places the most disproportionate weight on the criterion ranked number 1. They ranked the four selection criteria 1 through 4, as previously mentioned.

This method is a good way to establish the relative weights of trade-off criteria whenever all agree that the number one objective is clearly the most important

[Componation, 2005]. The LTA team, with direction from the EVA team and ISS management, established the timeline as the most important criterion. Table 6-17 summarizes the rank reciprocal model's analysis and evaluation results.

6.7.5   Choose an Alternative and Evaluate

This decision analysis offered enough differentiation to recommend a specific design alternative: Alternative III—installing the LTA power cable underneath Node 3's protective shields before launch and using extravehicular activity (EVA) to connect each end of the cable. The analysis focused attention on the work-performance criteria for EVA and produced new insights and a clearer recommendation for the ISS program manager's trade-study decision.

Eventually, the team must stop analyzing and report their results. A key question the manager must ask is: "Do we have enough insight for a sound recommendation?" This question is often difficult to answer. Unlike other forms of modeling, validating that decision-analysis models accurately reflect reality is difficult.

TABLE 6-17. **Overall Scoring Results Using a Rank Reciprocal Trade Model.** This supports the comparison of the three alternatives. The alternative with the highest total score (Alternative III) seems to be the best candidate based on the criteria and assessments. (EVA is extravehicular activity.)

Criteria	Rank Reciprocal				Node 3 LTA Cable Alternatives					
	(A) Rank	(B) 1/A	Weight Calculation (B) / Σ (1/ (A))	Weight	Alt I EVA Installed Evaluation	Score (Weight × Eval)	Alt II Ground Installed Above Shields Evaluation	Score (Weight × Eval)	Alt III Ground Installed Below Shields Evaluation	Score (Weight × Eval)
EVA Time	1	1.00	1/2.083	0.48	Fair (2)	0.96	Very Good (4)	1.92	Excellent (5)	2.4
Fatigue	2	0.50	0.5/2.083	0.24	Fair (2)	0.48	Good (3)	0.72	Excellent (5)	1.2
Task Complexity and Training	3	0.33	0.33/2.083	0.16	Fair (2)	0.32	Good (3)	0.48	Very Good (4)	0.64
Access/Reach	4	0.25	0.25/2.083	0.12	Very Good (4)	0.48	Fair (2)	0.24	Good (3)	0.36
	Sum 2.083			1	Alt I Total Score	2.24	Alt II Total Score	3.36	Alt III Total Score	4.60

References

Adams, Douglas. 1979. *The Hitchhiker's Guide to the Galaxy.* New York, NY: Harmony Books.

Alenia Spazio. December 2004. Node 3 Launch-To-Activation Report. N3-RP-AI-0180 Turin, Italy.

Cialdini, Robert B. 2006. *Influence: The Psychology of Persuasion.* New York, NY: William Morrow and Company, Inc.

Clemen, Robert T., and Terence Reilly. 2001. *Making Hard Decisions with Decision Tools.* Pacific Grove, CA: Dusbury.

Columbia Accident Investigation Board (CAIB). August 2003. Columbia Accident Investigation Board Report. Washington, DC: U. S. Government Printing Office.

Componation, Paul. 2005. ISE 734, Decision Analysis Lecture Notes. University of Alabama in Huntsville, Alabama.

Crocker, Lori, Stephanie E. Barr, Robert Adams, and Tara Jochim. May 2003. EVA Design Requirements and Considerations, EVA Office Technical Document, No. JSC 28918. Houston, YX: NASA Johnson Space Center.

Dobson, Michael, and Heidi Feickert. 2007. *The Six Dimensions of Project Management,* Vienna, Virginia: Management Concepts, pp. 20–34.

Freivalds, Andris, and Benjamin Niebel. 2003. *Methods, Standards, and Work Design.* Eleventh Edition. New York: McGraw-Hill Companies, Inc.

Goodwin, Paul and A. George Wright. 2005. *Decision Analysis for Management Judgment.* West Sussex, England: John Wiley & Sons, Ltd.

Habitation Extension Module (HEM) home page. c.a. 2008. www.aer.bris.ac.uk/.../hem /hem_and_node_3.jpg

Hammond, John S., Ralph L. Keeney, and Howard Raiffa. 1999. *Smart Choices: A Practical Guide to Making Better Life Decisions.* Boston, MA: Harvard Business School Press.

Harless, D. 2005. Preliminary Node 3 LTA WSA. 2005. Houston, TX: NASA Johnson Space Center.

Knebel, Fletcher. n.d. Original source unavailable. http://thinkexist.com/quotes/fletcher_ knebel/

NASA. n.d. https://node.msfc.nasa.gov/geninfo/PhotoGallery/Node3/CleanRoom/ DSCN1609.JPG

Node 3 Program Management Review (PMR). January 2005. Briefing for Node 3 LTA, no document number. Huntsville, Alabama: Marshall Space Flight Center.

Shaw, George Bernard. 1898. Caesar and Cleopatra, Act II.

US Marine Corps. October 1996. MCDP6 - Command and Control. Washington, DC: Department of the Navy.

Chapter 7

Lifecycle Cost Analysis

Joseph W. Hamaker, *SAIC*
Douglas Morris, *NASA Langley Research Center*
Michael B. Nix, *Qualis Corporation*

Cost estimating and analysis have always been part of defining space systems. We often think of cost analysis as a separate discipline, and in many organizations a dedicated functional unit estimates costs. But the trend today is to consider cost as one of the engineering design variables of systems engineering. These days, most aerospace design organizations integrate cost analysis into concurrent engineering. An initiative in systems engineering called "cost as an independent variable" strongly emphasizes this approach.

In concurrent engineering, cost analysts quantify the cost effects of design decisions. This chapter provides an effective cost model, which we use to evaluate the acquisition costs of technical performance metrics, technology level, and degree of new design for various launch systems and spacecraft. We offer other ways to estimate cost for the systems' operations, so we can evaluate the total lifecycle cost.

Table 7-1 displays the process to estimate the lifecycle cost of space systems. In the first step, we develop the project's work breakdown structure (WBS), which delineates the systems and subsystems that constitute the flight hardware as well as the project's functional elements, such as project management or systems engineering and integration.

TABLE 7-1.   **Estimating Lifecycle Costs of Space Systems.** We must iterate this process many times to get a credible estimate. Here we use a launch system (space launch and transportation system) as an example.

Step	Description	Where Discussed
1	Develop a work breakdown structure (WBS)	Section 7.1
2	Gather space system characteristics	Section 7.2, Chaps. 5, 6, 9, 10, and 11
3	Compute the system development cost	Section 7.3
4	Estimate operations and support costs	Section 7.4
5	Estimate cost of developing the launch site's infrastructure	Section 7.5
6	Estimate total lifecycle cost and assess cost risk	Section 7.6
7	Conduct cost-risk analysis	Section 7.7

We also gather the system characteristics for estimating the lifecycle cost. We collect the cost estimating inputs such as technical data, ground rules, assumptions, and enough information to allow us to understand the project's inheritance and complexity. The following are key principles in estimating costs:

- Use cost models and other estimating techniques to estimate cost for system acquisition, operations, and support, and for infrastructure development

- Aggregate the total lifecycle cost phased over time using beta distributions and the project schedule

- Do a sensitivity or cost-risk analysis to evaluate the cost estimate's sensitivity to the major assumptions and variables, weigh the major project risks, and quantify our confidence in the estimate

- Employ, as needed, economic analyses to compare alternatives and to access the project's metrics for return on investment

There are two distinct approaches to estimate a space system's lifecycle cost: parametric cost estimating and detailed engineering estimating (other methods, such as analogy estimating, are similar to these). *Parametric cost estimating* mathematically relates cost to engineering variables of the system. *Detailed engineering estimating* uses functional estimates of the labor hours and materials we expect to need for designing and fabricating items in the WBS or vendor quotes on those items. Early in project definition, planners commonly prefer parametric cost estimating, moving to detailed engineering estimating once the design is relatively mature or manufacturing has begun. Because we focus on conceptual approaches, this chapter illustrates parametric techniques.

The premise of parametric estimating is that we can predict cost with variables analogous to, but not wholly the cause of, final cost. The Rand Corporation invented this technique just after World War II, when they needed a way to estimate military aircraft cost rapidly and early in development. Today many industries, including aerospace, the chemical industry, shipbuilding, building construction, mining, power plants, and software development, use parametric estimating. The technique relies on statistically derived mathematical relationships called *cost estimating relationships (CERs)*. Cost is the (predicted) variable, and engineering variables are the independent (input) ones. We usually derive CERs using data from records of past projects and often do regression analysis to find the best statistical fit among cost and engineering variables. In space launch systems, dry mass is one of the most common engineering parameters, but the best CERs also take into account other variables, including those for management.

In the example presented here, parametric cost estimating helps gauge essential elements of the lifecycle cost for developing, producing, launching, and operating space launch and transportation systems (SLaTS). A space system's lifecycle cost divides into three main phases. The *design, development, test, and evaluation (DDT&E)* phase includes designing, analyzing, and testing breadboards, brassboards, prototypes, and qualification units. It also encompasses proto-flight units and one-time infrastructure costs, but not technology development for system components. The *production* phase includes the cost of producing the system. One concept in modeling costs is the *theoretical first unit (TFU)*, the cost of the first flight-qualified vehicle or system off the assembly line. For multiple units, we estimate the production cost using a *learning curve* factor applied to the TFU cost, which we discuss later. Design, development, test, and evaluation, together with production cost, form *acquisition cost*. The *operations and support (O&S)* phase consists of ongoing operations and support costs, often called *recurring costs*, including the vehicle or system.

We use a SLaTS as an example throughout this chapter. But the cost model is appropriate for most space systems, including remote spacecraft, crewed spacecraft or space stations, launch systems (crewed or not), and the supporting infrastructure. This chapter is not so much about the SLaTS cost model as it is about the issues for any cost model. In using one we must consider the following:

- Impact of the selected cost model
- Definitions in the work breakdown structure
- Cost estimating relationships, variables, maturity, range of values, and applicability
- Development and operations as part of total lifecycle cost
- Models useful for estimating the cost of system test operations, systems engineering and integration, and project management
- Assessment of the technology's readiness
- Time phasing of money for budget development
- Cost risk

Project managers and systems engineers should be very concerned about the lifecycle cost model for their projects, because it greatly affects both the project and their professional life. Ideally, they estimate their project's lifecycle cost with the estimating package and estimator by which others will judge the project in the future. A project's lifecycle cost dangles between two extremes: the estimate is too high, so the project gets killed or de-scoped, or the estimate is too low, gets accepted, and makes the project manager's and systems engineer's life excessively difficult. Our goal is to avoid both extremes.

Selecting the proper estimating package and approach for lifecycle cost is vital to project success. The package should be validated and verified, just like other models in the project. We have to track the model's maturity and how well its results align with the known costs of previous projects. Does the current project estimate exceed the range of values for any of the CERs? If so, the estimate may be unusable. Is the cost estimating package truly appropriate for the type of project and the environment within which the project is developed and operated? Many cost estimating packages provide an estimate of development costs, but very few estimate operations and support costs for space systems.

Below, we show one approach to both types of cost, with many details to help understand the complexities of such models. The development cost model also tells how to estimate the cost of systems engineering and integration, a particularly difficult task. What's more, it illustrates cost estimating for project management and detailed areas, such as system test operations.

Projects often get into trouble through optimism about the technology readiness level (TRL) of their hardware and software. In the example below, we describe an excellent way to assess these levels.

To prepare or review the project budget, or certain elements of it, the section on time phasing project cost is important. It helps us think through the strategy for budgeting a project. The last section offers some useful thoughts about identifying and resolving cost risk.

7.1   Develop a Work Breakdown Structure (WBS)

Figure 7-1 shows the costing WBS for our space launch and transportation system (SLaTS) example. It includes an accounting framework for the major systems associated with a SLaTS and a number of functional system-level elements, sometimes referred to as "wraps." It also contains elements for contingencies, such as allowances for risk and unknowns; fee-contractor profit; and program support. These allowances go beyond the prime contract's scope to support the buying organization. For more details on developing and assessing a work breakdown structure, see *Applied Project Management for Space Systems* (APMSS) [Chesley et al., 2008].

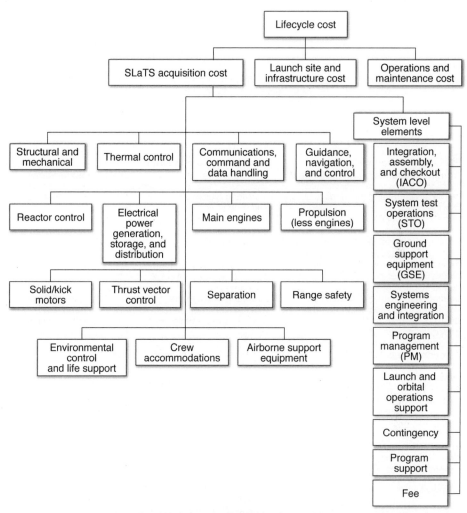

FIGURE 7-1.   Work Breakdown Structure (WBS) for Estimating the Cost of the Space Launch and Transportation System (SLaTS). These elements constitute the WBS. See APMSS [Chesley et al., 2008] for general details on work breakdown structures.

7.1.1   Cost Estimating Tool

This chapter sets out the space systems cost model for estimating acquisition costs. Parametric cost models work at various levels of the WBS. As an example, for structural and mechanical systems, we can develop CERs that cover individual structural components, such as beams, trusses, frames, skin panels, fairings, cargo bay doors, tanks, adapters, or thrust structures. Alternately, we can combine structural items into groupings and develop CERs at a higher level. The space systems cost model takes the latter approach, developing acquisition CERs at the group level. Table 7-2 exemplifies the dictionary for WBS elements in the model, including their functional description, typical components, content, and other considerations.

TABLE 7-2. **Example of Work Breakdown Structure (WBS) Definitions for the Space Systems Cost Model.** The table deals only with the structural and mechanical element. Good practice for WBS definitions demands three basic elements: functional description, typical components and content, and other considerations. (RCS is reaction control system; CER is cost estimating relationship.)

Functional Description of the Structural and Mechanical Element	Typical Structural and Mechanical Components and Contents	Other Considerations
The structural and mechanical group provides the load-bearing, containment, and attachment functions for all the other systems during ground processing, launch, on-orbit operations, re-entry, and landing. The space systems cost model for the structural and mechanical element includes propellant and oxidizer tankage, load-bearing and nonload-bearing, but excludes reaction control tanks, which the cost model includes in the RCS. Primary and secondary structures are included. The primary structure carries the major loads, while secondary structure supports lighter loads associated with smaller components. On crewed spacecraft, the structural and mechanical system includes the pressurized structure that provides a shirtsleeve environment for the mission crew (in which case the variable *Human* in the CER equals 2). This CER also encompasses mechanisms and other assemblies of moving parts.	Among typical structural and mechanical components are beams, trusses, frames, skin panels, struts, fairings, skirts, wings, control surfaces, cargo bay doors, tanks, spacecraft or stage adapter structures, thrust structure, deployment mechanisms, docking mechanisms, landing gear, and such secondary structural elements as appendages, component boxes, brackets, fasteners, fittings, and others.	The space systems cost model group level structural and mechanical CERs may not yield appropriate results at lower levels of the WBS. For example, the SLaTS cost model structural and mechanical CER might yield anomalously high cost for a beam but anomalously low results for a complex mechanism. Cost estimators must sometimes introduce judgmental complexity factors into the calculation to adjust a CER to apply in an instance where it may not be totally suitable.

7.2   Gather Space System Characteristics

Architectures that government and industry are studying for SLaTS today include the Earth-to-orbit launch segment and the in-space orbital segment. They address transferring crews and cargo from orbit to orbit, orbital maneuvering, stationkeeping, docking, extending power, extensive down-linking of data, and other functions. Thus, many launch systems—especially reusable launch vehicles (RLVs)—take on the attributes of spacecraft in orbit. Consistent with this approach, our space systems cost model's CERs often include spacecraft-like parameters that account for these expanded requirements. Examples are deployable structures, power sources such as fuel cells, active thermal control, and three-axis stationkeeping.

We derived the space systems cost model's acquisition CERs using data from NASA's REDSTAR database: a limited-access repository of historical data on most space missions flown since the 1960s. For each group in the WBS, the model provides two acquisition CERs, one for design, development, test, and evaluation (DDT&E) and another for the theoretical first unit's (TFU) production cost. These two costs are the basic building blocks of the model. In addition, the model provides CERs for the system-level costs that result when we sum costs for the DDT&E and TFU groups.

Table 7-3, the SLaTS cost model's CER variables, lists the system characteristics by the elements that the model requires. The global variables (affecting all WBS elements) are first, followed by variables for specific elements in the WBS. The first column gives the element variable name with the corresponding range of values in the second column. Dry mass is a required input to all CERs for the flight-hardware systems. Each system and element CER also uses other independent variables specific to it. Descriptions of global variables follow:

- *Human-rated* is an indicator variable that allows the model user to specify whether the flight hardware is human-rated. The space systems cost model's database contains projects rated and not rated for humans. Human-rated space hardware costs much more. Human = 1 excludes the cost of this requirement, whereas Human = 2 includes the added cost for human rating.

- *Mission type* is an indicator variable that allows us to specify whether the system we're estimating is an Earth-orbiting spacecraft mission, an expendable launch vehicle, a reusable launch vehicle, or a planetary spacecraft system.

- *XYZ factor* is an indicator variable that allows users to specify characteristics of the project's developmental environment in terms of the stringency of the specifications, documentation, traceability, and requirements for testing. Many of the historical projects in the model's database were higher-risk experimental missions to some degree (such as X vehicles or low-cost satellites). Others were demonstration-type projects (called Y vehicles) with higher-quality requirements and less risk than an experimental project, yet not under full-scale development standards (some of NASA's "faster, better, cheaper" missions). Most of the database represents full-scale development projects with normal aerospace standards, labeled "Z vehicles." Model users should set the XYZ Factor to 1 (X vehicle), 2 (Y vehicle), or 3 (Z vehicle).

- *Year of technology freeze* is a quantitative variable that specifies the system's estimated year of technology, usually the year of authority to proceed. Thus, a project with an expected 2005 proceed year would have year of technology freeze set equal to 2005. The cost model's database includes projects from the early 1960s through 2000. Regression analysis shows that development costs for space projects have steadily decreased in many systems, holding all else constant. This variable accounts for these trends.

- *Pre-development* is an indicator variable that measures the amount of study a project has had before full-scale development. Projects that have had more pre-development study cost less for full-scale development (FSD) because their program definitions are more mature. We set pre-development from 1, meaning very low (increasing the cost estimate), to 5, meaning very high (decreasing the cost estimate).

- *Engineering change activity* is an indicator variable similar to pre-development but measuring how volatile requirements are during FSD. Requirement changes are internally driven to meet the design specifications or externally driven to meet customer requests. Change activity has contributed heavily to space projects' costs, so it's an important variable in this cost model. We can set change traffic from very low (1) to very high (5).

- *Number of organizations* is an indicator variable that allows us to describe how much we expect the cost estimate to track with the number of organizations involved in design, development, and production. The model allows it to be as low as 1 for projects involving only one organization and as high as 5 for projects involving more than six. We should count customer and provider organizations in this total.

- *Team experience* is an indicator variable that allows us to describe how we expect organizations to perform in terms of their experience on similar projects. We expect teams with more experience to be more cost effective. Team experience can be from 1, very low (a junior team) to 5, very high (a highly experienced team).

- *Subsystem technology readiness level (TRL)* is an indicator variable we can set from TRL 1 to TRL 9, depending on the technology's maturity. (See the relationship between TRL and percent new design in Table 7-5.) Lower TRL settings dramatically increase the cost estimate. In general, we expect to use subsystems that are at least TRL 6 before starting FSD.

- *Percent new design* allows us to adjust the estimated development cost for inheritance from previous projects. Because this parameter is a multiplier in the model's design and development (D&D) CERs, it strongly affects cost, so we should choose it with care.

Other independent variables for the system or its elements appear in the CERs, as defined in Table 7-3.

Just having a cost estimating relationship doesn't mean the estimated cost is correct. The CER may exclude certain variables of interest that affect cost. It does the best it can with the information given to estimate cost, but it often doesn't contain all pertinent cost drivers. So when we say a CER is valid, we're just saying it does a decent job of explaining variances between the historical database and included variables. Other variables may be at work, and they may change over time. A further caveat: the input values fed into the cost estimating relationship may be wrong—CERs don't correct for bad input.

TABLE 7-3.   Space Systems Cost Model CER Variables. We must be precise in defining variables, since we use them to develop our estimate. See the discussion in the text for expanded definition of some of the variables. (Al is aluminum.)

System or Element/Variable Abbreviation	Variable Range of Values	
Globals and Other Inputs Common To All CERs		
Human-rated	1 = no 2 = yes	
Mission type	1.0 = Earth-orbiting spacecraft 1.3 = expendable launch vehicles	1.8 = reusable launch vehicles 1.9 = planetary spacecraft
XYZ factor	1 = X vehicle 2 = Y vehicle 3 = Z vehicle	
Year of technology freeze	Generally, year of authority to proceed (ATP)	
Pre-development	1 = very low or none 2 = relatively low (0%–2% of DDT&E) 3 = nominal (2%–4% of DDT&E)	4 = relatively high (4%–5% of DDT&E) 5 = very high (>5% of DDT&E)
Engineering change activity	1 = very low 2 = relatively low 3 = nominal	4 = relatively high 5 = very high
Number of organizations	1 = very low (in-house) 2 = relatively low (single customer, single prime) 3 = nominal, (~3–4 organizations involved)	4 = relatively high, (~5 or 6 organizations involved) 5 = very high (>6 organizations involved)
Team experience	1 = very low (junior team) 2 = relatively low (low to mixed, 2nd quartile)	3 = nominal (mixed to high, 50th percentile) 4 = relatively high (3rd quartile) 5 = very high (4th quartile)
Subsystem TRL	1 = TRL 1 2 = TRL 2 3 = TRL 3 4 = TRL 4 5 = TRL 5	6 = TRL 6 7 = TRL 7 8 = TRL 8 9 = TRL 9
Percent new	Degree of new design: Example: 0.7 for 70% new design	
Structural and Mechanical		
	Dry mass	
Material complexity	1 = mostly Al 2 = Al + other well-characterized metallics 3 = metallics + minor composites or exotic metallics	4 = significant composites or exotic metallics 5 = mostly exotic metallics or composites

TABLE 7-3.   Space Systems Cost Model CER Variables. (Continued) We must be precise in defining variables, since we use them to develop our estimate. See the discussion in the text for expanded definition of some of the variables. (Al is aluminum.)

System or Element/Variable Abbreviation	Variable Range of Values	
Deployables complexity	1 = no deployables 2 = simple or minor 3 = nominal	4 = moderately complex 5 = complex
Parts count	1 = very low (extensive use of net casting/machining to avoid assemblies of parts; minimal number of different parts 2 = low parts count	3 = nominal parts count 4 = high parts count 5 = very high parts count
Thermal Control		
	Dry mass	
Coatings, surfaces, multi-layer insulation (MLI), tiles, reinforced carbon-carbon (RCC), etc.	1 = none 2 = coatings, reflective surfaces 3 = MLI, blankets, reusable flexible surface insulation, etc.	5.5 = low temperature tiles 7 = RCC
Cold plates, heaters, etc.	1 = none 2 = cold plates without radiators, electric heaters, etc.	
Radiators	1 = none 2 = radiators with passive cold plates 2.2 = radiators with louvers	
Pumped fluids, heat pipes, etc.	1 = none 2 = radiators with pumped fluid cold plates	2.2 = radiators with louvers 2.5 = heat pipes
Stored cryogenics or refrigeration	1 = none 2 = stored cryogenics 2.5 = active refrigeration	
CC&DH (Communication, Command, and Data Handling)		
	Dry mass	
Communication downlink	1 = very low (~1 Kbps) 2 = low (~10 Kbps) 3 = moderate (~100 Kbps)	4 = high (~1 Mbps) 5 = very high (~10 Mbps) 6 = ultra high (~100 Mbps)
Redundancy level	1 = single string 2 = dual string 3 = triple string	4 = quadruple redundancy 5 = quintuple redundancy
Electronics parts class	1 = Class B 2 = Class B+ 3 = Class S	

TABLE 7-3.   Space Systems Cost Model CER Variables. (Continued) We must be precise in defining variables, since we use them to develop our estimate. See the discussion in the text for expanded definition of some of the variables. (Al is aluminum.)

System or Element/Variable Abbreviation	Variable Range of Values	
Calculated nonredundant CC&DH mass	Total mass reduced by redundancy level	
GN&C (Guidance, Navigation, and Control)		
	Dry mass	
Stabilization type	1 = spin/gravity gradient 2 = despun 3 = 3-axis	4 = momentum wheels 5 = control moment gyros
Redundancy level	1 = single string 2 = dual string 3 = triple string	4 = quadruple redundancy 5 = quintuple redundancy
Sensor types	1 = none 2 = Sun sensors	3 = Earth/horizon sensors 4 = star trackers
Degree of autonomy	1 = none 2 = very limited 3 = limited	4 = some 5 = significant 6 = totally autonomous
Calculated nonredundant GN&C mass	Total mass reduced by redundancy level	
Electrical Power		
	Dry mass	
Generation type	1 = none 2 = silicon solar cells 3 = gallium-arsenide (GaAs) solar cells	4 = fuel cells 5 = radioisotopic thermoelectric generators (RTGs)
Redundancy level	1 = single string 2 = dual string 3 = triple string	4 = quadruple redundancy 5 = quintuple redundancy
Battery type	1 = none 2 = nickel-cadmium (NiCd) 3 = silver-zinc (AgZn)	4 = nickel-hydrogen (NiH) 5 = lithium-ion (Li-Ion)
Calculated nonredundant electrical power mass	Total mass reduced by redundancy level	
Reaction Control		
	Dry mass	
Propellant type	1 = monopropellant 2 = biprop monomethyl hydrazine (MMH) 3 = bipropellant mode	
Liquid Rocket Main Engines		
Mass (kg) - 1 engine	Dry mass	

TABLE 7-3.   Space Systems Cost Model CER Variables. (Continued) We must be precise in defining variables, since we use them to develop our estimate. See the discussion in the text for expanded definition of some of the variables. (Al is aluminum.)

System or Element/Variable Abbreviation	Variable Range of Values	
Vacuum thrust (kg) - 1 engine	Vacuum thrust for a single engine	
Chamber pressure (kPa) - 1 engine	Chamber pressure, P_c, for a single engine	
Vacuum I_{sp} (s) - 1 engine	Specific impulse, I_{sp}	
Engine cycle	1 = pressure-fed 2 = gas generator	3 = staged combustion, single preburner 4 = staged combustion, dual preburner
Fuel	1 = hydrogen peroxide (H_2O_2) 2 = rocket propellant 1 (RP1) 3 = liquid hydrogen (LH_2)	
Expendable vs. reusable	1 = expendable 2 = reusable	
Calculated thrust-to-weight ratio	Calculated thrust-to-weight ratio	
Propulsion Less Engines		
	Dry mass	
Solid/Kick Motors		
	Dry mass	
Thrust Vector Control and Control Surface Actuation		
	Dry mass	
Type	1 = hydraulic 2 = electro-mechanical	
Separation System		
	Dry mass	
Range Safety		
	Dry mass	
Environmental Control and Life Support		
	Dry mass	
Crew days	Number of crewmembers multiplied by days on orbit	
Crew Accommodations		
	Dry mass	
Crew days	Number of crewmembers multiplied by days on orbit	

TABLE 7-3.   **Space Systems Cost Model CER Variables. (Continued)** We must be precise in defining variables, since we use them to develop our estimate. See the discussion in the text for expanded definition of some of the variables. (Al is aluminum.)

System or Element/Variable Abbreviation	Variable Range of Values
Airborne Support Equipment	
	Dry mass
Electrical and data requirements	1 = no 2 = yes
Fluid requirements	1 = no 2 = yes
Jet Engine Package	
	Dry mass
Air start qualification	1 = no 2 = yes

7.3   Compute the System Development Cost

Table 7-4 provides examples of the cost-model CERs we use to estimate the development cost of the proposed SLaTS. Combining the individual group costs for design and development (D&D) and the system-level cost in the D&D column yields total cost for DDT&E. Similarly, summing individual group costs for the theoretical first unit (TFU) and the system-level costs in the TFU column gives total TFU costs. To learn how TFU cost and learning help us obtain total production cost, see Section 7.3.2.

7.3.1   Percent New Design

A key parameter in the CERs is the percent of new design. At lower levels of the work breakdown structure for a launch vehicle's systems, some of the hardware or software may be off-the-shelf. This means that designers intend to use hardware or software identical to, or modified from, those in previous projects. Typical examples are structural brackets, fasteners, common mechanisms, insulation, heaters, transmitters, computers, antennas, navigation sensors, guidance instruments, batteries, voltage converters, engines, thrusters, or tanks. We also sometimes reuse software subroutines with only minor changes. Using such inherited components saves design hours, cuts lead time, and reduces component qualification and testing. But it may come at the expense of not having a component or software code best designed for the current application or not based on the latest technology.

Some systems have more opportunity to use existing hardware than others. For example, primary load-bearing structures are usually new. Some secondary structures or minor mechanisms may be off-the-shelf or modifications, but they probably don't involve much money. Thermal control, communication, guidance,

TABLE 7-4. Space Systems Cost Estimating Relationships (CERs) for Development and Theoretical First Unit (TFU) Costs. Here we give CERs for selected categories from Table 7-3. We need these CERs to estimate development and TFU costs. See the note below for details. Values for the inputs and exponents are in Table 7-8B. (D&D is design and development; TRL is technology readiness level; MLI is multi-layered insulation; RCC is reinforced carbon-carbon; CC&DH is communication, command and data handling; TFU is theoretical first unit; ELV is expendable launch vehicle; RLV is reusable launch vehicle.)

Design and Development (D&D) CERs	Theoretical First Unit (TFU) CERs

Structural and Mechanical

D&D = 1.603 (mass in kg)$^{0.40}$ (Material Complexity)$^{0.10}$ (Deployables Complexity)$^{0.20}$ (Parts Count)$^{0.30}$ (Human Rated)$^{0.40}$ (Mission Type) (XYZ Factor)$^{0.70}$ (Year of Technology Freeze)$^{0.015}$ (Pre-Development)$^{0.20}$ (Engineering Change Activity)$^{0.35}$ (Number of Organizations)$^{0.20}$ (Team Experience)$^{0.25}$ (Subsystem TRL)$^{1.10}$ × %New × Year Dollars

TFU = 0.060 (mass in kg)$^{0.70}$ (Material Complexity)$^{0.20}$ (Deployables Complexity)$^{0.10}$ (Parts Count)$^{0.25}$ (Human Rated)$^{0.30}$ (Mission Type) (XYZ Factor)$^{0.30}$ (Year of Technology Freeze)$^{0.010}$ (Pre-Development)$^{0.05}$ (Engineering Change Activity)$^{0.05}$ (Number of Organizations)$^{0.10}$ (Team Experience)$^{0.15}$ (Subsystem TRL)$^{0.30}$ × Year Dollars

Thermal Control

D&D = 0.705 (mass in kg)$^{0.40}$ (Coatings, Surfaces, MLI, Tiles, RCC, etc.)$^{0.10}$ (Cold Plates, Heaters, etc.)$^{0.10}$ (Radiators)$^{0.15}$ (Pumped Fluids, Heat Pipes)$^{0.20}$ (Stored Cryogenics and/or Refrigeration)$^{0.25}$ (Human Rated)$^{0.50}$ (Mission Type) (XYZ Factor)$^{0.80}$ (Year of Technology Freeze)$^{0.015}$ (Pre-Development)$^{0.20}$ (Engineering Change Activity)$^{0.35}$ (Number of Organizations)$^{0.20}$ (Team Experience)$^{0.20}$ (Subsystem TRL)$^{1.10}$ × %New × Year Dollars

TFU = 0.078 (mass in kg)$^{0.70}$ (Coatings, Surfaces, MLI, Tiles, RCC, etc.)$^{0.05}$ (Cold Plates, Heaters, etc.)$^{0.05}$ (Radiators)$^{0.10}$ (Pumped Fluids, Heat Pipes, etc.)$^{0.15}$ (Stored Cryogenics and/or Refrigeration)$^{0.20}$ (Human Rated)$^{0.40}$ (Mission Type) (XYZ Factor)$^{0.40}$ (Year of Technology Freeze)$^{0.010}$ (Pre-Development)$^{0.05}$ (Engineering Change Activity)$^{0.05}$ (Number of Organizations)$^{0.10}$ (Team Experience)$^{0.15}$ (Subsystem TRL)$^{0.15}$ × Year Dollars

CC&DH

D&D = 5.93 (Calculated Nonredundant CC&DH mass in kg)$^{0.40}$ (Comm Kilobits Down)$^{0.20}$ (Redundancy Level)$^{0.20}$ (Electronic Parts Class)$^{0.08}$ (Human Rated)$^{0.60}$ (Mission Type) (XYZ Factor)$^{0.90}$ (Year of Technology Freeze)$^{0.040}$ (Pre-Development)$^{0.20}$ (Engineering Change Activity)$^{0.35}$ (Number of Organizations)$^{0.20}$ (Team Experience)$^{0.25}$ (Subsystem TRL)$^{1.10}$ × %New × Year Dollars

TFU = 0.198 (Calculated Nonredundant CC&DH mass in kg)$^{0.70}$ (Comm Kilobits Down)$^{0.10}$ (Redundancy Level)$^{0.10}$ (Electronic Parts Class)$^{0.12}$ (Human Rated)$^{0.30}$ (Mission Type) (XYZ Factor)$^{0.50}$ (Year of Technology Freeze)$^{0.030}$ (Pre-Development)$^{0.05}$ (Engineering Change Activity)$^{0.05}$ (Number of Organizations)$^{0.10}$ (Team Experience)$^{0.15}$ (Subsystem TRL)$^{0.30}$ × Year Dollars

In above CERs: Mission Type Multiplier Coefficient = 1.0 for Earth-orbital spacecraft
= 1.3 for ELVs
= 1.8 for RLVs
= 1.9 for planetary spacecraft

Year of Technology Multiplier Coefficient = 1/(input − 1960)$^{(exponent)}$

Pre-development Multiplier Coefficient = 1/(input)^{exponent}
Team experience Multiplier Coefficient = 1(input)^{exponent}
Subsystem TRL Multiplier Coefficient = (1/(input)^{exponent})/0.1393

navigation, power, propulsion, and environmental control and life support systems may use inherited materials or components. But they're usually configured into a new geometry, which engineers must design, analyze, and test (and this is where most of the expense is anyway). Reusing software isn't very practical for launch systems because most are unique and were developed many years apart. For example, Ariane V's first flight failed because it reused software from Ariane IV.

Off-the-shelf hardware and software also may not meet specifications as the project's definition matures and engineers cope with demands to optimize vehicle performance. These issues have been major reasons for cost growth during concept formulation and while transitioning to implementation. We have to judge carefully the reasonableness of early estimates regarding how much to use off-the-shelf elements. We should document assumptions and rationale for inheritance, examine the rationale carefully, and carry enough cost reserves to cover new hardware requirements that naturally creep into projects.

By dividing the actual historical cost by an estimated percent new factor from Table 7-5, we've adjusted the D&D-group CERs for our space systems cost model to yield 100% new design costs. The *percent new design* factor included as a multiplier in each of these CERs enables us to account for specific inheritance situations when estimating the cost of a new system. We should select the new-design factor in the SLaTS cost model based on the guidelines below.

TABLE 7-5.   **Percent New Design Definitions.** We should choose new design factors carefully to correspond to the design effort's expected scope and the technology's maturity.

New Design Factor	Basis
0.0 to 0.1	Existing design • The system is an off-the-shelf design requiring no modifications and no qualification. • All drawings are available in required format • All engineering analyses apply • All elements are within the current state-of the-art • Needs no new standards, specifications, or material adaptations or requires only very minor interface verification • Requires no qualification testing of the system • The system is in production or has been produced recently • Current team fully understands the design
0.1 to 0.2	Requires very minor modifications to a design • A few components or features require very minor modification • Must verify interfaces and integrate new components or features • Needs no—or very limited—qualification testing of the system • The system is in production or has been produced recently • Current team fully understands the design
0.2 to 0.3	Requires only minor modifications to a design • Several components or features require modification • Must verify interfaces and integrate new components or features • Requires minor requalification analysis and minor testing of the system • Current team fully understands the design

TABLE 7-5.   **Percent New Design Definitions. (Continued)** We should choose new design factors carefully to correspond to the design effort's expected scope and the technology's maturity.

New Design Factor	Basis
0.3 to 0.4	Requires significant modifications to a design • A significant number of components or features require modification • Must verify interfaces and integrate new components or features • Requires significant requalification and testing of the system • Current team fully understands the design
0.4 to 0.5	Requires very significant modifications to a design • Many of the components or features require modification • Must verify interfaces and integrate new components or features • Requires major requalification analysis and major testing of the system • Current team fully understands the design
0.5 to 0.6	Requires major modifications to a design • An adaptable design exists but requires major modifications to many of the components or features • Current team fully understands the design
0.6 to 0.7	Requires very major redesign or new design with considerable inheritance • An adaptable design exists but requires major modifications to most of the components or features • Requires a new design but one that can reuse many components, engineering analyses, or knowledge • Current team fully understands the design
0.7 to 0.8	New design with significant inheritance • Requires new design, but some components, engineering analysis, or knowledge exists—or advanced development work has brought the system into state-of-the-art
0.8 to 0.9	Almost totally new design with minor inheritance or advanced development work
0.9 to 1.0	Totally new design but within current state-of-the-art • Mostly new components • TRL[*] 6 at authority to proceed (ATP)
1.0	New design requiring minor advancement in state-of-the-art • Significant advanced development work has occurred • TRL[*] 5 at ATP
1.0–1.5	New design requiring some advancement in state-of-the-art • Some advanced development work has occurred • TRL[*] 4 at ATP
1.5–3.0	New design requiring significant advancement in state-of-the-art • Some advanced development work has occurred • TRL[*] 3 at ATP

[*] TRL refers to NASA's Technology Readiness Level scale (Chapter 13). In general, no new NASA project should enter the implementation phase before achieving TRL 6 or higher.

7.3.2   Learning and Rate Effects

Learning curve theory predicts that the unit production cost of repetitively manufactured or assembled products decreases as the number of produced units increases. Learning curves have helped predict this decrease in cost since the 1930s. Although the phenomenon is called "learning," we attribute it not only to direct labor learning but to many other economy-of-scale factors, such as quantity buys of materials and parts or more efficient use of indirect costs and overheads. Some analysts distinguish between learning and rate effects. They say learning is a function of total production quantities over many years, whereas rate effects result from annual production quantities. Other analysts emphasize that in highly automated environments, no real learning occurs—that is, automated machines don't learn. But as production rates increase, industry uses more efficient methods, whatever the source. This viewpoint combines all these effects into the single concept of learning.

Most actual cost data on production runs indicate that cost decreases at a rate that follows a power curve $y = ax^b$, where y represents cost, a is the first unit's cost, x is the number of units produced, and b is the logarithm of the learning curve fraction divided by the natural logarithm of 2. We have two ways to compute the effects of learning: the cumulative-average method (first published by J.R. Crawford and called Crawford curves) and the unit method (first published by P.T. Wright and often called Wright curves).

These two methods differ in the way they define the y variable. In the cumulative average method, y is the **average** unit cost over x units. The unit method defines y as the cost of the **xth** unit. For example, assume we have an item's first-unit cost $a = \$100$, with $x = 10$ units being built, and a 90% learning curve. To get either the predicted cumulative average over 10 units or the cost of the 10th unit (depending on the method), we have

$$y_{10} = \$100(10)^{\ln 0.90/\ln 2} = \$70.47 \qquad (7\text{-}1)$$

In space applications we recommend using the unit (Wright) method learning curve because it's more conservative. With this method and given the same example conditions as above, cost decreases from a TFU of $100 to $90 for the second unit, then to 90% of that cost by the fourth unit ($81). To compute the cumulative average cost of n units with this method, we sum the production costs for all n units and divide by n. The cumulative average for 10 units turns out to be $79.94, vs. $70.47 with the Crawford method. A more computationally convenient formula for calculating cumulative average is

$$y_{avg} = \frac{a}{x(1+b)}[(x+0.5)^{1+b} - (0.5)^{1+b}] \qquad (7\text{-}2)$$

where y_{avg} is the average unit cost over x units, and a and b are defined as above. Thus, the tenth average unit cost using a TFU of $100 and assuming 90% learning is

$$y_{avg10} = \frac{\$100}{10\left(1 + \dfrac{\ln 0.90}{\ln 2}\right)}\left[(10 + 0.5)^{1 + \frac{\ln 0.90}{\ln 2}} - (0.5)^{1 + \frac{\ln 0.90}{\ln 2}}\right] = \$80.06 \qquad (7\text{-}3)$$

The resulting average unit cost of $80.06 from the computationally convenient formula is acceptably close to the actual average unit cost of $79.94 calculated the hard way (and shown in Table 7-6).

TABLE 7-6.   Comparison of Unit and Cumulative Average Theories. Results from the learning curve depend on which of the two types of learning-curve theories we apply.

Core formula*: $y = ax^b$

	Unit Method			Cumulative Average Method		
Unit #	Cost/Unit†	Average Cost	Total Cost	Cost/Unit†	Average Cost	Total Cost
1	100.00	100.00	100.00	100.00	100.00	100.00
2	90.00	95.00	190.00	80.00	90.00	180.00
3	84.62	91.54	274.62	73.86	84.62	253.86
4	81.00	88.91	355.62	70.14	81.00	324.00
5	78.30	86.78	433.92	67.49	78.30	391.49
6	76.16	85.01	510.08	65.46	76.16	456.95
7	74.39	83.50	584.47	63.81	74.39	520.76
8	72.90	82.17	657.37	62.44	72.90	583.20
9	71.61	81.00	728.98	61.26	71.61	644.46
10	70.47	79.94	799.45	60.23	70.47	704.69

First Unit Cost: $a = 100$
Number of units: $x = 10$
Learning Curve Fraction: $b = 0.90$
† To find the unit cost of the nth unit with this method, calculate y_n and y_{n-1}, and subtract $(n-1)y_{n-1}$ from ny_n.

Space launch and transportation systems have exhibited learning curve rates between 80% and 100% (100% means no learning). We should choose a learning-curve percentage based on our best judgment about the economies of scale being contemplated for the program as production ramps up. Learning doesn't affect the cost of RLVs much (because not many units are likely to be built in the foreseeable future), but it's a very important factor in ELVs at their higher production quantities.

7.3.3   Cost Estimate for the SLaTS Example Vehicle

We're now ready to estimate the development cost of the expendable launch vehicle (ELV) for our reference SLaTS. We have some—but not all—of the cost inputs needed for a space systems cost-model estimate (Table 7-7). This situation is normal in cost estimating; we often must surmise some of the required information based on our knowledge of the project.

TABLE 7-7.   Mass Statement and Other Information For Costing the Reference SLaTS Vehicle. Cost analysts often do not receive all the information required for cost estimating, so assumptions are made and verified at a later date. (CC&DH is communication, command, and data handling; GN&C is guidance, navigation, and control; LH_2 is liquid hydrogen; LOX is liquid oxygen.)

	Booster (kg)	2nd Stage (kg)	Fairing (kg)
Structural and mechanical	14,774	4319	2700
Thermal protection system	1477[*]	216[*]	136[*]
CC&DH	88[†]	--	--
GN&C	38	--	--
Electrical power	82	179	--
Reaction control	--	589[**]	--
Main engines	3589	761	--
Main propulsion system	1061	191	--
Other information			
Number of engines	1	1	--
Vacuum thrust (N)	3020	302	--
Vacuum I_{sp} (s)	430	456	--
Chamber pressure (kPa)	--	--	--
Fuel/oxidizer	LH_2/LOX	LH_2/LOX	--
Engine Cycle	Gas generator	Gas generator	

[*] Not provided; calculated as 10% × total structural and mechanical weight for booster, 5% for 2nd stage and fairing

[†] Moved to 2nd stage for costing; obtained from the given avionics mass of 126 kg by alloting 70% to CC&DH and 30% to GN&C

[**] Not provided; calculated as 10% × total other stage mass

We begin by specifying the cost model's global variables in Table 7-8. (For brevity, we show only the inputs for the second stage here, excluding a similar set of inputs required for the booster and for the payload fairing). The first global variable reflects the fact that our vehicle isn't human-rated (human rated = 1). Next we set mission type to 2, corresponding to an ELV. Our configuration is a fully operational vehicle (XYZ factor = 3) and has an authority to proceed date of 2003 (implying a year of technology freeze = 2003). We estimate that an average amount of work has defined the configuration before the project start and therefore set pre-development = 2. We expect no more than the usual amount of changes to occur in development (engineering change activity = 2) and no more than three or four customer and performing organizations to participate in the program (number of organizations = 3). The team has normal experience, so team experience = 3.

We want the cost to include one system test article, one rocket engine on the second stage, and no jet engines or solids. The model's cost outputs must be in 2003 dollars. Finally, we enter the fractions we want the model to use in calculating cost contingency, project support, and fee for the design and development (D&D) and theoretical first unit (TFU) parts of the estimate. We specify contingency values of 30% and 15% for D&D and TFU, respectively—values typical for a definition-study level of understanding (but reduceable as the project becomes more mature). Project support gets values of 10% and 5% for D&D and TFU (again, typical values for costs beyond the prime contractor's scope). Finally, we input a fee percentage of 8% for both phases of the project, which is about average for contractor profit margins.

TABLE 7-8A. Global Inputs for the Space Systems Cost Model. These are for the second stage of the reference vehicle. Global inputs are the same for all CERs. (D&D is design and development; TFU is theoretical first unit.)

Global Inputs	D&D Input Values	TFU Input Values
Human rated	1	1
Mission type	2	2
XYZ factor	3	3
Year of technology freeze	2003	2003
Pre-development	2	2
Engineering change activity	3	3
Number of organizations	3	3
Team experience	3	3
Number of systems test articles	1	1
Number of rocket engines/stage	1	1
Number of jet engines/stage	0	0
Number of solid rockets	0	0
Year dollars (enter 4 digit year)	2003	2003
Model calculated inflation from 1999	1.118	1.118
Cost contingency %	0.30	0.15
Project support %	0.10	0.05
Prime fee %	0.08	0.08

TABLE 7-8B. Subsystem Inputs for the Space Systems Cost Model. Results here are only for structural and mechanical, thermal control, communication, command, and data handling (CC&DH) and GN&C elements. (D&D is design and development; TFU is theoretical first unit; TRL is technology readiness level; MLI is multi-layered insulation; RCC is reinforced carbon-carbon.) For "percent new" see Table 7-5.

Cost Estimating Relationships (CER) Variables	Inputs	D&D Exponent (b)	TFU Exponent (b)	D&D Co-efficients (a)	TFU Co-efficients (a)	Computed ($ Millions) D&D	TFU
Structural and Mechanical				1.603	0.060	$162.0	$13.8
Mass (kg)	5420	0.40	0.70	--	--		
Material complexity	1	0.10	0.20	1.00	1.00		
Deployables complexity	2	0.20	0.10	1.15	1.07		
Reserved	1	1.00	1.00	1.00	1.00		
Parts count	3	0.30	0.25	1.39	1.32		
Reserved	1	1.00	1.00	1.00	1.00		
Human rated	1	0.40	0.30	1.00	1.00		
Mission type	2	--	--	1.30	0.40		
XYZ factor	3	0.70	0.30	2.16	1.39		
Year of technology freeze	2002	0.015	0.010	0.54	0.66		
Pre-development	2	0.20	0.05	0.87	0.97		
Engineering change activity	3	0.35	0.05	1.47	1.06		
Number of organizations	3	0.20	0.10	1.25	1.12		
Team experience	3	0.25	0.15	0.76	0.85		
Subsystem TRL	6	1.10	0.30	1.00	1.00		
Percent new	1.00			1.00	1.00		
Thermal Control				0.705	0.078	$22.2	$3.0
Mass (kg)	544	0.40	0.70	--	--		
Coatings, surfaces, MLI, tiles, RCC, etc.	3	0.10	0.05	1.12	1.06		
Cold plates, heaters, etc.	1	0.10	0.05	1.00	1.00		
Radiators	1	0.15	0.10	1.00	1.00		
Pumped fluids, heat pipes, etc.	1	0.20	0.15	1.00	1.00		
Stored cryogenics and refrigeration	1	0.25	0.20	1.00	1.00		
Human rated	1	0.50	0.40	1.00	1.00		
Mission type	2	--	--	1.30	0.40		
XYZ factor	3	0.80	0.40	2.41	1.55		
Year of technology freeze	2002	0.015	0.010	0.54	0.66		
Pre-development	2	0.20	0.05	0.87	0.97		
Engineering change activity	3	0.35	0.05	1.47	1.06		
Number of organizations	3	0.20	0.10	1.25	1.12		

TABLE 7-8B. Subsystem Inputs for the Space Systems Cost Model. (Continued) Results here are only for structural and mechanical, thermal control, communication, command, and data handling (CC&DH) and GN&C elements. (D&D is design and development; TFU is theoretical first unit; TRL is technology readiness level; MLI is multi-layered insulation; RCC is reinforced carbon-carbon.) For "percent new" see Table 7-5.

Cost Estimating Relationship (CER) Variables	Inputs	D&D Exponent (b)	TFU Exponent (b)	D&D Co-efficients (a)	TFU Co-efficients (a)	Computed ($ Millions)	
						D&D	TFU
Team experience	3	0.25	0.15	0.76	0.85		
Subsystem TRL	6	1.10	0.30	1.00	1.00		
Percent new	1.00			1.00	1.00		
CC&DH				5.93	0.198	$39.9	$1.8
Mass (kg)	88	0.40	0.70	--	--		
Comm kilobits down	3	0.20	0.10	1.25	1.12		
Redundancy level	2	0.20	0.10	1.15	1.07		
Electronics parts class	2	0.08	0.12	1.06	1.09		
Reserved	1	1.00	1.00	1.00	1.00		
Reserved	1	1.00	1.00	1.00	1.00		
Human rated	1	0.60	0.30	1.00	1.00		
Mission type	2	--	--	1.30	0.40		
XYZ factor	3	0.90	0.50	2.69	1.73		
Year of technology freeze	2002	0.040	0.030	0.19	0.29		
Pre-development	2	0.20	0.05	0.87	0.97		
Engineering change activity	3	0.35	0.05	1.47	1.06		
Number of organizations	3	0.20	0.10	1.25	1.12		
Team experience	3	0.25	0.15	0.76	0.85		
Subsystem TRL	6	1.10	0.30	1.00	1.00		
Percent new	1			1.00	1.00		
Calculated nonredundant CC&DH mass (kg)	53	--	--	--	--		
GN&C				3.55	0.164	$29.5	$1.3
Mass (kg)	38	0.40	0.70	--	--		
Stabilization type	1	0.10	0.20	1.00	1.00		
Redundancy level	2	0.20	0.10	1.15	1.07		
Sensors type	1	0.20	0.10	1.00	1.00		
Degree of autonomy	1	0.10	0.20	1.00	1.00		
Reserved	1	1.00	1.00	1.00	1.00		
Human rated	1	0.90	0.50	1.00	1.00		
Mission type	2	--	--	1.30	0.40		
XYZ factor	3	1.10	0.70	3.35	2.16		
Year of technology freeze	2002	0.025	0.020	0.35	0.44		

TABLE 7-8B. Subsystem Inputs for the Space Systems Cost Model. (Continued) Results here are only for structural and mechanical, thermal control, communication, command, and data handling (CC&DH) and GN&C elements. (D&D is design and development; TFU is theoretical first unit; TRL is technology readiness level; MLI is multi-layered insulation; RCC is reinforced carbon-carbon.) For "percent new" see Table 7-5.

Pre-development	2	0.20	0.05	0.87	0.97		
Cost Estimating Relationship (CER) Variables	**Inputs**	**D&D Exponent (b)**	**TFU Exponent (b)**	**D&D Co-efficients (a)**	**TFU Co-efficients (a)**	**Computed ($ Millions)** D&D	TFU
Engineering change activity	3	0.35	0.05	1.47	1.06		
Number of organizations	3	0.20	0.10	1.25	1.12		
Team experience	3	0.25	0.15	0.76	0.85		
Subsystem TRL	6	1.10	0.30	1.00	1.00		
Percent new	1			1.00	1.00		
Calculated nonredundant GN&C mass (kg)	22	--	--	--	--		

Next we need to feed information on each subsystem and other cost elements in our WBS into the space systems cost model in Table 7-8. Some of this information comes from Table 7-7, but we need to supply other inputs based on our knowledge of the project and our experience. For example, to estimate structural and mechanical cost, the space systems cost model requires inputs on material complexity, deployables complexity, parts count, and so on. In this case, we've generated these inputs ourselves, as shown in Table 7-8, using our best engineering judgment. We specify a material complexity of 1.0 corresponding to "mostly aluminum" structure because design engineers tell us this material is dominant. We use a deployables complexity of 2.0 corresponding to simple or minor deployables because the only mechanism required is the staging mechanism that separates the booster from the second stage. The assigned parts-count value is 3, which corresponds to a normal number of parts, because the design uses some net casting and other parts-count strategies but not more than usual for a project of this type. We get several other inputs for structural and mechanical parts from the global inputs. The final two structural and mechanical inputs are the subsystems' technology readiness level (TRL), which we set at 6 because all technology issues related to this element have been demonstrated in a relevant environment.

We set percent new design to 1.0 to reflect our assessment that the design represents all new effort. This last input requires more explanation. Table 7-5 implies that, in general, a TRL level of 6 also implies a percent new design factor less than 1.0 (the table gives values of 0.5 to 0.8 for a TRL of 6). But we must sometimes depart from the cookbook instructions and use our own judgment. In this case, even with a technology readiness level of 6, we judged the design to be all new with essentially no inheritance from previous designs, so it warrants a new design factor of 1.0.

Now we must provide the cost model's required inputs for the reference vehicle's other subsystems and elements, either by estimating them ourselves or by interviewing design engineers or project managers. In Table 7-8, for brevity, we show only three of these additional subsystems: thermal control (TCS); communication, command, and data handling (CC&DH); and guidance, navigation, and control (GN&C).

After we insert inputs for all WBS elements of each stage of the reference vehicle, the space systems cost model computes each element's cost (based on all the CERs). Table 7-8B gives the resulting estimated costs for the second stage. The estimates for the structural and mechanical element for the SLaTS vehicle's second stage are a D&D cost of $162.0 million and a TFU cost of $13.8 million. Similarly, the table shows costs for the TCS, CC&DH, and GN&C elements. All these cost elements accumulate to a DDT&E cost of $1200 million and a TFU cost of $85 million for the second stage. (We use the term D&D cost for the subsystem and the term DDT&E for total stage cost because the stage-level costs include those for system-test hardware and operations. We combine these costs with the stage-level integrate and check-out, ground support equipment (GSE), systems engineering and integration and project management, to bring the D&D cost of the subsystems up to full DDT&E.)

Though we don't discuss it here, we employ the same process to generate costs for the booster stage and the payload shroud. These costs are, in total, a booster-stage cost of $2240 million DDT&E and $189 million TFU and a payload shroud cost of $330 million DDT&E and $13 million TFU.

Table 7-9 summarizes these results. We derived the average unit costs of the vehicle stages and shroud in the table by using the TFU from the space systems cost model and the learning curve method we described in Section 7.3.2. To apply learning, we multiplied the TFU costs for the booster, the second stage, and the shroud by the average unit cost factor—assuming a production run of 12 units per year over 10 years (120 total units). We assume the fairly standard 90% learning curve here. Using the learning curve equations from Section 7.3, we find that the average unit cost decreases to 56% of the TFU cost over the production run of 120 units. This computation is

$$\left(y_{120} = \frac{100}{120\left(1 + \frac{\ln 0.90}{\ln 2}\right)} \left[(120 + 0.5)^{1 + \frac{\ln 0.90}{\ln 2}} - (0.5)^{1 + \frac{\ln 0.90}{\ln 2}} \right] \right) \approx 56\% \quad (7\text{-}4)$$

As we see in Table 7-9, applying a 0.56 learning factor to the TFU costs results in a total average unit cost of $161 million for the vehicle stage.

7.4   Estimate Operations and Support Costs

The recurring cost for operations and support (O&S) of new space or launch systems is the largest cost element, as Figure 7-2 indicates. But until recently, this area has been underemphasized when formulating and developing new launch

TABLE 7-9.   Summary of the SLaTS Reference Vehicle's Average Unit Costs for Design, Development, Test, and Evaluation (DDT&E). The table gives the costs in 2003 dollars. (TFU is theoretical first unit.)

	DDT&E ($M)	TFU ($M)	Learning Factor (90% Over 120 Units)	Average Unit Cost With Learning (Over 120 Units) ($M)
Booster	2240	189	0.56	106
Second stage	1200	85	0.56	48
Shroud or payload carrier	330	13	0.56	7
Total vehicle	3770	287	0.56	161

systems. Operations data and the development of assessment and costing methods have particularly suffered from neglect. As a result, we've lacked the information needed to help develop an analysis process, and the aerospace community hasn't agreed on the approaches to use. More recently, emerging analysis methods have addressed space transportation O&S from several perspectives and various levels of detail to give us better insight. Many of these tools were developed to assess the operability of new concepts, but they now offer the best ways to get information needed to assess operations and support costs.

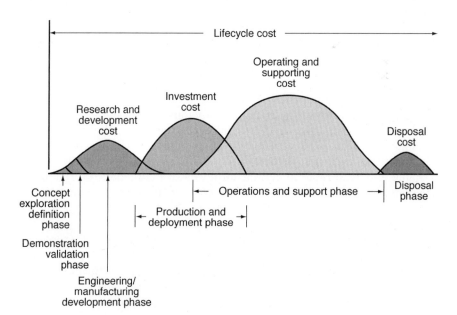

FIGURE 7-2.   Lifecycle Phases. Here we show that the operations and support costs are usually the most significant cost element over a program's lifecycle. [DoD, 1992]

Because reusable launch systems are unique and relatively new to operations, we haven't developed simple rules of thumb for estimating O&S costs based on their development or production cost. We also haven't had enough systems to develop parametric relationships of cost-to-engineering variables for O&S, as is the case for DDT&E estimates. Unlike aircraft systems, which have a common operational scenario, the launch systems' low flight rates and unique support structures require a more involved approach to developing O&S cost estimates. Possible launch concepts are crewed and uncrewed; horizontal and vertical takeoff and landing; reusable and expendable rocket systems, aircraft-like systems with air-breathing engines, and others.

Because no two systems are alike, our analysis must reflect the specialized nature of each system's operation and support needs that may also vary with flight rate. Such needs include facilities, specialized support equipment, processes, levels of support, and so forth. The analysis process and models must fit the complexity of these distinctive concepts with their different operating requirements. Costing for O&S requires us to realistically estimate the resources needed to achieve the flight rates required over the system's operating life. In this section, we:

1. Explain what elements are part of O&S cost

2. Identify the O&S cost issues we must address

3. Outline a process for analyzing a new launch system

4. Exemplify the types of results expected and desired

7.4.1   Operations and Support (O&S) Cost Elements—What to Consider

As part of the lifecycle, O&S addresses a vehicle system's operational phase throughout its useful lifetime. Operations and support costs comprise all costs of operating, maintaining, and supporting a fully deployed system. In Figure 7-3 we see a representative operational flow that repeats for each flight over a launch system's lifetime. We need to capture the costs of these functions with all pertinent off-line functions over the system's useful life. These costs include personnel; consumable and repairable materials; all levels of maintenance (organizational, intermediate, and depot); facilities; and the costs of sustaining the investment. We often organize the results into personnel, hardware, material (including consumables), and facility costs.

Figure 7-3 illustrates the concept for the launch processing flow, plus many off-line functions we must also account for in O&S costs. The operational flow includes the steps required for launching, retrieving, and returning to flight a reusable launch vehicle, as well as the process for expendables. For reusable concepts, the process defines the system's turnaround ability, which is a major factor in its economic viability. The analytical scope must include all functions pertaining to the launch system's operating scenario.

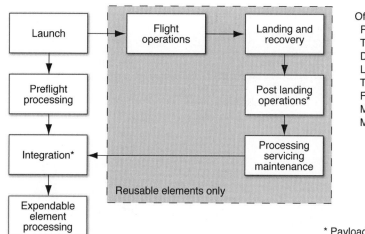

FIGURE 7-3. **Online Processing Flow.** The on-line processing flow, along with the vehicle and facilities definition, drives the operations and support cost.

Part of the O&S cost is the nonrecurring startup costs (incurred while preparing for operations) and the costs of support when the system is fully operational. The *nonrecurring* cost elements usually cover initial costs of designing, developing, and building the support facilities, as well as the cost of acquiring and training people to operate and support the system before it deploys. The *recurring* costs include the cost of people needed to process and maintain the launch vehicle, together with the launch system's support equipment and facilities. We usually express them as an annual cost that depends on flight rate and the hardware's characteristics. Costs for O&S also encompass those for material, consumables, and expendables used for operations. (We normally account for the cost of O&S software and its maintenance as a part of the cost element that the software supports.) Costs to replace components that exceed limits on hardware life, such as an engine, also fall under O&S. Table 7-10 summarizes typical O&S costs.

In addition, recurring costs have fixed and variable components, as Figure 7-4 illustrates. In economic terms, fixed costs are those not affected by changes in output or activity for a specified system capability, whereas variable costs change with the output level. We express a space launch system's capability in terms of flight rate. Fixed costs are for maintaining the system's ability to meet a specified flight rate at a given time, whether or not the specified number of flights is actually flown.

Among fixed costs are staff salaries and infrastructure upkeep required to sustain the specified flight rate (users may add interest, fees, insurance, and so on). A fixed cost may change with major changes in the system's capability. Variable costs apply to individual flights. Examples are the cost of spares, expendable hardware, consumables, and extra people needed to meet a higher flight rate. No specific conventions govern how to classify O&S costs, so we should compare carefully with other O&S costs to be sure we include similar cost elements in each study.

TABLE 7-10. Operations and Support (O&S) Cost Elements. These divide into nonrecurring costs to build the infrastructure for initial operation and recurring costs for ongoing operations. The costs include facilities, personnel, hardware, and software. Many of these items are discussed in Chapter 9 of APMSS [Chesley et al., 2008]. (GSE is ground support equipment.)

Cost Element	Nonrecurring	Recurring
Site	Acquisition, development	O&S
Facilities	Design, build, facility GSE	O&S
Launch and flight operations	Equipment, facilities, personnel	O&S
Vehicle hardware		O&S, replacement
GSE (vehicle-unique)	Initial	O&S
Maintenance		O&S (organizational, intermediate, depot)
Spares	Initial	Replenishment
Consumables	Initial	Replenishment
Expendables	Initial	Replenishment
Training	Initial	Refresh and replacement
Documentation	Initial	Updates
Transportation	Initial	On-site, return to site
Management	Initial	Replacement
Insurance, fees, taxes		Per flight

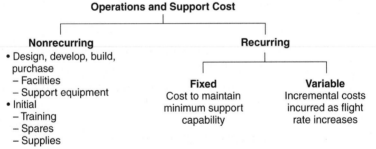

FIGURE 7-4.   Elements of Operations and Support (O&S) Costs. The figure illustrates the taxonomy we use to distinguish O&S cost elements.

7.4.2   Operations and Support Issues

The cost of operating and maintaining a launch system over its useful life has five main drivers:

- System design (vehicle, flight, and ground support)
- Reliability, maintainability, and supportability characteristics
- The operations concept
- Support policy
- Flight rate

Because decisions that affect these issues occur early in the design process, we must capture their effects on the O&S costs during the conceptual phase. Designers can modify a concept early in design to meet its cost goals when changes are least costly. Next, we look at the effect of each issue on new launch concepts.

System Design Factors

A new launch system must meet performance goals. Design features and technology choices play an important role in the support burden in terms of people, time to do tasks, support equipment, and facilities. The technology choices made during initial concept development heavily influence the design layout of a new concept. An example is selecting propellants for the reaction control system and orbital maneuvering system that are different from the main propulsion system's propellants, and require separate tankage and feedlines. A host of technology choices confronts a designer in space systems today. Choices made solely to save money over the near term, or program drivers such as the availability of funds, can increase future O&S costs by orders of magnitude more than the original savings. On the other hand, technology solutions to make O&S affordable must be consistent with up-front realities, such as budget and the physical and environmental demands of having to achieve orbit with the desired payload.

Multiple studies [NASA, 1997 and Nix et al., 1998] have documented improvements required in future designs if we want to reduce O&S costs and increase overall productivity. Five key guidelines have emerged for early design and technology choices:

- Increase the supportability of servicing systems for launches by reducing the numbers of different fluids, commodities, gases, and electrical or communication requirements and accompanying interfaces
- Eliminate environmental and safety hazards by dropping hazardous operations, such as using highly toxic propellants
- Eliminate hazardous areas and operations introduced by vehicle design choices, such as vehicle purge requirements or confined spaces
- Select components of proven high reliability
- Increase readiness levels of new technologies by properly demonstrating and understanding whole flight and ground system technologies

By incorporating these choices, a design requires less workforce, support equipment, and processing time. For example, the Shuttle and its support infrastructure use more than eight toxic fluids. They demand special handling, isolated work environments that sometimes halt concurrent work, and special training and equipment. Design changes that limit the number of different fluids used on a launch concept significantly reduce the required support.

Reliability, Maintainability, and Supportability

Technologies in a reusable design affect its reliability and maintainability, which determine the maintenance burden we must handle before reflight. We have to replace or repair items that have failed during a mission or ground processing (or have reached life limits) before the next flight. And many systems require periodic servicing, regardless of their condition. Servicing also has to account for the technology's maturity and complexity. Support plans and policies attempt to balance maintenance and servicing needs to ensure reliability and safety, yet repair any items that don't work properly. For example, Shuttle processing often has to deal with more than 500 problem reports after each flight (excluding thermal tile work) besides about 1700 routine maintenance tasks. Systems that have low functional reliability, or are difficult to access and maintain, require higher levels of support. Estimates for these parameters have to come from engineering judgment based on historical records of similar systems or on test data for newer components.

Operations Concept

An operations concept is integral to assessing O&S costs. Chapter 3 and *Cost Effective Space Mission Operations* (CESMO) [Squibb et al., 2006] describe how to develop an operations concept. Suppose a concept includes vehicles that land away from the launch site or arrive in segments to be mated at the pad. These features affect O&S costs. Typical choices for a launch vehicle operations concept are an integrate-transfer-launch or assemble-on-pad scenario, vertical or horizontal processing and integration, where to install the payload, and how much processing to place in a processing station versus a launch pad. Horizontal or vertical processing and integration affect O&S differently for any concept, but the ultimate driver—the amount of work—influences it the most.

Support Policy

Repairs and servicing must take place before reflight, but support policy affects the environment under which they occur. Support policy encompasses such decisions as the level of repair at the launch site, the number of available spares, shift and staffing for a facility, and the level of inspection required to ensure safe and reliable flight. These policies often result from trying to meet mission flight-rate goals as safely and economically as possible. Except for shift policy, many of these decisions are implicit in conceptual studies. We should be aware of a study's assumptions and stay consistent with them. For instance, the planned work and level of inspection necessary before flight are a matter of policy because they depend on the system's maturity, technology, and design confidence.

Developmental and crewed systems receive more extensive servicing and inspection to ensure the safety of flight, so they require more staffing and longer turnaround times between flights. That effort reduces a vehicle's availability and increases O&S cost. The change in minimum turnaround times observed after the

Challenger accident best exemplifies the effect of a change in support policies. The shortest turnaround time before the Challenger accident had been 46 workdays (STS 61-B), but new policies for flight safety have lengthened the best time since then to 81 workdays (STS 94) for what is essentially the same vehicle system.

Role of Flight Rate

A launch system needs certain types of equipment and facilities (support infrastructure) regardless of the flight rate. The fixed cost of the core infrastructure usually differs only incrementally as the flight rate increases and as we add facilities, equipment, and people needed to support the increased capability. So although we would expect overall O&S cost to rise as flight rate increases, the increase isn't linear. In fact, the per-flight cost increases for flight rates that don't fully use the support facilities. In general, higher flight rates tend to use people, facilities, and equipment more effectively and thus lower the cost per flight.

A system's ability to ready a launch vehicle for reflight is crucial to its productivity. It depends on the inherent reliability, maintainability, and supportability of the system element and the support environment. This environment includes flight hardware, support equipment, and facilities, as well as the philosophy (culture) used to maintain and launch the system. Designers often set flight-rate goals that they assume a system can meet based on economic needs entirely independent of the design. Although this practice may work for "what-if" parametric studies, it's not appropriate for assessing new concepts. For these assessments, we must consistently connect goals with the system's capabilities. For example, the design, technologies used, and support concept—in combination—must meet the mission's turnaround goals. Establishing this capability for new designs isn't an exact science. It involves estimating with a mixture of techniques, such as engineering estimates, analogies, models, and simulations. These methods are the essence of estimating O&S cost.

7.4.3   Process for Operations and Support Cost Analysis

Table 7-11 lists the steps required to analyze O&S costs on new launch systems. Although listed sequentially, many of these steps occur concurrently. Explanations of the main issues for each step follow Table 7-11.

Step 1.   Understand the System and the Purpose of the Analysis

Understanding the system to be analyzed is critical to a successful analysis. Knowing why the analysis is being done helps to determine scope and depth: we need to grasp the total project and its elements to ensure completeness. If the purpose of the analysis is to cost O&S for a new system, all critical elements must be in the study. If we compare competing designs or operating scenarios, we can omit elements not affected by the design choices if this doesn't affect the relative results, thus saving time and effort. We need to capture all pertinent costs for comparison, not necessarily all costs.

TABLE 7-11. Steps Required to Estimate Operations and Support (O&S) Cost. We use these steps to estimate the cost for operations and support. (APMSS is *Applied Project Management for Space Systems* [Chesley et al., 2008].)

Step	Task	Where Discussed
1. Understand the system and purpose of the analysis	Define purpose, scope, and depth of analysis	Chaps. 1–6
2. Gather information on the conceptual design	Determine characteristics of the proposed launch system	Chaps. 3, 6, and 10
3. Develop an operations concept	Provide a general description of the operating flow and major facilities required for the launch system to function	Chap. 5
4. Determine the ground rules and assumptions	Derive from mission needs, requirements, and constraints. Limit scope and depth.	Chaps. 1 and 2
5. Choose a cost element structure (CES)	Develop a CES to assure that results capture all the cost elements needed for analysis	APMSS Chap. 11
6. Select an appropriate method or model	Select a method or model appropriate to the goals, scope, and depth of analysis	Chap. 11
7. Perform a study based on the selected method or model	Run models	Chap. 11
8. Evaluate results	Compare estimates with nearest comparable system	Chap. 11
9. Document results	Prepare documentation	Chap. 11

We should make sure that the study's depth matches its purpose and the available information. If the purpose is to trade the number of stages, we may not need to estimate cost to the subsystem level. If the purpose is to evaluate the effect of a new subsystem technology, we must estimate subsystem support costs so we can assess that effect. Not all parts of the cost analysis require the same level of detail for analysis.

Step 2. Gather Information on the Conceptual Design

We need to define the conceptual launch vehicle system at least to a level that agrees with the cost model's input requirements. We usually describe the major elements that constitute the launch system and the overall dimensional data, masses, and technologies of the launch vehicle. The design process covered in Chapters 3–6 produces this information, which we should gather on all elements or stages that need support and facilities. The definition should identify the technologies used in the system design along with the type of materials, propellants, and so forth. We also need to know the mission for which the concept was defined, called the design reference mission. Technology changes can result in resizing the concepts, so we must ensure that the concepts and reference mission remain consistent. Otherwise, we may cost a system that can't meet the mission requirements. If we do resize the concept, we should recheck it for consistency with the mission goals, ground rules, and assumptions.

Step 3.   Develop an Operations Concept

The *operations concept* describes how the launch system is deployed, processed, maintained, and operated to achieve the desired mission flight rate. It details the types of facilities, ground support equipment, and skill levels required, but not necessarily the number of each because this may vary with flight rate and is captured in the analysis step. The operations concept should define the order for normal operations and options for abnormal ones. Examples include what to do after a mission scrub or delay and provisions for abort scenarios (landing options, how to return to a launch site, and so on). The operations concept can be as simple as a flow chart similar to Figure 7-3, or it can be more complex, with a detailed written description. The concept guides analysis and gives us confidence that we've addressed all elements. (Trade studies require alternative concepts and scenarios.)

Step 4.   Determine the Ground Rules and Assumptions

This work establishes the ground rules and assumptions that govern the O&S cost estimate. They help us limit the depth and scope of the analysis. For this step, we need information such as whether the support uses a "clean-sheet" approach (assumes that no required facilities exist) or uses existing facilities where possible in the analysis. We also need to establish schedule information such as date of first flight, the initial operating capability, and the operational life. Here we identify assumptions about the number of developmental flights, the value of a learning effect to reduce processing time and workforce, the vehicle's design life, the flights per year required to support the mission, the location of manufacturing and support suppliers, and alternate launch and landing sites. We specify the year to use as the basis for test analysis and state any constraints on flight and processing requirements.

Step 5.   Choose a Cost Element Structure

The *cost element structure* (CES) establishes a standard vocabulary for identifying and classifying a system's costs. (Our choice of CES here is part of the operations and maintenance cost WBS in Figure 7-1.) We should design the CES to capture as many relevant costs as practical so we can meet the assessment's goal and not inadvertently omit major O&S costs. Several CESs are in use for launch systems, but not all are populated with complete reference data. Table 7-12 gives some examples, but we don't have to restrict ourselves to them. To reflect the unique support required, we may have to change the basic structure to fit the system's characteristics.

Each cost element structure has organized the O&S information relevant to the data source and the intended use of the analysis. We need to be aware of the data sources in making this selection because, for the costs to fit, we have to interpolate and estimate data not collected in the chosen CES. That adds a lot of work and uncertainty to the analysis.

TABLE 7-12. **Examples of Cost Element Structures (CES).** Column A reflects a CES based on the support of ground and launch operations for the Shuttle. Column B represents a more general type of cost structure used by the military for logistics support. Column C lists the major cost elements for new Spaceport launch systems. (GFE is government-furnished equipment; GSE is ground support equipment; JSC is Johnson Space Center; KSC is Kennedy Space Center; LPS is launch-processing system; O&M is operations and maintenance; ILS is integrated logistic support; R&D is research and development.)

A. Access to Space Cost Estimating Structure	B. General CES	C. Spaceport Operations Modules and Facility Functions
External Tank	Operations	Traffic and flight control facilities
Solid Rocket Motor (SRM)	Refurbishment	Launch facilities
Solid Rocket Booster (SRB)	Organizational maintenance	Expendable element facilities
Engine (Sustaining Engineering)	Processing operations	Vehicle assembly and integration facilities
Orbiter and GFE (JSC)	Integration operations	Payload and cargo processing facilities
Orbiter logistics and GSE (KSC)	Payload operations	Landing and recovery facilities
Propellant (KSC)	Transfer	Vehicle turnaround facilities
Launch operations (KSC)	Launch operations	Vehicle depot maintenance facilities
Shuttle processing	Mission operations	Spaceport support infrastructure facilities
Systems engineering and support	Landing, recovery, and receiving operations	Concept-unique logistics facilities
Facility operations and maintenance	Non-nominal operations	Transportation system operations planning and management facilities
LPS instrumentation and calibration	Logistics Support	Community infrastructure
Modifications	Depot maintenance	
Technical operations support	Modifications	
Program operations support	Spares	
Communications	Expendables	
Base operations contract	Consumables	
Launch support services	Inventory management and warehousing	
Weather support	Training	
Payload operations	Documentation	
Mission operations	Transportation	
Mission operations facilities	Support equipment	
Mission planning and operations	ILS management	
Program and document support and management	System Support	
Crew operations (JSC)	Support	
Crew training and medical operations	Facility O&M	
Program office/headquarters	Communications	
Institutional	Base operations	
Program management support	Program Support	
Network support	R&D	
Systems engineering		

Step 6.   Select an Appropriate Method or Model

Based on the goals, scope, and depth of our analysis, we choose an analysis model or develop a method consistent with the goals. Though we would prefer to have high-level parametric models during concept development because of the uncertainty in concept definition, they aren't available. Lacking parametric relationships, we need to use accounting methods that base the estimate on the aggregate of each cost element. Here we develop the cost from simple relationships or direct input. We can use simulation models to study the effects on cost of such issues as operational constraints, basing concepts, and different maintenance, sparing, and support policies. The needs of the analysis may require some combination of these.

Operations and support cost models are under development and should soon be available, but none is complete. Among them are the operations cost model (OCM) [NASA, 1993] and NROC (iNtegrated RMAT and OCM); the Reliability and Maintainability Analysis Tool/Life Cycle Analysis (RMAT/LCA) [Morris, 1995; NASA, 2000]; and the Architectural Assessment Tool-enhanced (AATe) and the Vision Spaceport Strategic Planning Tool [NASA, 2001]. We encourage contact with these sources to determine the development status and to search for other models that may have become available.

Step 7.   Perform a Study Based on the Selected Method or Model

We do the study to meet overall goals or requirements, usually by exploring and quantifying alternate design, support, and policy options. For this purpose, we must identify and estimate all relevant O&S costs. Future O&S costs are very uncertain, so we should search out elements that drive uncertainty. Sometimes a sensitivity analysis of key cost drivers helps identify how much uncertainty affects costs and places the estimate into perspective. To do this analysis, we must estimate the effect of the high and low uncertainty ranges for significant input factors, treating each one independently or in combinations as part of a Taguchi analysis.

Step 8.   Evaluate Results

We should compare model or analysis cost results with known O&S costs for comparable systems to assess credibility. Unfortunately, cost information isn't always available for comparison because it's proprietary for many systems, but Table 7-13 can help. It gives the estimated price for launching various expendable systems, plus a per-flight cost estimate for the U.S. Space Transportation System (STS) (Shuttle). The price includes any profit margin for commercial ventures. Profit isn't part of the O&S cost; it's added to it. Therefore, any comparison should allow for this difference.

Concepts whose O&S costs exceed the price values shown should be suspect. Similarly, for values significantly lower than those of comparable systems, we need to find out where the cost savings are. Also, we must keep the system's functional capability in mind. For example, the Shuttle and Titan IVB have similar size payloads, but only the Shuttle can carry crews. These differences in capability contribute to differences in O&S costs.

TABLE 7-13. **Comparing Per-flight Prices.** Here we show the estimated price ranges for current launch systems. Values come from Isakowitz, Hopkins, and Hopkins [1999] and aren't converted to 2003 dollars.

Launch Vehicle		Type*	O&S Price/Flight ($M) (Cost of expendables included)	Payload (kg)
Shuttle		R	375–600†	24,400
Atlas	IIA	E	75–85	7316
	IIAS	E	90–105	8600
	IIIA	E	90–105	8641
	IIIB	E	90–105	10,720
	V900	E	75–90	12,500
	V500	E	85–110	20,050
	VHVL	E	140–170	--
Delta II	7320	E	45–55	2870
	7920	E	45–55	5140
Delta III		E	75–90	8290
Delta IV		E	140–170	25,800
Titan IVB		E	350–450	21,700

R is Reusable, E is Expendable
† Cost based on 5 to 8 flights per year

Step 9. Document Results

Although no specific convention exists, we recommend that the documentation state the purpose of the analysis and describe the concept, ground rules, assumptions, and costs included in the study. Depending on the study's purpose, we may present the results in a number of formats. Examples are a total lifecycle cost that includes all phases, a time-phased estimate that illustrates the cost distribution over the program's life, or an annualized cost. For annualized costs, we should assume mature system operations. The documentation should contain enough information for the reader to understand the scope and character of the results.

7.4.4   Example Case—Estimating Operations and Support Cost for the SLaTS Reference Vehicle

To illustrate our analysis, we assume the following concept: an expendable vehicle requiring two stages to orbit contains an uncrewed payload that will go into low-Earth orbit. Single LH_2/LOX engines power the first and second stages. The payload* has a solid kick motor to place it in orbit. We integrate the payload

---

* Here "payload" refers to the entire satellite that is placed in orbit, not just to the part of the satellite that carries out the mission.

on the pad, and no elements will return for recovery. The first stage's characteristics are 4.6 m diameter, 30.6 m height, 145,600 kg of propellants. The second stage's characteristics are 4.6 m diameter, 10.3 m height, 30,400 kg of propellants. The payload shroud is 14.7 m in height. The flight vehicle's total stacked height is 55.5 m. (We assume a mixture ratio of 6:1 for oxidizer: fuel.)

Step 1.   Understand the System and Purpose of the Analysis

The scope of the analysis includes the launch site and mission support functions only, capturing the direct cost for the design and its support infrastructure. This allows us to compare designs and support concepts. We don't address secondary (but not inconsequential) costs for off-site support such as sustaining engineering and testing, off-site depot work, supporting off-site landing sites, and improving or managing products. (We would need to address these costs if the purpose were to develop a cost for budgeting or if we wanted to establish a commercial launch price.)

Step 2.   Gather Information

This step organizes the information we have obtained so we can ascertain whether we are missing any critical data.

- Design—Uncrewed, expendable vehicle requires two stages to orbit
- Technologies—The first and second stages (boosters) each require a single LH_2/LOX engine. The payload has a solid kick motor (upper stage) to place it in orbit.
- Dimensions—Assembled flight vehicle has 4.6 m diameter, 55.5 m stacked height
- Propellant masses—Propellant includes 151,000 kg LOX and 25,200 kg LH_2 based on an assumed mixture ratio of 6:1
- Mission—Requires payload delivery to orbit, 12 flights per year

Step 3.   Operations Concept

We deliver launch elements to the site and erect them directly on the launch pad. Teams check each stage independently before adding the next stage. They integrate the payload stage on the pad, enclose it with the payload shroud, and do an integrated check-out. A separate support crew provides mission support from an onsite firing room. We assume launch-day operations require one day. An off-site organization handles on-orbit payload support.

Step 4.   Ground Rules and Assumptions

- "Clean-sheet" approach: estimate the cost of major facilities
- Ten-year lifecycle for operations
- Flight rate of 12 flights per year
- Missions split between NASA and commercial enterprises (50% each)
- No learning curve

- No developmental flights
- Constant flight rate (no growth over lifecycle)
- Results in FY2003 dollars

Step 5. Choose a Cost Element Structure

In this case, the COMET/OCM model in Step 6 determines the cost element structure and thus defines the cost data needs.

Step 6. Select Method or Model for Analysis

The preceding sections in this chapter describe elements we should consider in developing O&S costs. Computer-based models that address them are still being developed. To exemplify how to compute recurring costs for O&S, we've developed a high-level modeling process based mainly on information extracted from the Conceptual Operations Manpower Estimating Tool/Operations Cost Model (COMET/OCM, or OCM). We use it for rough order-of-magnitude cost estimates using parameters typically available during concept development. But the model can't directly reflect all the benefits or disadvantages of using newer technologies, processes, or designs—all of which we should consider in a cost estimate. For that level of estimate, we suggest reviewing the models discussed in Section 7.4.3, Step 6.

Step 7. Perform Analysis

We trace costs for space launch operations to the tasks required to prepare a vehicle for launch, conduct mission operations, and provide support. These cost elements focus on people, facilities, infrastructure, ground support equipment, consumables, expendables, spares, and so forth. But detailed information required to determine accurately the number of people or the time needed for specific tasks usually isn't available for a conceptual system. So OCM uses historical data and a technique called ratio analysis to estimate costs for new concepts. Its main parameters are the number of people required to process vehicles and plan flights, which then helps estimate the staffing needed to do the additional launch and flight operations (Table 7-14 and Figure 7-5). Most of the other cost elements depend on these staffing levels. We also allow for cost elements not related to staffing levels, such as consumables and spares. Below, we describe that process for the example concept.

We've developed the total recurring O&S cost for ground and mission operations. As presented here, the results reflect the support environments in the mid-1990s for the Shuttle and expendable launch systems such as Atlas and Titan. We have to adjust these values based on engineering judgment on how new designs, technologies, and processing changes affect the support of new concepts.

Step 7.1. Describe Vehicle and Mission Concepts

We begin by using Tables 7-15 and 7-16 to describe the vehicle and the mission. The parameters we need to cost these functions are based on concepts defined in that format. We use the following terms: booster is any stage that provides lift to orbit but doesn't achieve orbit or participate in on-orbit maneuvers. A low-Earth-orbit (LEO) stage does on-orbit maneuvers but need not be reusable. An upper

TABLE 7-14. Process to Estimate Recurring Costs for Operations and Support (O&S). These five steps describe how to estimate recurring support costs based on information from the Operational Cost Model.

Step	Source
7.1. Describe vehicle and mission concepts	Tables 7-15 and 7-16
7.2. Compute primary head count for vehicle processing and flight planning • Launch operations (vehicle processing) • Flight operations (flight planning)	 Tables 7-17 through 7-19 Table 7-20
7.3. Compute total head counts • Head count for basic flight rate of 8 per year • Head count for desired flight rate	 Eqs. (7-12) and (7-13)
7.4. Compute costs • Labor costs • Compute supplies and materials cost from labor costs • Compute propellant cost from concept definition • Compute GSE spares from nonrecurring GSE cost • Network support from mission description • Wraps from labor costs • Expendable hardware from Chapter 7, Step 3	Tables 7-21 and 7-22 and Eq. (7-16)
7.5. Show results—total recurring O&S costs per flight and per year for a given flight rate	

stage is the final element that either places a payload in orbit or transfers it from LEO to a higher orbit. It may be carried in the bay of a LEO vehicle such as the Shuttle, or atop a booster stage such as Titan or Ariane. An upper stage may act as a cargo-transfer vehicle or orbital-maneuvering vehicle if we have no LEO stage. We must complete the process in Table 7-14 for each booster element of the concept (we show only the stage-one element). A major purpose for filling out Tables 7-15 and 7-16 is to define the system in terms that match the format needed in Tables 7-17, 7-18, and 7-20.

Step 7.2. Compute Primary Head Counts for Vehicle Processing and Flight Planning

To estimate the number of people for vehicle processing, we consider them by operational phase (Table 7-17) and type of element being processed (Tables 7-18 and 7-19). Then we use Equation (7-5) to estimate the head count for vehicle processing. We estimate the number of people for flight planning in a similar way, using Equations (7-7) through (7-11) plus Table 7-20.

$$
\begin{aligned}
\text{Ground operations cost} = &\ \text{cost of expendable hardware} \\
&+ \text{labor cost to support launch operations} \\
&+ \text{cost of propellant and other consumables} \\
&+ \text{cost of supplies and materials} + \text{cost GSE and facilities spares} \\
&+ \text{cost of wraps, fees, taxes, and other miscellaneous charges}
\end{aligned}
\tag{7-5}
$$

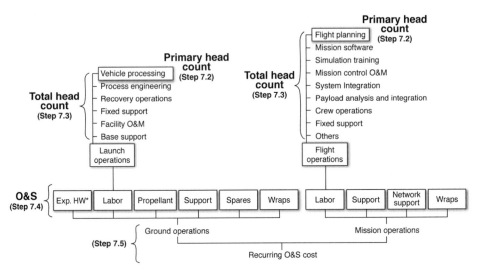

* Exp. HW = Expendable hardware, assumed to be a
   production cost pulled from another area

FIGURE 7-5.   The Contribution of Head Count to Total System Costs. Head counts for vehicle
processing and flight planning drive the total head count, which in turn feeds into the
aggregate cost of the system. Step 7.1, "Describe vehicle and mission concepts", is
omitted in this figure. (O&M is operations and maintenance; O&S is operations and
support.)

TABLE 7-15. Concept Definition for Operations and Support (O&S). The first step in estimating
O&S costs is to describe the concept and vehicle using this table and others that follow.
As shown here, we fill out the table for the first-stage element (first and second stages are
boosters), the upper stage (with the solid kick motor) which contains the payload, and the
total propellant requirements for the launch vehicle. (CTV is crew transfer vehicle; OMV is
orbital maneuvering vehicle; TPS is thermal protection system.)

Concept Description

Booster Definition:

For each booster type:

1) Enter the number of boosters: [1]

2) Enter the type of boosters: Solid: [   ]
                                            Hybrid: [   ]
                                            Liquid: [X]

   a. If solid or hybrid, what type?   Monolithic: [   ]   Segmented: [   ]
   b. Enter the number of engines per [1]
      booster:

3) Are the boosters reusable?   Reusable: [   ]   Expendable: [X]
   a. If so, enter recovery type: Parachute/water: [   ]
                                            Parachute/land: [   ]
                                            Flyback/land: [   ]

TABLE 7-15. Concept Definition for Operations and Support (O&S). (Continued) The first step in estimating O&S costs is to describe the concept and vehicle using this table and others that follow. As shown here, we fill out the table for the first-stage element (first and second stages are boosters), the upper stage (with the solid kick motor) which contains the payload, and the total propellant requirements for the launch vehicle. (CTV is crew transfer vehicle; OMV is orbital maneuvering vehicle; TPS is thermal protection system.)

Concept Description

Upper Stages Definition

For each upper stage:

a. Enter propulsion type:		Solid:	X		
		Hybrid:			
		Liquid:			
b. Enter the number of engines:	1				
c. Are CTV/OMV functions performed		Yes:		No:	X
d. Reusable or expendable?		Reusable:		Expendable:	X

LEO Stage Definition

1) Is there a LEO stage?	Yes:		No:	X
2) Does the LEO stage return?	Yes:		No:	
a. Is so, enter return method:	Parachute/water:			
	Parachute/land:			
	Flyback/land:			
b. Is this LEO stage reusable or expendable?	Reusable:		Expendable:	
3) Is this LEO stage crewed or un-crewed?	Crewed:		Uncrewed:	
4) Is there a main propulsion system (MPS)?	Yes:		No:	
a. How many engines?				
5) The Orbiter has approx. 27,500 tiles and blankets for reusable TPS. What fraction of this, if any, does the LEO stage have?				

Cargo Integration

1) Is there a payload other than man?	Yes:	X	No:	
2) What integration method is used?	Off-line encapsulation:	X		
	Pad encapsulation:	X		
	Payload bay:			

Type	Mass (1000s kg)
Liquid oxygen (LOX):	151
Liquid hydrogen (LH$_2$):	25.2
Methane (CH$_4$):	
Propane (C$_3$H$_8$):	
Kerosene (RP1):	
Nitrogen tetroxide (N$_2$O$_4$):	
Unsymmetrical DiMethyl Hydrazine (N$_2$H$_4$-UDMH):	
MonoMethyl Hydrazine (MMH):	

TABLE 7-16. **Mission Description.** The table helps define the mission parameters in enough detail to facilitate operations and support O&S cost estimation. The events are based on liftoff, engine ignition, cutoff (both stages), and separation for each stage plus jettison of the payload fairing. The table uses a standard design and mission type and splits mission payloads evenly between commercial and NASA launches. (OMS/RCS is orbital maneuvering system/reaction control system; ET is external tank; EVA is extravehicular activity.)

Mission Description			
Mission Profile:		Factor	Factor
Number of events during:			
	Ascent:	8	See Events Table*
	On-orbit	2	See Events Table*
	Descent		See Events Table*
Type of mission:	Standard	X	Unique
Maturity of payload design:	Mature	X	First flight
Trajectory and post-flight analysis will be:	Manual	X	Automated
Crew/Payload Planning:			
Will the vehicle have a flight crew?	No	X	Yes
Average mission length (days):			
Average crew size:			
Mission man-days:		*Mission length x crew size*	
Will there be EVA operations?	No	X	Yes
Mission Model:			
Percent of missions that are:			
Commercial		50	
Civil/NASA		50	
DoD		0	

Events Table

Ascent Maneuvers or Events	On-Orbit Maneuvers or Events
Main engine start	Orbit change OMS/RCS ignition
Booster engine start	Orbit change OMS/RCS cut-off
Liftoff	Alignment to spacecraft separation attitude
Booster engine burnout	EVA attitude adjustments
Main engine cut-off	Spacecraft separation
1st stage separation	Rendezvous with docking platform
Payload module separation	Docking maneuver
2nd-stage engine ignition	Separation from docking platform
2nd-stage engine cut-off	
2nd-stage separation	**Descent Maneuvers or Events**
OMS ignition	Deorbit OMS/RCS ignition
OMS cut-off	Deorbit OMS/RCS cut-off
Insulation panel jettison	Pre-reentry attitude alignment
Payload fairing jettison	Post-reentry attitude alignment
Alignment to spacecraft separation attitude	Parachute deployment
Spacecraft separation	Final approach landing alignment
Upper stage collision avoidance maneuver	Runway landing
Booster or payload module parachute deployment	Splashdown
ET separation	Flotation device deployment

For launch operations, we calculate the personnel head count (HC) for each element in the concept (booster, LEO vehicle, upper stage, cargo, and integrated vehicle) for each applicable vehicle-processing phase (process, stack, integrate, countdown, and recover) by multiplying the number of elements of each type by the HC required to perform the phase activity for the element. Table 7-17 gives guidance on matching the necessary head-count parameters to our concept description, while Table 7-18 provides the basic HC values required to perform each function.

TABLE 7-17.   Vehicle Processing Manpower Requirements by Phase. For each element type, the table identifies the corresponding head-count parameters to compute the vehicle processing head count. Table 7-18 defines the terms and head-count values. Table 7-19 summarizes the results. (R is reusable; E is expendable; other terms are defined in Table 7-18.)

Element Type[1]	Process	Stack	Inte-grate	Count-down	Recovery
Booster-solid monolithic (R)	Mono				PR, GR
Booster-solid segmented (R)	SegP	SegS			PR, GR
Booster-hybrid (R)	SegP	SegS			PR, GR
Booster-liquid (R)	LCBase, 1Eng, XEng		LMate		PR, GR
Booster-liquid/no eng	ET				
LEO crewed (W/engine[2], R, TPS[3])	BLEO, C, 1Eng, XEng, RLEO, TPS		LSI		PR, GR
LEO crewed (W/engine[2], E)	BLEO, C, 1Eng, XEng		LSI		
LEO uncrewed (W/engine[2], R, TPS[3])	BLEO, 1Eng, XEng, RLEO, TPS		LSI		PR, GR
LEO uncrewed (W/engine[2], E)	BLEO, 1Eng, XEng		LSI		
Upper stage solid (on-orbit functions[4], R)	SUS, COF, RUS		SUS		PR, GR
Upper stage liquid (on-orbit functions[4], R)	USB, 1eng, XEng, COF, RUS		USB		PR, GR
Cargo—payload bay	PBP		PBI		
Cargo—on-pad encapsulated			OnEP		
Cargo—off-line encapsulated	OffEP		OffEP		
Integrated vehicle operations (crewed, uncrewed)				CDM, CDU[5]	

NOTES:
[1] Use terms in ( ) only if applicable; e.g., drop PR and GR terms if not reusable.
[2] Engine support head count is (1Eng + ((#of Engines on stage-1) × XEng))
[3] TPS term is adjusted as a percentage of the new concepts TPS coverage relative to Shuttle; e.g., for 60%, (0.6 × 300) = 180.
[4] Performs on-orbit functions, e.g., crew transfer vehicle (CTV) or orbital maneuvering vehicle (OMV).
[5] In some cases (Atlas, etc.), the same personnel that support processing may support countdown, so we enter zero to avoid double bookkeeping.

$$\text{Vehicle processing HC for "n" elements} = \text{sum of}$$
$$[\text{\# of elements of type n} \times (\text{HC/element for processing} + \text{HC for stacking}$$
$$+ \text{HC for integrating} + \text{HC for countdown} + \text{HC for recovery})] \qquad (7\text{-}6)$$

TABLE 7-18. Vehicle Processing Manpower Requirements by Element Type. We use these values to define head counts (HC) for each element identified in Table 7-17. (CTV is crew transfer vehicle; OMV is orbital maneuvering vehicle.)

Element Abbreviation	Head Count per Element	Element Description or Function
1Eng	30	Basic engine processing head count per stage
BLEO	770	Basic head count required for vehicles such as an Orbiter, CTV, etc. performing operations in low-Earth orbit (LEO)
C	62	Head count for a crewed system
CDM	288	Crewed countdown head count
CDU	98	Uncrewed countdown head count
COF	10	Head count for upper stages performing CTV/OMV functions
ET	60	External tank processing head count
GR	100	Glider recovery head count
LCBase	10	Liquid rocket booster and core base head count
LMate	34	Liquid rocket booster mating head count
LSI	230	LEO stage integration head count
Mono	4	Monolithic processing head count
OffEP	9	Head count contribution for off-line encapsulated processing and integration
OnEP	55	Head count to encapsulate and integrate cargo on the launch pad
PBI	136	Head count contribution to integrate cargo into the payload bay
PBP	70	Head count contribution to payload bay processing
PR	17	Parachute recovery head count
RLEO	200	Head count required for reusable LEO vehicle processing
RUS	15	Reusable upper stage head count
SegP	36	Segmented processing head count
SegS	48	Segmented stacking head count
SUS	10	Solid upper stage processing base head count term
TPS	300	Thermal protection system processing head count (Shuttle value)
USB	10	Liquid upper stage processing base head count
XEng	15	Head count for processing each additional engine on a stage

Manpower sharing is sharing staff between activities or elements by having the same people do different tasks based on the operational phase. It makes more efficient use of people and resources. We reflect any assumption of sharing by adjusting the head-count values defined by the model for each element. Table 7-19 shows the estimated staffing required to process our example vehicle.

We can now estimate the number of people required for flight planning in a similar manner, by using Equation (7-7) to calculate the cost of mission operations:

$$\text{Mission operations cost} = \text{cost of labor to support flight operations}$$
$$+ \text{cost of supplies and materials} + \text{cost of network support}$$
$$+ \text{cost of wraps, fees, taxes, and other miscellaneous charges} \qquad (7\text{-}7)$$

TABLE 7-19. Estimated Manpower to Process the Reference Vehicle. The table summarizes the results from using Tables 7-17 and 7-18 by element type and processing phase. The total is the head count required to support vehicle processing. We use this value later to define the total head count to support launch operations as part of defining recurring costs for ground operations. Numbers in this table are based on eight flights per year.

Vehicle Processing Manpower Worksheet Example Results						
Element	Process	Stack	Integrate	Count-down	Recovery	Total/ Element
Booster 1-liquid (1eng) (LCBase + 1eng)	40 (10+30)	-	-	-	-	40
Booster 2-liquid (1eng) (LCBase + 1eng)	40 (10+30)	-	34	-	-	74
Solid upper stage (SUS)	10	-	10	-	-	20
Cargo (OnEP)	0	-	55	-	-	55
Integrated vehicle	-	-	-	98	-	98
Total head count	90	0	99	98	0	287

We derive the *cost of labor to support flight operations* from the number of people required to support flight planning. This primary flight-planning head count comes from the mission description in Table 7-16. We compute flight planning from the number of people needed to design and plan the mission, plus the number required to plan the crew activities and payload handling while on orbit. All these calculations are adjusted by factors that account for the change in management requirements based on the type of organization launching the payload.

Flight planning head count =
[(HC for flight profile design and mission planning) +
(HC for crew or payload activity planning)] ×
(program management, document control, data reporting, testing, etc. factor)   (7-8)

where

HC for flight profile design and mission planning =
{[(10 × ascent phase difficulty level × ln(# events during ascent phase +1))
+ (10 × on-orbit phase difficulty level × ln(# events during on-orbit phase +1))
+ (10 × descent phase difficulty level × ln(# events during descent phase + 1))]
× crewed mission multiplier × payload design multiplier
× automated post-flight analysis (PFA) multiplier
× mission-peculiar multiplier}   (7-9)

Table 7-20 defines the terms in Equation (7-10). The *difficulty level* ranges from 1.0 to 8.0, with values for the reference system equal to 4.5 for the ascent phase, 7.5 for the on-orbit phase, and 6.0 for the re-entry phase. We derive these values from model calibrations using flight-planning histories for the Shuttle and expendable launch vehicles. The *crewed mission multiplier* is 1.0 for uncrewed and 1.9 for crewed flights. The *payload design, automated PFA,* and *mission-peculiar multiplier* values range from 1.0 to 1.2, 1.0 to 0.97, and 1.0 to 1.2, respectively.

The next term in Equation (7-7) is

HC for crew or payload activity planning =
[10 × crew activity difficulty level × ln(# man-days in space + 1)]
× EVA factor   (7-10)

where *man-days in space* = mission length in days × crew size and where the *crew activity difficulty level* is similar to that in the flight planning difficulty level (Table 7-20) with a value range from 1–8. *EVA factor* is 1.2 for a mission with EVA and 1.0 for no EVA. We find the final term in Equation (7-7) using

Program management, document control, data reporting, testing, etc. =
(% commercial payloads × commercial payload factor)
+ (% civil/NASA payloads × civil/NASA payload factor)
+ (% DoD payloads × DoD payload factor)   (7-11)

TABLE 7-20. Mission Operations Factors. This table contains typical factors and ranges needed to estimate the flight-planning head count [Eqs. (7-7) through (7-11)]. (PFA is post-flight analysis; EVA is extravehicular activity.)

Mission Operations Factors					
Function		**Range of Values**			**Example SLaTS System**
Mission Profile:					
Ascent phase difficulty level	Low	1.0	High	8.0	4.5
On-orbit phase difficulty level	Low	1.0	High	8.0	7.5
Descent phase difficulty level	Low	1.0	High	8.0	6.0
Mission-peculiar multiplier	Standard	1.0	Unique	1.2	1.0
Payload design multiplier	Mature	1.0	First flight	1.2	1.0
Automated PFA multiplier	Manual	1.0	Automated	0.97	1.0
Crewed mission multiplier	Uncrewed	1.0	Crewed	1.9	1.0
Crew/Payload Planning:					
Crew activity difficulty level	Low	1.0	High	8.0	6.0
EVA factor	No EVA	1.0	EVA Ops	1.2[*]	1.0
Mission Model:					
Commercial payload factor = 1.0					
Civil/NASA payload factor = 1.4					
DoD payload factor = 1.6					

[*] Represents 50% commercial and 50% NASA.

Staffing for the flight design and crew activity functions depends strongly on the customer. Commercial missions in general don't require the same level of effort for these tasks as, for instance, a DoD or NASA mission. NASA missions normally require many briefings to report the results of analysis, as well as more program management. These add a lot of staff to mission planning for flight design and crew activities. Values for these three types of missions are

- Commercial missions (baseline) = 1.0
- Civil/NASA missions = 1.4
- DoD missions = 1.6

For our example problem, the head count for flight planning equals 176 for the flight profile design and mission planning and 0 for the crew or payload activity planning. The program management factor of 1.2 means the total head count is 211 for launching 50% commercial and 50% NASA payloads at a rate of eight flights per year.

Step 7.3.   Compute Total Head Count

We based the head counts for vehicle processing (VPHC) and flight planning (FPHC)—computed in the previous steps—on a rate of eight flights per year, because this rate was typical when we collected the support data. We must adjust these head-count values to the desired flight rate (R) using

$$VPHC_{Flight\ Rate\ R} = 0.440 \times VPHC_8 \times R^{0.394} \qquad (7\text{-}12)$$

for the vehicle-processing head count and

$$FPHC_{Flight\ Rate\ R} = 0.756 \times FPHC_8 \times R^{0.135} \qquad (7\text{-}13)$$

for the flight-planning head count. (Note: In estimating the O&S costs of conceptual launch systems, we have to be cautious using Eqs. (7-12) and (7-13). They represent Shuttle operations and are usually accurate only for relatively low flight rates—fewer than 17 per year. Above this rate they tend to underestimate the head counts.)

For the example case, the vehicle-processing head count is 287 for eight flights per year and 337 for 12 flights per year. The flight-planning head count is 211 and 223 respectively. From the primary vehicle-processing and flight-planning head counts for launch and flight operations, we use a ratio analysis with historical data as a basis to calculate the HC for other launch and flight-operation activities. The analysis is based on Table 7-21 and the following equation:

$$SOP = Base\_SOP + K1 \times (1 - \exp[K2 \times Flt/yr^b]) \qquad (7\text{-}14)$$

Here the *standard operating percentage (SOP)* represents the fractional contribution of each cost element for the launch and flight-operations head counts respectively. The parameters Base, K1, K2, and b come from the table values. Because the distribution of people varies with flight rate and the type of architecture, we need to base the parameters from Table 7-21 on one of these four types: crewed-reusable, uncrewed-reusable, crewed-expendable, and uncrewed-expendable. The distributions that this equation computes capture the way of doing business in the early 1990s and for that combination of characteristics (manning, reusability, etc.) and flight rate. We need engineering judgment to adjust these values and assess the effects of changing technologies, designs, or support policies. For the example concept, we apply the uncrewed-expendable portion (Set 4) of the table.

Using ratio analysis, we calculate the head counts for the added launch and flight-operations elements, such as process engineering or fixed support for launch operations. First, we compute the total *launch-operations head count (LOHC)* using the *VPHC* for 12 flights per year ($VPHC_{12} = 337$). Then we divide by the fraction that it represents of the total ($SOP = 0.512$) as calculated by Equation (7-13).

$$SOP = Base\_SOP + K1 \times (1-\exp[K2 \times (Flt/yr)^b])$$

$$= 0.3980 + 0.3903 \times (1-\exp[-0.0288 \times 12^{1.0}])$$

$$= 0.512$$

Ratio analysis helps us find the total number of people required to support launch operations: $LOHC_{12}$ equals $337/0.512 = 658$. We then determine the head counts for the rest of the elements in launch operations:

$$HC_{element} = (LOHC_{12}) \times SOP_{Element} \qquad (7\text{-}15)$$

We calculate similarly to determine the *total flight-operations head count*; the result is 606. Table 7-22 shows these results.

Step 7.4. Compute Costs

We multiply the head count for each launch- and flight-operations element by an annualized labor cost. For vehicle processing, this is $98,600; for flight planning, it's $110,900. As required, we add the costs for propellants and other consumables, spares, network support, fees, and contingencies. Table 7-23 shows these costs. We drew the values needed for ground support equipment spares and propellant calculations from the concept definition.

Step 7.5. Show Results

The total recurring operations costs summarized above include ground and mission costs and represent an annual cost for 12 flights per year. The value in FY2003 dollars is $193M annual and $16M per flight. This illustration covers only the recurring cost. We haven't accounted for learning in this example, but we can do so with more parameters using the modeling software. We should find a model that best matches the analysis needs. The recurring O&S cost then equals the annual cost summed over the lifecycle using Equation (7-16):

$$O\&S_{Recurring} = \sum_{i=1}^{n} (Cost/flight)(Learning_i)(Inflation_i)(Flight\ rate_i) \qquad (7\text{-}16)$$

TABLE 7-21. Coefficients (Base, K1, K2, and b) for Standard Operating Percentage (SOP) Equations. This table provides the coefficients to compute the fraction of support personnel required (out of the total) to support each element of launch and mission operations. Four combinations exist for manning and reusability: crewed/reusable (M/R), uncrewed/reusable (UM/R), crewed/expendable (M/EX), and uncrewed/ expendable (UM/EX). (O&M is operations and maintenance.)

Launch Operations								
	Set 1 (M/R)		Set 2 (UM/R)		Set 3 (M/EX)		Set 4 (UM/EX)	
	$K2 = -0.1813$ $b = 0.7882$		$K2 = -0.0593$ $b = 0.9576$		$K2 = -0.1508$ $b = 0.8306$		$K2 = -0.0288$ $b = 1.0000$	
Launch operations	K1	Base	K1	Base	K1	Base	K1	Base
Vehicle processing	0.1211	0.2529	0.3364	0.3690	0.1749	0.2819	0.3903	0.3980
Process engineering	−0.0070	0.0988	−0.0576	0.1064	−0.0196	0.1007	−0.0702	0.1084
Recovery operations	−0.0245	0.0323	−0.0049	0.0065	−0.0196	0.0259	0.0000	0.0000
Fixed support	−0.0050	0.2627	−0.1149	0.2283	−0.0325	0.2541	−0.1424	0.2197
Facility O&M	−0.0280	0.2104	−0.0665	0.1360	−0.0376	0.1918	−0.0761	0.1174
Base support	−0.0566	0.1428	−0.0925	0.1538	−0.0656	0.1455	−0.1015	0.1565

Flight Operations								
	Set 1 (M/R)		Set 2 (UM/R)		Set 3 (M/EX)		Set 4 (UM/EX)	
	$K2 = -0.0426$ $b = 0.9563$		$K2 = -0.0959$ $b = 0.8697$		$K2 = -0.0559$ $b = 0.9346$		$K2 = -0.1093$ $b = 0.8481$	
	K1	Base	K1	Base	K1	Base	K1	Base
Flight operations	0.0759	0.2013	−0.0244	0.3579	0.0508	0.2405	−0.0495	0.3970
Mission software	−0.0464	0.0957	−0.0014	0.0849	−0.0352	0.0930	0.0099	0.0822
Simulation and training	−0.0588	0.0765	−0.0395	0.1410	−0.0540	0.0727	−0.0347	0.0573
Mission control O&M	−0.1295	0.1743	0.0985	0.1410	−0.0725	0.1660	0.1554	0.1327
System integration	−0.0469	0.1786	−0.0922	0.1660	−0.0582	0.1755	−0.1035	0.1628
Payload analytical integration	0.1766	0.0292	0.1300	0.0393	0.1649	0.0317	0.1184	0.0418
Crew operations	0.0538	0.0595	0.0108	0.0119	0.0430	0.0476	0.0000	0.0000
Fixed support	0.0103	0.1043	−0.0747	0.1218	−0.0110	0.1087	−0.0960	0.1262
OTHER	−0.0348	0.0806	−0.0070	0.0161	−0.0279	0.0645	0.0000	0.0000

TABLE 7-22. Operations and Support Cost Estimate for the Reference Mission and Vehicle. We use this table in conjunction with Table 7-23 to summarize the cost elements for ground and mission operations' annual costs. (Dist is distribution; MP is manpower; HC is head count; prop is propellant mass; GSE is ground support equipment.)

Cost Element	Fractional Dist of MP	HC	$M/Yr	
Ground Operations				
Vehicle processing	0.50	337	$33.2	
Process engineering	0.09	58	$5.7	
Recovery operations	0.00	0	$0	
Fixed support	0.18	117	$11.5	
Facility operations and maintenance	0.10	63	$6.2	
Base support	0.13	83	$8.2	
Labor	1.00	658	$64.8	= Operations HC × annual cost factor of $98,600
Supplies and material			$6.5	= Labor cost × operations factor
Propellants			$3.2	= Flight rate × [(prop $/kg) × (kg prop) × (1 + %boil-off)
GSE spares			$0.9	= Cost of nonrecurring GSE* × GSE spares factor
Wraps			$22.7	= (Fee + contingency + government support) × labor cost
Total cost/year for vehicle processing			$98.2	
Cost/flight			$8.2	
Flight planning	0.37	223	$24.7	
Mission software	0.09	53	$5.9	
Simulation and training	0.04	22	$2.4	
Mission control ops and maintenance	0.22	136	$15.1	
System integration	0.10	61	$6.8	
Payload analytical integration	0.11	68	$7.5	
Crew operations	0.00	0	$0.0	
Fixed support	0.07	42	$4.7	
Other	0.00	0	$0.0	
Labor	1.00	605	$67.1	= Operations HC × annual cost factor of $110,900
Supplies/material			$3.4	= Labor cost × operations factor
Network support			$1.1	= (Average hours/flight)† × (flights/ year) × (network support factor)
Wraps			$23.5	= (Fee + contingency + government support) × labor cost

**TABLE 7-22. Operations and Support Cost Estimate for the Reference Mission and Vehicle.
(Continued)** We use this table in conjunction with Table 7-23 to summarize the cost
elements for ground and mission operations' annual costs. (Dist is distribution; MP is
manpower; HC is head count; prop is propellant mass; GSE is ground support
equipment.)

| | Ground Operations | | |
Cost Element	Fractional Dist of MP	HC	$M/Yr
Total cost/year for flight planning			$95.1
Cost/flight			$7.9
Total operations			Note: This excludes the cost of the expendable vehicle
Total cost/year			$193.3
Cost/flight			$16.1

* Obtain "Cost of nonrecurring GSE" from infrastructure computations in Chap. 8, Step 5, *Space
Launch and Transportation Systems* [Larson et al., 2005].
† "Average hours /flight" is the amount of time we need the range and mission networks to support the
flight.

TABLE 7-23. General Rates and Factors. We use these values in Table 7-15 to compute costs.
(LOX is liquid oxygen; LH_2 is liquid hydrogen; CH_4 is methane; C_3H_8 is propane; RP1 is
rocket propellant 1; N_2O_4 is nitrogen peroxide; N_2H_4-UDMH is hydrazine-
unsymmetrical dimethylhydrazine; MMH is monomethyl hydrazine.)

| Propellants | | | General Rates and Factors | | | |
Type	$/kg	Boiloff				
LOX	0.082	74.5%	**$M / (HC* yr)**		**Wraps**	
LH_2	4.40	122.9%	Launch operations	$0.0986	Fee	10%
CH_4	0.734	30.6%	Flight operations	$0.1109	Contingency	15%
C_3H_8	0.516	0.0%	**Supplies and materials**		Government support	10%
RP1	1.22	0.0%	Launch operations	10.0%	Total Wrap Factor	35%
N_2O_4	7.47	0.0%	Flight operations	5.0%	**GSE spares**	
N_2H_4-UDMH	35.30	0.0%	**Network support**		% of Non-recurring	1.17%
MMH	27.20	0.0%	$M/hr	$0.0111		

where n = total years in the lifecycle
      i = year of the lifecycle
      Cost/flight = from analysis ($M/flt)
      Learning = reduction in cost/flight anticipated from launch experience
      Inflation = adjustment for annual inflation
      Flight rate = number of flights per year

Although new launch systems usually have a development or ramp-up period in which the flight rate gradually increases to an operational flight rate, we don't account for this growth here. Nor do we assume any increase in flight rate over the lifecycle. We assume learning has no effect, which is consistent with the low flight rates of launch vehicles. Instead, we assume an operational flight rate of 12 flights per year for the 10-year lifecycle.

$O\&S_{Recurring}$ = $1933 M for the lifecycle, excluding the costs of expendable elements, and inflation.

Step 8.   Evaluate Results

We should now compare the answer in Step 7 with a benchmark value such as the Atlas IIAS in Table 7-13. (Section 7.4.3, Step 8.)

Step 9.   Document Results (Section 7.4.3, Step 9)

7.5   Estimate Cost of Developing the Launch Site's Infrastructure

Development costs for the launch site's infrastructure are a significant part of a new system's lifecycle costs. They include the facilities and ground support equipment (GSE) needed to receive or recover, process, and launch a space vehicle, as well as to support it in flight. The type and number of facilities depend on the mission's design, technologies, support concept, and flight rate. Besides this nonrecurring cost, the launch site's infrastructure also contributes to the recurring cost for maintaining the infrastructure and replacing depleted supply and stock.

To estimate costs of the launch site's infrastructure, we start with the new system's operations concept, which grew out of the logical process required to prepare the vehicle concept for flight, launch it, do the mission, and recover it, if necessary (Section 7.4.3, Step 3). We must

1. Judge effectively based on the design, technologies used, and support concept, plus knowledge of how similar concepts have been supported

2. List the facilities required (we discuss their number later)

3. Compute each facility's costs from the dimensions of the individual elements and integrated concept (using the costing relationships in Table 7-24)

4. Calculate the GSE cost from the facility cost and use it to compute the initial spares for GSE support

Table 7-24 captures the building's "brick and mortar" costs plus that of equipment to support what it does. Among these are the costs to design and develop the systems, as well as a contingency allowance, site inspection and engineering studies, and activation. All values are expressed in FY 2003 dollars, which we must convert to the time frame of interest.

We extracted the cost estimating relationships in Table 7-24 from a modeling analysis conducted at Kennedy Space Center in the 1980s. They come from the Shuttle and Saturn facilities, so they reflect a heavy influence from those concepts and support needs. But we've added notes to help adjust calculations for the required dimensions. We must choose facilities whose functions best match the needs in the new operations concept. Then, using the dimensions of the element or integrated vehicle, we compute the cost of the facility, associated GSE, and initial spares from the cost coefficients for each facility in the operations concept. These facilities mainly support the online processing flow. Although many ancillary buildings at a launch site are important to the support, those in Table 7-24 are of primary interest and generally capture the main facility cost requirements.

TABLE 7-24. **Facility Cost Table.** We use these cost coefficients to estimate the cost of facilities that support new concepts. Estimates are usually based on concept dimensions. (GSE is ground support equipment; COF is cost of facilities.)

Major Facility Type	Cost Coeff FY2003 $/m³	GSE %COF	Initial Spares %GSE	Notes	Suggested for: R- Reusable E- Expendables
Solid rocket motor processing facility	185	95	6.5	1	R, E
Solid rocket motor assembly and refurbish facility	685	278	6.5	1	R, E
Solid rocket motor recovery facility	277	112	6.5	1	R
Solid rocket motor stacking facility	211	55	6.5	1	R, E
Tank processing facility	458	80	6.5	1	R, E
Horizontal vehicle processing facility	590	447	6.5	1, 6	R
Horizontal vehicle maintenance and refurbish facility	293	200	6.5	1, 6	R
Main engine processing facility	241	63	6.5	8	R
Payload integration facility	704	206	6.5	1, 4	R, E
Hazardous payload processing facility	298	57	6.5	1	R, E
Vehicle assembly and integration facility	207	57	6.5	1	R, E
Mobile launch platform (MLP) without launch umbilical tower (LUT)	5670	72	6.5	7	R
Launch umbilical tower for a mobile launch platform	1570	88	6.5	2	E
Launch pad with mobile service structure (MSS)	336	142	6.5	5	E

TABLE 7-24. **Facility Cost Table. (Continued)** We use these cost coefficients to estimate the cost of facilities that support new concepts. Estimates are usually based on concept dimensions. (GSE is ground support equipment; COF is cost of facilities.)

Major Facility Type	Cost Coeff FY2003 $/m³	GSE %COF	Initial Spares %GSE	Notes	Suggested for: R- Reusable E- Expendables
Launch pad with fixed and rotating service structure	538	142	6.5	1	R
Mission control and firing room	1011	1,339	6.5	3	R, E
Landing facility	62	142	6.5	9	R

General: Use element or stack dimensions in determining facility dimensions of length, width, and height.

1. Add about 11 m clearance around and above the concept dimensions for support equipment, cranes, etc.

2. Use stack height plus 9 m. Choose base dimensions based on Saturn type LUT (12.2 m × 12.2 m × 122 m).

3. For Shuttle, 37 m × 27 m × 5 m is usual for a firing room.

4. Shuttle used 3.5 times the payload bay length for payload processing.

5. Used for expendables, based on stack height (SH) plus MLP height. Add 12.2 m to all sides and above SH for MSS dimensions.

6. Include landing gear height if used.

7. Multiply stack dimensions (wingspan, depth, SH) by 2.12, 1.76, and 0.26 respectively to obtain a concept MLP proportional to the STS MLP. Multiply stack dimensions (Dia., Dia., SH) by 4.85, 4.09, and 0.13 respectively to obtain a concept MLP proportional to the Saturn V MLP.

8. A single Shuttle main engine processing cell measures 19.5 m × 11.3 m × 14.0 m. The Space Shuttle main engine dimensions are 4.3 m long × 2.3 m diameter. Multiply the appropriate dimensions by the number of cells desired to obtain facility size.

9. This cost coefficient is per square meter and used for total runway, taxiway, and apron areas. Shuttle runway is 91.4 m wide by 4572 m long.

Information derived primarily from ground-operations cost model (GOCOM).

Cost depends heavily on the number of facilities required, on how long the launch element occupies the facility, and on the necessary flight rate. Computing these values is beyond this book's scope, but Table 7-25 has representative times for successive launches of several ELVs from the same launch pad [Isakowitz et al., 1999]. We use these values to guide our assumptions for the required launch-pad time. For integrate-transfer-launch (ITL) systems, it encompasses readying the integrated vehicle for launch as well as preparing the payload. For build-on-pad (BOP) concepts, it's how long it takes to stack and integrate the vehicle and integrate the payload.

Although we emphasize time-on-pad (TOP) as a driver for the number of launch pads required to support a flight rate, we can apply the same process to other facilities used in processing. We define BOP or ITL as a type of operations concept, but a clear distinction usually doesn't exist. Preprocessing, stacking, and

TABLE 7-25. Representative Times Selected Vehicles Need to Use the Launch Pad. To establish a time-on-pad (TOP) estimate for new launch concepts, select a system with characteristics similar to the new concept and use the TOP value as a starting point. (Cal Days are calendar days; BOP is build-on-pad; ITL is integrate-transfer-launch.)

Launch Vehicles Model	Model	Stages	Boosters	Stack Height Up to _ (m)	Gross Liftoff Mass (×1000 kg)	Payload (kg)	Support Type	Time-on-Pad (TOP) Cal Days
Atlas	I	2	-	44	163	3911	BOP	130
	II	2	-	48	187	4420	BOP	218
	IIA	2	-	48	187	4916	BOP	84
	IIAS	2	4	48	234	5800	BOP	79
Delta II	6925	3	4-9	39	218	2675	BOP	40
	7925	3	4-9	42	232	3450	BOP	54
	7920	2	3-9	42	232	3450	BOP	26
Titan	II	2	-	36	154	2083	ITL	59
	IVA	2	2	57	82	12,000	ITL	115
	IVB	2	2	62	926	14,600	ITL	188
Shuttle		2	2	56	2040	16,400	ITL	25

integrating of some elements occur off the pad. We must estimate the number and mix of facilities the new concept needs to support the desired flight rate. Picking a length of time similar to one in Table 7-25 implies a similar type of launch vehicle and support concept. Assuming TOP values different from the ranges shown implies new technologies or support procedures. To determine the number of launch pads needed to support a given flight rate, we multiply the annual flight rate desired by the assumed time-on-pad (in days) for that concept, divided by 365 days. A similar process applies for the other facilities based on the time they're occupied (plus refurbishment time) and not available for the next element.

Table 7-25 shows the Shuttle's TOP for comparison of a reusable concept. In addition to the 25 workdays on the pad, this well-defined processing flow usually has a dwell time of 84 workdays in the Orbiter Processing Facility and five workdays in the Vertical Assembly Building to integrate with the external tank and solid rocket boosters.

For the example system, which will be built on the pad, the facilities are a launch pad with a mobile service structure for assembly, a launch umbilical tower, and a firing room for launch operations. Based on Table 7-25, we choose a TOP of 60 days for the new concept, as its assembly, number of stages, stack height, and payload characteristics are similar to the Atlas IIAS. The estimate for the example is lower because it uses no solid boosters. Using the required flight rate of 12 flights/year, we need two launch pads to meet this demand ((12 flt/yr × 60 d/flt)/ (365 d/yr)) ≈ 2 pads.

From Table 7-24, we use a cost coefficient of $336 per cubic meter to estimate the cost of a launch pad with a mobile service structure. As the note indicates, we add 12.2 meters of work space on all sides of the vehicle and to the stack height (for an overhead crane) to estimate the size of the structure used to compute the costs: $[(12.2 + 4.6 + 12.2)\ (12.2 + 4.6 + 12.2)\ (55 + 12.2)]m^3 = 57,400\ m^3$, for a cost of $19.3 million each. Because the flight rate requires two pads, the total cost including GSE and initial spares is $75 million in FY2003 dollars.

The cost coefficient for the launch umbilical tower is $1568 per cubic meter based on the Saturn's tower (as the note indicates). Because the new concept is about half the stack height and doesn't support a crewed launch, we assume the tower dimensions chosen for the example concept are 9 m × 9 m × 65 m, resulting in a facility cost of $8.3 million and a total of $24 million for two towers including the GSE and spares.

Again from Table 7-24, we use a cost coefficient of $1011 per cubic meter to estimate cost for the firing room. Based on the Shuttle firing room's size of 36.6 m × 27.4 m × 4.6 m and the fact that this system isn't crewed, we assume for the example concept a smaller room of 18 m × 14 m × 3 m. This concept yields a facility cost of $0.76 million. With the GSE and initial spares included, the cost is $11 million. So we estimate the total nonrecurring cost for O&S infrastructure is $110 million in FY2003 dollars.

7.6   Estimate Total Lifecycle Cost and Assess Cost Risk

Finally, we're ready for Step 6: assembling the lifecycle cost of our SLaTS reference vehicle by aggregating costs for design, development, test, and evaluation (DDT&E); production (from Step 3); operations and support (from Step 4) and facilities (from Step 5). Table 7-26 shows this total lifecycle cost. The DDT&E costs are directly from Table 7-9. Table 7-23 summarizes the O&S costs. The vehicle's total DDT&E cost is $3770 million, plus the facility's nonrecurring cost of $113 million, which gives us a grand total DDT&E of $3880 million. We calculate production cost by multiplying the average unit costs from Table 7-9 by the 120 production units planned, to obtain a total production cost $19,270 million. The operations and support cost from Step 8 of $193 million per year for 12 flights is included in Table 7-26 as a total 10-year cost of $1930 million. All this gives us a 10-year total lifecycle cost estimate for the SLaTS reference vehicle of about $25,100 million.

7.6.1   Time Phasing of Cost

After we estimate the basic building blocks of lifecycle cost, we time-phase these costs using the expected project schedule. If a detailed project schedule were available, we could spread the cost out at lower levels of the WBS. But for our purposes, we allocate by year the costs of the major acquisition cost elements: stage, engine, launch site, and infrastructure, plus O&S. Time phasing often uses a standard beta distribution, which spreads cost given a start date, an end date, and the distribution's shape. We use a distribution with shape parameters that spread

TABLE 7-26. **Summary of SLaTS Reference Vehicle's Lifecycle Costs.** The table gives costs in millions of 2003 dollars.

	Design Development Test and Evaluation	Production Cost (120 units over 10 years)	Operations and Support (120 flights over 10 years)	Totals
Booster	$2240	$12,700	--	$14,940
Second stage	$1200	$5700	--	$6900
Shroud or payload carrier	$330	$870	--	$1200
Vehicle subtotal	$3770	$19,270	--	$23,040
Facilities nonrecurring	$110	--	--	$110
Total DDT&E Cost	**$3880**			**$3880**
Total production cost		$19,270		**$19,270**
Total operations and support cost (10 yrs)	--	--	$1930	**$1930**
Total lifecycle cost				**$25,080**

60% of the cost in 50% of the time. This shows fairly accurately the funding requirements by time for space projects. If we need very accurate spreading, the time periods can be quarters or even months. Here, we spread cost by year.

Figure 7-6 gives the spreading percentages for a 60% cost in 50% time beta distribution for up to 10 periods of time. In our case, this is 10 years, so we use the 10th row to obtain the cost allocation by year.

	Beta Distribution—60% Cost in 50% Time for Schedules of 1 to 10 Periods										
	Number of Periods Over Which to Phase Funds										
Number of Periods	1	2	3	4	5	6	7	8	9	10	Total
1	100.00%										100.00%
2	60.00%	40.00%									100.00%
3	25.41%	62.58%	12.01%								100.00%
4	12.59%	47.44%	35.18%	4.79%							100.00%
5	6.83%	32.36%	39.15%	19.45%	2.20%						100.00%
6	4.29%	21.12%	34.62%	27.97%	10.78%	1.22%					100.00%
7	2.42%	16.09%	26.96%	27.87%	18.77%	7.29%	0.60%				100.00%
8	1.93%	10.66%	22.51%	24.93%	22.83%	12.36%	4.45%	0.34%			100.00%
9	1.16%	7.75%	16.50%	22.16%	23.99%	16.44%	8.93%	2.83%	0.24%		100.00%
10	0.87%	5.96%	13.32%	19.05%	20.83%	18.32%	12.83%	6.63%	2.03%	0.17%	100.00%

FIGURE 7-6. **Cost Spreading for Budgeting Analysis.** Spreading percentages represent a 60% cost in 50% time beta distribution for one to ten periods of time. Start and end dates for the time phasing come from the project schedule.

We should take the start and end dates for the time phasing from the project schedule. The Shuttle required about nine years from authority to proceed to first flight. Because the SLaTS vehicle is expendable and uncrewed, we assume seven years is enough time for total DDT&E. This can be time-phased using the beta distribution percentages from Figure 7-6 for seven years and applying those percentages against the total DDT&E cost of $3880 million from Table 7-27. For example, the first year of DDT&E is 0.0242 × $3880 million = $94 million. We spread production costs over two or three years using the beta distribution (two or three years being the likely production span for one vehicle). This approach increases production costs in the first two or three years before reaching steady state. But we assume production is a steady-state cost of $1927 million per year (1/10 of the total ten-year production cost of $19,270 million from Table 7-27). Similarly, O&S costs are $193 million per year (1/10 of the total ten year O&S cost of $1930 million).

TABLE 7-27. Time Phased Lifecycle Cost (2003 Dollars in Millions). These numbers can be used to establish budget requirements for the reference vehicle. Negative years are before launch. (DDT&E is design, development, test and engineering; O&S is operations and support.)

Year	−7	−6	−5	−4	−3	−2	−1	1	2	3	...	10	Total
DDT&E	$94	$624	$1046	$1082	$728	$283	$23				...		$3880
Production								$1927	$1927	$1927	...	$1927	$19,270
O&S								$193	$193	$193	...	$193	$1930
Total	$94	$624	$1046	$1082	$728	$283	$23	$2120	$2120	$2120	...	$2120	$25,080

7.7   Conduct Cost-Risk Analysis

Cost estimating isn't an exact science. Uncertainty arises from three basic sources (only two of which risk analysis can deal with). First, the cost estimating tools we use contain statistical uncertainly. Parametric models are derived using a best-fit CER through the scatter of historical data. There's some variance between actual and predicted cost in any equation (a statistic called the standard error measures this variance). Second, cost estimating suffers from uncertainty in the input variables. Even with a perfect cost model, bad input yields bad output. After all, we estimate costs while the project is still conceptual. Most inputs for cost estimating are uncertain. Third, cost estimates must contend with "state of the world uncertainties" or "unknown unknowns." No pro forma cost estimate and risk analysis can—or should be expected to—account for random, unpredictable events.

A cost-risk analysis is meant to inform managers and decision makers about how uncertainties in cost estimating and project definition may affect the project's cost. A risk analysis can be as simple and unsophisticated as expert judgments by an experienced project manager, chief engineer, or a team. It also can be a relatively straightforward sensitivity analysis, in which we vary the cost estimate's major parameters that we believe are uncertain over their expected reasonable range to see how they affect cost. Usually, sensitivity analyses vary only one parameter at a time. Project cost simulations can gauge the effects of numerous variables that change simultaneously.

Figure 7-7 shows one approach for cost-risk analysis, in which we "dollarize" safety, technical, schedule, and cost risk using CERs similar to a deterministic cost estimate. But the cost-risk process accounts for uncertainty in the CERs, project schedule, and technical specifications. We define these uncertainties with probability distributions and model the entire cost estimating process in a Monte Carlo simulation to build a statistical sample of cost outcomes over a large number of simulations.

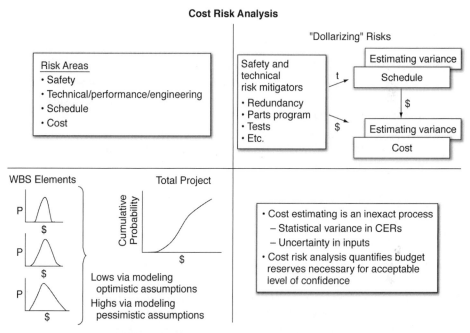

FIGURE 7-7.   **Cost-Risk Analysis Using CERs in a Deterministic Estimate.** Here, we define uncertainties with probability distributions and model the entire cost estimating process in a Monte Carlo simulation to build a statistical sample of cost outcomes over many simulations.

We order the sample of cost outcomes and plot it as a probability distribution. Plotted (or tabulated) cumulatively, the results give a probability for any given cost over the outcome range. This approach gives decision makers the probability that a given cost won't be exceeded, so we can select a cost based on perceived risk. Most analysts recommend using at least the 50% confidence level, and many say a higher confidence level (perhaps 75%) is more prudent because analysis, at best, still excludes unknown unknowns.

Summary

The uncertainties in cost estimating—especially for such complex entities as space systems—entail the need for cost-risk analysis. Still, we do have reliable and tested cost models to help determine what the project outlay will be, even over a lifecycle spanning many years. This chapter has discussed at length how the space systems cost model predicts several cost elements for a space launch and transportation system (SLaTS). The judicious use of cost estimating tools, applied to a well-defined operational concept, reduces the number and severity of nasty budgetary surprises.

References

Chesley, Julie, Wiley J. Larson, Marilyn McQuade, and Robert J. Menrad. 2008. *Applied Project Management for Space Systems.* New York, NY: McGraw-Hill Companies.

Department of Defense. 1992. *Cost Analysis Improvement Group, Operating and Support Cost-Estimating Guide.* Washington, D.C.: Office of the Secretary of Defense.

Isakowitz, Steven J., Joseph P. Hopkins Jr., and Joshua B. Hopkins. 1999. *International Reference Guide To Space Launch Systems.* Washington, DC: American Institute of Aeronautics and Astronautics.

Larson, Wiley J., Robert S. Ryan, Vernon J. Weyers, and Douglas H. Kirkpatrick. 2005. *Space Launch and Transportation Systems.* Government Printing Office, Washington, D.C.

Morris, W. Douglas et al. 1995. *Defining Support Requirements During Conceptual Design of Reusable Launch Vehicles.* Presented at the AIAA Space Programs and Technologies Conference, Sept. 1995, Paper No. AIAA 95-3619.

NASA. 1993. *Transportation Systems Analysis. Operations Cost Model,* User's/Analyst's Guide (NAS8-39209). Hunstville, AL: NASA Marshall Space Flight Center.

NASA. 1997. *A Guide for the Design of Highly Reusable Space Transportation.* Cape Canaveral, FL: NASA Space Propulsion Synergy Team, Kennedy Space Center.

NASA. 2000. *The NASA Reliability and Maintainability Model (RMAT 2001), User's Manual (NAS1-99148).* Hampton, VA: NASA Langley Research Center.

NASA. 2001. *Spaceport Concept and Technology Road mapping, Investment Steps to Routine, Low Cost Spaceport Systems.* Final Report to the NASA Space Solar Power Exploratory Research and Technology (SERT) Program, by the Vision Spaceport Partnership, National Aeronautics and Space Administration, John. F. Kennedy Space Center, and Barker-Ramos Associates, Inc., Boeing Company, Command and Control Technologies Corp., and Lockheed Martin Space Systems, JSRA NCA10-0030.

Nix, Michael, Carey McCleskey, Edgar Zapata, Russel Rhodes, Don Nelson, Robert Bruce, Doug Morris, Nancy White, Richard Brown, Rick Christenson, and Dan O'Neil. 1998. *An Operational Assessment of Concepts and Technologies for Highly Reusable Space Transportation.* Hunstville, AL: NASA Marshall Space Flight Center.

Squibb, Gael, Daryl Boden, and Wiley Larson. 2006. *Cost-Effective Space Mission Operations.* New York, NY: McGraw Hill.

Chapter 8

Technical Risk Management

Robert Shishko, *Jet Propulsion Laboratory*

Technical risk management is an organized means of collecting and controlling project risks. This chapter describes the technical risk management process and summarizes some methods and techniques the risk manager may employ. Technical risk management uses both qualitative and quantitative methods. Probability and statistics play a fundamental role in the quantitative methods. A previous grounding in these disciplines is useful in understanding the quantitative methods described in this chapter.

Technical Risk Management Basics

Risk has many definitions in space systems engineering. Some have said that risk is the "potential for a negative future reality that may or may not happen." [DoD, 2000]. NASA NPR 8000.4 states: "Risk is characterized by the combination of the probability that a program or project will experience an undesired event...and the consequences, impact, or severity of the undesired event, were it to occur." [NASA, 2002 (1)]. This chapter uses the latter as a working definition and therefore treats risk as a vector consisting of a likelihood measure and a consequence measure.

Risk is an inherent part of space missions. Risk management is important because the programs and projects by which we seek to accomplish some space objective have limited resources (usually budget) or must meet strict return-on-investment (ROI) constraints. In a resource-constrained environment, a project has a better chance of succeeding if it deals with its risks systematically and objectively. Even then we have no guarantees that the project will remain viable, but technical risk management encourages or even forces the project team to collect and analyze risk information and control the risk within its resources. (For an in-depth treatment of risk management from a project management viewpoint, see Chapter 15 of *Applied Project Management for Space Systems* (APMSS) [Chesley et al., 2008.)

So in technical risk management, the project risk manager and project team need to ask first: "What are the project risks and what is the overall project risk position?" and second: "What can and should we do about the risks?" The activities involved continue throughout the lifecycle and this chapter explains them in greater detail.

To determine the project's risk position, the risk manager works to identify and collect the "knowable" risks in the project. This step is intended to be comprehensive, covering all types of risks, project elements, and project phases. The risk manager, project manager, and some project team members supported by others with specialized skills in risk assessment work together to assess the collected risk items. Ideally, this activity results in consistent and useful assessments, finding the "tall poles" of the risk tent. Some of these tall poles may warrant in-depth analysis to confirm or reject likelihood or consequence estimates. For the overall risk position, the risk manager aggregates the risks into an integrated picture of the project's risk exposure and communicates that picture to stakeholders.

In deciding what can and should be done, the risk manager and risk owners try to identify, before the risk is realized, fruitful mitigations and other pro-active actions to reduce the risk exposure. As part of this thinking ahead, they also determine if and when to trigger such actions. Lastly, they track the risk and the effects of any mitigation actions to see if these are working as intended. Throughout, the risk manager communicates to stakeholders any changes in risk exposure.

Risk as a Vector

To identify and characterize a project risk in terms of its likelihood and consequences, we plot the risk in a three-dimensional space. One axis represents the likelihood (probability) and the other represents the consequence (e.g., dollars, reduced mission return). Unfortunately, we may not know either of these with certainty. The real-world complexity of risks coupled with the practical limitations of even the best engineering models results in some uncertainty in one or both measures. To a risk manager, a risk item looks like Figure 8-1 (left).

When we apply risk reduction mitigations and they work, the risk likelihood, consequences, or both are lowered. The uncertainty may also show less of a spread. To a risk manager, a successful risk reduction action looks like Figure 8-1 (right).

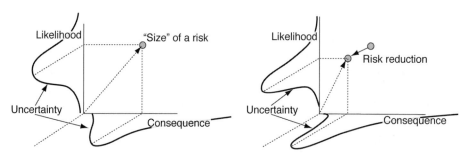

FIGURE 8-1.   **Risk as a Vector.** The size of a risk is characterized by its likelihood and consequence, though considerable uncertainty may exist in both measures. Uncertainties are plotted in the third dimension as probability distributions. The left shows the size of the risk before risk reduction actions are taken. The right shows how the size of the risk (along the axes) and its uncertainty decrease (narrower probability curves) after such actions.

Types of Risks

Categorizing risks is useful if it communicates something important about the source, timing, and consequences of the risk and possible mitigations. Table 8-1 shows some categories, but there appears to be no formal taxonomy free of ambiguities. One reason for this is that one type of risk can often be traded for another, for example, exchanging a schedule risk for a cost risk.

TABLE 8-1.   **Risk Types and Examples.** Most risk items involve more than one type of risk.

Risk Type	Example
Technical performance risk	Failure to meet a spacecraft technical requirement or specification during verification
Cost risk	Failure to stay within a cost cap
Programmatic risk	Failure to secure long-term political support for program or project
Schedule risk	Failure to meet a critical launch window
Financial or market risk	Failure of demand to materialize at the specified price
Liability risk	Spacecraft de-orbits prematurely, causing damage over the debris footprint
Regulatory risk	Failure to secure proper approvals for launch of nuclear materials
Operational risk	Failure of spacecraft during mission
Safety risk	Hazardous material release while fueling during ground operations
Supportability risk	Failure to resupply sufficient material to support human presence as planned

It's often useful to distinguish between implementation risks and mission risks. Implementation risks deal with getting to the launch pad with a spacecraft that meets the technical performance requirements, on time and within budget. Mission risks deal more with mission return after launch and whether it meets stakeholder expectations.

Sources of Risks

While each space project has its unique risks, the underlying sources of risks include the following:

- Technical complexity—many design constraints or many dependent operational sequences having to occur in the right sequence and at the right time
- Organizational complexity—many independent organizations having to perform with limited coordination
- Inadequate margins and reserves
- Inadequate implementation plans
- Unrealistic schedules
- Total and year-by-year budgets mismatched to the implementation risks
- Over-optimistic designs pressured by mission expectations
- Limited engineering analysis and understanding due to inadequate engineering tools and models
- Limited understanding of the mission's space environments
- Inadequately trained or inexperienced project personnel
- Inadequate processes or inadequate adherence to proven processes

Overview of Technical Risk Management Process

Figure 8-2 illustrates the risk management process for a typical NASA robotic mission. The remainder of this chapter provides a more detailed description of each step and high-level examples that suggest some best practices. Table 8-2 shows where these steps are discussed.

8.1   Prepare a Technical Risk Management Strategy

In this task we craft our overall risk management philosophy, decide how to implement it in the technical effort (i.e., the risk management strategy), and document that strategy in the risk management plan.

The project manager articulates the project's risk management philosophy after consulting the stakeholders. This philosophy is important, because it affects the project's implementation approach, design reference missions, engineering designs (e.g., whether single-point failures are allowed), and the system operational concept (e.g., whether the system is monitored continuously).

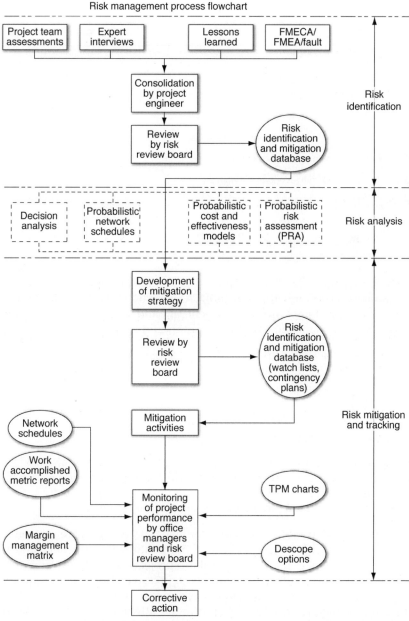

FIGURE 8-2.   Risk Management Process Flowchart. This summarizes the project risk management activities, methods, and interfaces discussed in this chapter. The dashed lines for risk analysis indicate options; the analyst may choose one or more of the techniques shown. (FMECA is failure modes, effects, and criticality analysis; FMEA is failure modes and effects analysis; TPM is technical performance measurement.)

TABLE 8-2.   **Risk Management Process.** These steps are for managing risk in a space systems engineering project. [NASA, 2007]

Step	Description	Where Discussed
1	Prepare a technical risk management strategy	Section 8.1
2	Identify and document technical risks	Section 8.2
3	Conduct technical risk assessments	Section 8.3
	Qualitative assessment methods	
	Quantitative assessment methods	
4	Determine an initial risk handling approach	Section 8.4
5	Plan risk mitigations and triggers	Section 8.5
6	Monitor risk status	Section 8.6
7	Implement risk mitigations	Section 8.7
8	Capture technical risk management products	Section 8.8

To help managers develop a risk management philosophy, NASA uses a risk classification scheme with four levels [NASA, 2004 (1)]. The following criteria apply to space systems projects in general, and may help determine how aggressive the risk management philosophy should be:

- Priority (criticality to long-term strategic plans)
- Acceptable risk level
- National significance
- Mission lifetime (primary baseline mission)
- Estimated project lifecycle cost
- Launch constraints
- In-flight maintenance feasibility
- Alternative research opportunities or re-flight opportunities

Clearly, NASA projects involving human space flight are critical to the organization, and fall into the category of high national significance and project lifecycle cost. NASA's risk tolerance in these projects is very low. Consequently, the risk management philosophy is highly aggressive, and system requirements (e.g., for fault tolerance) are levied on the design to avoid or reduce potential risks. A project's risk management philosophy boils down to how much the project (reflecting the stakeholders'—either government or commercial investors— preferences) is willing to spend, not just in dollars, but in other factors like schedule and technical margins, to protect against unacceptable outcomes. The degree of risk aversion is a key descriptor of any risk management philosophy.

Even so, risk-averse design practices cannot eliminate all risks. We still need a technical risk management strategy. This need translates the received philosophical direction into an action plan. It identifies the technical risk management products to be created, by whom (at least notionally), on what schedule, with what degree of depth, and against what standards. The work breakdown structure and mission operations phases (launch, proximity operations, entry, descent, and landing) usually serve as organizing structures for these products. Generally projects with a greater lifecycle cost spend more on technical risk management activities than smaller ones, although the percentage of the lifecycle cost may be less.

Whatever the resources assigned to technical risk management, the risk manager should create a strategy that maximizes its benefits (or return-on-investment). We document the strategy formally in the project risk management plan. This plan also describes other essential technical risk management activities:

- An overview of the technical risk management process (see technical risk management basics at the beginning of this chapter)
- The organization within the project responsible for technical risk management activities
- The communication paths for reporting risk issues and status to review boards
- The risk tracking and reporting frequency and methods to use during formulation and implementation (including during operations)
- General risk handling responses based on the risk classification (Section 8.3)
- Training needed for technical risk management personnel
- Integration with other cross-cutting processes such as technology insertion, risk-based acquisition, earned-value measurement, IT security, and knowledge capture

8.2   Identify and Document Technical Risks

This task identifies project risks from a variety of sources and documents them with enough information for future resolution.

8.2.1   Identify the Risks

A useful starting point for identifying risks is to review project documentation, including:

- Project requirements documentation
- Work breakdown structure (WBS)
- Proposal documents
- Project plan, including schedule and cost baselines
- Architecture and operations concept descriptions
- Design documentation

As the availability and maturity of such documentation changes over the project lifecycle, the risk manager may have to look elsewhere early in the project. Other documentation, outside of the current project, may also provide insights, including lessons learned databases, problem and failure reports for heritage hardware, previous projects' significant-risk databases, and risk templates (essentially a structured series of suggested issues culled from many past projects).

The most effective source of risks, however, is from individual interviews or brainstorming sessions with key project personnel and independent experts. Experienced members of the project team are an essential source for identifying technical risk. First, we should assemble a good significant-risk list (SRL) from the top down. (Many good engineers like to focus on detailed design issues too soon.) It's a good idea to look at major trade studies—completed or planned—and initial system and subsystem sizing analyses. A top-down approach should also consider the number of organizational interfaces, including supply-chain relationships.

Later in the project cycle, when acquisition contracts have been let, an *integrated baseline review* (IBR) may be a risk identification source. An IBR is a joint assessment by the contractor and government to verify the technical content and the realism of the related performance budgets, resources, and schedules. It should provide a mutual understanding of the risks inherent in the contractor's plans and underlying management control systems.

The proactive identification of risks continues throughout the project lifecycle. During implementation, monthly status assessments, project independent reviews, development test data, and unanticipated events provide new sources of risks. Specialized studies such as a hazard analysis, failure modes and effects analysis (FMEA), and failure modes, effects, and criticality analysis (FMECA) may also point to new risks in the engineering designs.

8.2.2   Develop the Significant Risk List (SRL)

The SRL is a key product in technical risk management. It represents the expert judgment of risk owners, risk managers, and project manager on risks that pose potential threats to the project and require further attention. The SRL goes beyond a mere risk description. At a minimum, it includes the following attributes for each identified risk:

- Title or ID number
- Description or root cause
- Possible categorizations
  - System or subsystem
  - Cause category (technology, programmatic, cost, schedule, etc.)
  - Resources affected (budget, schedule slack, technical margins, etc.)
- Owner

- Assessment
  - Implementation risk
    - Likelihood
    - Consequences
  - Mission risk
    - Likelihood
    - Consequences
- Mitigation options
  - Descriptions
  - Costs
  - Reduction in the assessed risk
- Significant milestones
  - Opening and closing of the window of occurrence
  - Risk change points
  - Decision points for implementing mitigation effectively

In filling out the SRL, we must list as many specifics as we can to describe the concern and identify the context or condition for the risk. We should try to envision and describe the feared event as specifically as possible. We need to describe the effect of the event in terms that other project team members understand. This may involve thinking ahead to characterize the effort (time, money, mass, power, etc.) needed to recover from the event (implementation risk) and the penalty in mission success that would result (mission risk). The risk manager assigns ownership of a risk to the individual or organization responsible for its detailed management. This is often the one that first identified the risk.

Sometimes the risk management plan includes an initial SRL, i.e., a list of those risks that have been identified early in the systems engineering process. It's good practice to maintain a single, configuration-managed data repository for the project's significant risks that is kept current. That's why on large projects the operational SRL is kept in some form of database or dedicated risk management tool.

8.3   Conduct Technical Risk Assessments

The SRL contains information attributes that may not be available until we've completed technical risk assessments. These characterize each risk in terms of the severity of consequences and likelihood of occurrence of the risk event. Severity may be measured by performance, cost, and schedule effects on the project. Likelihood is best quantified as the probability of occurrence.

Early in the project cycle, we may have only qualitative assessments of a risk's severity and likelihood and of the effect of a proposed mitigation. Later, when more design information is available, quantitative assessment may be possible. Quantitative assessment is more difficult and generally more costly, but also more valuable for determining how to allocate scarce mitigation resources and project reserves. It's also sometimes the only way to verify risk-related requirements.

8.3.1   Qualitative Assessment Methods

A simple, yet effective, method of qualitative risk assessment relies on expert judgment (generally from the project team, but sometimes independent experts) to classify risks using relative scales for severity of consequences and likelihood of occurrence. The consequences scale ranges from "minimal" to "mission failure" for mission risk or "exceeds project termination limits" for implementation risk. The likelihood scale ranges from "very low" to "very high", though sometimes we supplement these descriptions with actual probability ranges. One difficulty with relative scales is that verbal descriptions are subjective. Different people interpret them differently.

Project practices differ, but five levels for consequences and five levels for likelihood are common. Tables 8-3 and 8-4 list levels for consequences and likelihood, respectively. Figure 8-3 shows a 5 × 5 risk classification matrix. In this risk assessment method, each risk is placed in one of the boxes and classified as high, medium, or low risk.

TABLE 8-3.   Consequence Levels. This table provides definitions to guide the classification of each risk in terms of consequences. We use separate definitions to distinguish mission risk from implementation risk.

Level	Mission Risk Level Definitions
5	Mission failure
4	Significant reduction in mission return
3	Moderate reduction in mission return
2	Small reduction in mission return
1	Minimal or no impact to mission

Level	Implementation Risk Level Definitions
5	Overrun budget and contingency, cannot meet launch
4	Consume all contingency, budget, or schedule
3	Significant reduction in contingency or launch slack
2	Small reduction in contingency or launch slack
1	Minimal reduction in contingency or launch slack

For high or medium risks, the systems engineer may decide whether the project can benefit from a quantitative analysis. If an unplanned quantitative analysis is deemed advantageous and the resources are available, we amend the risk management plan accordingly.

TABLE 8-4.   **Likelihood Levels.** This table classifies risk—here, cost risk—in terms of likelihood. For other types of risk, we assign different values to the levels.

Level	Likelihood Definitions	Probability (Pr)	Qualitative Description
5	Very high	Pr ≥ 70%	Better than 2:1 odds for
4	High	70% > Pr >50%	More likely than not
3	Moderate	50% ≥ Pr >30%	Less likely than not
2	Low	30% ≥ Pr >1%	Better than 2:1 odds against
1	Very low	Pr ≤ 1%	Very unlikely

Tools like 5×5 matrices and likelihood tables have limitations. They allow only a few subjective levels, which must cover probabilities from zero to one. They also pose problems with their imprecise language. Qualitative terms like "low", "medium", and "high" mean different things to different people, and the meaning changes depending on the context. For example, Table 8-4 provides typical values for likelihood definitions of cost risk, perhaps that of a cost overrun. We routinely accept a 30% probability of a cost overrun—regrettable, but that's the reserves target for NASA projects. Acceptable mission risks, however, are considerably lower, and for human spaceflight, a loss-of-mission risk of 0.5% (1 in 200) may be unacceptably high.

8.3.2   Quantitative Assessment Methods

Based on early, qualitative classifications of risks, we determine which risks warrant additional analysis to refine the severity of consequences and the risk's probability. Several engineering methods are available. In this section we describe the following methods[*] with some examples of their application to space systems:

- Decision analysis and decision trees
- Probabilistic network schedule analysis
- Cost risk analysis
- Probabilistic risk assessment (PRA)
- Availability, maintainability, and supportability analysis

For example, at the end of Pre-phase A, FireSAT engineers and managers assessed the top five risks to the project and formed the risk matrix shown in Figure 8-4.

---

[*] Common to these methods is Monte Carlo simulation techniques [Morgan et al., 1990].

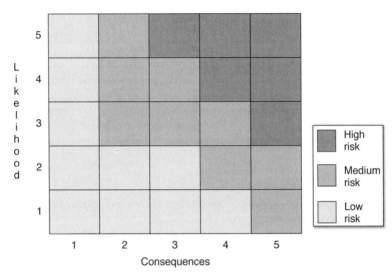

FIGURE 8-3. **A Typical 5×5 Risk Classification Matrix.** Risks are subjectively assigned a level of consequences and a level of likelihood using expert judgment. Each is classified as high, medium, or low risk based on the box it's in.

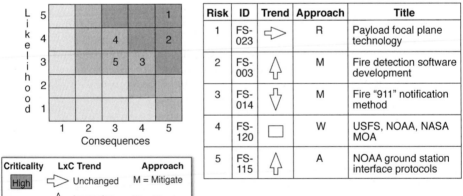

Risk	ID	Trend	Approach	Title
1	FS-023	⇨	R	Payload focal plane technology
2	FS-003	⬆	M	Fire detection software development
3	FS-014	⬇	M	Fire "911" notification method
4	FS-120	▢	W	USFS, NOAA, NASA MOA
5	FS-115	⬆	A	NOAA ground station interface protocols

Criticality	LxC Trend	Approach
High	⇨ Unchanged	M = Mitigate
Med	⬆ Increasing (worsening)	W = Watch
	⬇ Decreasing (improving)	A = Accept
Low	▢ New since last report	R = Research

FIGURE 8-4. **A 5×5 Risk Classification Matrix for FireSAT in Phase A.** These top five risks for FireSAT are the focus of the risk mitigation planning for the project. (USFS is United States Forest Service; NOAA is National Oceanic and Atmospheric Administration; MOA is memorandum of agreement.)

Decision Analysis and Decision Trees

Decision analysis helps us deal with a complex set of uncertainties by using the divide-and-conquer approach common to much of systems engineering. We divide a complex uncertainty into simpler ones, which we then treat separately. The decomposition continues until either hard information is brought to bear or expert judgment functions effectively. We represent it graphically as a decision tree (Figure 8-5). The tree branch points (nodes) represent either decision points or chance events. Endpoints of the tree are the potential outcomes (consequences).

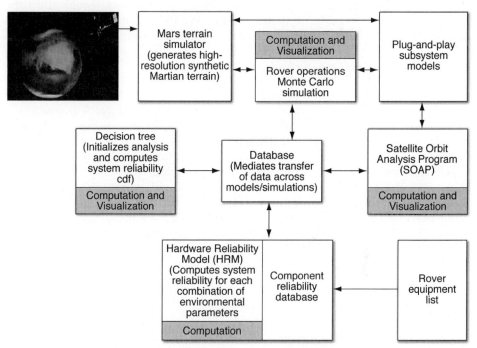

FIGURE 8-5.   A Decision Tree for Sojourner Surface Operations. This decision tree was constructed using commercial software to capture formally the uncertainty in the atmospheric opacity and surface roughness at the Mars Pathfinder landing site. Project scientists supplied the probabilities. A limited probabilistic risk assessment (PRA), using hardware failure models, calculated the outcomes.

In most applications, especially when the consequences are monetary, we assign dollar values to these outcomes. From the probabilities assigned at each chance node and the dollar value assigned to each endpoint, we derive the distribution of dollar values (consequences) for the tree. A decision tree allows a systematic enumeration of uncertainties and encoding of their probabilities and consequences.

This technique was part of a quantitative risk assessment of the Mars rover, Sojourner [Shishko, 2002]. In that assessment, we faced major uncertainties regarding the conditions the rover would find when it reached the Martian surface. These uncertainties concerned the atmospheric opacity, which had implications for anticipated near-surface temperatures, and the rock size distribution in the immediate vicinity of the Mars Pathfinder lander. Would the frequency distribution of sufficiently large rocks impede rover movements and would temperatures be too low for the rover to survive its mission? We captured these uncertainties in the decision tree shown in Figure 8-5 and encoded probabilities. While the environmental conditions and the attendant uncertainties were of interest by themselves, our concern was to estimate how such conditions would affect the ability of the rover to perform its full mission. We did a limited probabilistic risk assessment (PRA), described below, to deal with this loss-of-mission issue.

Probabilistic Network Schedule Analysis

This method estimates a probabilistic completion date for a complex set of interrelated activities in a project's integrated master schedule. A probabilistic network schedule allows us to describe each activity's duration using a probability density function (PDF), usually a triangular distribution so that only three points are needed—minimum, maximum, and most likely durations. (A beta distribution is also common.) We use this information to determine, for example, the chances that a project, or any set of activities in the network, will be completed by a given date. In this probabilistic setting, however, we may not have a unique critical path. Some risk managers have also cited difficulties in eliciting meaningful input data, especially in large network schedules involving hundreds of activities. A simpler alternative to a full probabilistic network schedule is to focus on a Monte Carlo simulation of activity durations along the deterministically computed critical path. While computationally much easier, this method may not catch all schedule risks.

Cost Risk Analysis

Several methods of estimating a project's cost risk have been proposed and vigorous debate continues regarding their efficacy. The objective of a cost risk analysis is a credible project cost S-curve—that is, the cumulative distribution function for the project cost without reserves (Figure 8-6). The S-curve provides far more information than a single cost number and helps us choose a defensible level of reserves. The curve depends on the cost estimating methodology we use and how much augmenting risk information the cost analyst secures within the allotted time and money.

Cost Risk Analysis for Parametric Estimation. Any cost estimating relationship (CER) estimated value has some uncertainty associated with the statistical properties of the CER; these uncertainties are not indicators of the inherent project risks. By the time the project does a parametric estimate, some analytical work has probably been done in the form of a conceptual study or a

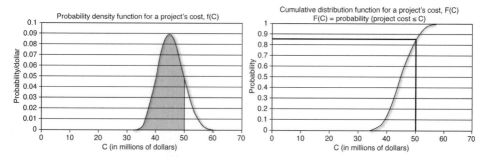

FIGURE 8-6.   **The Project Cost S-Curve.** A project cost probability density function (PDF) is shown on the left and its cumulative distribution function (CDF) or S-Curve on the right. The line on the CDF indicates the 85th percentile of the estimated cost distribution corresponding to the shaded area on the PDF.

design reference mission. But we have yet to understand the detailed technical and schedule or programmatic risks. As a proxy for these risks, we commonly place probability distributions on the continuous inputs (such as mass) in the estimating relationship, and use Monte Carlo simulations to develop the cost S-curve.

These probability distributions are often subjective. The usual source for them is the project team, though the risk manager should be aware of the potential for advocacy optimism here. Any probability elicitation and encoding should follow established protocols and methods such as those in Morgan et al. [1990].

Cost Risk Analysis for Analogy Estimation. Even with analogy estimation, we can capture cost risk and build a project cost S-curve. As with parametric estimation, we often do analogy estimation before we understand the detailed technical and schedule or programmatic risks. In analogy estimation, each estimator, usually a discipline expert with substantial project experience, scales an analogous project's actual cost, accounting for changes in requirements, technology, and other project implementation factors. As a proxy for project risks, we represent each scaling factor by a subjective probability distribution, thus turning a point estimate into a probability distribution. We then use Monte Carlo simulation to develop the cost S-curve. As with any subjective probability elicitation, we should use established protocols and methods.

Cost-Risk Analysis for Grass-Roots Estimation. A cost-risk analysis for grass-roots estimation requires an understanding of the sources of cost risk. A thorough risk analysis of the project should have already identified those elements of the WBS that have significant technical and schedule or programmatic risk. These risks typically arise from inadequacies in the project definition or requirements information, optimistic hardware and software heritage assumptions, assumptions about the timeliness of required technological advances, and overestimating the performance of potential contractors and other implementers. Two methods are available for performing a cost risk analysis for a grass-roots estimate. We should identify both the method and the analysis data.

In the first method, we elicit a cost distribution from the individuals responsible for each WBS element identified as a significant risk. Common cost distributions used in this elicitation are triangular, beta, truncated normal, or lognormal. In practice, we usually ask for three points: minimum, maximum, and most likely costs, and convert that information into the parameters of the pre-selected distribution. A Monte Carlo simulation combines these individual distributions into a project cost S-curve. Some risk managers assume a positive covariance between costs in certain WBS elements, which has the effect of increasing the variance of total project cost [USAF, 2007].

In the second method, worst case (actually, 95th percentile) costs are elicited instead for each WBS element identified as a significant risk. These values are conditional on the proposed budget (without reserves), performance attributes, and schedules specified in the grass-roots estimates. To obtain the conditional cost S-curve, they are combined, as in the first method, using Monte Carlo simulation. This method, however, is based on different behavioral assumptions and uses a different mathematical approach (a constrained Wiener process). This process, a type of random walk, simulates cost growth over T periods using the stochastic equation shown here.

$$dC(t) = \mu\, C(t)\, dt + \sigma C(t)\, dw \qquad\qquad (8\text{-}1)$$

$$\text{subject to } dC(t) \geq \mu\, C(t)\, dt \text{ for } t \in [0,\, T]$$

where
$C(t)$	= predicted WBS element cost at time t in year t dollars
μ	= inflation rate per period
σ	= WBS element volatility parameter
T	= WBS element duration
dw	= random variable that is normally distributed with mean zero and variance dt

Each WBS element has a characteristic volatility parameter (σ) derived from the 95th percentile elicitation and the element's duration [Ebbeler et al., 2003]. Because the cost growth process is stochastic, we have to do many runs for each WBS element to generate a cost PDF like the one shown in Figure 8-6 (left).

Probabilistic Risk Assessment (PRA)

A PRA seeks to quantify the likelihood—and the uncertainty in that likelihood—of formally defined end states (outcomes) in complex engineering systems. These end states usually involve low-probability but high-consequence or catastrophic events such as the loss of a crew, mission, or spacecraft.

Probabilistic risk assessment is a structured quantitative analysis methodology that draws upon probability theory, reliability analysis, and decision analysis. The focus is on understanding the contributors to system risks beyond just hardware failures so that project decisions are risk-informed and risk reduction mitigations and resources are more wisely applied.

The principal effort in PRA is developing scenarios, which are series of key events. Each scenario begins with an initiating event, which propagates through the system, and ends with a particular outcome. Non-quantitative failure modes and effects analyses (FMEAs) often serve as precursor analyses for a PRA. In developing scenarios, a PRA analyst also attempts to capture any subtle interactions among system elements (hardware, software, and operator) that might together lead to undesired end states. Scenarios are developed and documented in a variety of diagrams and formal PRA computational tools such as the Systems Analysis Programs for Hands-on Integrated Reliability (SAPHIRE) and the Quantitative Risk Assessment System (QRAS). Diagrams help organize the information needed for the formal computational tools in a way that's "human-readable" so that the underlying logic may be verified.

Probabilistic risk assessments come in a variety of "flavors," depending on the scope and nature of the problem, the time, resources, and data available. NASA characterizes these flavors as full, limited, and simplified [NASA, 2004 (2)]. In the following paragraphs we describe applications of these three types of PRAs.

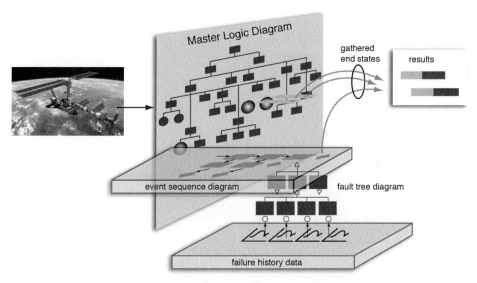

FIGURE 8-7.   The International Space Station (ISS) Probabilistic Risk Assessment (PRA) Top-down Logic. The architecture of the ISS PRA is depicted, showing the flow from the master logic diagram (MLD) through the event sequence diagrams, fault trees, and component reliability data. Though not shown in the figure, the event sequence diagrams (ESDs) are translated into event trees for computational data entry. Event trees are essentially decision trees with chance nodes only.

International Space Station (ISS). The PRA conducted for the ISS provides a particularly good didactic example of a full assessment [Smith, 2002]. The principal end states of concern were those involving the loss of the station, loss of crew, and evacuation of the station resulting in a loss of mission (LOM). Figure 8-7 shows the top-down logic of this model. The master logic diagram (MLD) represents a decomposition of the ISS into functions and systems needed for a working station. At the bottom are initiating events, each of which kicks off an event sequence diagram (ESD). One such ESD for the Russian CO_2 removal assembly, Vozdukh, is illustrated in Figure 8-8.

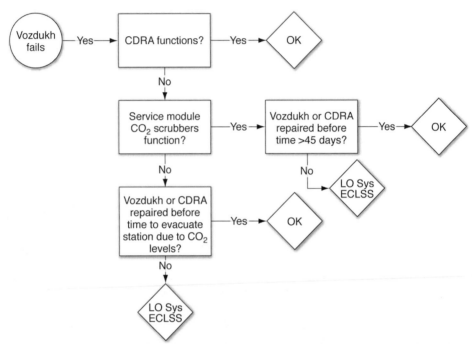

FIGURE 8-8.   **Event Sequence Diagram (ESD) for Vozdukh Failure.** In this ESD, the Vozdukh assembly failure is the initiating event. Other events depicted involve whether backup systems (CDRA and Service module CO_2 scrubbers) work and whether spares or maintenance resources to repair either Vozdukh or backups are available in time to avoid the loss of the environmental control and life-support system (LO Sys ECLSS). (CDRA is Carbon Dioxide Removal Assembly.)

For each key event (called pivotal events in this application) in the ESD, a fault tree (FT) diagram is associated to help determine the event's likelihood of occurrence. At the bottom of each FT are the basic hardware, software, and human failure events, whose probabilities of occurrence come from actual or estimated failure data. Because of the uncertainty in such failure data, that uncertainty is allowed to propagate (via Monte Carlo simulation) up to the top-level PRA results.

The likelihoods of occurrence and their uncertainty for the end states of concern for the ISS were computed and aggregated using the SAPHIRE software. The principal value in building the logical and data structures for performing a PRA is the ability to exercise it in trade studies and other "what if" scenarios. Although it takes a great deal of effort, once built, the PRA model and supporting data become part of the risk management process for the remainder of the project lifecycle.

Mars Rover Surface Operations. A full PRA in the style described above was performed for the entry, descent, and landing of the Mars Pathfinder spacecraft. A PRA for the surface operations of its principal payload, the Sojourner rover, required a more limited approach focused on the major mobility and electronic hardware items and the risks identified by the project team.

The first step was to identify a quantified performance objective for the Sojourner rover during its surface mission. A performance requirement for the rover's total travel distance during the mission was set at 100 meters of geodesic (straight-line) travel. The project-level systems engineer working with a risk manager established the hardware and software configuration of the rover in the analysis, the Mars site to be explored, and with the help of the mission designers, the surface mission start time.

Next, the risk manger and project scientist established a probabilistic description of the environmental conditions at the selected landing site. They used the decision tree tool to elicit the scientist's beliefs about surface, atmospheric, and near-surface thermal conditions, and their probabilities. The scientist used whatever hard data were available from actual observations. The results of this step were a set of quantitative values and concomitant probabilities describing the surface terrain (including rock-size frequency distribution, average slope, and surface roughness), atmospheric opacity, and diurnal temperature minimums and maximums (including an estimate of their statistical dispersion). Atmospheric opacity and diurnal temperatures are causally connected, so these values and probabilities were elicited conditionally. Figure 8-5 shows part of the user interface of the decision tree tool used in the elicitation.

One of the PRA's uncertainties was how far the Sojourner rover would have to travel on the odometer to meet the 100-meter goal. The actual travel distance is affected by the rover's physical size, navigation and hazard avoidance algorithms, and the rock-size frequency distribution. This question was important because actual travel distance was the main failure driver for the rover's mobility system. (The failure driver was really the wheel motor revolutions, which is nearly proportional to distance traveled.)

To produce an estimate of this random variable, a simulation was performed in which a virtual Sojourner rover, complete with actual flight software, was run on computer-generated patches of Martian terrain. This was no small feat since it required specially developed parallel processing software running on the Caltech supercomputer. The characteristics of the synthetic Martian terrain could be parametrically varied to match the different values represented in the decision tree. For each terrain, the simulation captured the actual travel distance. By randomizing the initial location of the virtual rover relative to its target, and running the simulation many times, the risk manager generated enough data to produce a Weibull probability density function (PDF) that described the actual travel distance needed to complete the 100 meters geodesic distance. Figure 8-9 shows the resultant PDFs for two Martian terrains.

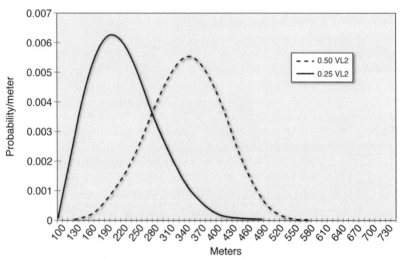

FIGURE 8-9. **Results of the Sojourner Surface Operations Simulations.** The two curves represent different terrains; the dashed line, a terrain with 50 percent of the rock-size frequency of the actual Viking Lander 2 site; and the solid line, a terrain with 25 percent. Each probability density function was produced from multiple simulations of a virtual Sojourner moving across a computer-generated terrain.

The results of each simulation (time, distance traveled, and on-off cycles) were also passed to another model, the hardware reliability model (HRM). For each simulation run, the HRM computed the likelihood of no critical failures (system reliability). It based its calculations on data provided by the simulation of failure drivers and on component-level reliability data provided by reliability engineers. The HRM user interface served mainly to input these component-level data, namely the failure modes for each component, the associated failure driver, failure density functional form (exponential, lognormal, Weibull, etc.), and the parameters of the distribution. Data collection was a challenge for the reliability engineers, since quantitative data of this sort are very scarce.

To complete the limited PRA, the risk manager used the original decision tree to merge the system reliability results with the probability estimates for each combination of Mars environmental parameters. Gathering successful end states resulted in a cumulative distribution function (CDF) for Sojourner's reliability at the Mars Pathfinder landing site. The PRA results in Figure 8-10 capture the uncertainty in the estimate.

FireSAT. The methodology for full and simplified PRAs is the same, but the latter usually contains a reduced set of scenarios designed to capture only the major (rather than all) mission risk contributors. A simplified PRA may be

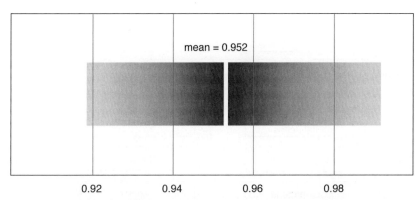

0.92   0.94   0.96   0.98

FIGURE 8-10.   Estimated Reliability of the Sojourner Rover at the Mars Pathfinder Landing Site. A limited probabilistic risk analysis (PRA) was performed to obtain the reliability of the rover at 100 meters of geodesic travel. The PRA used a Monte Carlo simulation of the rover traversing Mars terrains combined with component reliability models and a decision tree to determine the uncertainty in the estimate. The shaded band approximates the 75 percent confidence interval.

appropriate early in the project when less detail design data are available and as a precursor to a full PRA later. During trade studies, for example, it may provide the systems engineer with some early quantification of the top-level risks inherent in alternative designs. The next few paragraphs illustrate a simplified PRA for the FireSAT mission and describe some of the issues involved.

One FireSAT success criterion requires the constellation of satellites to be operational for at least five years, with a goal of seven years. The simplified PRA master logic diagram (MLD) for FireSAT, which addresses the probability of failing to meet that criterion, might look like the one in Figure 8-11.

Understanding the risks due to the space environments depends on scenarios in which micro-meteoroid and orbital debris events, charged-particle events, and solar proton events lead to the loss of a satellite. We analyze each of these events to determine the likelihood of their occurrence using models (some standardized) of such phenomena[*]. The systems engineer needs to validate these models to be sure that they apply to FireSAT and to account for the FireSAT orbit (700 km circular, 55° inclination), spacecraft shielding, and parts types.

We analyze spacecraft hardware reliability early in the project based on subsystem block diagrams showing the degree of redundancy. And we estimate subsystem failure rates by applying parts reliability and uncertainty estimates from several standard databases. The systems engineer may wish to augment random failure rate estimates by taking into account infant failures, which occur

---

[*] In the case of solar proton events, these models generally take the form of predictions for the probability of a particular fluence (particles/m^2) during the period of solar activity maxima [Larson and Wertz, 1999].

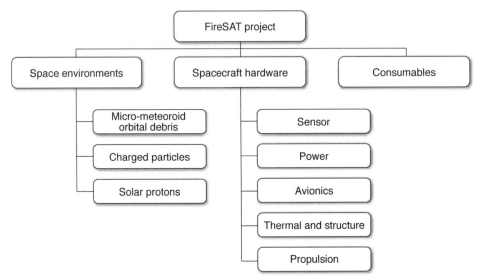

FIGURE 8-11.   **Master Logic Diagram (MLD) for the FireSAT Project.** A simplified probabilistic risk assessment (PRA) does not include risks that are deemed less significant. Here, software failures and operator mistakes do not appear in the MLD. Later in the project cycle, these risk sources should be included in a full PRA.

early in operations, and wear-out failures, which occur with greater likelihood as the system ages.

The exhaustion of consumables (primarily propellant) is another risk to the FireSAT mission success. The systems engineer can determine the likelihood of such scenarios from assumptions made in developing the project's ΔV budget. For example, what assumption was made regarding atmospheric density during FireSAT's operational period? What is the probability that the actual density will exceed the assumed density and how does that translate into propellant consumption rates? The latter depends on the FireSAT spacecraft effective frontal area, orbital parameters, and phase in the solar cycle.

To illustrate the results of a simplified PRA, we assume that the estimated five-year loss of mission (LOM) probability from space environments effects is 7×10^{-4} ±15%, from spacecraft hardware failures 1×10^{-3} ±10%, and from exhaustion of consumables 5×10^{-5} ±10%. Then using the laws of probability, assuming stochastic independence for convenience, the overall ten-year LOM probability is approximately 1.75×10^{-3} ±8%, or roughly one chance in 570. It is perhaps more interesting to view these results in terms illustrated in Figure 8-12, which shows the relative contribution of each top-level factor in the MLD. To the systems engineer, the simplified PRA suggests that the spacecraft hardware reliability and space environments effects contribute over 97% of the risk. We should confirm this result by further analyses as better design information becomes available.

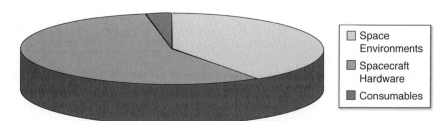

- Space Environments
- Spacecraft Hardware
- Consumables

FIGURE 8-12.   Loss of Mission (LOM) Contributors. The figure shows how the top-level factors in the master logic diagram contribute to the total LOM probability for one particular project architecture and spacecraft design. Changes in the design or orbit parameters affect the relative contributions as well as the total probability.

Availability, Maintainability, and Supportability Analysis

As humans expand their permanent presence in space, we need to address new risks regarding space system availability, maintainability, and supportability. While PRAs tend to focus on high-consequence mishaps, these longer-term system risks demand other quantitative analysis methods that deal with reparable systems with periodic, but uncertain, resupply. Availability models relate system design (reliability and repair parameters) and integrated logistics support (ILS) characteristics to the probability that the system is operational[*]. Maintainability models focus on the ability of maintenance resources (spares, shop equipment, and repair personnel) to keep pace with the demand for corrective and preventive maintenance. We usually use them to estimate the total maintenance effort, or the amount of some critical maintenance resource, needed to support a particular level of system availability. We may conduct this analysis in conjunction with an optimal level-of-repair determination. Supportability models encompass aspects of both and may incorporate additional features of ILS, such as echeloning spare parts inventories and maintenance capabilities and the responsiveness of transportation services to operational sites.

System reliability block diagrams and component reliability and repair parameters such as mean time to failure (MTTF) and mean time to repair (MTTR) are basic data for all these types of models. The usefulness of such models is in quantifying the effects of providing various levels of support resources in terms of the risks to crew and mission. These risks are due to the stochastic nature of failures, uncertain reliability and repair parameters, and unanticipated (but not unpredictable) operational events. Some models deal with a single supportability resource at a single operational site, and some deal with multiple resources at multiple sites. Here we

---

[*] Instantaneous availability $A(t)$ is formally defined as the probability that the system is performing its function at a prescribed time, t. Average availability in the interval $[t_1, t_2]$ is defined as $(t_2 - t_1)^{-1} \int_{t_1}^{t_2} A(t)dt$. Steady-state availability is defined as $lim_{t \to \infty} A(t)$ when that limit exists. Analytic expressions for these availability functions can be derived for simple systems that can be modeled as Markov chains with constant hazard and repair rates.

describe two NASA-sponsored models—one analytic and one a discrete event simulation—representing examples at both ends of this spectrum.

LMI (Logistics Management Institute) Logistics Assessment Model for Human Space Missions. This model is a single-echelon sparing-to-availability model. It computes the minimum orbital replacement unit (ORU) spares needed to achieve a given level of system end-of-mission availability, $A(T)$, within a mass, volume, or cost constraint. The model provides the risk manager with the entire availability versus resource curve, so we can determine the system availability effects of incremental changes in the resource constraint. The model is compact enough to write down its equations and discuss its solution.

The end-of-mission system availability is a function of the quantity of various ORU spares, s_i, carried with the spacecraft. The availability relationship is shown as Equation (8-2). The availability function is maximized subject to a resource constraint, Equation (8-3). The resource constraint could be, for example, the total mass of the selected ORU spares. The quantity EBO_i is the number of expected back orders for ORU type i, which is simply computed under the assumption made in the model of a Poisson failure process, as in Equation (8-4). The Poisson distribution parameter is $\lambda_i T$, where T is the mission duration and q_i is the total number of applications of ORU type i.

$$\max A(s_1, s_2, \ldots, s_m) = \prod_{i=1}^{m} A_i = \prod_{i=1}^{m} \left(1 - \frac{EBO_i(s_i, \lambda_i T)}{q_i}\right)^{q_i} \qquad (8\text{-}2)$$

$$\text{subject to } R(s_1, s_2, \ldots, s_m) \le \overline{R} \text{ and } s_i \ge 0 \qquad (8\text{-}3)$$

$$\text{where } EBO_i(s_i, \lambda_i T) = \sum_{x = s_i + 1}^{\infty} (x - s_i) p(x, \lambda_i T) \qquad (8\text{-}4)$$

We find the solution by first taking the natural log of the availability function, transforming the product into a weighted sum, and then applying the Kuhn-Tucker Theorem for constrained optimization. The solution requires for each ORU type the calculation of the left-hand side of Equation (8-5), which is approximated by the right-hand side. If the total mass of the spares is the resource constraint, then the denominator is just the mass of ORU type i, m_i.

$$\frac{\partial \ln A / \partial s_i}{\partial R / \partial s_i} \cong \frac{\ln A_i(s_i + 1) - \ln A_i(s_i)}{m_i} \qquad (8\text{-}5)$$

We find the optimal spares set by ranking ORU selections from highest to lowest value according to the value of the right-hand side of Equation (8-5). The cut-off ORU is determined when the constraint mass is reached. The resulting availability then is computed from Equation (8-2) [Kline, 2006].

We can embellish this model by adding features that allow indentured parts, usually shop replaceable units (SRUs), to be included in the set of potential spare parts in the availability function. This allows presumably lighter and smaller SRUs to be carried on human space missions rather than their parent ORUs. Repairs using SRUs, however, may engender increases in crew maintenance time that are not valued in the model, so the optimality of the solution is necessarily local. Other sparing-to-availability models have been proposed that also allow for multiple echelons of spares inventories [Sherbrooke, 2004].

SpaceNet. This model is a discrete event simulation of an interplanetary supply chain, developed at MIT and JPL [de Weck et al., 2006]. The scope of the simulation is comprehensive, from launch to Earth return, and models exploration from a logistics perspective. SpaceNet simulates individual missions (e.g., sortie, pre-deploy, or resupply) or campaigns (sets of missions). It allows evaluation of alternative exploration scenarios with respect to mission-level and campaign-level measures of effectiveness (MOEs) that include supply chain risk metrics.

The basic building blocks of the simulation are nodes, elements, and supplies, along with two concepts that tie these together: the time-expanded network and processes for movement through the time-expanded network. We explain each of these below. Collectively, these building blocks allow a complete description of the demand and the movement of all items in a logistics scenario.

- *Nodes*—Nodes are dynamic spatial locations in the solar system. Nodes can be of three types: surface locations, orbits, or Lagrange points.

- *Supplies*—Supplies are any items that move through the network from node to node. Supplies include all the items needed at the planetary base, or during the journeys to and from the base. Examples include consumables, science equipment, surface vehicles, and spares. To track and model the extraordinary variety of supplies that could be required, they are aggregated into larger supply classes. Each supply class has a unique demand model.

- *Elements*—Elements are defined as the indivisible physical objects that travel through the network and in general hold or transport crew and supplies. Most elements are what we generally think of as "vehicles," such as the Orion Crew Exploration Vehicle and various propulsion stages, but they also include other major end items such as surface habitats and pressurized rovers.

- *Time-expanded network*—Nodes in the simulation are linked to other nodes to create a static network. The time-expanded network adds the "arrow of time" by allowing only those trajectories that reflect astrodynamic constraints.

- *Processes*—Five processes are modeled in SpaceNet: each has its particular parameters (e.g., ΔV) that create relationships among the other building blocks and time:

  - *Waiting*—remaining at the same physical node

- *Transporting*—moving to a new physical node along an allowable trajectory
- *Proximity operations*—rendezvous/docking, undocking/separation, and transposition of elements
- *Transferring*—transferring crew and supplies from one element to another co-located element
- *Exploration*—science and exploration EVA operations at a node

Risk-based MOEs in SpaceNet are of two types: process-associated and network-associated. Each process above can fail in a variety of ways initiated by hardware, software, or operator failures and the consequences may indeed be catastrophic. One example is the proximity operations mishap at Mir*. Identifying initiating events and accident sequences and calculating the probability and consequences of accidents is properly the domain of a PRA. SpaceNet doesn't perform that function, but it aggregates PRA results from the process level to the mission and campaign level.

Network-associated risks are hard to capture in a traditional PRA. SpaceNet addresses such risks with two MOEs based on similar considerations facing terrestrial supply chains. These MOEs address two significant questions. The first is: "How well does an interplanetary supply chain strategy fare under uncertain demand estimates?" Uncertain demand estimates can be the result of poor demand parameter forecasts such as the Poisson distribution parameter, $\lambda_i T$, in Equation (8-4) above. Is the interplanetary supply chain strategy, which may involve pre-positioning, carryalong safety levels, resupply, and in-situ resource utilization (ISRU) reliance, robust enough to handle less likely, but plausible, future scenarios?

The second question is: "How well does an interplanetary supply chain strategy fare when parts of the chain fail?" Is the supply chain strategy robust enough to cope with launch, proximity operations, or landing failures, or missed or delayed resupply missions? To address these complex risk issues pragmatically, SpaceNet employs new techniques from computational, multi-scenario simulation approaches and scenario-based planning. From hundreds of runs covering widely divergent future scenarios, SpaceNet computes robustness metrics (e.g., maximum regret) to identify those candidate strategies that perform satisfactorily under conditions of multiple uncertainties. More information is available on decision making when the wide range of plausible future scenarios makes it difficult to employ traditional decision analysis methods [Bankes, 2002 and Kouvelis and Yu, 1997].

---

* The proximity mishap of Mir happened in June 1997, when there was a collision between a Progress resupply spacecraft and the Mir space station, damaging a solar array and the Spektr module.

8.4   Determine an Initial Risk Handling Approach

This task involves assigning an appropriate initial risk handling approach. Four risk handling alternatives are:

- *Mitigate*—We mitigate risk by applying methods to eliminate or reduce its likelihood or consequences. We do this through engineering, schedule, or budgetary changes to designs, processes, or procedures; adding parallel approaches (hedging) to supplement the project's main approach; or transferring the risk through contractual mechanisms and insurance.

- *Accept*—The project manager accepts the risk and documents the rationale

- *Research*—This alternative includes collecting additional information and doing analyses to refine the likelihood and consequences estimates, and sometimes, to reduce the uncertainty surrounding these estimates. These analyses provide a better basis for future risk handling decisions.

- *Watch*—The project manager decides not to take immediate action, but to track, monitor, or watch the trends and behavior of risk indicators over time

8.5   Plan Risk Mitigations and Triggers

In this task, the risk owner thinks ahead about how to mitigate a risk item. Risk mitigation planning is the proactive search for cost-effective ways to deal with a risk that the project is unwilling to accept. Table 8-5 shows some ideas. The result of this task is a plan for each such risk that includes a set of actions with triggering events or dates.

Sometimes the best risk mitigation option is obvious, but many risks have several options with different costs and potential effectiveness. In these cases, the risk owner should conduct trade studies to determine the best mitigation approach. Quantitative models like PRAs are very useful in these situations, when they're available. Even when they're not, the risk owner still needs to estimate how much each option will mitigate the risk and what each option costs (in terms of dollars, mass, power, and other scare resources). To complete the trade studies, the risk owner must select the option that best balances the marginal reduction in the size of the risk against its marginal cost.

Experienced space systems engineers and designers already carry a certain amount of margin to cover normal uncertainty in the final requirements and in the design process. When we develop contingency mitigation plans to cover risks beyond that, the project as a whole must carry reserves to meet the project's needs, while providing the customer with reasonable assurance that the budget will not be exceeded.

We should tie project reserves to project-specific risks and to the cost of contingency risk mitigation plans. A fixed percentage of estimated project costs (used in some cost models) to cover contingency risks meets neither the needs of the customer nor of the project. A careful analysis of all contingency risk mitigation costs

TABLE 8-5. **Typical Mitigations.** This table shows some actions to consider in risk mitigation planning. (FTA is fault tree analysis; FMEA is failure modes and effects analysis; CPU is central processing unit.)

Category	Subcategory	Mitigations
Safety risks	Ground handling	• Develop and test procedures • Certify that personnel have adequate training • Minimize or limit overtime
	Ground environments	• Proof-test facilities, ground support equipment, test equipment
Space environment risks		• Provide adequate radiation design margins and shielding • Provide adequate micro-meteoroid and orbital debris shielding • Have robust thermal margins • Have adequate mechanical factors of safety
Operational risks	In-flight space vehicle reliability	• Conduct FTAs, FMEAs, reliability analyses
	Space vehicle performance degradation	• Include adequate redundancy, fault protection software • Provide robust performance margins for power, memory, CPU cycles, ΔV, telecommunications • Test, test, test
	Ground systems degradation	• Provide more training for personnel • Verify that measurement units are consistent • Hold rigorous review and approval for commands; do error checking for sequences and flight rule violations • Perform spacecraft performance trend analysis • Plan backup downlink stations; leave time for redundant downlink playback
Cost risks	Cost estimation	• Verify and validate all cost models used; understand the estimate's prediction interval • Use several models or independent estimates • Perform Monte Carlo cost risk analysis
	Design margins	• Provide mass, power, and other design margins early
	Workforce application	• Control the tendency to maintain "standing armies" (large groups of workers) during development and operations
	Monitor cost trends	• Use earned value management to track the estimate at completion (EAC)
Schedule risks	Schedule estimation	• Know and schedule all deliverables; understand integrated schedule risk; monitor the critical paths
	Schedule margins	• Plan schedule margins at key points in the project cycle (e.g., major reviews, system tests)
	New technology insertion	• Provide hedges for new technology developments; monitor new technology
	Spares application	• Buy enough spares to support early testing

is an important step in deriving appropriate project reserves and reducing overall project cost risk. In practice, we activate contingency mitigation plans and commit risk resources when the triggering events associated with these risk items occur. However, there's always a danger of running out of project reserves before we've handled all risks (retired, mitigated, or accepted).

The effect of committing some project reserves against a contingency risk is a shift in the cost S-curve to the right and a reduction in its spread (variance), as shown in Figure 8-13. As a general principle, the cost of a contingency risk mitigation plan should exceed the expected marginal reduction in risk only if the project is risk-averse (the insurance principle).

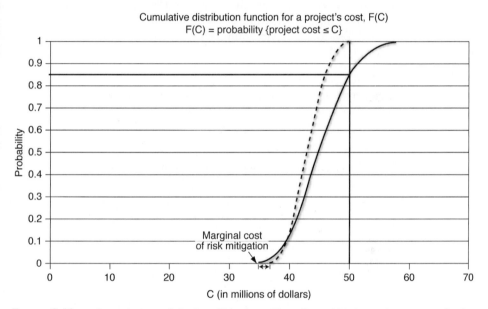

FIGURE 8-13.   Commitment of Project Reserves. The effect of implementing a cost-effective contingency risk mitigation plan is a change in the project cost S-curve that improves the project's overall risk position. Here, the project has committed reserves equal to the marginal cost of a risk mitigation, and the probability of meeting the $50M budget has increased from about 85% (the original solid line) to nearly 100% (the dashed line). In general, projects typically face complex situations in which a more detailed calculation is needed to determine if the overall risk position (probability and consequences) has improved.

8.6   Monitor Risk Status

This task involves monitoring the status of each technical risk periodically. We track the status of each risk item to determine whether the risk has changed, and in particular, whether the current risk status requires that we activate mitigation plans. The risk manager reports to the project manager when a threshold has been triggered.

8.6.1   Risk Early Warning Tools

Several recognized systems engineering tracking tools and techniques provide early warning signals for technical risk management. These include:

- Implement technical performance measurement
- Perform earned value management
- Track cost and schedule against plan
- Track receivables and deliverables against plan

Technical Performance Measurement (TPM). This tool gives insight into the performance, physical, and supportability attributes that determine whether the delivered systems will meet performance requirements. A rapid rise in the measured value of a TPM such as spacecraft mass (as in Figure 8-14) after the critical design review may signal an impending breach of a margin requirement. The risk manager may then generate a new risk item or increase the size of an existing risk. If the breach actually occurs, the project manager may activate mitigation or descope plans to restore the TPM to an acceptable level, thereby reducing or eliminating the risk item.

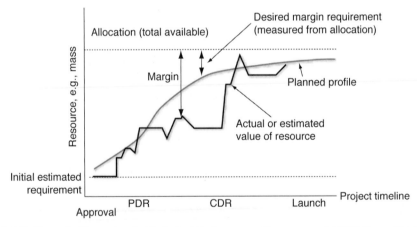

FIGURE 8-14.   An Example of a Technical Performance Measurement (TPM) Reporting Chart. This chart shows a typical plot of a TPM such as spacecraft mass as a function of project maturity. The chart contains the measured (or estimated) value of the mass, its allowed upper bound (allocation), the margin requirement, and the available margin. When the available margin is less than the margin requirement, a mitigation action may be triggered.

Earned Value Management (EVM). Earned value management complements technical performance measurement by giving insight into the project's progress against the baseline cost and schedule plan. In DoD and NASA, contracts above a certain dollar threshold require EVM.

The cost variance (CV) at a given point in the project is the difference between what the project budgeted for the work elements completed to that point (BCWP) and the actual cost for that work (ACWP). A large negative CV (as in Figure 8-15) suggests, in the absence of corrective actions, a project cost overrun. Schedule variance (SV), is the difference between what the project budgeted for the work elements accomplished up to that point (BCWP) and what the project budgeted for the work elements planned up to that point (BCWS). A large negative SV suggests, in the absence of corrective action, a project schedule slip.

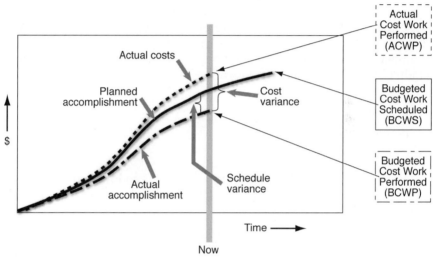

FIGURE 8-15.   Earned Value Management (EVM). EVM combines actual cost, schedule, work planned, and work completed data to produce the estimate at completion (EAC).

One important use of EVM is to calculate the estimate at completion (EAC). We can compute this at any point in the project. The appropriate formula depends on the underlying reasons associated with any variances. If a variance is due to a one-time event, such as an accident, then Equation (8-6) applies. If a variance exists for systemic reasons, such as a general underestimation of schedule durations or a steady flow of rework, then we assume the variance will continue to grow over time and Equation (8-7) applies.

$$EAC = BCWS_{\text{at completion}} + ACWP - BCWP = BCWS_{\text{at completion}} - CV \qquad (8\text{-}6)$$

$$EAC = BCWS_{\text{at completion}} \left(\frac{ACWP}{BCWP}\right) = BCWS_{\text{at completion}} \left(1 - \frac{CV}{BCWP}\right) \qquad (8\text{-}7)$$

A growing number of liens, action items, requirements redefinitions, or other problems that increase the difficulty of future work, may cause the EAC to grow faster than Equation (8-7) implies. We should not use a rote formula as a substitute for understanding the underlying causes of variances.

Actuals against plans. Earned value management requires specialized software and considerable time and effort to develop appropriately detailed sets of work elements. Smaller projects may not have the resources to do full EVM. One approach for a low-cost project is to track actual costs against planned costs and to track actual receivables against the planned schedule of deliverables in a cumulative receivables/ deliverables chart [Jorgensen et al., 1998].

8.6.2   Reporting the Project Risks

Here the risk manager prepares periodic technical risk status reports for the project manager and project team. These reports usually include the status of the top risk items in the significant risk list (SRL), highlighting any major changes. During formal project technical reviews, risk item status reports usually contain more in-depth information. Tools for communicating the project's risk position and individual risk item status include:

- 5×5 risk matrix
- Risk database or SRL
- Problem/failure reports (P/FRs)
- Critical items list (CIL)

Figure 8-16 shows the 5×5 risk matrix on the left and the top ten project risk items on the right. Each of these risk items is tagged with the current risk handling approach described in Section 8.4.

8.7   Implement Risk Mitigations

Implementing a risk mitigation plan is the responsibility of the risk owner. The risk owner first reviews the plan to ensure that it will have the desired effect, then provides revised budgets, schedule, and technical performance inputs to the project manager.

We must also monitor the results of implementing the mitigation plan. This task involves: 1) determining how effective the actions are in stopping or reversing adverse trends or in recovering from a realized risk, and 2) tracking the resources needed to complete the mitigation. The risk manager reports to project management on both of these.

8.8   Capture Technical Risk Management Products

This task results in the capture of all technical risk management products as part of the project archives. Main products include the risk management plan, significant risk lists over the project lifecycle, quantitative models (e.g., probabilistic risk assessment) developed or used and analysis results; risk mitigation plans; risk action items; and risk review board and project manager decisions. Finally, we should capture technical risk lessons learned in a formal manner to serve future projects.

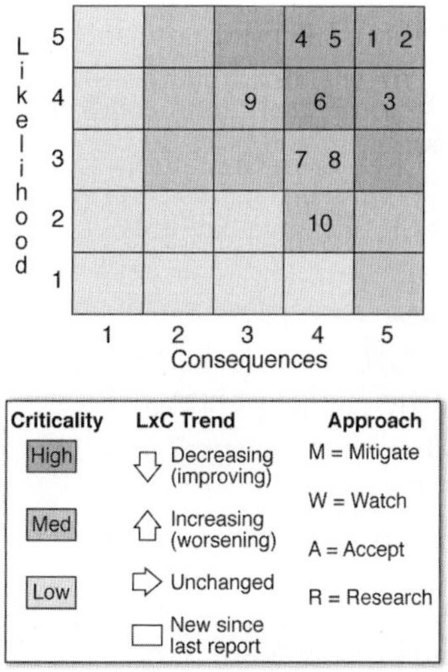

	Rank and Tread	Risk ID	Approach	Risk Title
⇧	1	SLI-P1-02	M	Thermal vacuum/acoustic test
⇨	2	SLI-P2-06	W	On-orbit propellant transfer
⇩	3	ISS-P1-03	M	Aggressive schedule
⇧	4	ISS-P4-08	M	Government furnished property (GFP)
□	5	SRB-P1-05	M	Hot fire test schedule slip
⇨	6	UG-P1-01	M	Design launch weight exceeds the Shuttle capacity
⇩	7	SRB-P5-04	R	Returnability requirement
⇧	8	HST-P9-09	W	12 year life certification
□	9	SLI-P4-07	M	Reboost engine/thruster lives
□	10	ISS-P1-15	M	Unachievable thruster technology

FIGURE 8-16. **Project Risk Position Reporting.** In this combination chart, the project's top ten risks are in a 5x5 risk matrix on the left. The right side identifies each risk, shows the risk's trend since the last risk report with block arrows, and displays codes for the risk handling approach. This chart is used often in high-level presentations, because it conveys a large amount of risk information in a compact form.

References

Bankes, Steven C. May 14, 2002. "Tools and Techniques for Developing Policies for Complex and Uncertain Systems." *Proceedings of the National Academy of Sciences.* 99:7263–7266.

Chesley, Julie, Wiley J. Larson, Marilyn McQuade, and Robert J. Menrad. 2008. *Applied Project Management for Space Systems.* New York, NY: McGraw-Hill Companies.

Department of Defense. 2000. *Systems Engineering Fundamentals.* Ft. Belvoir, VA: Defense Acquisition University Press.

Department of the Army. May 2002. *Cost Analysis Manual.* Arlington, VA: U.S. Army Cost and Economic Analysis Center.

de Weck, Olivier L., David Simchi-Levi, et. al. 2006. "SpaceNet v1.3 User's Guide." NASA/TP—2007–214725. Cambridge, MA: Massachusetts Institute of Technology and Pasadena, CA: Jet Propulsion Laboratory.

Ebbeler, Donald H., George Fox, and Hamid Habib-agahi. September 2003. "Dynamic Cost Risk Estimation and Budget Misspecification." *Proceedings of the AIAA Space 2003 Conference.* Long Beach, CA.

Frankel, Ernst G. March 31,1988. *Systems Reliability and Risk Analysis (Engineering Applications of Systems Reliability and Risk Analysis),* 2nd edition. Dordrecht, The Netherlands; Kluwer Academic Publishers.

Jorgensen, Edward J., Jake R. Matijevic, and Robert Shishko. 1998. "Microrover Flight Experiment: Risk Management End-of-Mission Report." JPL D-11181-EOM. Pasadena, CA: Jet Propulsion Laboratory.

Kline, Robert C., Tovey C. Bachman, and Carol A. DeZwarte. 2006. *LMI Model for Logistics Assessments of New Space Systems and Upgrades.* LMI, NS520T1. McLean, VA: Logistics Management Institute.

Kouvelis, Panos and Gang Yu. 1997. *Robust Discrete Optimization and Its Applications.* Dordrecht, The Netherlands: Kluwer Academic Publishers.

Larson, Wiley J. and James R. Wertz, eds. 1999. *Space Mission Analysis and Design,* 3rd edition. Torrance, CA: Microcosm Press and Dordrecht, The Netherlands: Kluwer Academic Publishers.

McCormick, Norman J. 1981. *Reliability and Risk Analysis.* New York: Academic Press.

Morgan, Millett G., and Max Henrion. 1990. *Uncertainty: A Guide to Dealing with Uncertainty in Quantitative Risk and Policy Analysis.* Cambridge, UK: Cambridge University Press.

NASA. March 26, 2007. "Systems Engineering Processes and Requirements." NPR 7123.1. Washington, DC: Office of Safety and Mission Assurance.

NASA. April 25, 2002. "Risk Management Procedural Requirements." NPR 8000.4. Washington, DC: Office of Safety and Mission Assurance.

NASA. August 2002. "Probabilistic Risk Assessment Procedures Guide for NASA Managers and Practitioners." Washington, DC: Office of Safety and Mission Assurance.

NASA. June 14, 2004. "Risk Classification for NASA Payloads." NPR 8705.4. Washington, DC: Office of Safety and Mission Assurance.

NASA. July 12, 2004. "Probabilistic Risk Assessment (PRA) Procedures for NASA Programs and Projects." NPR 8705.5. Washington, DC: Office of Safety and Mission Assurance.

Sherbrooke, Craig C. 2004. *Optimal Inventory Modeling of Systems: Multi-Echelon Techniques,* 2nd edition. Norwell, MA: Kluwer Academic Publishers.

Shishko, Robert. August 2002. "Risk Analysis Simulation of Rover Operations for Mars Surface Exploration." *Proceedings of the Joint ESA-NASA Space-Flight Safety Conference.* ESA SP-486. Noordwijk, The Netherlands: ESTEC.

Smith, Clayton A. August 2002. "Probabilistic Risk Assessment for the International Space Station." *Proceedings of the Joint ESA-NASA Space-Flight Safety Conference.* ESA SP-486. Noordwijk, The Netherlands: ESTEC.

United States Air Force (USAF). April 2007. Air Force Cost Risk and Uncertainty Analysis Handbook. Arlington, VA: U.S. Air Force Cost Analysis Agency.

Chapter 9

Product Implementation

David Y. Kusnierkiewicz,
Johns Hopkins University Applied Physics Laboratory

Product implementation produces specified system elements [ISO, 2002]. The process transforms performance specifications, interfaces, and implementation constraints into fabrication and coding to create system components and assemblies. We integrate these elements to form a system, as Chapter 10 discusses. We must verify that the system satisfies design requirements, and validate that it satisfies stakeholder requirements, as described in Chapter 11.

Product implementation involves a combination of buying, making, and reusing. We may employ any of these methods at any level in the product realization phase. The prime contractor, subcontractors, and their suppliers purchase piece-parts and raw materials to fashion into components, and ultimately, into the end product. For systems engineers from the implementing organization, the process comprises the following:

1. Prepare to conduct product implementation

2. Participate in

    a. purchasing components and end products

    b. reusing components and end products

    c. making components and end products

3. Capture the work products

Table 9-1 depicts the technical planning process, organized around these major activities.

TABLE 9-1. **The Product Implementation Process.** The process includes not only physical acquisition, but preparation, documentation, and communication. Not all steps apply to all acquisitions.

Step	Description	Where Discussed
1	Prepare for implementation	Section 9.1
2	Participate in buying the product	Section 9.2, Chaps. 11 and 12
3	Participate in acquiring the reuse end product	Section 9.3
4	Evaluate the readiness of items that enable product implementation	Section 9.4
5	Make the product	Section 9.5, Chaps. 10, 11, 14, and 16
6	Prepare product support documentation	Section 9.6
7	Capture product implementation work products	Section 9.7, Chaps. 16 and 17
8	Ensure effective communication	Section 9.8

Figure 9-1 shows how the process steps interact; all participants in manufacturing do the activities shown in the dotted box. Inputs to the implementation process fall into three main categories:

- Documentation (design and interface specifications and configuration documentation)
- Raw materials (metallic, non-metallic, electronic piece-parts, etc.)
- Things that enable product implementation (computer-aided design, modeling, and simulation tools, machine tools, facilities, etc.)

Outputs include:

- The product
- Documentation (end-item data package, user's manuals, etc.)
- Work products (procedures used, decision rationale and assumptions, corrective actions, and lessons learned)

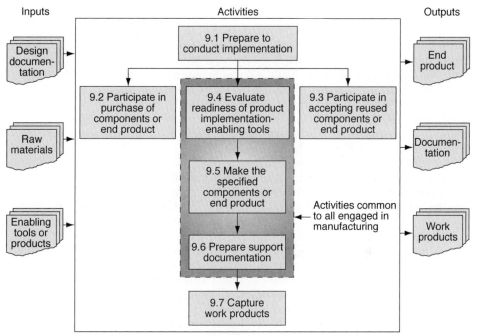

FIGURE 9-1.   **Implementation Process Interactions.** Manufacture requires more steps than either purchase or reuse.

9.1   Prepare for Implementation

An implementation strategy includes partnering considerations, acquisition strategy (build, buy, or reuse), and the type of procurements (single-source versus competitive, and government-furnished equipment (GFE)). Partnerships occur between government organizations, between government and industry, and with foreign organizations. The latter may even contribute equipment, such as an instrument, at no cost to the project, in exchange for a reciprocal service from a US government agency. We must address any required fabrication processes, tools, equipment, and facilities. The cost estimates and schedule have to identify and account for any necessary development activities.

Product implementation strategies vary between two extremes: wholesale procurement of an entire system on one extreme, to complete design, fabrication, integration, verification, and operational deployment by one organization on the other. The latter extreme is less common today; most systems embody a combination of building and buying. The strategy directly affects specific systems engineering responsibilities and implementation details, but the activities are similar in any case; the differences lie in who does what.

The FireSAT example implementation strategy is typical of many space systems today. The prime contractor, Acme Space Systems Enterprises, is responsible for design, integration, test, and delivery of the completed satellite based on the requirements of NASA's Goddard Space Flight Center (GSFC). Acme builds part of the system, including the primary mechanical structure, the flight avionics stack, and the spacecraft harness. They also subcontract some of the other system components: the solar arrays, batteries, the entire attitude determination and control subsystem (AD&CS), and the propulsion module. The suppliers are directly responsible to Acme. The AD&CS and propulsion suppliers are foreign companies; therefore, International Traffic in Arms (ITAR) Regulations apply, requiring government approval for technical information exchange. IREyes Corporation is under contract to NOAA to implement the payload module, which is then GFE to Acme through NOAA. IREyes is directly responsible to NOAA, not Acme.

Putting together the final system requires communication and coordination among all participants; they must all understand the chain of command among them. For instance, Acme may give technical direction to their subcontractors (as well as to the Acme members of the FireSAT team), but not to IREyes. Technical direction to them goes through NOAA, who holds their subcontract. Acme must manage the programmatic impacts of technical changes on their team members and subcontractors, but impacts on the payload require agreement from NOAA. And although Acme and IREyes must communicate directly on interfaces between the spacecraft and the payload, payload performance requirements come from NOAA.

The procurement authority (GSFC) provides high-level (mission- or system-level) objectives or requirements—with few quantified parameters—to Acme, and for the payload, to NOAA. These requirements go into a "Level 1 Requirements" document (usually after negotiating and iterating with the prime contractor, and in the FireSAT case, with NOAA and IREyes). The Level 1 requirements determine the contractual requirements that Acme must fulfill.

Acme decomposes the Level 1 requirements into lower-level requirements that define the system in successively greater levels of detail (system, subsystem, and component). They then flow them down accordingly, to their project team and subcontractors (and to IREyes, through NOAA). As at the system level, subsystem implementation often consists of a combination of direct manufacture, reuse, and procurement.

In procuring subsystems or components, the contractor does much of what the sponsor does in procuring the system. The requirements for procured subsystems or components are quantitative and more detailed than the high-level objectives the sponsor sends to the prime. But the relationships and activities between successive tiers in the supplier chain are comparable (Figure 9-2). A component may be an off-the-shelf product with performance specifications that meet or exceed system performance requirements, a modified version of an off-the-shelf product, or a totally new design.

If the component is off-the-shelf, we need to review details of hardware and software interfaces, environmental qualification, and parts quality to ensure compatibility with the current application. We may need to modify the product, or develop a new, custom design to meet project technical, environmental, or reliability requirements.

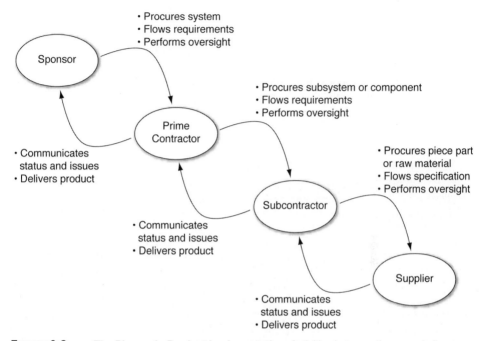

FIGURE 9-2.   The Players in Product Implementation. Activities between tiers are similar.

9.1.1   Determine Build, Buy, or Reuse Strategy

The build, buy, or reuse decisions account for a number of factors. The best value to the sponsor should be a driving consideration. Reusing existing components is cheaper than procuring or manufacturing new ones. But we must evaluate thoroughly the performance of the available component, the levels to which it was qualified, and the electrical (hardware and software) and mechanical interface requirements (how easily does the new system accommodate the existing component's interfaces?).

Many components are easier to buy from vendors instead of having the prime contractor design and fabricate them. Star trackers, propulsion systems, solar arrays and battery cells frequently fall into this category. As with reused components, we must evaluate the performance, interface, and qualification specifications of candidate purchased components against the current application. The impact of modifications may become part of the trade space. Wholesale subcontracting of the entire product may be the most effective way to implement (e.g., the FireSAT AD&CS and propulsion subsystems). Where a vendor has a standard spacecraft bus that meets mission requirements, we might buy the entire system. But even in these cases, the spacecraft vendor usually relies on procured subsystems or components. The availability of personnel and such physical resources as manufacturing tools and facilities may also factor into these decisions.

9.1.2   Understand Sponsor, Prime, and Subcontractor Interactions

Figure 9-3 depicts the activities and interactions between the sponsor, prime, and subcontractors in the course of a system development (reviews are not shown). Because a subcontractor has suppliers, and perhaps subcontracts as well, these relationships extend to additional levels. Table 9-2 summarizes the communication between sponsor, prime, and subcontractors.

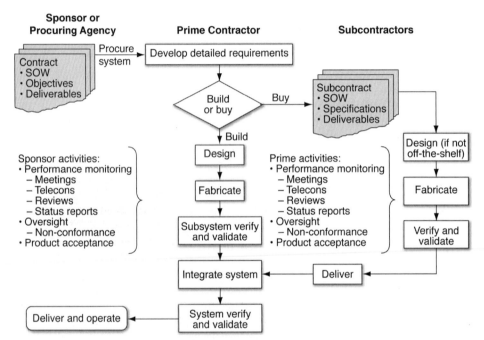

FIGURE 9-3.   **Implementation Activities and Interactions.** Sponsor, prime, and subcontractor activities and interactions are similar between tiers. Activities for reuse are much the same as for purchase. (SOW is statement of work.)

We need to establish relationships and lines of communication between the organizations involved in product implementation. For example, we must define policies regarding notification (people and timing) of failures during test. Nonconformances involving the failure to meet a sponsor's or prime's requirement usually need their approval. We must also obtain agreement on the timing and format of regular communication via telecons (often weekly), written progress reports (monthly) and formal reviews (e.g., preliminary design review). Such communication provides the means for progress monitoring, management, and problem resolution.

TABLE 9-2. **Communication for Implementation.** Sponsor, prime, and subcontractor communications must be timely and thorough throughout the project.

	Input to Lower Tier	Output Received	Communication and Oversight

Sponsor	• Contract to prime
	  – Statement of work
	  – Requirements or objectives
	  – Deliverables (operational system, documentation)
		• Status reports
		• Waivers and deviations
		• Deliverables
			• Meetings
			• Emails
			• Telecons
			• Reviews
			• Non-conformance reporting from prime

Prime	• Contract to subcontractor
	  – Statement of work
	  – Specifications
	  – Deliverables (qualified subsystem or component ready for system integration, documentation)
	• Purchase requisition to suppliers
		• Status reports
		• Waivers and deviations
		• Deliverables from subcontractors
		• Piece parts from suppliers for components fabricated by prime
			• Meetings
			• Emails
			• Telecons
			• Reviews
			• Non-conformance reporting from subcontractors
			• Review of certificates of compliance from suppliers

Subcon-tractors	• Purchase requisition to supplier	• Piece parts from suppliers	• Review of certificates of compliance from suppliers

9.1.3 Plan Long-lead Procurements

During the conceptual design phase of a project, we may need to begin procurement early on some components and materials to meet the project schedule. Some of these long-lead procurements may be developmental, as with enabling technologies, without which the system can't meet its performance goals. In other cases, the time to deliver a standard component may exceed the time available in the schedule due to inherent processes or supplier workload. For example, a standard star tracker that is sufficiently accurate for a three-axis stabilized spacecraft bus may need hardware or software modifications to perform adequately on a spinning bus. Or the electrical piece-part performance in an off-the-shelf design may degrade to unacceptable levels in the spacecraft's radiation environment.

After identifying such a problem, we decide on a risk mitigation strategy. Several alternate paths may be available, which we might pursue in parallel. For instance, in the star tracker modification case, we might place contracts with multiple subcontractors to study the scope of necessary enhancements, produce a working prototype, and provide a detailed cost estimate for delivery of flight units. We may also need an environmental qualification program for the prototypes, to select the eventual provider. If development work isn't needed, but subcontractor workload dictates procurement early in the project, we simply begin the procurement early, with the appropriate allocation of funds. These funds may be liens against budget reserves.

9.2   Participate in Buying the Product

Many of the activities discussed above are common to buying a system, subsystem, or component. The main differences are in who is on the giving end, and who is on the receiving end. In this discussion, the prime is the provider of the system—hardware and software, and a subcontractor is responsible for a subsystem or component (a spacecraft "box", which may also contain software). Suppliers are providers of electrical or mechanical piece parts, or software.

9.2.1   Manage the Sponsor-to-Prime Contractor Interface

The contract between the sponsor and prime specifies the statement of work (SOW) that the prime is to perform, a high-level statement of system requirements, the deliverables, and the cost and schedule requirements. The high-level requirements from the sponsor are usually referred to as system requirements. At first they may not be quantitative. System requirements are often detailed and quantified after the prime has been selected, are negotiated and iterated between the sponsor and the prime, and eventually incorporated into the contract.

Regular communication is necessary during product implementation. The sponsor may appoint a team to oversee the prime. For instance, the sponsor may appoint someone to be responsible for oversight of budgetary and schedule matters, plus a technical authority, who communicates primarily with the project's systems engineer. Members of the sponsor team and the prime's project team may communicate directly, but technical or programmatic direction should only be through an established and agreed-to chain of communication (usually through the project manager).

Telecons and email are familiar tools for everyday correspondence. More formal means include regular status or progress reviews covering technical and programmatic status, written reports to the sponsor from the prime (often monthly), and milestone project reviews such as preliminary and critical design reviews attended by the sponsor. A sponsor may also attend meetings between the prime and the subcontractors that supply major system elements.

The sponsor must approve any changes to contractual requirements. Sponsor approval is also required for non-conformances, waivers, and deviations from Level 1 requirements. Other project documentation, such as project plans, systems engineering management plans, and performance assurance plans, are often contract deliverables; the sponsor reviews these, but may not have to directly approve them (the project plan may need joint approval from both the sponsor and the prime contractor).

The prime is usually responsible for implementing the system, with sponsor oversight. This organization manages the allocation of technical resources, such as mass, power, data rates and volume, RF communication links, and pointing budgets, and it monitors use trends to maintain appropriate margins. The sponsor formally accepts the system upon delivery or deployment for operational use, but may not take physical possession. The prime may also perform operations and deliver contracted data products.

9.2.2   Insight into the Prime-to-Subcontractor Interface

The interface between the prime and the subcontractors is similar to the sponsor-to-prime relationship (Table 9-2). A contract contains the SOW for the subcontractor, detailed performance requirements or specifications (including environmental and reliability requirements), deliverables to the prime (including documentation and support equipment), and cost and schedule requirements. The subcontractor receives technical resource allocations, such as mass, power, and others as appropriate.

An allocation normally starts with a current best estimate of the resource, to which we apply a growth estimate, and assign a contingency reserve. The subcontractor is allowed to grow the resource within this allocation; it can only be exceeded with the consent of the prime by releasing some of the unallocated margin held at the system level. The amount of contingency reserve is a function of the component maturity. Typical initial contingency reserves are 15–30% for new designs, 10–20% for modified designs, and 5–10% for "build-to-print" designs.

The prime performs many of the same oversight functions as the sponsor does. The sponsor may also attend technical interchange meetings between the prime and subcontractor, as well as milestone reviews with the subcontractor to obtain detailed insight into project status. The project systems engineering team needs to keep current with subcontractor progress through the oversight function, to facilitate resolving issues that have system impact, track technical resource use trends, and keep project management informed.

Technical interfaces between the system and the subcontracted subsystem or component are documented in an interface control document (ICD). The detailed information in an ICD is usually the result of a negotiation between the prime and subcontractor, in much the way that system requirements are the result of negotiation between the sponsor and the prime. The ICD contains requirements that the subcontractor must verify before delivery to the prime. Information in an ICD often includes:

- Electrical interfaces
    - First circuit interfaces
    - Electrical connector identification and pin-out assignment
    - Power source characteristics
- Mechanical interfaces
    - Mounting footprint
    - Mechanical envelope
    - Clear field-of-view requirements
- Thermal interfaces
    - Interface type (thermally conductive versus isolated)
    - Interface temperature range

- – Active or passive thermal management methods to be employed
- Technical resource allocations (mass, power, data rate, data volume, etc.)
- Software command and telemetry definitions and formats
- Definition of electromagnetic, mechanical, thermal, and radiation environments
- Contamination control measures
- Shipping and handling methods

Interface development and control are fully addressed in Chapter 15.

In addition to the regular communication between prime and subcontractor, the prime may require physical inspection of the items at several points before delivery, such as before a board is installed in a housing, or before final box close-out. The prime may also require a witness for subsystem or component qualification testing.

9.2.3   Monitor Subsystem or Component Verification and Validation

Subsystems and components undergo a verification and validation (V&V) program before delivery for system integration. Procurement documentation normally establishes and documents V&V responsibility. The final compliance and verification matrices must explicitly identify unverified requirements, and requirements that were verified but not complied with. We must pay particular attention to system-level requirements that can only be validated at the subsystem level. Section 9.5 discusses details of subsystem V&V. Chapter 11 addresses system verification and validation, and these processes scale by analogy to subsystem V&V.

At the end of the subsystem verification and validation program, the prime formally accepts delivery of the subsystem or component, and takes physical possession of the hardware and any required support equipment (Section 9.2.5 describes the buy-off review). The subsystem is then integrated into the system, along with other subcontracted items and items fabricated by the prime before system-level V&V.

9.2.4   Procure Products

The build or buy trade studies that the prime performs deal with the technical performance of the candidate purchased products, as well as the cost and schedule. They also assess the qualifications of the potential subcontractor to supply the product. If the subcontractor is not certified to AS9100 or has been audited by the Joint Audit Planning Group (JAPG) or International Aerospace Quality Group within the last three years, a quality audit is advisable. Other conditions may warrant an audit, such as a change in corporate ownership or structure, or indications of poor performance, such as industry or Government-Industry Data Exchange Program (GIDEP) alerts. If the subcontractor is certified or has undergone a JAPG audit, we should review a copy of the audit report to be sure no outstanding findings exist.

Example: Acme Space Systems Enterprises, the FireSAT spacecraft supplier, has elected to buy a star tracker through a competitive procurement. They awarded the contract to a relatively new company on the Approved Supplier List (the company having passed an initial audit). A larger corporation certified to ISO9001, but not the more stringent AS9100, recently acquired the company. After negotiation with the subcontractor, the Acme supplier quality assurance manager conducts an on-site audit of the quality management system and practices. The audit focuses on the following areas:

- Configuration management process
- Corrective and preventive action system
- Equipment calibration records
- Electrostatic discharge (ESD) and material handling processes
- Inspection process
- Test process
- Limited-life (e.g., shelf-life) material tracking and handling process
- Documentation and disposition of non-conformances
- Manufacturing process control
- Quality records control and retention
- Training and certification
- Self-audit process

The auditor reviews process documents and as-built records, inspects facilities, and interviews personnel. The audit results in two findings:

1. The solder training is conducted to military/DOD standards, but not to NASA-STD-8739.3 (Soldered Electrical Connections), which the contract requires

2. Some of the ESD workstations had prohibited materials that produced excessive static fields, as determined by the auditor's hand-held static meter

The audit recommendation is to "conditionally accept the supplier." The subcontractor receives the results of the audit, and agrees to address the findings through their corrective action system. They train to the required NASA soldering standard, and forward copies of the certification records. They also retrain and certify staff to ESD handling and housekeeping practices, and increase the frequency of self-audits.

A kick-off meeting commonly follows contract award, normally at the subcontractor's facility. This is usually the first face-to-face meeting between project personnel and the subcontractor's team. Depending on the scope of the procurement, project representatives may include the project manager, the lead systems engineer, the primary technical point of contact for the project (often the lead subsystem engineer), a performance assurance engineer, and others as desired, such as the sponsor.

Part of this kick-off may be the heritage review for an off-the-shelf purchase. The review focuses on confirming the component's suitability for the application. It encompasses assessments regarding technical interfaces (hardware and software) and performance, and the environments to which the unit has been previously qualified, like electromagnetic compatibility (EMC), radiation, and contamination. It also assesses the design's compatibility with parts quality requirements. We have to identify and document all non-compliances, and address them either by modifying the components to bring them into compliance, or with formal waivers or deviations for accepted deficiencies.

We must carefully review and analyze all modifications to an existing, qualified design before incorporating it. How will these changes affect the performance and qualification of the product? Are the changes driven by requirements? A variety of factors may drive modifications:

- Quality requirements—do the existing parts in the design comply with the program parts quality requirements?
- Environmental qualification requirements (EMC and mechanical, which includes random vibration or acoustics and shock, thermal, radiation, and contamination)
- Parts obsolescence issues
- Product improvement

We must take care when buying even simple off-the-shelf components or mechanisms with previous flight heritage. Manufacturers may make seemingly minor changes for reasons of materials availability. They may consider such changes minor enough not to mention in the documentation normally reviewed by a customer. We have to be conscientious in inquiring into such changes and work with the vendor to ensure the changes are acceptable, and that the modified design still qualifies for the application.

The vendor and the project team members must communicate problems, failures, and non-conformances. It's usual to report failures within 24 hours. The project should participate in the disposition of all non-conformances. Any changes to requirements must be agreed to and documented via revisions to requirements documents, contracts, or formal waivers and deviations as appropriate.

The statement of work may specify formal milestone reviews. Reviews to determine readiness to enter the environmental qualification program, and a pre-ship or buy-off review at the completion of the test program, are common.

9.2.5   Conduct Buy-off Review

The vendor prepares an end-item data package, and makes copies available to the project representatives accepting delivery of the product. The data package usually contains the as-built documentation, completed non-conformance and discrepancy reports, problem and failure reports, waivers and deviations, a requirements verification and compliance matrix, and test reports detailing the product performance through the acceptance qualification and environmental test program. Life-limited item logs, such as connector mate-demate logs, should also be included where appropriate. Other required documents, such as a user's manual, complete with operating constraints and contingency procedures, should come with delivery of the flight product, if not before.

9.2.6   Support Equipment

In addition to the ICD, we have other ways to help ensure interface compatibility and aid the development of purchased products. Interface emulators are useful development tools on both sides of the interface. Engineering model processors may be needed to support software development. Early interface tests between breadboards, brassboards, or engineering model hardware help verify hardware compatibility. Such tests are most beneficial before the fabrication of flight articles begins so that we can incorporate any necessary changes.

Some products require additional equipment to provide stimulation as part of subsystem- and system-level verification. Examples are illumination sources for optical instruments, or radiation sources for particle detectors. We may need more sophisticated software models to provide dynamic inputs (e.g., star tracker quaternions) to guidance and control processors to verify algorithms.

Many projects maintain a hardware-in-the-loop simulator for post-launch support. These simulators are usually a combination of engineering model hardware and testbeds that run software models to provide closed-loop simulations of in-flight operations for verifying before executing. We also use them to verify command sequences and software loads post-launch, and they provide a suitable training environment for mission operations personnel, for nominal and contingency operations. In some cases, a full engineering model may not be necessary or cost-effective. The command and telemetry interface—perhaps with canned data streams—that emulates the behavior of the flight system, may suffice. We should specify the need for engineering models or flight spares and simulators along with the flight product.

We should also consider the need for post-launch support contracts. Diagnosing in-flight anomalous behavior may require the vendor's expertise. International Traffic in Arms Regulation (ITAR) requirements may dictate the need for a technical assistance agreement (TAA) with foreign vendors to share in-flight data for anomaly resolution. We should plan well in advance for the possibility that we might need to return the product to a foreign vendor after delivery for repair, refurbishment, or maintenance. This planning minimizes schedule disruptions due to export control requirements.

9.3   Participate in Acquiring the Reuse End Product

End product reuse involves many of the same considerations as purchasing a new product. A heritage review should cover all technical interface and performance specifications, parts quality, and environmental qualification history to determine the product's compatibility with the current application. This review should occur well before the system product implementation phase, so that we can make any necessary changes in a timely manner, and address any issues that might arise during retesting.

We need to assess the parts in the reused product against the new radiation environment. We should have previously documented the operational hours (ground test hours) accumulated on the product, and made an assessment of any impact on expected operational life, particularly if the product contains any known life-limited items. We should do an audit of GIDEP alerts against the as-built parts lists to determine if any affected parts require change-out. We should also review the as-built documentation with particular attention to non-compliances to ensure they are acceptable in the new application. We must determine command and telemetry interface compatibility. We have to decide what performance testing and environmental requalification tests are required, and address fatigue concerns if there's a possibility of over-testing for the new application.

In addition to the flight article being considered for reuse, we must also evaluate the availability and suitability of supporting and enabling equipment, such as ground support equipment, calibration equipment, and simulators. We have to plan for accommodations during the integration and test phase. We must similarly consider engineering models and emulators as for purchased products. We need to procure a user's manual, constraints documents, and other documentation similar to those required of a purchased product. A formal buy-off review marks the official transfer of ownership. Depending on the source of the reused product, ITAR regulations may require a TAA or other export control documents.

9.4   Evaluate the Readiness of Items that Enable Product Implementation

Planning ensures the correct design and operation of a system or product under development. Implementing it entails a significant infrastructure that also requires planning. We should start well ahead of the implementation phase to be sure the infrastructure is ready to support product realization.

Items that enable the product implementation consist not just of products such as machine tools and electronic assembly pick-and-place machines, but also the software that controls them, facilities such as clean rooms, and emulators and simulators. The software development environment must be ready to support software production, including verification. Other tools such as scheduling and configuration management tools must be in place well before implementation begins. We should upgrade currently stable tools or new product roll-outs only when we've demonstrated they will be sound in their operational environment.

9.5   Make the Product

Manufacturing is a complex series of interacting and interdependent steps that occur in sequence to create the product. Assembling a complicated box requires a number of inputs to be brought together in a timely fashion, with many parallel processes occurring near-simultaneously. For instance, an electronics box may consist of several different circuit cards that are developed and tested in parallel, then integrated into the box. Software development also proceeds in a planned sequence in parallel with hardware manufacture, and is eventually integrated on the target hardware. The process must be planned well, with established mechanisms for resolving problems and controlling changes to the product and the process. Configuration control boards control these changes, while material review boards plan how to resolve technical problems that arise during manufacture, integration, and test.

The primary inputs to the implementation phase are the drawing packages ready for release to fabrication, and software specifications. The drawings reflect the current requirements flowing from the specifications, ICDs, and engineering model evaluations. Parts kits and raw materials are ready to support the start of implementation, while additional kitting and procurements may be ongoing throughout this phase as the work proceeds. Requirement verification plans, test plans, and procedures should be developed thoroughly to ensure readiness during and after fabrication. Technical and programmatic resource estimates should be kept up to date to support system-level budgeting.

Subsystems and components are designed, fabricated, and verified roughly in parallel, depending on the availability of fabrication and test facilities. We assemble components in stages, and test them at successively higher levels of integration into completed assemblies. Figure 9-4 illustrates the typical integration of an electronics box. The completed component then proceeds through a verification process. Any software that runs on the component should be installed before testing. Software is usually delivered in builds of increasing functionality, and the final software build is often delivered sometime during system integration. But it must be complete enough to support component-level verification before delivery to system integration. Chapter 10 treats system integration in detail.

9.5.1   Control the Configuration

Requests for changes (hardware and software) go before a configuration control board, sometimes referred to as a change control board, which is often chaired by the project manager or the systems engineer. We must determine if the requested change is needed to meet a non-waiverable requirement, or is simply meant to improve the product. We have to evaluate how making—or not making—the proposed change will affect product performance, cost, and schedule. Changes that directly affect a top-level (Level 1) requirement often require concurrence from the customer or sponsor. The contract should reflect this policy.

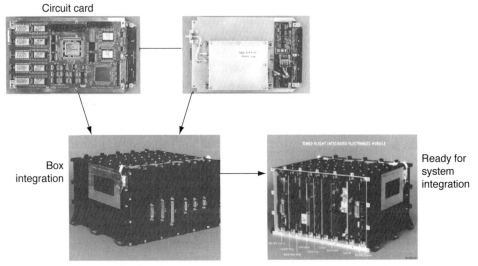

FIGURE 9-4.   Electronics Box Integration. The integration progresses in steps, from component through subsystem.

9.5.2   Plan Manufacturing and Assembly

Some organizations don't produce a manufacturing and assembly plan. These organizations are usually small enough that communication is regular and thorough, and they have well-established roles and routines for manufacturing, assembly, and problem resolution. In any case, the activities are generally the same, and the project schedule accounts for the steps. The main purpose of a manufacturing and assembly plan is to explain how the product will be made. It helps identify conflicts we must resolve to minimize the impact to project cost and schedule. The plan may contain:

- Definitions of roles and responsibilities
- Points of contact and communication flow
- External and internal resources, facilities, and processes required
  - Examples of bare board manufacture, machine shop facilities, assembly manpower; we may need to subcontract for specialized services not found in-house for some pieces, or to relieve internal resource conflicts
- Fabrication master plan
- Priorities

9.5.3   Conduct Manufacturing Reviews

We should hold several package design (manufacturing) reviews during the design and implementation phases. Minimum attendance should consist of manufacturing engineering, fabrication personnel, parts engineering, the lead engineer responsible for the product, and performance assurance (quality) personnel. Systems engineers should attend, particularly to evaluate the need for—and impact of—suggested changes.

A recommended review before the preliminary design review (PDR) and the start of the detailed design phase is a package concept review of such high-level requirements as access issues for tailoring performance, repair, rework, thermal interfaces, connector locations, etc. For example, a subsystem design might be based on a previous design that was flown once. Assemblers point out that the previous design required a significant amount of disassembly if a repair was needed. The decision at that time was to accept that feature, because only one unit was being made, and the cost and schedule impacts of changing the design were prohibitive. But the FireSAT system will build multiple units, and the systems engineers determine that early incorporation of this lesson learned will reduce schedule risk later in the project. They advise the project manager to accept this change. The project manager determines that the cost impact at this stage is minimal, and approves the change.

A review of the more detailed design—a detailed packaging review—should occur before the critical design review (CDR) and drawing sign-off. In our review of the detailed drawing package, we must analyze, identify, and understand board and electrical part maximum and minimum operating temperatures; mechanical stresses on boards; materials compatibility and compliance with contamination requirements and fabrication sequences, methods, and processes. The mass estimate should be updated. Systems engineers need to be aware of proposed changes; are they driven by the need to meet requirements, or simply by desire?

Two to four weeks before the start of fabrication, a manufacturing or production readiness review is held. It should be far enough in advance of the planned start of manufacturing to incorporate any necessary changes and replanning into the schedule without delaying the start date. Most manufacturing drawings should be near completion and ready for release. The goal of the review is to ensure that everyone understands the manufacturing sequence, procedures, and processes. Sequences should include inspection test points and in-process tests. We should address any issues from previous reviews and identify resulting changes. We need to determine the readiness of parts, materials, fixtures, facilities, and personnel. Systems engineers must resolve conflicts over manufacturing resources to establish priorities, or decide if alternate manufacturing sources are necessary.

9.5.4   Establish Material Review Board

Progress monitoring and problem solving are key systems engineering activities in the implementation phase. Daily communication to resolve issues promptly helps to avert schedule impacts. During fabrication, we must continually resolve product non-conformances, often with project office concurrence. A best practice is to schedule a daily session of the material review board (MRB) to handle non-conformances. Additional ad hoc sessions may be needed to keep the effort flowing. Board membership should consist of a manufacturing engineer, systems engineer, and performance assurance (quality) engineer at a minimum. Other participants (e.g., lead engineer for an affected product) are added as needed.

Increasingly prevalent is the use of electronic routing and sign-off for non-conformance documentation. While this alleviates the scheduling burden of face-to-face meetings, it can also reduce the frequency of valuable dialogue. Failure review boards (FRBs) are also necessary as testing starts on sub-assemblies. The make-up of the FRB is similar to the MRB.

We should review periodically the status of non-conformances (including problems and failure reports) and their causes, and trend the data to identify systemic issues requiring corrective action. For instance, are workmanship errors at an unacceptably high level? This problem can indicate a need for training or retraining. A prevalence of design errors may point to deficiencies in the design review process. Software problems and failures may be a sign of inadequate testing.

Example: During the fabrication of electronic boards for FireSAT, Acme Space Systems Enterprises quality control inspectors start rejecting solder workmanship associated with the attachment of surface-mounted capacitors using tin-lead solder. The solder is not adhering properly to the capacitor end-caps. Materials engineers determine that the end-cap finish is silver-palladium, and these parts are intended for use with silver epoxy for terrestrial applications, and not tin-lead solder. The manufacturer, who has eliminated lead from their processes for environmental reasons, discontinued the tin-lead end-cap finish. Parts are available with a pure tin end-cap finish that's compatible with tin-lead solder, but parts with a pure tin finish are prohibited due to concerns over tin whisker growth.

The Acme systems engineer and project office inform GSFC of the situation, and ask to engage GSFC materials engineers in resolving the issue. Between the two organizations, they determine that pure tin end-caps can be tinned with tin-lead solder before being attached to the boards, mitigating the tin whisker issue. They modify the part attachment process for these parts, and replace all affected surface-mount capacitors with pure tin end-caps, which are readily available from the part supplier. Schedule and cost impacts to the program are minimal.

9.5.5   Create Software

Systems engineering responsibilities for software development are analogous to those for hardware development. The project systems engineering team sets the top-level software requirements after determining the system functional partitioning between hardware and software. They then derive and review detailed software requirements before starting software design. A software systems engineer responsible for all aspects of system software (flight, ground, test, data analysis, etc.) is part of the systems engineering team. This person is responsible for ensuring interface compatibility between all software and hardware elements (e.g., the flight software and flight processor and memory), software-to-software interfaces (e.g., command and data handling software to guidance and control software), and system element-to-element (flight system to ground system).

We use detailed schedules to track progress during development. Software is normally delivered in staged builds of increasing functionality, and may not be final until well into system integration and test. We use various metrics to track progress and issues. For example, change requests or problem reports provide an indication of software stability during development and verification; measurements of processor memory and throughput use and margin indicate the adequacy of selected hardware resources and the ability to handle post-launch upgrades.

A common problem in software development is underestimating resources, such as manpower and development platforms. Flight processors that have commercial analogues allow us to make inexpensive development platforms available to all developers. The software architecture can also affect the ability to use inexpensive commercial platforms. The systems engineer needs to be alert early to ensure adequate development resources. Manpower staffing issues sometimes dictate that functionality not required until after launch and commissioning be deferred to a post-launch software build. In these cases we need to be sure that we have adequate resources for post-launch validation and verification (such as a hardware-in-the-loop simulator).

During the implementation phase, we have to resolve issues, evaluate change requests, determine a disposition, and monitor and manage the software verification. Software change requests go before the configuration control board for approval before we implement them in the software design. They may be needed to meet the software requirements, or may enhance the operability of the system, reducing the burden on, and risk associated with, operations. The systems engineer must weigh the impact of such changes, and make recommendations to the project manager.

System-level software verification and validation continue through integration and test. They begin with testing against requirements by the developer, followed by acceptance testing—that may include stress testing—by a team independent of the developers, and then testing and review by an independent verification and validation (IV&V) facility. The latter activities usually begin early in the project with a review of requirements, and continue through the system verification of the final build. The IV&V team participates in software reviews, witnessing tests, and reviewing test artifacts. They present an independent assessment of software quality at major project reviews. The software system engineer is the primary interface to the IV&V team.

9.5.6   Integrate Hardware and Software

Developing hardware and software follows roughly the same flow:

1. Requirements definition
2. Design
3. Build
4. Test

The build processes are analogous in that construction starts with lower levels of assembly that are tested individually, and then integrated into a more complete subsystem with increasing complexity and functionality. That is, an electronics box may consist of a number of boards that are tested individually before being integrated into the box, and several boxes may constitute a single subsystem. Similarly, software may consist of a number of individual modules that are developed separately and then integrated into a "build" for integration with the target hardware. Subsequently, the completed hardware subsystem undergoes a qualification (or acceptance) test program, after which modification is limited to rework due to failures or incompatibility issues. An incompatibility may be a requirement or interface non-compliance that we can't remedy with a software fix.

Software development is often a series of successive builds of increasing functionality that extends after the first build delivered for hardware integration. Several software builds may be planned during the system (spacecraft) integration phase. The schedule may reserve a slot for delivery of a software build intended solely to fix bugs identified through testing and use. The final software build is usually delivered during system-level integration and testing. Figure 9-5 illustrates this process.

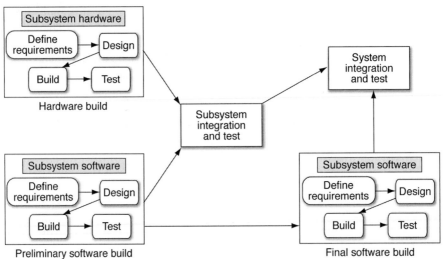

FIGURE 9-5.   System-level Integration and Testing. Increasingly mature software is progressively integrated during the hardware lifecycle.

We use one of at least two approaches to phasing hardware and software reviews during development. One approach is to schedule software development reviews to lag the hardware and system reviews, which allows system and subsystem issues to stabilize, providing more stability in the software requirements. The other approach keeps the hardware and software development in lock-step. This approach helps drive the software effort to stay closer to the overall system development schedule, and minimizes late deliveries. Standard hardware and software interfaces that maximize the reuse of existing software help mitigate the risk of significant disconnects that can severely affect the schedule.

9.5.7   Verifying the Subsystem or Component

A typical subsystem- or component-level verification process flow involves:

- Determining final mass properties
- Full electrical performance testing at ambient pressure and temperature (including measurement of power, data rates, and volumes)
- Thermal testing at ambient pressure
- Electromagnetic compatibility (EMC) testing
- Vibration, shock, and acoustic testing (shock testing is often not done at the component level, while acoustic testing is only for sensitive components such as thin foils)
- Thermal vacuum testing

Some organizations, to shorten the schedule and reduce the cost, elect not to perform EMC, vibration, and thermal-vacuum testing at the component level. But this increases the risk of costly disruption late in the system integration and verification process.

At the end of the verification program, we review the results to be sure we've identified all non-conformances for proper disposition, and then prepare the end-item data package. This package should include a subsystem verification matrix. The matrix should capture all the requirements (performance, interface, environmental, etc.), state the verification method (test, analysis, inspection, simulation/modeling, or demonstration), and provide references to the test or analysis report or verification event. It notes non-compliances, with reference to the appropriate waiver or deviation. After satisfactory disposition of all items, the unit or subsystem is then accepted for system integration.

Procured subsystems follow a similar flow, and are delivered for system integration. We may stage subsystem deliveries to accommodate the integration process. And we can accommodate late deliveries by substituting engineering model hardware to allow integration to proceed.

9.5.8   Conduct Buy-off Review

As with the purchase or reuse of products, subsystems may undergo a formal acceptance (buy-off review) to confirm readiness for delivery for integration into the system (Section 9.2).

9.6   Prepare Appropriate Product Support Documentation

Product support documentation consists of a user's manual, complete with constraints definitions and contingency procedures, if required. The manual should include command and telemetry definitions and descriptions of modes of operation. A system description, including block diagrams and simple schematics, is useful. Other supporting documentation, such as the manufacturing drawing package—including detailed schematics—should be maintained and easily accessible. These documents aid problem and anomaly resolution during ground test and flight operations. Test reports and other data from the product acceptance or qualification test provide helpful reference. The product support documentation also includes items necessary for operating and maintaining the operational-phase ground system.

9.7   Capture Product Implementation Work Products

Modern product lifecycle management tools effectively and automatically capture work products in progress. Such products may constitute records whose retention is required by contract or AS9100 compliance for years after contract termination. Implementation work products may include end-item data packages consisting of manufacturing drawing packages, as-built documentation, non-compliance reports, plans, procedures, test reports, corrective actions taken, lessons learned, and even engineering notebooks.

Implementation Worry List

Space mission success requires attention to detail. Literally thousands of details need attention, and getting any one of them wrong can result in a catastrophic loss of mission. As if that weren't enough to worry about, here are some things that can make life difficult during the implementation phase:

- Interface incompatibilities—In spite of the best attempts to document and implement compatible interfaces, it's not unusual to discover an error or oversight during implementation. Early check-out of subsystem and system interfaces via brassboard or engineering model integration helps mitigate this risk, at least at the hardware level. It's especially important to assess changes implemented after this step for any effects on interfaces.

- Intermittent failures or anomalies, or one-time anomalous events. Intermittent failures or anomalies are inherently troublesome due to their lack of repeatability. This makes it extremely difficult to isolate the problem and implement effective corrective action. Is the problem in hardware, software, ground support equipment, or operator error? Is the problem induced by environmental conditions? One-time observations are no less vexing. All such anomalies must be recorded and tracked. Those for which we can't identify or implement any corrective action must be assessed for the impact on mission success if the condition persists or worsens in flight.

- Late discovery of parts quality or process issues — A parts quality or process issue usually surfaces during the implementation phase, or later. Such issues can arise in several ways, such as in-flight failures that occur on other missions or unacceptably high failure rates from life tests conducted by the project to qualify a particular part for flight. Results from such life tests may not be available at the time flight hardware fabrication must start to support the schedule, so fabrication proceeds at risk pending life test results. A parts issue may be confined to specific lots, or may result from a particular application. Process issues can arise from "minor" changes in materials (example in Section 9.5). The later in the project we discover such issues, the greater the impact on cost and schedule.

9.8   Ensure Effective Communication

Effective communication is essential for project success. But communication has also been the bane of many organizations throughout time. Incomplete or erroneous communication can be extremely costly to a project, and even result in mission failure. The proliferation of modern communication methods is both a help and a hindrance. Email, voicemail, videoconferencing, and various document-sharing systems are great communication tools, particularly with geographically diverse teams, but can easily create an overwhelming amount of information. Modern engineering and communication tools have increased workforce efficiency and productivity, but the greater complexity of many projects and systems increases the amount of communication required for mission success. And with a smaller, more efficient project team, the demand on the engineer's time to support meetings, prepare for reviews, and answer emails and action items can leave little time for engineering.

Given this ease of communication, everyone must manage the flow and volume of information, particularly between a contractor and subcontractor. Lines of communication must be respected, particularly when giving directions. Otherwise, a project or task can become unmanageable. Contractors need insight into subcontracted work and are obligated to provide oversight, but they must establish boundaries and adhere to them; efforts must be managed, not micro-managed. This balance is difficult to achieve.

Weekly team status meetings are a useful way to communicate status and issues. In addition to keeping the systems engineering team informed of progress and significant events and issues, they generate interaction among team members. Representatives from manufacturing (fabrication) should also attend, as well as those responsible for monitoring work done by subcontractors or partners. A summary schedule from each sub-product showing milestones achieved and anticipated is presented, as well as updates to technical resource requirements (power, mass, data rate, etc.).

Communication with the sponsor or customer is a prime responsibility of systems engineers. Weekly telecons, where inputs are gathered from the weekly team meetings, are common means of communication. Formal written monthly status or progress reports are also the norm. Depending on sponsor requirements, immediate (e.g., within 24 hours) notification of major problems may be necessary.

Summary

Product implementation involves a mix of buying, making, and reusing. A number of factors drive the decision as to which method to employ. What approach results in the best value to the sponsor? Considerations include cost, schedule, performance, reliability, and availability of existing system elements, manufacturing and test facilities, and even personnel (manufacturing and design). All participants in the building process have a set of activities in common, whether the product is a component (star tracker), a subsystem (attitude determination and control system), or the entire end product (spacecraft). Oversight of all these activities comes from successively higher levels in the project organization, starting with the sponsor, the prime contractor, and flowing down through the layers of subcontractors and suppliers.

References

International Organization for Standardization (ISO). November 2002. *Systems Engineering—System Life Cycle Processes*, ISO/IEC 15288. ISO/IEC.

Chapter 10

System Integration

Gerrit Muller, *Buskerud University College*
Eberhard Gill, *Delft University of Technology*

Systems engineering is a process initiated by a set of mission needs or business opportunities and resulting in a system that satisfies the customer's needs. Increasing pressure to reduce costs in the public and private sectors forces such systems to be commercially viable as well. In this chapter we describe the system integration process as shown in Table 10-1.

10.1   Define the Role of System Integration within Systems Engineering

System integration is one of the most significant activities within systems engineering. It combines capabilities of platforms, systems, operators, and infrastructure to satisfy customer needs. We often use the term in conjunction with testing and verification. In space engineering, system integration comprises such diverse activities as assembling space sensor components, functional integration of a spacecraft subsystem, and integrating the spacecraft to the launch vehicle. From a systems point of view, the final integration of a mission may occur in space,

TABLE 10-1. **Chapter Structure.** This table summarizes the topics of this chapter.

Step	Description	Where Discussed
1	Introduction to System Integration	Section 10.1 and Chap. 1
2	System Integration Strategies and Approaches	Section 10.2, Chaps. 11, 15, and 19
3	Integration and Testing	Section 10.3 and Chap. 11
4	Integration Plan	Section 10.4
5	System Integration Roles and Responsibilities	Section 10.5 and Chap. 16
6	Configuration Management	Section 10.6, Chaps. 4 and 14
7	Special Topics	Section 10.7 and Chap. 11
8	Lessons Learned	Section 10.8 and Chap. 19

following the deployment and check-out phase of a spacecraft. In this phase the space segment operates together with the ground and control segments to realize the mission objectives.

Figure 10-1 depicts the role of system integration within system and use contexts. Here, components and functions come together to achieve a reproducible and supportable system working in its environment and fulfilling its users' needs. Detailed knowledge of integration on the technical, organizational, and reflection levels serves as a framework for the chapter.

10.1.1   System Integration Definitions

System integration is an ongoing flow of activities during the systems engineering process. The nature of these activities changes with program phase and increasing system maturity. In Phase D, for example, we typically integrate technologies or lower-level components to build and verify higher-level subsystems. Full system integration occurs in Phase E following the operational readiness review.

System integration is a process of physically and functionally combining lower-level products (hardware or software) to obtain a particular functional configuration [ECSS, 2008]. The Defense Acquisition Guidebook defines integration as "the process of incorporating the lower-level system elements into a higher-level system element in the physical architecture" [OUSD, 2006]. While these definitions emphasize the bottom-up character of system integration, other definitions focus on interfaces. The International Council on Systems Engineering (INCOSE) defines system integration as a function "to establish system internal interfaces and interfaces between the system and larger program(s)" [INCOSE, 2004]. NASA [2007] defines product integration as the process used to transform

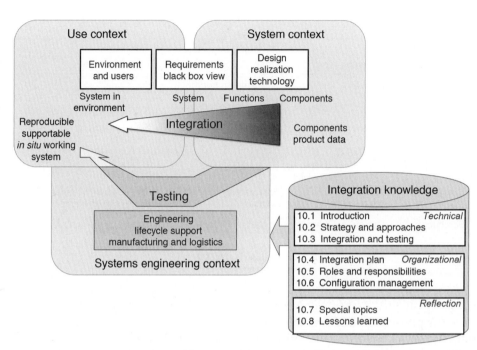

FIGURE 10-1.　**Map of System Integration.** The systems engineering context ties together the system integration effort.

the design solution definition into the desired end product of the work breakdown structure (WBS) model through assembly and integration of lower-level validated end products in a form consistent with the product-line lifecycle phase exit criteria and that satisfies the design solution definition requirements.

These definitions of system integration vary substantially in emphasis, but they all share the idea of combining less complex functions to achieve a system satisfying its requirements. Everyone on a space project must have a clear and unambiguous understanding of what system integration is and what it comprises. Its processes and activities depend on the scope and complexity of the space mission, and on the customer. Large-scale integration for manned space flight, for example, is far more complex than integrating a technology demonstration mission using a micro-satellite. Table 10-2 lists typical engineering activities related to the system integration tasks within the development of a space mission.

TABLE 10-2. **Space System Levels with Their Associated System Integration Tasks.** The tasks depend not only on system level, as shown here, but on mission complexity.

System Level	Example	Typical Integration Task	Task Example
Part	Resistor	Integrate on board	Functional test
Component	Printed circuit board	Integrate component	Soldering
Subassembly	Actuator electronics	Integrate hardware and software	Software upload
Assembly	Reaction wheel	Integrate actuator	Measurement of exerted torques
Subsystem	Attitude and orbit control subsystem	Integrate functional sensor	Replacement of sensor model by physical breadboard
Segment	Ground segment	Implement interfaces	Setting up the communication network
System	Satellite mission	Integrate final system	In-orbit check-out

10.1.2   Objectives and Challenges of System Integration

The primary goal of system integration is a system that includes all functional and physical elements needed to meet the stakeholders' requirements and expectations. Closely related to this objective is the reduction of risk, be it cost, scheduling, functional, or qualification risk. During integration, the project team tries to identify unforeseen problems as early as possible to analyze and solve them in a timely manner. The word *unforeseen* indicates the main challenge of integration: How do we find problems when we don't even know they exist?

Problems can be unforeseen due to the project team's limited knowledge or invalid assumptions. Assumptions we make early in the design to cope with inherent uncertainties limit how well we see their consequences. Examples are assumptions about the clock drifts of oscillators, which may cause failure in tasking processes in the on-board computer, or unforeseen inrush currents that threaten electronic equipment.

Another common source of unforeseen problems is interference between functions or components. For example, two software functions running on the same processor may perform well individually, but running concurrently may be too slow due to cache pollution. The contents of the cache are replaced by interrupting activities, slowing the interrupted thread much more than expected. These page reads and writes are several orders of magnitude slower than ordinary memory reads and writes. This also causes a tremendous slow-down of the software. Physical interference of radio frequencies between an actively transmitting satellite payload, such as an active synthetic aperture radar, and spacecraft subsystems might severely degrade system performance.

10.1.3   Integration within Systems Engineering and Architecture

A universal organizational instrument in systems engineering is *decomposition*. Decomposition enables the logical allocation and distribution of work concurrently. Its complement is *integration* [NASA, 1995]. Every system or process that has been decomposed into smaller parts or steps must be integrated again to obtain the desired outcome [Forsberg and Mooz, 1992]. Figure 10-2 illustrates the decomposition and integration processes.

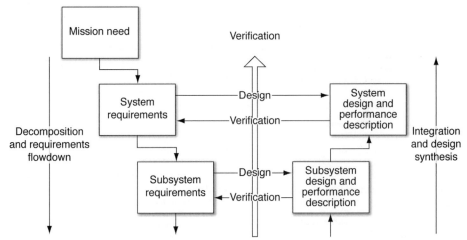

FIGURE 10-2.   **System Decomposition and Integration within the Systems Engineering Framework.** Decomposition follows a top-down approach, which is complemented by the bottom-up integration process. The vertical verification arrow indicates verification during final integration; the horizontal verification arrows indicate verification at the respective requirement levels.

We may encounter a number of problems from decomposition. When a system is decomposed, the subsystems may perform well, but crosscutting functionality and characteristics suffer. The root causes of such problems are lack of ownership, attention, or communication across organizational boundaries. Constant attention to system integration helps counter these difficulties.

10.1.4   The Process View

Systems engineering is a sequence of phases with increasing levels of realization and decreasing levels of risk. This may be a useful high-level mental model, but we should keep in mind that most activities overlap each other. The pure waterfall model, where requirements, design, integration, and test are sequential phases, is not practical. Much more practical is an approach with a shifting emphasis, as shown in Figure 10-3b. We note especially the long ramp-up of the integration activity.

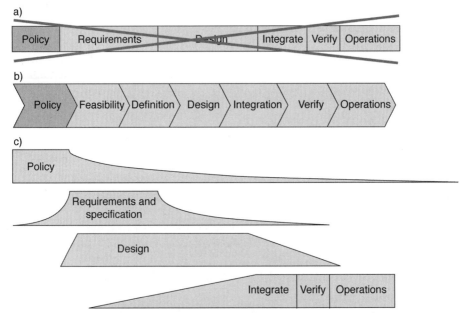

FIGURE 10-3.   **Typical Phases of a Systems Engineering Process. a)** The top diagram depicts a simplistic unidirectional waterfall scheme, which is inefficient. **b)** The middle process is acceptable, but suboptimal, as it limits integration activities. **c)** The bottom shows a shifting focus of activities, which is preferable from an integration perspective.

10.2   Select a System Integration Approach and Strategy

10.2.1   Interface Specifications

Interfaces play a crucial role in space system integration (Chapter 15). This role comes mainly from system complexity, which involves the interaction of many people and things. We must document interfaces properly, usually with interface control documents (ICDs). A good way to define and maintain interfaces is through interface control working groups (ICWGs), which carry the definition, specification, and maintenance of interfaces along the integration process [DoD, 2001]. Interfaces cover a broad range of types, as Table 10-3 shows.

Developing a clear interface specification that covers the interface completely and unambiguously is a challenge. A model of the interface definition, usually described in the ICD, may cover much of it but neglect aspects that raise problems during integration. An example of an incomplete interface is one that accounts for the steady-state current of an electronic component but neglects the inrush current that occurs when the component is switched on. Another challenge to writing good ICDs is often the authors' differing backgrounds. Engineers and scientists, for

TABLE 10-3. **Types of Interfaces.** The examples here are typical of space systems.

Interface Type	Examples
Mechanical and structural	Mounting, mass, dimensions, material
Electrical	Voltage, connector, and pin allocation
Optical and radio-frequency	Wavelength, bandwidth, modulation
Environmental	Thermal, radiation
Human	Operations effort, man-machine interface
Software	Calling list, performance, parameter specification

example, may lack a common language and a sufficient knowledge of the others' area. Beyond the conventional ICDs, a system integration point of view demands that we consider all interfaces between subsystems, assemblies, etc. A matrix of all elements at a particular system level is a good way to indicate and trace interfaces.

10.2.2   Integration from Components to Systems

The integration of a system starts bottom-up with provisional testing of individual components. The purpose of this approach is to identify, isolate, and recover from faults in a sufficiently small and controllable scope to allow for diagnoses of such faults. If we simply bring thousands of components together into a single system, the system will certainly fail. Moreover, identifying the source of this failure is impossible, because of the multitude of unknowns, uncertainties, and ill-functioning parts.

The focus of activity shifts during the integration phase. Figure 10-4 shows the bottom-up levels of integration over time. Quality engineering begins to integrate the higher levels before the lower levels are finished, so the different levels overlap. During early integration, we emphasize the functionality and behavior of parts, components, and assemblies. Typical questions are: "Do the items fulfill their functional and performance requirements and constraints?" and "How do we handle unexpected behavior resulting from specific test conditions?" As the integration proceeds, we concentrate more on subsystem- and system-level functionality. Here we ask such questions as, "Are the interfaces described in the interface control document correct and complete?" and, "Do the combined subsystems operate correctly?" The last step in integration stresses system qualities, such as performance and reliability.

We must often test system qualities in a realistic environment where other systems provide inputs or consume outputs. The environment of a system may comprise check-out equipment, electrical and mechanical ground support equipment, and integration facilities. Examples of integration facilities for a large and a small space mission are shown in Figures 10-5 and 10-6. An example of a realistic system environment for a spacecraft in low-Earth orbit equipped with a

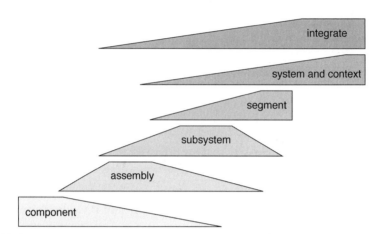

FIGURE 10-4.   Integration Over Time. Integration takes place in a bottom-up fashion with large overlaps between the integration levels.

GPS receiver is a GPS signal simulator (GSS). Since a terrestrial receiver experiences much lower Doppler shifts than a receiver moving at about 7500 m/s, a representative dynamical environment must simulate GPS signals as the spacecraft would receive them in space. The simulated environment of the system has to be realized in parallel with the system, and the two are integrated stepwise.

The discussion thus far has emphasized system integration as a process that integrates hardware or software to achieve the required functionality. But we may apply system integration to any physical or abstract end-to-end model. In the initial phase of a space project, we may start with a very simple model of a sensor. As the project matures, we integrate an increasingly detailed model into an end-to-end model, which finally allows for sophisticated sensitivity analysis. We may formulate environmental conditions, such as cleanliness for the spacecraft development, as high-level requirements. During the development and integration process, we develop and integrate models of the cleanliness of test facilities into a complex end-to-end environmental model.

10.2.3   Integration Approaches

We try to integrate subsystems or functions as early as possible, looking for unforeseen problems. So integration occurs while most of the new components, subsystems, and functions are still being developed. We normally use partial systems or modified existing systems in the early phases of integration as substitutes for the not-yet-available parts. Figure 10-7 shows this transition from using partial and existing subsystems to systems based on newly developed parts.

Now we describe the integration approach for a specific subsystem of the FireSAT spacecraft. This provides a good understanding of applied integration in practice, which translates readily to other subsystems. We focus on FireSAT's attitude

FIGURE 10-5.   **NASA's AURA Earth Observation Platform with a Mass of about 1800 kg.** Here we show the completely integrated AURA spacecraft during acoustic tests in 2003.

FIGURE 10-6.   **Integration of the Delfi-C3 Nano-satellite with a Mass of 3 kg.** Here the components come together to form subassemblies, subsystems, and the systems in the clean room at the Delft University of Technology.

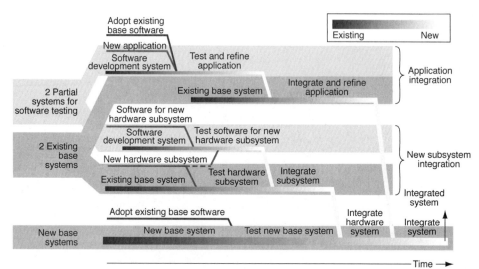

FIGURE 10-7.   A Process that Integrates a New Application and a New Subsystem for Final Integration into a New System. During integration, we transition from previous systems and partial systems to the new system configuration. The integration does not necessarily have to wait for new hardware but can apply simulated hardware instead.

and orbit control subsystem (AOCS), with the goal to verify its correct functional operation. We selected the AOCS because such subsystems are notoriously complex. This complexity is not due primarily to the dynamics or the control law, but because it involves many sensors and actuators, and we must accommodate many modes, safety precautions, and operational constraints. This scenario considers a set of four gyroscopes to measure the spacecraft's rotation rates, a coarse Earth-Sun sensor to determine the spacecraft's orientation with respect to Earth and the Sun, a set of two star trackers for determining the precise inertial spacecraft orientation, and fine Sun sensors for precision orientation with respect to the Sun.

The actuators for the FireSAT AOCS are a set of four reaction wheels to orient the spacecraft in all three axes, two magnetic torquers to support wheel saturation, and a thruster system for orbit control. Typical attitude modes are: Sun acquisition mode where the solar panels orient toward the Sun for charging the batteries; Earth acquisition mode for the payload, such as a camera, pointing toward Earth; and safe mode with a limited set of sensors and actuators to assure survival in critical situations. As the AOCS requires micro-second timing, we must prove the performance of the subsystem with flight software and hardware. Thus, in addition to computer simulations, we also need closed-loop hardware-in-the-loop tests. The relevant elements of the FireSAT environment include the spacecraft orbit, torques acting on the spacecraft, the position of the Sun, and other spacecraft subsystems, all of which are software simulated. The AOCS sensor and actuator units and the attitude control computer (ACC) are first software simulated and then replaced step-by-step by hardware.

Figure 10-8 shows a test setup for the FireSAT attitude orbit and control subsystem unit. These tests usually happen at the unit contractor to test the interfaces of their unit against the standardized check-out equipment (COE). They also test the physical performance of the sensor with specialized equipment.

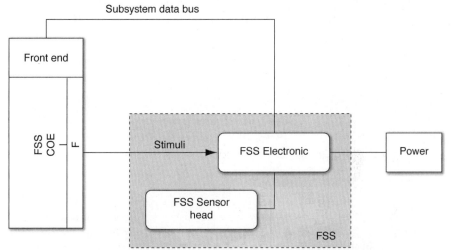

FIGURE 10-8.   Integrated FireSAT Attitude and Orbit Control System (AOCS) Closed-loop Test with the Integrated Attitude Control Computer (ACC). Here, the electronics of the fine Sun sensor (FSS) are stimulated by check-out equipment (COE). The loop is closed by a serial data bus between the FSS electronics and the COE.

The integrator establishes a closed-loop test bed for the FireSAT AOCS. The attitude control computer is tested on interfaces and is then the first unit to be integrated into the subsystem. At this stage, the subsystem check-out system (AOCS check-out equipment), which simulates the spacecraft's environment, is integrated as well (Figure 10-9). After establishing and testing this setup, engineers enhance it for new functionality. Software models for the sensors and actuators then are integrated into the ACC. The prescribed integration setup is rather flexible. If, for example, the attitude control computer is not yet available in hardware, the subsystem COE could substitute for it.

The goal of the integration test is to integrate as much FireSAT AOCS hardware as possible to verify the system's performance. To this end, the software models are replaced step-by-step by the COE (Figure 10-7) and by their sensor and actuator counterparts (based on Figure 10-8). Figure 10-10 shows a fully integrated closed-loop test setup for the FireSAT AOCS. The stimuli for the sensors and actuators, generated by the COE, are electrical. A physical excitation is in most cases not possible in the test environment.

In the figures, spacecraft hardware or engineering models are indicated as boxes with rounded corners. In Figure 10-10, actuators are indicated with bold

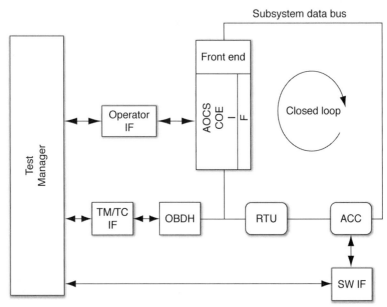

FIGURE 10-9.   **Closed-loop Integration Test Setup for the FireSAT Attitude and Orbit Control System (AOCS).** The attitude control computer (ACC) connects to the AOCS check-out equipment (COE) via a remote terminal unit (RTU). The ACC connection through the onboard data handling (OBDH) and the telemetry and telecommand interface (TM/TC IF) allows us to achieve an integrated operation at early stages. We note that the software interface (SW IF) is available to provide the means to upload software or debug output generated by the ACC. *(Adapted from the Infrared Space Observatory [ISO] AOCS.)*

boxes. The setup comprises assembly check-out equipment providing stimuli for gyroscopes, a coarse Earth-Sun sensor, star trackers, a fine Sun sensor, magnetic torquers, reaction wheels, and thrusters, which are connected through a data bus to an attitude control computer (ACC). The ACC connects to the simulator through a remote terminal unit. Communication happens through an operator interface to the AOCS simulator and through the onboard data handling system and a telemetry and telecommand interface.

In closed loop AOCS test setups, such as the one depicted in Figure 10-10, we have to verify the correct functional performance of all the attitude modes and the transitions between them. We must also test all of the safeguard functions. This amount of testing often challenges engineers because it may require massive manipulation of the test configuration to activate these functions, which are not activated in normal operations.

The discussion so far has focused on the functional aspects of integration testing. But another important role of environment simulation is the removal of the terrestrial environment. Earth's gravity in a laboratory is a completely different environment from the free-fall of objects in space. The deployment mechanisms

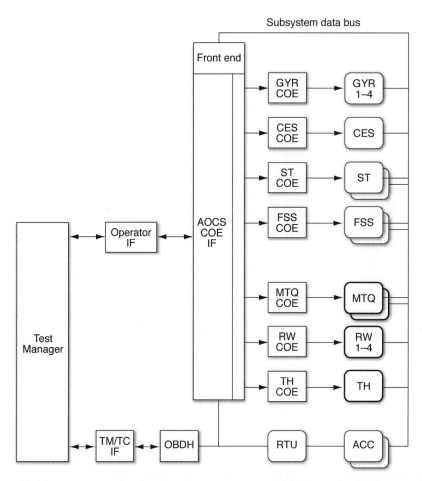

FIGURE 10-10. Closed Loop Integration Test Setup for the FireSAT AOCS Subsystem. Multiple interfaces present special challenges in testing. (AOCS is attitude and orbit control system; COE is check-out equipment; GYR is gyroscope; CES is coarse Earth-Sun sensor; ST is star trackers; FSS is fine Sun sensor; MTQ is magnetic torquers; RW is reaction wheel; TH is thrusters; TM/TC IF is telemetry and telecommand interface; Operator IF is operator interface; RTU is remote terminal unit; OBDH is onboard data handling; ACC is attitude control computer.) *(Adapted from Infrared Space Observatory attitude and orbit control system.)*

and drives for large solar arrays require special test setups to eliminate or compensate for the effects of gravity. The magnetic environment in an integration facility must be known and possibly changed to test the functional behavior of AOCS sensors like magnetic coils.

The challenge for the project team is to determine which intermediate integration configurations are beneficial. Every additional configuration costs money to create and keep up-to-date and running. An even more difficult matter is that the same critical resources, an imaging expert for instance, are needed for the different configurations. Do we concentrate completely on the final product, or do we invest in intermediate steps? Finally we have the difficulty of configuration management for all the integration configurations. When hundreds or thousands of engineers are working on a system, most are busy with changing implementations. Strict change procedures for integration configurations reduce the management problem, but in practice they conflict with the troubleshooting needs during integration. Crucial questions in determining what intermediate configurations to create are:

- How critical or sensitive is the subsystem or function to be integrated?
- What aspects are sufficiently close to final operation that the feedback from the configuration makes sense?
- How much must we invest in this intermediate step? We have to pay special attention to critical resources.
- Can we formulate the goal of this integration system in such a way that it guides the configuration management problem?
- How do contractual relations (e.g., acceptance) affect and determine the configurations?

Table 10-4 details a stepwise integration approach based on these considerations. The first step is to determine a limited set of the most critical system performance parameters, such as image quality, attitude control accuracy, or power consumption. These parameters are the outcome of a complicated interaction of system functions and subsystems; we call the set of functions and subsystems that result in a system parameter a chain. In the second step we identify the chains that lead to critical performance parameters. For the FireSAT AOCS example of Section 10.2.3, the chain comprises the star trackers as attitude determination sensors, the reaction wheels as attitude actuators, and the algorithms implemented in the attitude control computer.

Step three specifies these chains as the baseline for integration configurations. We test components and subsystems to verify compliance with the specification. The integration focuses on the aggregate performance of the elements. In this way we verify that the composition behaves and performs as it should. We start to define partial system configurations as integration vehicles after we identify critical chains. These chains serve as guidance for the integration process.

Performance parameters and their related chains are critical for different reasons: a parameter may be critical per se, or we may know (perhaps based on historic data) that the parameter is very sensitive, vulnerable, or uncertain. An example is the performance of the attitude control algorithm, composed of many low-level functions. Experience shows that individual functions tend to perform well, but the composition of many functions performs badly, due to low-level resource conflicts.

We should identify the critical system performance parameters as early as possible. In Table 10-4, this occurs in steps 4 and 5, starting with a focus on typical

TABLE 10-4. Stepwise Integration Approach. This approach guides the use of resources to the intermediate integration of the most important system parameters.

Step	Description
1	Determine the most critical system performance parameters
2	Identify subsystems and functions involved in these parameters
3	Work toward integrating configurations along these chains of subsystems and functions
4	Show system performance parameters as early as possible; start with "typical" system performance
5	Show worst-case and boundary system performance
6	Rework manual integration tests in steps into automated regression tests
7	Monitor regression results with human-driven analysis
8	Integrate the chains: show system performance of different parameters simultaneously on the same system

performance. Once the system gets more stable and predictable, we have room to study worst-case and boundary performance, e.g., a reduced number of star trackers. In steps 6 and 7, we replace preliminary integration test setups with mature systems, which allow, depending on the test case, automated regression testing. We then evaluate, analyze, and iterate the test results until they show compliance with the requirements. The final step integrates the chains and enables a simultaneous performance demonstration of the system.

10.3 Set Up a Test and Integration Plan

10.3.1 Assembly, Integration, and Verification

Space engineering normally uses the terms assembly, integration, and verification (testing) together. While *assembly* emphasizes the mechanical mating process of a complex system following manufacturing, *integration* encompasses the entire process of combining less complex functions to achieve a system satisfying its requirements. With integration, we identify unknowns in the system and resolve uncertainties. Furthermore, integration allows us to identify erroneously specified or incomplete interface definitions with potential changes in the interface control documents (ICDs) and system requirements. *Testing* is operating a system or one of its parts in a predefined way to verify its behavior. The system passes the test if the result fits the specified behavior and performance; otherwise it fails. *Verification* provides proof of the compliance of the product with the user requirements. In general, we develop a verification concept based on the system level (e.g., component or subsystem) and the applicability, using a set of verification methods.

10.3.2   Model Philosophy

Analyzing cost, risk, and development time is crucial to efficiently implementing a space mission. Because verification is costly and time-consuming, a sequential system integration of a single physical model is inefficient and quite risky. We deal with this problem by using a *model philosophy*. This defines an optimum number of physical models required to gain a high confidence in the product verification with short planning and suitable trade of costs and risks [ECSS, 1998]. Thus, we don't necessarily have to test fully representative hardware and software models. As part of a mission's model philosophy, test models only need to be representative as far as required for the test purposes. This streamlining allows early testing of certain characteristics, e.g., mechanical stability or thermal compliance, at the cost of making several models of the product. For example, we may build an exact mechanical model of the spacecraft to verify its proper accommodation and integration with the launcher's payload fairing.

The number of physical models we define for a mission depends primarily on the mission characteristics and program constraints. Manned space projects generally follow a *prototype* philosophy, which involves an extensive model development to minimize risk. This allows concurrent activities on the different models, but it's costly. In contrast, a *proto-flight* philosophy qualifies the design of a single flight model by replacing critical assemblies during the integration process. This approach is attractive if we use mostly already space-qualified hardware; it costs less but is riskier. Most common are hybrid model philosophies that combine advantages of both approaches tailored to the specific mission needs.

Table 10-5 shows the most commonly used physical verification models, together with descriptions of their major objectives and representations. Additional models may be required depending on the mission needs. The model terminology differs among agencies, industries, and countries. Chapter 11 describes a general suite of models that includes graphical, mathematical, and software models.

Additional or intermediate models may include the flight spare or the proto-flight model (PFM). Combining or upgrading models according to their qualification standard is a common practice. For example, we often combine structural and thermal models to represent a structural-thermal model, or a PFM becomes the flight model after passing necessary qualification steps. And the engineering and qualification model combines the separate engineering and qualification models.

Models differ significantly in the use of redundancy and highly reliable parts. An engineering model (EM) normally has neither. Instead, it uses the same items (except not space qualified), often from the same supplier. These parts suffice for the EM's purpose and they cost much less.

TABLE 10-5. **Physical Models for Space Systems Integration and Verification.** Almost no space mission uses all of these models. The project must identify the relevant ones based on the mission characteristics and the constraints. (FEM is functional engineering model.)

Model Name	Model Objective	Representation
Mock-up	• Interface optimization • Validation of interface process • Accommodation control • Architectural analysis	• Geometrical configuration • Layout • Interfaces
Development model, breadboard	• Development support • Proof of concept	• Selected functions
Integration model	• Functional and interface tests • Failure mode analysis • Validation of procedures	• Full functionality using commercial equipment
Structural model	• Qualification of structure • Validation of FEM	• Structural flight standard • Structure dummy
Thermal model	• Qualification of thermal design • Validation of thermal model	• Thermal flight standard
Engineering model	• Functional qualification • Failure tolerance	• Flight representative in form and function
Qualification model	• Design qualification	• Flight design and standard
Flight model	• Flight use	• Full flight design and standard

10.3.3   Test Types

Space missions employ high-tech and complex systems exposed to a hostile environment. High-tech and innovative components and subsystems require extensive testing of their functions before we can consider them for a space mission. The complexity of space missions requires a staggered testing approach as the system integration advances. The space segment must withstand extreme conditions such as temperature variations when moving from sunlight to umbra, charged-particle bombardment when crossing the South Atlantic Anomaly, and mechanical loads during launch.

Because space systems must be highly reliable, extensive testing is key to space engineering. A variety of test methods have been developed, which we group into three main categories:

- Functional tests

- Integration tests

- Environmental tests

These test methods may be applied to different physical models at different system levels.

Functional tests verify the system's functional requirements at all levels. They may include a variety of test types depending on the functions of the product. Examples are deployment tests of a solar array, closed-loop control tests of an attitude control system with real hardware in a simulated environment, radio frequency tests of a communication system, or the test of an algorithm on the onboard computer.

Integration tests normally comprise mechanical, electrical, and software integration. In mechanical integration tests, we test the assembly of lower-level systems, check mechanical interfaces, verify envelopes and alignments, and determine mass properties. Electrical integration tests check the electrical and power interfaces, power consumption, data transfer, grounding, and soldering and may include electromagnetic compatibility tests as well. Software tests check the correct functioning of software on simulated hardware or in a simulated environment.

Environmental tests verify that items will withstand the conditions they're exposed to during their operational lifetime. Included are tests related to radiation exposure and thermal and mechanical loads. Electromagnetic compatibility demonstrates that the electromagnetic emission and susceptibility of the equipment won't lead to malfunction. Total dose tests verify that the equipment can tolerate the ionizing space environment. Thermal testing demonstrates the equipment's ability to withstand worst-case conditions in a thermal-vacuum environment. These tests predominantly address orbital conditions, but mechanical tests, for such matters as linear acceleration or vibration and shock, focus on transport, handling, launch, and reentry conditions.

Some items may require other test methods, such as humidity tests if the equipment can't be environmentally protected, or leakage tests for pressurized equipment. Extraterrestrial space missions must meet protection standards to prevent biological contamination of planetary surfaces. Human space missions require tests for toxic off-gassing and audible noise.

Environmental tests are usually conducted sequentially and combined with a number of functional tests as shown in Figure 10-11. Testing the quality of workmanship is also important. Besides a visible or electrical inspection, a test under harsh environmental conditions, especially thermal-vacuum and mechanical ones, may uncover problems with workmanship, like soldering errors or loose screws.

Testing is a significant cost factor in space systems engineering. Costs come from more than planning, defining, conducting, analyzing and documenting tests. Testing also entails the necessary infrastructure, particularly costly check-out equipment (COE). The equipment is sometimes specific for a particular test or mission and developing it is often a project in itself. Examples of COE are electronic ground support equipment (EGSE) or the GPS signal simulator (GSS). We use the EGSE during ground tests to connect to the space segment via the space link and to effectively replace the ground station. The GSS simulates, based on a predefined user satellite trajectory and GPS ephemerides, the radio frequency signals that the antenna of a GNSS receiver would receive in space, including the time-varying Doppler shift. Mechanical ground support equipment should eliminate or compensate for gravity effects during the deployment tests for booms, antennas, or solar arrays.

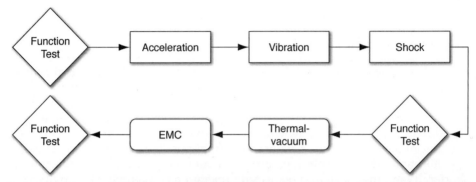

FIGURE 10-11.   Typical Sequence of Environmental Tests Combined with Functional Tests.
Environmental tests are depicted as mechanical (rectangles) and radiation (rounded boxes) tests. The sequence varies depending on the tested item and the mission requirements. The embedded functional tests vary depending on the effort of functional tests and specific environmental tests and the risks associated with it. (EMC is electromagnetic compatibility)

10.3.4   Test Methods

During early phases of development, we often test single functions at a low system level. We execute the function, observe its behavior, and measure the performance. But many problems only appear under certain circumstances. For instance, performance degradation may be noticeable only under heavy loads or after a long time. Another characteristic of early testing is a limited or ad hoc set of test data where, again, problems arise only under specific circumstances. The selection of a test configuration therefore bears heavily on integration effectiveness. This section discusses several types of testing methods:

- Testing under stress
  - Load testing
  - Lifecycle testing
- Test configuration
  - Testing using systematic data
  - Testing using random data
  - Regression testing
- Interface testing

Testing Under Stress

Load testing—This testing puts the system under heavy load conditions to verify performance under stress on system resources and algorithms. We must create or emulate such conditions early and at low cost. For example, load testing can use software simulators generating system stimuli, or intentionally stimulate items physically beyond their specification limits. Typical examples are thermal testing of electronic equipment at temperatures close to or exceeding the component specifications, and vibration tests that simulate the stresses on the spacecraft structure during launch.

Lifecycle testing—This testing denotes system tests under conditions that emulate the lifetime of the system. It's comparable to load testing with time as the critical parameter. In software dominated areas, we target effects with long time constants, such as drift or resource fragmentation. For hardware dominated areas, we address lubrication, fatigue, wear, and creep. We have to design the load and vary the inputs so as to find as many problems as possible with long time constants in a short time. The available integration and test time is often orders of magnitude shorter than the system's expected operational lifetime. Creativity and engineering ingenuity is required to "accelerate" system aging to find the effects of a very long time constant. Risk analysis helps decide the need for costly lifecycle tests. Lifecycle testing is a good example of the link between operational and functional analysis.

Test Configuration

Testing using systematic data—This involves letting parameter sets vary systematically within a predefined parameter space. An example is the power input generated by solar arrays to the spacecraft's power subsystem. Systematic test sets are desirable for their reproducibility and ease of diagnostics. We systematically test the boundary of the parameter space to verify that the system is well-behaved. This last conclusion, however, is based on the assumption that the design is smooth and continuous, without any resonances or hot spots. But in many complex systems this assumption isn't valid. For instance, mechanical systems have resonance frequencies that cause significantly higher deformations when excited than do other frequencies. So we must include test configurations at pre-determined or suspect parts of the parameter space.

Testing using random data—The previous paragraph discussed the potential system failure due to resonances or hot spots. During system design and analysis, we identify and try to eliminate these problems as far as possible. But we must also be alert for unknown singularities in the parameter space. Because systematic test data is based on a priori knowledge, it won't catch all these problems. Random test data complements systematic data. It may unveil problems that would otherwise stay invisible.

Regression testing—Since many parts of the system are still changing, we have to monitor system performance regularly. Common problems with software updates include systems that may not be executed any longer or that deviate from nominal performance. Regression tests allow us to monitor and identify regression bugs—one way is by redoing previously successful test runs. Early integration tests are usually done manually, because the system is still premature, and because integrators have to respond to many unexpected problems. In time, the chain and surrounding system get more stable, allowing automated testing. The early manual integration steps then transition into automated regression testing. Systems engineers must monitor and analyze the results of regularly performed regression tests. This analysis isn't pass or fail, but looks for trends, unexplained discontinuities, or variations.

Interface Testing

The objective is to uncover inconsistencies or errors in design of subsystems or interfaces. This is often done on the mechanical, electrical, and software levels. Interface testing allows us to identify and resolve these problems before a subsystem or instrument undergoes final production and verification. It's an important tool to reduce schedule risk during integration.

Later during integration, we integrate the chains and show the simultaneous performance of the critical performance parameters. We must document and archive all test results and test conditions. This process creates a set of historical system data, which helps troubleshooting. It answers such questions as, "Did we see this effect before?" and "How and when did the effect occur?"

10.3.5   Test Levels

As the integration process advances, different test levels are involved: *development testing* in the early phase, *qualification testing* proceeding to *acceptance testing* in the late development phase. Other test levels such as proto-flight or in-orbit testing may be introduced as well [ECSS, 2002]. In Chapter 11 we describe product verification approaches.

Development tests validate new design concepts or the application of proven concepts and techniques to a new configuration. We normally use them to confirm things like performance margins, manufacturability, testability, maintainability, reliability, failure modes, and compliance with requirements. The procedural, control, and documentation effort for development tests is usually moderate since development testing still depends on the design maturity. Nevertheless, we must maintain adequate records of test configurations and results to support other parts of the verification program. Development tests should occur over a range of operating conditions exceeding the design limits to identify marginal design features. Chapter 11 describes qualification testing and acceptance testing.

10.3.6   Test Recommendations

With the variety of test methods, types, and levels, we need a clear concept of testing, which should follow these general recommendations:

- Properly distinguish the mindset of integration testing from that of verification and validation
- Don't allow the reuse of test data

- Do allow qualification data from the integration process to be used for verification and validation

Even when testing the same system performance parameters, e.g., on attitude control accuracy, the mindset of testing in integration is completely different from that in verification and validation (V&V). During integration, we focus on identifying still unknown problems in the system. Testing during verification and validation emphasizes the system performance with respect to predefined requirements.

In general, reusing data from one test in another test is unwise due to an increased complexity of the testing concept by linked tests. Reused data is hard to overlook, laborious to properly document, and often difficult to reproduce. An exception is proper when such a test is explicitly "pre-declared" such that it receives the proper review and allows us to maintain the oversight in the test concept. Under rigorously defined conditions, a test that serves two or more purposes can accelerate integration and reduce cost.

Integration covers most of the development process for a space mission. Near the end of the development cycle, the emphasis shifts toward verification and validation. The data we've acquired during testing constitutes the link between integration and V&V.

10.3.7   Concurrent Integration

Integration within space projects is not a sequential task but a concurrent process. It shortens development time and allows us to identify and solve critical issues early in development, reduces risk, and lets us verify the adequacy and completeness of interfaces.

Concurrent integration is largely based on the applied model philosophy and on simulating and emulating functionality. The model philosophy lets us test subsystems independently of each other and in parallel, as discussed above. Table 10-6 depicts the relation of verification efforts to the model philosophy.

Another important aspect of concurrent engineering is the use of simulation or emulation models early in the development. The underlying idea is that software, to a large extent, simulates the performance of an assembly or subsystem long before the physical model is available. To this end, a model of the entire or partial segment may be built early in the development process, which allows us to simulate its functionality based on simulated subsystems. Thus, concept validation and interface verification is already taking place while the system is gradually becoming more and more mature through the successive replacement of mathematical models by their physical counterparts (Figure 10-7).

TABLE 10-6. Verification and Test Levels Related to the Model Philosophy. Test levels are indicated by demonstration (D), qualification (Q), and acceptance (A). The relation shows a typical space scenario, which may be adapted depending on mission characteristics.

Model	Verification Level			Test Level
	Assembly	Subsystem	Segment	
Mock-up	X			D
Development model	X	X	X	D
Integration model	X	X	X	D
Structural model		X		Q
Thermal model		X		Q
Engineering model	X	X	X	Q
Qualification model	X	X		Q
Flight model	X	X	X	A

10.4   Schedule in the Presence of Unknowns

Space project scheduling is tricky, even when we know what to anticipate. But previously unknown difficulties invariably crop up, and we have to deal with them.

10.4.1   Integration Planning and Scheduling

The approach described in the previous section requires lots of logistics support. The project leader, therefore, must prepare integration schedules in close cooperation with the systems engineers. Integration schedules have two conflicting attributes:

- Predictability and stability to ensure timely availability of resources
- Flexibility and agility to cope with the inherent uncertainties and unknowns

The starting point to creating a schedule is determining a detailed order of integrating components and functions to measure the desired critical system performance parameters. Figure 10-12 shows an example of a specific order of functions required to determine the performance parameter for the image quality system of a waferstepper (semiconductor equipment to expose wafers). Such a diagram often starts at the right-hand side, the desired output parameter. Next, we recursively ask the question, "What do we need to achieve this output?" This partial diagram is still highly simplified. In reality many of these functions have multiple dependencies.

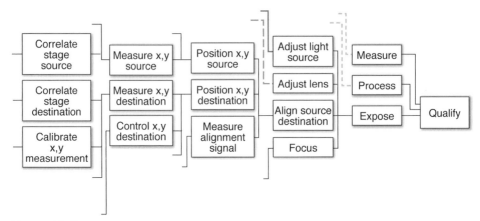

FIGURE 10-12.   **Simplified Subset of Functions Required for the Performance Parameter of an Image Quality System of a Waferstepper.** At the left hand side, elementary (local) functions are shown. As we move to the right, these local functions are combined into system-level functions. Combining functions continues until the desired integral system function is achieved (an exposure that qualifies in the semiconductor factory context).

Circular dependencies complicate the process. For instance, to align source and destination we need to be in focus, and to find the focal plane we need to be aligned. These dependencies are known and solved at design time. A frequently used design pattern is a stepwise refinement of coarse and fine alignment, and coarse and fine focusing. A detailed integration schedule provides worthwhile inputs for the design. Making the integration schedule specific forces the design team to analyze the design from an integration perspective and often results in the discovery of many unresolved implicit assumptions.

10.4.2   Reviews

Integration is a continuous process extending through most phases of a space mission. The systems integration review (SIR) marks the end of the final design and fabrication phase (Phase C) and its successful completion indicates readiness to proceed to the system assembly, integration and test, and launch phase (Phase D) [NASA, 2007]. The objective of the SIR is to ensure that the system is ready to be integrated. To that end all subsystems and components, integration plans and procedures, and integration facilities and personnel must be prepared. The review's success is determined by criteria for accepting the integration plan, identifying and accepting the risk level, and defining and documenting the work flow. The process also includes defining consequences and recovery strategies in the case of a less than fully successful review.

Given the close relation between integration and testing, a test readiness review (TRR), conducted early in Phase D, follows the SIR. Here, the objective is to ensure that the article to be tested, the required test facilities, and means for data handling are ready. A successful TRR provides approved test plans, including the coordination of required test resources and facilities. It doesn't mark the beginning of the project test program but provides a formal initiation of tests to be conducted at a higher integration level.

10.4.3   Dealing with Disruptive Events

A well-defined integration schedule is valuable for the design and for understanding the integration. But the integration process isn't very predictable; it's an ongoing set of crises and disruptions, such as late deliveries, broken components, non-functioning configurations, missing expertise, wrong tolerances, interfering artifacts, etc. Improvising and troubleshooting are crucial.

The integration schedule is a volatile and dynamic entity. We don't formalize the integration heavily, nor keep it updated in all details. Formalizing and extensive updating take a lot of effort and yield little benefit. It's better to use the original integration schedule as a reference and to apply short cyclic planning steps to guide the integration process. Meetings during peaks of integration activity occur as often as every day; attendees discuss results and problems, required activities and resources, and the short-term schedule.

Despite the inevitable disruptions, we should follow the integration schedule. Lessons learned from previous projects help plan integration phases that are long enough to be realistic. We must carefully identify risks to the integration process and how to mitigate them. We have to adhere to national and international export, import, and shipping regulations. We need to identify alternate suppliers for space qualified hardware early in the process. Ordered unit numbers should account for all testing efforts and possibly some overhead. Available spares add flexibility to the integration and test schedule. We minimize the consequences of scheduling conflicts or risks by decoupling critical steps.

10.5   Form the Integration Team

Integration involves many project team members in different roles and responsibilities:

- Project leader
- Systems engineer and architect
- System tester
- Engineers
- System owner
- Logistics and administrative support personnel

Figure 10-13 shows these roles in relation to their responsibilities. The names of these roles depend on the organization, so this chapter uses generic labels.

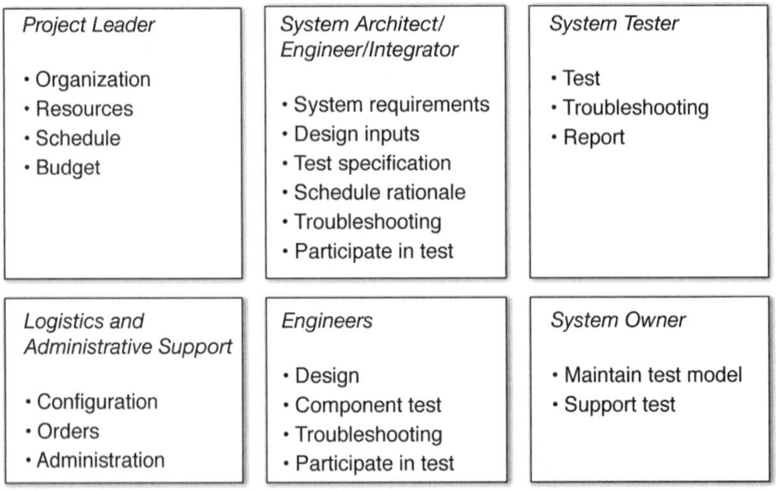

Project Leader	System Architect/ Engineer/Integrator	System Tester
• Organization • Resources • Schedule • Budget	• System requirements • Design inputs • Test specification • Schedule rationale • Troubleshooting • Participate in test	• Test • Troubleshooting • Report
Logistics and Administrative Support	Engineers	System Owner
• Configuration • Orders • Administration	• Design • Component test • Troubleshooting • Participate in test	• Maintain test model • Support test

FIGURE 10-13.   **Roles and Responsibilities during the Integration Process.** These roles and responsibilities aren't isolated. Communication and coordination among team members are crucial for successful integration of a complex system.

The *project leader* takes care of managing resources, schedule, and budget. Based on inputs from the systems engineer, the project leader claims and acquires the required resources, and also facilitates the integration process. This contribution is critical for project timing.

The *systems engineer's* role is in fact a spectrum of roles comprising architecture, integration, and systems engineering. Depending on their capabilities, one or more persons may take on this role—a systems engineer might be a good architect and a bad integrator or vice versa. This role depends on content, relating critical system performance parameters to design and test. It determines the rationale of the integration schedule; the initial integration schedule is a joint effort of project leader and systems engineer. The integral perspective of this role results in a natural contribution to the troubleshooting. The *systems architect* or engineer is responsible for the integration tests' success, and is also in charge of evaluating the tests.

The *system tester* actually performs most of the tests. During the integration phase, this person spends a lot of time in troubleshooting, often taking care of trivial problems. More difficult problems should be referred to engineers or the system integrator. The system tester documents test results in reports for the systems engineer to review and evaluate.

The *system owner* is responsible for maintaining a working up-to-date test model. This is a challenging job, because many engineers are busy making updates and doing local tests, while system integrators and system testers need undisturbed access to a stable test model. Explicit ownership of one test model by one system owner increases the test model stability significantly. Projects without such a role lose a lot of time to test model configuration problems.

Engineers deliver locally tested and working components, functions, or subsystems. But their responsibility continues into the integration effort. Engineers participate in integration tests and help in troubleshooting.

Many kinds of support personnel supplement the project team. For integration, *logistics and administrative support* is crucial. This support comprises configuration management of test models and of the products to be manufactured. Integration problems may lead to changes in the formal product documentation and the logistics of the final manufacturing. These changes in turn have significant financial consequences due to the concurrency of development and preparation for production. The logistics support staff also has to manage and organize unexpected but critical orders for parts of test models.

10.6   Manage a Changing Configuration

Configuration management documents the functional and physical properties of a product, controls changes to these properties, and documents change processing and implementation status of the product. This effort allows all parties to verify compliance with agreements, requirements, and specifications for the product. Furthermore, configuration management enables everyone to use identical data during the project lifecycle, provided under a controlled process. It's vital for verifying contractual obligations, allows a system replication at any point in the process, and enables us to trace anomalies to the design or implementation. Chapter 16 discusses configuration management in considerable detail.

10.6.1   Concurrency of Integration Testing and Repeated Production

Configuration management is complicated because it generally involves more than the company that develops the product. The company acquires items from suppliers and delivers a product to a customer. Figure 10-14 depicts this situation. The company is responsible to the customer for the supplier's configuration management activities. Integration testing occurs concurrently with the start-up of the logistics flow of components that have to be produced in large quantities.

While configuration management applies to processes with repeated assembly, integration, and testing, space missions usually involve one-of-a-kind systems. This tends to simplify configuration management, because the system used for integration is the same as the one delivered. We must still take care to properly apply configuration management in the framework of the model philosophy. An advantage of one-of-a-kind systems is that the production chain omits the difficult transfer step from creation to repetition.

We have to bear in mind that within the single system a repetition of subsystems still requires repeated production, with all its integration consequences. Examples are the charged-coupled device lines for stereoscopic pushbroom scanners, L-band antenna arrays on GPS spacecraft, or mechanisms for bearing assemblies that connect the spacecraft body and the solar arrays.

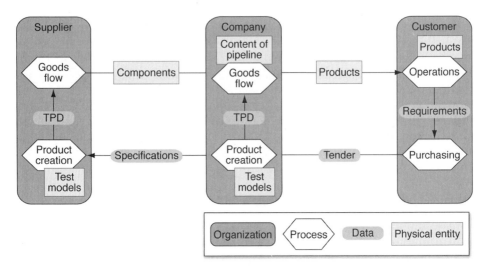

FIGURE 10-14. **Simplified Process Diagram That Shows Processes That are Relevant from a Configuration Management Perspective.** Here we show that the supplier fills orders from the company that is filling orders from the customer. All orders must meet specifications. All data and physical entities shown in the diagram are under configuration management. The company is responsible to the customer for the supplier's configuration management activities. (TPD is technical product documentation.)

10.6.2 The Role of Testing Environments and Analysis Techniques

Destructive or potentially damaging integration steps are especially harmful to one-of-a-kind systems. Alternative environments, as shown in Figure 10-10, become even more important. The virtual nature of most of the integration environment demands a lot of attention to the potential differences between it and reality. Techniques such as failure modes and effects analysis or fault tree analysis applied in the broader context of the environment help identify potential problems early. Integration requires support for configuration management of testing environments and additional information such as failure scenarios.

10.6.3 Impact and Propagation of Changes

Many design and implementation changes happen during integration. Problems arising from the integration are one of the main sources of these changes. The theory of requirements traceability [Ramesh et al., 1997] suggests that we should predict and manage the impact of change (see Chapter 4). Unfortunately, the situation in practice is much more difficult. The left-hand side of Figure 10-15 shows the formal situation as defined by the traceability graph. The effect of a low-level change is traceable to the impact on system-level requirements. The right-hand side of the figure shows the real situation: a number of undocumented relations between design choices pop up, causing an avalanche of additional affected requirements. Every additional relationship triggers its own avalanche of affected higher-level requirements. The counteraction to repair the design triggers a new avalanche in the top-down direction.

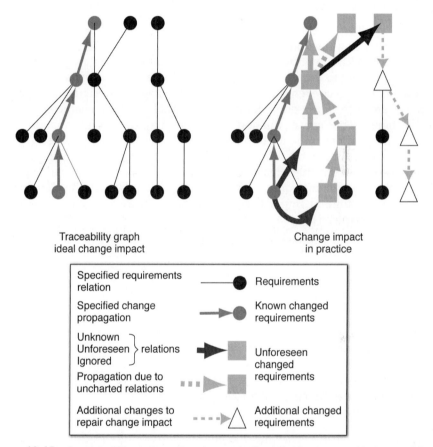

Traceability graph
ideal change impact

Change impact
in practice

Specified requirements relation	——●	Requirements
Specified change propagation	⟶▶●	Known changed requirements
Unknown ⎤ Unforeseen ⎬ relations Ignored ⎦	▬▬▶■	Unforeseen changed requirements
Propagation due to uncharted relations	▪▪▪▶■	
Additional changes to repair change impact	- - -▶△	Additional changed requirements

FIGURE 10-15.   **Impact of Changes on Requirements.** The left side shows an ideal case while the right side exhibits a much more complicated situation as encountered in practice.

These effects turn integration into a barely predictable and controllable activity. The system is constantly changing while being integrated. These changes continually change subsystem and system performance, undoing the previous integration progress. Sufficient regression testing helps us cope with this problem.

10.7   Adapt Integration to the Project Specifics

This section discusses special topics related to system integration. The references cited in each subsection provide more extensive information.

10.7.1   Integrating Commercial-off-the-shelf (COTS) Products

Small satellite projects especially make extensive use of COTS products. Unit cost may be a tiny fraction of fully space-qualified units. Often, however, these products are neither mature nor stable. Production processes may change from one production lot to the next without notice and qualification procedures may be modified or interfaces even changed. The disadvantage for design and integration is that many times we don't know the internal product characteristics and specifications. The space mission designer may need information about functionality and performance beyond the black box specification. Designers are tempted to make implicit assumptions about the product without being aware of the implications. A feasible alternative is to document assumptions and to provide a basis for these assumptions by characterizing the product. The challenge for integration is to find the implicit assumptions as early as possible to facilitate the verification (Chapter 11).

Characterizing a product means measuring and modeling it to capture the required information. Ideally, this information feeds back into the supplier's specification. But in the COTS market, suppliers are often unwilling to modify their specifications. If we can't anchor this information at the supplier, the system design is vulnerable. The supplier might change the product implementation, without changing the specification, but still affecting the earlier characterization. For example, the operating system supplier might update the network protocol stack for security reasons. The update doesn't change any functionality, but some operations might be slowed down considerably.

10.7.2   Subcontracting and Outsourcing

The complexity of space projects often calls for subcontracting and outsourcing to realize a space mission efficiently [NRC, 2004]. Government space projects, in contrast to purely commercial ones, may include subcontracts to universities, research institutions, and knowledge institutes. In outsourcing, the same issues play a role as for COTS technology. The integration strategy depends on the chosen outsourcing model, which may be a black-box supplier, a co-designer, or a co-manufacturer. In general, the main challenge is to agree on early deliverables for integration. A common mistake is to wait until the production and qualification process for the final delivery. This mistake results in late discovery of integration problems, with late rework as a consequence. Even in the case of black box outsourcing, intermediate (grey box) deliverables must be organized for integration purposes.

10.7.3   Integration for Small Space Projects

Budgetary constraints and a rapid evolution of miniaturization technology have led to space missions employing smaller and smaller spacecraft. Such missions have far fewer suppliers and actors than large space missions, but may still be highly complex. So we must implement a clear systems engineering approach for these projects as well [Meijers et al., 2004]. The project manager is responsible for adapting the analysis, design, integration, and test processes to fit within the scope of the project [Hamann and van Leeuwen, 1999]. Integration and test facilities are more easily accessible for small than for large space projects.

10.8   Solve Typical Problems

Having integrated a complex system, we face a good chance that the system will not behave as expected but will exhibit unforeseen problems. Table 10-7 provides a typical sequence of integration problems. NASA [2001] has archived a host of lessons learned from the process of designing and integrating a launch vehicle, which apply to other space missions as well. The following paragraphs discuss these integration problems in detail along with typical examples from space projects. In theory, none of these problems should occur, but in practice all of them **can** occur, despite mature processes.

TABLE 10-7.   **Typical System Integration Problems, Their Occurrence, and How to Avoid Them.** In spite of the project team's best efforts, and adherence to correct processes, unforeseen difficulties can and do arise.

Problem	When it Occurs	How to Avoid
System cannot be integrated	At the beginning of the integration process	• Check implicit assumptions on the subsystems and the integration environment
System does not function	After the system has been integrated	• Check implicit assumptions • Check the interfaces
System is too slow	After the system functions	• Check test specification • Check concurrent processes and timeline • Check resources due to monitoring
System performance is not achieved	After the system operates	• Review the test conditions • Analyze the performance under simplified conditions
System is not reliable	After the system operates nominally	• Analyze late design changes • Identify conditions not tested before

System cannot be integrated. The failure to integrate the system (or to build it when software integration is concerned) often arises from the use of implicit know-how. An example is a relatively addressed data file that resides on the engineer's workbench, not working in the independent test environment. Network connections shorten software change cycles and help in troubleshooting, but they don't uncover these problems.

System does not function. The next level of problem is when individual functions work, but cease to function when combined. The source of the trouble may be a violation of implicit assumptions. This problem is often related to interface errors. It could be missing or unclear specifications in the interface control documents (ICDs), or implementations that don't adhere to the ICD specifications. Another type of problem comes from the implicit assumptions themselves. For example, the use of a calling subsystem depends on the called subsystem functioning properly. Unexpected behavior on the part of the called subsystem causes problems for the caller. These difficulties are often not visible at the interface specification level, because none of the subsystem designers realized that such behavior is relevant at the interface level.

System is too slow. Once the system reaches a basic operational functionality, the nonfunctional system properties become apparent. Integrators usually encounter deficiencies in the system's speed or throughput. Individual functions and components in isolation perform well, but when everything runs concurrently, sharing the same computing resources, performance suffers. The mismatch of expected and actual performance comes not only from concurrency and sharing, but possibly also from an increased load due to more realistic test data. On top of this, non-linear effects may worsen the overall performance even more. After some redesigns, the performance problems tend to go away, but we should continue to monitor the system. Performance might degrade further during integration, due to added functionality of the integration environment.

System performance is not achieved. When the system is functional and performs well, then its core functionality, the main purpose of the product, is tested extensively. In this phase, the application experts are closely involved in integration. They may use the system differently and look differently at the results. Problems in critical system functionality are discovered in this phase. Although present in the earlier phases, they stayed invisible because other integration problems dominated and the testers and the application experts had different perspectives. We should also review the test conditions to be sure they're fully applicable. Finally, it may be worthwhile to simplify some test conditions to trace down the problem.

System is not reliable. During the last integration phase we use the system more and more intensively. Such uses touch on the less robust parts of the design, causing flaws to become manifest. A common complaint in this phase is that the system is unreliable and unstable. This is partly due to continual design changes triggered by earlier design phases, with the designers not realizing that every change may also trigger new problems.

Summary

Systems integration is critical to the entire creation process. All unknowns and uncertainties left over from previous activities show up during this phase. In space systems the final integration takes place after launching the spacecraft, which increases the importance of all types of tests during integration. We've discussed how to set up tests and an integration plan, and have elaborated the roles in the integration team and the complexity of configuration management. Integrators always have to troubleshoot and solve problems, despite the smooth interplay of process, people, and plans.

References

Department of Defense (DoD). January 2001. System Engineering Fundamentals. Fort Belvoir, VA: Defense Acquisition University Press (DAU).

European Cooperation for Space Standardization (ECSS) Secretariat. April 1996. *Space Project Management*. ECSS-M-40A. Noordwijk, The Netherlands: ESA Publications Division.

ECSS Secretariat, ESA-ESTEC Requirements and Standards Division. November 17, 1998. *Space Engineering—Verification*. ECSS-E-10-02A. Noordwijk, The Netherlands: ESA Publications Division.

ECSS. February 15, 2002. *Space Engineering—Testing*. ECSS-E-10-03A. Noordwijk, The Netherlands: ESA Publications Division.

ECSS. 2008. *System Engineering—Part 1: Requirements and Process*. ECSS-E-ST-10 Part 1B. Noordwijk, The Netherlands: ESA Publications Division.

Forsberg, Kevin and Hal Mooz. September 1992. "The Relationship of Systems Engineering to the Project Cycle." *Engineering Management Journal*, Vol. 4, No. 3, pp. 36–43.

Hamann, Robert J. and Wim van Leeuwen. June 1999. "Introducing and Maintaining Systems Engineering in a Medium Sized Company." *Proceedings of the 9th Annual International Symposium of the International Council on Systems Engineering*.

International Council on Systems Engineering (INCOSE). June 2004. INCOSE *Systems Engineering Handbook*, Version 2a. INCOSE-TP-2003-016-02.

Meijers, M.A.H.A., Robert J. Hamann, and B.T.C. Zandbergen. 2004. "Applying Systems Engineering to University Satellite Projects." *Proceedings of the 14th Annual International Symposium of the International Council on Systems Engineering*.

National Aeronautics and Space Administration (NASA). June 1995. *NASA Systems Engineering Handbook*. NASA SP-610S.

NASA. 2001. *Launch Vehicle Design Process Characterization, Technical Integration, and Lessons Learned*. NASA TP-2001-210992. Huntsville, AL: Marshall Space Flight Center.

NASA. March 26, 2007. *NASA Systems Engineering Procedural Requirements*, NPR 7123.1a.

National Research Council (NRC) Committee on Systems Integration for Project Constellation. September 21, 2004. *Systems Integration for Project Constellation: Letter Report*. Washington, DC: The National Academies Press.

Office of the Under Secretary of Defense (OUSD) for Acquisition, Technology and Logistics. July 2006. *Defense Acquisition Guidebook*, Version 1.6. Fort Belvoir: DAU Press. http://akss.dau.mil/dag.

Pisacane, Vincent L., ed. 2005. *Fundamentals of Space Systems*, 2nd Edition. Oxford: Oxford University Press.

Ramesh, Balasubramaniam et al. January 1997. "Requirements traceability: Theory and practice." *Annals of Software Engineering*, no. 3 (January): 397–415. Dordrecht, The Netherlands: Springer Netherlands.

Chapter 11

Verification and Validation

Jerry Jon Sellers, *Teaching Science & Technology Inc.*
Riley M. Duren, *Jet Propulsion Laboratory*
William H. Arceneaux, *NASA Johnson Space Center*
Edward B. Gamble Jr., *Jet Propulsion Laboratory/*
California Institute of Technology

In this chapter we turn our attention to the evaluation processes of the systems engineering "engine" as shown in Figure 11-1. In Table 11-1 we show the structure of our discussion. Inadequately planned or poorly executed verification and validation (V&V) activities have been at the heart of too many high-profile space failures. Table 11-2 summarizes a few of the major mission failures that more rigorous system V&V might have prevented. Project verification and validation encompasses a wide variety of highly interrelated activities aimed at answering several key questions throughout the mission lifecycle in roughly the following order:

1. Requirements validation—Are these the right requirements?

2. Model validation—Are the models that support design and verification activities correct?

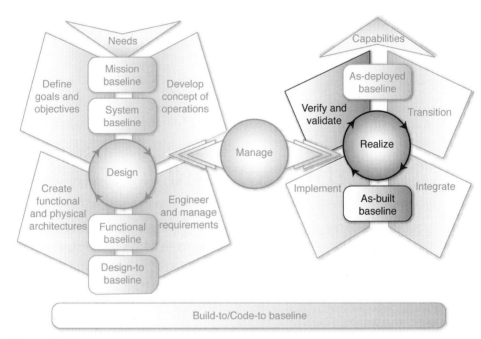

FIGURE 11-1.   We Are Here! Verification and validation activities fit on the right side of the systems engineering "V." However, much of the planning begins considerably earlier, on the left side of the "V."

TABLE 11-1.   Overview of Verification and Validation Activities. This table summarizes the major steps and key activities that fall under the broad umbrella of verification and validation.

Step	Activity	Where Discussed	Related Chapters
1	Validate requirements	Section 11.1	Chapter 4
2	Validate models	Section 11.2	Chapter 10
3	Verify products • Prepare for product verification • Implement product verification	Section 11.3	Chapter 10
4	Validate products • Prepare for product validation • Implement product validation	Section 11.4	Chapters 9 and 12
5	Certify products for flight	Section 11.5	Chapter 12
6	Address special challenges of commercial off-the-shelf and non-developmental items product verification	Section 11.6	Chapter 10
7	Verify and validate software	Section 11.7	Chapter 10
8	Document and iterate	Section 11.8	

3. Product verification—Does the system we built or coded meet the requirements we wrote?
4. Product validation—Did we build or code the right system (sanity check on all of the above)?
5. Flight certification—Is the system ready to fly?

TABLE 11-2. **Excerpts from NASA Mishap Reports.** This table lists several examples of inadequate verification and validation on deep space missions. Quotations are from the missions' mishap reports.

Mission Year	Mishap	Verification and Validation Contributing Factors
Genesis 2004	G-switch installed backwards • Parachute not deployed • Hard landing	"...tests used a bypass of the G-switch sensors and focused on higher-level verification and validation..." "...inappropriate faith in heritage designs..."
Columbia 2003	Debris damaged thermal tiles • Loss of crew and vehicle	"...current tools, including the Crater model, are inadequate to evaluate Orbiter Thermal Protection System damage..." "Flight configuration was validated using extrapolated test data...rather than direct testing"
Comet Nucleus Tour (CONTOUR) 2002	Overheating of spacecraft by solid rocket motor plume • Vehicle lost	"Project reliance on analysis by similarity"
Wide Field InfraRed Explorer (WIRE) 1999	Electronic startup transient • Early cover jettison • Cryogen boil-off • Science mission lost	"The anomalous characteristics of the pyro electronics unit were not detected during subsystem or system functional testing due to the limited fidelity and detection capabilities of the electrical ground support equipment."
Mars Polar Lander 1998	Software flaw • Descent engine shut off too soon • Vehicle lost	"Employed analysis as a substitute for test in the verification and validation of total system performance...tests employed to develop or validate the constituent models were not of adequate fidelity..."

Table 11-3 shows the top-level process flow for each of these activities. Because systems of interest vary from the part level to the complete system of systems level, or, in the case of software, from the code level up to the installed full system level, the terms *product verification* and *product validation* apply at any level.

Verification and validation occur for products we use every day. From shampoo to ships, extensive V&V has occurred (we hope) to ensure these products meet their intended purpose safely. What makes space systems so different? Unlike shampoo, space systems are usually one-of-a-kind. We don't have the luxury of lot testing to verify production processes. Unlike ships, space systems don't go on "shakedown cruises" to look for flaws, then sail back to port to be fixed; we have only one chance to get it right the first time. So space system verification and validation must be creative, rigorous, and unfortunately, often expensive.

TABLE 11-3.   Top-level Process Flow for Verification, Validation, and Certification Activities. Many of the sub-process outputs serve as inputs for succeeding sub-processes. For instance, validated requirements are an output of the requirements validation sub-process, and serve as an input to the product verification sub-process. (FCA is functional configuration audit; PCA is physical configuration audit.)

Inputs	Verification, Validation, and Certification Activities	Outputs
• Unvalidated requirements	**Requirements validation**	• Validated requirements
• Unvalidated mission critical models • Validated model requirements	**Model validation**	• Validated models • Model uncertainty factors • List of model idiosyncrasies
• Validated requirements • Unverified end products • Verification plan (including incompressible test list) • Verification enabling products (e.g., validated models)	**Product verification**	• Verification plan (as implemented) • Verified end products • Verification products (data, reports, verification completion notices, work products)
• Verified end products • Customer expectations (e.g., measures of effectiveness and other acceptance criteria) • Operations concept • Validation plan • Validation enabling products	**Product validation**	• Validated products • Validation products such as data test reports and work products
• Verified and validated products • Verification and validation products • Real-world operational context for end products	**Flight certification**	• Certified product • Certification products (e.g., signed DD250–Material Inspection and Receiving Report; completed FCA and PCA; mission rules)

Let's start with some basic definitions. NPR 7120.5d [NASA (1), 2007] defines *verification* as "proof of compliance with design solution specifications and descriptive documents." It defines *validation* as "proof that the product accomplishes the intended purpose based on stakeholder expectations." In both cases, we achieve the proof by some combination of test, demonstration, analysis, inspection, or simulation and modeling. Historically, the space industry has developed one-of-a-kind products to fulfill a single set of requirements that encompass the design and manufacturing specifications, as well as the customer requirements. This heritage has merged verification and validation into one process in the minds of most practicing aerospace engineers. But products developed by the space industry are becoming increasing varied, and we need to fit new space systems into larger, existing system architectures supporting a greater diversity of customers. So we treat the two processes as having two distinct objectives. This chapter describes product verification and validation separately. It also includes a separate description of certification, and highlights how V&V supports this important activity. For any mission ultimately to succeed, we carry out all five of the activities listed in Table 11-3.

Figure 11-2 shows the basic flow of V&V activities. The emphasis of the different types of validation and verification varies throughout the project lifecycle, as shown in Figure 11-3. An appreciation for the activities that constitute V&V, and their shifting emphasis throughout the project, better prepares us to develop good requirements and models, verified products, and ultimately systems that work the way the customer intended.

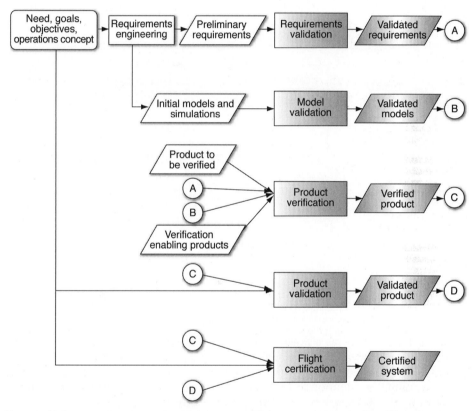

FIGURE 11-2.   Verification and Validation Activities in the System Lifecycle. This diagram depicts the logical sequence of these activities and their interdependencies. Shaded blocks indicate topics discussed at length in this chapter. Parallelograms indicate inputs or outputs of activities, which are shown in rectangular boxes.

This chapter examines each of the five activities of project V&V, starting with a brief review of the importance of having properly validated requirements as the basis. Next we discuss model validation as a critical but somewhat separate activity that supports other V&V and operational planning activities. We then address product verification activities, the most crucial, and typically most expensive, aspect of all V&V. After this, we examine the issues associated with product validation

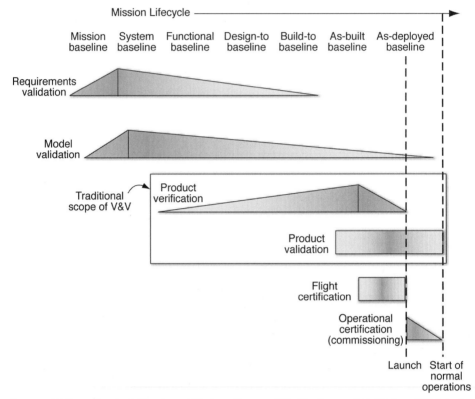

FIGURE 11-3.   Nominal Phasing of Various Types of Verification and Validation (V&V) in the Project Lifecycle. This figure illustrates the relative level of effort expended on each of the V&V activities throughout the lifecycle. The traditional scope of V&V activities centers on product verification and validation. But the broader scope includes requirements and model validation, and flight certification. For completeness, we include operational certification, also known as on-orbit commissioning.

before turning our attention to the capstone problem of flight certification. Finally, we review unique aspects of using commercial-off-the-shelf (COTS) or non-developmental items (NDI) for space applications followed by an examination of software V&V. The chapter concludes with case studies and lessons learned followed by a look at the FireSAT example.

This chapter uses the terms "system" and "product" interchangeably. A product or system V&V program includes flight hardware, flight software, ground support equipment (GSE), ground facilities, people, processes, and procedures (i.e., everything we use to conduct the mission). As Figure 11-3 illustrates, these activities span the entire project lifecycle. We categorize V&V activities as follows:

1. Pre-launch V&V leading to certification of flight readiness—Phases A–D

2. Operations V&V such as on-orbit check-out and commissioning—Phase E

We focus primarily on pre-launch V&V activities, but most of these also apply to operational and calibration mission phases. Let's review the supporting elements that make V&V possible. These include:

- Requirements
- Resources, such as:
    - Time—put V&V activities in the schedule on day one
    - Money
    - Facilities—GSE, simulation resources
    - People—subject matter and verification experts who have "been there and done that" to avoid reinventing the wheel and repeating old mistakes
- Risk management
- Configuration management
- Interface management
- Technical reviews

The first item is requirements. Without good requirements we have nothing to verify the product against and can't even build the product. Therefore, good, valid requirements form the basis for all other V&V. The next section addresses this first of the five activities.

11.1   Validate Requirements

Clear, well-written, and stable requirements are critical to the success of any program. Chapter 4 describes the process of requirements engineering that leads us from customer expectations to detailed technical requirements, and lays out aspects of "good" requirements. This process relies heavily on requirements analysis tools such as functional flow block diagrams. *Product verification*, in Section 11.4, depends totally on the quality of requirements and specifications defined early in the project and continually updated as the project evolves. *Requirements validation* confirms that the mission objectives and stakeholders' needs have been crisply captured in requirements necessary to design, build, verify, and operate the system. Figure 11-4 shows the feedback loop between customer expectations and technical requirements.

We have to establish from the outset a team culture that strives to ensure all requirements are VALID. We must constantly ask, "Are the requirements..."

- Verifiable
- Achievable
- Logical
- Integral
- Definitive

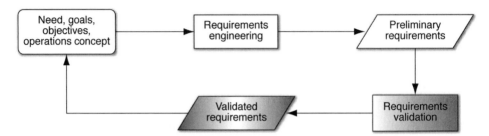

FIGURE 11-4.   **Requirements Validation Flow Chart.** This figure illustrates the feedback between customer expectations and technical requirements. Requirements validation determines whether a system built to meet the requirements will perform to the customers' expectations. Shaded blocks indicate topics discussed at length in this chapter. Parallelograms indicate inputs or outputs of activities, which are shown in rectangular boxes.

Verifiable—We must express the expected performance and functional utility so that eventual verification is objective, and preferably quantitative. Such words as "excessive," "sufficient," "resistant," and other non-quantifiable terms, are not easily verifiable. The verification activity must also be practical. For example, a mean time between failure requirement of two years for an ion rocket engine may not be practical to verify by demonstration if we have only a few months available for development.

Achievable—A requirement must be realistic. Early project assessments should focus on technology readiness levels to ensure that we have enough resources to mature enabling technologies in pace with overall system development. This normally involves iteration and negotiation; the project objectives must be appropriately balanced with available resources and risk posture. Detailed assessments of achievability require extensive modeling and simulation and other requirements-engineering techniques as described in Chapter 4. A detailed technology development plan must be part of technical planning as described in Chapter 13.

Logical—Top-level requirements should fit within the context of the system. Lower-level requirements should follow logically from and support higher-level requirements. This order is consistent with a well-defined requirements traceability process that supports the interrelationship between and among requirements.

Integral—The requirements set should be complete and all-encompassing. Missing requirements lead to missing functionality.

Definitive—Each requirement must precisely address one and only one aspect of the system design or performance. It must be unambiguous, with only one possible meaning. It must be complete and contain all mission profiles, operational and maintenance concepts, environments, and constraints. All information necessary to understand the customer's need must be there. It must be expressed in terms of need, not solution; that is, it should address the "why" and "what" of the need, not how to meet it.

Good requirements may have attributes that go beyond these (e.g., "needed" as described in Chapter 4). However, VALID serves as a good starting point and provides succinct guidance for anyone charged with reviewing and validating requirements.

We usually complete formal validation of the preliminary or top-level system requirements before the system requirements review (SRR) early in Phase A. Subsequently, we should constantly review all new, derived, or evolving requirements for validity as the system design matures. Requirements we initially thought to be achievable and verifiable, for example, may turn out to be neither, due to unavailability of technology, facilities, time, or money. Therefore, we should periodically revalidate the set of requirements for each system of interest as part of each technical baseline.

Table 11-4 shows an example of FireSAT requirements. We start with a mission-level requirement to maintain 12-hour revisit of the coverage area and see how this flows down through a series of mission analyses and allocation to component-level performance specifications. We then judge each requirement against the VALID criteria with methods and issues presented in the discussion. Such a matrix is useful to systematically assess each requirement. The earlier we identify flaws in requirements, the less it costs to fix them.

Chapter 4 addresses the challenge of crafting well-written requirements that are clear and concise. Here we focus more on their achievability and verifiability. While precise wording is essential to communicating requirements, it's equally important that they be the right ones. Thus, to truly validate a requirement, we must dig into the analyses or other models used to justify it. The next section examines the issue of model validation in greater detail.

Poorly defined or poorly allocated requirements lead to mission failure. This occurred on the Mars Polar Lander mission. Premature shutdown of the descent engines was the probable cause of the failure. The mishap report attributed the root cause of this failure to poor allocation of system-level requirements to subsystems, in this case, software-level requirements. Figure 11-5 illustrates the mapping from system to software requirements. The system requirements specified not using sensor data until reaching 12-meters altitude. They set further criteria to protect against premature descent engine thrust termination in the event of failed sensors and possible transients. However, the requirements did not specifically state failure modes. So software designers did not include protection against transients, nor think of testing for such transients. The lack of a complete software specification meant failure to verify that requirement during unit-level tests. In Section 11.5, we describe how to use product validation to uncover such problems.

A key part of requirements validation, and product verification and validation, is conducting and reviewing the results of the analyses we did to develop requirements. We often think of analyses as models that predict or describe system behavior. A fundamental challenge in any system development is keeping in synch all the myriad assumptions used in multiple, parallel analysis exercises.

System complexity often defies the "black-box" model. Real-world systems exhibit tight coupling and interdependencies between the subsystems and the

TABLE 11-4.   Example of FireSAT Requirement Validation Exercise. A compliance matrix such as this helps us to systematically review the validity of each requirement and includes the validation thought process for it. Requirements must embody the VALID characteristics. (USFS is US Forest Service; N/A is not applicable.)

Level	Requirement	Discussion (V A L I D)
Mission	The FireSAT constellation shall revisit the coverage area at least once per 12-hour period. *(Rationale: The USFS determined that a maximum 24 hour revisit period would meet mission needs.)*	✔ Verifiable—by analysis. Orbit simulations give precise estimates of revisit time for a given constellation configuration. ✔ Achievable—by analogy. Small constellations of satellites currently have similar revisit times. ✔ Logical—by engineering judgment. This requirement is reasonable given the goals of the project. ✔ Integral—by N/A. Constellation revisit supports overall mission goals. ✔ Definitive—by engineering judgment. This requirement is concise and unambiguous.
System	The FireSAT spacecraft shall station keep* in a 700 km (+/– 5 km), 55 deg (+/– 0.01 deg) inclined orbit, separated 180 deg (+/– 1 deg) in true anomaly *(Rationale: The size of the station keeping* box was determined through analysis [Sellers, 2005] based on maintaining a 12-hr revisit by a two-satellite constellation. (All three parameters are contained in the same requirement, reflecting their strong interdependence.))* * Station keeping means using maneuvers to keep a spacecraft in its required orbit.	✔ Verifiable—by analysis. Orbit maneuver simulations give precisely the "box" size needed to maintain the constellation to provide the requisite coverage. ✔ Achievable—by analogy. Small constellations of satellites currently achieve similar station keeping requirements. ✔ Logical—by engineering judgment. This requirement is reasonable given the goals of the project. ✔ Integral—by N/A. Maintaining the constellation configuration supports overall mission goals. ✔ Definitive—by engineering judgment. This requirement is concise and unambiguous.
Subsystem	Propulsion subsystem shall deliver a total of at least 513 m/s ΔV over the mission lifetime. *(Rationale: The ΔV requirement was determined by analysis [Sellers, 2005] including maneuver to mission orbit, station keeping over mission lifetime, and end-of-life de-orbit.)*	✔ Verifiable—by analysis. Orbit maneuver simulations give precise estimates of the total mission ΔV needed, first to move from the launch vehicle deployment orbit to the final mission orbit, then to keep within the defined station keeping box over the life of the mission given reasonable assumptions about perturbations. ✔ Achievable—by analogy. Small satellites currently meet similar ΔV requirements. ✔ Logical—by engineering judgment. This requirement is reasonable given the goals of the project and is derived from the station keeping requirement. ✔ Integral—by N/A. Achieving this ΔV is necessary to attain all mission goals. ✔ Definitive—by engineering judgment. This requirement is concise and unambiguous.

TABLE 11-4.   **Example of FireSAT Requirement Validation Exercise. (Continued)** A compliance matrix such as this helps us to systematically review the validity of each requirement and includes the validation thought process for it. Requirements must embody the VALID characteristics. (USFS is US Forest Service; N/A is not applicable.)

Level	Requirement	Discussion (V A L I D)
Component		
	Orbital maneuver thruster specific impulse (I_{sp}) shall be at least 210 seconds *(Rationale: Rocket performance was derived by system trade-off analysis [Sellers, 2005] using ΔV requirement, spacecraft dry mass.)*	✔ Verifiable—by test. Component-level tests of the propulsion subsystem thruster provide evidence of this performance level. ✔ Achievable—by inspection. Mono-propellant thrusters with this I_{sp} are widely available and represent state of the industry. ✔ Logical—by engineering judgment. This requirement is reasonable given the goals of the project and follows logically from the mission ΔV requirement. It's based on a reasoned trade-off of propulsion technology options. ✔ Integral—by N/A. Achieving this I_{sp} is necessary to support the mission ΔV requirement. ✔ Definitive—by engineering judgment. This requirement is concise and unambiguous.
Component		
	Usable propellant tank volume shall be 65 liters *(Rationale: Propellant volume derived by analysis (Ref: FS-TR-105) using ΔV requirement, spacecraft dry mass and assumed mono-propellant rocket performance)*	✔ Verifiable—by test. Component-level tests of the propellant tank produce evidence of this performance level. ✔ Achievable—by inspection. Propellant tanks with this volume are widely available. ✔ Logical—by engineering judgment. This requirement is reasonable given the goals of the project and follows logically from the mission ΔV requirement and rocket engine I_{sp} specification. ✔ Integral—by N/A. This volume is necessary to support the mission ΔV requirement. ✔ Definitive—by engineering judgment. This requirement is concise and unambiguous.

components. Many interfaces are porous or amorphous. Furthermore, real systems are highly multi-dimensional; interdependencies often cross dimensional boundaries and are not always orthogonal. In addition, requirements come in two basic types: 1) characteristics (e.g., mass, volume, color) and 2) capabilities (e.g., produce power, collect images). We have several useful tools to aid in this aspect of requirements validation, including:

- *Merit function*—A model showing sensitivity of the ultimate mission objective to the key mission design parameters (e.g., "goodness" of a global climate investigation as a function of spatial and spectroscopic resolution, which in turn depends on spacecraft pointing precision)

- *State analysis*—A behavioral model that simulates how a system responds to stimuli and commands

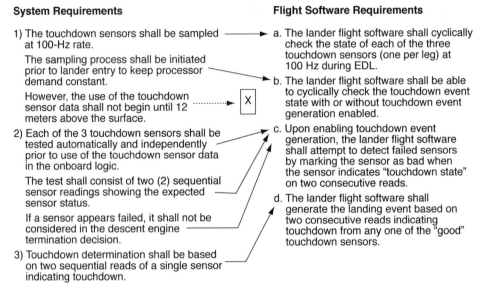

System Requirements

1) The touchdown sensors shall be sampled at 100-Hz rate.

The sampling process shall be initiated prior to lander entry to keep processor demand constant.

However, the use of the touchdown sensor data shall not begin until 12 meters above the surface.

2) Each of the 3 touchdown sensors shall be tested automatically and independently prior to use of the touchdown sensor data in the onboard logic.

The test shall consist of two (2) sequential sensor readings showing the expected sensor status.

If a sensor appears failed, it shall not be considered in the descent engine termination decision.

3) Touchdown determination shall be based on two sequential reads of a single sensor indicating touchdown.

Flight Software Requirements

a. The lander flight software shall cyclically check the state of each of the three touchdown sensors (one per leg) at 100 Hz during EDL.

b. The lander flight software shall be able to cyclically check the touchdown event state with or without touchdown event generation enabled.

c. Upon enabling touchdown event generation, the lander flight software shall attempt to detect failed sensors by marking the sensor as bad when the sensor indicates "touchdown state" on two consecutive reads.

d. The lander flight software shall generate the landing event based on two consecutive reads indicating touchdown from any one of the "good" touchdown sensors.

FIGURE 11-5.   **Incorrect Allocation of Requirements for the Mars Polar Lander Mission.** Part of system requirement #1 isn't allocated to any flight software requirement, and so the final system doesn't incorporate it. (EDL is entry, descent, and landing.) [Feather et al., 2003]

- *Trace analysis* — An independent mapping of low-level requirements back to top mission objectives

- *Fault trees and failure mode analyses* — Tools that identify potential soft spots in the design and highlight where to require mitigations

In addition to using these models to develop requirements, we may later depend on them — or models derived from them — to verify requirements, either solely or as part of a test campaign. Models of system behavior play a crucial role in the space business, during requirements development and validation, and verification and eventual system validation. Therefore, we must validate them.

11.2   Validate Models

Verifying and validating space missions is different from almost any terrestrial systems engineering effort because of the heavy reliance on analysis, simulation, and modeling for verification. Efforts to duplicate the space environment on Earth by test frequently yield crude approximations. And many aspects of that environment, e.g., free-fall and its associated effects on attitude dynamics, are impossible to recreate accurately here on the ground. Thus, in many cases, we must rely almost totally on models.

Because we depend so much on models, we must demonstrate that they're trustworthy. Model validation is particularly important when an analysis takes the place of a test as a critical link in the "verification chain"; errors in the models that support these analyses are sometimes mission-ending. We must also be sure that our models are trustworthy where a feel for system behavior is absent, or when serious consequences result from inaccuracy. The purpose of *model validation* is to develop objective evidence that the models we use to analyze mission systems and products reflect the real world as accurately as necessary to support critical decision making. We say "as accurately as necessary," **not** "as accurately as possible", because as with all V&V efforts, there's a point of diminishing returns beyond which more and more resources are required to obtain miniscule improvements in model certainty.

Models, and the resulting analyses they support, are of many different types, ranging from simple "back-of-the-envelope" calculations to full-blown integrated hardware-in-the-loop simulations. A model is a simplified representation of the real system of interest (spacecraft, instrument, ground segment, subsystem, or full-up mission). We emphasize "simplified" as a reminder that no model—even a validated one—is a perfect representation of the real thing. Table 11-5 lists the four categories of models that we use, along with examples of each. Some of the models that need validating in a space program include:

- Structural finite element models (FEMs) (e.g., NASTRAN)
- Optical prescription model (e.g., CodeV)
- Thermal control system model (e.g., SCIDA)
- Antenna (electromagnetic) model (equations)
- Pointing control system model (e.g., control algorithms)
- Monte Carlo simulation of trajectories (custom tools)
- Instrument performance model (e.g., spreadsheet performance trade-off model)
- Flight software behavioral model (e.g., Simulink software resource model)

Most reasonably complex systems defy effective representation by simple black-box models. Real-world systems don't compartmentalize easily into neat little boxes; interfaces and boundaries are often fuzzy. For this reason, we must ensure that requirements address functional capabilities (in a cross-cutting sense) as opposed to simply stating isolated specifications for individual hardware and components. This reasoning underlies the importance of logical decomposition as addressed in Chapter 5. Allocation of requirements to physical solutions too soon usually leads to missed opportunities for robust design.

To further complicate things, real-world systems exhibit multi-dimensional complexity. In Figure 11-6, width is end-to-end data flow, depth is top to bottom performance, and length is operational evolution in time. While different models address these dimensions separately and hence are validated separately, we

TABLE 11-5.   Types of Models. We employ a wide range of model types in varying fidelities for systems analyses throughout the system lifecycle.

Physical Models	Graphical Models	Mathematical Models	Statistical Models
• Wind tunnel model • Mock up (various degrees of fidelity) • Engineering model (partial or complete) • Hangar queen • Testbed • Breadboards and brassboards • Prototype • Mass/inertial model • Scale model of section • Laser lithographic model • Structural test model • Acoustic model • Trainer	• Functional flow charts • Behavior diagrams • Plus function flow charts • NxN charts • PERT charts • Logic trees • Document trees • Time lines • Waterfall charts • Floor plans • Blue prints • Schematics • Representative drawings • Topographical representations • Computer-aided drafting of systems or components	• Dynamic motion models • Structural analysis, either finite element or polynomial fitting • Thermal analysis • Vibrational analysis • Electrical analysis as in wave form or connectivity • Finite element • Linear programming • Cost modeling • Network or nodal analysis • Decision analysis • Operational or production throughput analysis • Flow field studies • Thermal model • Work flow analysis • Hydrodynamics studies • Control systems modeling	• Monte Carlo • Logistical support • Process modeling • Manufacturing layout modeling • Sequence estimation modeling

should recognize the need for opportunities to cross-validate models between the different dimensions. In fact, we ought to seek out such opportunities.

Table 11-6 outlines the model validation activities. The starting point for model validation depends on when in the lifecycle it occurs. In concept it begins with an understanding of at least preliminary mission, system, or product requirements, depending on the system of interest. But models of individual product-level behavior, especially in the space environment, may depend highly on mission-level definitions of that environment as reflected in mission-level models. How we develop these models in response to project needs is beyond the scope of this discussion. Industry-standard references such as *Space Mission Analysis and Design* [Larson and Wertz, 1999] or *Space Launch and Transportation Systems* [Larson et al., 2005] contain the detailed technical processes for modeling spacecraft and launch vehicles respectively at the system and subsystem levels. Depending on the type of model needed, and its required fidelity, modeling can cost a lot of time and money. The next section, on product verification, describes the factors driving project model philosophy. The specific requirements for these models, as well as the timing of their delivery, is a central aspect of technical planning reflected in Chapter 13. With requirements and unvalidated models in hand, model validation begins.

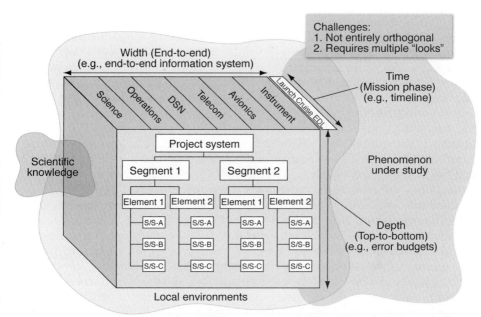

FIGURE 11-6. **Three-dimensional Elements of System Definition.** We illustrate the complexities of real-world systems using three dimensions. For scientific missions, our knowledge is expanded by studying one or more phenomena. Here the "depth" represents the physical decomposition of the physical system from segment to element to subsystem. The "width" spans all elements of the system architecture. Finally, the "length" represents the mission lifecycle from launch through EDL. (EDL is entry, descent, and landing; S/S is subsystem.)

TABLE 11-6. **Model Validation Inputs, Outputs, and Associated Activities.** Model validation activities transform unvalidated models into validated models and provide insights into system interdependencies and behavioral characteristics.

Inputs	Model Validation Activities	Outputs

• Unvalidated mission models
• Preliminary product requirements
	• Prepare for model validation
	  – Identify critical models
	  – Develop model validation plan
	• Implement model validation
	  – Develop model validation data
	  – Validate models
	• Document and iterate
	  – Update model validation matrix
	  – Document and correct differences between model and mission article
	  – Identify and document model uncertainty factors (MUFs)
	  – Iterate
		• Validated model
		• Model uncertainty factors (MUFs)
		• List of model idiosyncrasies

11.2.1   Prepare for Model Validation

We prepare for model validation by first identifying mission-critical models and simulations, then developing a validation plan to tackle each one. Models often play a critical role in bridging the gap between tests in a verification program (which we'll call verification by analysis). Model validation resources must focus on those models most critical to mission success. We should consider models mission-critical if errors in such areas could mask problems leading to failures in the operational phase of the project. One finding from the Mars Polar Lander Mishap Investigation was: "[T]he propulsion system...employed analysis as a substitute for test in the verification and validation of total system performance... end-to-end validation of the system through simulation and other analyses was potentially compromised in some areas when the tests employed to develop or validate the constituent models were not of an adequate fidelity level to ensure system robustness." [JPL, 2000] To avoid these types of failures, we must rigorously validate the models we depend on for mission design and analysis.

Identifying critical models begins early as part of the V&V planning process. One useful method is to develop verification and system validation "storyboards" to show where and how models and simulations fit into the overall scheme relative to other design and V&V activities. As an example, consider verification of the FireSAT payload angular resolution requirement:

FireSAT payload shall have an angular resolution $< 4.0 \times 10^{-5}$ rad at 4.3 μm

Cost and physical constraints usually preclude a true flight-like test of such an instrument. Instead, we verify lower-level requirements by test, using that information to validate various models and then rely on analysis to combine the results to verify the higher-level requirement. Figure 11-7 shows this process. We first measure the instrument's optical prescription, producing a test-validated model. We do the same for reaction wheel jitter, the spacecraft's structural finite element model (FEM), and its thermal model. The resulting test-validated models feed a simulation of the closed-loop pointing control of the instrument. This simulation, when convolved with the instrument optical prescription, attempts to predict the ultimate instrument angular resolution in operation.

Due to cost and physical limitations, neither the model used in the pointing control simulation, nor the final model of instrument response have been validated by test (short of hanging the integrated spacecraft and instrument in a free-fall configuration in a vacuum chamber, we don't do true flight-like closed loop imaging tests on the ground). Instead, we accept the risk of using other methods to validate those models (e.g., peer-review of the physics and computer code).

The verification storyboard is useful in two respects: 1) It highlights where we don't have test-validated models, and 2) It highlights where we have "test as you fly" exceptions. In both cases, this information guides risk assessment, and where warranted, mitigation (such as additional independent peer-review and cross-checks). The results of this effort also help define, refine, and validate the requirements levied on these mission-critical model functions, capabilities, and accuracies.

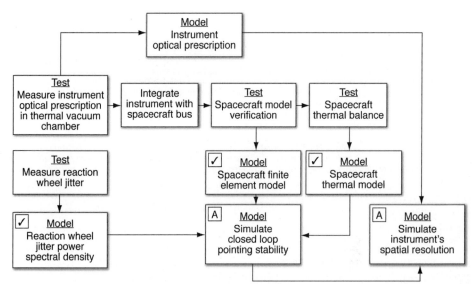

FIGURE 11-7.   Verification Storyboard (FireSAT Example). A verification storyboard such as this one helps to define a combined sequence of tests and analyses to verify payload performance requirements. Here a "check" in the box implies that a model has been or will be test validated. An "A" indicates the model will be validated by analysis only.

We now turn to developing the model validation plan. Most projects include it as part of the overall system verification or validation activities. But it's useful to at least conceptually separate this activity given the criticality and long shelf-life of some models (the Space Shuttle program relies on models for operational planning at the end of the program that were initially created a generation earlier during system development). Table 11-7 gives an example of the information in the model validation matrix. We must also show the specific requirements for each model. The information generated by test, inspection, or separate analysis is then compared against each requirement for each model. In some cases, we use models to validate other models. It's crucial to get at the underlying assumptions behind each model, and to perform a sanity check against reality before using the results generated by the model.

11.2.2   Implement Model Validation

Veteran analysts admit that model validation is not straightforward in a quantitative, statistical sense and that they place great emphasis on fairly subjective, qualitative techniques [Arthur et al., 1999]. A complementary mix of model validation techniques reduces the risk of errors. Such techniques include [Youngblood et al., 1995]:

1. *Face validity*—Experts inspect model results on the system, or similar systems, as a sanity check to confirm that the outputs are reasonable and represent the required functional capabilities. They may use simple hand calculations or results from previously validated models as a comparison.

TABLE 11-7.   **Examples of FireSAT Model Validation Planning.** During FireSAT system design and development, we need to validate different models using different approaches at different points in the system lifecycle.

Model	Model Description	Validated By...	Validated When...
Orbit planning model	Commercial-off-the-shelf (COTS) mathematical model used to predict orbital parameters for operational planning	Inspection of vendor data showing previous validation activities	Before mission baseline
Spacecraft structure finite element model	Mathematical model used to predict structural behavior (strength, stiffness, and stability)	Inspection of code to validate that assumptions match structural model Vibration test campaign (modal survey, static load, and random vibration tests)	Developmental testing before design-to baseline
Spacecraft structure engineering development unit	Physical model emulating the predicted mass properties used for validating finite element model	Inspection of design to validate design and construction as per drawings	Developmental testing before design-to baseline
Spacecraft attitude determination model	Statistical model used to predict attitude determination error using multi-cycle Monte Carlo	Inspection of code to validate that assumptions match flight hardware configuration and predicted noise profiles Empirical validation comparing predictions with actual flight data from similar missions	Before build-to baseline

2. *Peer review*—Independently review the theoretical underpinnings and examine in detail the model's internal components

3. *Functional decomposition and test*, also called *piece-wise validation*—Inject test data into individual code-modules and compare actual output with predicted output

4. *Comparison or empirical validation*—Compare performance of the model against performance of the physical system (or a similar system) being modeled

Comparison or empirical validation is the preferred technique and ideally provides quantitative estimates of model credibility or accuracy via cost functions such as Theil's Inequality Coefficient (TIC) and multivariate statistics [Smith et al., 1998; Balci, 1997]. However, in practice this technique has several limitations. Chief among them is that the final form of the system being modeled does not yet exist in the early phases of the project when we're using the model for requirements validation. We often remedy this situation later in the project when the real system is undergoing test and we use that to support model evolution. When making comparisons we must isolate noise induced by test artifacts from inherent system noise. In early project phases, we sometimes compare models with real systems by modifying the model's capabilities to describe a similar existing system and analyzing the differences, making adjustments to the model as necessary.

We manage project costs by limiting formal validation efforts to those aspects of models deemed mission-critical. The systems engineer must decide what level of credibility is sufficient to meet the needs, and balance that against model cost and utility requirements. We may decide that we're better off investing project resources in more comprehensive testing than in spending more to obtain an incremental improvement in the credibility and utility of a particular model. The space industry unfortunately has numerous examples of overdependence on what turned out to be insufficiently accurate models. The failure during the maiden flight of the Ariane V launch vehicle, for example, illustrates a case of guidance, navigation, and control models not fully accounting for real-world hardware-in-the-loop issues. When the navigation code originally developed for the much different Ariane IV rocket was used on the Ariane V, the guidance system failed, resulting in a total loss of the vehicle less than 40 seconds after launch. [Ariane 501 Inquiry Board, 1996]

11.2.3   Document Model Validation Results

Model validation has several potential outcomes:

- It may reveal that the model behaves as expected, with negligible errors

- It may reveal that the model warrants modification to more accurately reflect the real system (followed by revalidation to confirm the correction)

- It may reveal differences between the model and real system that can't be corrected but are important to quantify (knowledge versus control)

We must understand and document these differences as part of an archive of model uncertainty factors (MUFs). We capture these MUFs in a model validation matrix and in summaries of the individual analyses. Model uncertainty factors are effectively the error bars around the model results and help guide the need for design or other margin. But documentation activities must go beyond capturing MUFs. We must also manage the model configuration (as with any other mission-critical software) to ensure that the model's future users understand the evolution of changes and idiosyncrasies exposed by the V&V program.

Consider a model that predicts communication links between Earth and a deep space mission to Pluto, for example. Such a model would be essential early in the project to guide system design and analysis efforts. We would need it again 12 to 15 years later for operational planning after the spacecraft reaches its destination on the edge of the solar system. By that time, it's unlikely that any of the original RF engineers who developed the model would still be readily available. To make matters worse, software and hardware platforms will have gone through many generations, making it more difficult to maintain legacy software models. All this underscores the critical need to first validate and then meticulously document the results to guide mission design, analysis, and operation throughout the lifecycle.

11.3   Verify Products

The previous two activities, requirements validation and model validation, prepare us to begin the next, and likely most expensive and time-consuming activity, product verification. Most general definitions of verification address the activities of *product verification*. NPR 7123.1a defines this type of verification as "proof of compliance with design solution specifications and descriptive documents [NASA (2), 2007]." JPG 7120.3 calls verification "the methodical development of evidence of system of interest compliance with requirements ("shalls"). It is accomplished at each level of the system architectural hierarchy." [NASA, 2004] The European Cooperative for Space Standardization ECSS-P-001 says that satisfactory verification is typically the basis "for a contractual acceptance of the product by the customer." [ECSS, 2004]

Because product verification forms the heart of the V&V effort, and typically consumes the bulk of the effort and funding, we describe it in considerable detail. The purposes of product verification are of two types:

- Routine product verification—Performed at numerous points throughout the assembly, integration, and test (AIT) phase, up to and including launch vehicle integration. It verifies basic functionality and lack of damage from transportation or other events.

- Formal product verification—Performed at certain points during AIT to verify that the product meets a given design requirement or specification.

The processes we use to plan for and execute both types of product verification are similar, so we focus largely on the second type, formal product verification. Figure 11-8 gives a big-picture look at the overall flow. Table 11-8 lists the top-level activities that constitute product verification. These activities fall into two broad categories, *preparing for product verification* and *implementing product verification*. We start with validated technical requirements, along with enabling products such as validated models, test equipment, personnel, and other resources (time and money being the most important). From there, we develop verification requirements based on the project verification philosophy. We then use the verification requirements to guide definition of more detailed test and verification requirements (TVRs). These, together with detailed plans and procedures as well as facilities, equipment, and trained personnel, come together at verification events (tests, demonstrations, analyses, inspections, or modeling and simulation, or some combination of these) along with the end product to be verified.

Verification delivers a verified end product, one that has been shown objectively through one or more methods to meet the defined requirements. This result is supported by an as-executed verification plan and a variety of documentation products including verification completion notices, test reports, and other critical evidence.

Figure 11-9 illustrates how we implement verification activities throughout the project lifecycle. At every phase, the project manager (PM) and systems engineer

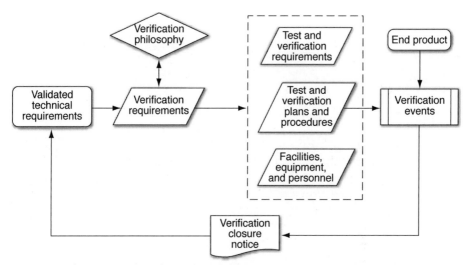

FIGURE 11-8.   Product Verification Process Flow Chart. This flow chart illustrates how technical requirements are ultimately verified and closed out.

TABLE 11-8.   Product Verification Activities. The verification activities provide objective evidence that the end product meets the system's requirements.

Inputs	Verification Activities	Outputs
• Validated requirements • Verification enabling products (e.g., validated models)	1. Prepare for product verification 2. Implement product verification	• Verified end product • Verification plan (as executed) • Verification products (data, reports, verification compliance matrix, verification completion notices, work products)

(SE) must see that product verification activities are performed in accordance with the verification plan and defined procedures. These procedures collect data on each requirement, giving special attention to measures of effectiveness (MOEs).

All military leaders learn that "no plan survives contact with the enemy." Problems are inevitable when product verification moves into high gear. Together with the stakeholders, the PM and SE must repeat as necessary those steps that aren't compliant with planned product verification procedures or the planned environment, including equipment, measurement, or data capture failures.

Throughout the assembly, integration, and testing phase, we perform periodic system integration reviews (SIRs) at each major step as identified by the integrated verification fishbone, described later. The SIR is a formal review to certify that flight elements and systems are ready to be integrated, including confirmation that facilities, support personnel, plans, and procedures are ready to support integration (see Chapter 10). The SIR focuses on the integration of flight elements and systems following acceptance by the customer. Most missions have several SIRs.

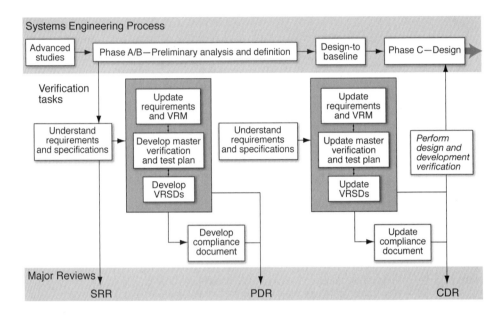

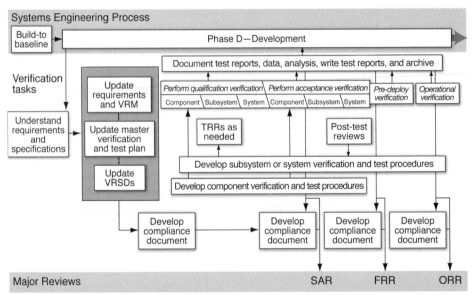

Figure 11-9.   Verification and Validation in the Project Lifecycle. This figure illustrates how verification activities fit within the project lifecycle, beginning immediately after the system requirements review (SRR). (VRM is verification requirements matrix; VRSD is verification requirements and specifications document; PDR is preliminary design review; CDR is critical design review; SAR is system acceptance review; FRR is flight readiness review; ORR is operational readiness review; TRR is test readiness review.) (Adapted from NASA (4), [2007].)

11.3.1 Prepare for Product Verification

A successful product verification program depends on good preparation. Chapter 13 emphasizes the importance of detailed technical planning for all systems engineering processes. This is especially true when it comes to verification. We have to plan adequately, starting in Pre-phase A, with continual updates until verification begins in earnest in Phase D. Otherwise the project may be left with inadequate resources to deliver a fully verified product, i.e., one that meets the intended requirements. Lack of adequate planning also causes more difficulty dealing with the inevitable setbacks that occur throughout the assembly, integration, and verification phase. Table 11-9 summarizes the inputs, activities, and outputs for the preparation phase of product verification.

TABLE 11-9. **Inputs, Outputs, and Activities Supporting Verification Preparation.** Early verification planning is important to assure readiness of test articles, facilities, and other long-lead items needed for system verification. (GSE is ground support equipment.)

Inputs	Verification Preparation Activities	Outputs
• Validated requirements • Verification enabling products (e.g., validated models, test facilities)	1. Establish the project or program verification philosophy 2. Develop verification requirements (identifying verification methods, specific verification events, and success criteria) for each requirement 3. Derive test and verification requirements (TVRs), plans and procedures, and support requirements 4. Write master verification plan	• Verification requirements (e.g., verification matrix) • Derived enabling product requirements (e.g., GSE, test/emulators) • Verification plans • Detailed test and verification requirements • Detailed test and verification procedures (e.g., verification logic networks, master schedules) • Master verification plan capturing detailed verification support requirements (e.g., facilities, equipment, product availability, personnel, time, money)

The inputs for verification preparation are as described above. The intermediate outputs all aim at defining what methods and events we need to fully verify each technical requirement along with detailed plans, procedures, and supporting requirements. The following sections step through each of the verification preparation activities.

Establish the Project Verification Philosophy

The philosophy of the verification effort reflects the fundamental approach to defining how much and what kind of evidence is **enough** to objectively verify a given requirement. This philosophy determines:

• The method or methods chosen for verification

- The number and types of models developed for verification and flight
- The number and types of tests, and the appropriate levels
- The project phases and events during which we implement verification

The program verification philosophy is driven by risk tolerance. Less verification implies (but doesn't necessarily create) more risk. More verification implies (but doesn't guarantee) less risk. Chapter 8 addresses risk management in the broad sense. Here we're concerned with risk management in developing the verification program. In deciding how this program will look, risk managers and verification engineers must account for:

- System or item type
  - Manned versus unmanned
  - Experiment versus commercial payload
  - Hardware versus software
  - Part, component, subsystem, complete vehicle, or system of systems
- Number of systems to be verified
  - One of a kind
  - Multiple, single-use
  - Multiple, multiple re-use
- Re-use of hardware or software
- Use of commercial-off-the-shelf (COTS) or non-developmental items (NDI)
- Cost and schedule constraints
- Available facilities
- Acquisition strategy

When we consider all these factors, two broad types of verification programs emerge, *traditional* and *proto-flight*, with lots of variations on these basic themes. The basic difference between them is that the traditional approach makes a clear distinction between verification activities for qualification and verification for acceptance. The proto-flight approach normally does not. This difference is reflected in the number and types of models developed. Before we implement either approach, we often conduct some level of developmental testing. Developmental testing is essential for early risk reduction and proof-of-concept. However, because it's done under laboratory conditions by developers absent rigorous quality assurance, it normally doesn't play a role in final close-out of requirements.

In the traditional approach, *qualification* activities verify the soundness of the *design*. Because requirements focus mostly on the design, evidence collected during a qualification program is usually defined to be the formal close-out of the requirements, documented in something like a verification completion notice (VCN). Verification requirements for qualification are associated with the design-to performance or design-to development specifications. When verified, qualification requirements establish confidence that the design solution satisfies the functional or performance specifications (compliance with the functional baseline). We typically complete flight qualification activities only once or when modifications are made that may invalidate or not be covered within the original verification scope.

Qualification activities are conducted using dedicated *flight-like* qualification hardware with basic flight software as needed to conduct the verification. The hardware subjected to qualification should be produced from the same drawings, using the same materials, tooling, manufacturing process, and level of personnel competency as flight hardware. For existing COTS or NDIs, the qualification item would ideally be randomly selected from a group of production items (lot testing). A vehicle or subsystem qualification test article should be fabricated using qualification units if possible. Modifications are permitted if required to accommodate benign changes necessary to conduct the test. For example, we might add instrumentation or access ports to record functional parameters, test control limits, or design parameters for engineering evaluation.

For qualification, we set test levels with ample margin above expected flight levels. These conditions include the flight environment and also a maximum time or number of cycles that certain components are allowed to accumulate in acceptance testing and retesting. However, qualification activities should **not** create conditions that exceed applicable design safety margins or cause unrealistic modes of failure. For example, at the vehicle level, qualification test levels for a random vibration test are typically set at **four times** (+6 dB) the anticipated flight levels induced by the launch vehicle for two full minutes.

In the traditional approach, we follow a formal qualification program with separate acceptance of flight hardware. *Acceptance verification* ensures conformance to specification requirements **and** provides quality-control assurance against workmanship or material deficiencies. Acceptance verifies workmanship, **not** design. Acceptance requirements are associated with build-to or product specifications and readiness for delivery to and acceptance by the customer. They often include a subset and less extreme set of test requirements derived from the qualification requirements. When verified, acceptance requirements confirm that the hardware and software were manufactured in accordance with build-to requirements and are free of workmanship defects (compliance with certified baseline). We normally perform acceptance verification for each deliverable end item. "Acceptance" encompasses delivery from a contractor or vendor to the customer as well as delivery from a government supplier (e.g., NASA or DoD) to the project as government furnished equipment.

Acceptance activities are intended to stress items sufficiently to precipitate incipient failures due to latent defects in parts, materials, and workmanship. They should **not** create conditions that exceed appropriate design safety margins or cause unrealistic modes of failure. Environmental test conditions stress the acceptance hardware only to the levels expected during flight, with no additional margin. Continuing the example above, at the vehicle level, flight article acceptance test levels for a random vibration test typically are set **at the anticipated flight level** (+0 dB) induced by the launch vehicle for only one minute.

An added, and often under-appreciated, benefit of the traditional approach is that the qualification activities serve as a valuable dress rehearsal to validate the subsequent acceptance program. This includes test techniques, procedures, training, logistics, equipment, instrumentation, and software.

In contrast to the traditional approach, with a pure *proto-flight* (a.k.a. flight-proof) strategy, all flight items are subjected to enhanced acceptance testing only. No dedicated qualification items are developed. The risk is that no formal demonstration occurs for the remaining life of the flight items. This risk is alleviated somewhat by the fact that each flight item has met requirements under acceptance testing at higher than normal levels. The test levels are mostly less than those for qualification, but are never less than those for acceptance. The test durations for the proto-flight test strategy are usually the same as those for acceptance. When choosing to use the proto-flight strategy, we should conduct extensive developmental testing to gain confidence that adequate margin, especially for fatigue or wear, remains after the maximum allowed accumulated acceptance testing at the enhanced levels. Using the same example as above, at the vehicle level, proto-flight test levels for a random vibration test might be set at **two times** (+3 dB) the anticipated flight level induced by the launch vehicle for one minute as a compromise between the qualification and acceptance levels.

Most common these days for robotic spacecraft is a hybrid approach that combines some items developed using the traditional approach, such as the spacecraft structure, and other items using the proto-flight approach, such as payloads or individual subsystems. Any of these may be based on some degree of flight heritage, mitigating risk at the component level.

Regardless of the approach, the result is a verified item that has been qualified to meet or exceed all design specifications, formally closing out contractual or other requirements; and is free from latent defects in parts, materials, and workmanship and has thus been deemed **acceptable** for flight. The large body of evidence collected during this process then moves forward with the spacecraft to the final series of project reviews to determine a go/no-go for flight. We describe these final activities under the topic of certification in Section 11.6.

Product verification activities span the project lifecycle. During planning, we decide when different verification events will occur. In the space business, final vehicle configurations (e.g., satellite integrated with solid rocket motor upper stage) are not usually achieved until shortly before integration with the launch vehicle. This means we may not be able to fully verify some interface or functional requirements until just before launch. Other performance requirements, especially those involving new capabilities or that depend on the interaction with the operational environment, may be verified only during on-orbit test or demonstration.

Additional activities focus on assembly and check-out requirements. These activities are associated with integrating flight elements and systems following acceptance in preparation for flight. When verified, these requirements confirm readiness to proceed to the next higher level of system integration.

Finally, we have *operational requirements*, those specifically associated with in-flight operations. Normally they're incorporated by detailed flight planning activities and get at the heart of the mission. For example, FireSAT has an operational requirement for on-orbit sensor calibration before moving to normal operations. These calibration activities verify that the sensor survived launch and is operating well enough in the space environment to fulfill the mission. When verified, operational requirements confirm readiness to proceed to the next phase of the flight (i.e., normal operations).

We now envision the three-dimensional problem posed by developing a robust product verification program as illustrated in Figure 11-10. The depth of the program encompasses the top-to-bottom system breakdown structure. The length of the program includes the verification activities during different program phases to build a "portfolio" of verification evidence throughout the product lifecycle. The width encompasses the various methods of verification at our disposal, with different methods being more or less appropriate for different parts of the system at different points in the lifecycle.

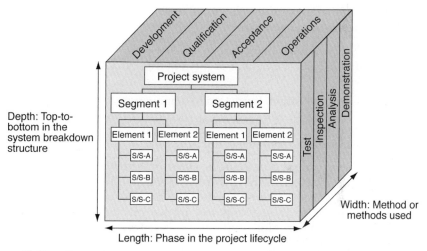

FIGURE 11-10.   Three-dimensional Aspect of a Product Verification Program. Verification activities address the entire system over the lifecycle. (S/S is subsystem.)

Develop Verification Requirements

Verification planning begins with a set of validated requirements. These are typically contained in a system or segment specification document (e.g., as per MIL-SPEC-961c) and perhaps electronically managed in a database that forms the basis for writing verification requirements. However we define and manage the requirements, a useful output is a summary requirements matrix (RM) that we use as a starting point for verification planning.

While the design or performance requirement states what to do and how well, the *verification requirement (VR)* states how we will know if the system does in fact meet that requirement. Ideally, we should write VRs at the same time as the source requirement. But several practical reasons often preclude this. Usually the pressing schedule leads planners to defer the detailed consideration of verification. Furthermore, the person or organization writing the design or performance requirement may not be an expert in the best verification methods to employ. Persons representing both sets of expertise—design and verification—should write the VRs as early as practical in the program.

Let's examine verification requirements in more detail. As Chapter 4 describes, writing good requirements is challenging enough. Writing good VRs can be even more taxing. They define the method and establish the criteria for providing evidence that the system complies with the requirements imposed on it. One technique for writing a verification requirement is to break it into three parts:

1. The method of verification (inspection, analysis, simulation and modeling, test, or demonstration)

2. A description of the verification work

   - For simulation and modeling, the item to be modeled and which characteristics to simulate

   - For an inspection, the item to be reviewed and for what quality

   - For an analysis, the source of the required data and description of the analysis

   - For a demonstration, the action to demonstrate

   - For a test, the top level test requirements

3. The success criteria that determine when the verification is complete

Writing a verification requirement starts by reviewing the originating requirement and stating concisely what we're verifying. If we can't, the original requirement may not be valid and may need to be renegotiated or rewritten! At this point we mustn't try to converge too quickly on the verification methods. Simply writing the description and success criteria helps define the most reasonable methods.

Next, we describe what work or process (at a high level) we need to perform the verification activity. We state a single top-level procedure for each activity, though we choose different procedures for different activities. Some example procedures are:

- Be performed in accordance with x (standard, another requirement, constraint)

- Include x, y, and z (given partial objectives or sub-methods)

- Using xyz (a specific tool, mode or data, or hardware or software)

The success criteria should define what the customer will accept when we submit the product with the formal verification completion notice. In the rationale, we have to capture any additional thoughts about the verification activity and why we selected certain methods. Then we go back to the first sentence and determine the methods to use at this verification level. We don't have to use every method for every VR. We'll probably use additional methods as we decompose the requirement to lower levels.

Figure 11-11 provides a simple template for writing a verification requirement. After we write and document them, we summarize the VRs in a verification requirements matrix (VRM). This allows us to see the source requirement side-by-side with the verification requirement. It may also include verification levels and sequencing.

- The _____ shall be verified by _____.
        function or quality being verified   method
  - Concisely state what is being verified and by what methods. In stating what is being measured, the original requirement may have to be updated for clarity.
  - Methods (inspection, analysis, demonstration, or test).
    - Example: "Interface C steady-state voltage shall be verified by test."
    - Not: "This requirement shall be verified by analysis and test."
- The _____ shall _____.
       method   description of the verification work to be done
  - Provide a top-level description of what is to be done. May be multiple verification objectives. Provide significant detail only if we know the methodology is a standard and is unlikely to change.
  - Is data needed for a higher system analysis? If so, state what is needed.
  - If appropriate, define what fidelity unit is needed and the test conditions.
- The verification shall be considered successful when the _____ shows _____.
                method   success criterion
  - Example: The verification shall be considered successful when the coupled loads analysis shows that the components mounted to ISPR maintain positive margins of safety.
- The rationale for the VR should also be captured using CORE or another requirements tracking tool.

FIGURE 11-11. **Writing Verification Requirements.** A good verification requirement defines the requirement to be verified and includes the verification method and the success criteria for determining when verification is complete.

Verification requirements should not be blindly "thrown over the wall" from designers to verification engineers. Rather, they should reflect a meeting of the minds between the two legs of the SE "V." And we must be careful not to imbed implied design requirements or contractor statement of work direction in the verification requirements. An example of a disconnect between design and verification is, "The test shall connect diagnostic equipment to the test port..." when no test port is specified in the design requirements. An example of a disconnect between engineering and contracting is, "...verified using a qualification model subjected to flight random vibration levels plus 6 db..." when the SOW places no contractual requirement on the vendor to produce a qualification model. To avoid this tendency, some organizations use more generic verification requirements that simply state the method to use and rely on the VRM for the acceptance criteria. Table 11-10 shows a partial VRM from the FireSAT mission. Verification matrices are useful for summarizing source requirements, verification requirements, and other information in a compact form.

TABLE 11-10. **Excerpts from the FireSAT Verification Requirements Matrix (VRM).** A VRM displays the originating requirement, the verification requirement ("what," "when," and "how well"), and the level at which verification will occur.

Requirement	Verification Requirement	Method	Levels
Space vehicle first-mode natural frequency shall be greater than 20 Hz	The space vehicle first-mode natural frequency shall be verified by analysis and test. The analysis shall develop a multi-node finite element model to estimate natural modes. The test shall conduct a modal survey (sine sweep) of the vehicle using a vibration table. The analysis and test shall be considered successful if the estimated and measured first mode is greater than 25 Hz.	Analysis and test	Vehicle
Structural components shall be marked with (1) axes orientation, and (2) component identification	The appropriate markings on all system structural components shall be verified by inspection. The inspection shall determine if axes and identifications are properly indicated. The verification shall be considered successful if all structural components are properly marked.	Inspection	Part, component, subassembly, assembly
The attitude determination system shall estimate vehicle attitude to within 0.01 deg (3σ).	The accuracy of the attitude determination system estimates shall be verified by analysis. The analysis shall use Monte Carlo simulations of expected sensor accuracy, plus noise, to determine statistical distribution of error. The analysis shall be considered successful if the predicted error is less than or equal to 0.01 deg (3σ).	Analysis	Subsystem
Battery charging GSE shall display current state of charge.	Battery charge GSE state of charge display shall be verified by demonstration. The demonstration shall show that state of charge is indicated when connected to a representative load. The demonstration shall be considered successful if state of charge is displayed.	Demonstration	System
Mechanical interface between structural assemblies shall use 4-40 stainless steel fasteners.	Fastener type shall be verified by inspection. The inspection shall review the vendor's records to look for the type and size of fasteners used. The inspection shall also review the documentation on fastener material. The verification shall be considered successful if all interface fasteners are 4-40 in size made from stainless steel.	Inspection	Part

We now turn to the verification methods identified by the verification requirements. We can describe four fundamental methods for verifying any requirement: inspection, analysis, demonstration, and test. (Some organizations, e.g., European Cooperative for Space Standardization (ECSS), divide these methods into slightly different categories, but they're equivalent to the categories used here. Others add simulation and modeling, which we categorize with analysis.) Table 11-11 describes them and gives examples. Figure 11-12 shows a decision tree for selecting the correct method for a given situation.

A major source of design and verification requirements is the induced launch and natural space environments. These impose requirements unique to space systems, and they warrant special attention here. Launch usually imposes the most

TABLE 11-11.   Verification Methods. The description of each method includes typical applications.

Verification Method	Description	Typical Language
Inspection (including design review)	Engineering inspection determines conformance to requirements by standard quality control methods in reviewing drawings and data. Inspection is appropriate whenever we have drawings, documents, or data to verify that the product incorporates the item's physical characteristics requirements.	The kind of language used in system or item requirements that usually indicates verification by inspection is: "… shall be 27 cm long…" "… shall be part marked per…" "… shall have a nameplate in accordance with…"
Analysis (including verification by similarity)	The techniques may include engineering analysis, statistics and qualitative analysis, computer and hardware simulations, and analog modeling. Analysis is appropriate when: (1) rigorous and accurate analysis is possible, (2) testing is not cost effective, and (3) verification by inspection is not adequate. Verification by similarity analyzes the system or item requirements for hardware configuration and application to determine if it is similar in design, manufacturing process, and quality control to one that has previously been qualified to equivalent or more stringent requirements. We must avoid duplication of previous tests from this or similar programs. If the previous application is similar, but not equal or greater in severity, additional qualification tests concentrate on the areas of new or increased requirements.	The kind of language used in system or item requirements that usually indicates verification by analysis is: "… shall be designed to…" "… shall be developed to …" "… shall have a probability of …"
Demonstration	Demonstration determines conformance to system or item requirements through the operation, adjustment, or reconsideration of a test article. We use demonstration whenever we have designed functions under specific scenarios for observing such characteristics as human engineering features and services, access features, and transportability. The test article may be instrumented and quantitative limits or performance monitored, but check sheets rather than actual performance data are recorded. Demonstration is normally part of a test procedure.	The kind of language used in system or item requirements that usually indicates verification by demonstration is: "… shall be accessible…." "… shall take less than one hour…" "… shall provide the following displays in the X mode of operation…"
Test	Testing determines conformance to system or item requirements through technical means such as special equipment, instrumentation, simulation techniques, and the application of established principles and procedures for evaluating components, subsystems, and systems. Testing is the preferred method of requirement verification, and we use it when: (1) analytical techniques do not produce adequate results, (2) failure modes exist that could compromise personal safety, adversely affect flight systems or payload operation, or result in a loss of mission objectives, or (3) for any components directly associated with critical system interfaces. The analysis of data derived from tests is an integral part of the test program and should not be confused with analysis as defined above. Tests determine quantitative compliance to requirements and produce quantitative results.	The kind of language used in system or item requirements that usually indicates verification by test is: "… shall be less than 2.5 mW at DC and less than 100 mW at 1 MHz…" "… shall remove 98% of the particulates larger than 3 μm…" "… shall not be permanently deformed more than 0.2% at proof pressure…"

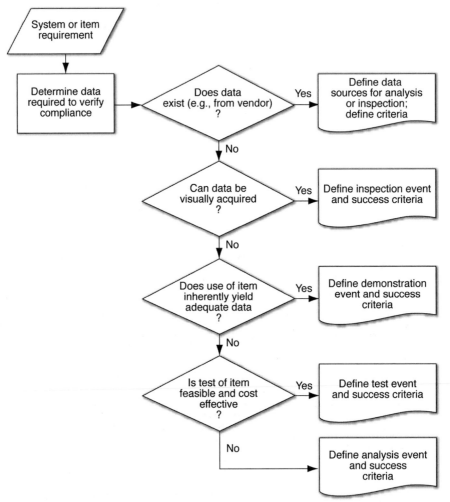

FIGURE 11-12.   Generalized Verification Method Selection Process. We determine the preferred verification method on the basis of data availability—can we acquire the data visually, by operating the system, by testing the system, or by analysis of system models?

severe mechanical stresses a spacecraft will ever see. However, about 25% of all recorded spacecraft anomalies are traceable to interactions with the space environment. Famous space environment problems include:

- Boeing 702 satellite: contamination of solar array collectors due to outgassing
- Stardust comet flyby: spacecraft went silent for four days after solar radiation burst caused the spacecraft to go into standby mode

- Tempo 2 and PanAmSat 6: Gallium arsenide solar cells experienced short circuits from plasma charging
- Global Positioning System (GPS): deposited contamination on solar arrays
- Pioneer Venus: memory anomalies related to galactic cosmic rays
- Skylab: reentered because of atmospheric drag

The anticipated launch environments—and associated design-to and test-to requirements—are normally in the launch vehicle user's guide. Key launch issues that drive design and verification include:

- Mechanical and electrical interfaces
- Mechanical environments
- Thermal environments
- Radio frequency and electromagnetic environment
- Contamination and cleanliness

Table 11-12 summarizes the requirements imposed by each of these along with typical verification methods and techniques they drive.

TABLE 11-12. Summary of Launch Environment Issues, Design Requirements, and Verification Methods. The payload-to-launch-vehicle interface requirements and verification requirements are in an interface control document (ICD) as described in Chapter 15. Interface verification assures that the spacecraft and launch vehicle work together. (CAD is computer-aided design; EMC is electromagnetic compatibility; EMI is electromagnetic interference; RF is radio frequency; N_2 is nitrogen.)

Launch Environment Issue	Design Requirements Imposed	Verification Methods
Mechanical and electrical interfaces	The interface between the spacecraft and launch vehicle is typically through a separation system such as a Marmon clamp or light band. Design requirements arising from the launch vehicle interface include: • Mechanical interface (bolt hole pattern) • Static and dynamic envelopes • Center-of-gravity (CG) and moment of inertia (MOI) ranges • Electrical interface (connector types, number and assignment of pins, power conditioning, telecommanding, grounding) • Deployment ΔV and tip-off torques (spring stiffness and misalignment) • Accessibility	• Analysis (clearance, tip-off torques) • Inspection (of design using CAD or paper mechanical drawings) • Demonstration (fit check) • Testing (separation or deployment test, end-to-end electrical tests, CG and MOI spin table test)
Mechanical environments	• Steady-state acceleration • Low-frequency vibration • Acoustic vibration • Shock	• Analysis (strength, stiffness, coupled loads) • Testing (static load, shock, random vibration, acoustic, modal survey)

TABLE 11-12.   Summary of Launch Environment Issues, Design Requirements, and Verification Methods. (Continued) The payload-to-launch-vehicle interface requirements and verification requirements are in an interface control document (ICD) as described in Chapter 15. Interface verification assures that the spacecraft and launch vehicle work together. (CAD is computer-aided design; EMC is electromagnetic compatibility; EMI is electromagnetic interference; RF is radio frequency; N_2 is nitrogen.)

Launch Environment Issue	Design Requirements Imposed	Verification Methods
Thermal and other environments	• Temperature (under fairing, pre-launch, and in-flight) • Aero-thermal flux at fairing separation • Humidity • Venting Rates	• Analysis (venting, thermal models) • Inspection (interface drawings, part of launch vehicle interface deliverables)
Radio frequency and EM environment	• Electromagnetic (EM) environment (range-to-spacecraft, launch vehicle-to-spacecraft)	• Analysis (RF flux) • Inspection (RF inhibit chain) • Testing (EMC/EMI)
Contamination and cleanliness	• Cleanliness class (e.g., Class 10,000) • Need for or availability of N_2 purge	• Inspection (of spacecraft storage requirements)

Once in space, a satellite endures a whole range of other stresses. We divide these into six basic areas:

- The *free-fall* environment causes real challenges to Earth-bound verification
- Earth's atmosphere, known as *atmospheric neutrals* to differentiate from charged particles, affects a spacecraft even in orbit, primarily due to drag. Atomic oxygen is also an issue at low altitudes.
- The *vacuum* in space above the atmosphere presents outgassing, thermal control, and other challenges
- *Electromagnetic radiation* from the Sun provides light for electrical power, but also causes heating on exposed surfaces, and long-term degradation from ultraviolet radiation
- *Charged particles* from the Sun and galactic cosmic rays that cause surface charging, single event phenomena, and long-term degradation of solar arrays and electronics due to total dose effects
- The *micrometeoroid and orbital debris* (MM/OD) environment, while a threat, has such a low probability that it is practically ignored for all but human missions. Thus it has little or no impact on verification planning. However, more recently, concerns about general proliferation of space debris have led to additional design and operational requirements for end-of-life disposal. These capabilities must be verified.

Table 11-13 summarizes the design requirements and associated verification methods for the significant challenges from the space environment.

TABLE 11-13.   **Space Environment Issues and Associated Design Requirements and Verification Methods.** Different aspects of the space environment place various constraints on the system design in the form of requirements that we must verify. (UV is ultra violet.)

Space Environment Issue	Design Requirements Imposed	Verification Methods
Free-fall	• Liquids management—use positive expulsion methods or wicking mechanisms • Mechanism dilemma—should be as light as possible to work in free-fall (or reduced g) but strong enough to test on Earth. Because of the complexity in setting up the demonstration, usually done only a few times (maybe only once).	• Analysis of loading and free-fall stresses • Test or demonstration
Atmospheric neutrals	• Materials—choose materials with ideal thermal properties and that resist atomic oxygen erosion • Configuration—shield sensitive surfaces to minimize the effects of atomic oxygen. Design the vehicle to have a low drag coefficient. • Operations—fly the side of the vehicle with the lowest cross-sectional area into the ram direction to minimize aerodynamic drag. Fly at higher altitudes to minimize interactions.	• Analysis to support material selections, configuration choices, orbit dynamics • Inspection of records and design to verify materials used, configuration
Vacuum	• Materials selection—choose materials and coatings that are stable under prolonged exposure to vacuum and solar UV and that generate little contamination • Configuration—design to ensure that water dumps, gas purges, and spacecraft venting are away from sensitive surfaces • Margin—allow for degradation in thermal and optical properties of sensitive surfaces on orbit • Materials pre-treatment—perform vacuum bake-out of contamination-generating materials during manufacturing and before installation in the spacecraft • Pre-launch handling—dry nitrogen purges for sensitive optics • Flight and ground operations—provide time for on-orbit bake-out of contaminants during early operations; provide cryogenic surfaces the opportunity to warm up and out-gas contaminant films	• Analysis of thermal model, to support material selections and configuration • Inspection of records and design to verify materials used • Testing in thermal/vacuum chamber to validate thermal model and characterize system behavior
Charged particles (plasma)	• Prevent differential charging of surfaces by ensuring that entire surface is of uniform conductivity (bonding). Use conductive surface coatings. • Tie all conductive elements to a common ground • Shield all electronics and wiring physically and electrically • Use electric filtering to protect circuits from discharge-induced currents • Make all exterior surfaces at least partially conductive and thick enough to withstand dielectric breakdown	• Analysis of conductivity and grounding scheme • Inspection of bonding scheme, shielding, and filtering design • Test of continuity

TABLE 11-13. **Space Environment Issues and Associated Design Requirements and Verification Methods. (Continued)** Different aspects of the space environment place various constraints on the system design in the form of requirements that we must verify. (UV is ultra violet.)

Space Environment Issue	Design Requirements Imposed	Verification Methods
Radiation	• Shielding—place structures between the environment and sensitive electronics (and crew) to minimize dose and dose rate effects • Parts specifications—use parts with sufficient total dose margin and that resist latch-up and upset • Software implementation—implement error detection and correction (EDAC) algorithms capable of detecting errors and of recovering the system from single event upsets	• Analysis (of total mission dose) • Review of design (to verify shielding, software architecture, part selection) • Inspection (of records and design to verify parts used) • Testing (at part- or component-level for total dose limits, software stress testing)

Derive Test and Verification Requirements, Plans, and Procedures

If the verification requirements represent the **strategic** plan for obtaining the objective evidence that a requirement has been met, then the test and verification requirements (TVRs) represent the **tactical** plan. For each TVR we capture the following information:

- TVR revision or identification number
- Identification of parent requirement and source (e.g., Section 4 VR)
- Hardware and software assembly level and fidelity (e.g., flight, emulator) to support TVR execution—should include part number when appropriate
- Organization responsible for defining and controlling the TVR
- Organization responsible for implementing the TVR
- Input products and conditions required to support execution of the TVR (e.g., analytical data from model, test data from separate test)
- Output products associated with TVR execution (e.g., test plan, reports, analytical model), and detail about how we will use those products to support verification
- Success criteria for satisfying the TVR

The test and verification requirement lists the details for the necessary test plans and test reports, and any support data we need to verify the requirement. The test plan tells how to implement the TVR and the test report shows the results. Test plans should incorporate by reference, or directly document, the following:

- The environmental specifications or lifecycle environmental profiles for each of the environmental test zones

- The identification of separate states or modes where the configuration or environmental levels may be different (such as during testing, launch, upper-stage transfer, on-orbit, eclipse, or reentry)
- Required test equipment, facilities, and interfaces
- Required test tools and test beds, including the qualification testing planned for them, to demonstrate that they represent an operational system environment and to verify that simulated interfaces are correct
- Standards for recording test data on computer-compatible electronic media to facilitate automated accumulation and sorting of data
- The review and approval process for test plans and procedures, and for making changes to approved ones
- Overall schedule of tests showing conformance with the program schedules, including the scheduled availability of test articles, test facilities, special test equipment, and procedures

The TVR leads us systematically from the source requirement to describe how we will achieve the verification approach identified in the verification requirement. Test and verification requirements help us organize the plans and data needed to support all the verification events. A *verification event* is any specific, formal activity (test, inspection, analysis, demonstration, or combination) whose purpose is to provide objective evidence that a requirement has been met. A single TVR is normally associated with a single, or several, verification events. However, we often satisfy multiple TVRs at each event. Section 19.4 gives a detailed example of a TVR for a set of FireSAT requirements.

Several key events are common to most space missions during the assembly, integration, and test phase. These are referenced in Chapter 10 and include, in the order typically performed:

1. Pressure/leakage test (repeat after vibration/acoustic and modal survey)
2. Electromagnetic compatibility (EMC) and susceptibility test
3. Mass properties—center of gravity and moments of inertia measurements
4. Fit check (may be combined with separation test)
5. Modal survey (repeat after each level of shock, random vibration, or static load test)
6. Shock test
7. Static load test
8. Acoustic or random vibration test
9. Separation test
10. Thermal cycle (required if thermal cycling acceptance test conducted)
11. Thermal balance (may be combined with thermal vacuum test)
12. Thermal vacuum test

Figure 11-13 gives an example of the coupled planning that must go into decisions on the number and types of units to be developed and the testing and levels on each for the FireSAT spacecraft.

Engineering development units are low fidelity and support system design. Qualification units are high-fidelity models or prototypes used for system verification. Flight units are subjected to acceptance tests to verify workmanship and demonstrate fitness for mission operations. Few systems in the real world rely purely on qualification or proto-flight models. Most systems are hybrid combinations of some components developed using dedicated qualification hardware and others relying more on proto-flight hardware.

During and between each of the above events, we also do hardware inspections and functional tests. These routine check-outs verify that nothing has been broken during a verification event or while moving or configuring an item for an event. Table 11-14 summarizes the major environmental tests, their purpose, needed equipment or facilities, and the basic processes to use.

Figure 11-14 shows a hypothetical environmental test campaign sequence for the FireSAT spacecraft at the vehicle level (assuming all events occur at the same test facility). Each event uses a specific system or item configuration (hardware or software) being verified along with some support equipment, and implements detailed procedures designed to perform the event precisely. The result of each event is captured in a verification report and other documentation that serves as a formal closeout (e.g., a verification completion notice) of that requirement.

Maximizing the number of TVRs for each event is more efficient, so careful planning is important. One useful planning tool is to prepare a logical flow of all the activities to better group them into events. The *integrated verification fishbone (IVF)* defines and documents the integration-related TVRs (e.g., project-to-project, project-to-program) associated with the different stages of mission integration. It uses a "top down" and "leftward" approach. The IVF is similar to a PERT diagram in that it illustrates logical relationships between integration TVRs. It may also include additional information not usually included on a PERT diagram, such as program milestones, facility locations, etc. IVFs help planners focus on the end item and encompasses all activities associated with final mission integration, such as:

- TVRs associated with the qualification requirement category
- TVRs associated with accepting flight hardware and software and for verifying its proper assembly and check-out
- TVRs associated with the acceptance, check-out, and operational requirement categories

The IVF also does not illustrate timing per se; rather it shows the logical order of things. All qualification and acceptance activities must be complete before an end item is fully verified. Figure 11-15 illustrates an example IVF for FireSAT, with emphasis on the logical flow of events for the propulsion module.

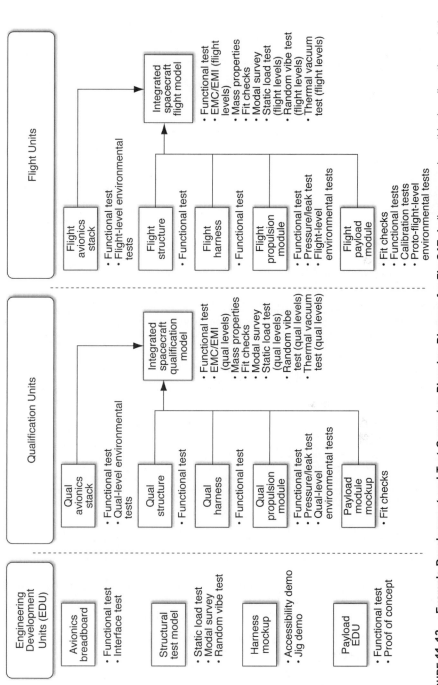

FIGURE 11-13. **Example Development and Test Campaign Planning Diagram for FireSAT.** A diagram such as this describes the evolution and pedigree of the myriad models developed throughout a project. (EMC/EMI is electromagnetic compatibility/electromagnetic interference; EDU is engineering development units.)

TABLE 11-14. **Summary of Major Space Environmental Tests.** We do this testing at the system component or box level, and on higher-level assemblies within facility capabilities. Environmental verification at the system level is usually done by analysis on the basis of component or box testing results. (EM is electromagnetic; EMC/EMI is electromagnetic compatibility/electromagnetic interference.)

Test	Purpose	Equipment or Facilities Required	Process
Vibration and shock testing	• Make sure that product will survive launch • Comply with launch authority's requirements • Validate structural models	• Vibration table and fixture enabling 3-axis testing • Acoustic chamber	• Do low-level vibration survey (modal survey) to determine vibration modes and establish baseline • Do high-level random vibration following profile provided by launch vehicle to prescribed levels (qualification or acceptance) • Repeat low-level survey to look for changes • Compare results to model
Thermal and vacuum (TVAC) testing	• Induce and measure outgassing to ensure compliance with mission requirements • Be sure that product performs in a vacuum under extreme flight temperatures • Validate thermal models	• Thermal or thermal/vacuum chamber • Equipment to detect outgassing (e.g., cold-finger or gas analyzer) as needed • Instrumentation to measure temperatures at key points on product (e.g., batteries)	• Operate and characterize performance at room temperature and pressure • Operate in thermal or thermal vacuum chamber during hot- and cold-soak conditions • Oscillate between hot and cold conditions and monitor performance • Compare results to model
Electromagnetic compatibility/ electromagnetic interference (EMC/EMI)	• Make certain that product doesn't generate EM energy that may interfere with other spacecraft components or with launch vehicle or range safety signals • Verify that the product is not susceptible to the range or launch EM environment	• Radiated test: sensitive receiver, anechoic chamber, antenna with known gain • Conduction susceptibility matched "box"	• Detect emitted signals, especially at the harmonics of the clock frequencies • Check for normal operation while injecting signals or power losses

Next, we develop detailed procedures for each event, along with a list of needed equipment, personnel, and facilities. The test procedure for each item should include, as a minimum, descriptions of the following:

- Criteria, objectives, assumptions, and constraints
- Test setup—equipment list, set-up instructions
- Initialization requirements
- Input data

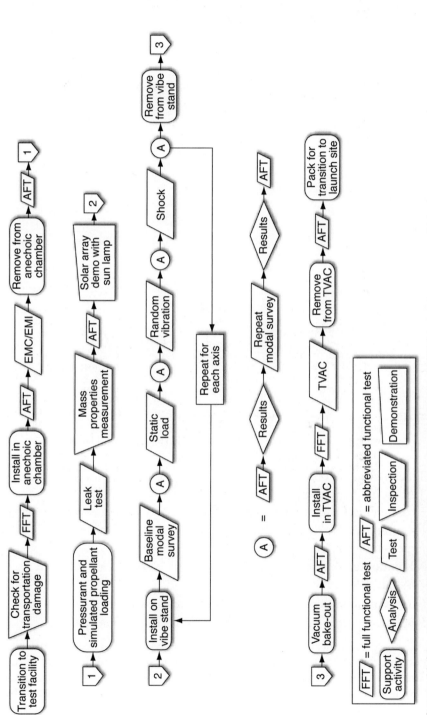

FIGURE 11-14.   FireSAT Flight-Model Environmental Test Campaign Sequence. This flow chart shows the notional sequence for major events in the FireSAT environmental test campaign. (EMC is electromagnetic compatibility; EMI is electromagnetic interference; TVAC is thermal vacuum.)

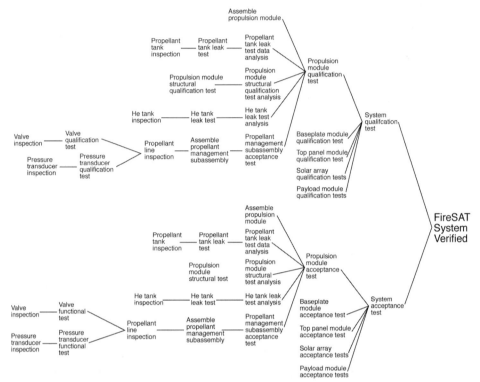

FIGURE 11-15. Example Integrated Verification Fishbone (IVF) for FireSAT. This diagram illustrates the use of an IVF for planning the sequence of assembly, integration, and verification activities for the FireSAT propulsion module. (He is helium.)

- Test instrumentation
- Test input or environment levels
- Expected intermediate test results
- Requirements for recording output data
- Expected output data
- Minimum requirements for valid data to consider the test successful
- Pass-fail criteria for evaluating results
- Safety considerations and hazardous conditions
- Step-by-step detailed sequence of instructions for the test operators

We then organize the TVRs into individual events that tie back to the project's integrated master schedule (IMS). This tells what to verify, how to verify it, and when. The next logical question is how much. Knowing how much testing is "good enough" is a major part of the planning process. The best answer is, "It depends." For example, the test levels and durations for a random vibration test depend on:

- The item being tested (unit or system level)
- Its model type (qualification, proto-flight, or acceptance)
- The anticipated launch environment (from the launch vehicle user's guide)
- Organizational standards or guidelines

For the FireSAT example, complete system verification depends on a qualification model and an acceptance model. For each model, we first inspect the components and then test them before integrating them into higher-order assemblies. Table 11-15 compares test levels for three types of models for some of the major environmental tests.

TABLE 11-15. **Representative Environmental Test Levels and Durations at the Vehicle Level.** Depending on the model, industry standards recommend more or less severe test levels. These levels are representative only. For actual test levels, each project should refer to its own governing standards. (Adapted from NASA/GSFC, [1996], ECSS, [1998], and DoD, [1998].) (MEOP is maximum expected operating pressure.)

Test	Model Design Qualification Test Levels	Proto-Flight Acceptance Test Levels	Flight Model Acceptance Test Levels
Structural loads	1.25 × limit load	1.25 × limit load	1.0 × limit load
Acoustic level (duration)	Limit level + 6dB (2 minutes)	Limit level + 3dB (1 minute)	Limit level (1 minute)
Random vibration level (duration) each axis	Limit level + 6dB (2 minutes)	Limit level + 3dB (1 minute)	Limit level (1 minute)
Mechanical shock	2 actuations 1.4 × limit level 2× each axis	2 actuations 1.4 × limit level 1× each axis	1 actuation 1.0 × limit level 1× each axis
Combined thermal vacuum and thermal cycle	10° C beyond acceptance temperature extremes for a total of 8 thermal cycles	10° C beyond acceptance temperature extremes for a total of 4 thermal cycles	0° C beyond acceptance temperature extremes for a total of 4 thermal vacuum cycles
Pressure/leak	1.5 × MEOP 5 minutes (3 cycles)	1.5 × MEOP Sufficient to establish leakage	MEOP Sufficient to establish leakage

Write the Master Verification Plan

The preceding discussion details the requirements for people, processes, and equipment to support the verification effort. In parallel with writing the verification requirements and TVRs, systems engineers develop the master verification plan (MVP), also known as the test and evaluation master plan (TEMP, in DoD). Along with the SEMP (Chapter 14), the MVP provides the philosophical and practical guidance for all product verification aspects of the systems engineering effort. Table 11-16 shows an example table of contents for a typical MVP.

TABLE 11-16.   Master Verification Plan (MVP) Outline. The MVP describes the roles and responsibilities of project team members in the project verification activities and the overall verification approach.

1.0 Introduction	5.0 Systems qualification verification
1.1 Scope	5.1 Tests
1.2 Applicable documents	5.2 Analyses
1.3 Document maintenance and control	5.3 Inspections
2.0 Program or project description	5.4 Demonstrations
2.1 Program or project overview and	6.0 Systems acceptance verification
verification master schedule	6.1 Tests
2.2 Systems descriptions	6.2 Analyses
2.3 Subsystems descriptions	6.3 Inspections
3.0 Integration and verification (I&V)	6.4 Demonstrations
organization and staffing	7.0 Launch site verification
3.1 Program or project management	8.0 On-orbit verification
offices	9.0 Post-mission and disposal verification
3.2 Base or field center I&V organizations	10.0 Verification documentation
3.3 International partner I&V organizations	11.0 Verification methodology
3.4 Prime contractor I&V organization	12.0 Support equipment
3.5 Subcontractor I&V organizations	12.1 Ground support equipment
4.0 Verification team operational relationships	12.2 Flight support equipment
4.1 Verification team scheduling and	12.3 Transportation, handling, and other
review meetings	logistics support
4.2 Verification and design reviews	12.4 Tracking station and operations
4.3 Data discrepancy reporting and	center support
resolution procedures	13.0 Facilities
	14.0 Master verification schedule

The MVP is supported by the detailed implementation plans reflected in the relationships between the VRs, the TVRs, and their associated plans and procedures. We should develop it in draft form by the system requirements review and update it with each baseline as the project matures. The final version approved at the critical design review guides the critical assembly, integration, and test (AIT) phase. A key product of this planning process is inputs to the project integrated master schedule showing details for the AIT phase. Figure 11-16 shows a schedule based on the sequential test layout given in Figure 11-14.

Test engineers have to try to estimate the duration of what are somewhat uncontrolled events. How long do X-axis vibrations take from start to finish? What if the planned test facility isn't available or needs to be reconfigured? The space test business has all too many enemies of schedule. These include:

- Launch slips—an inevitable part of any mission planning. These slips lead to widely known, but not officially recognized, changes in required delivery dates that involve last-minute schedule perturbations.

- Uncontrolled processes—such as set-up in an unknown test facility, or getting data acquisition systems to talk to each other.

- Hazardous operations—such as propellant and pyrotechnic loading or moving large, expensive equipment. Schedule doesn't take priority over safety.

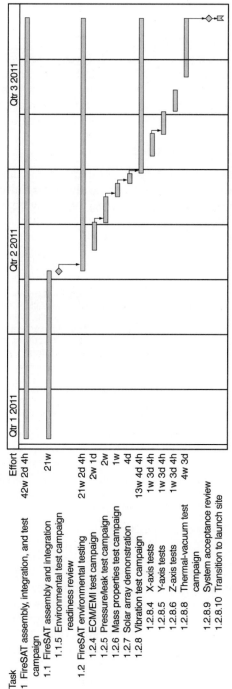

FIGURE 11-16. Top-level FireSAT Assembly, Integration, and Test (AIT) Schedule. This Gantt chart illustrates a hypothetical schedule for the FireSAT AIT phase. We budget about three months for system integration and baseline testing, followed by about four months of environmental testing. The environmental test sequence follows the flow chart shown in Figure 11-14. (EMC is electromagnetic compatibility; EMI is electromagnetic interference.)

- Dependence on infrastructure—if the facility is not ready, we wait or find another one
- Failures—we hope for the best, but plan for the worst

Systems engineers and program managers must rely on expert experience, and build in ample margin, in laying out the detailed project verification schedule. Planning, in some ways, never ends. Continuous replanning is needed in response to the inevitable schedule impacts that arise. No plan is perfect, but at some point schedule pressures dictate that we turn our attention from planning to execution.

11.3.2   Implement Product Verification

Chapter 13 stresses the need for good technical planning in any project. Systems engineers must be ready to improvise on the fly if carefully laid test sequences don't go according to plan due to late delivery of hardware or facilities failures. This flexibility pays dividends as the engineering team adapts and overcomes with a clear understanding of the ultimate goals, trade-offs, and overall verification philosophy driven by the project risk posture. Table 11-17 summarizes the inputs, outputs, and activities associated with implementing verification.

TABLE 11-17.   **Inputs, Outputs, and Activities Associated with Implementing Verification.** We employ the verification methods (analysis, inspection, demonstration, simulation and modeling, and test) on analytical and physical models for qualification and on flight articles for acceptance. All the outputs from the previous section serve as inputs to this phase. The results are verified end products along with the body of evidence that formally confirms to stakeholders that the item complies with the validated requirements.

Inputs	Verification Implementation Activities	Outputs
• Verification requirements (e.g., verification matrix) • Master verification plan (as-planned, including incompressible test list) • Test and verification requirements (TVR), with associated plans and procedures • Unverified end products • Verification enabling products (e.g., validated models, test facilities)	• Execute TVR plans and procedures • Track completion of (or exception to) each TVR • Audit (spot-check) lower-level verification events • Identify regression test needs • Identify test uncertainty factors • Document results	• Verified end product • Verification plan (as-executed) • Verification products (data, reports, verification compliance matrix, verification completion notices, work products)

During execution of test and verification requirement (TVR) plans and procedures, the project systems engineer or verification lead must track completion of each TVR, noting any exceptions or anomalies. However, when the verification activities flow down to the parts level, project-level engineers don't get involved in the potentially thousands of individual events. Responsibilities must be delegated carefully and overseen by appropriate audits and spot checks.

As Section 10.3.4 describes, *regression testing* involves repeating a test already run successfully, and comparing the new results with the earlier ones to be sure that something didn't break in the meantime. It's essential to build inspections and functional tests into the AIT sequence to serve as regression tests at each stage of integration or after a system has been moved to a new facility (or any other change that may have introduced faults). Routine regression tests uncover problems in three categories: 1) local issues where the change has introduced some entirely new problem, 2) unmasked issues, where the changes uncovered existing problems, and 3) remote issues where changing one part of the item introduces a fault in another part. When we encounter any of these issues, we have to develop and implement additional regression tests "on the fly" to troubleshoot the problem.

A vital part of implementing the verification plan is learning its limits. The aim of verification is to learn as much as possible about the system's behavior. The evidence collected during verification documents that behavior under a variety of conditions. These are the "known knowns" of the system. Also important are the test uncertainty factors (TUF) or "known unknowns," those aspects of the system that were not or could not be measured or demonstrated. Establishing this boundary of knowledge is crucial to flight certification. (Section 11.5 discusses "unknown unknowns.")

After we collect product verification data, we must analyze it to determine if the desired result was achieved. If it was, we can use the evidence of the data to formally close out the requirement(s) involved. In addition, we should capture as much other information as possible on the system behavior, which will be useful during later regression testing and ultimately during operations. If the results obtained were not as desired, in other words, if the product failed to pass the verification success criteria, then we face some important program decisions. Depending on the severity of the anomaly, anything from minor changes to verification procedures to a complete redesign of the product may be needed.

No job is complete until the paperwork is done. Following the event, and any subsequent analysis, we must prepare a detailed *verification report*. This should include:

1. The source paragraph references from the baseline documents for derived technical requirements

2. Technical requirements and stakeholder expectations

3. Bidirectional traceability among these sources

4. Verification methods employed

5. Reference to any special equipment, conditions, or procedures used

6. Results of verification

7. Variations, anomalies, or out-of-compliance results

8. Corrective actions taken

9. Results of corrective actions

We can summarize the information in this report in a verification compliance matrix (VCM). We usually establish and maintain the VCM once we've initiated requirements traceability after obtaining stakeholder commitment to the set of stakeholder expectations.

The final step is to capture the work products from the product verification. These include verification outcomes; records of procedural steps taken against planned procedures; any failures or anomalies in the planned verification procedures, equipment, or environment; and records citing satisfaction or non-satisfaction of verification criteria. They should also list the:

- Version of the set of specification and configuration documents
- Version or configuration of the verified end product
- Version or standard for tools and equipment used, together with applicable calibration data
- Results of each verification, including pass or fail declarations
- Discrepancy between expected and actual results
- Idiosyncrasies or behavior anomalies observed that, while within specifications, may be important to know during operations
- Lessons learned

11.4   Validate Products

Product validation answers the following questions:

- Did we build the right system?
- Does the end-to-end system (including hardware, software, people, processes, and procedures) meet the operational objectives?
- Does the system behave in a predictable and robust fashion?

Figure 11-17 illustrates the subtle difference between verification and validation. Verification confirms that the product meets the technical requirements while validation confirms that the product meets the original customer expectations. Done right, validation gets at the heart of these expectations and shows that "this thing will work in space as intended."

In execution, system validation events often occur concurrently with (and are sometimes confused with) product verification events. In fact, we can make the following argument:

IF...all of the following are true during the project:

- All driving needs are captured by requirements
- All requirements are validated, with no gaps or errors
- All requirements are iteratively and correctly updated as the design evolved

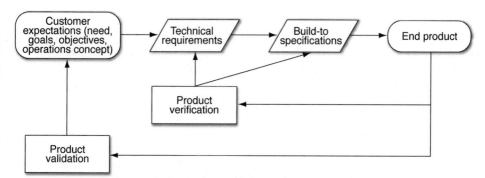

FIGURE 11-17. Product Validation Versus Product Verification. This flowchart illustrates the main differences between product verification and product validation (adapted from Perttula [2007]). While product verification closes the loop between product and requirements, product validation closes the loop between product and customer expectations.

- All models are validated, with no remaining model errors or uncertainties
- All requirements are verified without error
- No unexpected system behaviors or design features exist

THEN…system validation is almost automatic

Unfortunately, in the real world these conditions are never true. To begin with, real-life systems are organic. No matter how many requirements we write, we're producing only an approximation of the real thing (say at the 95% confidence level). A perfect description of a design would require an infinite number of requirements. But it's the missing 5% (often the ones someone assumes without communicating the assumption) that can spell the difference between mission success and mission failure.

Furthermore, producing a completely defect-free system is practically impossible for many reasons. The process of translating customer needs into verifiable requirements is imperfect, because stakeholders often don't have completely clear ideas of their true needs, developers don't grasp fully what stakeholders' expectations truly are, or some combination of both. And it's the nature of any project that requirements inevitably change as we refine the system definition and as needs evolve. As this happens, the loop between developer and stakeholder is not fully closed.

Another issue is that some customer expectations are ultimately subjective (e.g., astronaut comfort or work-load) no matter how exactly we try to translate them into design requirements. They can't be fully validated until exercised during flight-like operational scenarios as an integrated system. For example, automobiles undergo validation when test-driven by members of a target market or customer advocates. Human space systems are subjected to extensive man-in-the-loop test scenarios before operational deployment.

Another challenge is that as systems become more complex, designing and analytically predicting their behavior a priori becomes increasingly difficult. Software in general, and sequence engines and fault protection in particular, are prone to exhibit behaviors the designer never expected. We expose flaws only by rigorously exercising the integrated system in a full-up flight-like fashion.

A final matter relates to the imperfect nature of testing and other verification methods. An improperly designed test or calibration errors in test equipment may provide the illusion of the desired result. To complicate matters, testing only for correct functional behavior doesn't always show anomalous behavior under non-ideal circumstances. The Mars Polar Lander mishap report, for example, found that "the flight software was not subjected to complete fault-injection testing. System software testing must include stress testing and fault injection in a suitable simulation environment to determine the limits of capability and search for hidden flaws" [JPL Special Review Board, 2000]. Unfortunately, validation programs designed to inject errors and look for hidden flaws are difficult to design and are usually expensive to implement.

The following section discusses validating the project system within the context of flight systems, ground systems, and launch systems working together to execute the mission. Some of these concepts scale to other systems of interest. Table 11-18 shows the fundamental activities associated with product validation.

TABLE 11-18. **System Validation Inputs, Outputs, and Associated Activities.** System validation begins with a verified end product and documented stakeholder expectations. The activity affirms that the system will meet stakeholder expectations.

Inputs	Product Validation Activities	Outputs
• Verified end products • Customer expectations (e.g., need, goals, and objectives, mission requirements) • Validation enabling products (e.g., concept of operations, mission timelines, ops procedures, flight system and ground system interface control documents, command and telemetry handbook, fault trees, fault protection book)	• Prepare for product validation • Implement product validation	• Validated products • Validation products (e.g., test reports, work products)

11.4.1   Prepare for Product Validation

As Table 11-18 illustrates, product validation activities start with a verified end product (although it may be verified only partially depending on the concurrency of verification and validation activities). Validation success depends on knowing the true customer expectations, especially measures of effectiveness and success criteria. We refine these criteria further with enabling products that detail how the product will be used in the real world (e.g., the concept of operations as described in Chapter 3) and how it will work. Figure 11-18 shows the FireSAT concept of operations.

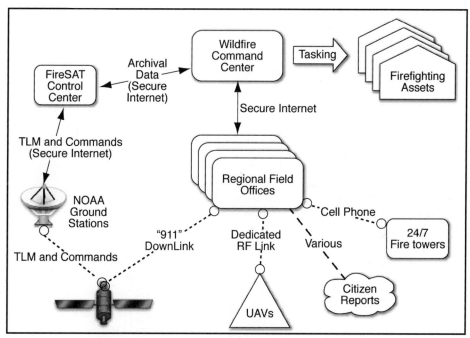

FIGURE 11-18. **FireSAT Concept of Operations.** The mission concept of operations is one of several key inputs into the validation planning process. We can't "test like we fly" if we don't know how we're going to fly! (TLM is telemetry; UAV is unmanned aerial vehicle; NOAA is National Oceanic and Atmospheric Administration; RF is radio frequency.)

From this operational context, engineers derive crucial information to support validation planning such as fault trees (or success trees, their conceptual complement). Fault trees help the validation planner develop clever ways to stress the system under realistic, but often extreme conditions. We don't know if we have a valid product unless we know the customer's definition of success. Figure 11-19 shows an example of a top-level fault-tree for the FireSAT mission.

With these inputs in mind, the systems engineering team develops a detailed validation plan aimed at achieving the primary output—a validated product, supported by a body of objective evidence. This evidence, like the test reports generated as part of verification, is invaluable during flight certification and later operations to give users insight into the real behavior of the system. Table 11-19 shows the inputs, outputs, and activities associated with preparing for product validation.

System validation sometimes appears to duplicate a product verification event, but in fact, it often goes well beyond it. Verification applies at many levels ranging from component to system level and may employ a variety of methods including test, demonstration, analysis, simulation and modeling, or inspection. In contrast, validation normally focuses on the highest level of the system of interest

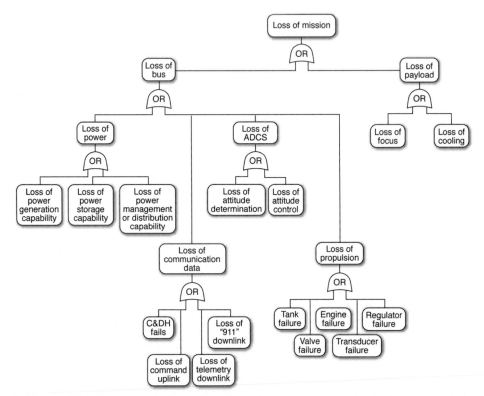

FIGURE 11-19.   **FireSAT Fault Tree.** We derive a fault tree from an understanding of the mission concept of operations and the underlying system functionality. The concept of operations, together with the fault tree, guides overall validation planning. (ADCS is attitude determination and control subsystem; C&DH is command and data handling subsystem.)

TABLE 11-19.   **Prepare for Product Validation.** Validation planning determines how to objectively ensure that the system will meet customer expectations. (FS/GS ICD is flight system/ground system interface control documents; GSE is ground support equipment.)

Inputs	Validation Preparation Activities	Outputs
• Verified end products • Customer expectations (e.g., need, goals, and objectives, mission requirements) • Validation enabling products (e.g., concept of operations, mission timelines, ops procedures, FS/GS ICD, command and telemetry handbook, fault trees, fault protection book)	• Establish the project or program validation philosophy • Develop validation requirements (identifying validation methods, specific validation events, and success criteria) for each acceptance criterion • Write master validation plan	• Validation requirements (e.g., validation matrix) • Derived enabling product requirements (e.g., GSE, test/emulators) • Validation plan (as-planned) • Detailed validation support requirements (e.g., facilities, equipment, product availability, personnel, time, money)

and is limited almost exclusively to test and demonstration (even if we do so using a simulator or engineering model). The principal difference here is that validation looks at the overall system, with the intent to discover unexpected behaviors. So models and analyses aren't useful here, because they're limited to what we know when we create them.

The focus during product validation then, is not so much on the requirements, but on what needs to happen in practice. This is an organic approach, not an algorithmic one. The system validation plan should include the following techniques:

- End-to-end information system testing—Shows compatibility of the project information systems (command, data, timing, etc.)

- Mission scenario tests—Demonstrate that the flight hardware and all software execute the mission under flight-like conditions (nominal and contingency), but do not necessarily follow the exact flight timeline

- Operational readiness tests (ORTs)—Demonstrate that all elements of the ground segment (software, hardware, people, procedures, and facilities) work together to accomplish the mission plan, using real timelines. These timelines should include the flight segment for at least one ORT that validates the interfaces between flight and ground segments during the entire launch sequence (nominal and contingency). Cross-system compatibility tests or "scrimmages" can serve as useful precursors and to complement ORTs.

- Stress testing and simulation—Assess system robustness to variations in performance and fault conditions. The example region of robust operation in Figure 11-20 depicts the space over which we should test system performance. Likewise, the results from system fault-tree analysis and probabilistic risk analysis should guide the definition of mission scenario tests that include fault injection.

- Analysis—everything not covered by the above

Stress testing validates the design, demonstrating system robustness by exercising it over a broad parameter space—beyond nominal. It determines capability boundaries and demonstrates robustness of designs, to assure health and safety, and provide confidence in project validation. It must also consider single faults that cause multiple-fault symptoms, occurrence of subsequent faults in an already faulted state, etc. We should do it on the integrated flight segment whenever we can do it safely and cheaply enough; otherwise we should do it on system test beds that duplicate the flight segment to high fidelity. However, stress testing does **not** imply physically stressing the flight hardware. The emphasis is on hardware/software interactions under normal and extreme (but not unreasonable) flight conditions.

Stress test planning requires rank ordering (balancing risk versus resources) to determine which tests to run. We have two basic methods for determining this order: (1) organic—brainstorming sessions with cognizant engineers and subject matter experts to identify "what-if cases"; and (2) algorithmic—results of analyses

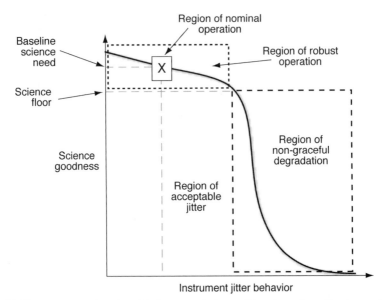

FIGURE 11-20. Example Relationship between Science "Goodness" and Instrument Pointing Jitter. Knowing where the "cliffs" are in system performance is a useful guide to validation planning. (Science floor is the minimum science return for mission success.)

to identify "soft" spots using such techniques as fault-tree analysis, state analysis, or performance sensitivity analysis.

The importance of stress testing and simulation was cited in the WIRE failure report: "Testing only for correct functional behavior should be augmented with significant effort in testing for anomalous behavior" [WIRE Mishap Investigation Board, 1999]. Some projects can't do full fault-injection testing with the flight system without incurring excessive risks or costs. The solution is software testbeds. An early verification and validation (V&V) plan identifies the needs for such testing and the required testbed capabilities.

The Jet Propulsion Laboratory has amassed several decades of lessons learned in deep-space missions, in a set of design principles and flight project practices. Besides the V&V principles already discussed in this chapter, they address the following:

- Perform a mission design verification test to validate the end-to-end flight-ground system by covering mission-enabling scenarios (e.g., science operations, trajectory correction maneuver, safing, cruise)

- Validate the launch sequence and early flight operations on the flight vehicle using the launch version of flight software

- "Test as you fly and fly as you test" (e.g., using flight sequences, flight-like operating conditions, and the same software functionality)

- Carry out stress testing involving single faults that cause multiple-fault symptoms, occurrence of subsequent faults in an already faulted state, etc.
- Perform system-level electrical "plugs-out" tests using the minimum number of test equipment connections
- In addition to real-time data analysis, perform comprehensive non-real-time analysis of test data before considering an item validated or verified
- Regard testing as the primary, preferred method for design verification and validation
- Arrange for independent review of any V&V results obtained by modeling, simulation, or analysis
- Reverify by regression testing any changes made to the system to address issues found during testing
- Perform verification by visual inspection, particularly for mechanical clearances and margins (e.g., potential reduced clearances after blanket expansion in a vacuum), on the final as-built hardware before and after environmental tests
- Include full-range articulation during verification of all deployable or movable appendages and mechanisms
- Validate the navigation design by peer review using independent subject matter experts
- Employ mission operations capabilities (e.g., flight sequences, command and telemetry data bases, displays) in system testing of the flight system

Tools for product validation are similar to those for product verification planning. A validation matrix (VaM) maps validation methods against customer expectations and key operational behaviors. However, unlike verification planning, which goes back to the design requirements, validation must reach back further and tap into the original intent of the product. This reaching back highlights the critical importance of getting early buy-in on the mission need, goals, and objectives as well as mission-level requirements as described in Chapter 2. These measures of effectiveness (MOEs) or key performance parameters (KPPs), along with the operations concept and failure modes and effects analysis, guide validation planners in preparing a cogent "test as you fly" plan. This plan aims at discovering unintended or unwanted product behaviors in the real-world environment. Table 11-20 shows a partial VaM for the FireSAT mission with a cross-section of originating intent.

Because detailed planning for validation so closely parallels that of verification planning (and may be almost indistinguishable in all but intent), we do not address it to the same level of detail here. The same tools and techniques easily apply. Because validation focuses primarily on system-level versus component or subsystem behavior, we usually don't need to develop IVFs as we do for assembly, integration, and test planning. But we must attend to configuration control

TABLE 11-20.   **Example FireSAT Validation Matrix.** This excerpt from the FireSAT validation matrix shows the information we need in preparing for product validation.

Capability	Mission Scenario Test	Operational Readiness Test	End-to-End Information System Test	Stress Test	Other System Test
Detect wildfires and communicate location and intensity to end-users	X	X	X		X
Optimize payload optical focus	X	X	X		X
Execute, launch, and reach a safe, power-positive, commandable state in the presence of a failed primary avionics string	X			X	X
Maintain safe spacecraft state during an extended solar flare event (24 hrs of spacecraft computer resetting and scrambled memory) with a simultaneous gyro failure				X	

(Chapter 15), because not all flight elements (especially software) are available during a ground-based validation event.

A major part of this strategic planning is developing the project's *incompressible test list (ITL)*. Project resources are scarce commodities. As the launch date looms, schedules get compressed and budgets get depleted so we must make tough decisions on how best to allocate resources to V&V activities. This resource reallocation often creates overwhelming pressure to reduce or even eliminate some planned testing. Recognizing this eventuality, the project leaders should define an ITL that identifies the absolute minimum set of tests that must be done before certain project milestones (such as launch) occur, regardless of schedule or budget pressures. Systems engineering and project management must approve any changes to this list. We should resist the temptation to immediately reduce the entire program to this minimum list under the "if the minimum wasn't good enough, it wouldn't be the minimum" philosophy. Instead, if properly presented and agreed to at the highest project levels early in the lifecycle, it's useful ammunition against pressures to arbitrarily reduce or eliminate verification or validation activities. Test objective considerations for all tests on the ITL generally fall into one or more of the following:

- Safe, reliable launch
- Flight segment health and safety
- Successful commissioning
- Successful completion of post-launch mission-critical activities
- Successful science data collection and return operations

Major test categories in the ITL include:

- Environmental tests (e.g., vibration, shock, electromagnetic compatibility, electromagnetic interference, and thermal vacuum)
- Integration, functional, and performance tests, including baseline tests used to establish initial system behavior for comparison with later regression testing
- Fault protection testing
- Sequence and scenario testing, both nominal and contingency
- Operations testing and training
- End-to-end information system testing
- Stress testing

Table 11-21 gives an example of the critical items included in the FireSAT ITL.

TABLE 11-21.   Example FireSAT Incompressible Test List (ITL). Subject to the demands of schedule and hardware safety, the flight segment (FS) for the FireSAT spacecraft is the preferred venue for testing. Testing may be done on the system test bed (STB), or during flight software flight qualification test (FSWFQT) on the software test bench (SWTB) as long as the fidelity of the test environment is adequate to demonstrate the required function or characteristic. (Adapted from the Kepler Project Incompressible Test List, courtesy of JPL). (GS is ground system; KSC is Kennedy Space Center; LV is launch vehicle; NOAA is National Oceanic and Atmospheric Administration; EMC/EMI is electromagnetic compatibility/electromagnetic interference; RF is radio frequency.)

NO.	Title	Description	Venue	Comment
1	Integration and functional verification of all FireSAT spacecraft subsystems	Verification of interfaces and functional requirements of each subsystem	FS, STB	--
2	Validation of all FireSAT spacecraft telemetry measurements	Ensure validation of telemetry function and calibration for each telemetered engineering data point	FS, STB, SWTB, GS	Does not include all permutations of telemetry subscription packets and does not include calibration of science data (done in flight)
3	Validation of generation and function of all commands	Ensure validation of commands from translation in GS through function on the FS. Include variation in critical parameter settings.	FS, STB, SWTB, GS	Validation of commands in the flight command dictionary

TABLE 11-21.   Example FireSAT Incompressible Test List (ITL). (Continued)

NO.	Title	Description	Venue	Comment
4	System-level environmental design verification and workmanship tests, including plugs-out test	Per the environmental test matrix, with demonstration of acceptable margins against flight predicted environments, including: thermal-vacuum, vibration, acoustic, separation shock, and EMC/EMI testing (radiated susceptibility and emissions). Self-compatibility testing shall be done as plugs-out.	FS	--
5	Phasing verification of all actuators, sensors, thrusters, gimbals, heaters, and telecom systems in flight configuration on the FireSAT spacecraft	End-to-end verification of functional to/from physical phasing for all devices. At least one phasing verification of actuators, attitude determination and control system (ADCS) sensors, and thrusters using the launch version of flight software. Phasing shall be verified in sufficient cross-strap configurations to confirm that phasing is correct in all configurations. Phasing of the telecom systems shall be verified by a combination of test and analysis.	FS	--
6	Dust cover ejection tests	Verify dust cover ejection to first motion on the integrated flight segment, after environmental testing	FS	--
7	Flight segment functional (baseline) tests	FS functional tests that demonstrate function of the spacecraft bus and photometer subsystems and devices. To be performed before and after environmental tests, and at the launch site.	FS	--
8	End-to-end information system (EEIS) tests	Functional verification of UHF and S-band telecom systems' compatibility with NOAA ground station equipment. Includes command and telemetry data flow end-to-end test between FS and GS via RF link and NOAA equipment. At least one test includes loading of any multi-mission resources.	FS, GS, KSC	--
9	Launch vehicle interface tests	Verification of capability to operate the FireSAT spacecraft in the launch configuration, to include all launch vehicle interfaces and launch pad interfaces including telemetry data flow	FS, LV, KSC	--

TABLE **11-21.**   **Example FireSAT Incompressible Test List (ITL). (Continued)**

NO.	Title	Description	Venue	Comment
10	Flight software reload test	Verification of the capability to update or reload flight software on the ground and in flight	FS, STB	--
11	Alignment verification tests	Verification of ADCS sensor alignments, thruster alignments, and payload alignment before and after system environmental tests	FS	Payload to spacecraft alignment only. Internal payload alignment checked before and after lower-level environmental test.
12	Test bed and simulation fidelity validation	Validation of the fidelity of test beds and simulations against the flight segment, particularly any software simulation of hardware missing in the STB	FS, STB	--
13	Fault test cases	Demonstration of successful mitigation of mission faults identified in the mission fault trees (for verification by test items only)	FS, STB	--
14	Contingency operations procedure validation	Validation of all Category A contingency procedures for launch and commissioning phases and health- and safety-related contingency procedures for science operations	FS, STB, GS	--
15	GS integration and test	Completion of GS integration and test for the launch version of GS	GS	--
16	FSW long duration test	A long duration test (at least 169 hours uninterrupted) that demonstrates freedom from unanticipated counter or register overflow conditions, unanticipated errors or halts, or other software interactions with the FSW performing normal mission operations (through at least one missed ground station contact)	STB	--
17	Independent flight workmanship verification	A verification walk-down of flight system workmanship by an independent team, including but not limited to blanket dress, harness dress, service loops, critical clearances, etc. Performed at the launch site before encapsulation.	FS	--

11.4.2   Implement Product Validation

Table 11-22 summarizes the product validation activities. We emphasize here one key difference between a verification event and a validation event. For a verification activity to count toward requirement close-out, it must be witnessed by an independent quality assurance team. At a validation event, one additional witness (or collection of witnesses) is also crucial—the customer. Product validation closes the loop with customers and their expectations as embodied in the need, goals, and objectives and further refined by measures of effectiveness and key performance parameters, and aims to show (by test, analysis, simulation and modeling, inspection, or demonstration) that these expectations have been met. The customers or their cognizant representatives should be present during major validation events.

Just as for verification events, we must fully document the validation results as part of the as-executed plan along with test reports and other data. The now verified and validated end product, along with the large body of data collected during all these events, moves forward to the final phase of V&V: flight certification.

TABLE 11-22.   Input/Output and Activities Associated with Implementing Product Validation. Using the master validation plan as a guide, the validation implementation activities transform the verified end product into a validated end product with the objective rationale underlying that transformation.

Inputs	Validation Implementation Activities	Outputs
• Validation requirements (e.g., validation matrix) • Master validation plan (as-planned) • Unvalidated end products • Validation enabling products (e.g., validated models, test facilities)	• Execute validation plans and procedures • Track completion of (or exception to) each item on the VaM • Identify regression test needs • Identify test uncertainty factors • Document results	• Validated end product • Validation plan (as-executed) • Validation products (data, reports, validation compliance matrix, work products)

11.5   Certify Products for Flight

General discussions of verification often blur the subtle differences between it and flight certification. For that reason, we address these unique activities separately. Product verification usually culminates in a verification completion notice (VCN) or equivalent, a formal closeout of the design-to, build-to, and test-to requirements agreed to by stakeholders and system developers. In addition, it meets or exceeds all build-to specifications, that is, it's free from latent defects in parts, materials, and workmanship and has thus been deemed acceptable for flight.

The large body of evidence collected during this process serves to indicate the item's pedigree and overall flight worthiness. Figure 11-21 illustrates the main differences between verification, validation, and flight certification. Verification and validation feeds back to the technical requirements and customer expectations, while flight certification feeds forward to support the effective conduct of mission operations. It's the technical hand-off between the product's known characteristics and behavior and the mission operations community, who must live with them.

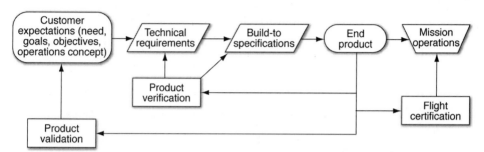

FIGURE 11-21.   **Flight Certification in the Big Picture of Verification and Validation Activities.** This figure illustrates the relationship between flight certification and product verification and validation.

Flight certification comprises those activities aimed at sifting through the large body of V&V data to formally—and usually independently—determine if the fully integrated vehicle (hardware, software, support equipment, people, and processes) is ready for flight. A *certificate of flight readiness* after the flight readiness review confirms that the integrated system has been properly tested and processed so that it's ready for launch. That is, the system has been successfully integrated with other major components and performance has been verified; sufficient objective evidence exists to verify all system requirements and specifications; and all anomalies or discrepancies identified during product verification have been satisfactorily dispositioned. Final flight system certification nomenclature differs widely between organizations. The terminology used here is adapted from DoD and NASA projects. In any case, the basic intent is universal—formally review the objective evidence indicating that the flight system is ready to fly.

Flight certification closes the loop between the end product and the operators or users of the product. It draws upon all the data collected throughout product verification and validation, as well as a detailed understanding of the mission operations concept. Table 11-23 shows the primary activities that support flight certification, along with their inputs and outputs.

Every stage of mission integration generates important data. Data products include verification plans, data, and reports; problem reports and history; variances (waivers and deviations); material review board dispositions; acceptance data packages; etc. Other important build data is also recorded, such as cycle counts and run time for limited life items, maintenance history, conforming but "out-of-family" anomalies, etc. For any project to effectively and efficiently plan, execute, and be sustained throughout its lifecycle, we must maintain this materiel history and keep it readily retrievable, especially when mission lifetimes span entire careers.

TABLE 11-23. Flight Certification Inputs, Outputs, and Associated Activities. The flight certification activities transform the verified and validated products into products that are certified to successfully perform mission operations. (DCR is design certification review; FCA is functional configuration audit; PCA is physical configuration audit; SAR is system acceptance review.)

Inputs	Flight Certification Activities	Outputs
• Verified and validated products • Verification and validation products • Real-world operational context for end products	• Confirm system integration activities completed • Confirm product verification activities completed • Confirm system validation activities completed • Review V&V products to derive the certification baseline • Conduct final reviews (DCR, FCA, SAR)	• Certified product • Certification products (e.g., certification baseline, signed DD250 (Material Inspection and Receiving Report), completed FCA, PCA, mission and flight rules, lessons learned, product idiosyncrasies)

The documented materiel history plays a critical role in supporting the investigation and disposition of problems that arise during the mission. Therefore, the project must require delivery of materiel history for various levels of assembly from its prime and sub-contractors so that it's readily stored, accessed, and used. A complete materiel history is far too expensive and impractical to collect. But deciding much later that something is needed involves even greater expense and difficulty. So the project must conservatively identify up-front (before the preliminary design review) what materiel history it needs to support operations and sustainment and clearly indicate these products as formal contract deliverables. Detailed flight certification includes the following (adapted from JPL's Certification of Flight Readiness Process):

1   Documentation of completion of critical activities

1.1   System and subsystem design reviews, including action item closures, are complete

1.2   System and subsystem environmental design and test requirements are documented and have been met, and test reports have been released

1.3   System and subsystem design analyses (fault trees, failure modes and effects criticality analyses, reliability, timing margin, functional models, mass properties, error budgets, etc.) have been completed, updated with test results, and reviewed

1.4   Hardware drawings (ICD's, parts, assemblies, schematics, circuit data sheets, etc.) and design review documents, including action item closures, are complete

1.5   Software design description, source code, command and telemetry dictionary, and design review documents, including action item closures, are complete

1.6   Ground data system and mission operations system design reviews (including mission design and navigation), through the operations readiness review, including action item closures, are complete

1.7   Hardware and software certification reviews, acceptance data packages, inspection reports, log books, discrepancy reports, open analysis items, and problem reports are complete and all open items closed out

2   Documentation of residual risks to mission success

2.1   Functional and performance requirements for complete and minimum mission success (including planetary protection) are documented and are being met

2.2   Matrices showing compliance with institutional requirements, design principles, and project practices have been audited and approved by the cognizant independent technical authority

2.3   Verification and validation requirements compliance matrix, including calibration, alignment and phasing tests and as-run procedures and test/analysis reports are complete and have been reviewed by the cognizant independent technical authority

2.4   Testbed certification of equivalence to flight system is complete and all differences documented and accounted for

2.5   Incompressible test list (ITL) tests, including operational readiness tests with flight software and sequences, are complete and have been reviewed, and any deviations have been approved by the cognizant independent technical authority

2.6   Test-as-you-fly exception list is complete and has been reviewed by the cognizant independent technical authority

2.7   All safety compliance documents have been approved

2.8   Commissioning activities, flight rules, launch/hold criteria, idiosyncrasies, and contingency plans are complete, reviewed and delivered to the flight team

2.9   Waivers and high-risk problem reports have been audited, are complete, and have been approved

2.10 All external interface (e.g., communication networks, launch vehicle, foreign partners) design and operational issues have been closed

2.11 Flight hardware has been certified and any shortfalls for critical events readiness, to allow post-launch development, have been identified, reviewed, and approved by the cognizant independent technical authority

2.12 All post-launch development work has been planned, reviewed, and approved

2.13 All work-to-go to launch activities have been planned, reviewed, and approved

2.14 Residual risk list has been completed, reviewed, and approved by the cognizant independent technical authority

In addition to focusing on the prima facie evidence derived from model validation and product verification efforts, flight certification dictates how the system should be operated. Early in the project, stakeholders define their expectations and develop an operations concept. Developers derive detailed performance specifications from this intent that ultimately flow down to every system component and part. But users and operators must live with the system as it actually is, not as it was intended to be.

One example is temperature limits. A mission design decision to operate the spacecraft in low-Earth orbit largely fixes the thermal conditions the system will experience. With this in mind, a good systems engineer strives to obtain the widest possible operational margins by defining survival and operational temperature range specifications for every component. Verification events, such as thermal node analysis and thermal-vacuum testing, determine if a given component will operate reliably in the anticipated thermal environment, plus margin.

This example uses a simple pass/fail criterion. Either the component works within specifications during and following thermal-vacuum testing, or it doesn't. For the designer and test engineer, this criterion is perfectly acceptable. But for the operator (one of the major active stakeholders), this result may not suffice for optimal operations. A transmitter, say, may perform within specifications between –20° and +40° C. But optimal performance may be between 5° and 15° C. Such behavior may even be unique to a single component. The same model on a different spacecraft may have a slightly different optimal range. Space hardware, like people, often exhibits different personalities.

Failure to account for model uncertainties when selecting parameters such as fault-protection alarm limits can result in a spacecraft unnecessarily initiating safe-mode response during critical early operations due to modest temperature excursions. Validation testing—particularly stress and scenario testing—plays a valuable role in selecting and certifying the set of key operational parameters. Flight certification involves translating observed system behaviors into flight or mission rules to guide the operator in squeezing the best performance out of the system once it's on orbit.

Other roles involve interactions of subsystems and payloads. System validation should uncover the "unknown unknowns" of system behavior. For example, we may find during system validation that for some reason, Payload A generates spurious data if Payload B is simultaneously in its calibration mode. Instead of redesigning, rebuilding, and reverifying the entire system, we may decide that an acceptable operational work-around is to impose a flight rule that says, "Don't collect data from

Payload A if Payload B is in calibration mode." The principle investigators who own the two payloads may not be happy with this, but it's likely to be more acceptable than a cost increase or schedule delay to fix the interaction problem.

Flight certification is a practical handoff between the developers and operators that tries to achieve the best results with the system we actually launch. The above examples involved limiting operational performance. More often than not, though, operators working with developers discover ways of enhancing performance or even creating entirely new capabilities far beyond what the designers or stakeholders ever envisioned. For instance, the Global Positioning System is an amazing example of a system that delivers services beyond the wildest dreams of the initial designers.

The as-built baseline (referred to as a *certification baseline* especially if multiple copies of the system are to be produced), along with supporting verification data, defines the certification limits to which the hardware or software was qualified. It addresses certification at several levels of assembly, from line-replaceable units to whole vehicles. It also places constraints on the appropriate level of assembly to ensure that performance is as expected and as qualified. Examples of these constraints include:

- Operational limits (e.g., thermal, pressure, duty cycle, loads), which may be lower or higher than the design-to performance specification limits

- Preventive maintenance limits on which qualification was based (e.g., filter change frequencies, lubrication frequencies, inspections)

- Production or fabrication process limits upon which qualification was based

The as-built baseline may be a separate document from the design-to or build-to performance specifications. Its limits may initially be the same as defined by the original specifications, but only because the verification results support those limits. The as-built baseline may also define "derived" constraints that aren't in the design or performance specifications. Engineering drawings and models, analytical tools and models (e.g., thermal, loads), and variances (with associated constraints and expansions) also constitute part of the as-built baseline. Expansion of a certification limit or change to the product configuration baseline (e.g., fabrication process) requires rigorous examination of all of the components that make up the certification to determine where additional verification activities, such as delta qualification or revised analysis, are warranted. Any approved changes are made to the as-built baseline, not necessarily the design or performance specification. We only revise hardware or software specifications when we identify changes to the production or fabrication processes, acceptance processes, or deliverables associated with acceptance.

Normally each "box" in a given subsystem has a design or performance specification. For example, the radiator and heat exchanger in a thermal control subsystem may be certified to different pressure and temperature limits. Sometimes, specific fabrication processes (e.g., braze welding, curing) for a box would, if changed, invalidate its certification, requiring requalification. As we assemble the boxes into a larger assembly, we may need additional operational constraints at the subsystem or higher levels. For example, the radiator and heat exchanger may have upper operating temperature limits of 200° C and 180° C, respectively. But their temperatures may be measured by a single sensor somewhere between them. To protect subsystem components, we may need to limit the maximum subsystem operating temperature to 150° C based on testing or thermal modeling showing that a 150° C temperature sensor reading equates to a radiator temperature of 200° C during operation. The subsystem certification limit for temperature thus becomes 150° C, while the radiator and heat exchanger certification limits remain unchanged. The thermal model we use to define the constraint is part of the certification baseline. We also may need to establish vehicle-level operational constraints (e.g., attitude and orientation) to avoid exceeding system, subsystem, or lower-level certification limits.

The design certification review (DCR) focuses on the qualification and first-use requirements identified during development of the VRs and TVRs. It's a formal review chaired by the customer, with other stakeholders represented, to certify that a configuration item (CI), contract end item (CEI), or computer software configuration item (CSCI) has met its specified design-to performance requirements and that the design is ready for operational use or further integration in the mission flow. We may hold a DCR for any such item, including software, line-replaceable unit (LRU), assembly, subsystem, element, system, vehicle, and architecture.

The DCR is an expansion of the functional configuration audit (FCA) process described in the next paragraph. Its primary objectives are to ensure the following:

- All qualification requirements for the item have been satisfied and verified or variances approved
- All operating constraints for the item have been baselined and placed under configuration management (CM) control
- All production or fabrication constraints associated with the item that could affect qualification and acceptance have been baselined and placed under CM control
- All maintenance requirements associated with the item's qualification have been baselined and placed under CM control
- All anomalies and non-conformances have been successfully dispositioned
- All hazards have been dispositioned and successfully mitigated
- The required materiel history associated with the product and certification baseline has been delivered to the customer and placed under CM control
- Customer and suppliers acknowledge, through a formal certification process, that all of the above steps have been completed and that the item being certified is ready for operational use and integration into the next phase of mission assembly

In contrast, the customer performs the FCA before delivery of an item from a contractor (after the critical design review (CDR) but before the system acceptance review (SAR)) to establish confidence that the design satisfies applicable functional or performance requirements. This is one of the last steps in the qualification process and is typically done once for each CI, CEI, CSCI, or significant modification. It differs from a DCR in several respects:

- The FCA does not require delivery of materiel history data to the customer as part of a certification package; the DCR does

- When only an FCA is performed, data retention responsibility often resides with the supplier

- The FCA is an audit of requirements (0% to 100%), whereas the DCR requires customer confirmation that each requirement has been satisfied and that data supporting compliance has been provided

- The FCA is tied to delivery of an item from a contractor to the customer and doesn't encompass review of design adequacy for subsequent assembly and integration work that the customer performs. The DCR evaluates each stage of mission integration following delivery of the item to the customer.

Depending on the project's scope and nature, it may require both types of review. Functional configuration audits are essential for the customer and contractors to ensure that their lower-level components (e.g., LRUs, assemblies) satisfy design-to requirements before integrating them into deliverable elements and systems. The customer also needs them to gain confidence that lower-level components designed and fabricated by its contractors comply with design-to (performance) requirements flowed down from the architecture- and system-level requirements. These audits give the customer insight into the effectiveness of contractor processes and confidence that they'll discover no significant issues (e.g., performance issues, materiel history deficiencies) during the DCR.

The system acceptance review (SAR) occurs near the end of verification, validation, and certification. It's a formal review to certify that a CI, CEI, or CSCI has been fabricated in accordance with and conforms to build-to (product-baseline) requirements, is free of workmanship defects, and is ready for operational use or further integration in the mission flow. We may hold an SAR for any such item, including software, LRU, assembly, subsystem, element, system, vehicle, and architecture. The SAR focuses on acceptance requirements identified as part of the product verification planning process. We might complete multiple SARs for every element in a mission. However, we normally don't do one for items to be integrated into higher-level assemblies before transferring responsibility for those items to the government, program, or project. The SAR constitutes the customer's acceptance review. They're conducted for subsystem or lower assemblies and components only when they are stand-alone deliverables to the customer. The primary objectives of the SAR are to confirm the following:

- All acceptance and assembly and check-out requirements have been satisfied and verified or variances approved

- All anomalies and non-conformances have been successfully dispositioned

- The required materiel history associated with the product baseline has been delivered to the customer and placed under CM control
- The customer and the contractor acknowledge, through a formal certification, that all of the above steps have been completed and that the item being certified is ready for operational use and integration into the next phase of mission assembly

An SAR focuses on completing the manufacturing and assembly processes; it isn't meant to replace or repeat day-to-day customer oversight and integration. It's also unlikely that all V&V activities will be complete by the SAR. Many of them, especially those involving final configuration changes, occur just before launch (e.g., installation of thermal blankets, or propellant loading). Chapter 12 addresses these as part of system transition.

Until now, our discussion approached the challenges of V&V from the perspective of developing nominally new hardware items. But space systems are increasingly composed largely of existing items. Furthermore, software differs from hardware when it comes to V&V. The next section explores the challenges of commercial-off-the-shelf (COTS) and non-developmental items (NDI). Software V&V issues are then examined in the section that follows.

11.6   Address Special Challenges of Commercial Off-the-Shelf and Non-Developmental Items Product Verification

A special challenge to space system product verification is the ubiquitous application of items that don't require development as part of the design process; they're either available on the commercial market, or the design exists in house. We use the following definitions:

- COTS: commercial-off-the-shelf (we buy it)
- NDI: non-developmental item (we already have it)

An existing design modified to a limited degree is often considered a COTS or NDI item. It's important to realize that *all* spacecraft include COTS or NDI items. Even entire spacecraft, such as the Boeing 702 commercial satellite bus that forms the backbone of the US Air Force's Advanced Extremely High Frequency satellite is nominally a COTS item. Spacecraft data handling systems are developed from existing microprocessors. Attitude determination sensors, thrusters, and other actuators are available off-the-shelf from many vendors. And at the parts level (resistors, capacitors, op-amps, wiring, connectors, etc.) spacecraft electronics is virtually all COTS.

Failure to fully account for the extent of verification needed for COTS and NDI items has led to many high-profile failures. The European Space Agency's Ariane V rocket exploded on its maiden launch as it veered off course due to navigation errors caused by using the NDI guidance, navigation and control system from the older

Ariane IV launch vehicle. NASA's Lewis spacecraft was lost soon after launch due to an attitude control problem [Harland and Lorenz, 2005]. Both of these missions relied on COTS or NDI hardware or software with proven flight heritage. This section highlights some of the verification challenges that COTS and NDI items pose, in the hope of preventing similar problems in the future.

From the standpoint of product verification, the widespread availability and use of COTS and NDI can be either good or bad. The benefit is that they already exist—we don't have to invent them. They likely have proven performance, possibly with demonstrated flight heritage. But these facts don't obviate the need for some level of verification, especially for new applications. This verification may be complicated by the unwillingness or inability of a commercial vendor to supply the level and type of detailed design or previous verification data we need. A further complication is the flow-down requirements that come from the decision to use a given COTS or NDI item. Any item comes with a pre-defined set of interfaces that the rest of the system may have to adapt to, entailing more verification activities.

We may insert COTS into a design anywhere during the flow-down side of the system design effort. But doing so before PDR is far more likely to maximize savings while minimizing system redesign efforts. If we use an item as is, we must still understand the requirements for it in the context of the new application. If we modify a COTS item, the requirements flow-down drives new requirements for the modified item to meet.

After we select a COTS or NDI item, derived requirements need to flow to the other subsystems, e.g., electrical power subsystem, thermal control subsystem, data handling subsystem. A *heritage review (HR)*, as described in Chapter 9, is advisable to identify issues with the reuse of a particular component in the new design. Table 11-25 summarizes a set of HR topics. The purpose of the review, best done before the preliminary design review (PDR), is to:

- Evaluate the compatibility of the inherited or COTS item with project requirements

- Assess potential risk associated with item use

- Assess the need for modification or additional testing

- Match the requirements of the inherited design against the requirements of the target subsystem. A compliance matrix is very useful for this activity.

- Identify changes to current design requirements

The heritage review forces a systematic look at the item's history and applicability to a new project. At the end of it, an item initially thought to be technology readiness level (TRL) 9 may be only TRL 7 or 8 for a new application, necessitating further verification before flight.

The outcome of the HR helps determine the need for, and level of, additional verification for the item. As we describe in Section 10.7.1, a further complication with COTS is that we can't apply the same rigor for V&V at the lowest levels of assembly simply because the vendor probably can't provide the required traceability, nor the rigor defined in the specifications. Proprietary issues may exist as well. We must often settle for higher-level verification of the unit and accept some risk (e.g., run some stressing workmanship, vibration, thermal-vacuum, radiation, and other tests on some sacrificial units to determine if they're good enough, even if we don't know exactly what's inside the box). The main point here is that tougher verification standards at higher levels, such as worst-case analysis; parts-stress analysis; or failure mode, effects, and criticality analysis, may be necessary to provide sufficient margins to offset the lack of lower-level knowledge.

By far the biggest risk in the use of COTS and NDI is the decision to fly terrestrial hardware in space. The launch and space environments pose unique challenges that otherwise perfectly suitable terrestrial components may not handle well. And even if they do, we determine this capability only after an extensive environmental qualification program. After careful analysis, we may decide that the additional time, cost, and complexity of such a program negate any advantage that COTS and NDI offer in the first place.

System Heritage Review Topics

Table 11-24 provides a way to categorize the heritage of a component to determine the need for and extent of a qualification program. The criteria for avoiding a qualification program of some type are quite stringent. But such a critical review of COTS items is essential to thoughtful inclusion in any system. The space industry is at the whim of the commercial market when it chooses to use COTS components. This means that replacement components may be difficult, if not impossible, to obtain if needed (some programs resort to searches on eBay to find replacement items). Even if the item is available, suppliers are not compelled to provide an identical one, even if the same part number is used. The change in one pin definition on an item with the same part number resulted in the need to completely relay the power board on the FalconSAT-2 satellite. The use of COTS components in space systems is compelling and inevitable, but it must be carefully approached with rigorous verification in mind.

11.7   Verify and Validate Software

Software V&V poses unique challenges that warrant a separate discussion. This section explores these challenges and how to address them.

In modern space missions, software is ubiquitous. It's used onboard the spacecraft, in ground data systems, for operations planning and control systems, within testbeds and their simulators, in ground support equipment, for behavior modeling, in analytical tools, and others. Figure 11-22 illustrates the software architecture for the FireSAT mission. Software is necessary for every function from mundane, routine data archiving to mission-critical fire detection.

Because of this breadth in functionality, we tailor the software development process, including the V&V phase, based on the software's criticality. We

TABLE 11-24.   Categorizing Commercial Off-the-shelf (COTS) and Non-developmental Items (NDI) Qualification Requirements. We use this simple checklist to categorize items based on their heritage to determine the necessity of a modified or full flight qualification program. A "delta" qualification program refers to something less than a full qualification program that fills in the gaps, or "deltas," between what the item was previously qualified for and the new application. (Adapted from ECSS [1998].)

Category	Description	Qualification Program
A	• Item is an off-the-shelf product without modifications, **AND** • It has been subjected to a qualification test program at least as severe as that imposed by the project specifications (including environments), **AND** • The product is produced by the same manufacturer or supplier, using identical tools, materials, and manufacturing processes	None
B	Item is an off-the-shelf product without modifications, **BUT** it has been... • Subjected to a qualification test program less severe or different from that imposed by the project specifications (including environment), **OR** • Produced by different manufacturer or supplier, or using different tools and manufacturing processes, **OR** • Made with substitution of parts and materials with equivalent reliability	Delta, or adjusted, qualification program, decided on a case-by-case basis
C	Item is an off-the-shelf product with design modifications	Delta or full qualification program, decided on a case-by-case basis depending on the impact of the modification
D	Item is a newly designed and developed product	Full qualification test program

determine the criticality early in the project lifecycle, when we specify the top-level system design. Thereafter, software classified as safety or mission critical is developed with a high degree of process rigor. NPR 8715.3b [NASA (3), 2007] defines mission critical and safety critical as follows:

- *Mission critical*—Any item or function that must retain its operational capability to assure no mission failure

- *Safety critical*—Any condition, event, operation, process, equipment, or system that could cause or lead to severe injury, major damage, or mission failure if performed or built improperly, or allowed to remain uncorrected

These definitions are clearly based on mission success criteria. Different companies and government agencies use different criticality or classification systems but each system is fundamentally related to a mission success criterion.

We base software criticality on a project-level software failure analysis that includes all project systems as well as human operator procedures. Because the additional process rigor required of safety and mission critical software is onerous,

Table 11-25.　System Heritage Review (SHR) Topics. This table lists a series of topics that can be addressed as part of a SHR. A formal system heritage review provides an opportunity to address the specific design and construction of the item in question and fully vet its suitability for a new application.

Topic Description
Description and prior history • The inherited item was developed to certain requirements, workmanship issues, and conditions • Original design, if available • Performance history • Failure history and failure trends • Testing performed and test results, analyses performed in lieu of test, and waivers for noncompliance • Problem/failure report (PFR) system used, summary of all PFRs and red flag PFRs, waivers for noncompliance, and adequacy of product family representation and redesign framework closures
Intended application in the target project: • Level of redundancy in application • Single-point failure philosophy • Reliability analysis results and extent of independent review of these results
Compatibility with project requirements: • Design, qualification, and environmental requirements • Extent of changes required for project use • Parts classification, parts list of inherited hardware or design, specifications for parts used, nonstandard parts used, nonstandard parts authorization requests, and waivers • Material and process requirements, standards and controls used, packaging design, and conformal coating • Operating system interfaces • Programming language • Compatibility with host machine • Support from provider • Operating environment • Configuration control: design compared to "as built," change control requirement, when begun, waivers • Design changes made since qualification, and changes made or planned since last flown • Cost

early system design trades seek to reduce this critical software. For modern non-human rated space systems, examples of safety and mission critical software include:

Flight Software

- *Executable images*—The executable binary image derived from the software source code and persistently stored onboard a spacecraft

- *Configuration parameters*—Onboard, persistently stored parameter values used to adjust flight software functionality

- *Command blocks*—Persistently stored sequences of commands embodying common software functionality and often used for fault protection responses

Ground Software

- *Uplink tools*—Any software used to generate, package, or transmit uplink data once we determine that the data is correct

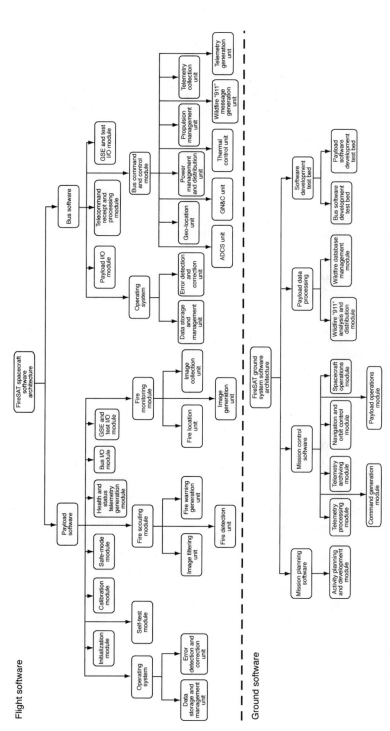

FIGURE 11-22.   FireSAT Flight and Ground Software Architecture. The complexity of the FireSAT mission software (flight and ground) illustrates the pervasiveness of software in every aspect of the mission. (GSE is ground support equipment; I/O is input/output; ADCS is attitude determination and control system; GN&C is guidance, navigation, and control.)

- *Spacecraft health and safety tools*—Any software used to downlink, unpack, and analyze data used to determine the health and safety of the spacecraft
- *Navigation tools*—Any software used to determine or to project the position and trajectory of the spacecraft
- *Data archival*—Any software that preserves mission or safety critical engineering or science data
- *Critical ground support equipment (GSE)*—Any GSE software required for mission success
- *Models*—Any software that models a behavior critical to mission success and without which we can't properly assess spacecraft behavior

The focus here is on space mission software classified as either safety critical or mission critical and which thus demands the highest level of process rigor and corresponding V&V.

11.7.1   Software-specific Validation and Verification (V&V) Challenges

The unique challenges presented by mission software arise from

- The great breadth of software functionality
- The ever-growing size and complexity of software
- The unique qualities required in space applications

We start with issues raised by the breadth of software functionality. Software provides the primary user interface to the flight and ground systems and interacts with nearly every component in both systems. Nearly every hardware interface, user interface function, and all of the software-internal interfaces contribute to this complexity. This breadth makes comprehensive testing infeasible because of the complex combinations of the functionality space. The user interface functions, such as a flight system command and sequencing, are designed with flexibility to support the unknown space and operational environment. This flexibility, by its nature, limits comprehensive testing. The functional complexity challenges the system and software engineers to devise a manageable set of processes and tests to ensure mission success within the larger project constraints.

The exponential increase in software complexity also presents many challenges. This increase is due, at least in part, to the complications associated with environmentally embedded (*in-situ*) missions, with the scope of fault protection, and with the desire to maximize the science return of the spacecraft. Certain aspects of the software design, such as multi-threaded software task-control architectures and implementation language, also increase the complexity. Table 11-26 lists source lines of code for ten JPL-managed space missions between 1989 and 2004. The increased software complexity affects the size and scope of the software V&V activities, along with the requirement for supporting resources.

TABLE 11-26.   **Source Lines of Code over Time in JPL-managed Space Missions.** Here we illustrate a general trend toward increasing lines of source code in more recent missions. (From a private communication with G. Holzmann, JPL, 2008.)

Year	Mission	Source Lines of Code
1989	Galileo	25,000
1992	Mars Observer	25,000
1996	Mars Global Surveyor	50,000
1996	Mars Pathfinder	100,000
1997	Cassini	125,000
1998	Deep Space One	175,000
2001	Mars Odyssey	110,000
2003	SIRTF	180,000
2003	Mars Exploration Rovers	325,000
2004	Deep Impact	140,000

The uniqueness of space mission software is due largely to several generic software qualities. These qualities are a response to the harsh nature of the space environment, the extreme inter-system distances with their associated navigation and communication challenges, and the autonomy required to sustain functionality over the mission life. Some of these qualities have analogues with mission hardware; some are purely software concerns. The foremost generic qualities required of space mission software include:

- *Reliability*—Spacecraft need to operate reliably for long periods in the harsh space environment. We need to manage software resources, such as dynamic memory, to ensure the health and safety of the spacecraft. Fault protection, attitude control, and other autonomous software activities need to maintain and to recover the spacecraft to a safe, self-preserving state.

- *Operability*—The software needs to provide an operator interface, generally through commanding and sequencing, that safely, efficiently, and consistently allows for meeting mission objectives even under changing objectives, space environment, and spacecraft state

- *Visibility*—The software needs to afford insight into all spacecraft systems to support reconstruction of the spacecraft state to enable remote debugging of hardware or software

- *Maintainability*—The software needs to support operational updates to its functionality. For spacecraft, such updates need to be made across the space link and without threatening the spacecraft's health and safety.

These generic qualities in space mission software usually result from the software development process or with software engineer experience, both of which are inadequate, although accepted, for safety and mission critical software. They never have a comprehensive set of requirements defined and rarely have more than a few requirements defined that are critical to mission success. This means we don't have a comprehensive set of tests analogous to hardware testing.

Without adequately defined requirements, traditional verification of these qualities is problematic. So we must rely on software validation, at multiple levels, in multiple test venues, and for multiple development phases, to demonstrate the suitability of the software for its mission application.

11.7.2   Development Aspects of Software Validation and Verification (V&V)

We perform V&V activities throughout the software development lifecycle and more generally as an integral part of the software development process. This *in-situ* V&V is a software development best practice, with the fundamental goal to identify defects as early as possible.

We model typical software development lifecycle phases as a waterfall as illustrated by Path A in Figure 11-23. Regardless of the path, all software development has the following phases:

- *Requirements analysis*—This phase defines the scope of a software development based on customer elicitation of product capabilities, features, and requirements. Requirements analysis normally ends with the successful completion of a software requirements review. Typical software defects introduced during this phase are: requirements subject to multiple interpretations, untestable requirements, and unnecessary requirements that over-constrain the software design.

- *Architectural design*—Here we specify the highest-level design of the complete software development, the architectural design. The software architecture must accommodate the required functionality as well as all the qualities and other constraints. Software defects often introduced in this phase are: performance bottlenecks, complex and unmaintainable module interfaces and dependencies, over-constrained architectural principles that don't allow for a detailed design solution, under-constrained architectural principles that leave too much variability in the detailed design, inconsistent resource handling and conflicts, lack of visibility into the internal software state, and limited testing interfaces.

- *Detailed design*—Here we further decompose the architectural design into ever smaller software abstractions. This phase normally ends with a software design review. Software defects introduced in this phase include: multiprocessing conflicts, poor algorithm selection leading to performance degradation, inadequate resource management schemes, excessive tasking overhead, excessive memory allocations for code or data space, and interface misinterpretations.

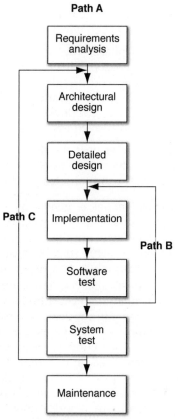

FIGURE 11-23.   Waterfall Paths in Software Development. Path A (top to bottom with no loops) illustrates a traditional single-pass waterfall lifecycle for software development. Path B shows a lifecycle appropriate for a development with confidence in the software requirements, architecture, and detailed design but without a delivery into system test for any increments. Path C shows a lifecycle with multiple, detailed design and system test phases. This path is common for space mission software development.

- *Implementation*—This phase encompasses the creation of software component code, the unit test of that code, and the integration of multiple components into an executable build. Implementation frequently occurs within a software-specific simulation venue. It normally includes a code review and ends with a test readiness review. This phase introduces such software defects as: multiprocessing race conditions and deadlocks, data corruption to either persistent or dynamically maintained data, memory leaks or other resource management errors, and boundary condition omissions.

- *Software test*—In this phase, the software team verifies and validates an executable software build relative to the customer-defined requirements, features, and capabilities. It normally ends with a test completion review.

Software defects often overlooked in this phase are: requirement errors, hardware interface errors, hardware performance issues, limited longevity, and correctness of off-nominal behaviors.

- *System test*—During this phase, the software development customer performs acceptance testing of the delivered software functionality. System testing may occur in multiple venues, with each venue designed for a unique aspect of the system functionality. The ultimate test platform is the space system. Typical software defects overlooked in this phase are: requirement errors, limited longevity, and correctness of off-nominal behaviors.

- *Operations*—Here the customer uses the delivered software functionality. The software development team considers this phase the start of software maintenance. They discover residual software defects during this phase.

A simple waterfall lifecycle model is generally inadequate for safety and mission-critical software development largely because it doesn't allow early visibility into the quality of the software nor the performance of the software team relative to the development plan. Problems may be uncovered relatively late in the schedule, creating significant cost, schedule, and performance risks.

A modification to the waterfall lifecycle is to iterate through the process multiple times, as Paths B and C in Figure 11-23 illustrate. This leads to an iterative delivery schedule with functionality added incrementally or reworked if necessary. With each delivery, project management assesses the product quality and the software team's performance with respect to the project requirements. This approach mitigates significant risks. And we tailor the functionality assigned to each software build to address project risk—whether those risks are in the software or elsewhere within the project (e.g., software needed to test the functionality of critical hardware). An iterative and incremental development approach, though, is a challenge to integrate with other project developments, such as hardware development.

With an iterative software development lifecycle, we perform V&V activities multiple times. For example, we verify the software requirements each time we perform the software test phase. Validation occurs iteratively in all phases. A well-defined software development process is designed to intercept defects introduced during a lifecycle phase or propagated through multiple phases. It thereby reduces the number of defects in the delivered software. This section describes software V&V mechanisms that reduce defect insertion rates and increase defect detection rates systematically. Table 11-27 summarizes these techniques. Not every technique is performed during the lifecycle phase for which the technique exposes errors.

Some V&V techniques apply to any phase. One of these is formal technical reviews. During lifecycle reviews, such as a software requirements review, software stakeholders assess the software development artifacts. Peer reviews, such as code reviews, make use of software experts, independent of the development, to assess the development artifacts. Success criteria involve completing a checklist that embodies software best practices.

TABLE 11-27.   Verification and Validation (V&V) Techniques Applicable for Each Lifecycle Phase. Different phases employ different V&V techniques. This table highlights the most useful ones. Some techniques are appropriate during multiple phases.

Lifecycle Phase	V&V Technique
Requirements analysis	• Early test case generation • Bidirectional traceability • Prototyping with customer feedback • Incremental delivery • Stress testing • Hardware in the loop
Architectural and detailed design	• Reuse • Design constraints • Prototyping for qualities • Complexity measures • Viability analysis • Formal verification
Implementation	• Reuse • Source language restrictions • Continuous integration • Static analysis • Auto-code generation • Software operational resource assessment
Software and system test	• Full-path coverage • Independent testers • Differential testing • Boundary conditions • Random reboot injection • Hardware off-line faults • Longevity assessment • Multiple, simultaneous fault injection • Multi-tasking overload
Operations	• Post-launch software check-out • Software upgrade demonstration • Executable image validity checks

Another lifecycle-independent software V&V approach is increased rigor in software development artifacts. One problem with the traditional software development process is that many design artifacts are captured in natural language that's subject to misinterpretation. If, however, the design artifacts are captured in a machine-readable form instead, based on rigorous semantics, then we can analyze the artifacts objectively using automated tools. A second benefit of this semantically rigorous approach is that it allows us to generate more detailed design artifacts directly from the source form. For example, test sequences, used during the software or system test phases, may be generated from rigorous requirement statements. We also feed those same requirements into the high-level design process to allow for systematic verification of the soundness of design decisions. Let's now look in greater detail at the techniques listed in Table 11-27.

Requirements Analysis

Flaws in requirements analysis, including elicitation and elucidation, often propagate through the majority of the software development lifecycle only to be exposed as residual defects during the system test phase or during operations. Such errors are the most expensive to fix because of their late detection and the resulting full lifecycle rework needed to resolve them.

In Section 11.1 we discuss requirements validation. Applying the VALID criteria described there, we have several useful techniques for identifying flaws in software requirements. These include:

- *Early test case generation*—Forces system, software, and test engineers to understand requirements statements so as to codify test cases and thus uncover weaknesses in a requirement's description

- *Bidirectional traceability*—Ensures that all requirements and their implementing components are interrelated. The interrelationships allow us to identify requirements that don't have any implementation process or conversely, requirements or implementations that are unneeded.

- *Prototyping with customer feedback*—Enables the customer to assess the correctness of the system early in the software development lifecycle. The prototype should stress end-to-end, customer-facing behavior.

- *Incremental delivery*—Allows us to perform the full scope of V&V activities, including system testing, early in the software development schedule. A root cause analysis of flaws in requirements during the early increments sometimes reveals systematic errors in the requirements analysis phase. Subsequent incremental software builds should be able to eliminate these errors.

- *Stress testing*—Examines the 'boundary conditions' of requirement statements that identify misunderstandings and weaknesses in the statements

- *Hardware-in-the-loop*—Prevents defects in the software development simulation environment from suffering identical or related requirement errors and thus from hiding flaws in the software product

We don't apply all of these techniques during the requirements phase. Some may be applied only after the software has reached a certain stage of development. Numerous software development tools are available to facilitate requirements analysis, documentation, and V&V. (Tools suitable for requirements analysis include DOORS by Telelogic, CORE by Vitech Corp, and Cradle by 3SL.) Increased rigor in requirement specification is particularly useful in eliminating requirement flaws and fostering correct design.

Architectural and Detailed Design

After the valid requirements are defined, architectural and detailed design leads to a decomposition of the software into modules. The modules should be recursively structured and designed depending on their size or complexity. The primary results of a design are these modules, their interconnectivity, and their interface specifications. The design must be shown formally to satisfy the requirements. Techniques to avoid flaws in design include:

- *Reuse*—Exploits prior designs, particularly architectural but detailed design if possible, to reduce and to simplify V&V. Adopting an architecture or design that's unsuitable for the current product introduces many of the design problems that reuse is meant to avoid. Careful systems engineering is required to adopt a reuse approach that reduces the project risk.

- *Design constraints*—Limits the space of design solutions to facilitate other aspects of the development. Facilitation could support formal analysis of the design or implementation, reduce the complexity of testing, adapt to limitations in software engineer skills, ensure suitable visibility for remote debugging, or others. A common example is to forbid preemptive multiprocessing systems and dynamic memory allocation and thus the types of implementation errors resulting from such systems.

- *Prototyping for qualities*—Provides early feedback on the architecture's strengths and weaknesses relative to the product's generic qualities (reliability, for example). This kind of prototype focuses not on customer-facing behavior but on the driving product qualities. For example, a prototype helps assess a design's ability to meet strict temporal deadline requirements during the mission's critical events, such as a planetary landing or orbit insertion.

- *Complexity measures*—Assesses the design relative to its modular complexity. A complex design, based on a prescribed measure, either warrants a design rework or additional software development resources for later lifecycle phases. Common measures are based on the nature of the interconnectivity of the modules within the software design.

- *Viability analysis*—Using high-level scenarios derived from the requirements ensures that the design meets its requirements. The scenarios are, at least conceptually, run through the design to confirm the ability to implement them. This analysis often identifies overlooked modules and module interrelationships. If the design is executable, we execute the scenarios directly to yield definitive viability results.

- *Formal verification and logic model checkers*—If we express requirements in a semantically rigorous way, we can apply logic model checkers to verify designs in an implementation-independent manner. (The Spinroot website has a software tool that can formally verify distributed software systems.)

These design phase V&V activities result in a software design with fewer residual errors, giving us greater confidence that the implementation phase will produce correct software code.

Implementation

During the implementation phase the detailed design is converted into a binary executable code based on software source code compiled for a target platform. Errors in design and requirements easily propagate into the executable code. The source code, expressed in a computer language, is prone to syntactic and semantic errors. Of these, the semantic errors are harder to eliminate. Common software source languages include C, C++, and Java, although growing numbers of higher-level languages exist. For example, The Mathworks and Telelogic websites have more information about their capabilities. Techniques to avoid flaws during implementation include:

- *Reuse*—Use of a proven software executable or library directly avoids implementation errors as the implementation is already complete. This approach is not a panacea, however, because the reused software may be used incorrectly or introduce subtle errors into the rest of the software executable. Trade-offs for this type of implementation reuse are based on risk assessments.

- *Source language restrictions*—Some software source languages offer coding constructs that are prone to error and to misuse. An effective way to eliminate such errors is simply to forbid such constructs and to enforce the prohibition during code reviews, for example with code checklists, or during compilation with tool or compiler support.

- *Continuous integration*—A large software development involves integrating many individual components. If we integrate components all at once or late in a development, we won't identify interface or behavioral errors until late, and by then they're usually on the critical path. Instead, if we do integration as soon as individual components are complete, the errors will surface earlier.

- *Static analysis*—Source code static analyzers check source code for common semantic errors. They are increasingly required in the development process to augment compilation of the source code. The sophistication of an analysis varies, but many errors, such as those related to array bounds checking, variable initialization, and multiprocessing locks, are identified statically. A related step, short of a full-blown static analyzer, is to fine tune the source code compiler's warnings. Simply increasing the compiler's verbosity is beneficial in identifying possible errors. At the time of writing, the leading examples of these types of tools are Coverity, KlocWork, and CodeSonar, each with a website for information.

- *Auto-code generation*—Use of tools to generate source code is a higher form of software reuse. These tools are suitable for repetitive source code constructs such as for telemetry or commanding, among others. We should treat the result of the auto-generated code as if it were manually created and require it to undergo full software V&V.

- *Software operational resource assessment*—Runtime software resource use increasingly requires monitoring code, given the complexity of resource use in the *in-situ* operational environments. These internal monitors, like any system monitors, identify anomalous situations based on deviations from an expectation and initiate recovery actions where appropriate. Deviations arise from the operational environment or from errors in the software design or implementation.

Software V&V activities are effective at catching errors from the implementation phase and also errors from earlier phases that have propagated undetected through to implementation. The level of detail required for implementation and the options available for implementation analysis often force a reexamination of the early lifecycle software artifacts. This exposes latent errors, while software and system tests expose others.

Software and System Test

The objective of the software and system test phases is to identify all differences between the desired behavior and the delivered behavior. The desired behavior is rarely fully documented and thus should not be confused with the documented software requirements. Verification focuses on the software and higher-level requirements. Validation focuses on two key aspects: the validation of the complete system functionality with software as one component and the validation of the generic software qualities.

Validating the generic software qualities is akin to hardware environmental testing. Hardware systems are decomposed into multiple components with hierarchies from parts to boards to boxes. We test the hardware components with a thoroughness that we don't achieve in system testing—longevity and stress testing being two examples. Thorough testing of each software component is a major aspect of software validation.

Validation of the system functionality helps catch errors in software requirements. By its very nature, software verification testing does not identify requirement errors. Only validation, particularly stress and scenario testing, can do so. Techniques to identify flaws during software and system test include:

- *Full-path coverage*—Software source code includes branching constructs that introduce multiple paths through the code. Common branching constructions include 'if-then', 'if-then-else' and 'switch-case' statements that introduce one, two, or N paths respectively. With full-path coverage, every path through the code is executed at least once within the complete test suite. This testing goes a long way toward exposing errors.

   But full-path coverage is rarely applied correctly. Consider two back-to-back 'if-then' statements. Full path coverage testing requires four test cases: both 'if' conditionals are 'true,' both are 'false,' and the two cases where one conditional is 'true' and the other is 'false'. The normal interpretation of 'full path coverage' only requires two test cases: both 'true' or both 'false'. Worse, tools to track this type of coverage lead to a false sense of confidence in the implementation—the missed paths are surely the paths with latent errors. Furthermore, normal flight code has considerably more conditionals and non-deterministic external interrupts and multiprocessing task schedules.

In practice, we do true full-path coverage only on subsets of space mission software. The focus is on the safety critical software or the most important mission critical software—typically fault protection, resource management, and attitude control. Other techniques include:

- *Independent testers*—Software tested by the software developer tends to contain errors based on the developer's incorrect assumptions. For example, if a developer assumes an interface contains values from 0 through 7, then the source code and the test code will handle only 0 to 7, even if the actual values are from 0 through 9. Independent testers bring different assumptions and thus produce test cases that expose developer errors.

- *Differential testing*—If we have a provably correct gold standard implementation, such as previously existing baseline data or results from software being reused, then we do differential testing by running test cases through the gold standard and the software under test and then compare the results. Such an implementation only exists for a small subset of the full software system.

- *Boundary conditions* lie at the extreme edges of the anticipated flight conditions. Test cases that test all boundary conditions improve the chance of catching latent errors.

- *Random reboots injection*—Spacecraft computers, because of space environmental conditions (primarily charged particles and hardware interactions), tend to shut down without giving the software a chance to prepare. Such spontaneous shutdowns leave the software's persistent storage in an inconsistent and thus corrupted state. Testing that involves random reboot injections helps identify storage corruption errors.

- *Hardware off-line faults*—Spacecraft software interacts with a large variety of hardware sensors, actuators, and controllers. These devices sometimes fail or have their communication paths interrupted. Testing that simulates hardware off-line faults at random times identifies software errors related to hardware trouble.

- *Longevity assessment ("soak tests")*—Operating the software for long periods tends to expose subtle resource management bugs. A common error is a memory leak in which a finite memory resource is eventually exhausted. Particularly insidious types of resource management errors are those involving persistent storage; these errors manifest themselves only after a very long time and perhaps multiple reboots. Unfortunately, longevity assessments are difficult to implement in practice due to the durations required and the limited availability of test beds. *Long* is generally defined to be a significant fraction of the expected time between computer reboots; however, for the insidious persistent storage errors it's even longer.

A useful technique for longevity testing is to compress the mission activities into a reduced period of time and then to repeatedly execute those activities while monitoring the internal software resources. After multiple iterations, we should compare the software resource use with expected behavior and identify discrepancies. For example, if system memory is monitored and expected to remain constant across iterations, any growth indicates a memory leak.

- *Multiple, simultaneous fault injection ("acid tests")*—An extension of the boundary condition technique whereby multiple faults are injected into the software. Watching how the software attempts to respond in an overwhelming situation is often very illustrative of weaknesses in its ability to handle nominal functions.

- *Multi-tasking overload*—Another extension to a boundary condition technique, whereby one or more tasks are intentionally bogged down. We then assess the effects of the slowed tasks through the rest of the software. Tasks get slowed by unanticipated conditions or parameters that lengthen their required CPU time. The delay then ripples through the system to the point where, for example, critical command sequences might not execute.

Not all of these V&V activities for software and system tests are done on the spacecraft or even hardware-in-the-loop testbeds. In particular the tests of the generic software qualities, while critically important for correct software behavior, likely involve test bed configurations or scenarios that are inappropriate when hardware is involved. So we must do many of the tests within a software-specific test environment. We have to be careful that the testing does not introduce errors into the system under test. Two characteristic software examples are:

- *Excessive use of hardware with a limited life span*—If running tests causes the hardware to accumulate large numbers of cycles, then the hardware reliability is compromised. Switches and flash memory are two types of hardware devices that wear out if stressed during software or other testing.

- *Persistent storage changes*—Sometimes persistent storage is modified specifically to facilitate a test. A common example is to narrow fault protection monitor thresholds to ensure that a fault protection response is tripped in a test case. The persistent storage must be restored following the test. Such a change is a "test as you fly" exception and is thus symptomatic of a more general test plan problem.

Operations

Testing software in operations is best viewed as "system check-out." This entails exercising software functionality in the operational space environment but before its critical need. Plenty of hardware check-out analogies to this software approach exist. We generally perform them when data rates, and therefore visibility, into the spacecraft are high. Techniques to identify flaws during operations, before they manifest themselves in potentially mission-ending ways, include:

- *Post-launch software check-out*—Exercise operational scenarios before they're needed while monitoring internal software resources and other states to confirm expected behavior

- *Software upgrade demonstration*—Demonstrate the ability to patch the software in flight during a quiescent period when anomalous spacecraft or software behaviors are not demanding the patch

- *Executable image validity checks*—Validate the binary, executable images to confirm that they are uncorrupted and thus usable in the event of a spacecraft computer reboot

Catching software errors during a controlled period of operations is better than uncovering errors during a critical spacecraft event.

11.7.3   System-level Aspects of Software Validation and Verification (V&V)

After the software is delivered into a system test environment, either for a flight or ground system, the nature of the testing changes dramatically. While the software-specific V&V techniques described above for the system test phase still apply, the focus moves to the full system within which software is one aspect. The next two sections describe the nature and impact of the system focus and the system-level test venues.

Software-in-the-system Context

From a software V&V perspective, the following differences and their effects characterize the system testing context:

- *Real hardware*—The system context uses flight, engineering model, or ground support equipment hardware. This hardware is irreplaceable, and thus the consequences of a software error become mission critical. Simulations in a system context are generally hidden behind real hardware interfaces, whereas in the software development environment, simulations are the primary interface to anything outside the software. Consequently the delivered software sees the true data rates, interrupt rates, register layouts, command protocols, and hardware interactions—this reality often exposes errors in the software development platform. Furthermore, the real hardware and the associated simulations and environments tend to have true closed-loop behaviors; latent software errors surface after the system exhibits closed-loop behavior.

- *Absent software engineers*—After the software is in the system context, the software developers no longer operate "their" system. So all sorts of assumptions incorporated into the software behavior get exposed and latent errors identified. Without the software engineers the documentation of the software design and use becomes critical, but sufficient documentation is often lacking for software developments. The developers disappear either when explicitly rolled off the project or simply by restricted access to the system and its test environment.

- *Personnel and procedures*—The system context uses real test venue support personnel, operations personnel, and associated operational procedures. The software developers' assumptions no longer apply. The consequences of a software error may now be safety critical from the perspective of human life. Operational procedures may have errors that use the software in off-nominal, even dangerous, ways.

To avoid these issues we need additional tests within suitable test venues. The software development team's completion of the software based on their own qualification testing is only a first step in achieving a qualified software executable for flight or ground operations.

System Test Venues

There are several venues for performing software validation at the system level. We introduced these earlier during our discussion of product validation but offer some software-specific considerations here:

- *End-to-end information system (EEIS) tests*—Exercise end-to-end command and data flow to prove that all the software products in the flight and ground system are compatible, in terms of the basic data transport capability and performance (latency, throughput, etc.), and that the appropriate combinations of end-items are exercised. A big lesson from past missions is that we must carefully exercise software associated with multi-mission assets (e.g., communication relay satellites and ground stations). Tests of these assets should include exercising them in realistic loading scenarios, including simultaneous downlink/uplink by other users.

- *Mission scenario tests (MSTs)*—Ensure that the spacecraft, using the final launch-version of flight software, commands, parameter tables, and other products generated by the ground software, properly executes the mission objectives under nominal and off-nominal conditions. Following the mission timeline exactly is less important than exercising mission-critical events such as launch and deployments. We must be especially careful in this venue to exercise the fault detection and responses and the critical parameters.

- *Operational readiness tests (ORTs)*—Ensure that all flight and ground mission hardware, software, people, processes, and procedures properly execute crucial segments of the official mission timeline under nominal and off-nominal conditions. We should conduct at least one ORT with the flight hardware, in conjunction with an MST, but should do most of them with testbeds and hardware emulators that allow appropriate injection of simulated anomalies into the telemetry to exercise the ground software and operators. Software provides the user interface to the spacecraft, so to exercise it properly, we must exercise the operators too.

- *Stress tests or risk reduction tests*—Ensure that the mission and safety-critical software is robust and has no hidden behaviors not exposed in prior testing. The complexity of modern software makes it impossible to completely test all permutations. The V&V techniques applied in a software context, while necessary, are not sufficient. Stress-testing at the system level, on the launch version of the software, helps catch any remaining flaws. Particularly appropriate test phase techniques are: boundary conditions; longevity assessment; multiple, simultaneous fault injection; and multi-tasking overload.

11.7.4   Other Verification and Validation Topics

We summarize here a number of other important software-centric topics in systems engineering and the overall V&V program, each of which could fill a chapter.

- *Independent V&V (IV&V)*—The V&V software techniques and test efforts described so far are performed under the auspices of the project organization using software, system, and test engineers. This type of V&V, informally known as iV&V (or "little i" V&V), raises questions about the independence of the V&V effort when faced with programmatic pressure applied by the project. An alternative is to use a fully independent organization, referred to as IV&V (or "big I" V&V), that is not controlled by the project. For example, NASA employs a dedicated IV&V facility, headquartered in West Virginia. This facility is ideally financially, managerially, and technically independent and conducts software verification and validation for certain classes of space missions based on the criticality. However, NASA IV&V efforts are of a considerably smaller scope than project-directed iV&V efforts, so it's an augmentation to, not a replacement for, project activities.

- *Software reuse*—The reuse of software design and implementation has advantages but is not without pitfalls. In many cases, we realize substantial cost savings by reusing heritage software (from the component level up to wholesale flight images for multi-spacecraft programs). But heritage software carries risks that the assumptions built into the software, and even its prior V&V program, do not apply for the new use. Reuse is an example of "verification by similarity," and thus carries significant risk. The systems engineering team needs to factor in that risk. Many of the issues for commercial-off-the-shelf and non-developmental items (COTS/NDI) apply here as well.

- *Regression testing*—Either by design or as a consequence of "bug fixes," software is delivered multiple times within a space project. With each delivery a fundamental question is the scope of the necessary regression testing. Ideally a regression test consists of the complete test suite. In practice, though, we rarely have the resources to do this. The V&V effort must assess what is affected by the software changes and then prune the test suite appropriately to fit into the resource constraints. We must test carefully, because a missed test case might obscure latent errors that lead to loss of mission.

- *Incomplete launch loads*—A fundamental advantage to flight software is that it is changeable in flight whereas hardware is not. Taking this advantage to an extreme allows for launching a spacecraft without the full required functionality and then uploading the software sometime in the future. In some applications, notably deep space missions with cruise phases lasting from months to years, the project might exploit this time by deferring software development. However, the future builds need to be developed and tested with all the rigor of the launch build.

- *Post-launch software builds*—A fundamental trade-off in the decision for a post-launch build is whether post-launch V&V activities only on testbeds are sufficient. It's crucial for this trade to cross-validate the testbed against the spacecraft before launch, and as part of that cross-validation to quantify differences and idiosyncrasies that could otherwise impair the in-flight V&V. Failing to complete this cross-validation before launch leads to high risks later in the mission after the spacecraft is gone!

- *System safety*—Safety critical software is one aspect of overall system safety. A safe system generally requires multiple inhibits, with the software counting for one inhibit at most. Other system inhibits may include, for example, firing pins in propulsion systems. During final stages of launch preparations, a spacecraft's propulsion system is usually filled with hazardous material and its deployment ordinances (pyrotechnic initiators) are installed and armed. From this point onward, any powered test of the spacecraft carries a risk to personnel and vehicle safety in the form of accidental firing of the propulsion system or pyrotechnic initiators. By this time, safety-critical software, including flight, ground, and GSE software, must be complete and fully qualified so that the software's defined inhibits preclude accidents.

Let's look at the example test flow for the FireSAT flight software whose top-level architecture was shown in Figure 11-22. Figure 11-24 illustrates a nominal test flow for this software from the unit, to module, to integrated functional level. It also includes place holders for integrated functional tests, end-to-end information system (EEIS) tests, and mission scenario tests. Like the IVF presented in Section 11.4, this diagram helps systems and software engineers integrate their test planning.

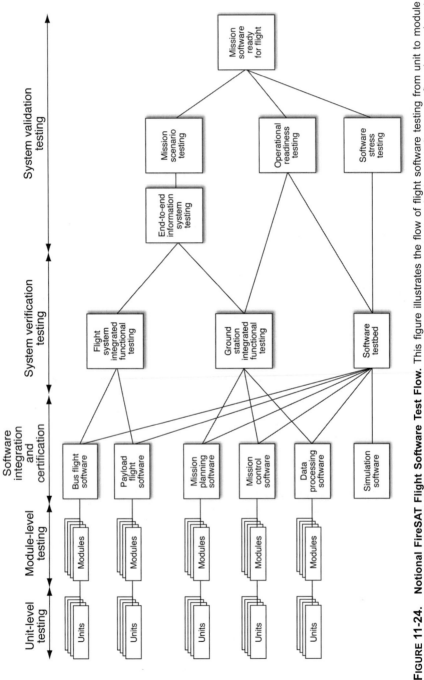

FIGURE 11-24.   **Notional FireSAT Flight Software Test Flow.** This figure illustrates the flow of flight software testing from unit to module to integrated software deliverable items. Included in the diagram are placeholders for integrated spacecraft and ground system software tests, use of software testbeds, end-to-end information system (EEIS) tests, operational readiness tests, and software stress tests.

11.8   Document and Iterate

We carry out V&V to convince ourselves that we built the system right **and** that we built the right system. In the space business, we typically don't get "do overs." We get one chance to get it right. Besides sheer luck, the only way to be sure that we should give the "go" to launch, committing millions if not billions of dollars and possibly human lives, is to rely on the results from rigorous V&V. No certainties exist in this business; there will always be risks. Verification is a risk mitigation and management tool to help reduce the number of known unknowns. Validation builds confidence that the remaining unknown unknowns will not be mission-ending.

To this end, V&V activities must amass a body of objective evidence indicating that the product was built according to its requirements and provides a capability the customer needs. This same evidence is crucial to realistic operations planning as part of flight certification. In that respect, the documentation generated by V&V activities is a singularly important output. This step places a heavy burden on engineers to plan meticulously and document rigorously. A poorly written procedure likely will execute poorly. A poorly documented test (or one with inadequate data-review) essentially did not happen.

This chapter has focused on the interrelated activities that fall under the broad umbrella of V&V. Along with a verified and validated product, these efforts produce master verification plans, verification requirements matrices, and numerous other artifacts. These documents are much more than simply stacks of paper to add to the pile; they are the tools that systems engineers use to define and guide the V&V process and ultimately justify the decision to launch. Organizing and managing that documentation is the purpose of the technical data management process described in Chapter 17. Hand in hand with the technical data is configuration management as described in Chapter 16. Verification and validation activities begin at the lowest level and progress through increasingly more complex assemblies until reaching the final integrated state. However, schedule and availability of hardware and software often entail the product not being in its exact, final configuration. Understanding and managing the risks associated with these differences requires rigorous configuration control at every level. If we don't know what we tested, we won't know what to do with the results.

On a final note, the majority of practices, procedures, and techniques used in space system verification and validation were developed by sometimes painful trial and error since the 1950s. At the dawn of the space age, few of the challenges of the launch and space environment were well understood. The rest had to be learned through the school of hard knocks.

Sometimes systems fail. Many of these failures are the result of design flaws. The purpose of V&V is to detect those design flaws. These results are the largely unsung "diving catches" that discover potentially mission-ending problems before launch. But some of those failures are the result of inadequate V&V. These failures are the ones that cause us to wish "if only we'd done this one simple test..." Because so much of V&V is based on experience, V&V engineers must learn the lessons from previous missions, the good and the bad, in planning and executing their activities.

Therefore, a final V&V responsibility is to carefully review the lessons learned, good and bad, from each project to avoid repeating the same mistakes and to preclude reinventing best practices on the next project. An internet search on "spacecraft verification" returns over 200 hits. NASA maintains a large database of lessons learned on their website. Table 11-28 gives just one example from this website that stresses the "test as you fly, fly as you test" philosophy we emphasize throughout this chapter.

TABLE 11-28.   **Lesson Learned Example.** The NASA Lessons Learned website captures thousands of lessons such as this one on a vast array of topics. This example illustrates just one of hundreds on the subject of spacecraft verification. All systems engineers, especially those involved in verification and validation (V&V), have a duty to learn from these lessons so as not to repeat previous mistakes.

Public Lessons Learned Entry: 1196

Lesson Info:
Lesson Number: 1196
Lesson Date: 2002-01-24
Submitting Organization: JPL
Submitted by: David Oberhettinger

Subject: Test as You Fly, Fly as You Test, and Demonstrate Margin (1998)

Abstract: Mars Polar Lander had deficiencies in compliance with the principle of "test-as-you-fly" that requires ground tests and simulations accurately reflect the planned mission profile, plus margin. Enforce the system-level test principle of "test as you fly, and fly as you test," and carefully assess any planned violations. When using simulations for system-level verification, validate models and include sufficient parametric variations in the simulations to ensure that adequate margins exist.

Description of Driving Event: The principle of "test-as-you-fly" means that ground tests and simulations should accurately reflect the planned mission profile, plus margin and the appropriate off-design parameters. Although the system-level test and verification process for Mars Polar Lander (MPL) was well planned and executed, there were deficiencies in the test program for the parachute, terminal descent, and touchdown phases. MPL test deficiencies included:

- The touchdown sensing software was not tested with the lander in the flight configuration, leading to the probable cause of the MPL mission loss. (See Lesson #0938.)
- Fault-injection testing of the flight software was not thorough enough to detect all logic errors in post-landing fault-response algorithms. (See Lesson #0939.)
- The thermal design of the propulsion subsystem was incorrectly characterized in the system thermal-vacuum test due to an error in the thermal model, causing major errors in the propulsion thermal design that went undetected until after launch.
- The test and verification process for the Deep Space 2 (DS2) probe that was co-launched with MPL and did not survive Mars encounter, also departed from the test-as-you-fly principle in end-to-end testing and analysis.
- A decision was made to not conduct a system-level impact test of the probe with aeroshell. The risk of structural failure from the dynamic interaction between the aeroshell and the probe was recognized and accepted.
- Though DS2 was designed to strike the Martian surface at a velocity of 200 meters per second, there was no impact test of an electrically powered, complete system.
- The flight battery lot was not subjected to impact tests. Testing performed on an 8-cell predecessor flight-like lot did not provide statistical confidence, as it resulted in one structural failure.

Adequate system margins were not demonstrated during the MPL terminal descent control system testing. These margins were subject to erosion from propulsion system dynamics (impulse variations due to water hammer or thermal effects), propellant center-of-mass migration, the lack of a high-fidelity fuel slosh model, and nonlinear pulse-width modulation effects. The true margins of the system were not fully characterized in the presence of these effects.

TABLE 11-28.   **Lesson Learned Example. (Continued)** The NASA Lessons Learned website captures thousands of lessons such as this one on a vast array of topics. This example illustrates just one of hundreds on the subject of spacecraft verification. All systems engineers, especially those involved in verification and validation (V&V), have a duty to learn from these lessons so as not to repeat previous mistakes.

References:
1. Report on the Mars Polar Lander and Deep Space 2 Missions, JPL Special Review Board (Casani Report), JPL Internal Document D-18709, 22 March 2000, Sections 3.4 and 5.2.
2. JPL Corrective Action Notice No. Z69164, Mars Program Investigation Results: "System Engineering/Risk Management/Error Detection," 1 May 2000.
3. JPL Corrective Action Notice No. Z69165, Mars Program Investigation Results: "Verification and Validation Process," 1 May 2000.

Additional Key Words: Entry, Descent, and Landing (EDL), Environmental Test, Fault Protection, Integration and Test, Risk Assessment, Robust Design, Simulation Accuracy, Software Testing, Spacecraft Test, Technical Margins, Test and Evaluation, Test Errors, Test Fidelity, Test Planning

Lesson Learned: The process of end-to-end system verification (either through testing, simulation, or analysis) may be compromised when it is not consistent with the mission profile (plus margin and the appropriate off-design parameters).

Recommendations:
1. Enforce the system-level test principle of "test as you fly, and fly as you test." Carefully assess any planned violations of this principle; if they are necessary, take alternate measures such as independent validation. Departures from this principle must be reflected in the project risk management plan, communicated to senior management for concurrence, and reported at reviews.
2. When using simulations for system-level verification, models must have been validated (e.g., supported by test); and sufficient parametric variations in the simulations must be performed to ensure that adequate margins exist.

If our final responsibility is to capture lessons learned, our first responsibility is to be aware of these hard-won lessons. It behooves all systems engineers, especially those responsible for V&V, to apply best practices based on previous experience. Verification and validation is a vast subject that could easily fill a book. With a successful V&V process, a project proceeds with confidence to finally transition the product to the user.

References

Ariane 501 Inquiry Board. 1996. Ariane 5 Flight 501 Failure—Report by the Inquiry Board. URL: sunnyday.mit.edu/accidents/Ariane5accidentreport.html

Arthur, James D. et al. 1999. "Verification and Validation: What Impact Should Project Size and Complexity have on Attendant V&V Activities and Supporting Infrastructure?" *Proceedings of the IEEE: 1999 Winter Simulation Conference*, Institute of Electrical and Electronic Engineers, P.A. Farrington, ed. pp. 148–155.

Balci, Osman. March 1997. "Principles of Simulation, Model Validation, Verification, and Testing." *Transactions of the Society for Computer Simulation International*, 14(1): 3–12.

Department of Defense (DoD). April 10, 1998. MIL-STD-340(AT)—*Process For Coating, Pack Cementation, Chrome Aluminide*. Washington, DC: DoD.

European Cooperative for Space Standardization (ECSS). July 14, 2004. ECSS-P-001B—Glossary of Terms. Noordwijk, The Netherlands: ESA-ESTEC.

ECSS Requirements and Standards Division. November 17, 1998. ECSS-E-10-02-A—Space Engineering: Verification. Noordwijk, The Netherlands: ESA-ESTEC.

Feather, Martin S., Allen P. Nikora, and Jane Oh. 2003. NASA OSMA SAS Presentation. URL: www.nasa.gov/centers/ivv/ppt/172561main_Feather_ATS_RDA_v7.ppt

Harland, David Michael and Ralph D. Lorenz. 2005. *Space Systems Failures: Disasters and Rescues of Satellites, Rockets and Space Probes.* Berlin: Springer.

JPL Special Review Board (JPL). March 22, 2000. "JPL D-18709—*Report on the Loss of the Mars Polar Lander and Deep Space 2 Missions.* URL: spaceflight.nasa.gov/spacenews/releases/2000/mpl/mpl_report_1.pdf

Larson, Wiley J., Robert S. Ryan, Vernon J. Weyers, and Douglas H. Kirkpatrick. 2005. *Space Launch and Transportation Systems.* Government Printing Office, Washington, D.C.

Larson, Wiley J. and James R. Wertz (eds.). 1999. *Space Mission Analysis and Design.* Dordrecht, The Netherlands: Kluwer Academic Publishers.

National Aeronautics and Space Administration (NASA (1)). March 6, 2007. NPR 7120.5d—*NASA Space Flight Program and Project Management Requirements.* Washington, DC: NASA

NASA (2). March 26, 2007. NPR 7123.1a—*NASA Systems Engineering Processes and Requirements.* Washington, DC: NASA.

NASA (3). April 4, 2007. NPR 8715.3b—*NASA General Safety Program Requirements.* Washington, DC: NASA.

NASA (4). December 2007. NASA/SP-2007-6105, Rev 1. *NASA Systems Engineering Handbook.* Washington, DC: US Government Printing Office.

NASA. March 2004. NASA/JPG 7120.3 *Project Management: Systems Engineering & Project Control Processes and Requirements.* Washington DC: NASA.

NASA Goddard Space Flight Center (NASA/GSFC). June 1996. GEVS-SE Rev. A—*General Environmental Verification Specification For STS and ELV Payloads, Subsystems, And Components.* Greenbelt, MD: GSFC.

Perttula, Antti. March 2007. "Challenges and Improvements of Verification and Validation Activities in Mobile Device Development." 5th Annual Conference on Systems Engineering Research. Hoboken, NJ: Stevens Institute of Technology.

Sellers, Jerry Jon. 2005. *Understanding Space.* 3rd Edition. New York, NY: McGraw-Hill Companies, Inc.

Smith, Moira I., Duncan Hickman, and David J. Murray-Smith. July 1998. "Test, Verification, and Validation Issues in Modelling a Generic Electro-Optic System." *Proceedings of the SPIE: Infrared Technology and Applications XXIV,* 3436: 903–914.

WIRE Mishap Investigation Board. June 8, 1999. *WIRE Mishap Investigation Report.* Available at URL http://klabs.org/richcontent/Reports/wiremishap.htm

Youngblood, Simone M. and Dale K. Pace. "An Overview of Model and Simulation Verification, Validation, and Accreditation", *Johns Hopkins APL Technical Digest,* 16(2): 197–206. Apr–Jun 1995.

Chapter 12

Product Transition

Dr. Randy Liefer, *Teaching Science and Technology, Inc.*
Dr. Katherine Erlick, *The Boeing Company*
Jaya Bajpayee, *NASA Headquarters*

Transition, as a systems engineering process, means delivering an operational product to an end user. Key questions in planning and executing transition are:

- What's the product?
- Where does it need to go?
- How should we get it there?
- When is it "finished"?
- When does transition begin?
- When is transition complete?
- What happens after transition?

Any space project involves designing, building, testing, and then transitioning a wide range of products, either to the next level up for integration or to the end user for launch or operations. A product might be documentation, a database, a software program or module, a physical model, a demonstration prototype, a ground station, a piece of flight hardware, or an entire spacecraft. Chapter 9 addresses the transition of a subsystem or supporting assembly up to the next level in the work breakdown structure; Chapter 10 discusses the succeeding integration of those components into higher-level assemblies, up to and including the system of interest.

This chapter focuses on transitioning a finished product to the end user. Transition begins at the end of the satellite integration and test program and ends with handing the satellite over to the mission operators following early-orbit check-out. At this point, day-to-day operations begin. Hence, the product usually needs to go into orbit—around Earth or another body in the Solar System. It normally gets there aboard a launch vehicle for at least part of the journey. We deliver it to the end user when it's placed into an operational orbit, tested, and certified ready for routine operations. Table 12-1 provides the framework for transition activities.

TABLE 12-1.   **A Framework for Organizing Transition Activities.** A general process for product transition is shown here along with which section discusses each step.

Transition Activities		
Step	**Description**	**Where Discussed**
1	Plan the transition	Section 12.1
2	Verify that the product is ready for transition	Section 12.2
3	Prepare the product for transportation to the launch site	Section 12.3
4	Transport to the launch site	Section 12.4
5	Unpack and store the product	Section 12.5
6	Integrate with the launch vehicle	Section 12.6
7	Roll vehicle out to the launch pad	Section 12.7
8	Integrate with the launch pad	Section 12.8
9	Conduct launch and early operations	Section 12.9
10	Transfer to the end user	Section 12.10
11	Document the transition	Section 12.11

Transition also involves a series of formal milestones consisting of an operational readiness review, pre-ship review, flight readiness review, and ultimately a post-launch assessment review. Subsequent sections, and further discussion in Chapter 18, describe these events with recommended entry and exit criteria.

Transitioning a satellite to the end user occurs late in the product lifecycle (Figure 12-1, [NASA, 2007 (1)]). It begins toward the end of Phase D after the satellite is integrated and tested and is ready to begin launch and on-orbit operations. Transition is complete at the post-launch assessment review, which marks the beginning of Phase E. Depending on mission complexity, this portion takes about three to six months. NPR 7123.1a [NASA, 2007 (1)], maps transition as a set of inputs, activities, and outputs as shown in Figure 12-2.

12.1   Plan the Transition

Transition applies to a wide variety of products and documentation, entailing many different activities. Besides the space product, it involves transferring responsibilities for the logistics elements from the developers to the end users. These elements include, as a minimum, the formal delivery of the logistics support analyses (LSA) report, the packing, handling, storage, and transportation (PHS&T) instructions, the facilities requirements for operations and storage, the spares and inventory management system, the technical manuals, any required support equipment, and personnel requirements for operations and management. Developing these elements provides the basis for transition planning.

12.1.1   Define the Transition Process

To begin transition planning, we must answer the following: what product are we transitioning, why do we believe it's ready for transition, to whom are we delivering the product, where will it be located, and in what configuration should the product be when transition is complete? Major steps in this process include:

- Verify that the satellite meets requirements at its level of assembly
- Ensure that the satellite is ready for interfacing to the launch vehicle
- Pack and transport the satellite to the launch site
- Unpack at the launch site and put into storage if necessary
- Integrate to the launch vehicle
- See that spares and procedures are available if necessary
- Conduct launch and early-orbit check-out
- Transfer operational responsibility to the end user

Once we integrate the spacecraft with its payload (remote sensing sensors, communication payload, etc.), this higher level of assembly is called an observatory or a satellite. We must select the launch vehicle early in the product lifecycle because this choice levies many requirements on the product's design and testing. The satellite must survive the launch environment, function after deployment in the operational orbit, and communicate with the ground stations during operations. The transition process begins only after the satellite undergoes detailed testing to verify that it meets all requirements at this level of assembly, and that it's ready for interfacing to the launch vehicle.

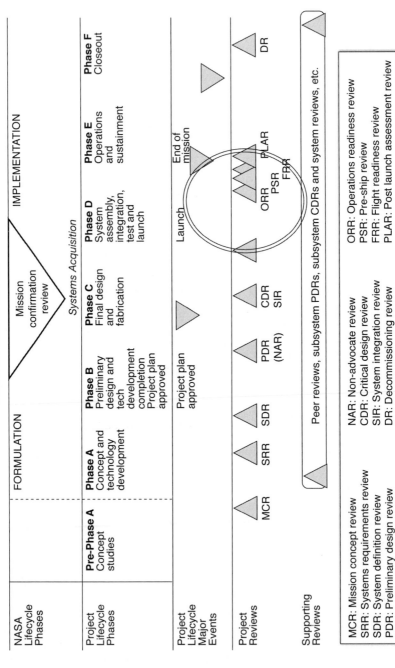

FIGURE 12-1.   Project Lifecycle. Here we depict the NASA project lifecycle. Commercial and other government organizations divide the lifecycle differently and use different names, but the essence is the same. The oval indicates the part of the lifecycle that this chapter discusses.

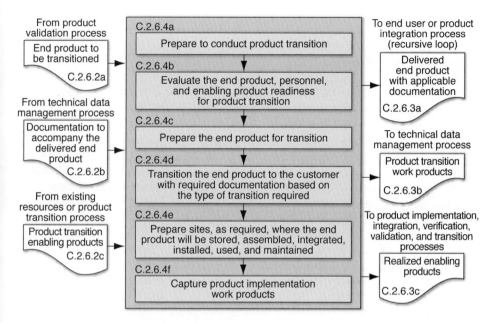

FIGURE 12-2.   The Product Transition Process. At the end of this process, the system begins routine operations. [NASA, 2007 (2)]

Early in transition, the transitioning organization should identify all the relevant stakeholders and formally document their requirements and expectations. Stakeholders include the customer, the integrator, the designer, the builder, the shipper, the launcher, the maintainer, and the operator. Requirements from these people, requirements derived from the product, and the outputs of the logistics support analysis (LSA) help define any special transition procedures, including packing, storage, and handling requirements.

12.1.2   Logistics of Transition

Ideally, all stakeholders have recognized early-on the need to plan and budget for logistics support of the space and ground segments of the mission. During transition, this planning—and procuring of facilities, spares, and personnel—culminates in the formal delivery of the following items:

LSA reports—The backbone of the logistics suite of products, LSA is an analysis tool that develops the requirements for logistics. Off-the-shelf computer-based modeling tools are available. The input data for the LSA modeling is all of the product's engineering analyses (such as reliability, survivability, test data, trades, dimensions of the spacecraft, etc.) and design drawings. The LSA model results include spares profiling; design drawings, software builds, reliability data, maintenance tasks; training tasks; packaging, handling, storage, and transportation (PHS&T) profiles; facilities requirements; support equipment requirements; procedures for maintenance and operations; and personnel requirements. The deliverables to the customer are numerous LSA records. The LSA analyzes all deliverable hardware, including the ground station, for its effects on the mission. The reports produced by the LSA evolve throughout the design and test phase and are periodically updated throughout the operations and support phases. Guidelines for LSA are in MIL-STD-1388-2B, although best commercial practices are preferable.

PHS&T instructions—These directions detail how all items can be handled and moved efficiently by available modes of transportation, with considerations for safety and hazardous material. They also document all long-term and short-term storage requirements as well as requirements for maintenance during storage.

Facilities requirements for operations and storage—These requirements define types of facilities or facility improvements, locations, capacity needs, environmental requirements, and equipment needed during transition and operations. Requirements analysis and determination mostly identify facilities needs for after delivery of the spacecraft. This analysis includes all ground station and mission operations events. However, during the spacecraft build and test phase, we need to consider facilities for storage of spare parts (for use if a component fails) and for assembling, testing, and processing the spacecraft.

Spares and inventory management—The mission requires a supply system to provide data and interfaces for acquiring spares, and a planning process to ensure that the right items, in the right quantities, are at the right places, at the right times, to enable cost efficiencies. Source data comes from the LSA and provides the foundation for spares modeling, which results in spares profiles to use in operations and maintenance. We must also manage and administer the spare parts inventory. In general, this inventory applies to the ground network and mission operations. During the build and test phase of the spacecraft, though, adequate spares need to be on hand in case a component fails.

Technical manuals (or procedures for operations and maintenance)—All documents that instruct the maintainers on how to operate, install, maintain, and service the product, including support equipment, must be developed, transitioned to the users, and subjected to their configuration control. For consistency, most projects adhere to a series of MIL standards, such as MIL-STD-38784, that guide the writing style and content of technical manuals.

Training manuals and course work—We use these documents to train, certify, and qualify personnel to maintain the product being transitioned. The source data are the LSA reports. Trained specialists lead the planning process for identifying training requirements and skilled personnel. They also develop the appropriate curriculum, simulation, and training courses for these personnel.

Support equipment—This includes all equipment needed to maintain and operate the product, system, or facility, including aerospace ground equipment and any unique support equipment—hardware and software. We may need to design equipment to support a product's unique maintenance requirements.

Personnel requirements for operations and maintenance—We must identify, hire, and train people with the skills to operate and support the system over its lifetime. While formal delivery of these logistics products occurs during product transition, we should make preliminary or draft versions of the documents available at major technical milestones such as preliminary design review (PDR) and critical design review (CDR) as a part of the overall product technical design disclosure. Typically, the logistics products begin to mature following CDR.

12.1.3   Identify Product Transition Inputs

The product—This chapter focuses on the transition of an end product (e.g., the FireSAT spacecraft) to the user. The input product is the "as-built" baseline that results from successfully completing the systems acceptance review as described in Chapter 11.

Documentation—An acceptance data package constitutes part of the as-built baseline. No product is complete and validated until the documentation has been written and validated for proper procedural conduct. Some subset of the product's documentation accompanies the product during transition. These items include manuals and procedures for unpacking, storing, transporting, operating, maintaining, and integrating the product (we define these more completely in Section 12.5). This data matures progressively as the design matures until final check-out and test. Following the testing phase is a logistics validation phase with regard to the operations and maintenance of the product. We validate the technical manuals or procedures by using the procedural documentation and validating that the process flow is correct. We do some of this during testing but must complete it before delivering the product. The documents are then ready for transition to the end user. The documentation usually includes (see Subsection 12.1.1 for more detail):

- Logistics support analysis (LSA) reports
- Packaging, handling, storage, and transportation (PHS&T) instructions
- Facilities requirements for operations and storage
- Technical manuals (or procedures for operations and maintenance)
- Training manuals and course work

These products, while part of the as-built baseline approved at the system acceptance review (SAR), are considered transition products.

Support equipment—This equipment may include packing material, handling equipment, containers for transportation and storage, and equipment for environmental control and instrumentation. Much of it is existing or off-the-shelf but some is designed and built just for the product. Examples include the barge to float the Shuttle external tank from the Michoud Assembly Facility in Louisiana to Kennedy Space Center in Florida, or plastic wrap to encapsulate a microsatellite before shipping it. In any case, equipment needs to go through the same process for validation and verification as the product. The support equipment requires its own technical manuals and spares to ensure seamless operations. We must treat these as "end items," with the same calibration and maintenance requirements as the main product. Because transition requires a supportive infrastructure, support elements for this equipment must also be transitioned to the end user, including:

- Spares and inventory management
- Personnel requirements for operations and maintenance
- Facilities requirements for operations and storage

12.1.4   Identify Product Transition Outputs

The operational product—After transition, the product is in the possession of the end user and is ready for normal operations. However, we must carefully negotiate and clearly define the interfaces between delivering and receiving organizations. Important issues include:

- When do deployment and check-out operations conclude?
- When does legal liability shift from the delivering to the receiving organization?
- How do we determine if anomaly resolutions during deployment and check-out are sufficient?
- When do routine operations begin?
- What constitutes routine operations?

Documentation—All the documentation prepared for transition also goes to the end user. Additional documentation of the transition event should record such information as send and receive dates, environmental conditions and mechanical loads the product experienced during ground transportation and launch, configuration changes, etc. The sending and receiving organizations must negotiate data formats and media for all documentation. We must address as early as possible any issues involving the transfer and protection of classified, proprietary, or International Traffic in Arms Regulations (ITAR) restricted data. By the time of transition most of this documentation should be in the technical manuals. After these manuals have been validated and verified by the customer, the customer "owns" them and is responsible for their configuration control. Any future changes or updates require analysis, upgrade, validation and verification, and buy-off.

Support equipment—All the support equipment that arrives at the launch site with the product must be stored, returned for maintenance and reuse, or disposed of. Normally, when the product with its associated support equipment and operations and support capability have been accepted by the end user, this end user owns the equipment and becomes responsible for its maintenance and disposal.

12.2   Verify that the Product is Ready for Transition

The first step in the transition process is making certain that the product is ready. So, how do we know when a satellite is ready? It's when satellite development is complete, rigorous system-level testing has verified and validated that it meets all of its requirements (direct and derived), end-to-end data flow tests confirm that it can communicate with the ground system, and meticulous documentation attests that it interfaces properly with the launch vehicle. Indeed, the satellite is ready for transition when it's ready for shipment to the launch facility for pre-launch procedures.

12.2.1   Pre-ship Review

Before packing the product with its support equipment and documentation for delivery to the launch site, we conduct a pre-ship review. In this review, the mission systems engineers present facts to a review team showing that satellite development is complete, the transportation logistics (container and plan) are in place, the range is prepared to receive the satellite, and facilities are ready to begin the next phase of operations upon its arrival. A successful pre-ship review affirms that:

- The product is ready for delivery to the launch site or integrator
- The appropriate support equipment, whether uniquely designed, developed, or purchased off-the-shelf, is available and functional
- Necessary spare items are available
- Technical manuals are complete, validated, and verified against product and support equipment
- Trained personnel are available
- Facilities are ready to receive the product, support equipment, spares, personnel, and documentation
- Appropriate PHS&T materials and procedures are on hand and have been validated and verified against the product, support equipment, and spares

In some missions, the end user (e.g., NASA) takes responsibility for integrating the satellite to the launch vehicle. In this case, we must define the point at which the end user accepts the satellite. The satellite vendor often tries to define the pre-ship review as the acceptance review. But the project managers and systems engineers at NASA's Goddard Space Flight Center always define the point of acceptance as the post-delivery functional tests at the launch site. This means the vendor is responsible for proper design of the transportation container, choosing the transportation method, and transporting the satellite. An end user that accepts the hardware at the pre-ship review is then responsible for its transportation. The satellite provider cannot be held responsible for activities outside the contract.

FIGURE 12-3.   **Testing a Satellite before Transition. a)** A DirecTV satellite goes into the compact antenna test range, a massive anechoic chamber at Space Systems/Loral where its broadcast capabilities will be tested. **b)** Satellite being lowered into thermal vacuum chamber for testing. *(Courtesy of Palo Alto Online)*

12.3   Prepare the Product for Transportation to the Launch Site

Packaging a delicate spacecraft or a piece of it for shipment across a facility or around the world is complex, time consuming, and expensive. Packaging process requirements come from the product and the selected transportation mode. The product specifications should define the spectrum of allowable environments and stress conditions that the product can tolerate inside the packaging. The transportation providers should be bound to the environmental and stress limits that the packaging may be exposed to during transit. The combination of these limits—what the product is allowed to experience and what the package may be exposed to—drives the design of the packaging materials and containers. We should give particular emphasis to:

- Physical damage to surfaces
- Corrosion
- Damage to wiring or cabling
- Shock or vibration damage
- Thermal stress
- Moisture and contamination that could damage sensors or interfere with moving parts

Other factors include economy and ease of handling and transport, accountability (by shrink wrapping to keep all parts and manuals together during shipping, for example), security, and safety of unpacking.

After packaging the product and accompanying documentation (technical manuals), the sender either moves them to the shipping location or gives access to their facility to the shipping provider. Interfaces here need to be carefully negotiated and defined. For example, does the transport provider assume responsibility for the product at the manufacturer's loading dock or at the aircraft loading ramp?

12.4   Transport to the Launch Site

Transition may involve the physical movement of large hardware items (a Shuttle external tank, for example) across many miles or delivery of a small electronic assembly for integration at the launch site. In any case, available transportation modes may affect how and where the product is designed, built, and tested. Figure 12-4 shows typical transportation modes.

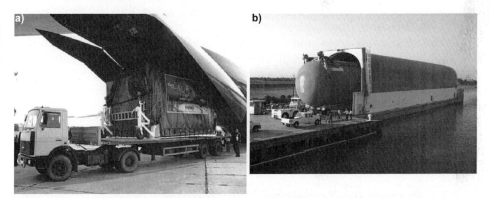

FIGURE 12-4.   Typical Transportation Modes. a) The first MetOp meteorological satellite arrives on 18 April 2006 in Kazakhstan, on board an Antonov-124 transport plane. It was launched on 19 October 2006 from the Baikonur Cosmodrome in Kazakhstan, on a Soyuz ST rocket with a Fregat upper stage. *(Source: ESA)* **b)** Lockheed Martin and NASA personnel at NASA's Michoud Assembly Facility in New Orleans load the Space Shuttle's external tank designated ET-118 onto its seagoing transport, the NASA ship Liberty Star. (Source: NASA/Michoud Assembly Facility)

12.4.1   Transportation of Space Systems

After we know all the stakeholder requirements, and have completed the analysis for PHS&T, we select the appropriate transportation mode for travel to the launch site. Candidate modes are airlift, sealift, land transport (road or rail) or some combination of these. We should choose a transportation method that provides a technically acceptable (mass, volume, environment) solution at reasonable cost and risk, and on a reasonable schedule. For large products, such as launch vehicles and propellant tanks, the manufacturing location may limit our transportation choices.

Conversely, access to transportation may influence where the product is built or integrated. For example, the Shuttle external tank (9.8 m diameter × 30.5 m length) can only be transported to Kennedy Space Center by barge. This fact influenced the choice of the Michoud Assembly Facility, located on a deep-water port on the Mississippi River, as the manufacturing facility for the external tanks.

Commercial carriers usually provide land transport, though specialized containers and vehicles are often needed. Careful route planning helps avert problems from low bridges, small tunnels, and tight turns. The limitations for air transport are size, weight, proximity to airfields, and availability of (sometimes specialized) large aircraft. For any transportation mode, we must provide for securing the product or its container to the transport vehicle during transition. Containers are typically instrumented to record the loads experienced during transport. A decision analysis tool such as the one in Table 12-2 helps clarify the choice of transportation mode. For more detailed information on the choice between air, sea, and land transport, we recommend *Space Launch and Transportation Systems*, Chapter 4 [Larson et al., 2005].

Before product delivery, the sending and receiving organizations must verify that the facilities at the receiving site are prepared, as well as any facilities for sustainment or storage; unpacking and installation procedures have been scripted, documented, and rehearsed; and responsibilities for each step are clearly defined.

The transition schedule should allow for testing at appropriate points in the process. Each time the product's configuration changes and each time it arrives at a different facility, some functional testing is needed to ensure that nothing has been damaged. This data is captured in the technical manuals.

Finally, we should plan for disposal or return of any transition support equipment, including, for example, shipping containers, packing materials, barges, or aircraft. This plan is also recorded in the configuration controlled technical manuals for the product and its associated support equipment.

12.4.2   Practice the Transition (Pathfinding)

The best way to ensure that all participants in the transition process understand their roles and are properly prepared and equipped is a practice run with something other than the actual hardware. These practice runs are called *pathfinders*, and we should plan, schedule, and fund them from the beginning of transition planning.

Examples of pathfinding activities include loading an engineering model into the shipping container to be sure it fits; sending a shipping container (weighted to full mass and center of gravity) from the manufacturing site to the launch facility; or driving an inert launch vehicle from the integration facility to the launch pad.

The Iridium® program employed pathfinders extensively to prepare the constellation's deployment. They used them to practice the transition of the spacecraft bus and subsystems to and through the manufacturing facility in Phoenix. FedEx practiced by flying an empty shipping container from Phoenix to the launch site at Baikonur. The program then shipped physical models of the

TABLE 12-2.   Sample Decision Matrix for Selecting Transportation Modes. The choice of transportation mode can be aided and documented by weighting the 17 criteria listed here and then scoring the competing transportation modes against these criteria. [Larson et al., 2005] (Wt. is weight of factor; Sc. is score; Wt.Sc. is Wt. x Sc.)

	Transportation Decision Support										
	Classify Objective		**Option A**		**Option B**		**Option C**		**Option D**		
	Must be		(Description)								
	Length xx m										
	Diameter xx m										
	Mass xxxx kg										
	Want – Desirable	Wt.	Sc.	Wt. Sc.	Sc.	Wt. Sc.	Sc.	Wt. Sc.	Sc.	Wt. Sc.	
1	Cost risk			0		0		0		0	
2	Schedule risk			0		0		0		0	
3	Performance risk			0		0		0		0	
4	Lifecycle cost			0		0		0		0	
5	Time to transport			0		0		0		0	
6	Transport availability			0		0		0		0	
7	Temperature—within acceptable range			0		0		0		0	
8	Humidity—within required range			0		0		0		0	
9	Pressure—within required range			0		0		0		0	
10	Acoustic level—less than requirement			0		0		0		0	
11	Acceleration—within acceptable levels			0		0		0		0	
12	Shock—within acceptable levels			0		0		0		0	
13	Speed—acceptable for requirements			0		0		0		0	
14	Door provides extra clearance			0		0		0		0	
15	Ordinance accommodation			0		0		0		0	
16	Propellant accommodation			0		0		0		0	
17	Permits			0		0		0		0	
	Total			0		0		0		0	

Iridium spacecraft to each of the launch sites in the US, China, and Russia. Finally, they used pathfinders at each launch facility to rehearse loading fuel into the satellites and loading the satellites onto the launch vehicles. This detailed planning, pathfinding, and practice contributed to a launch campaign that orbited 67 operational satellites with 15 launches in slightly more than 12 months.

12.5   Unpack and Store the Product

When the product arrives at the receiving facility, all personnel and equipment needed to unpack, store, assemble, integrate, install, test, use, and maintain it must be in place. Appropriate spares and repair parts must be available.

The transition of the Galileo spacecraft to Kennedy Space Center illustrates the need for careful planning, preparation, and training at the receiving facility. Because the spacecraft's radioisotope heaters were integrated before it was shipped to Kennedy, active cooling was necessary during the entire time the spacecraft was inside its shipping container. When Galileo arrived at Kennedy, the purge gas line was switched from the transporter to the building's supply, which ran at a different temperature. The abrupt temperature change caused condensation to form inside the container. Galileo stayed in its container (which was instrumented for temperature but not humidity monitoring) for several weeks over the Christmas holidays. When the container was opened, there were puddles of water on the floor of the container and some of Galileo's external surfaces had started to corrode [NASA, 1995].

We may also need to provide for storing the product during transition, perhaps for a long time. Solar arrays destined for the Space Station are a prime example; they were transitioned to KSC and stored there for more than five years, far longer than originally planned. Had this delay been anticipated, the transition process would undoubtedly have been different.

Operational satellites, such as weather or GPS satellites, often require storage during transition. The Polar Orbiting Environmental Satellites (POES) of NOAA are often built and then stored until called for launch. They must be ready to launch when needed to replace a failed or failing operational satellite. NOAA's constellation of Geosynchronous Operational Environmental Satellites (GOES), by contrast, includes an on-orbit spare, which can be moved to an operational slot upon call.

12.5.1   Receive and Check Out the Product at the Launch Site

Now we transport the product with its accompanying documentation, spares, and support products to the launch site. Trained personnel and support equipment should be available and prepared; facilities must be ready at the sending and receiving sites; procedures, roles, and responsibilities must be defined, negotiated, and practiced.

The launch site usually makes available an assembly and test facility, where the systems engineers test the product to be sure it wasn't damaged in shipment and that it still meets all performance requirements. After arrival at the launch site, the receive and check-out operations occur as follows:

1. Visually inspect the container

2. Examine the sensor data

3. Unpack

4. Visually inspect the satellite and the payload

5. Place the satellite in its environmentally controlled assembly and test facility

6. Perform functional tests

The tests begin with a visual inspection of the container to verify that no visible damage has occurred during transportation. This check is done almost immediately after dismounting from the transport carrier (ship, airplane, truck).

Next, we review the sensor data from accelerometers, temperature monitors, cleanliness monitors, etc., to determine the real environments that the satellite endured during transport. Early in the design process, we include the transportation environment in determining the satellite's design environments. Ideally, after arrival at the launch site, all the sensor data will show that the transportation environment was well within the satellite's design environment. If not, the systems engineer must assemble the engineering team to determine the impact of these environments on the satellite, what tests must be done to confirm the integrity of the product for launch, how much disassembly is needed, where to disassemble it, and whether it should it be shipped back to the vendor's facility.

The next step is the visual inspection after opening the container. The systems engineer hopes to find no dents, discoloration, or loose parts in the box. Otherwise we have a real headache. Now the systems engineering team must determine what happened, how it happened, how it affects the product, what needs to be done to ready the product for launch, and where this should be done.

After unpacking and conducting a more thorough inspection, we place the satellite in its environmentally controlled facility to begin its functional testing. A check-out procedure should be part of the acceptance data package, to give assurance that no damage occurred to the system during transit to the launch site. This functional test (versus a verification test) usually requires hookup to the ground support equipment (GSE) that accompanies the space system. If the check-out indicates anomalies, we formally record and track these problems in a problem and corrective action (PRACA) system, as described in the safety and mission assurance plan. Depending on the severity of the anomaly, launch site personnel will attempt resolution per the technical manuals (also in the acceptance data package) that accompany the product. Chapter 11 also discusses functional testing and problem resolution during test. Upon successful check-out, the space system begins integration to the launch vehicle. The GOES-N mission, shown in Figure 12-5, underwent two months of tests on the imaging system, instrumentation, communication, and power systems. At the end of this step, the product with its technical manuals and support equipment is at the launch facility in the configuration that meets all stakeholders' requirements.

FIGURE 12-5. **Geosynchronous Operational Environmental Satellite-N.** These images show the GOES-N satellite undergoing testing at Astrotech, KSC, during Spring, 2005.

12.5.2   Operational Readiness Review (ORR)

At some point during the launch processing operations, we hold an ORR to review the readiness of the space system for flight operations. This review occurs very late in the launch processing sequence, because we usually include the space system in live simulations with the mission operators. This could conceivably occur before shipment to the launch site (e.g., at the factory), although this practice is rare. It can also happen during integration to the launch vehicle, which is more common.

The ORR examines the system characteristics and the procedures used in the system's operation; ensures that all system and support (flight and ground) hardware, software, personnel, and procedures are ready for operations; and that user documentation accurately reflects the system as deployed. The ORR occurs when the system with its personnel and operational and support equipment are ready to undertake the mission. The conditions for a successful ORR are:

- **Science operations are ready**—All operational procedures are functional and contingency procedures are in place. We can process data from the satellite to generate science data products, share science data products with the science community, and archive data. For this, all hardware and software on the spacecraft and on the ground that are needed to perform science operations must be in place and ready. Also, all personnel must be trained for operations, including failure modes.

- **Ground systems are ready**—Data can be sent to the science operations center (SOC). For some missions, the satellite downloads all data into a mission operations center (MOC), which then routes the science data to the SOC. For this scenario, the network between the MOC and SOC must be verified operational by flowing simulated data. For other missions, the SOC receives data directly from the satellite; for this, we test the receive antenna and the data flow system. Ground system readiness also requires that a security plan, risk management plan, and contingency plans be in place and that personnel be trained to execute them.

- **Mission operations plans are in place and the team has practiced them** — The MOCs must be ready. Hardware and software must be compatible with that of the orbiting satellite and the mission operations team must be ready. Ensuring mission operations readiness includes performing simulations of each phase—launch, separation, deployment, acquisition, maneuver into operational orbit, normal operations, and contingencies. The MOCs often perform the maneuver simulations weekly and keep a training matrix to track each team member's progress. These simulations exercise the entire ground system and tracking networks. The MOCs perform multiple mission operations simulations to resolve all issues and to exercise the contingency and launch scrub procedures.

- **Ground tracking networks are ready to support the on-orbit operations** — A number of ground-based receive and tracking sites have been established around the world to support satellite operations. For example, NOAA operates Command and Data Acquisition ground stations in Virginia and Alaska to track satellites in low-Earth orbit. The Deep Space Network, which tracks and receives from interplanetary and other extremely distant spacecraft, has stations in California, Spain, and Australia.

  Early in the planning phase, mission architects must decide which ground stations to use based on the orbit and trajectory of the spacecraft and the availability of potential stations. As the spacecraft flies over a ground site, it establishes communication with the site, then downlinks data to it. The satellite controllers send commands to the spacecraft through antennas located at these sites. The tracking network is ready once we've scheduled the use of each site, all agreements are in place, and we've performed a communication test by flowing data to that site.

- **Space networks (if applicable) are ready**—If a space network, such as TDRSS, is required, the systems engineers must verify that it's ready to support on-orbit operations. This means we must have signed agreements documenting the use schedule and verifying communication through the space network asset by flowing data from the spacecraft.

- **Mission timeline is clear and well understood**—This plan includes documented timelines of pre-launch, ascent, and initial orbital operations. Often the satellite must be maneuvered after launch into its operational orbit. We must document the burn sequence for maneuver operations, develop the software scripts, and practice maneuver operations during end-to-end testing. The maneuver team documents and tests contingency procedures.

- **Contingency philosophy is clear and well understood for all mission phases, especially launch countdown and early-orbit check-out**—During these mission phases, the time for reacting to anomalies is short. Success depends on the robustness of contingency planning and the team's ability to execute these plans. Most teams develop a contingency philosophy from which they derive their contingency plans.

Inadequate contingency planning was a primary contributor to the loss of the Lewis mission, part of NASA's Mission to Planet Earth enterprise, in 1997. Here a series of spacecraft anomalies in the hours after launch caused the vehicle to go into a safe mode that, in retrospect, was poorly understood. At that point, controllers had reached the end of their duty day and had gone home. Upon returning nine hours later, they discovered the spacecraft in an unrecoverable spin with batteries virtually dead [Lewis, 1998]. Contingency planning guidelines that might have saved Lewis include:

- Develop contingency procedures for mission-threatening failures
- Have a flow chart for each contingency
- Have contingency plans to respond to failures (for manned missions, we consider two failures)
- Require ground involvement
- Augment onboard autonomous action to ensure total safety

12.6   Integrate With the Launch Vehicle

After the launch site functional tests have confirmed that the performance of the satellite is unchanged after transport, the satellite is ready for the next step—integration with the launch vehicle and pre-launch processing. For some projects, the satellite is integrated with the launch vehicle at the launch pad; on others it's integrated at a vehicle assembly building and then rolled out to the launch pad. Depending on the availability of launch vehicle interface emulators and simulators, much of the interface verification occurs at the factory before the system acceptance review, significantly simplifying integration activities at the launch site. The processes governing the verification are analogous to those described in Chapter 11. The launch vehicle to payload interface requirements document contains the requirements that must be verified, and we develop verification plans for them just like any other set of requirements. We normally summarize or reference them in the payload user's guide for the chosen launch vehicle.

Before integration, the satellite systems engineers inspect the satellite one last time. Integration begins with connecting the satellite mechanically and electrically with the launch vehicle, also called "mating". This includes carefully placing the satellite on the launch vehicle, affixing all the mechanical attachments, mating the connectors, and installing the mechanism that will separate the satellite from the launch vehicle in orbit. Extra care, and lifting equipment, is needed during the mating process so as not to bump the satellite into the launch vehicle.

We test each mechanical connection to verify its integrity. Electrical continuity checks verify that electrical connections have indeed been made and are functioning per design. If we use electro-explosive devices (EEDs) for the separation function, we test their electrical functionality using simulators. Technicians verify that the flight EED has the appropriate safing mechanism installed with "remove before flight" safety flags.

We conduct every step in this process using a carefully scripted procedure and in the presence of quality assurance (QA) personnel. Completion of each procedure step requires the signatures of at least two people—the technician or engineer performing the step and the QA representative. Repeated functional checks are done at each step, consistent with the integration and test processes described in Chapters 10 and 11.

The last step of integration is placing the payload fairing onto the integrated launch vehicle to cover the payload. After the fairing is placed, access to the satellite becomes limited. Strategically placed doors allow access to safety critical items (e.g., arming EEDs, venting high-pressure vessels, disposing of hazardous materials such as hydrazine) and allow monitoring of satellite temperature and pressure to ascertain that proper environmental conditions have been maintained.

The series of photos in Figure 12-6 shows ESA's Venus Express Satellite undergoing launch vehicle integration. In this example, the payload with the fairing installed is attached to the Soyuz-Fregat launch vehicle in the MIK-40 integration building.

Figure 12-7 shows GOES-N, which was integrated with the launch vehicle on the launch pad. For this integration, only the payload module was rolled out to the launch pad, then attached to the launch vehicle.

12.7   Roll Vehicle Out to the Launch Pad

Following integration with the launch vehicle, the satellite is ready to roll out to the launch pad. This movement is normally done on a carrier that safely transports the launch vehicle and places it upright on the launch pad. Some rockets (Soyuz, Pegasus) are transported horizontally, while others (Shuttle, Ariane 5) are transported vertically. The carrier contains air-conditioning equipment to maintain temperature, humidity, and cleanliness inside the payload fairing. It crawls very slowly to the launch pad. For the Space Shuttle, the total time for the 6.8-km crawl to the launch pad is around 7 hours and 45 minutes. The transport carrying the Soyuz with ESA's Venus Express took 45 minutes to travel a couple of hundred meters.

12.8   Integrate With the Launch Pad

After arrival, the launch vehicle is placed upright on the launch pad. It's attached to the launch tower with sufficient accuracy for the tower to guide the vehicle through the initial seconds of launch. The type of launch tower and accuracy of attachments depends on the launch vehicle. The systems engineer managing this portion of the project should get a copy of the user's guide for the launch vehicle immediately after launch vehicle selection and be familiar with

FIGURE 12-6.   **Venus Express.** Here we show the Venus Express during launch vehicle integration, Fall 2005. **a)** 30 Oct 2005: Venus Express is inspected before integration. **b)** 30 Oct 2005: Inspection of the fairing. **c)** 30 Oct 2005: The Venus Express spacecraft and Fregat upper stage in the MIK-112 building. **d)** 4 Nov 2005: Final integration of the Soyuz launch vehicle with the Venus Express Spacecraft has taken place in the MIK-40 integration building. The launch vehicle and its payload are now fully assembled and preparations begin for rollout to the launch pad. *(Source: European Space Agency website)*

these processes. In Figures 12-8 and 12-9, we show several instances of transporting the launch vehicle and connecting it to the launch pad.

The Space Shuttle crawlers have a laser-guided docking system that allows drivers to position the massive load accurately to within 6.4 mm at the launch pad. Standing up to 8 meters high at the tallest, NASA's two crawler vehicles are each 34 meters wide and 40 meters long. They carry a team of 25 engineers and technicians under full operations and have a mass of up to 8 million kilograms when capped with a Shuttle launch stack and launch platform.

The Soyuz vehicle is transported in a horizontal position, situated on the launch pad, then raised as shown in Figure 12-9.

FIGURE 12-7.   **Geosynchronous Operational Environmental Satellite-N (GOES-N).** The GOES-N spacecraft is prepped for integration onto its launch vehicle. **a)** June 8, 2005: Sitting atop a transporter, the encapsulated GOES-N spacecraft arrives at Launch Complex 37 on Cape Canaveral Air Force Station in Florida. The spacecraft will be lifted up into the mobile service tower and mated with the waiting Boeing Delta IV. **b)** June 8, 2005: At the upper level of the mobile service tower on Launch Complex 37, the spacecraft is lowered toward the Delta IV second stage. **c)** Boeing workers attach the spacecraft to the Delta IV second stage. *(Source: NASA Kennedy Space Center website.)*

FIGURE 12-8.   **Roll-out to the Pad.** Here we show various approaches for transporting the launch vehicle to the pad. **a)** Nov 5, 2005: ESA's Venus Express is transported to the launch pad horizontally (left). After arriving on the launch pad, it's raised to the upright position for launch. *(Courtesy of European Space Agency)* **b)** May 26, 2005: NASA's Crawler hauls the Space Shuttle Discovery's launch stack back to the Vehicle Assembly Building. The external tank is swapped with another, which has an additional heater to minimize ice debris that could strike Discovery during launch. *(Courtesy of NASA Kennedy Space Center)* **c)** March 9, 2007: Ariane 5 rolls out with INSAT 4B and Skynet 5A. *(Source: EFE, the Spanish News Agency website.)*

FIGURE 12-9.   Soyuz on the Launcher. The Soyuz launcher is integrated horizontally, then erected on the pad. **a)** November 5, 2005: The integrated Soyuz launcher being prepared to tilt the rocket to vertical position on the launch pad. **b)** November 5, 2005: Tilting launcher to vertical position. **c)** November 5, 2005: The fully integrated Soyuz FG-Fregat vehicle carrying Venus Express stands upright on the launch platform, secured by the four support arms. *(Courtesy of European Space Agency)*

12.8.1   Flight Readiness Review

After the spacecraft is mated with the launch vehicle, systems engineers perform tests to make sure that the newly integrated system and its support equipment are ready for flight. They confirm that interfaces are compatible and functioning. Technical manuals (procedures) have been rehearsed and validated, and operator training is completed. The systems engineers then review, confirm, and document these steps in the flight readiness review (FRR).

The FRR updates the mission status, closes out actions from previous reviews, and certifies readiness to proceed with the launch countdown. It occurs approximately three days before launch at the launch site after the mission is in launch configuration.

The system at this point consists of the space segment (launch vehicle and satellite); launch range; spacecraft's telemetry, tracking, and command (TT&C) system; and the communication system, including the network of receive sites around the world. The system functional test ensures that the interfaces between the launch range, the space segment, the communication system, and the command and control system are working smoothly. The space segment functional test must:

- Confirm that each stage of the launch vehicle can ignite and separate at the right time. We also test the flight termination system to be sure that it responds to the range safety officer's commands from the control center. Of course, no ordnance is actually activated during these tests. It's placed in a safe position by mechanical safe and arm devices, which isolate all ordnance initiators from the circuit by mechanically rotating them out of line. And safe plugs are placed on the circuit to short it, which allows for continuity checks.

- Validate blockhouse functionality by establishing that it can send commands, for example to charge batteries, safe a circuit, turn power off, and receive data.

- Prove that the mission communication system is operational. As the spacecraft flies over a receive site, it establishes communication with the site, then dumps data to it. The controllers must be able to send commands to the spacecraft through their antennas. We verify communication system performance by flowing data from the satellite to all receive sites around the world—usually before the satellite is mated to the launch vehicle. Spacecraft controllers also send commands to each of these sites to see that the site transmits them.

The ground segment is ready for launch when the TT&C is configured for launch, and personnel are trained to handle nominal and anomalous situations. In ground segment functional tests:

- We confirm the satellite TT&C functionality as part of the communication test. The TT&C receives data from the satellite and sends commands to it.

- To validate personnel and operational procedures, we present personnel with nominal, off-nominal, and anomalous conditions and assess their reactions. When things go wrong during launch operations, many things go wrong simultaneously, the time to react is short, and mistakes are costly. Therefore, planning is crucial to avoid mistakes during this stressful time. We must develop comprehensive procedures and see that the team can execute them smoothly. This attention to detail directly affects the quality of the launch operation.

Table 12-3 defines the entrance and success criteria for a flight readiness review.

12.9   Conduct Launch and Early Operations

The launch and early operations phase is the period beginning with live spacecraft activities during countdown, through the launch event, and concluding with transition to normal operations on orbit (Figure 12-10). For a NASA science mission, *normal operations* start when primary science data collection begins. For a crewed mission to the moon there will likely be no such formal transition to "normal operations."

TABLE 12-3.   **Flight Readiness Review (FRR) Entrance and Success Criteria [NASA, 2007 (1)].** Here we show a useful checklist to decide when a system is ready for its FRR and afterwards, to determine if it is, indeed, ready for flight. (DGA is Designated Governing Authority.)

Flight Readiness Review	
Entrance Criteria	**Success Criteria**
1. Certification has been received that flight operations can safely proceed with acceptable risk	1. The flight vehicle is ready for flight
	2. The hardware is deemed acceptably safe for flight (i.e., meeting the established acceptable risk criteria or documented as being accepted by the project manager and DGA)
2. The system and support elements have been confirmed as properly configured and ready for flight	
3. Interfaces are compatible and function as expected	3. Flight and ground software elements are ready to support flight and flight operations
4. The system state supports a launch "go" decision based on go/no-go criteria	4. Interfaces are checked and found to be functional
5. Flight failures and anomalies from previously completed flights and reviews have been resolved and the results incorporated into all supporting and enabling operational products	5. Open items and waivers have been examined and found to be acceptable
	6. The flight and recovery environmental factors are within constraints
7. The system has been configured for flight	7. All open safety and mission risk items have been addressed

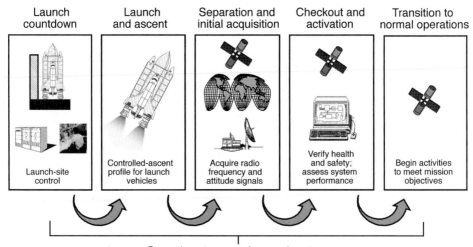

Launch countdown	Launch and ascent	Separation and initial acquisition	Checkout and activation	Transition to normal operations
Launch-site control	Controlled-ascent profile for launch vehicles	Acquire radio frequency and attitude signals	Verify health and safety; assess system performance	Begin activities to meet mission objectives

Operations team and ground system

FIGURE 12-10.   **Phases of Launch and Early Operations.** These images depict the launch and early operations from countdown to the beginning of normal operations [Boden and Larson, 1996].

We separate launch and early-orbit operations into distinct phases: countdown; ascent; separation and initial acquisition; activation and check-out; and transition to normal operations (Figure 12-11). Table 12-4 describes the steps associated with each phase.

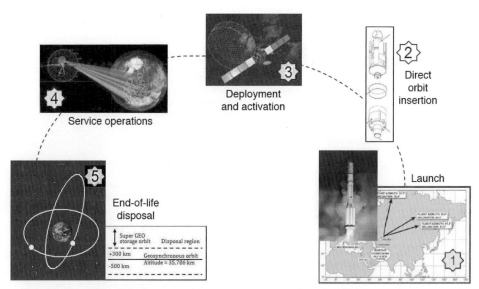

FIGURE 12-11.   Pictorial Depiction of the Launch and Operational Phases. Handover to the end user occurs at item 4 in the picture. *(Source: NASA)*

12.10   Transfer to the End User

After the satellite is launched and placed in the orbital slot to perform its mission, it undergoes an early-orbit check-out phase. During this phase, the engineers for the spacecraft and payload (scientific instrument, remote sensor, communication system, etc.) perform detailed tests of the system in its operational environment to validate that it meets all requirements. The early-orbit check-out phase normally ranges from one to three months, depending on mission complexity. The development team then hands the satellite over to the operations team and routine operations begin. This handover happens at the post-launch assessment review.

12.10.1   Post-launch Assessment Review (PLAR)

This is the final step in the transition process, and also signals the end of lifecycle Phase D. Now the spacecraft is functioning on orbit and all products and documentation necessary for normal operations have been tested and delivered. Users and operators are trained and ready to assume control.

A PLAR is a post deployment evaluation of the readiness of the spacecraft to proceed with full, routine operations. Its purpose is to validate that the satellite system (spacecraft and payloads) is functioning in its operational environment as planned, that it's meeting all requirements, that contingency plans are valid, that the operations team has successfully operated the satellite through all modes

TABLE 12-4.   Description of Launch and Early-orbit (L&EO) Operations Phases. This table lists the major steps involved in each phase of launch and early operations [Boden and Larson, 1996].

Subphase	Steps	Description
Launch countdown	• Power up spacecraft • Test spacecraft's functions • Test ground system's readiness • Determine go/no-go • Monitor system	Involves the final launch readiness tests of the end-to-end system (short aliveness tests) to support a final go/no-go determination by project management. After the launch is a go, places the systems in their final configuration for lift-off.
Launch and ascent	• Lift-off • Ascend through the atmosphere	Uses the pre-programmed, powered-flight profile to achieve orbit. Spacecraft operations during this phase are mostly passive, with some automated operations controlled by timers.
Separation and initial acquisition	• Separate spacecraft • Start activities immediately after separation • Acquire initial attitude • Acquire initial signal with ground system • Transfer to operational orbit	Involves the spacecraft's separation from the launch vehicle and the critical operations associated with acquiring an initial attitude, settling into a power positive mode, acquiring routine communications with the ground system, and deploying appendages. Timers, separation switches, or stored commands often control the separation activities automatically.
Activation and check-out	• Activate subsystems • Activate instruments • Check system performance	Activate the spacecraft's subsystems and instruments; configure for orbital operations; adjust biasing, settling, and drift rate; do check-out. Check-out entails verifying modes and configurations and doing initial system calibrations. Verify overall system performance against predictions, adjust ground models, modify operating plans, and modify procedures.
Transition to normal operations	• Configure for normal operations • Begin normal activities	Configure the spacecraft bus and instruments for routine operations to meet mission objectives.

(normal and contingencies), and that the system is now ready to begin the operations phase. Below are the conditions needed for a successful PLAR:

- Launch and orbital insertion phase has been completed and documented, especially variances between planned and actual events. The systems engineers must have an explanation of the impact of each variance on the overall mission. For example, if additional maneuvers were required to get into orbit, the systems engineers must explain why these maneuvers were required, how much propellant they used, and how much propellant remains.

- The spacecraft bus and payload have successfully passed all functional tests, including all operational modes. Here again, the systems engineers must explain all variances in test results and procedures. A list of each test performed, its results, variances from expected, impact of each variance on operations, and how lessons learned from each variance are incorporated into the overall operational plan. An excellent example of such a variance is described in the FireSAT case study (Section 19.1.3).

- Contingency plans are in place and understood, with special emphasis on changes from the pre-launch phase. The operational team is ready to accurately recognize a contingency and execute the appropriate plan.

- We've reviewed project plans to assess the team's ability to conduct the operational phase of the mission. In preparing for this part of the review the systems engineers must emphasize near-term operations and mission-critical events.

Table 12-5 defines the entrance and success criteria for a post-launch assessment review.

TABLE 12-5.   Post-launch Assessment Review Entrance and Success Criteria. This table gives a useful checklist to determine that a spacecraft is ready to commence normal operations. [NASA, 2007 (1)]

Post-launch Assessment Review	
Entrance Criteria	**Success Criteria**
• The launch and early operations performance, including the early propulsive maneuver results, are available • The observed spacecraft and science instrument performance, including instrument calibration plans and status, is available • The launch vehicle performance assessment and mission implications, including launch sequence assessment and launch operations experience with lessons learned, are completed • The mission operations and ground data system experience, including tracking and data acquisition support and spacecraft telemetry data analysis is available • The mission operations organization, including status of staffing, facilities, tools, and mission software (e.g., spacecraft analysis and sequencing), is available • In-flight anomalies and the responsive actions taken, including any autonomous fault protection actions taken by the spacecraft or any unexplained spacecraft telemetry, including alarms, are documented • The need for significant changes to procedures, interface agreements, software, and staffing has been documented • Documentation is updated, including any updates originating from the early operations experience • Future development and test plans are developed	• The observed spacecraft and science payload performance agrees with prediction, or if not, is adequately understood so that future behavior can be predicted with confidence • All anomalies have been adequately documented, and their impact on operations assessed. Further, anomalies affecting spacecraft health and safety or critical flight operations are properly disposed. • The mission operations capabilities, including staffing and plans, are adequate to accommodate the actual flight performance • Liens, if any, on operations identified as part of the operational readiness review have been satisfactorily disposed

12.10.2   Post Transition Support

Upon the successful conclusion of the PLAR, the spacecraft now belongs to the user. However, the developer's involvement in its operation will likely continue throughout the mission. Depending on the relationship between the developing organization and the operational users, the developers may be involved in routine, recurring functions such as trend analysis and mission planning. Certainly they will need to participate in unscheduled activities such as anomaly resolution and incident investigations. These sustaining activities are often underestimated or neglected in the developers' planning and budgeting processes and can be a significant burden when missions last far longer than originally anticipated. Voyager operations and the Mars rovers Spirit and Opportunity are classic examples of this "nice to have" problem.

12.11   Document the Transition

Finally, we must capture and archive all the work products from the transition event. These include a description of the procedures or technical manuals (as planned, executed, and configuration controlled), rationale for decisions made, problems encountered, corrective actions taken, lessons learned, and recommendations for future transition events. Each significant step in the transition process (loading, unloading, mating, fueling, etc.) should be photographed or filmed. These visual records often serve as final evidence of how we executed procedures and installed hardware. This documentation is important for feeding back into the LSA and may also become a useful resource for planning the transition of future products.

References

Boden, Daryl G. and Wiley J. Larson. 1996. *Cost Effective Space Mission Operations.* New York: McGraw-Hill.

Larson, Wiley J., Robert S. Ryan, Vernon J. Weyers, and Douglas H. Kirkpatrick. 2005. *Space Launch and Transportation Systems.* Government Printing Office, Washington, D.C.

Lewis Spacecraft Mission Failure Investigation Board. 12 Feb 1998. Final Report. Washington, DC: NASA.

National Aeronautics and Space Administration (NASA). January 30, 1995. Lessons Learned website, Public Lessons Learned Entry 0376.

NASA (1). March 26, 2007. NPR 7123.1a, *NASA Systems Engineering Processes and Requirements.* Washington, DC: NASA.

NASA (2). March 6, 2007. NPR 7120.5D, *NASA Space Flight Program and Project Management Requirements.* Washington, DC: NASA.

Chapter 13

Plan and Manage the Technical Effort

L. Dale Thomas, *NASA Marshall Space Flight Center*

"It's not the plan, it's the planning." *Dwight D. Eisenhower*

"The battle plan is obsolete once the first shot is fired." *Norman Schwartzkoff*

A model for the systems engineering process provides a common framework for technical planning. Although many organizations have developed standards for this process, they have much in common. Any of the popular standards will work for technical planning, but we use NASA Procedural Requirement (NPR) 7123.1a for this book. It's closely related to International Organization for Standardization/ International Electrotechnical Commission (ISO/IEC) 15288 [ISO, 2002].

This procedural requirement discusses 17 processes grouped into three general categories: system design, technical management, and product realization. The process of technical planning prepares us to apply and manage the other 16 processes in an integrated, consistent way. It produces a key document called the systems engineering management plan (SEMP), which describes how these 17 processes apply to a system development. This chapter explains technical planning and how we apply it: information crucial to getting any project started down the right path and—even more important—being able to recover from the unforeseen. Table 13-1 shows the technical planning process; for a broader project management-oriented treatment of planning, we refer the reader to Chapter 3 of *Applied Project Management for Space Systems* (APMSS) [Chesley et al., 2008]. In Chapter 14 we describe how to develop a SEMP based on this planning information and how to manage its changes throughout the system development.

TABLE 13-1.   **The Technical Planning Process.** Process steps address planning the work, creating the plans, committing stakeholders to the plans, and implementing them.

Step	Description	Where Discussed
1	Prepare for technical planning	Section 13.1
2	Define the technical work	Section 13.2
3	Schedule, organize, and cost the technical work	Section 13.3; Chap. 7
4	Prepare the systems engineering management plan and other technical plans	Section 13.4; Chaps. 5, 8, 9, 11, 12, 14, 16, and 17
5	Get stakeholder commitments to technical plans	Section 13.5; Chaps. 2 and 14
6	Execute the technical plan	Section 13.6; Chap. 18
7	Document and iterate	Section 13.7; Chaps. 16 and 17

Any plan includes **what** to do, **who** will do what, **when** the work must be complete, and **how many** resources will be available to do it. Thus, we have four main objectives in technical planning, as shown in Table 13-2.

TABLE 13-2.   **Matching Technical Planning's Four Main Objectives to Key Activities.** These activities produce the plan, which we revisit as the need arises.

Objective	Description	Activity	Section
1	What to do	Define technical work	13.1–13.2
2	Who will do what	Establish roles and responsibilities	13.3.1–13.3.3
3	When the work must be complete	Set up phasing and scheduling	13.3.4
4	How many resources will be available to do it	Specify technical work	13.3.5–13.3.6

Our ultimate objective isn't the plan but completing the technical work—for this book, developing a space system and transitioning it to mission operations. Technical management activities apply the plan and complete the work, including adjusting the technical plans to respond to the inevitable deviations between plans and reality. Figure 13-1 shows these activities, and Sections 13.4 through 13.6 describe them.

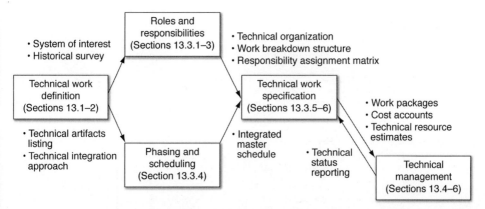

FIGURE 13-1.   Progression of Technical Planning Activities. This chapter discusses the four main activities that constitute technical planning, plus the technical management that implements the planning. The diagram includes the planning products and shows how these activities depend on each other.

13.1   Prepare for Technical Planning

Technical planning begins with understanding the implications of viewpoint. For example, planning for a thruster development as part of FireSAT's attitude control system differs from that for the FireSAT spacecraft. Section 13.1.1 addresses the implications of viewpoint by introducing the "system of interest" concept. Section 13.1.2 then reviews similar system developments to establish a point of reference for technical planning These two sections prepare us to define the technical work, covered in Section 13.2.

13.1.1   Identify the System of Interest

Both NPR 7123.1a and ISO/IEC 15288 establish the *system of interest* nomenclature to describe the technical effort's scope. How we perceive and define a system, its architecture, and its system elements depend on our interests and responsibilities. One person's system of interest can be another's system element in a different system of interest or just part of the operational environment for yet another's system of interest. Figure 13-2 exemplifies the many perceivable systems of interest in an aircraft and its operational environment. It shows how we can view an entity at any level in a hierarchical structure as a system, an element within a system, or a system's operational environment. The 17 systems engineering processes apply to any system of interest. Figure 13-2, for instance, shows

- Global positioning receiver system operating within the environment of other navigation system elements
- Navigation system operating within the environment of the aircraft system elements

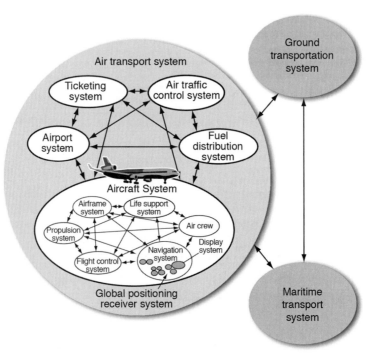

FIGURE 13-2.   **Typical System View of an Aircraft in Its Environment of Use [ISO, 2002].** We can view an entity at any level in a hierarchical structure as a system, a system element, or a system's operational environment.

- Aircraft system operating within the environment of the air transport system elements
- Air transport system operating within the environment defined by the ground transportation and maritime transportation systems

Identifying the system of interest helps to establish the scope of the technical work, and we must understand it clearly before starting to plan that work. The work includes four main phases:

1. Characterize the influences of the system context, where the context (such as the natural environment) can affect the system but the system doesn't affect it

2. Characterize the system's interactions with external elements, such as external interfaces (an example is a satellite's ground communications network)

3. Characterize the system performance to meet the objectives for which we're developing the system

4.   Design, develop, and integrate a system to meet the required performance within the proper environment and context

By establishing the system of interest, we limit the scope of this work at any location because technical planners need to consider only one level up and one level down in the system hierarchy. For example, in Figure 13-2, technical planners for the navigation system work within the context, environment, and system performance set only by the aircraft system; they needn't consider the air transport system's environment and context. Likewise, technical work for the navigation system defines the context, environment, and system performance for the navigation system's elements: the GPS receiver, display, and so forth. But its scope focuses only on integrating these elements, rather than exhaustively describing each element's technical work.

We consider also the FireSAT case study. Figure 13-3 shows various system of interest perspectives, including the overall system (the project), the FireSAT system (integrated spacecraft and payload), the spacecraft; and the spacecraft's electrical power subsystem. Indeed, the system of interest for the FireSAT SEMP is the FireSAT system; other elements of the project, such as the ground systems, provide context and environmental considerations. Also, the SEMP describes only one level of refinement in the spacecraft and payload systems. For instance, whereas it identifies the spacecraft and payload as system elements, it doesn't address **components** of the electrical power subsystem. The SEMP's technical plan encompasses the overall technical work for the electrical power subsystems, but it defers their details to technical planning.

One consequence of this approach is that we understand more as we step through levels of refinement in the systems hierarchy. Presumably, a good understanding of the FireSAT project from Figure 13-3 precedes understanding of the FireSAT system, consistent with the discussion in Chapter 2. This progression isn't a strict waterfall, as any systems engineer knows that system definition is iterative. Rather, technical planning moves sequentially from higher to lower levels in the system hierarchy. Although concurrency is desirable in that it leads to a more effective system, it comes at the price of iteration.

13.1.2   Review Similar Developments

Preparation is a key ingredient in technical planning, which in turn is vital to system development. We begin preparing for technical planning by gathering information on successful and unsuccessful technical plans that planners have previously applied to similar systems. These plans come from our own or other business units, most often from their systems engineering management plans (SEMPs). In looking at past efforts, we try to find technical planning precedents that will help us conceive a skeleton technical plan for our system development.

Perhaps the most informative outcome of the historical survey is a preliminary listing of technical products and the time phasing of their production during the lifecycle. These deliverable work products are the technical effort's content, and

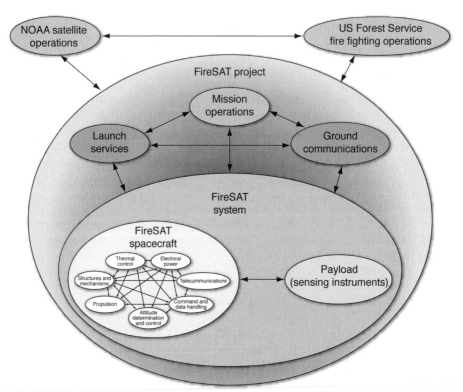

FIGURE 13-3.   **System-of-Interest Diagram for FireSAT.** Here we depict the differing system environments and elements for the FireSAT project, the FireSAT system, and the FireSAT spacecraft.

the phasing sets its pace. (We define these technical products fully in Section 13.2.) This historical survey also helps us scope other elements of the technical effort:

- Technical reporting requirements, including measures of effectiveness (MOEs), measures of performance, and technical performance measures (Chapter 2 discusses MOE development)

- Key technical events, with technical information needed for reviews or to satisfy criteria for entering or exiting phases of system lifecycle management (see our brief discussion in Section 13.2 and full discussion in Chapter 18 concerning technical data packages for technical milestone reviews)

- Product and process measures to gauge technical performance, cost, and schedule process. Given the preliminary definition and time phasing of technical products, we develop resource requirements as described in Chapter 7 and may then track technical progress against the technical plan, as described in Section 13.6.

- The approach for collecting and storing data, as well as how to analyze, report, and if necessary, store measurement data as federal records (Chapters 16 and 17 discuss configuration and data management respectively)
- How we should manage technical risks in planning (Chapter 8)
- Tools and engineering methods the technical effort will use (Chapter 14 tells how they fit into SEMP development)
- The technical integration approach for integrated systems analysis and cross-discipline integration (Section 13.2.3)

Though not exhaustive, this list typifies the foundational technical planning we can complete based on a historical survey. Reviewing technical plans for recently developed Earth-observing satellites, for example, gives technical planners significant insight for the FireSAT project. We save time and effort by refining these basic plans, as described in Section 13.2, rather than generating all the planning information from scratch.

13.2   Define the Technical Work

Once we have a foundation for planning, we scope the technical work.

13.2.1   Identify the Deliverable Products

The technical work for a system development focuses on systems engineering but requires involvement with almost every part of the project [Blanchard, 2004]. For instance, forming a budget isn't technical work but it influences, and is influenced by, the latter's phasing and content. Likewise, the project's schedule development and acquisition strategy depend heavily on the systems engineering results. Although planners commonly segregate a project's development work into program and technical camps, it's not a clean break. Hence, we must define and adjust technical work in terms of its relationship with other work elements.

In addition to the operational system, the deliverable products for a project include plans, specifications, drawings, and other technical artifacts from systems engineering processes. These products vary with each system development. For example, different acquisition strategies, such as making something versus buying it from a supplier, lead to distinct deliverable products. A standard process, such as a corporate method for configuration management, can eliminate or significantly reduce the scope of deliverable products, especially plans.

The scale of the system of interest likewise influences the deliverables. A very large project, such as NASA's Constellation Program, uses many plans to describe the technical work. One function of the Constellation SEMP [NASA, 2006 (1)] is to integrate these plans cogently, but the SEMP for each project within the program focuses primarily on its own technical planning. As the system's scale decreases, the number of technical products and distinct plans also decreases. Some NASA projects have included technical planning in the project plan and omitted a standalone SEMP. These small projects used standard NASA processes or umbrella planning by a parent program, so systems engineering planning required few distinctions and didn't warrant separate plans.

We use the 63 outputs for 17 technical processes in NPR 7123.1a Appendix C to list the technical deliverables for a system development, but we have to adapt this generic listing for a particular project. The listing includes technical artifacts typically incorporated in system developments, such as an operations concept. It also includes products normally integrated to form a single technical artifact but identified separately because different processes produce them. For example, the NPR lists measures of effectiveness, measures of performance, and technical performance measures as three distinct outputs from three separate processes. But we normally integrate them into a single technical artifact.

The NPR also lists some categorical outputs consisting of multiple technical artifacts. For example, one output of technical planning is the "SEMP and other technical plans," with the other plans depending on the project's scale. Still, these process outputs are a good point of departure for identifying technical artifacts for a given project, as illustrated in Figure 13-4. Table 13-3 lists all of FireSAT's technical artifacts, and Figure 13-4 shows how we cross-reference this listing to process outputs.

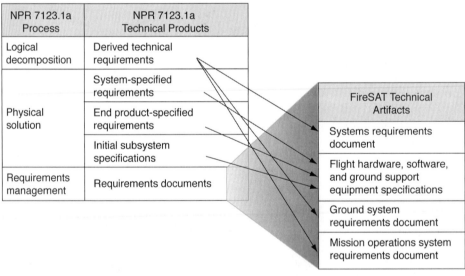

FIGURE 13-4.   Using NASA Procedural Requirement (NPR) 7123.1a Process Outputs as a Reference. The process outputs in NPR 7123.1a Appendix C are a point of departure for identifying FireSAT's technical artifacts.

Tailoring acquisition to fit various strategies is the rule rather than the exception. A project's acquisition strategy strongly influences its technical artifacts. The technical products in Table 13-3 would satisfy the technical work for FireSAT's system elements, if we were making everything. The bolded items in the table are what they'd be if we were buying everything. In this case, we would scope the technical work to get the bolded products from suppliers and then oversee the prime contractors, who would produce the non-bolded items. Acquisition uses pure "make" or "buy" only in exceptional circumstances. For example, TRW was the prime contractor on NASA's Chandra X-ray Observatory to:

- Develop the spacecraft and optical bench

- Integrate the upper stage with the Space Shuttle for launch

- Integrate instruments from the Smithsonian's Astronomical Observatory under a separate NASA procurement

TABLE 13-3.   FireSAT's Technical Products. Deliverables for the FireSAT Project include plans, specifications, drawings, and other technical artifacts produced by systems engineering processes in addition to the operable system.

• **Need, goals, and objectives**	• Disposal plan
• **Mission requirements**	• Technical review plans
• **Operations concept document (including ground systems and mission operations strategies)**	• Technology development plan
	• Launch operations plan
• Technical measurement plan and reports	• Payload-to-carrier integration plan
• **Systems requirements document**	• Technical work directives
• Flight hardware, software, and ground support equipment specifications	• Engineering change requests and notices
	• Waivers
• **Ground system requirements**	• **External interface control document (ICD)**
• **Mission operations requirements**	• Spacecraft to payload ICDs
• Natural environment definition document	• Spacecraft to launch vehicle ICDs
• Requirements validation matrices	• Spacecraft to ground system ICDs
• Logical decomposition models	• Launch vehicle to ground system ICDs
• Functional analysis document	• Ground system to mission operations ICDs
• **Master verification plan**	• Instrumentation and command listing
• Operations procedures	• Interface change requests
• Activation and inflight check-out plans	• Technical risk reports
• User manuals	• Hardware and software configuration items list
• Operational limits and constraints	• Baseline configuration document
• Integrated schematics	• Configuration management reports
• Spares provisioning list	• Specification tree
• **Integrated master plan**	• Drawing tree and engineering drawing list
• Integrated master schedule	• Technical data electronic exchange formats
• **SEMP**	• Design disclosure
• **Systems analysis plan**	• Technical measurement reports
• **Software management plan**	• Mass properties reports
• **Safety and mission assurance plan**	• Design and analysis cycle reports
• Configuration management plan	• Technical reports
• Data management plan	• Trade study and decision support reports
• Mass properties control plan	• Verification compliance reports
• Manufacturing and assembly plan	• Validation compliance reports
• Electromagnetic compatibility/interference control plan	• Integrated spacecraft and payload
	• Integrated spacecraft and payload acceptance data package
• Spacecraft systems analysis plan	• Mission operations system
• **Supportability plan (including training plan)**	• Ground system
• Spacecraft design and analysis cycle plans	

13.2.2   Mature the Technical Baseline

Blanchard [2004] points out that "system design is an evolutionary process, progressing from an abstract notion to something that has form and function, is fixed, and can be reproduced in specified quantities to satisfy a designated consumer need." As in nature, the project environment drives survival of the fittest, and the mortality rate of projects is very high. In the early 1990s, only one product development in four was successful in US industry, the rest being cancelled before they could make it to market or having too little market appeal to warrant mass production.

Industry responded by adapting *stage-gate systems* to develop products more effectively, getting appealing products from concept to markets much sooner. Stage gates divide development into stages consisting of prescribed, related, and often parallel activities—each usually more expensive than the preceding one.

Stage-gate systems typically have four-to-seven stages and gates, as Figure 13-5 shows, although no hard and fast rule exists. The entrance to each stage is a gate that functions like a quality control checkpoint on a manufacturing process. Senior managers staff these gates, with authority to approve needed resources and decide whether the product meets quality standards. In other words, they decide whether to go, kill, hold, or recycle based on quality criteria appropriate for that stage [Cooper, 1990].

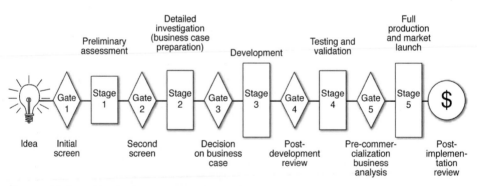

FIGURE 13-5.   The Stage-Gate Process for System Development [Cooper, 1990]. A typical stage-gate process models system development like manufacturing, with quality control checkpoints, or gates, between each process stage.

Stage-gate development dates to at least the 1960s in industry and the 1940s and 1950s in government, particularly the US Department of Defense (DoD). Space systems development today uses various stage-gate processes. NASA employs four different ones, depending on the type of system development; for example, the one for flight systems has seven stages or phases. The DoD's stage-gate process for system acquisitions has five phases, and IEEE-1220 specifies six phases [IEEE, 2005]. Although the number of phases appears arbitrary, it always depends on the system under development, just as it would for manufacturing. We wouldn't expect Proctor and Gamble's five rigorous stages for developing new consumer goods to be the same as NASA's stage-gate process for space systems.

Space system design evolves from the abstract notion of a mission concept to an operational system. As described in Section 13.2.1, technical work encompasses a broad array of artifacts, some of which must obviously precede others. For instance, engineering drawings must come before fabrication. So a technical artifact must be complete or mature enough to become the basis for succeeding artifacts and thus be part of the technical baseline. Once we baseline a technical artifact, it can change only through formal change control procedures (Chapter 16) because it influences technical work products under development.

At any point in the system's lifecycle, a *baseline* divides the finished work from the work in progress. The baselining of each technical artifact moves a project toward completing the technical baseline. Thus, a project may have tens or even hundreds of technical baselines throughout the development cycle as we baseline technical artifacts one by one. Still, a space project's technical baseline usually corresponds to the system's development phases and therefore consists of seven discrete parts:

- **Mission baseline**—fully articulated and stakeholder-affirmed mission need and objectives, as well as the concept for meeting those objectives

- **System baseline**—functional and performance requirements characterizing systems that perform the mission concept

- **Functional baseline**—the system design, with functional and performance requirements allocated to its elements

- **Design-to baseline**—the fully specified system design, with functional and performance requirements and design or construction standards fully allocated and flowed down to the configuration item (hardware and software) level

- **Build-to or code-to baseline**—the complete system design as described by engineering drawings or computer-aided design (CAD) models, software design documents, and plans for fabrication, assembly, integration, and test

- **As-built or as-coded baseline**—the realized system as verified, validated, and certified, and ready for deployment

- **As-deployed baseline**—the operational system as launched or otherwise deployed, activated, and checked out including calibration, and ready to proceed with full, routine operations

Figure 13-6 illustrates how the technical baseline evolves. We establish each baseline following comprehensive review at a key technical event, making sure the event is compatible with the existing baseline. Such events occur at the end of a lifecycle phase. At that point, managers assess progress against the plan, consider technical risks, and decide whether to proceed or to remain in the current phase for a certain time—or possibly to end the program because of poor performance.

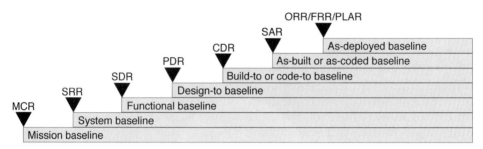

FIGURE 13-6.   **How the Technical Baseline Evolves.** Each successive technical baseline builds on the preceding ones and is a step toward the operational system. (See Figure 13-7 for acronym definitions.)

Figure 13-7 shows the key technical events within the system's lifecycle, based on NPR 7123.1a. Appendix G of the NPR describes review objectives, as well as entry and success criteria. (Chapter 18 gives a full description of technical review and assessment.)

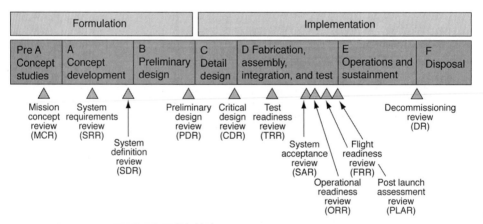

FIGURE 13-7.   **Key Technical Events during the Flight System's Lifecycle.** The project must pass these reviews successfully to continue through the lifecycle.

Entrance and success criteria for reviews help us refine the technical baseline by specifying artifacts at each key technical event. For instance, consider the system requirements review (SRR) for NASA's Ares 1 launch vehicle. Table 13-4 lists the success criteria, adapted from NPR 7123.1a Appendix G, with the technical artifacts reviewed as objective evidence that the criteria are fully satisfied.

Figure 13-8 shows how to develop notional technical baselines in terms of their technical artifacts. The as-deployed baseline has three reviews associated with it, but this depends strongly on the type of system. For instance, Space Shuttle

TABLE 13-4.   Technical Work Products for a System Requirements Review of NASA's Ares 1 [NASA, 2006 (2)]. The first column lists success criteria for the system requirements review. The second column lists the technical artifacts offered as objective evidence of having met the criteria.

Success Criteria	Review Data
1. The resulting overall concept is reasonable, feasible, complete, responsive to the mission requirements, and consistent with available resources, such as schedule, mass, or power	Requirements validation matrix, Ares system requirements document, requirements traceability matrix, integrated vehicle design definition, constellation architecture, and requirements document
2. The project uses a sound process for controlling and allocating requirements throughout levels and has a plan to complete the system definition within schedule constraints	Systems engineering management plan, project plan, configuration management plan, data management plan
3. The project has defined top-level mission requirements, interfaces with external entities, and interfaces between major internal elements	Ares system requirements document, interface requirements documents, functional flow block diagrams, constellation architecture and requirements document, design reference missions
4. Planners have allocated requirements and flowed down key driving requirements to the elements	Traceability matrix, functional flow block diagrams, Ares system requirements document
5. System and element design approaches and operational concepts exist and are consistent with requirements	Operations concept document, integrated vehicle design definition, design reference missions
6. Planners have determined preliminary approaches for verifying requirements and validating them down to the element level	Master verification plan, Ares system requirements documents, interface requirements documents
7. Planners have identified major risks and ways to handle them	Risk management plan, Ares risk list, functional failure mode and effects analyses
8. The project's technical maturity and planning are adequate to proceed to preliminary design review	Pre-board and board assessments, technical plans

missions combine the three. Conversely, for a satellite on an expendable launch vehicle, we may conduct three reviews in sequence:

- The operational readiness review addresses the satellite's readiness for launch, although not necessarily for operations, because we may load software or calibrate instruments after launch

- The flight readiness review covers launch readiness of the integrated payload and launch vehicle

- The post-launch assessment review covers the operational readiness elements reserved for after deployment

Applying success criteria for each review yields the technical baseline's phased evolution, as shown in Figure 13-9. For example, baseline releases of all six

Baseline	Representative Technical Artifacts	Established at
Mission	Need, goals, and objectives / Mission requirements	MCR
System	Concept of operations / System requirements / SEMP / IMP / Natural environment definition / Safety and mission assurance plan	SRR
Functional	Baseline configuration / Interface requirements / Allocated requirements / Test and evaluation master plan / Software management plan / Technology development plan	SDR
Design-to	Component specifications / Software specifications / Ground support equipment specifications / Interface control documents / Verification plans / Manufacturing plan	PDR
Build-to	Engineering drawings / Operational limits and constraints / Manufacturing process requirements / Acceptance plans and criteria / Verification procedures / Integration and assembly plan / Test plans / Training plan	CDR
As-built	End products / Enabling products / Acceptance data package / Operations procedures / In-flight checkout plan	SAR
As-deployed	Integration and test anomaly resolutions / Mission support training and simulation results / Scientific instrument calibration results / System activation results	ORR FRR PLAR

FIGURE 13-8.   **Progressive Development of the Technical Baseline.** The technical baseline matures incrementally as a project baselines groups of technical products following key technical events. (MCR is mission concept review; SRR is system requirements review; SDR is system definition review; PDR is preliminary design review; CDR is critical design review; SAR is system acceptance review; ORR is operational readiness review; FRR is flight readiness review; PLAR is post-launch assessment review.)

documents for the system baseline must be ready to support the system requirements review. If approved following their respective reviews, these technical artifacts then become part of their respective baselines.

This synchronizing of discrete technical baselines, development phases, and associated key technical events isn't coincidental; it has developed over the years and converged around logical breaks in the nature of technical activities during a system's development. The concept of a discretely evolving technical baseline is important for two reasons:

- The baseline provides a logical basis for sequencing technical products — sequencing we must understand to plan and schedule the technical work (Section 13.3)

- Examining entry and success criteria for the reviews reveals organizational expectations about the technical artifacts a development team must produce, when to produce them, and how to assess the artifacts' quality.

Again, this knowledge proves useful in detailed planning and scheduling of the technical work (Section 13.3).

Technical plans and reports	Mission concept review	System requirements review	System definition review	Preliminary design review
Concept of operations		Baseline		
System requirements document		Baseline		
Systems engineering management plan		Baseline		
Software management plan		Draft	Baseline	
Flight hardware, software, and ground support equipment specifications		Draft	Update	Baseline
Ground system specifications		Draft	Update	Baseline

FIGURE 13-9.   **Evolution of FireSAT's Technical Baseline.** Analyzing the success criteria of each key project milestone against the technical work products enables us to begin identifying the technical artifacts that will make up the technical baseline. Here we show only the first four milestone reviews.

13.2.3   Determine the Technical Integration Approach

Space systems require a lot of interaction between the system elements. This coupling lies at the heart of "rocket science," and gives rocket science its reputation for difficulty. Daily coordination among the engineering disciplines produces a broad array of informal or second-tier engineering products to help us develop the formal technical products. How we integrate the informal products profoundly affects the quality of the formal ones, and thus the effectiveness of the entire project.

Coordination among engineering specialties and disciplines leads to technical integration of the space system. The subsystems involved differ with the disciplines: structures and mechanisms, electrical power, propulsion, communications, and so forth. The interactions among these subsystems constitute the workings of the system itself. For example, the propulsion subsystem generates a motive force, which the structure subsystem distributes throughout the spacecraft. We characterize interfaces between project elements, such as between a spacecraft and the payload, in terms of how their respective subsystems interact.

The following scenario illustrates the rich interaction between subsystems in a space system. Satellite telecommunication requirements depend on the amount of information we must transmit, the time available to do so, and how far the signal must travel. Given a required data rate, designers of communication subsystems must trade off antenna gain and broadcast power. Higher gains imply narrower beam widths that demand more accurate pointing knowledge and stability from the antenna. That, in turn, requires structural stiffness that drives structural mass, material selections, and so forth.

Alternatively, increased broadcast power takes more electrical power, which means more solar cell area, which drives the electrical system's mass, etc. Increased electrical power consumption also affects the thermal subsystem because electricity produces heat that the system must dissipate [Griffin and French, 2004]. So space subsystems tend to be tightly coupled, necessitating integrated systems analysis.

To plan for technical integration, we must understand the task, beginning with the engineering disciplines for the project. The engineering disciplines for a launch vehicle form the main diagonal of an N×N matrix (Figure 13-10). The launch vehicle element in the upper-left corner allocates requirements, architectural elements, constraints, etc., to the engineering disciplines. The first row of the matrix captures this allocation. The outputs for a diagonal element are in its row, and inputs to it are given in its column. In Figure 13-10, the cell entry illustrates the outputs from aerodynamics needed to do the thermal analyses. Humphries et al. [1999] and Blair et al. [2001] contain a complete N×N matrix for a launch vehicle, which thoroughly describes how to technically integrate this subsystem.

A historical review of developments similar to FireSAT suggests that we must integrate the 13 engineering disciplines shown along the diagonal of the N×N matrix in Figure 13-11. Section 13.3.4 gives a detailed discussion of how this matrix works. The partial N×N matrix shown here illustrates the discipline interdependencies that underlie the design and analysis cycle.

By way of illustration, we consider the integrated master schedule (IMS) for a design and analysis cycle (DAC) planned to support FireSAT's preliminary design review. The N×N diagram in Figure 13-11 identifies the dataflow interdependencies between the engineering disciplines and establishes predecessor and successor relationships. For instance, the astrodynamics and environments element specifies FireSAT's orbit using the mission definition. The mission definition also allows the telecommunications analysts to determine the data rates. But they also need the orbital altitude and inclination definition from astrodynamics and environments to begin analysis of antenna configurations and power requirements to transmit that data rate. Moreover, the power requirements from all the subsystems are needed as inputs to the electrical power system analysis to determine the solar array area, battery sizing, and electrical power distribution schemes. And in turn, electrical power consumption parameters are needed for thermal analysis to assess the adequacy of the thermal control system to meet thermal dissipation needs. The thermal analysis results drive the choice for active or passive thermal control, thermal radiator area, and material selections.

These thermal system characteristics, along with the characteristics for all other subsystems, become inputs to the cost analysis, which estimates the system's cost and compares it to the budget. Later, it closes the loop with the mission

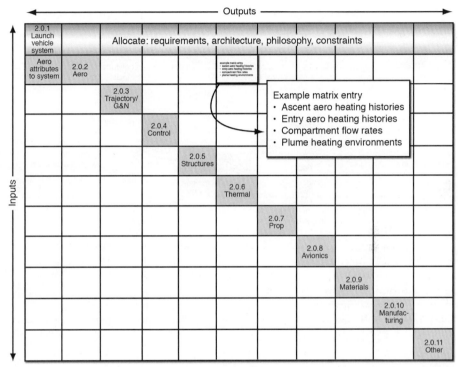

FIGURE 13-10.   NxN Matrix for Launch Vehicles [Blair et al., 2001]. For thermal analyses and design, the thermal discipline needs outputs from aerodynamic analyses, including histories of ascent and entry heating, compartment flow rates, and plume heating environments.

analysis, which considers it in determining the costs and benefits of mission objectives, particularly if the projected cost is larger than the budget.

Figure 13-2 shows the implications of the system-of-interest perspective for technical planning. Technical integration for the FireSAT system (spacecraft and payload) needs to address all 13 engineering disciplines in the list, whereas integration for the FireSAT project considers only those in boldface. The *Constellation Program System Integrated Analysis Plan* [NASA, 2006 (3)] and the *Exploration Launch Project Systems Analysis Plan* [NASA, 2006 (4)] are examples of actual systems analysis plans for a program and one project within that program.

FireSAT's system analysis plan must describe how to handle integrated systems analysis during an evolving spacecraft design, including the mechanics of design and analysis cycles. This analysis is necessary because the spacecraft's subsystems are highly interdependent, and technical resources such as mass and electrical power are severely constrained. In contrast, payload development doesn't usually require integrated systems analysis because analysis for the spacecraft or launch system includes the integrating parameter constraints, such as structural

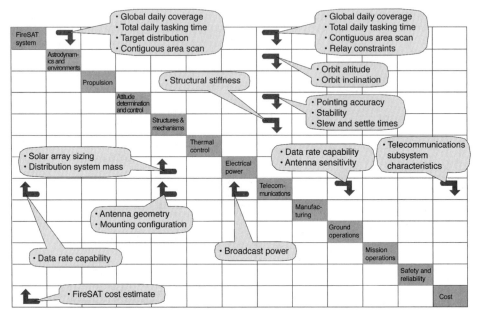

FIGURE 13-11.   NxN Diagram for the FireSAT System. The diagram includes selected interdisciplinary interactions for analysis and design of the telecommunication subsystem.

loads or electrical power consumption. But integrated systems analysis for the spacecraft requires data and information from the payload technical effort, so appropriate contracts or technical directives must include provisions for this data.

As the FireSAT example illustrates, the system of interest significantly influences technical integration, so we must account for it in technical integration planning. For instance, if the acquisition strategy has suppliers providing subsystems with relatively intense interactions, we must plan appropriate integrating mechanisms. We might include technical integration planning in the contract requirements. Or we might have to consider how much time the project schedule allows for a *design and analysis cycle* (DAC). The DAC iterates integrated engineering analysis across all disciplines, as illustrated by the NxN matrix in Figure 13-11. Planners typically schedule DACs to end with the project's various formal reviews, to best support evolution of the technical baseline.

Figure 13-12 depicts the planned progression of design and analysis cycles through system development of the Ares 1 launch vehicle. This progression supports the technical product deliverables for each key technical event, as described in Section 13.3.4. Brown [1996] describes the DAC process used to develop the International Space Station.

In summary, planning for technical integration identifies interdependencies between engineering disciplines and improves project success in three ways:

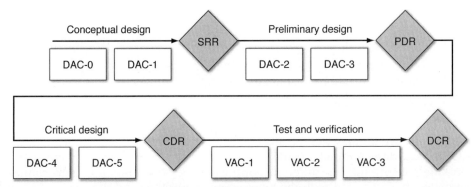

FIGURE 13-12.   **Design and Analysis Cycle (DAC) Support of Reviews for the Ares Launch Vehicle Project [NASA, 2006 (4)].** DACs and verification analysis cycles (VACs) support key technical events and provide objective evidence for decisions concerning the technical baseline. A VAC provides results of analytical verification. (SRR is system requirements review; PDR is preliminary design review; CDR is critical design review; DCR is design certification review.)

- It identifies the myriad informal data (second-order data) and information products that must pass between engineering functional organizations during the development of formal technical products. It thus ensures that we fully define the technical work and generate an adequate resource estimate.

- It traces interdependencies between engineering and management disciplines, enabling planners to develop and manage the engineering part of the integrated master schedule

- It delineates the second-order data and information interdependencies as a basis for the technical data architecture. This in turn is a foundation of the project's data management plan.

13.2.4   Plan Hardware and Software Integration

System integration involves putting together components from different suppliers to create the complete system, and we describe it in detail in Chapter 10. Integrating hardware and software is a particularly challenging aspect of space system development since the software is generally developed in parallel with the computer system on which it will execute. Furthermore, in most cases the software development parallels the maturing of the system hardware that the software is meant to monitor and control. Hence, software is developed concurrently in pieces, by different suppliers, in development environments that can only simulate the eventual operating environment. A space system such as FireSAT will have the software partitioned in rough alignment with the subsystem decomposition we describe in Section 13.2.3:

- Propulsion

- Attitude determination and control
- Structures and mechanisms
- Thermal control
- Electrical power
- Telecommunications, command, and data handling
- Ground and mission operations

The attitude and orbit control subsystem (AOCS) developer develops the AOCS software to carry out the attitude and control function in concert with the AOCS hardware. A project usually has multiple software development teams working concurrently on different portions of the software, all of which must ultimately be integrated with the other software portions and the spacecraft. An extreme example of spacecraft software development is shown in Figure 13-13, which illustrates the International Space Station Increment 14 software architecture. This complex architecture includes 21 software units distributed among 45 computers (accounting for redundancy) organized into three tiers. The program established and implemented a consistent, configuration-managed software stage plan for communicating past, present, and future on-orbit ISS software configurations. This was vital, because knowing exactly what software configuration is running onboard the vehicle is critical to crew safety and mission success. This example illustrates the challenge of integrating upgrades to software throughout the life of the vehicle.

The software management plan (Section 13.4.3) must address schedule considerations including the number of builds and releases, the hardware and software dependencies, and the ordering of the tasks. For instance, the AOCS flight software build testing cannot be performed without having the command and data handling (C&DH) software build as well. So the software schedule should show the C&DH build completed first. When we schedule the software development, higher complexity software should go into the early builds if possible; we should defer incomplete or unstable requirements to late builds, and try to have a full closed-loop build as early in the cycle as possible.

Furthermore, mature flight software is required for efficient flight hardware integration and test support, so the flight software development must be completed ahead of the flight hardware. The AOCS software must be sufficiently mature to enable the integration of the hardware components. (Section 10.2.3 goes into detail on the integration of the AOCS subsystem on FireSAT.) Likewise for the propulsion subsystem—if its software isn't ready, the propulsion subsystem verification cannot proceed.

The essence of hardware and software integration planning is reconciling how the hardware "wants" to be integrated with how the software "wants" to be integrated. An integrated verification fishbone such as the one in Figure 11-15 helps us visualize how to logically integrate the hardware components to successively higher levels of assembly, until ultimately the spacecraft is integrated. We also have sequencing dependencies and desirable practices for integrating the software components to

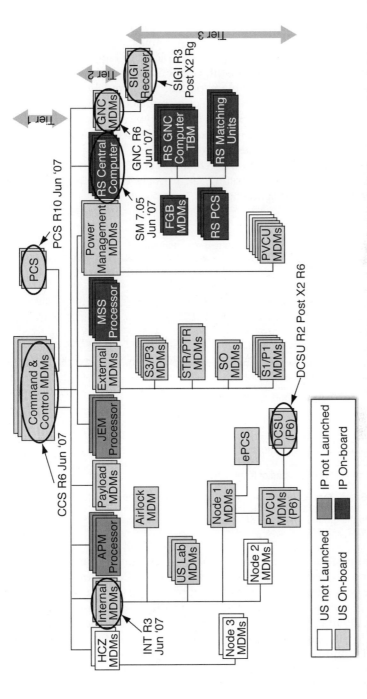

FIGURE 13-13.   International Space Station (ISS) Increment 14 Software Architecture [Schaefer, 2008]. The proper and timely integration of these disparate software components was vital to the safe and successful functioning of ISS. We don't define the acronyms here, since the point of this figure is to illustrate the complexity of the software architecture.

successively higher levels of assembly, until the full software complement is integrated and installed on the spacecraft. When the hardware integration and software integration sequences don't coincide, we must use simulations. A high-fidelity simulation of the spacecraft computers, depicted in Figure 13-14, provides a software development and test environment. This simulation supports both hardware and software integration. We can replace part of the simulation with hardware engineering units, yielding hardware-in-the-loop simulation configurations such as we describe for the AOCS example in Section 10.2.3. A typical spacecraft development incorporates several such simulations for various subsystems, and AOCS in particular may include multiple hardware-in-the-loop configurations. The degree to which these simulation configurations differ from the actual spacecraft determines the scope of effort (software patches, regression testing, etc.) involved in getting the software functional and validated on the spacecraft.

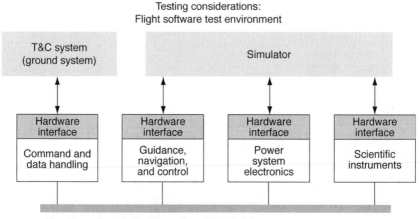

FIGURE 13-14.   Basic Test Environment for Flight Software [Pfarr and Obenschain, 2008]. Each software interface to the hardware requires a simulator, including the interface to the ground system. (T&C is telemetry and command.)

Each new simulation configuration—and each test run on that configuration—costs money, and higher fidelity simulations cost more. More and higher fidelity simulations are better, but how much is enough? Failure to plan sufficient fidelity testing can lead to mission failure. The maiden flight of Ariane 5 failed due to insufficient fidelity in the integration testing of the guidance and control software [Thomas, 2007]. But runaway costs can, and have, led to numerous project cancellations. Technical planners must balance cost and risk in developing a prudent test strategy; Chapter 8 furnishes several techniques and methods of risk management to help determine the appropriate balance.

13.3   Schedule, Organize, and Cost the Technical Work

The previous section affords a general understanding of technical work through broad descriptions and illustrative examples drawn from previous system developments. This section goes into detail on the technical work content for FireSAT.

13.3.1   Develop the Work Breakdown Structure (WBS)

The WBS organizes system development activities by apportioning systems and products into product architectures. Analysts organize and depict these architectures and their associated services, such as program management or systems engineering, in a hierarchical tree-like structure. Although the WBS is a non-technical product, it derives directly from the physical system architecture, as Figure 13-15 shows. A high-quality, cogent WBS benefits the technical work by organizing it more efficiently. The complete work breakdown structure, including refinements to three levels, is in the FireSAT systems engineering management plan.

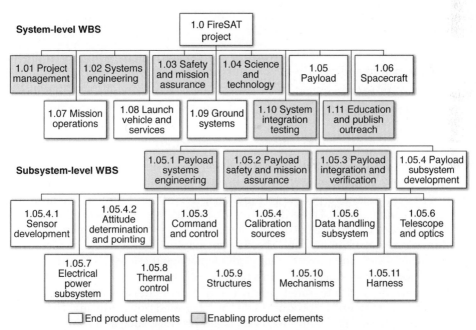

FIGURE 13-15.   Top Levels of the FireSAT Work Breakdown Structure (WBS). The WBS includes the end products and enabling products. Planners progressively refine it to precisely describe all work for the system development.

A project's WBS contains end product and enabling product elements. The end product part is based on the physical architectures developed from operational requirements. The enabling product part identifies the products and services required to develop, produce, and support the system throughout its lifecycle. The WBS is a foundation for all project activities, including planning for the project and its technical activities; defining the event schedule; managing the system's configuration, risk, and data; preparing specifications and statements of work; reporting status and analyzing problems; estimating costs; and budgeting [DoD, 2000].

To illustrate how the WBS organizes a project, we consider how to map the list of technical products in Table 13-2 to the WBS. The concept of operations and system requirements document are normally systems engineering products, so we identify them with WBS 1.02. The manufacturing and assembly plan maps to 1.05 and 1.06, so we have to decide whether to develop separate plans for the payload and spacecraft or a single plan that addresses both. If we assign a single plan to 1.05 or 1.06, the other WBS element has an input to the plan—identified as one of its technical products. Figure 13-16 shows part of the mapping of FireSAT's technical products from Table 13-3 to its WBS from Figure 13-15.

13.3.2   Organize the Technical Team

We place the many organizational schemes that government agencies and industry use into three basic categories:

- A project organization, in which everyone working on a project is directly assigned to and managed by this organization
- A functional organization, in which people work for departments within the agency or company that do specialized work, such as engineering or quality control
- A matrix organization, which is a hybrid of the other two

Smaller organizations more often use functional and project structures, whereas larger organizations employ matrix structures almost exclusively. A matrix structure assigns people to functional departments similar to those of a functional organization. These departments provide qualified people, tools, and standard procedures to project organizations. The latter then blend resources from the functional departments into effective cross-functional teams that focus on a particular product. The project organizations manage these teams [Grady, 1994].

This use of cross-functional teams has become the preferred approach in DoD, and is referred to as integrated product and process development (IPPD). The teams are called integrated product teams or integrated process teams. Their name depends on whether the team's focus is an end product, such as a communication subsystem for a satellite, or an enabling product element or process, such as a training program for satellite operators [DoD, 1998]. The approach has two key organizational principles: organizing around the work breakdown structure (WBS) and concurrent engineering.

The WBS shown in Section 13.3.1 provides a cornerstone for developing a project (and technical) organization. The organization should establish an effective

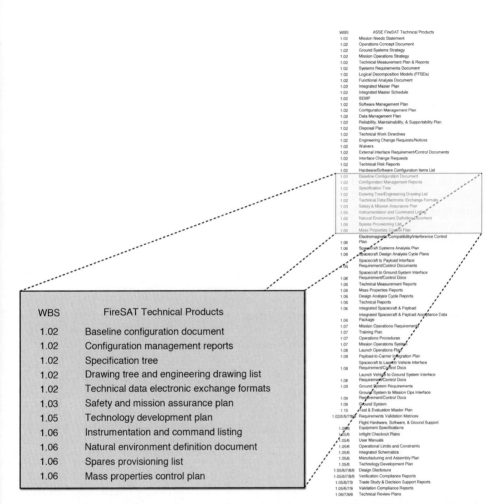

FIGURE 13-16. Categorizing the FireSAT Project's Technical Work Products by Work Breakdown Structure (WBS). The WBS logically groups technical products.

management structure for the products in the WBS. That's relatively straightforward for end products but more difficult for enabling ones that commonly address development and deliverable processes. A development process helps produce a balanced product but doesn't go to the customer. Examples are technical integration (Section 13.2.3), test and evaluation, software development, and production. Someone must apply them, and we must have a management structure to manage and improve them during the development. Its form depends on the size of the task and the importance of the process.

Deliverable processes are developed for and delivered to the customer. Typical examples are those for support, training, and maintenance. For some projects, a process is the only deliverable. As with those for development, someone must decide how to manage them.

It's usually appropriate to have both product- and process-oriented elements within the organizational structure. The most common management strategy for processes is to form one organizational element, such as an office, department, or branch, and then integrate the deliverable products into a coherent, effective system. A common name for this element is the "systems engineering and integration office." It integrates the efforts of the individual product teams, ensuring that they communicate and effectively apply accepted systems engineering principles to developing the product. The element delivers an integration process, whose effectiveness we measure by the integrated product's success or failure to meet total system requirements with the best balance of cost, schedule, and performance.

A project's acquisition strategy strongly influences organizational structure, which typically forms around major suppliers' products or processes. The more diverse the suppliers for end product and enabling product elements (processes), the more likely they will affect the project's organization. One example is the Chandra X-ray Observatory acquisition strategy we describe in Section 13.2. Although the acquiring organization may combine elements into a single organization to manage suppliers cleanly, the suppliers may refine that structure to reflect their own distinct elements.

To illustrate, let's consider the project organization for the FireSAT Project. From FireSAT's SEMP the acquisition strategy relies on a prime contractor, with NASA's Goddard Space Flight Center serving as the procuring organization. The prime contractor produces the integrated FireSAT System (spacecraft and payload) and support for launch service integration, and the government provides launch services. The prime contractor also supplies some elements of the mission's operations and command, control, and communication architecture in a joint development with the government. The government furnishes the ground element, including the communication and mission operations infrastructure and launch site, including launch processing and operations. The FireSAT system requires three distinct acquisition approaches:

- Procure deliverables from a prime contractor
- Get ground systems and launch services from the government
- Employ a prime contractor and the government for operations

Starting with the WBS in Figure 13-15 and using the prime contractor for WBS elements 1.05 and 1.06, it's appropriate to combine the two into an organizational element that covers the scope of the government's contract with the prime contractor. Likewise, with the government providing the ground systems (WBS 1.09) and launch services (WBS 1.08), it makes sense to form organizational elements to manage those activities. Goddard will acquire FireSAT for the US Forest Service's firefighting office, and the National Oceanic and Atmospheric Administration (NOAA) will operate it.

The joint approach for mission operations, including three government agencies (NASA, NOAA, and the US Forest Service) and the prime contractor suggests an organizational element for mission operations (WBS 1.07). Systems engineering, safety and mission assurance, science and technology, and systems integration and test (WBS 1.02, 1.03, 1.04, and 1.10) are enabling elements (development processes) for end products that a single systems engineering and integration office can handle. Similarly, a single management support office handles activities for project management and education and for public outreach (WBS 1.01 and 1.11). Figure 13-17 shows the resultant project office structure.

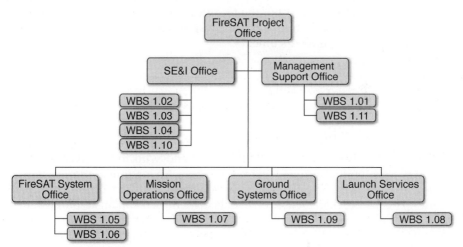

FIGURE 13-17.   Office Organization for the FireSAT Project. Each office's project responsibilities correspond to elements of the work breakdown structure (WBS). (SE&I is systems engineering and integration.)

13.3.3   Develop the Responsibility Assignment Matrix

A responsibility assignment matrix maps a project's technical work to the organizational elements. One type uses the RACI (responsible, accountable, consulted, and informed) format:

- Responsible—Does the work or is responsible for having it done. Each activity should have only one responsible organization. If an activity lists more than one, the work needs further division.
- Accountable—Approves the completed work
- Consulted—Has information or capability needed to complete the work, so it supports the responsible role
- Informed—Receives information on plans, progress, and results

These matrices are useful in estimating required resource levels for the organizational elements because they assign the organization's role for each activity [PMI, 2004]. We build them at a high level (WBS elements) or a detailed level (individual work products), and nest them in hierarchies. Developing them with participating organizations inevitably leads to clearer organizational roles and responsibilities, especially for process-oriented organizational elements.

Returning to the FireSAT case, we need to create a responsibility assignment matrix for the technical work products in Table 13-3. Here, we map them to the organization shown in Figure 13-17. Table 13-5 captures the results. Because this is a matrix for the technical work products, it shows limited roles for the management support office. The project office has the approval role for the major plans and for acceptance of the end products, but the systems engineering and integration office approves most of the technical work that other offices produce. The matrix reveals the scale of this office's responsibilities: it must produce most of the project's key planning documents.

As the procuring organization, Goddard turns to its own directorates for engineering and safety and mission assurance to support systems engineering. So these can use a refined responsibility assignment matrix to establish their matrixed team members' roles and responsibilities (Table 13-6).

13.3.4   Schedule the Technical Work

Preceding sections have defined the **what** and **who** for the technical work required to develop the system. Next is the **when**. An integrated master plan and an integrated master schedule serve to sequence and schedule the technical activities. The master plan focuses on events to identify the main accomplishments and show the system's maturation as it progresses through the development cycle. Its key technical events include milestone reviews, such as the system design or preliminary design review; system tests or test flights; prototype deliveries; and so forth. It defines each event precisely so that participants can

- Adequately plan their portions of the technical work for the event
- Agree on when the event has been completed

The master schedule flows directly from the master plan and adds detail, including the individual tasks for each event. Each task description includes that task's duration and relationship to predecessor or successor tasks. The former must be done before the task begins; the latter can't begin until the task is complete. Task durations and relationships enable us to develop a logic network of integrated tasks. When tied to an overall start date, they form the integrated master schedule. We must define the master schedule in enough detail to carry out the system's daily development. Figure 13-18 shows how the integrated master plan and schedule relate.

In developing an IMP, we use the key milestones from Figure 13-5. In a large project, each major end product may have its own such milestones. Concept and requirement reviews usually progress from higher to lower level, and design and readiness reviews from lower to higher. So the project's system requirements review (SRR) precedes the SRRs for the overall system, the ground system, and

TABLE 13-5.   Assignment Matrix for Technical Responsibilities on the FireSAT Project. This matrix assigns the role each organization is to play in developing the technical products. (SE&I is systems engineering and integration; R is responsible; A is accountable; C is consulted; I is informed.)

				Organizational Roles				
WBS	Technical Products	FireSAT Project Office	Management Support Office	SE&I Office	FireSAT System Office	Ground Systems Office	Mission Operations Office	Launch Services Office
1.02	Concept of operations	A		R	C	C	C	C
1.02	Systems requirements document	A		R	C	C	C	C
1.02	Systems engineering management plan	A	I	R	C	C	C	C
1.02	Software management plan			A/R	C	C	C	C
1.05/6	Flight hardware, software, and ground support equipment specifications			A	R	C	C	C
1.09	Ground system requirements			A	I	R	C	C
1.07	Mission operations requirements			A	C	C	R	C
1.03	Safety and mission assurance plan	A		R	C	C	C	C
.	.	.	.	.	.	.	.	.
.	.	.	.	.	.	.	.	.
.	.	.	.	.	.	.	.	.
1.06	Integrated spacecraft and payload	A		I	R			
1.07	Mission operations system	A		I			R	
1.09	Ground system	A		I		R		

TABLE 13-6. **Refined Responsibility Assignment Matrix from the Project Office for Systems Engineering and Integration (SE&I).** The shaded column depicts the responsible, accountable, controlled, and informed roles from Table 13-5. Differences with the unshaded column for the SE&I project office indicate delegations of responsibility to engineering or safety and mission assurance (S&MA).

		Organizational Roles				
WBS	Technical Products	SE&I Project Office	SE&I Project Office	Engineering	S&MA	Prime Center
1.02	Concept of operations	R	I	R		C
1.02	Systems requirements document	R	I	R	C	C
1.02	Systems engineering management plan	R	R	C	C	C
1.02	Software management plan	A/R	I	R	C	C
1.05/6	Flight hardware, software, and ground support equipment specifications	A	A	—	I	R
1.09	Ground system specifications	A	A	—	—	I
1.07	Mission operations specifications	A	A		C	C
1.03	Safety and mission assurance plan	R	I	—	R	C
•	•	•	•	•	•	•
•	•	•	•	•	•	•
•	•	•	•	•	•	•
1.06	Integrated spacecraft and payload	I	I	—	—	R
1.07	Mission operations system	I	I	—	—	•
1.09	Ground system	I	I	—	—	•

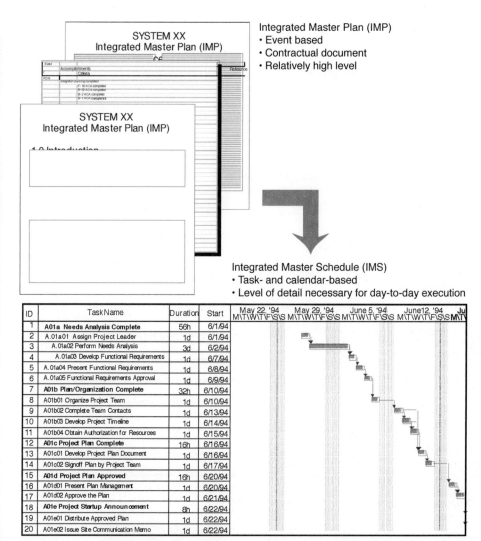

FIGURE 13-18.   **How the Integrated Master Plan (IMP) and Integrated Master Schedule (IMS) Relate [DoD, 2005].** The IMP consists of key events and their planned occurrence dates; the IMS details all the activities required to achieve the IMP.

mission operations. But the critical design reviews (CDRs) for these same elements come before the project CDR. Exceptions may be necessary. For example, requirement reviews for long-lead items may occur before those for the system into which they will integrate. When the sequencing changes, managers must identify and manage the risks (Chapter 8).

Figure 13-19 shows an integrated master plan for developing NASA's Ares 1 launch vehicle. Technical reviews dominate the milestones, but the project has other key events:

- Deliveries of test articles, such as development motors one and two for the first-stage element
- The cold-fire and hot-fire events for the main propulsion test article on the upper-stage element
- Deliveries of flight elements to support launch integration for test and operational flights
- Test flights and operational flights

Ares 1 Project Milestones

Sep 28, 2006

Name	FY06	FY07	FY08	FY09	FY10	FY11	FY12	FY13	FY14	FY15
Constellation level II milestones	RBR SRR — L1 DPMS	Oct	Pre NAR	PDR NAR — May	Ares 1 — Apr	CDR		Ares2 — Sep	Orion3 Orion4 Orion5 — Sep Jun Sep Dec	Orion6
136905.02 Vehicle integration		SRR SRB — Dec	PDR — Apr	CDR — Sep			Ares DCR — Jun			
136905.08.01 First stage	Pre SRR TIM — Aug	SRR — Dec	PDR — Dec	ICDR — Jul	CDR — Jul		DCR — Apr	DDCR — May		
Fabrication, integration, test, and delivery*				DM-1 DM-2 — Apr Oct		QM-1 — May Nov	Ares 1-2 QM-3 Orion3 Orion4 — Aug Feb Nov Feb		Orion5	
136905.08.05 Upper stage		SRR SDR PDR — Feb Jun Feb		CDR — Apr			DCR — Feb			
Fabrication, integration, test, and delivery*					MPTA MPTA CF HF — Sep Nov		Ares2 Orion3 — Jun Feb	Orion4 Orion5 — Sep Nov Sep		Orion6
Avionics		SRR — Feb	PDR — Feb	CDR — Apr		Ares 2 — May	Orion3 Orion4 — May Jan May	Ares2 Orion3 Orion4	Orion5	
Flight software		SW RR — Jun	PDR — Apr	CDR — Sep		Ares2 S/W — Mar	Orion3 S/W — Mar	Orion4 S/W — Jan Mar	Orion5 S/W	
136905.08.04 Upper stage engine	PRR — Jun	SRR PDR — Nov May	CDR — May				DCR — Mar			
Engine deliveries for integ. to upper stage				Ares 1 — Oct		MPTA — Oct	Ares2 — Mar	Orion3 — Nov	Orion4 Orion5 Orion6 — Jul Nov Jul	
136905.10 Flight integration and test ops		SRR PDR CDR — Sep Mar Sep		Ares 1 Flt — Apr						
Ares I-1										
Ares I-2				SRR — Jul	PDR — Aug	CDR — Jun	Ares2 Flt — Sep			

Note: All design review dates are board dates.

FIGURE 13-19.   Integrated Master Plan (IMP) for NASA's Ares 1 Launch Vehicle. The IMP identifies the milestones that managers use to schedule the technical work. (TIM is technical interchange meeting; RBR is requirements baseline review; SRR is systems requirements review; NAR is non-advocate review; PDR is preliminary design review; CDR is critical design review; MPTA CF is main propulsion test article cold flow test; MPTA HF is main propulsion test article hot flow test; SRB is standing review board; ICDR is initial critical design review; DCR is design certification review; DDCR is demonstration design certification review; DM-1 is development motor #1; QM-1 is qualification motor #1; SWRR is software requirements review.)

The FireSAT Project has a dedicated series of key milestone events, as will the FireSAT System, ground system, and mission operations. These events are scheduled to support each other and the integrated FireSAT Project. Figure 13-20 shows a multi-level IMP that captures acquisition milestones, technical reviews, launch dates, operational capability milestones, and system decommissioning.

Events in the integrated master plan set the pace for technical work. Because each event has defined expectations, mapping event descriptions to the associated technical work products is a good starting point for the IMS. Although the number of tasks often appears unmanageable, a well-constructed IMS effectively guides managers in controlling the daily technical work.

To develop an effective IMS, we rely on hierarchical tiers in the integrated master plan that correspond to the work breakdown structure (WBS). These are refined by the products for each WBS element. Each product requires tens or even hundreds of tasks, but the structure makes them clear. The WBS and product hierarchy also help us identify predecessor and successor relationships between tasks. For a given product, the lead or manager usually identifies these relationships. At the next level, managers of all products in a single WBS must collectively identify them, and so on through the WBS hierarchy.

Given a collection of discrete tasks, along with their precedence and succession relationships, we build a schedule network to organize the tasks and interrelationships. Figure 13-21 shows part of a schedule network for FireSAT's design and analysis cycle, derived from the N×N diagram in Figure 13-11.

By containing the durations for each task, the schedule network underpins the program schedule with one of several techniques: the program evaluation and review technique, arrow diagram method (also known as the critical path method), precedence diagramming method, or graphical evaluation and review technique. Several references cover network schedule methods [PMI, 2004; Kerzner, 1998; Hillier and Lieberman, 1995]. Figure 13-22 shows an example schedule for a design and analysis cycle on NASA's Ares 1 launch vehicle.

Finally, as with virtually all technical planning, developing the integrated master plan and master schedule is iterative: the pace of technical work (as reflected in the master schedule) influences the master plan's timing and content.

13.3.5   Define Work Packages

If we've followed the steps described in the previous sections, we now understand what the project must do, who is to do it, and when it's to be done. Now we're ready to define individual work packages. A work package is a distinct subset of the project work. It describes the organization's work and helps it monitor and report work. A prime contract is one example of a work package. Though a large one, it represents the part of a project a prime contractor must do as one designated team member.

Defining the work packages for a project enables us to develop and phase the resources required for the technical work. Work packages should be natural subdivisions of the overall project, so we should define them based on how the

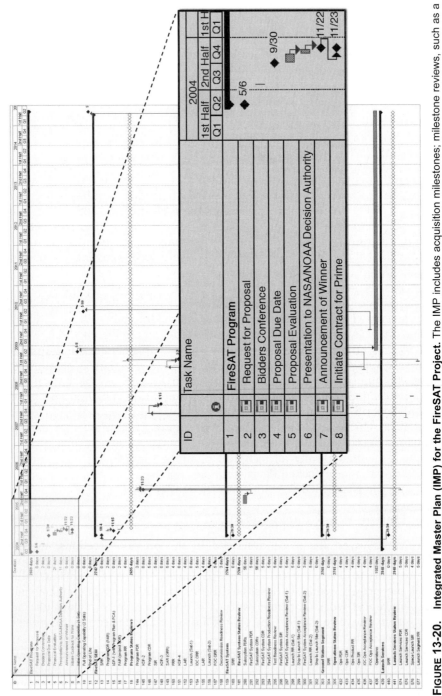

FIGURE 13-20.   Integrated Master Plan (IMP) for the FireSAT Project. The IMP includes acquisition milestones; milestone reviews, such as a system requirements review, preliminary design review, and so forth; system tests or test flights; and prototype deliveries.

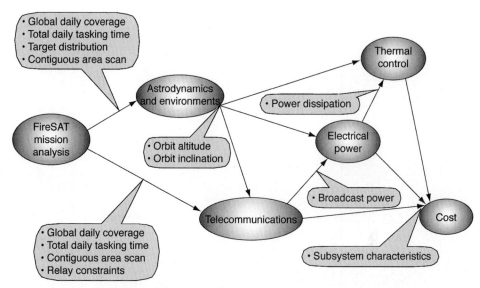

FIGURE 13-21.   Schedule Network for FireSAT's Design and Analysis Cycle. This partial schedule network supports the preliminary design review and depicts a subset of the task relationships illustrated in Figure 13-11.

work will be done. We get a first-order definition of the work packages for technical work from the responsibility assignment matrix. In the FireSAT example, Goddard Space Flight Center (GSFC) uses its Applied Engineering and Technology Directorate (AETD) and Safety and Mission Assurance Directorate (SMA) to do technical work for systems engineering and integration (SE&I), under the FireSAT project office's management. By refining this matrix, we clarify which organizations will do which work. Table 13-5 illustrates such a refined matrix.

In this example, the SE&I office has at least three formal work packages: one each for itself and the two GSFC directorates. We may also want to develop a work package for each of the four WBS elements assigned to the SE&I office, for a total of twelve. If some small ones combine with others for the same performing organization, we may end up with fewer work packages, as shown in Figure 13-23.

Although the refined matrix clarifies roles and responsibilities, we need more detail to define the work package. Continuing with the FireSAT case, we assume Goddard's organizations for engineering and for safety and mission assurance are experienced in developing scientific satellites like FireSAT. So engineering needs only the schedule to produce the system requirements document. On the other hand, we need to consider the informed role assigned to engineering for the mass properties reports. This role ranges from a casual review for errors and adherence to the mass properties plan, to independently analyzing and comparing results. Required resources vary a lot over this spectrum, so we must precisely describe the work package content. For the systems engineering management plan (SEMP),

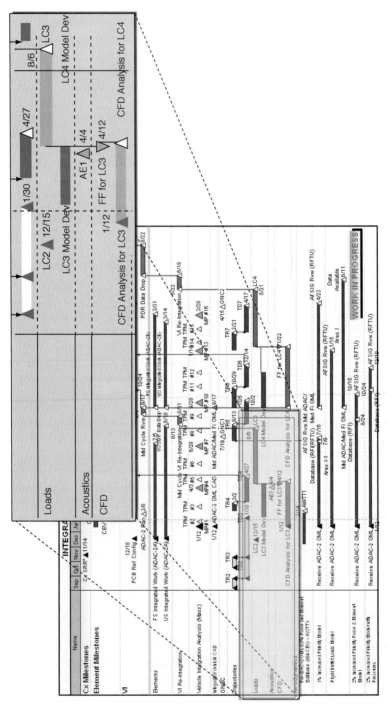

FIGURE 13-22.   **Schedule of the Design and Analysis Cycle for NASA's Ares Launch Vehicle.** This summary schedule illustrates durations and interdependencies between discipline analyses. The analysis for integrated flight loads is a critical-path dependency on the results from the computational fluid dynamics (CFD) analysis. The former includes structural loads from engine thrust and gimballing, sloshing propellant, operating the reaction control system, and so on. The latter is a numerical simulation of aerodynamics during ascent. (FF is final finite element model; LC is load cycle; AE is acoustic environment.)

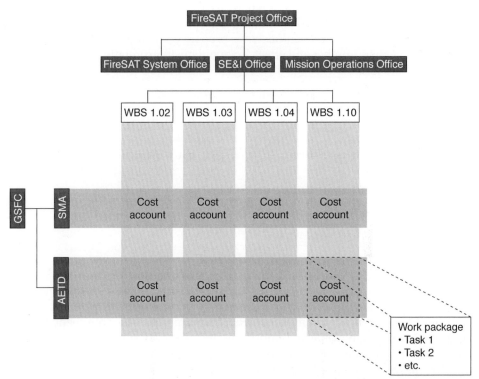

FIGURE 13-23.   **Work Packages for Systems Engineering and Integration (SE&I) on FireSAT.**
The intersections of the work breakdown structure (WBS) elements and the
functional organization elements define the work packages. (GSFC is Goddard
Space Flight Center; SMA is Safety and Mission Assurance; AETD is Applied
Engineering and Technology Directorate.)

engineering also has many potential roles, so project managers must provide
precise detail to make their intentions clear. To illustrate this point, we give the
following examples of misunderstanding on intent between the acquirer and
supplier [Kerzner, 2005]:

- The task description calls for at least fifteen tests to determine a new
  substance's material properties. The supplier prices twenty tests to include
  some margin, but at the end of the fifteenth test the acquirer demands
  fifteen more because the results are inconclusive. Those extra tests result in
  a cost overrun of $40,000.

- The task description calls for a prototype to be tested in "water." The
  supplier prices testing the prototype in a swimming pool. Unfortunately,
  the acquirer (the US Navy in this instance) defines water to be the Atlantic
  Ocean. The supplier incurs $1 million in unplanned expenses to transport
  test engineers and equipment to the Atlantic Ocean for the test.

As these examples illustrate, it's hard to give a rule of thumb for the detail needed in a work package description. Misinterpretations of task descriptions are inevitable no matter how meticulously we prepare them, but clarity and precision greatly reduce their number and effect. To avoid misinterpretation, we employ the following strategies:

- Avoid mixing tasks, specifications, approvals, and special instructions
- Avoid using imprecise language, such as nearly, optimum, minimum, maximum, or approximately
- Use a consistent pattern, structure, or chronological order for all task descriptions
- Define all tasks to have a similar scale
- Employ a consistent style for all the task descriptions
- Ask a third party for review

If an organization has plenty of experience, they'll easily understand a given responsibility. In this case, describing departures from business as usual is more important than details. The answer to the question "How much detail?" is quite simply "Enough so all participants consistently understand the technical work." We often just reference formally documented work processes for the needed detail. If the processes are informal and undocumented, the only rule of thumb to follow is: "Too much detail is better than too little."

The SEMP summarizes technical work packages as part of a narrative accompanying the responsibility assignment matrix. Eventually, these packages become part of technical work directives (Section 13.6) that formally document them. In other words, we don't formally document work packages as an end unto themselves. Still, we mark the types of information we should capture to characterize a work package:

- Reference WBS—the WBS element that encompasses the technical work package
- Requesting organization—the customer for whom the performing organization is doing the technical work and to whom they will deliver the technical work products
- Performing organization—the organization that must complete the technical work and deliver the technical work products to the requesting organization
- Task description—a narrative description, akin to a statement of work, that summarizes the work to be done and includes ground rules or assumptions to help interpret the task
- Deliverables—a list of the technical products deliverable within the work package

- Schedule—a list of delivery dates for each deliverable, including interim deliveries as appropriate to gauge technical progress

  Note: For interim deliveries, the schedule includes likely product maturity dates. In general, the delivery dates should be the appropriate key event milestones, such as the system requirement or preliminary design review, rather than calendar dates. Whenever projects revise the master schedule, this approach makes revising detailed schedules much easier.

13.3.6   Develop the Technical Resource Estimate

We use the work packages to estimate resources needed for the technical work, so we must build them carefully to make them complete but not overlapping. If we don't describe the needed technical work completely, our resource estimate won't cover what is missing. Conversely, overlapping work packages will result in an excessively large resource estimate because two or more providers will include resources to do the same work. They don't usually catch the overlaps because they expect work package definitions to divide the work effectively.

In Chapter 7 we describe how to estimate resources for technical work as part of the project's resource estimate. Estimates cover each top-level WBS element, as described in Section 7.1, and are sometimes necessary for lower levels of the WBS. For instance, we would develop technical resource estimates for the FireSAT spacecraft and instruments as elements of the FireSAT system, as well as for the spacecraft's subsystems. Thus, defining the technical work packages consistent with the WBS is essential to forming accurate technical resource estimates. The descriptions of technical work packages from Section 13.3.5 feed into the estimation process we describe in Chapter 7. If these descriptions are necessary and sufficient, the cost estimate will be unbiased and have little uncertainty.

We categorize the technical work for these work packages in terms of one or more of the 17 systems engineering processes in NPR 7123.1a, allowing us to approximate how the resource requirements will vary with time. For instance, in the very beginning of a project (before Phase A), the technical work consists almost entirely of defining stakeholders' expectations, as we describe in Chapter 2. After the mission has a baseline, we spend much less time on this area but increase efforts in defining the technical requirements and establishing processes for logical decomposition. We characterize the work packages that make up the technical work in this manner so that the relative resource requirements over time may be estimated. Conversely, if we analyze a given resource's phasing when the resources are arbitrarily spread over the lifecycle, we discover inadequate insight regarding the underlying rationale.

Figure 13-24 illustrates how the estimating effort varies over the system's lifecycle for each of the 17 systems engineering processes. So, if the estimate for definitions of stakeholder expectations is X person-years over a project's lifetime of Y years, we can establish the year-to-year budget for this process. This approach works as long as we have the milestones and dates in the integrated master plan that bound the respective phases.

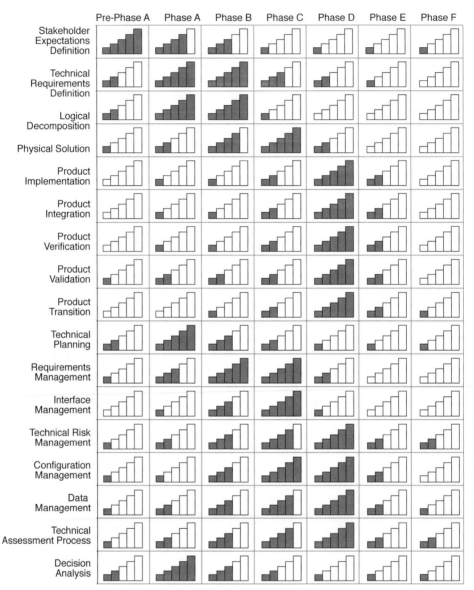

FIGURE 13-24. Intensity of Resource Use for Systems Engineering Processes by Project Phase. Analogous to a cellular phone signal, zero bars represents no or very low resource requirements, and five bars represent highest use. We must bear in mind that a number of bars for one process doesn't imply the same level of resources as that number of bars for a different process.

The resultant cost estimate invariably leads to changes in the scope of the technical work. Space systems engineers tend to avoid risk because the systems they build are complex, must work in an extreme environment, and have many routes to failure. For instance, given adequate resources, the leader of FireSAT's systems engineering and integration office wants to scope the "I" (informed) tasks for engineering and for safety and mission assurance with independent analysis across the board, even though not all the technical work is equally challenging. FireSAT's project manager should remember the conservative tendencies of all office leaders and provide enough resources for this comprehensive analysis. Risk management (Chapter 8) converges the scope and resource estimates, reducing the scope whenever necessary to keep the project's risk at acceptable levels.

For FireSAT, we assume we can reasonably develop the satellite using any of several satellite buses. This assumption greatly simplifies the integrated systems analysis and decreases the amount of work that engineering must do in reviewing its reports for consistency with the integrated systems analysis plan. But FireSAT's payload includes a sensor that may need cryogenic cooling, which none of the potential satellite buses has used. Therefore, we probably need full independent system analysis of the payload's cryogenic cooling system, including spacecraft thermal analysis, to be sure its thermal environment allows the satellite bus to work properly. If we don't include this detail in the technical work package for integrated system analysis, the resources needed to complete it won't be in the technical resource estimate for the appropriate work breakdown structure.

Another example involves the failure of NASA's Lewis spacecraft. The design for its attitude control system (ACS) derived from a heritage design for the proven Total Ozone Mapping Spacecraft. But Lewis's ACS design differed from the heritage design in subtle yet ultimately significant ways. The ACS analysis, based on the heritage spacecraft's tools and models, inaccurately predicted the control system's performance. Following a successful launch and orbital injection, the satellite entered a flat spin that caused a loss of solar power, fatal battery discharge, and ultimately mission failure [Thomas, 2007].

13.4   Prepare the Systems Engineering Management Plan and Other Technical Plans

Documented technical plans offer context for completing the technical work in the requirement assignment matrix, integrated master plan, and other technical planning artifacts described in this chapter. They state the methods, assumptions, and ground rules that underpin mutually consistent stakeholder expectations, technical work plans, budgets, and schedules.

13.4.1   Define the Discrete Technical Plans

The systems engineering management plan (SEMP) is the chief technical plan; it integrates subordinate plans, such as the software management plan and the master verification plan. The number of distinct subordinate plans depends on the project's scale and acquisition strategy. For example, NASA's Constellation program is very large. At the system requirements review, a key technical event, teams had developed 22 distinct technical plans, plus the SEMP. Although this number may appear unwieldy, the program's scale and duration suggest that 23 plans aren't too many. Constellation encompasses a launch vehicle project and spacecraft project to transport astronauts to the International Space Station and will include other launch vehicles, spacecraft, and habitats for lunar and Mars exploration through the 2020s. In contrast, the FireSAT program is much smaller in scale, requiring only 13 technical plans, including the SEMP. Table 13-7 lists these plans, and the chapters that address each one.

The number of technical plans for a project is a matter of style and preference. For example, we could easily merge the 23 technical plans for the Constellation Program into a single document, though a very thick one. Whether we separately document an aspect such as technical integration planning or include it in the SEMP isn't as important as doing the technical planning and doing it well.

13.4.2   Prepare the Systems Engineering Management Plan

The systems engineering management plan (SEMP) documents and communicates a project's technical approach, resources, and key technical tasks, activities, and events with their metrics and success criteria. It describes the technical effort to the technical team, managers, customers, and other stakeholders. The technical planning work products developed in this chapter—Section 13.3 in particular—are the basis for developing this plan. Chapter 14 fully describes how to develop, maintain, and apply the SEMP.

13.4.3   Prepare the Software Management Plan

Although most systems engineers emphatically assert that systems engineering encompasses highly interdependent hardware and software, most agree that we must develop these elements separately. A lot of research, such as that for model-based engineering methods, has resulted in smoother concurrent development and integration of system hardware and software [Chesley et al., 2008]. But virtually all projects with significant developmental content require a standalone software management plan. As Pfarr and Obenschain [2008] point out, this plan usually describes three main areas:

- System interfaces, including requirements, schedules, deliverables, acceptance criteria, operator training, customer-supplied elements, and maintenance after delivery

- How to manage software, including the work breakdown structure, organization and roles of the team lead and members, and project monitoring and control processes

TABLE 13-7.   Key Technical Plans for FireSAT. Entries briefly describe 13 discrete technical plans for the FireSAT program and indicate the chapters that discuss them. Some plans contain information that could split out to form a separate plan (e.g., the supportability plan includes the plan for training), whereas other documents might combine into one. (MOE is measure of effectiveness; MOP is measure of performance; TPM is technical performance measure.)

FireSAT Technical Plans	Relevant to
Electromagnetic Compatibility/Interference Control Plan—Describes design procedures and techniques to assure electromagnetic compatibility for subsystems and equipment. Includes management processes, design, analysis, and developmental testing.	Chap. 5
Mass Properties Control Plan—Describes the management procedures to be used for mass properties control and verification during the various development cycle phases.	Chap. 5
Risk Management Plan—Summarizes the risk management approach for the project, including actions to mitigate risk and program de-scope plans. Includes a technology development plan that may be split off as a stand-alone plan for technology-intensive system developments.	Chap. 8
Manufacturing and Assembly Plan—Describes the manufacturing strategy, facility requirements, organizational implementation, quality assurance, critical assembly steps, development of the master parts list, and manufacturing data collection, including problem reporting and corrective actions (PRACA).	Chap. 9
Master Verification Plan—Describes the approach to verification and validation for the assurance of project success. Addresses requirements for hardware and software verification and validation.	Chap. 11
Supportability Plan—Includes initial establishment of logistics requirements, implementation of supportability analysis activities, and requirements development for supportability validation. Typically integrates the reliability and maintainability plans. Also includes the training plan.	Chap. 12
Activation and Check-out Plan—Provides information needed for an engineering check-out of the system to determine readiness for full mission operational status. Information includes activation requirements, engineering test plans, and data evaluation tasks.	Chap. 12
Spacecraft Systems Analysis Plan—Approach to and phasing of integrated engineering analysis of functional/logical and physical system designs, decomposition of functional requirements and allocation of performance requirements, assessments of system effectiveness (MOEs, MOPs, and TPMs), decision support, and managing risk factors and technical margins throughout the systems engineering effort.	Chap. 13
Software Management Plan—Defines software management processes for the technical team, and describes responsibilities, standards, procedures and organizational relationships for all software activities.	Chap. 13
Safety and Mission Assurance Plan—Addresses the activities and steps to ensure mission success and the safety of the general public, the project workforce, and high value equipment and property used to develop and deploy the system.	Chap. 13

TABLE 13-7.   Key Technical Plans for FireSAT. (Continued) Entries briefly describe 13 discrete technical plans for the FireSAT program and indicate the chapters that discuss them. Some plans contain information that could split out to form a separate plan (e.g., the supportability plan includes the plan for training), whereas other documents might combine into one. (MOE is measure of effectiveness; MOP is measure of performance; TPM is technical performance measure.)

FireSAT Technical Plans	Relevant to
Systems Engineering Management Plan—The chief technical plan, which describes the project's technical effort and provides the integration framework for all subordinate technical plans.	Chap. 14
Configuration Management Plan—Describes the structure of the configuration management organization and tools. This plan identifies the methods and procedures for configuration identification, configuration control, interface management, configuration traceability, configuration audits, and configuration status accounting and communications. It also describes how supplier configuration management processes will be integrated with those of the acquiring organization.	Chap. 16
Data Management Plan—Addresses the data that the project team must capture as well as its availability. It includes plans for data rights and services, addressing issues that often require tradeoffs between the interests of various communities (e.g., acquirer versus supplier, performing versus managing, etc.).	Chap. 17

- The kind of technology to use, including the development process, the acquisition strategy, the verification and validation process, and product assurance

The software management plan is analogous to the SEMP in that it serves as the chief planning document for software. It often includes the planning to manage software configuration, risk, and security, although it may reference standalone documents for these topics depending on the project's size and complexity. Chapter 19 includes a table of contents for FireSAT's software management plan.

13.4.4   Prepare the Safety and Mission Assurance Plan

The safety and mission assurance (SMA) plan addresses the SMA functions and activities over the project's lifecycle. It documents the roles, responsibilities, and relationships in a mission assurance process map and matrix that's unique to the project, with appropriate support and guidance from the functional SMA organization.

This plan addresses such areas as procurement, management, design and engineering, design verification and test, software design, software verification and test, manufacturing, manufacturing verification and test, operations, and pre-flight verification and test. It includes such critical disciplines as

- Safety
- Quality assurance
- Compliance verification
- Audit, safety, and mission assurance reviews
- Safety and mission assurance processes
- Software safety and quality assurance

It also describes how the project will develop and manage closed-loop problem reporting and resolution—mainly while manufacturing, assembling, and integrating the space system. This approach defines a data collection system and process for reporting, analyzing, and correcting hardware and software problems or anomalies.

13.5   Get Stakeholder Commitments to Technical Plans

A stakeholder is someone interested in or affected by the system of interest. Section 2.2 identifies stakeholders as active or passive, with sponsors as a special subset of each.

To describe the stakeholders' commitment to technical plans, we must distinguish those for technical planning from those for the system. Chapter 2 addresses the latter; here, we describe the former. Stakeholders with roles in the project, as described in Section 13.3.3, are a good starting point for recognizing technical planning stakeholders. Organizations flagged with roles of responsible, accountable, or consulted tend to be active, whereas those flagged as informed tend to be passive.

Active stakeholders. *Active stakeholders* in technical plans are organizations that interact with those plans. They usually have to

- See that teams meet or manage to a plan

- Do the work within a plan's resources and timeline

- Materially contribute to or consume outputs produced by following the technical plan

- Participate in technical planning by developing the technical planning work products we describe in Section 13.3.

If an active stakeholder has a participating organizational role, a representative from that organization must coordinate the role. Otherwise we'll get a rude awakening when the project seeks approval for the technical work directives. Responsible managers, as active stakeholders, will refuse to approve the technical work directive pending concurrence from the right people (also active stakeholders) within their organization. So technical planning teams must secure the active stakeholders' commitment by engaging them while developing technical plans and getting their approval for technical planning documents and technical work directives.

Passive stakeholders. *Passive stakeholders* in technical plans, such as policy-making organizations, usually focus on **how** the technical work gets done. For instance, there's usually a standard procedure for calculating a spacecraft's mass properties, determining the mass margin, and managing the margin over time. A project's mass properties control plan often includes some provisions that deviate from the standard procedure. These deviations result from a project's peculiarities, such as the amount of commercial off-the-shelf hardware it includes, its technology risk, and so on—typically representing the most prudent, workable technical approach. Still, the organization that holds the standard procedure will have to review this plan and concur with it. Otherwise, the plan may still be approved and applied, but organizational controls such as ISO audits will eventually bring to light the disparity between policy and plan. Whereas active stakeholders must participate in developing the technical planning work products, passive stakeholders need only participate in reviewing and approving them.

Sponsors. The sponsor for a technical planning work product is usually the customer. Customers are mainly interested in a product that matches their expectations, on schedule, and within budget. Therefore, technical planners must manage their expectations from the beginning of a project through development and delivery of the product. Planners get a customer's commitment through mutual agreement about the deliverable item, schedule, and resources—in a formal acquisition, with a signed contract. But operational concepts and top-level requirements for space systems evolve as the system design matures, so we must maintain a customer's commitment throughout the development. Likewise, customers must regularly review schedule and budget status to concur with corrections to actual or anticipated deviations.

For FireSAT, the customer is the US Forest Service's firefighting office, so an interagency agreement between NASA and the Forest Service should detail how the customer will participate. To establish clear mutual expectations, the project manager must work closely with the customer and get approval for the concept of operations and systems requirements document. The customer should also help develop and approve the master schedule and budget.

Project managers should discuss with the customer the summaries of results from the design and analysis cycle to show how FireSAT's evolving design is expected to meet system requirements. Managers also must engage the customer in resolving technical issues that could affect system requirements, as well as in managing risk, so the customer knows about events that might change expectations. Customers also expect periodic status reports on the schedule and budget.

The active stakeholders for FireSAT include NASA's Goddard Space Flight Center, the National Oceanic and Atmospheric Administration (NOAA), and the prime contractor. The contract between NASA and the prime contractor commits both to the project. Formal interagency agreements secure mutual commitment from Goddard and NOAA to their roles and responsibilities. The Federal Communications Commission is a passive stakeholder because their regulations control how the FireSAT satellite uses the radio frequency spectrum to communicate with its ground systems. Because commands from the satellite to the ground will be encrypted, the National Security Agency is also a passive stakeholder.

Finally, organizations build roles and behaviors as they learn from and react to their successes and failures, as technology advances change processes and procedures, and as people come and go. Passive stakeholders for technical planning especially emerge and fade away over time. Changes in active stakeholders are less common, but stakeholders sometimes switch from active to passive roles. Therefore, we must continually confirm a stakeholder's commitment, not just assume we can "check the box" and be done.

13.6   Execute the Technical Plan

Eventually, the time comes to stop planning and begin doing. But space systems developments are characterized by their dynamism: opportunities for re-planning are so frequent that technical planning continues throughout. In this section we look at how we apply discrete technical plans and begin reporting progress. Chapter 14 of *Applied Project Management for Space Systems* [Chesley et al., 2008] affords an in-depth discussion of evaluation and monitoring from the project management point of view.

13.6.1   Issue Authorized Technical Work Directives

A technical work directive is organizational approval to start that work with the intent of delivering the product on schedule and on budget. Supplying and acquiring organizations must formally approve the work and jointly issue the directive.

A contract between an acquiring organization and a prime contractor is analogous to a technical work directive. But the latter commonly authorizes doing work and accruing costs within a given organization—for example, between a project office and a functional matrix organization. Issuing technical work directives means we've finished technical planning and are now realizing the work products. Thus, work directives include the work package descriptions as developed in Section 13.3.5 and cost estimates such as those in Section 13.3.6. The directive template used by NASA's Marshall Space Flight Center (see Appendix at the end of this chapter) is fairly typical of the sort used in large organizations. Information on the work package and resources fills out several fields in this form. The other fields include such items as

- Labor codes that require a visit to the organization's business office
- Names of people in the requesting and providing organizations that will be points of contact for the technical work
- Names of approving officials for the requesting and providing organizations

In general, the approving official's level depends on the resources associated with a technical work directive. Organizations (even small ones) have policies on who can approve technical work directives, so we must know the policies well in advance to keep stakeholders engaged. Also, changes to the work directive may occur at any time before approval to capture revisions to the technical work description, schedule, and resources. After approval, revisions require formal changes to the directive.

In the FireSAT example, the Applied Engineering and Technology Directorate (AETD) and Safety and Mission Assurance Directorate (SMA) at Goddard receive technical work directives to perform the work defined in their respective work packages. Goddard determines the approving officials, e.g., FireSAT's project manager and the directors of AETD and SMA.

13.6.2   Report Technical Status

After technical work begins, requesters and providers should track its progress. The project office establishes a periodic (often monthly) status review of work done versus work planned. Teams analyze variances and correct them when appropriate. More rigorous or formal reviews include key technical events, such as the system readiness or preliminary design reviews (Chapter 18) and other organizational management reviews.

Functional organizations, such as engineering, commonly use periodic status reviews to track progress of the technical work within their organization. They're usually monthly, but may take place more often for especially challenging technical work or as a key technical event approaches. These reviews tend to focus more on the technical issues and risks associated with completing the work on schedule. They consider resource issues only by exception, mainly when they need more resources to resolve a technical problem.

Technical status reviews differ from the project status reviews in that their purpose is to surface and **solve** technical problems before they lead to cost and schedule variances. In general, the leader of a functional organization is the expert in that discipline, and these status meetings make this expertise available if needed. A "quad chart" of the sort depicted in Table 13-8 commonly summarizes status, with backup material as necessary to elaborate on technical issues or risks. Astute functional managers also glean a lot of information from what's not on the quad chart. For example, sketchily planned accomplishments or a dearth of schedule milestones for the next status point may indicate a poorly conceived plan or one overcome by events—worthy of attention in either case.

TABLE 13-8.   Quad Chart for Technical Status. This type of chart helps functional managers stay on top of progress and problems with the technical work.

Project or task title Name or organization code	
Accomplishments • List accomplishments since previous status review	Issues and concerns • List any technical issues and concerns and provide additional information backup charts as necessary
Plans • Summarize planned accomplishments for next status review	Schedule milestones • List the key schedule milestones for the technical work, such as completion of draft documents, initiation of analyses, receipt of data, etc.

13.7   Document and Iterate

Chapter 17 describes technical data management, which incorporates technical planning work products. As we describe in Section 13.2.3, technical integration planning identifies the secondary data products that organizations must send one another to perform their technical work. In turn, secondary products influence the technical data architecture. Technical planners define work packages, which they incorporate into technical directives, and then archive both products. They should also archive the distinct technical plans, though most of these plans come from other systems engineering processes.

Let's consider the FireSAT case for illustration. The work package definitions, technical work directives, prime contract, and interagency agreements are technical planning work products that we must archive. Of the thirteen plans, only the SEMP and the plans for system analysis, software management, and safety and mission assurance are outputs of technical planning and available for archiving. Other systems engineering processes generate the rest—for instance, the configuration management plan results from the configuration management process (Chapter 16).

Archiving the work products, including revisions and updates throughout the system's development cycle, is vital to growing the technical planning knowledge base. Space systems developments are dynamic, so technical planning continues throughout system development. Thus, NPR 7123.1a calls for updating the SEMP as an entrance criterion for each key technical milestone review. Diligent archiving blazes a trail for updates, revisions, and future system developments.

Summary

This chapter has described a technical planning process for space systems. Any plan includes **what** to do, **who** will do what, **when** the work must be complete, and **how many** resources will be available to do it. Technical management focuses on the plans in action, comparing work done against the work planned. Technical planning and management correct any divergences between the two and update the work plan accordingly. In describing how to develop the plan for completing the technical work in a system development, this chapter has addressed the following points:

- A historical survey of analogous system developments produces a skeleton technical plan, including a preliminary list of technical artifacts and the time phasing of their production during the lifecycle
- The system of interest, its acquisition strategy, and systems engineering guides such as NPR 7123.1a help us refine the list of technical artifacts that make up the technical work
- Technical baselines for space systems evolve through a project's lifecycle in seven discrete steps that correspond to the development phases
- Technical integration planning focuses on identifying interdependencies between space systems engineering disciplines and underpins mutually consistent development of the technical artifacts
- The work breakdown structure (WBS) is a foundation for all project activities, including planning and organizing the project and technical work, defining the schedule, and estimating costs
- The integrated master plan consists of key project milestones, and the integrated master schedule details individual tasks needed to complete this plan
- Work package descriptions must be consistent with the WBS and must be necessary and sufficient to characterize the technical work
- The systems engineering management plan (SEMP) documents and communicates the technical approach; needed resources; and key technical tasks, activities, and events (with their metrics and success criteria). One of the SEMP's functions is to integrate subordinate technical plans.

- Technical work directives represent the end of technical planning and show how to apply the technical plan as defined in the technical planning work products

- Technical status reviews focus mostly on the technical issues and risks associated with completing the work on schedule, typically dealing with resource issues mainly when resolving a technical issue will require more resources

The technical planning methods described in this chapter have evolved and been adapted in response to lessons learned from glamorous successes and glaring failures. If planners use them pragmatically, they usually develop a credible technical plan that responds to the project team's needs.

References

Blair, James C., Robert S. Ryan, Luke A. Schutzenhofer, and William R. Humphries. 2001. Launch Vehicle Design Process: Characterization, Technical Integration, and Lessons Learned. NASA/TP-2001-210992, NASA Marshall Space Flight Center.

Blanchard, Benjamin S. 2004. *System Engineering Management,* 3rd edition. Hoboken, New Jersey: Wiley.

Brown, Gary L. 1996. Design Analysis Cycle Application to the International Space Station Design, in *Proceedings of the 6th Annual International Symposium of the International Council on Systems Engineering (INCOSE)*. Boston, MA.

Chesley, Bruce, Erik Daehler, Michael Mott, and L. Dale Thomas. 2007. Model Driven Systems Development for Space Systems, in *Proceedings of the 58th International Astronautical Congress* (ref. IAC-07-D1.3.04). Hyderabad, India.

Chesley, Julie, Wiley J. Larson, Marilyn McQuade, and Robert J. Menrad. 2008. *Applied Project Management for Space Systems.* New York, NY: McGraw-Hill Companies.

Cooper, Robert G. 1990. "Stage-Gate Systems: A New Tool for Managing New Products." *Business Horizons*, Volume 33, Issue 3, pp. 44–54.

Department of Defense (DoD). January 2001. *Systems Engineering Fundamentals.* Fort Belvoir, VA: Defense Acquisition University Press.

DoD. August 1998. *DoD Integrated Product and Process Development Handbook*, Office of the Under Secretary of Defense for Acquisition and Technology. Washington, D.C.: Government Printing Office.

DoD. October 2005. *Integrated Master Plan and Integrated Master Schedule Preparation and Use Guide*, Ver. 0.9. Washington, D.C.: Government Printing Office

Grady, Jeffrey O. 1994. *System Integration.* Boca Raton, Florida: CRC Press.

Griffin, Michael D., and James R. French. 2004. *Space Vehicle Design*, 2nd ed. Reston, VA: American Institute of Aeronautics and Astronautics.

Hillier, Frederick S., and Gerald J. Lieberman. 1995. *Introduction to Operations Research,* 6th edition. New York: McGraw Hill.

Humphries, William R., Wayne Holland, and R. Bishop. 1999. Information Flow in the Launch Vehicle Design/Analysis Process. NASA/TM-1999-209887, Marshall Space Flight Center (MSFC), Alabama.

Institute of Electrical and Electronics Engineers (IEEE). 2005. *IEEE Standard for Application and Management of the Systems Engineering Process*, IEEE Standard 1220–2005.

International Organization for Standardization (ISO). 2002. *Systems Engineering—System Life Cycle Processes*, ISO/IEC 15288.

Kerzner, Harold. 1998. *Project Management: A Systems Approach to Planning, Scheduling, and Controlling*, 6th ed. Hoboken, New Jersey: Wiley.

Monk, Gregg B. 2002. Integrated Product Team Effectiveness in the Department of Defense. Master's Thesis, Naval Postgraduate School.

NASA. March 2007. *Systems Engineering Processes and Requirements*. NPR 7123.1a. Washington, DC: NASA.

NASA. 2006 (1). *Constellation Program System Engineering Management Plan*, CxP 70013. NASA Johnson Space Center (JSC), Texas: Constellation Program Management Office.

NASA. 2006 (2). *Constellation Program System Requirements Review (SRR) Process Plan Annex 2.1 Crew Launch Vehicle*, CxP 70006-ANX2.1. MSFC, Alabama: Exploration Launch Office.

NASA. 2006 (3). *Constellation Program System Integrated Analysis Plan (SIAP) Volume 1*, CxP 70009. JSC, Texas: Constellation Program Management Office.

NASA. 2006 (4). *Exploration Launch Project Systems Analysis Plan (SAP)*, CxP 72024. MSFC, Alabama: Exploration Launch Office.

Pfarr, Barbara, and Rick Obenschain. 2008. *Applied Project Management for Space Systems*. Chapter 16, Mission Software Processes and Considerations. New York: McGraw Hill.

Project Management Institute (PMI). 2004. *A Guide to the Project Management Body of Knowledge*, 3rd ed. Newton Square, Pennsylvania: Project Management Institute.

Schaefer, William. 2008. Non-published presentation entitled "ISS Hardware/Software Technical Planning," NASA Johnson Space Center, 2008.

Thomas, L. Dale. 2007. "Selected Systems Engineering Process Deficiencies and Their Consequences," *Acta Astronautica* 61:406-415.

TECHNICAL TASK AGREEMENT (TTA)			1. TTA NUMBER	2. DATE	3. PAGE: 1 OF ___
TASK INFORMATION					
4. TASK TITLE			5. PROJECT PROGRAM		
6. POINT OF CONTACT (REQUESTING)	7. CENTER	8. MAIL CODE	9. EMAIL		10. PHONE NUMBER
11. POINT OF CONTACT (PROVIDING)	12. CENTER	13. MAIL CODE	14. EMAIL		15. PHONE NUMBER
Business POC					
16. PARTICIPATING ORGANIZATION(S)					
17. TASK DESCRIPTION (include specification descriptions and cite references where appropriate)					

19. DATA REQUIREMENTS:	20. GOVERNMENT FURNISHED DATA ITEM NO.	21. DELIVERY DATE
22. DELIVERABLE ITEMS:	21. GOVERNMENT FURNISHED EQUIPMENT ITEM NO.	24. DELIVERY DATE

MSFC Form4421 (September 2005) – Revised for EV

APPENDIX A-1.   Technical Work Directive for NASA's Marshall Space Flight Center (page 1 of 4).

TECHNICAL TASK AGREEMENT (TTA)			25. TTA NUMBER		26. DATE		27. PAGE 2 OF ___	
28. IMPACTS BY UPN & CENTERENTER	FY06	FY07	FY08	FY09	FY10	FY11	FY12	TOTAL
PROCUREMENTS	$K	$K	$K	$K	$K	$K	$K	$K
								0
29. IMPACTS BY UPN & CENTER	FY06	FY07	FY08	FY09	FY10	FY11	FY12	TOTAL
NON-PROCUREMENTS$	$K	$K	$K	$K	$K	$K	$K	$K
TOTAL	0.000	0.000	0.000	0.000	0.000	0.000	0.000	0.000
CS SALARY								0.000
CS TRAVEL								0.000
FACILITY UTILIZATION								0.000
SERVICE POOL								0.000
CENTER G&A								0.000

30. WORKFORCE BY	FY06		FY07		FY08		FY09		FY10		FY11		FY12		TOTAL	
UPN & CENTER	FTE	WYE	FTE	WYE	FTE	WYE	FTE	WYE	FTE	WYE	FTE	WYE	FTE	WYE	FTE	WYE
TOTAL	0	0	0	0	0	0	0	0	0	0	0	0	0	0	0	0
															0	0
															0	0
															0	0
															0	0

SCHEDULE ACTIVITIES/MILESTONES
31. CURRENT YEAR (ACTIVITIES/MILESTONES)
32. OUT YEARS (ACTIVITIES/MILESTONES ES):

APPROVAL SIGNATURES	*(Budget Signature From Each Organization REQUIRED)*	
REQUESTER		
33. TECHNICAL MANAGER/LEAD:	34. SIGNATURE TECHNICAL MANAGER/LEAD:	35. DATE:
36. BUSINESS MANAGER/LEAD:	37. SIGNATURE BUSINESS MANAGER/LEAD:	38. DATE:
39. DEPARTMENT MANAGER:	40. SIGNATURE DEPARTMENT MANAGER:	41. DATE:
42. PROJECT ELEMENT LEAD:	43. SIGNATURE PROJECT ELEMENT MANAGER:	44. DATE:
PROVIDER		
45. PROJECT MANAGER/LEAD:	46. SIGNATURE PROJECT MANAGER/LEAD:	47. DATE:
48. BUSINESS MANAGER/LEAD:	49. SIGNATURE BUSINESS MANAGER/LEAD:	50. DATE:

MSFC Form 4421 (September 2005) – Revised for EV

APPENDIX A-2. Technical Work Directive for Marshall Space Flight Center (page 2 of 4).

TECHNICAL TASK AGREEMENT (TTA)	44. TTA NUMBER	45. DATE	46. PAGE 3 OF ___
CONTINUATION BLOCK			

47. Type Block Number of Item to be continued followed by continuation information.

APPENDIX A-3.   Technical Work Directive for Marshall Space Flight Center (page 3 of 4).

TECHNICAL TASK AGREEMENT (TTA)		
INSTRUCTIONS FOR COMPLETION OF THE TTA FORM		
1	Task Title	Provide a brief title of the work to be performed.
2	TTA Number	Number assigned by the requesting organization.
3	Revision Number	Revision reference number assigned by the requesting organization.
4	Project	Program project for which the work is being performed.
5	Project WBS/UPN	Unique project number assigned by the requesting organization (code used to track resources expended).
6	*Labor Code*	Code used to track labor hours expended.
7	Requesting Center	Center where the work is being performed.
8	Requesting Organization	The center organization that has responsibility for determining requirements for the agreement and is the recipient of product or services from the provider.
9	Point of Contact	The point of contact for requesting organization.
10	Organization Code	Organization code where the TTA work is initiated.
11	Telephone Number	Telephone number of point of contact for the requesting organization.
12	Providing Organization	The center organization that has responsibility for providing products or services to the requester.
13	Point of Contact	The point of contact for the providing organization.
14	Organization Code	Organization code for providing organization.
15	Telephone Number	Telephone number for the provider point of contact.
16	Participating Organizations	Organizations other than the requester or the provider involved in the TTA.
17	Task Description	Brief description of the project and/or service being provided.
18	Reporting Requirements	List all reports to be delivered by the provider. Identify the frequency, delivery date, and specific requirements. If a specific format is required, a copy should be attached to the TTA.
19	Data Requirements	Identify all data requirements to be delivered by the provider.
20	Government Furnished Data Item Number	If applicable, provide an identification number for each data requirement that is also a government-furnished data (GFD) item.
21	Delivery Date	Date established by the contract containing the GFD item number for delivery by the government to the contractor.
22	Delivery Items	Specify products or services to be provided to the requester by the provider.
23	Government Furnished Equipment Item Number	If applicable, provide an identification number for each deliverable item that is also a government-furnished equipment (GFE) item.
24	Delivery Date	Date established by the contract containing the GFE item number for delivery by the government to the contractor.
25	Cost ($K) by Project	Cost ($K) by project reported by fiscal year.
26	Workforce by Project	Workforce by project reported by fiscal year.
27	Current Year	Identify specific milestones and activity schedules as determined by the requester and provider for the current year.
28	Out Year(s)	Identify specific milestones and activity schedules as determined by the requester and provider for the out year(s).

APPENDIX A-4. Technical Work Directive for Marshall Space Flight Center (page 4 of 4).

TECHNICAL TASK AGREEMENT (TTA)		
INSTRUCTIONS FOR COMPLETION OF THE TTA FORM		
29	Continuation Block	If more space is needed to provide sufficient details for any item, type Block number of item to be continued followed by continuation information.
	Requester Signature	The requester organization should enter the title of persons whose signature signifies acceptance of the TTA. Note that a "budget" signature is required for both organizations.
	Provider Signature	The provider organization should enter the title of the person whose signature signifies acceptance of the TTA and who has full signature authority for authorization of resources. If multiple organizations are used, more than one signature is required. Note that a "budget" signature is required for both organizations.

APPENDIX A-4. Technical Work Directive for Marshall Space Flight Center (page 4 of 4). (Continued)

Chapter 14

Technical Direction and Management: The Systems Engineering Management Plan (SEMP) in Action

Peter A. Swan, *Teaching Science and Technology, Inc.*
Eric C. Honour, *Honourcode, Inc.*

The primary purpose of the systems engineering management plan (SEMP) is to enable the technical team to understand, develop, and perform the activities necessary to acquire a complex space system. We should write this living document early (before the system requirements review in most cases), tailor it for the project, update it before all major reviews, and see that it reaches across the management and engineering disciplines.

14.1 Introduction to the Systems Engineering Management Plan

The SEMP must be flexible and complement all other plans over the project lifecycle. It provides the direction for engineering teams; specifies the technical efforts; describes the technical processes to use; and indicates the organization, schedule, and cost of the engineering efforts. Collaborative planning contributes to a better project in four major ways, as shown in Figure 14-1.

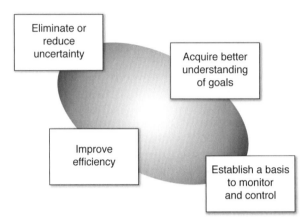

Figure 14-1.   Reasons for Planning. Planning contributes to a better project in four major ways. The SEMP planning team consists of the chief engineer, the lead systems engineer, and other members of the systems engineering team. *(Courtesy of Honourcode, Inc.)*

The SEMP designates the roles and responsibilities of the space systems engineering efforts as well as those of others on the technical team. It must explain communication activities, including at least vertical and horizontal integration, team communication, scope of decision making authority, and systems engineering staffing (when-which domain emphasis). Positive team-based communication is critical to success.

14.1.1   Purpose, Goals, and Objectives

The SEMP is multifaceted; it must support the engineering effort and also interface with all program management activities. It should be a single, integrated planning document covering the technical efforts throughout the lifecycle. The SEMP is the blueprint for planning and execution, including guidance for the contract, management, and control of the technical aspects. In particular, it:

- Specifies the technical efforts
- Gives the selection and description of the technical processes
- Describes the tasks to be done
- Defines the resources necessary (people, organization, money)
- Lays out the technical schedule with all its external interfaces
- Enables communication between engineering and management
- Establishes entry and exit criteria for each program phase

The SEMP emphasizes different efforts during each phase of the space program acquisition. An example of this is in Table 14-1, which lists the technical activities during the early phases of concept development and the systems level testing phase.

TABLE 14-1. **Systems Engineering Management Plan Activities Versus Program Phase.** The SEMP is key to understanding each phase of the program.

Early Phase of Concept Development	System Level Testing Phase
• Scope technical activities through projected lifecycle • Involve technical team in request for proposal (RFP) preparation • Involve technical team in source selection • Provide a preliminary list of standards	• Establish priority for testing • Establish time sequence for tests • Develop success criteria for each test • Define incoming configuration for each test

14.1.2 Timelines

All experienced space systems engineers know that early in a major acquisition program the master schedule dominates their lives. So the systems engineering team should emphasize an achievable technical schedule early. The integrated master plan should be compatible with the initial acquisition schedule, but not so rigid that the work can't be done in the allotted time. Chapter 13 discusses schedule management. Figure 14-2 shows an example of a schedule for the Constellation project.

14.2 Determine the Contents of the SEMP

Figure 14-3 shows the outline of the SEMP. This chapter describes what should and should not be in a space systems SEMP, which is the primary document for planning technical processes on a project.

14.2.1 General Structure

The general structure describes the major systems engineering processes that we need for a space systems acquisition. It allows the technical leadership team to explain how they intend to develop, launch, and operate the space system. For this chapter, the SEMP structure corresponds to NASA Procedural Requirement (NPR) 7123.1a, which lists 17 systems engineering processes.

14.2.2 Expanded Outline

Purpose and Scope. The SEMP describes the elements of the technical approach and helps the team answer the challenging questions that are always present. Developing the SEMP early enables the management and systems engineering teams to answer some of these hard questions, such as the ones in Table 14-2, which seem daunting at the early stages of a major program.

Applicable documents. This section is usually "copied and pasted" from previous SEMP documents in the same organization. If so, the lead systems engineer must make sure that each document is needed for the project because

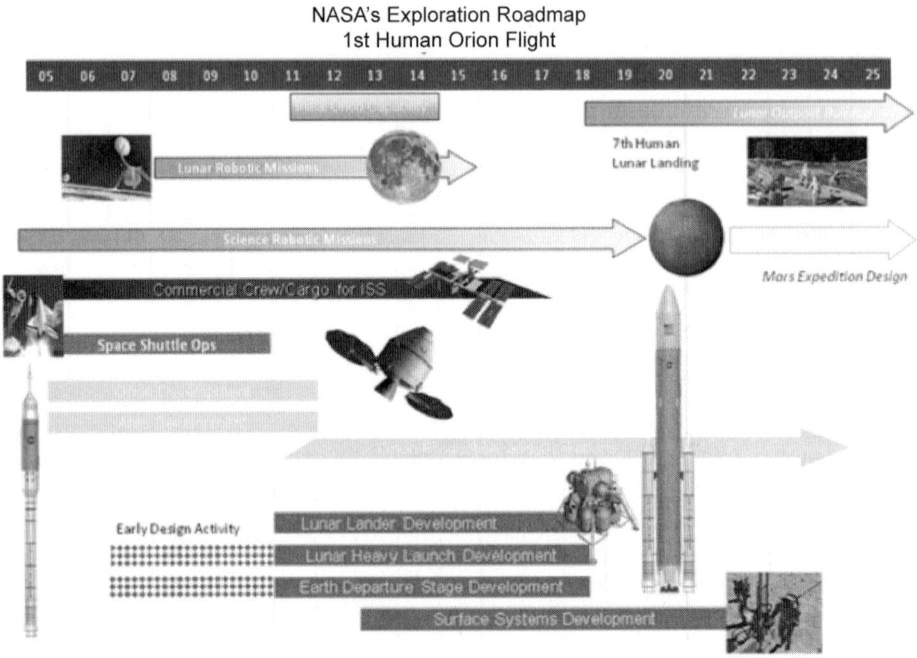

FIGURE 14-2.   **Constellation Sequence.** Early presentation of the program schedule enables
systems engineers to understand their long-term involvement.

everyone (government, contractor, subcontractor, parts suppliers, and operators)
follows this guidance. Nonessential documentation requirements slow the
acquisition, so we must tailor the list carefully.

Technical summary. This portion of the SEMP relies on the available data.
Before the request for proposal (RFP), the concept is soft; after launch, the technical
description is frozen. At the beginning, the technical summary should be a concise
statement of the problem we need to solve. One example is the IRIDIUM goal:
"Communications—anywhere—anytime—anyone!" As the system develops over
time, the systems engineers need to update the SEMP so that everyone on the
project views the design characteristics the same way. Once we establish the
design, configuration management controls the changes so that data needed to
update the SEMP is available before all major milestones.

System description. The system description should be at a high enough level
to depict all the interfaces while also showing the organization and technical
activities. An excellent method of describing the system is an architectural
systems-level picture (such as Figures 14-4 and 14-5) supported by a WBS. The lead
systems engineer must see that the systems description data is available for
everyone on the project, to permit a consistent approach to a solution. Flexibility
pays dividends early in the design. But freezing the design has a place and time
and we must be firm about it.

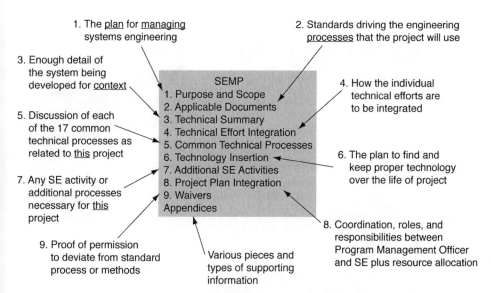

1. The <u>plan</u> for <u>managing</u> systems engineering

2. Standards driving the engineering <u>processes</u> that the project will use

3. Enough detail of the system being developed for <u>context</u>

4. How the individual technical efforts are to be integrated

5. Discussion of each of the 17 common technical processes as related to <u>this</u> project

6. The plan to find and keep proper technology over the life of project

7. Any SE activity or additional processes necessary for <u>this</u> project

8. Coordination, roles, and responsibilities between Program Management Officer and SE plus resource allocation

9. Proof of permission to deviate from standard process or methods

Various pieces and types of supporting information

SEMP
1. Purpose and Scope
2. Applicable Documents
3. Technical Summary
4. Technical Effort Integration
5. Common Technical Processes
6. Technology Insertion
7. Additional SE Activities
8. Project Plan Integration
9. Waivers
Appendices

FIGURE 14-3.   Systems Engineering Management Plan (SEMP) Outline. The SEMP is the primary document for planning technical processes on a project.

TABLE 14-2.   Programmatic Elements and Questions. Early assessment of key "daunting questions" lowers risk. *Note:* the items in the right column don't correspond one-to-one with those in the left column.

Elements of the Technical Approach	Daunting Questions
• Describe the space systems technical baseline • Identify systems engineering processes • Identify resources required (funding, organization, facilities) • Identify key technical tasks and activities • Identify success criteria for each major task • Schedule the technical development efforts • Identify reference documents • Identify specialty engineering tasks	• What are the technical issues and risks? • Who has responsibility and authority for executing the technical project? • What processes and tools should we use? • Which processes must be managed and controlled? • How does the technical effort link to outside forces? • How many test programs must the project have? • What are the major interfaces within each segment of the project? • Where will all the expertise come from?

System structure. This section describes the work breakdown structure (WBS) so that we see all technical efforts and recognize all interfaces. It gives the products of the system structure along with the specification tree and drawing tree. It also describes the interface specifications as the WBS is prepared to ensure that the design team covers them.

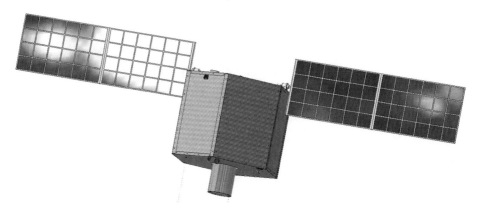

FIGURE 14-4.   Early FireSAT Systems Image [TSTI, 2006]. An early estimate of shape enhances the team's understanding of the problems.

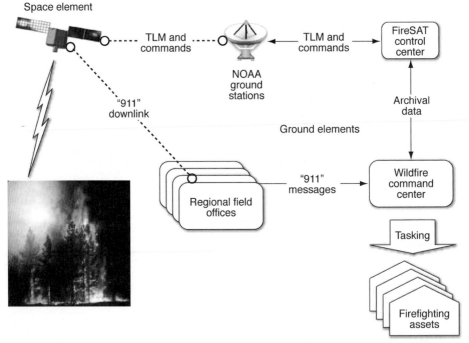

FIGURE 14-5.   FireSAT Concept of Operations. A high-level concept of operations provides the context within which the system will function and depicts system-level interfaces. (TLM is telemetry; NOAA is National Oceanic and Atmospheric Administration.)

Product integration. Product integration from parts, to assemblies, to subsystems, to systems, and then to systems of systems has to be shown in technical terms, organizational structures, product flow, and test schedules with validation and verification activities. This section describes the processes for program acceptance of integrated product flow.

Planning context. This section describes the lifecycle phases with entry and exit criteria and program constraints. We do this with respect to the: (1) milestone decision gates, (2) major technical reviews, (3) key intermediate events, (4) lifecycle phases, and (5) baseline product flows. But the real insight that the lead systems engineer contributes is through the technical inputs to the master schedule.

Boundary of technical effort. This portion of the SEMP defines where the program manager and chief engineer have authority. The boundaries are delineated so they can execute the project or escalate an issue to the appropriate level. A clear description of the system of interest as we discuss in Section 13.1.1 helps establish the boundary of technical effort.

Cross references. We need to be aware of any non-technical references that affect the technical activities. If necessary, we must also see that the contractor uses the same set of inputs as the project progresses.

Technical effort integration. A significant challenge in a major space acquisition is the timely meshing of schedules, product flow, test and verification activities, and facility availability. Technical effort integration tends to drive schedulers and planners while demanding herculean efforts of the individual teams, costing time and money to a program if a test facility is not ready when the product is delivered, or vice versa. For the Constellation Program, the tremendous complexity across the enterprise is evident in the architecture and technical effort integration shown in Figures 14-6 and 14-7.

Responsibility and authority. This section shows how the organization enables teams to complete the project. It specifies the combination of resource allocation (program manager and chief engineer responsibilities) and skill mix for the individual teams charged with the technical activities. It includes: (1) the roles, responsibilities, and authority for each technical activity, (2) the organization or panel that serves the decision maker, (3) the approach for multidisciplinary team work, (4) the breakout of technical staff planning by discipline, expertise level, and leadership role, and (5) the schedule of technical staff training.

Contractor integration. This section matures as the program develops. The initial challenge is to define the roles for contracts inside the technical arena. A NASA-executed project integrates the consultants and product developers into a program office under a normal contract. But if the contractor is responsible for the total system, the team integration depends on the customer's (NASA's) wishes. We must develop the contracts wisely so the contractor and government do what they do best. As the project progresses, a technical memorandum defines the relationship between contractor and government. Some items to discuss early are: (1) how to handle configuration management (CM), such as who chairs the board, (2) who has authority to implement or approve changes to CM, (3) who approves the final design, (4) who performs which of the 17 processes, (5) who approves

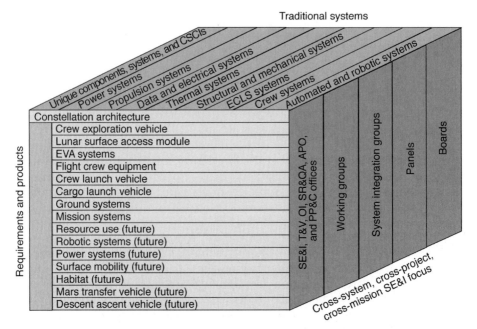

FIGURE 14-6.   Architecture Integration Model for Constellation [NASA (2), 2006]. Here we show how to divide a complex project into definable parts. (CSCI is computer system configuration item; ECLS is environmental control and life support; SE&I is systems engineering and integration; T&V is test and verification; OI is operations integration; SR&QA is safety, reliability, and quality assurance; APO is advanced projects office; PP&C is program planning and control; EVA is extravehicular activity.)

verification and validation reports, and (6) who signs off on test results. The real key is determining early who has the authority over which process (or portions of a process in case of shared responsibility).

Support integration. This portion of the SEMP describes the integrated support equipment that sustains the total effort. It includes data bases (common parts, etc.), computer design tools and manufacturing tools, planning-management information systems, and modeling and simulation setups.

Common technical processes implementation. The SEMP describes, with appropriate tailoring, each of the 17 common technical processes. Implementation includes (1) defining the outcomes to satisfy entry and exit criteria for each lifecycle phase, and (2) major inputs for other technical processes. Each process section contains a description of the approach, methods, and tools for the following:

- Identify and obtain adequate human and non-human resources to perform the process, develop the work products, and provide the process services

- Assign responsibility and authority to perform the process, develop the work products, and provide the process services

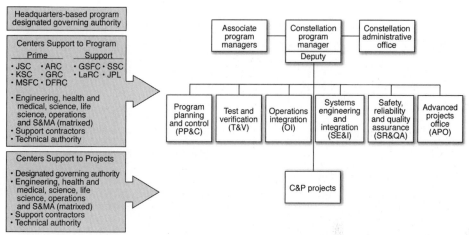

FIGURE 14-7.   Technical Effort Integration for Constellation [NASA (2), 2006]. Here we show how the Constellation Program leaders divided the program into more manageable tasks. (JSC is Johnson Space Center; ARC is Ames Research Center; GSFC is Goddard Space Flight Center; SSC is Stennis Space Center; KSC is Kennedy Space Center; GRC is Glenn Research Center; LaRC is Langley Research Center; JPL is Jet Propulsion Laboratory; MSFC is Marshall Spaceflight Center; DFRC is Dryden Flight Research Center; S&MA is safety and mission assurance; C&P is contracts and pricing.)

- Train the technical staff performing or supporting the process, as needed
- Place designated work products of the process under appropriate configuration management
- Determine and involve process stakeholders
- Monitor and control the process
- Objectively evaluate how well the process and its work products and services adhere to the applicable requirements, objectives, and standards; address noncompliance
- Review process activities, status, and results with appropriate levels of management; resolve issues

Technology insertion. The maxim, "No miracles required!" is a standard at the beginning of all space programs. But some technology insertion is inevitably necessary to achieve the mission. Table 14-3 gives examples of previous space programs where technology insertions entailed high risk.

This section of the SEMP describes each proposed technology within the project, including the associated risks and the criteria for accepting the technology. The NASA technology readiness levels (TRLs) help us evaluate the required technology and determine if it will be available or if we should develop an alternative in parallel. Table 14-4 compares the levels and shows the estimated TRL for many of FireSAT's necessary technologies. Significant technology

TABLE 14-3. **Miracles Required.** Risk reduction includes understanding the level of technological maturity of major system components.

Space Program	Miracles Required
Iridium	High-speed processor chips (Power PC3 chips)
DirecTV	Compression technology
Mars Rovers	Parachute and airbag technologies
Pioneer and Voyager	Radioisotope thermoelectric generators
Corona	Re-entry vehicles, film in vacuum
Space Transportation System	Re-entry tiles
Hubble Space Telescope	Fine guidance
Chandra X-ray Observatory	Large X-ray mirrors

development calls for a stand-alone technology development plan. The section also summarizes the technology development effort, with references to this standalone plan for details.

Additional SE functions and activities. This section is a catch-all for important processes not previously discussed. A common item is management or engineering professional development needs, such as team building or systems engineering training for the discipline engineers. This section may also include the following:

- **System safety**—Nothing stops project progress like an accident or safety violation. We must specify the safety approach and various techniques, and designate the person responsible for the safety program.

- **Engineering methods and tools**—This part covers the tools and methods not discussed previously, such as special payload handling needs.

- **Specialty engineering**—This part is usually a matrix dealing with engineering specialties and product descriptions based on the WBS. It could include safety, reliability, human factors, logistics, maintainability, quality, manufacturing, operability, and supportability.

Integration with the project plan and technical resource allocation. This section describes project management roles and responsibilities, emphasizing the technical efforts. It includes allocation of resources (time, money, personnel) and how we plan to coordinate changes to these resources.

Waivers. This section establishes the approach for obtaining waivers to current organization policies. It places special emphasis on standard process documents (e.g., NASA NPRs).

Appendices. This section includes the glossary, acronyms and abbreviation lists, information pertinent to multiple topics, charts or proprietary data, and the summary of technical plans.

The FireSAT SEMP is shown in Table 14-5.

TABLE 14-4.   **FireSAT Technology Readiness Levels (TRLs).** Early identification of maturity levels lessens risk. (Adapted from Table G-19 of NPR 7123.1a).

Technology Readiness Level	Description	FireSAT Example
9. Actual system flight proven through successful mission operations	In almost all cases, the end of the last 'bug fixing' aspect of true system development. This TRL does not include planned product improvement of ongoing or reusable systems.	Infrared payloads that have flown on space missions
8. Actual system completed and flight qualified through test and demonstration (ground or space)	This level is the end of true system development for most technology elements. It might include integration of new technology into an existing system.	Space system end-to-end prototype for testing of FireSAT concept
7. System prototype demonstration in a space environment	This is a significant step beyond TRL 6, requiring an actual system prototype demonstration in a space environment. The prototype should be near or at the scale of the planned operational system and the demonstration must happen in space.	Total system concept with payload supported by hardware in a communication network
6. System or subsystem model or prototype demonstration in a relevant environment (ground or space)	A major step in the level of fidelity of the technology demonstration follows the completion of TRL 5. At TRL 6, a representative model or prototype system or system, which goes well beyond ad hoc, "patch-cord," or discrete component level breadboarding, is tested in a relevant environment. At this level, if the only relevant environment is space, then the model or prototype must be demonstrated in space.	Payload optics tested on aircraft flying over forest fires
5. Component or breadboard validation in relevant environment	At this level, the fidelity of the component or breadboard being tested has to increase significantly. The basic technological elements must be integrated with reasonably realistic supporting elements so that the total applications (component level, subsystem level, or system level) can be tested in a simulated or somewhat realistic environment.	Payload optics tested on laboratory benches with simulated wildfires
4. Component or breadboard validation in laboratory environment	Following successful proof-of-concept work, basic technological elements must be integrated to establish that the pieces will work together to achieve concept-enabling levels of performance for a component or breadboard. This validation must support the concept that was formulated earlier, and should also be consistent with the requirements of potential system applications. The validation is relatively low-fidelity compared to the eventual system: it could be composed of ad hoc discrete components in a laboratory.	Payload optics for fire recognition based upon laboratory parts

TABLE 14-4.   FireSAT Technology Readiness Levels (TRLs). (Continued) Early identification of maturity levels lessens risk. (Adapted from Table G-19 of NPR 7123.1a).

Technology Readiness Level	Description	FireSAT Example
3. Analytical and experimental critical function or characteristic proof of concept	At this step, active research and development (R&D) is initiated. This must include both analytical studies to set the technology into an appropriate context and laboratory-based studies to physically validate that the analytical predictions are correct. These studies and experiments should constitute proof-of-concept validation of the applications and concepts formulated at TRL 2.	Constellation design with coverage timelines and revisit times
2. Technology concept or application formulated	Once basic physical principles are observed, practical applications of those characteristics can be invented. The application is still speculative and no experimental proof or detailed analysis exists to support the conjecture.	Needs analysis with satellite option
1. Basic principles observed and reported	This is the lowest level of technology readiness. Scientific research begins to be translated into applied research and development.	Physics of 1000° C fire being observed from space

TABLE 14-5. **FireSAT Systems Engineering Management Plan (SEMP).** This table describes what goes into a SEMP and gives a FireSAT example for each section. (SE is systems engineering; ICD is interface control document; CDD is common data dictionary; SSS is source selection statement; TEMP is test and evaluation master plan; ASAP is as soon as possible; DL is delivery; SRR is system requirements review; NPR is NASA Procedural Requirement; WBS is work breakdown structure; GrSysEngr is ground system engineer.)

	Section	Example	FireSAT SEMP Increment A (During Concept Development)
			FireSAT Subset Examples: Concept Development
1	Plan for managing SE	• Establish technical content • Provide specifics of technical efforts • Describe technical processes • Illustrate the project organization • Establish technical schedule of key events • Serve as the communication bridge between management and technical	**Purpose:** The purpose of the FireSAT Project is to save lives and property by providing near-real-time detection and notification of wildfires within the US **Mission objectives:** "The US needs the means to detect and monitor potentially dangerous wildfires." **Technical effort breakout:** See FireSAT preliminary architecture and ConOps **FireSAT organization:** Small program office (Two managers, systems engineer, satellite engineer, GrSysEngr, financial) **Master schedule:** Start ASAP, development 5 yrs, DL 6 yrs, SRR within 6 mos **Technically feasible:** Yes, within 5 year development schedule—no miracles required!
2	Documents driving the SEMP	• ICD • CDD • SSS • TEMP	At this time, draft documents are still in preparation
3	Technical summary	• System description • System structure • Product integration • Planning context • Boundary of technical effort • Cross references	See Figure 14-5. At this phase, understanding the approach is critical and buy-in from stakeholders is essential. We base the concept, the approach for operations, the project's breadth of reach, the schedule, and the financial estimates on a systems architecture layout with conops attached. We should do this before the SRR and the first full SEMP.
4	Integrate the technical effort	This section describes how we will integrate the various inputs into a coordinated technical effort that meets cost, schedule, and performance objectives. • 4.1 Responsibility and authority • 4.2 Contractor integration • 4.3 Support integration	Within the master schedule is an estimate of the technical effort. We have drafted the WBS and laid out the areas of responsibility for the team members. We have also addressed some major questions, such as: • Will we build or buy? • Which organization will develop the major subsystems? • Who will operate the ground segment? • Will we use on-going operations centers? • Which outside organizations must we coordinate with during development?

Table 14-5.   **FireSAT Systems Engineering Management Plan (SEMP). (Continued)**

		FireSAT SEMP Increment A (During Concept Development)
	Section	**Example**	**FireSAT Subset Examples: Concept Development**
5	Engineering processes	Each of the 17 common technical processes has a separate subsection that contains the plan for performing the process activities as appropriately tailored. Implementing the processes includes: (1) generating the outcomes needed to satisfy the entry and exit criteria of the applicable lifecycle phases and (2) generating the necessary inputs for other technical processes. The section contains a description of the approach, methods, and tools.	During this early phase, we list the 17 processes and the phases in which they dominate. We then draft a preliminary write-up of these processes to include: 1) Identifying and obtaining adequate human and other resources for performing the planned process, developing the work products, and providing the process services. We must know the peak activities across the master schedule so we can spread out the effort for ease of technical management. 2) Assigning responsibility and authority for performing the process, developing the work products, and providing the services (with dates of delivery for each of these, draft and final) 3) Training for the technical staff 4) Designating work products and placing them under configuration management across the master schedule 5) Identifying and involving stakeholders 6) Monitoring and controlling the processes
6	Technology insertion	This section describes the approach and methods for identifying key technologies and their associated risks. It also sets criteria for assessing and inserting technologies, including critical technologies from technology development projects.	We normally use the NASA technology readiness level (TRL) approach to decide whether a technology is ready for the project. We need to determine immediately if any "miracles" are required. For FireSAT at this preliminary stage, we assume that all parts of the system have been developed to at least TRL 8. Integrating the system is the key to the technology development. This integration is definitely a challenge to the System Program Office, but not a technological one.
7	SE activities	This section describes other areas not specifically included in previous sections but essential for planning and conducting the technical effort.	During the design concept development phase, determining necessary systems engineering tasks is of prime importance. The timing of the needs, the staffing requirements, and the tools and methodologies are required at least for: • 7.1 System safety • 7.2 Engineering methods and tools • 7.3 Specialty engineering

TABLE 14-5. FireSAT Systems Engineering Management Plan (SEMP). (Continued)

		FireSAT SEMP Increment A (During Concept Development)	
	Section	Example	FireSAT Subset Examples: Concept Development
8	Project plan integration	This section tells how the technical effort will integrate with project management and defines roles and responsibilities. It addresses how technical requirements will be integrated with the project plan to allocate resources, including time, money, and personnel, and how we will coordinate changes to the allocations.	At this early phase, the project plan is being put together with assistance from the chief systems engineer. This person is responsible for seeing that the project plan is consistent with the SEMP. Key to success is consistency across the management team and the engineering team with respect to the WBS, the master schedule, the staffing estimates, the breakout of contractor-internal efforts, and the basic financial estimates.
9	Waivers	This section contains all approved waivers to the SE implementation plan, required for the SEMP. It also has a subsection that includes any tailored SE NPR requirements that are not related and can't be documented elsewhere in the SEMP.	Ascertaining waivers early enables the process to proceed without undue overhead. We do this by reviewing programs similar to FireSAT (similar customers, similar orbits, similar timelines) and reviewing their waivers. We should list all waivers before the first draft of the SEMP because it drives all the processes and many of the project management activities.
A	Appendices	Appendices provide a glossary, acronyms and abbreviations, and information published separately for convenience in document maintenance.	Ascertaining the cross-program engineering issues early is essential to smooth operations. This effort includes information that references the customer, the needs, the vocabulary, the acronyms, etc. The categories of this information are: • Information that may be pertinent to multiple topic areas • Charts and proprietary data applicable to the technical effort • A summary of technical plans

14.3   Develop the Systems Engineering Management Plan (SEMP)

The best approach for writing the SEMP is to use as much of a successful past program's SEMP as possible. We may have to make significant alterations if it originates outside the parent organization. But if our space program is similar to one currently under development within the organization, sharing the document makes sense. This section describes how to write a SEMP, as depicted in Figure 14-8.

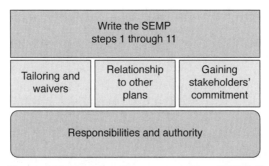

FIGURE 14-8.   **The Systems Engineering Management Plan (SEMP) Process.** The SEMP is built upon the foundation of responsibilities and authority, bolstered by tailoring and waivers to established processes, and tied to other technical plans and stakeholder commitment.

14.3.1   Responsibility and Authority

The SEMP is most efficient when the engineering team that will execute the program writes it. We may invite a specialist to start the writing, but the responsibility lies with the chief engineer or the lead systems engineer, depending on the organization. This responsibility, and authority, is inherent in the organizational layout and in guidance from the senior stakeholder or principle developer. After we develop the SEMP, the lead systems engineer or chief engineer must see that the processes are followed and the resources are allocated reasonably. This enforcement demands close association with program management so the right technical expertise is available at the right time.

14.3.2   Writing the SEMP

Getting started is paramount. We simply can't wait until all the information is in, especially for a new project with exciting challenges. We have to form a team, start writing, and persist until the SEMP has been approved. A suggested approach is as follows:

Step 1: Prepare to write. Because the SEMP must be in place before the systems requirements review (SRR), the ability to pull together diverse inputs is important. One good approach is where the chief engineer asks the team to determine who writes which section.

Step 2: Assign a lead and support members. The SEMP writing team is critical to a project's success. All 17 process owners plus the program managers need inputs. Deciding on the lead for the SEMP development is crucial.

Step 3: Schedule the SEMP. As in all project activities, setting the due dates is essential. This schedule includes dates for the outline, section inputs, preliminary draft, final version for the program manager and chief engineer, and the presentation to the whole project team, both management and technical.

Step 4: Identify technology insertion needs. This portion reviews the standard technologies and includes recommendations for inserting higher-risk technologies into the program. To mitigate risk, we should have a parallel development plan for the risky technologies.

Step 5: Develop technical schedule. In concert with the program management team, the technical team must decide when, where, and who will implement the technical activities. This input to the master schedule should allow flexibility for dealing with unforeseen problems as well as margin for technical issues. We must list all the processes and then spread resources across the schedule to support each one.

Step 6: Specify parallel plans. A major space systems development requires a whole series of plans. The SEMP should note how each plan works in concert with it and describe approaches to assure compatibility. This includes, as a minimum, the program management plan, the test plan, and the verification and validation plan.

Step 7: Determine needed technical expertise. We must ascertain early which specialties we need during the development program and then schedule them as necessary. Many times a specialty is needed throughout the program, but only periodically. Judicious planning and scheduling of specialist time is the best way to be sure the expertise is available when called for.

Step 8: Conduct draft reviews. Developing the SEMP requires many inputs from many sources inside and outside of the program office. These sources must review the SEMP many times during its development to make sure it contains all necessary inputs before we publish it.

Step 9: Submit the final SEMP. The space program's leadership team must review the draft SEMP. The final SEMP then goes to the leadership team for approval.

Step 10: Present the SEMP. We should present the final version of the SEMP during the month before the SRR. This timing lets all the participants of a technical program review it to ensure that it meets stakeholder needs.

Step 11: Implement continual improvement. Because the SEMP has many reviews and updates before all major milestones, we have to continually improve it. The proven approach for continual improvement entails a single point of contact responsible for refinements and updates to the SEMP. These changes arrive all the time from players inside and outside the program office. We insert the updates to the schedule in the SEMP and the master schedule in a timely manner; updates to the technical processes require approval from their owners. This approach allows all players to know where to get the latest version of the SEMP and how to help improve it. A separate annex should identify a process for SEMP inputs.

14.3.3   Tailoring and Waivers

The SEMP is important for tailoring and waivers because it's the document that formalizes them. Therefore, in writing the SEMP, the authors must plan to determine the necessary tailoring and waivers.

Tailoring concepts. *Tailoring* is adapting standard processes to make them directly applicable to project requirements. A *waiver* is a documented agreement releasing a project from meeting a requirement.

Systems engineering is a general process for engineering any complex product. We apply it to developing products as simple as a toaster or as complicated as the International Space Station. NPR 7123.1a Appendix C provides a full set of systems engineering processes that apply to the largest of space programs. But most projects are smaller. Tailoring allows us to define a set of technical processes for the project. This tailoring is essential to project planning (and writing the SEMP).

Tailoring is not just "picking the right processes," but a conscious decision on what levels of effort to apply to each task. The appropriate level of effort depends on many factors, such as those shown in Table 14-6. The planner must consider all these factors in evaluating each process step to determine the right level of effort. Tailoring is not a way to eliminate activities because of budget or schedule pressures, nor is it a way to throw out processes we don't like. It's not allowed on all processes. And it's not meant to undermine the intent of good systems engineering.

TABLE 14-6.   **Factors in Tailoring.** Tailoring should be applied with judgment and experience.

Organizational considerations	Programmatic considerations
• Organizational requirements for SE processes	• Goals and constraints of a project
• Baseline process and tailoring guidelines	• Essential success factors
System considerations	• Cost targets
• System scenarios and missions	• Acceptable level of risk
• Measures of effectiveness	• Directions and limitations of tasks
• Known constraints	• Technical events for demonstration and confirmation
• Key technologies and their maturity	• Level of detail
• Performance issues	
• Limiting technologies	

Tailoring adapts processes to better match the purpose, complexity, and scope of a space project. It varies from project to project based on complexity, uncertainty, urgency, and willingness to accept risk. If a large project misses any steps in the systems engineering processes, it may be extremely costly to recover or—as in the case of the failed 1999 Mars Climate Orbiter—it may not recover at all. On the other hand, a small project that uses the full formality of all process steps will be burdened unnecessarily and may founder in the details of analysis. Tailoring strikes a balance between cost and risk.

Tailoring for specific project types. NASA's NPR 7120.5 specifies several classes of space system programs and projects. Depending on the type of project, some tailoring guidance applies as shown in Table 14-7. We have to consider the impact of all these factors and more when tailoring, so that the processes are applicable to the project.

TABLE 14-7.   **Tailoring Guidance by Project Type.** By understanding which SEMP sections are most relevant to different types of space system projects, the chief systems engineer sees that appropriate effort is applied.

Project Type	SEMP Needed?	Sections to Emphasize	Sections to De-emphasize
Basic and applied research	Usually not	6-Technology insertion	Most
Advanced technology program	Yes	3-Technical summary 6-Technology insertion	5-Common tech process 7-Additional SE functions
Flight systems and ground support (new development)	Yes	4-Tech effort integration 5-Common tech process 8-Project plan integration	None
Flight systems and ground support (evolutionary acquisition)	Yes	4-Tech effort integration 5-Common tech process 6-Technology insertion 8- Project plan integration	None
Institutional (property)	No	Not applicable	Not applicable
Institutional (information technology)	Yes, if development	4-Tech effort integration 5-Common tech process 8-Project plan integration	6-Technology insertion 7-Additional SE functions
Institutional (other functional initiatives)	Yes, if development	4-Tech effort integration 5-Common tech process 8-Project plan integration	6-Technology insertion 7-Additional SE functions
Surveillance-type projects (work performed out of house)	Yes, if development	Emphasis per project type, plus 4—Tech effort integration, particularly technical selection and oversight	De-emphasis per project type

Sources for tailoring. Systems engineering processes vary from organization to organization and from project to project. We should base the SEMP on the processes specific to the organization, but we may find some process differences by reviewing the documentation from other organizations. We find other tailoring possibilities by observing patterns of similarity in other projects, and using them on ours. These projects should be similar to ours in scope, complexity, and risk tolerance. In NASA, an excellent archive of such sources is the NASA Lessons Learned web portal with its database of learned patterns. Figure 14-9 shows a typical entry page from this system. Each entry provides knowledge that may help tailor the SEMP of a similar project. Table 14-8 shows some lessons learned from DoD experiences with programs and SEMPs.

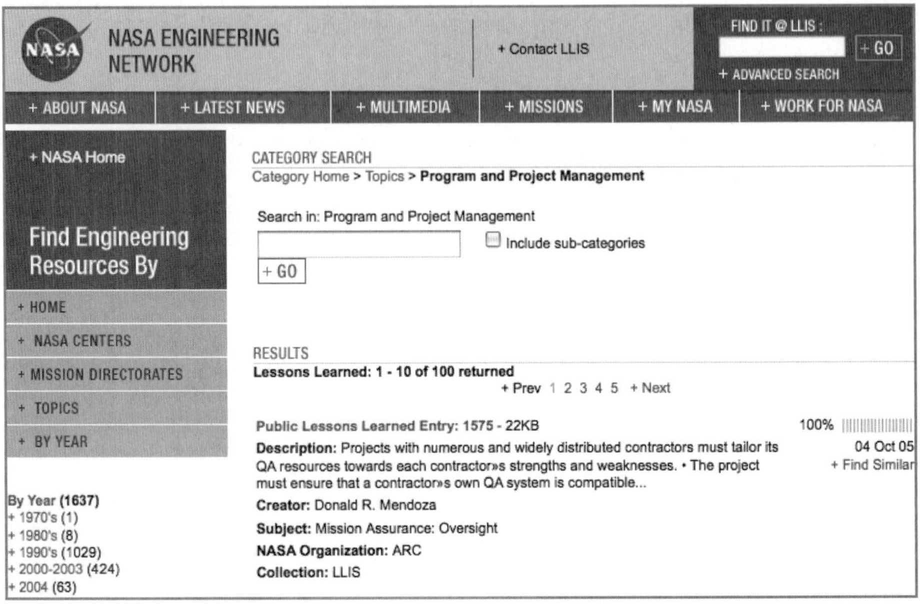

FIGURE 14-9.   **Example of NASA's Lessons Learned Information System (LLIS).** A careful search of the database helps keep common problems from afflicting the current project.

TABLE 14-8.   **Systems Engineering Management Plan (SEMP) Lessons Learned from DoD Programs [NASA, 1995].** The project leadership team should review this table before the systems requirements review.

Lesson	Lesson Learned
1	A well-managed project requires a coordinated systems engineering management plan that is used throughout the project cycle
2	A SEMP is a living document that must be updated as the project changes and kept consistent with the project plan
3	A meaningful SEMP must be the product of experts from all areas of the project
4	Projects with insufficient systems engineering discipline generally have major problems
5	Weak systems engineering, or systems engineering placed too low in the organization, cannot perform the functions as required
6	The systems engineering effort must be skillfully managed and well communicated to all project participants
7	The systems engineering effort must be responsive to the customers' and the contractors' interests

Systems engineering processes in many forms are all around us. While NASA has specified forms of systems engineering, other standards have alternate process forms that may be applicable to the project. Table 14-9 lists some useful standards documents. All of these references provide alternate processes for tailoring the plan.

TABLE 14-9.   Systems Engineering Standard References. Many organizations are sources of processes, standards, and approaches. (DoD is Department of Defense; FAA is Federal Aviation Administration; ISO is International Standards Organization; IEC is International Electrotechnical Commission; ANSI is American National Standards Institute; GEIA is Government Electronics and Information Association; IEEE is Institute of Electrical and Electronics Engineers.)

Government process standards	**International Council on Systems Engineering (INCOSE)**
• US DoD—*SE Fundamentals*, published by the Defense Acquisition University	• Systems Engineering Handbook
• FAA—Integrated Capability Maturity Model (iCMM)	• INCOSE Process Asset Library (IPAL)
Industry standards	• "Systems Engineering," the Journal of INCOSE
• Software Engineering Institute—Capability Maturity Model Integration (CMMI)	• INCOSE symposium proceedings
• ISO/IEC 15288	
• ANSI/GEIA 632	
• IEEE 1220	

Role of the SEMP in tailoring and waivers. Tailoring is important in writing the SEMP, because the SEMP describes and authorizes the tailoring. Any instance where the SEMP varies from standard organizational processes is a waiverable variance. By explicitly documenting the differences, the approval of the SEMP by the designated governing authority constitutes approval of the waivers it contains.

14.3.4   Relationship to Other Plans

The technical planning for a major space system is the core of the development approach. To minimize disruptions, it's critical to coordinate and meld a host of plans. Scheduling is key to smooth execution of a space system acquisition across the various organizations.

The integrated master schedule integrates all program elements into a manageable flow with few surprises. As the technical management lead document, the SEMP coordinates the technical schedule to ensure compatibility with other efforts. As the program matures to the middle of production, the schedule room (now a computer file on everyone's desk or laptop) becomes extremely important. The SEMP must coordinate with many plans (see list below), but the most critical are the project plan, the software management plan, and the risk management plan.

Project Plan

The project plan has the following characteristics:

- It's the overall project management lead document, to which the SEMP is subordinate
- It details how the technical effort will integrate with project management and defines roles and responsibilities
- It contains the project's systems engineering scope and approach
- It lists technical standards applicable to the project
- It's the result of the technical planning effort, which should be summarized and provided as input to the technical summary section of the project plan

Software Management Plan

The software management plan has the following characteristics:

- It's developed within the scope of the technical effort and subordinate to the SEMP
- SE planning ensures that software development is based on systems requirements
- System aspects represented or implemented in software are included in all technical reviews
- It describes
  - How the software activities are consistent with the systems engineering management plan
  - How the software activities are fully integrated parts of the technical effort

Risk Management Plan

The risk management plan has the following characteristics:

- It's developed by the project team, typically as a peer to the SEMP and subordinate to the project plan
- It lists technical risk sources and categories
- It identifies potential technical risks
- It characterizes and rank orders technical risks

Other technical plans may include: technical team organization; responsibility assignment matrix; configuration management plan; data management plan; electromagnetic compatibility or interference control plan; human factors or engineering plan; interface control plan; manufacturing and assembly plan; mass properties control plan; reliability plan; software development plan; reliability, maintainability, and supportability plan; systems analysis plan; design and analysis cycle plans; safety and mission assurance plan; technical measurement plan; training plan; in-flight check-out plan; test and evaluation master plan; disposal plan; technical review plans; technology development plan; launch operations plan; and payload-to-carrier integration plan. (See Chapter 13.)

The key to coordinating all these other plans is that the SEMP is the dominant plan for the program's technical processes. Every plan dealing with engineering should follow the SEMP for the sake of consistency. The SEMP engineering aspects:

- Control product requirements, product interfaces, technical risks, configurations, and technical data
- Ensure that common technical process implementations comply with requirements for software aspects of the system
- Provide the big picture for the technical view

14.3.5   Obtaining Stakeholder Commitments

An essential part of creating a systems engineering management plan (SEMP) is to gain acceptance among the project stakeholders; otherwise, the plan is useless. A *stakeholder* is any group or individual affected by or in some way accountable for the outcome of an undertaking. Chapter 2 discusses stakeholders in detail.

Stakeholder review of the SEMP. Technical plans are little more effective than the stakeholders' level of agreement allows them to be. Most stakeholders do not have signature-level approval of the SEMP, because this authority is reserved for the executive manager. But they can defeat a project by failing to buy into the plan. For instance, a developing contractor that opposes some of the technical processes may exhibit passive or active aggression against those processes, which in turn could defeat the entire technical plan. Or an executive manager whose expectations diverge from elements of the plan may impede progress repeatedly by asking us to explain again the plan's intent. Such political issues are often major obstacles to technical progress. The way to prevent them is to obtain stakeholder concurrence with the technical plan.

The best way to obtain this concurrence is through frequent stakeholder reviews of the plans. As the SEMP develops, we need to communicate its features through such means as face-to-face contact, telephone, and document reviews. Frequent small contacts are best, so we should look for opportunities to ask planning questions, to create workable relationships, and to consult with them on aspects of the developing plan. We must also remember that the stakeholders include the development teams that have to perform under the plan.

With this groundwork of relationships, a successful formal review of the final SEMP is usually smooth and easy. When the SEMP is near completion, we should provide it as a full document to the stakeholders for their reading and detailed review. If the project is large enough, we should schedule a stand-up presentation and formal review to allow interaction among the stakeholders and the planners. For some key stakeholders, an individual presentation may be appropriate. As part of the review, we find out which parts of the plan make the stakeholders uncomfortable and discuss possible changes with them.

SEMP negotiation and modification. Developing an effective technical plan is as much political as it is technical. The relationships we build during the planning effort determine how much support we get during the project.

The hardest way to try to gain stakeholder concurrence is to create a "perfect" technical plan before ever showing it to them. The best technical plan is not the one that has all elements perfectly devised, but rather the one that represents the best compromise between the technical needs and the desires of all relevant stakeholders. When asked after the fact about the D-Day plans, Gen. George S. Patton noted, "In the space of two days, I had evolved two plans, wholly distinct, both of which were equally feasible. The point I am trying to bring out is that one does not plan and then try to make circumstances fit those plans. One tries to make plans fit the circumstances."

The emphasis of systems engineering is on safely achieving stakeholder functional, physical, and operational performance requirements. We must negotiate and modify the technical planning to meet stakeholders' needs. Negotiation with stakeholders is a give-and-take process, usually involving the following steps:

1. We present an initial idea—in this case, some aspect of the technical plan

2. We discuss the idea until the stakeholders fully understand it

3. The stakeholders decide whether they agree with the idea

4. If not, they explain the objectionable parts

5. We discuss further to arrive at an alternate, satisfactory idea

This process creates effective agreement, in which both parties have the opportunity to influence the plan. It also fosters good relationships, in which each party learns to trust the other. Relationships are built on trust. Trust is built on fulfilled promises. Therefore, we must be careful about promises. When negotiating with one stakeholder, we have to remember and be honest about the impacts of other stakeholders. We must avoid over-promising, yet provide enough commitment to satisfy stakeholder needs. The relationships built during the technical planning process become the basis for much of the later technical work.

14.3.6   Continual Improvement During the Project

The engineering effort is only as good as the timely allocation of resources to the team. This effort means extensive planning on the engineering side. We must have a complete SEMP as soon as possible. As the program progresses, the issues become more complicated and the interfaces begin to dominate. So the transition between phases becomes more complex as we go from a big picture concept to requirements, to decomposition, to integration, to verification and validation, and finally to operations. The core team should be relatively consistent during many developmental phases while the ancillary engineering support rotates between programs. The SEMP has to allow these transient engineering teams to understand the chief engineer's recommended approach. It must continually improve.

14.4   Characteristics of Good and Bad Plans

A good SEMP has five main characteristics:

- **Clarity**—A good SEMP is easy to understand, so that those who join the project can quickly fit into the plan
- **Conciseness**—A good SEMP provides the necessary information without over-burdening the reader
- **Completeness**—A good SEMP covers all the necessary and specified topics
- **Currency**—A good SEMP is applicable to the current phase of work and contains valid plans for the upcoming phases
- **Correctness**—A good SEMP accurately captures and communicates the underlying technical plan

This section provides some descriptions and examples of good and bad approaches to documenting the technical plan. The FireSAT end-to-end case study in Chapter 19 also gives a good example.

14.4.1   Clarity

Anyone reading the SEMP must be able to understand it. A good SEMP, therefore, is written in language that is familiar and informative. The language and grammar should be appropriate to a plan, not a specification. Table 14-10 shows samples of good and bad writing styles. The SEMP is a formal document and should be fully reviewed for style. If written in haste with inadequate proofing, the plan probably won't convey the essential information. Project team members become confused, or simply don't read it, leading to technical confusion during execution. Grammar might seem to be a minor issue, but the impact of poor grammar is anything but minor. The SEMP writing style should feature fairly short sentences with declarative information in the active voice. Technical writers often get in the habit of using passive voice, explanatory phrases, and overuse of attempted precision. We need to avoid these.

Different grammatical tenses serve different purposes. When reciting the history of a project or the plans, we use past tense. When describing the product, the project, the plans, or the organization, we use present tense. We do so even when describing the **future** configuration of the product, as if that configuration were in front of the reader. Future tense is appropriate only when describing a set of actions that are part of the future. **Present** tense is usually preferable unless it would mislead the reader.

14.4.2   Conciseness

The SEMP is a working plan, documenting and providing essential information to guide the technical team. It should structure that information so readers quickly find and assimilate it. It should focus on the **deviations** from

TABLE 14-10.   **Samples of Writing Style.** After completing the first draft, we should compare it against this list.

Good Writing Style	Bad Writing Style	Why is This Bad?
The lead systems engineer is responsible for the content of the SEMP.	When necessary for management guidance, this SEMP shall be controlled, managed, implemented, and maintained by that individual assigned, at the time of control, to the responsibility of lead systems engineer.	This sample is too long, uses passive voice, adds explanatory phrases, and is overly precise.
The project completed the concept development phase in January 2006.	The project will follow the scheduled phase milestones in Figure 3-1.	Progression of time will invalidate the future tense as milestones pass.
The Earth Observing Satellite system provides geological data to a ground support station.	The Earth Observing Satellite system will provide geological data to a ground support station.	Extra words in future tense obscure the primary meaning. Even though the product, its delivery, and other plans may be in the future, future tense is not necessary to convey the intended meaning.
At each design review, the lead systems engineer is responsible for gathering and disseminating action items.	At each design review, the Lead Systems Engineer will be responsible for gathering and disseminating action items.	
The satellite will be mated with the launch vehicle during the vehicle integration phase.	The satellite is mated with the launch vehicle during the vehicle integration phase.	This is a specific action to be taken at a definite time in the future.

standard processes, paraphrasing them and invoking them by reference, instead of describing them completely. "Boilerplate" text is not appropriate. A common mistake is to create a project SEMP by copying material from a prior one. Section 14.3 mentions leveraging a successful project's past SEMP as the best approach for writing a new one. But the key is to leverage, not to copy. Copying leads to general descriptions that add little to the technical plan. Instead, we should take the time to plan first—then write the specifics of the plan into the SEMP.

The SEMP should contain sufficient detail to describe the technical work, but not so much as to override the latitude that the team members need. Insufficient detail comes from planning at too high a level, from inadequate consideration of technical risks, from the use of boilerplate, or from too little description. A plan written at the level of a slide presentation has insufficient detail. Too much detail, on the other hand, comes from over-control by the planner or too many task elements in the WBS. Table 14-11 describes some good and bad practices in level of detail.

The document's organization should also be concise. While standards such as NPR 7123.1a provide the primary outline, each SEMP must organize the subsections within the primary outline to fit the content. These subsections flow logically to clarify the relationship of each topic to other topics. Disorganized sections and poor flow are signs of a bad SEMP.

TABLE 14-11.   Samples of Level of Detail. After we write the draft, we edit the level of detail using samples such as these.

Good Level of Detail	Bad Level of Detail	Why is This Bad?
A technical summary of the project that details the major components, work efforts, and their relationships	A technical summary of the project that details the work effort and products of each person	Does not facilitate change
	A technical summary that only mentions the project and its purpose	Does not help the reader understand the scope of the work effort
Organizational charts that display the major contributing organizations and team groups, and assign leadership responsibility	Organizational charts that identify each individual by name	Does not facilitate change
	Organizational charts at the organization or center level without team groups depicted	Does not effectively assign responsibility
Technical staffing levels by organization, skill type, and month or quarter	Technical staffing levels by team group, by experience level, by week, or other detailed levels	Generates excessive tracking and management burden
	Technical staffing levels only by organization, or only by skill type, or by longer time periods	Insufficient detail for organizational managers to plan resources

14.4.3   Completeness

A good SEMP is complete; it covers all necessary topics. The outline in NPR 7123.1a identifies the topics. The SEMP should address each topic in the outline, with sufficient thought to the level of detail and tailoring necessary for the project.

Maintaining the complete primary outline is a good practice. NPR 7123.1a lists nine major sections with specific content. Each major section in the SEMP should have a title identical to that in the standard, so that others who read it understand the content. The major section should cover the topics specified, and in order. Where the standard specifies a second-level outline (as in Sections 3, 4, and 5), the SEMP second-level sections should also have titles identical to those in the standard. This compliance with standard is a sign of cooperation and experience, and it gives the reader confidence that the technical plan is worth reading and following.

If a section doesn't apply, or if the technical plan uses only standard processes for a topic, then the SEMP should document this situation. If the decision represents a waiver, then we must provide the rationale. Appropriate wording for such a section might be (1) "The [Name] project uses the standard processes for technical reviews as specified in MWI 8060.3." or (2) "This section is not applicable to the [Name] project because…"

Section 5 of the SEMP covers the 17 common technical processes. NPR 7123.1a Appendix C describes each process in sufficient detail to provide plan guidance. Authors might be tempted to gloss over this section, or to include duplicative detail from the standard; neither is appropriate. The section should cover all 17 processes, but be aimed at the unique differences and implementations.

The SEMP may also include appendices. A glossary of terms, acronyms, and abbreviations is essential. Technical staffing levels are often placed in an appendix due to the volume of information that may be necessary. Other appendices help meet the desired level of background detail for specific sections.

14.4.4   Currency

A good SEMP is current and relevant. It's a living document and should be kept alive. When first created, during the early formulation phase, it may be quite short—perhaps fewer than ten pages—and many sections may have only preliminary information. At the end of each development phase, we should update the SEMP in a "rolling wave" of planning for the upcoming phases. Every major project review addresses the sufficiency of plans for the next phase, so we need updates at least at the following reviews: system requirements review or mission definition review, system definition review, preliminary design review, critical design review, test readiness review or system acceptance review, flight readiness review, and operational readiness review. We should also update it in response to any significant project change, such as a technology breakthrough, funding change, or change in priority or emphasis.

Each SEMP revision should be documented and approved. A revisions page following the title page shows the revisions along with their date, reason for revision, and scope of revision. Approval of the revisions follows the same path as approval of the original SEMP, including obtaining concurrence from stakeholders (as described in Section 14.3.5) and signature approval from the designated governing authority.

14.4.5   Correctness

Finally, a good SEMP accurately reflects the technical plan. If it's not clear, concise, complete, and current as we describe in the preceding sections, it's a poor SEMP regardless of the quality of the technical plan. Conversely, it can be clear, concise, complete, and current, but mislead the reader because it obscures weaknesses or gaps in the technical plan. Although in general the adage "garbage in, garbage out" is true, it's possible to write a SEMP that looks better than the reality of the underlying technical plan. (Table 14-12)

The SEMP author must balance a desire for brevity (since it **is** a management plan) against the additional detail needed for correctness. In the words of Albert Einstein, "Make everything as simple as possible, but not simpler." The author must include sufficient detail to accurately communicate the technical plan, yet refer appropriate details to the other peer and subordinate technical plans. To the extent that the SEMP is correct, the continual improvement process (Step 11 in Section 14.3.2) focuses on and corrects the weaknesses in the technical plan over time. Conversely, a SEMP that glosses over weaknesses in the technical plan hinders and delays corrections.

TABLE **14-12.**   **The Systems Engineering Management Plan (SEMP) as an Indicator of Technical Planning Quality.** We should examine the SEMP critically to determine the quality of the overall technical planning activity.

The SEMP author and its critical readers should ask the following question: How do we know that the SEMP reflects good technical planning? A good technical plan yields the following answers:

- No technological surprises appear
- The scheduling of tests matches our verification and validation plan
- Other project documents mesh with the SEMP
- The modeling and simulation tools are available and validated when needed
- The technology cost estimates are not significantly beyond budget
- No one yells, "That's stupid!"

14.5   Necessary Elements for Executing a SEMP

Senior program managers and systems engineers have identified five elements of major system developments that lead to successful programs. They are: 1) leadership, 2) discipline in technical management, 3) collaborative environments, 4) scope control, and 5) stakeholder involvement. This section discusses how each element is also a necessary element for SEMP execution.

14.5.1   Leadership

In most projects, the systems engineer is not in a position of organizational authority, but does carry full responsibility for the technical performance of the system. Technical management must therefore be handled through leadership rather than through mandated authority. To implement a SEMP well, the lead systems engineer must help the technical team members cooperate and collaborate toward common goals. In technical teams, this task becomes more difficult by the nature of the people being led, who are typically engineers. Engineers tend to be highly intelligent and independent, confident in their own knowledge and expertise within their field. They also tend to be introverts, more comfortable working alone than in groups, and their confidence often leads to a technical arrogance; they're certain their way is the only correct way. These characteristics make leadership all the more important.

Leadership is more than just telling people what to do. The leader's role is to induce others to follow, as part of a team. It calls for an attitude of support rather than an attitude of power. The leader supports and encourages the rest of the full-time team. while the leader creates supportive relationships with others who may have to provide work for the project. The kinds of support vary and depend on circumstances, but usually include creating smooth interpersonal interfaces, resolving conflicts, and making others' jobs easier and more rewarding.

> Necessary element for SEMP execution # 1—The technical leader must lead the technical aspects of the project.

Figure 14-10 shows the relationship between technical management, project management, and system design. Project management focuses on cost and schedule control, corporate management, and stakeholder relations. System design focuses on the creative technical issues to develop a system that meets the requirements. Technical management focuses on the technical integration of the work efforts, so they fit together to create a working system.

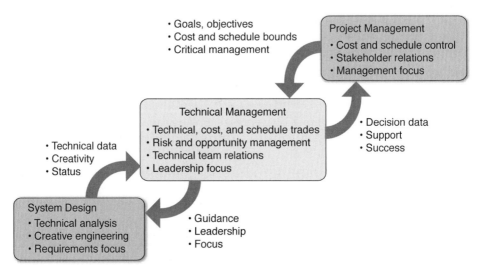

FIGURE 14-10. **The Discipline of Technical Management.** Excellent communication skills leverage the strengths of the management and the design teams.

14.5.2   Discipline of Technical Management

Technical management features in all systems engineering standards. Some standards dedicate more words to this discipline than to any other systems engineering activity. Without technical management, the creative design engineers all move in their own directions, making assumptions that may not be compatible with other engineers. The result can be disastrous during system integration and test, when the team first discovers the conflicting assumptions.

> Necessary element for SEMP execution # 2—Excellent technical management entails constant attention to the operational and technical goals, tempered by risk, to keep changes to a minimum.

Any of several different people may perform the technical management role. Figure 14-11 shows several possible configurations; the right one for any given project depends on the skill levels and interpersonal relationships of the project leaders. A strong project manager or a strong systems engineer may perform all technical management. They may share tasks, or others may be involved. A project engineer may be assigned specifically to this leadership role. Tools and techniques appropriate to technical management are: (1) general management methods such as communicating and negotiating, (2) product skills and knowledge, (3) status review meetings, (4) technical work directive system, (5) organizational processes, and (6) technical tracking databases. The technical manager communicates with technical experts through technical information. The architecture and design at the system or subsystem level provide the context within which the experts apply their expertise. Communication between the system and subsystem domains and the component and part domains determines the system's success. Team coordination and management are critical. Table 14-13 illustrates some of the strengths and shortcomings of teams. The technical leader, using the SEMP, must see that the teams work at their top potential.

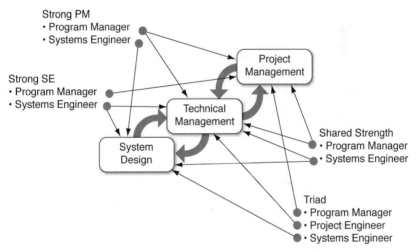

FIGURE 14-11.   Who Performs Technical Management. Any one of several relationships may succeed. *(Courtesy of Honourcode, Inc.)*

TABLE 14-13.   Traits of Effective and Ineffective Teams. A key to success is recognizing when a team is weak.

Effective Teams	Ineffective Teams
• A positive cooperative climate prevails • Information flows freely within team • No work is considered beyond an individual's job description • Interpersonal interactions are spontaneous and positive • The team's collective energy is high	• A climate of suspicion and distrust exists • Information is hoarded or withheld • Finger-pointing and defensiveness prevail • Counterproductive subgroups and cliques form • Fear of failure causes individuals to avoid or postpone making important decisions

14.5.3   Collaborative Environments

The Internet has changed forever the landscape of team collaboration. Today, teams collaborate over space and time in ways not possible even a few years ago. Because space systems are often international in scope, multi-cultural issues make collaboration more difficult. The nature of teams has changed rapidly over the last decades, with dynamic structures never before seen. In today's world of geographically dispersed, virtual teams, successful team development requires careful thought about the systems that support the team. Effective planning accounts for the technical processes and the social and relationship aspects of teams.

> Necessary element for SEMP execution # 3—The technical leader must excel within a collaborative environment, and enable it across the project.

Technical teams today have access to powerful tools that contribute to effective collaborative environments. Classes of collaboration tools include data warehouses, shared information tools, meeting support, and decision support.

14.5.4   Scope Control

The SEMP is useful for scope control, which means keeping the technical scope within bounds, across space, time, and technical concepts. Not only do requirements tend to grow, but so do the expectations of the stakeholders, investors, and members of the team. *Scope control* ensures that the project includes all and only the work required to complete the project.

> Necessary element for SEMP execution # 4—The SEMP and the technical leader must bound the scope of the project.

If properly planned, the SEMP defines the work required for the project. When we implement that work through negotiation, technical work directives, and leadership, the project should be successful. The challenges arise in the realizations that happen during implementation, because no plan is perfect. Nineteenth-century German general Helmuth von Moltke said, "No plan survives first contact," and this applies to technical plans as well. As the project proceeds, unplanned events occur and unexpected problems arise. Scope control aims to ensure that the project responds appropriately to these issues without overly affecting the original plan.

14.5.5   Stakeholder Involvement

Stakeholder relationships are a large part of effective scope control. Because different stakeholders have differing goals, they often have conflicting desires for the project. One stakeholder wants more time spent analyzing a possible risk, while another wants project costs reduced. These conflicts create the environment within which scope control is necessary.

> Necessary element for SEMP execution # 5—Effectively implementing the SEMP requires building and maintaining relationships with stakeholders.

With good relationships, the technical manager keeps stakeholders informed and handles changes in a controlled way. To build relationships, we need to understand the stakeholders' needs, put ourselves in their shoes, and see the project from their viewpoint. We must use their language and terminology. For example, to users we should present the operational impacts of a technical issue rather than presenting the depth of the issue. We should go only to the level that the stakeholders need. The relationship with each stakeholder should involve regular, quality communication. It should be ongoing, fed by positive contact. We should communicate any issues of interest to a stakeholder as they occur. One goal in the relationship is that there be no surprises. A common practice is the traffic light reporting technique as shown in Figure 14-12.

● Red Light — Major problem exists that may affect cost, schedule, scope, or performance. Sponsor involvement is necessary.

○ Yellow Light — Potential problem exists. Sponsor is informed but no action appears necessary at this time.

◐ Green Light — Work is progressing as planned. Sponsor involvement is not necessary.

FIGURE 14-12.   **Traffic Light Reporting.** A project manager and technical manager use the colored light scheme to report the project's status monthly to senior managers and other high-level stakeholders. *(Courtesy of Honourcode, Inc.)*

Project status "traffic light" reporting. Senior managers sponsor the project through their organization, but are not involved in its day-to-day workings. These sponsors feel ownership and keen interest in the project even if their participation may be minimal for months at a time while things are going well. Still, the technical manager and project manager must keep the sponsor informed about the project and its needs. A simple way is to use "traffic light" reporting, as shown in Figure 14-12. Each major element of the project is reported with a color that keeps the sponsor informed as to the status and the participation that the project team deems appropriate. Monthly traffic light reporting is common.

14.6   A Look at DoD's SEP

The systems engineering plan (SEP) that DoD uses for acquisitions is a very strong tool for the program office. "Systems engineering management is the technical function that ensures that SE and all other technical functions are properly performed. How the project will be managed to achieve its goals and objectives within the programmatic constraints is typically defined in a project plan. A government systems engineering plan (SEP) is an overarching document that defines how the project is to be technically managed within programmatic constraints." [DoD, 2004] This document shows how the government conducts the technical management. The oversight should require each contractor to have its own SEMP describing how they will manage according to direction from the top level SEP. This tie between the contractor and the program office must ensure consistency between them and timely updates of both documents. The stated approach for the SEP is as follows (all citations are DoD [2006]):

- "The SEP is the Program Manager's plan, but is often jointly developed by the government and its contractor(s)."
- "The SEP is a 'living' document that captures a program's current and evolving systems engineering strategy and its relationship with the overall program management effort. The SEP purpose is to guide all technical aspects of the program."

The SEP is developed around, and must answer, a series of questions:

- What systems capabilities, requirements, and associated design considerations must we address?
- What organizational integration is necessary to address these requirements, including the systems engineering organization and infrastructure? This integration includes staffing, individual and organizational responsibilities and authorities, and training needs.
- What engineering effort, work products, and schedule do we need to meet the requirements?
- Who will manage the technical effort and how? This encompasses technical baseline implementation and control, and technical reviews, including metrics, event-driven entry criteria, and exit criteria.
- How will the SEP link with other technical and programmatic planning efforts?

We also obtain the information needed on the technical side if the approach answers the following questions:

- What are the technical issues and risks?
- Who has responsibility and authority for managing the technical issues and risks?

- What processes and tools will we use to address the technical issues and risks?

- How will that process be managed and controlled?

- How does the technical effort link to the overall management of the program?

The multiple pages of specific questions are very valuable for any space development program [DoD (2), 2006]. Table 14-14 gives 5 of the 50 detailed questions in the SEP Preparation Guide that we use to develop the SEP for different phases of a program.

TABLE 14-14. **Systems Engineering Plan (SEP) Question Examples.** The approach to developing a government SEP uses questions that have been historically shown to result in a strong technical plan.

Program Phase	Focus Area	Question
Concept refinement	Program requirements	How well does the technical approach reflect the program team's understanding of the user's desired capabilities and concepts?
	Technical staffing	How well does the technical approach describe how technical authority will be implemented on the program?
Production	Technical management planning	How well does the technical approach describe who is responsible for managing the technical baselines?
	Technical review planning	How well does the technical approach describe who is responsible for overall management of the technical reviews?
Reduce logistics footprint	Integration with overall management	How well does the technical approach describe how the program manager will use the in-service reviews to manage the technical effort and overall operational and support (O&S) cost containment?

The outline for the SEP is very similar to the SEMP, but it emphasizes the program management side of the equation.

The Outline of the SEP [DoD, 2006]

Title and Coordination Pages

Table of Contents

1. Introduction

    1.1   Program Description and Applicable Documents

    1.2   Program Technical Status as of Date of this SEP

    1.3   Approach for SEP Updates

2.   Systems Engineering (SE) Application to Lifecycle Phases
    2.1   System Capabilities, Requirements, and Design Considerations
        Capabilities to be Achieved
        Key Performance Parameters
        Statutory and Regulatory Requirements
        Certification Requirements
        Design Considerations
    2.2   SE Organizational Integration and Technical Authority
        Organization of Integrated Project Teams (IPTs)
        Organizational Responsibilities
        Integration of SE into Program IPTs
        Technical Staffing and Hiring Plan
    2.3   Systems Engineering Process
        Process Selection
        Process Improvement
        Tools and Resources
        Approach for Trades
    2.4   Technical Management and Control
        Technical Baseline Management and Control (strategy and
          approach)
        Technical Review Plan (strategy and approach)
    2.5   Integration with Overall Program Management Control Efforts
        Acquisition Strategy
        Risk Management
        Integrated Master Plan
        Earned Value Management
        Contract Management

The list below shows the difference between technical planning in an SEP versus an SEMP. Both approaches rely on SEMPs at the lower levels of a project hierarchy. Both have tremendous strengths while emphasizing excellent management of the systems and discipline engineering aspects of the project. To achieve this high level of technical leadership, the project must have superb people and excellent processes. Completing an SEMP or an SEP early in the project is critical to seeing that the engineering elements are guided wisely.

DoD approach

- Dominant DoD SEP at the System Program Office (SPO) level
- Lower level contractor SEMPs supporting the government SEP

NASA approach
- Tiered SEMPs from government SPO level
- Contractor SEMPs supporting government leads

We must develop the SEMP early and continually improve it during the life of the program. This chapter illustrates some key elements of the process, along with a comparison with the DoD SEP methodology. Successful systems engineers focus on early involvement; concentrate on technical details (especially in requirements), uncover and resolve schedule issues early, conduct significant trade studies as appropriate; and pay constant attention to risk tradeoffs. A well-done SEMP enables the project to move forward smoothly while helping to uncover problems early enough to resolve them within project constraints.

References

Department of Defense (DoD). 2004. *Systems Engineering Plan Preparation Guide*, Version 0.85. OUSD(AT&L) Defense Systems/Systems Engineering/Enterprise Development.

DoD. 2006. *Systems Engineering Plan Preparation Guide*, Version 1.02, OUSD(AT&L) Defense Systems/Systems Engineering/Enterprise Development, ATL-ED@osd.mil.

National Aeronautics and Space Administration (NASA). 1995. *Systems Engineering Handbook*, SP-6105. Washington, D.C.: NASA Headquarters.

NASA. August 31, 2006. *Constellation Program Systems Engineering Management Plan*, CxP 70013. NASA.

Teaching Science and Technology, Inc. (TSTI). 2006. *FireSAT Systems Image.*

Chapter 15

Manage Interfaces

Robert S. Ryan, *Engineering Consultant*
Joey D. Shelton, *TriVector Services*

As many systems engineers have said: "Get the interfaces right and everything else will fall into place." Managing interfaces is crucial to a space system's success and a fundamental part of design. When we decide how to segment the system, we automatically choose the interfaces and what they must do. These interfaces take many forms, including electrical, mechanical, hydraulic, and human-machine. Our choice also includes communications (electromagnetic signals), as well as computer (software code) interfaces to other systems. Figure 15-1 shows typical space system components and their interfaces.

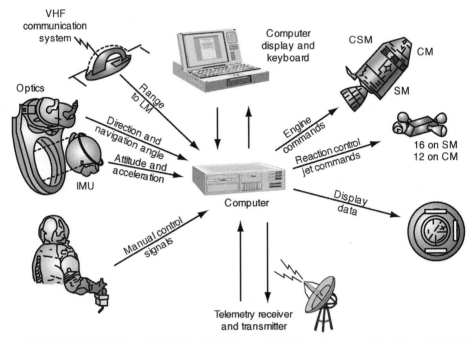

FIGURE 15-1.   Apollo's Top-level Interfaces. Here we show interfaces for human to machine; computations for guidance, navigation, and control; other controls; and telemetry. (CSM is command service module; CM is crew module; SM is service module; IMU is inertial measurement unit; LM is lunar module; VHF is very high frequency.)

Kossiakoff and Sweet [2002] say that managing interfaces consists of "(1) Identification and description of interfaces as part of system concept definition and (2) Coordination and control of interfaces to maintain system integrity during engineering development, production, and subsequent system enhancements." They go on to assert that defining "internal interfaces is the concern of the systems engineer because they fall between the responsibility boundaries of engineers concerned with the individual components." Similarly, according to the Department of Energy [DOE, 2006]:

> Interface management includes identifying system interfaces, defining the requirements at each system interface boundary, and managing (controlling) the interfaces during all steps of the systems engineering process. The interfaces are identified at the beginning of the task or project and continually managed throughout the life of the task.

To define and apply interfaces, we must often consider design trade-offs that affect both components. Implied in this idea are internal interactions, as well as external ones such as transportation, handling with ground support equipment (GSE), communications, human-machine interactions, and natural and induced environments.

The management process, outlined in Figure 15-1, is critical to product success and a fundamental part of systems engineering. We recommend NASA Procedural Requirement (NPR) 7123.1a for interface management requirements.

TABLE 15-1. **Process for Managing Interfaces.** Several other chapters describe topics similar to some of these steps, as noted in the "Where Discussed" column.

Step	Definition	Documentation	Where Discussed
1	Prepare or update interface management procedures	Work breakdown structure (WBS), organizational chart	Section 15.1
2	Decompose the system physically and functionally	Physical architecture diagrams, functional flow block diagrams	Section 15.2 and Chap. 5
3	List interfaces and prepare initial interface requirements documents (IRDs)	Interface list, IRDs	Section 15.3
4	Develop N×N and I×I diagrams	N×N and I×I matrices	Section 15.4 and Chap. 13
5	Develop sub-level WBS of the lifecycle interface design and verification tasks	WBS	Section 15.5
6	Develop an interface control document (ICD) for each interface	ICDs	Section 15.6
7	Manage interfaces during product integration	Interface control plan, interface control working group minutes	Section 15.7 and Chap. 10
8	Design interfaces, iterate, and trade	Analysis reports, IRD and ICD updates	Section 15.8
9	Build interfaces	Interface drawings, manufacturing plan	Section 15.9 and Chap. 9
10	Verify interfaces including integration with the system	Verification requirements, verification closure data, procedures documents	Section 15.10 and Chap. 11
11	Document, iterate, and control the configuration	Standards	Section 15.11
12	Develop operating procedures and training	Operational sequence diagrams, training procedures	Section 15.12

The following sections describe these steps for managing interfaces. We summarize them in Figure 15-2, where we start with requirements and then divide the system into elements as our first definition of potential interfaces. Using these elements and potential interfaces, we

1. Develop the interface function and define interfaces

2. Tailor induced environments so interfaces and concepts will meet requirements in their presence

3. Select and design the interface concept, using trade studies based on sensitivity and risk

4. Manufacture and verify the interfaces

5. Develop operational constraints, procedures, and plans

Standards, guidelines, criteria, philosophies, and lessons learned guide the process, which we manage and control with interface control documents.

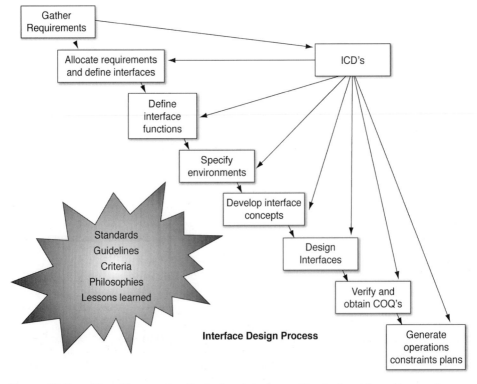

FIGURE 15-2.   **Flow Diagram for Designing Interfaces.** Standards, philosophies, and so on furnish guidance to the process, but we manage it by means of interface control documents (ICDs). (COQ is certification of qualification) (*Source: NASA*)

15.1   Prepare or Update Interface Management Procedures

From the top-level work breakdown structure (WBS), we develop interface management and control plans. Later, the WBS will describe the detailed design task or will refer to the interface management plan (IMP), which is generated when the systems engineering team selects an architecture and concept (Chapter 3). As we analyze the architecture's physical and functional components, we must define interfaces as simply as possible. Interface experts must participate in defining architectures and concepts, and this cooperation should be the first statement in the interface control plan. After this step, with all appropriate trades in place, we develop the IMP. Interface design is part of the design process discussed in Chapters 2–5.

Developing an interface management process includes interacting with the mission's entire support infrastructure. For example, a space launch vehicle has interfaces with transportation, assembly, processing, check-out, launch, and mission operations. Many of these interfaces involve human-machine connections, as well as electrical, power, hydraulic, mechanical, and other functions. A second set of interfaces involves the natural and induced environments—determined largely during technical systems integration (Chapter 3). The last set of interfaces—internal—is where components meet in at least four partitions: structural and mechanical, electrical, data and software, and fluid.

The IMP consists of project control (decision making authority) and management elements and tools. Typically, an interface control working group recommends interface control plans, changes, and so on to the project boards. This group coordinates and sees that interfaces are managed properly, usually through IRDs and ICDs that document the control of hardware and software. Preliminary interface revision notices are a formal way to suggest changes. The program level that governs an interface determines the final control. The interface management process dictated by the WBS is the triangle of requirements, processes, and designs under control of the working group and the project's chief management officer. It's based on the interface control plan, using interface requirements documents (IRDs), interface control documents (ICDs), databases, and certification documents.

Interface management helps control product development when efforts are divided among parties such as government, contractors, and geographically separated technical teams. It defines and maintains compliance between interrelated products. Figure 15-3, from NPR 7123.1a, depicts the process flow and identifies its inputs, outputs, and activities. Before beginning to manage interfaces, we need five main types of information:

- **System description**—Enables us to explore and examine the system's design, so we can determine where system interfaces are and include them in contractor arrangements

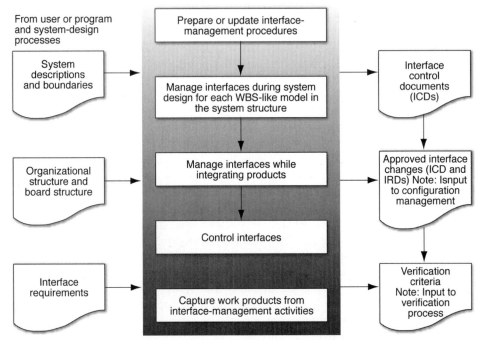

FIGURE 15-3. **Process Flow for Interface Management.** Here we define the necessary inputs and outputs for interface management. The key inputs come from the system, work breakdown structure, and requirements. The outputs focus on interface design, change control, and verification. (IRD is interface requirements document.) [NASA, 2007]

- **System boundaries**—Physical boundaries, components, and subsystems help us determine the interfaces

- **Organizational structure**—The mission organization must describe interfaces, particularly when groups need to agree on a system's shared interface parameters. The program and project work breakdown structures also describe interface boundaries.

- **Board structure**—The systems engineering management plan must reveal organizational interfaces and their locations

- **Interface requirements**—Defining a system's functional and physical requirements establishes the internal and external interfaces

While forming the concept of operations (Chapter 3), we analyze it to identify external and internal interfaces. This analysis establishes the origin, destination, stimuli, and special characteristics of the interfaces that we must document and maintain. As the system's structure and architecture emerge, we add or change interfaces to match them. At this stage, then, managing interfaces relates closely to such areas as defining requirements (Chapter 4) and managing system configuration (Chapter 16). An interface working group (IWG) normally establishes communication between team members, who must interface systems, end products, enabling products, and subsystems. The IWG makes sure that teams plan, schedule, and complete all interface activities. These are usually technical teams, with members from interfacing parties such as the project or contractor.

During product integration, interface management supports reviewing integration and assembly procedures to ensure that interfaces are properly marked and compatible with specifications and interface control documents. During this period, managing interfaces relates closely to integrating (Chapter 10) and verifying and validating (Chapter 11) products. We use interface control documents and approved interface requirement changes to verify and validate products, particularly whenever we need verification test constraints and interface parameters to set testing objectives and plans. Verifying interface requirements is critical to verifying the overall system.

15.2   Decompose the System Physically and Functionally

Chapter 5 describes decomposition analysis. The first step is to divide the hardware system into subsystems, elements, components, and parts based on industrial, academic, or other specialties, as well as by requirements resulting from the functional analysis (Chapters 3–4). We also divide these elements into design functions and discipline functions. Splitting the design task into compartments means we must strongly emphasize technical integration of interfaces and interactions, which must be properly managed and controlled. The "hows" and "whats" of decomposing a mission affect the interfaces. For example: in Project Apollo and now Ares 1, systems engineers decided the launch vehicle would have a guidance, navigation, and control (GN&C) system for the boost phase, whereas the command module would have a GN&C system for other mission phases. A simple interface between the two systems would provide critical information.

The command module could have supplied all GN&C, but this design would have greatly complicated the hardware and software, and routed all signals to subsystems on the launch vehicle. By separating the two functions, engineers could design and build simpler, more robust systems. They could also achieve more flexibility, because the launch vehicle could support payloads other than the command module. Adding the human-machine interfaces for the astronauts, as well as launch and mission control, increased the complexity of design, verification, and training. (Buede [1992] shows an example of segmenting functions and defining requirements for inputs, outputs, and interfaces.)

After decomposing the systems, we have to separate the hardware and electrical pieces at their interfaces and design special interface devices. For example, the electrical connections' failure to separate between the first and second stages of an Inertial Upper Stage resulted in the loss of a mission to send a satellite into geosynchronous orbit. Electrical bundles for the first and second stage connected using a pin connector, which required the parts to move toward each other to release a coupler that allowed them to separate. But teams wrapped insulation on each part, so the parts couldn't move, and the coupler never released. The unseparated cable kept the two stages coupled and thus destroyed the mission. Many types of stage interfaces are available to offer stiffness and load paths during one stage burn, as well as a way to separate the two stages. Decomposition produces various interfaces: from pad hold-down, to fluid or electrical, to human-machine. Table 15-2 lists the results of decomposing a project into systems, their requirements, and their interfaces.

TABLE 15-2.   Results of Decomposing the Project into Systems, Their Requirements, and Interfaces. In the left column we list the project-level requirements, specifications, standards, and constraints. In the right column we list where we must account for interfaces within the decomposition.

Project Specifications	Interface Requirements
• Allocated requirements for project functions and performance	• Allocated requirements for interface performance • Derived requirements for interface performance
• Constraints on end-item design	• Allocated constraints on interface design • Derived constraints on interface design
• Requirements of the natural environment • Requirements of the induced environment	• Interface requirements for induced environments
• Standards and specifications from industry, government, agency	• Industry, government, agency standards and agreements on how to apply specifications • Agreements on standards and specifications for derived component
• Software processing requirements	• Requirements for software inputs and outputs
• Allocated power consumption • Power standards	• Allocated requirements for interface voltages, currents, power quality
• Allocated physical constraints for end items	• Derived agreements on design constraints for physical interfaces
• Allocated requirements for heat generation and rejection	• Allocated requirements and constraints for heat transport • Agreements for heat transport

When we decompose the project, we get systems and their interfaces. Figure 15-4 shows a general interface plane and sample quantities that must pass through from one system to another. Table 15-3 lists the categories of interfaces, their description, and comments.

We illustrate how to define system interfaces using the FireSAT project. Figure 15-5 charts the FireSAT spacecraft's subsystems and shows how we divide the spacecraft into a number of configuration items for configuration management. The spacecraft's implementation plan calls for Acme Aerospace Inc. to be the

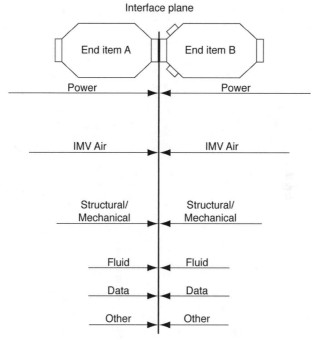

Part I ICDs define allocated and derived functional interface
performance requirements necessary to ensure integrated
performance at the element-to-element interface.

FIGURE 15-4.   **General Plane for Element-to-element Interfaces.** This generic interface plane passes power, air, data, and other quantities. It must also hold together structurally. (IMV is intermodule ventilation.)

primary spacecraft integrator. They also furnish the propulsion module as a single configuration item from their subcontractor, PN Test Systems Ltd. Thus, we must define carefully the interfaces between the propulsion module and the rest of the spacecraft. Figure 15-6 shows the location of key interfaces between the propulsion module and baseplate assembly. The structural-mechanical interface of the propellant module and base plate depend on the agreed-to design of the physical interface between the two items. Electrical interfaces activate the valves and pyrotechnic devices, and they measure pressure and temperature for telemetry. The fluid interface defines how the propellant (monomethyl hydrazine) and pressurant gas (helium) flow between the propellant module and the base plate (rocket engines). In the next few sections, we look at how to analyze and define these and other interfaces between the two.

TABLE 15-3. **Categories and Functions of Interface Types.** This table lists the types of interfaces, their functions, examples, and remarks. (EMI is electromechanical interference; EMC is electromechanical compatibility.)

Categories	Functions	Types/Examples	Remarks
I. Structural and mechanical	1. Structural integrity between elements, subsystems, and components • Load paths • Stiffness • Strength • Durability 2. Separation of elements as mission timeline dictates	1. a. Flanges b. Bolts c. Welds d. Links e. Fasteners f. Adhesives 2. a. Pyros b. Springs c. Hydraulics	1. Form and fit between interface mating parts is critical and a source of many problems. Verification of form and fit as well as structural capability is a major challenge. 2. Malfunctions in separation systems have created many problems. Verification of separation systems is a major activity and challenge.
II. Fluid and hydraulic	1. Propellant flow between elements 2. Air flow between elements 3. Control forces and separation forces	1. Duct and flanges 2. Duct and flanges 3. Actuators and links	1. Prevention of leaks with ability to separate as required 2. Prevention of leaks with ability to separate as required 3. Ability to handle point loads and varying dynamics
III. Electrical	1. Transmit power between elements 2. Communication between elements 3. Information flow between elements	• 1-, 2-, and 3-pin connectors • Wires • Busses • Transmission waves	1. Provide adequate power with the ability to separate 2. Provide clear communications with the ability to separate 3. Provide information with the ability to separate
IV. Environmental	1. EMI and EMC 2. Natural environments 3. Induced environments	1. Wire shielding and separation 2. System's ability to function in the surrounding environment 3. System's ability to control, manage, and function in the created environment	1. Design for integrity of electronic or electrical signals 2. System must function in the environment (examples: temperature, winds) 3. System must function in the environment it produces (examples: thruster plume, thermal-protection system, icing)

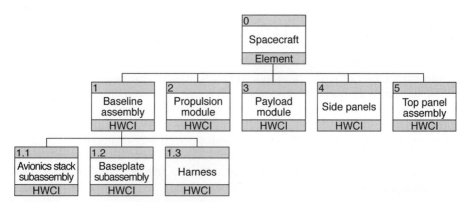

FIGURE 15-5.   Configuration Management Chart for the FireSAT Spacecraft. The diagram shows the FireSAT spacecraft's major elements as hardware configuration items (HWCI).

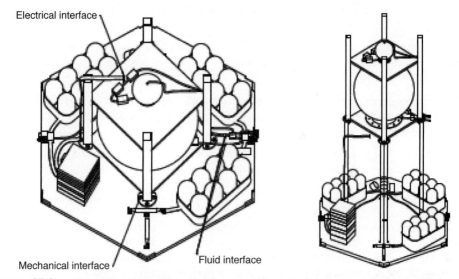

FIGURE 15-6.   FireSAT's Propulsion System and Its Interfaces. The electrical interface connects the batteries to electrical systems. Flow from the pressure vessels to the propulsion subsystem requires a fluid interface. The mechanical interface physically connects the structures.

15.3   List Interfaces and Prepare Initial Interface Requirements Documents (IRDs)

With physical and functional decomposition complete, our next step is to list all interfaces with their functions and requirements, which requires us to develop N×N and I×I diagrams in parallel. We start with a precise statement of the interface's functions, from which we derive all other requirements. The function describes what to do and how to do it. In addition, requirements must cover form and fit, operations (including sequencing), and the natural and induced environments. This step is critical to verifying requirements for the final product, including all system interactions across the interface. Starting with the list of interfaces, we define requirements in four main categories [DOE, 2006]:

- *Interface function*—Specifies what the interface is to do. Examples of this requirement are to provide stiffness and load carrying between two elements of a space vehicle, supply a way to separate the elements at the right time, and give mission control a way to abort the flight to save the crew.

- *Performance interface requirement*—Specifies how well the interface must maintain such parameters as pressure, temperature, flow rate, gas composition, batch amount, transfer frequency, voltage, power, purity, and water quality

- *Design interface constraint*—Specifies codes and standards that apply to the interface, a specific design, an operating or maintenance configuration, and essential features and materials

- *Physical interface requirement*—Specifies physically related characteristics for components at the interface boundary, such as materials, dimensions, tolerances, finish, size, weight, dynamic limits, equipment envelopes, footprints, layout locations, and orientation to plant reference marks

The human-machine interfaces require special consideration, for two main reasons. First, information flows across the interface in two ways—from the machine to the person and from the person to the machine. Second, the person must decide something based on the information received. This decision requires clear definitions for the interface, as well as system performance that enables the correct decision. These interfaces work in many ways. For example, they're continuous if we're driving a car, visual if we're assembling parts, and referential if we're drawing on queued performance information to decide how to abort or change a space vehicle for mission success or crew survival.

After assembling all the information, we develop, baseline, and control the requirements document. These requirements don't specify how to do the job, but they clearly define what the interface must do. Requirements include performance specifications and the environmental conditions under which the interface must work.

Opinions differ on whether interface requirements documents should be separate or part of the ICDs. Although having them in the ICDs offers advantages for traceability and control, we believe it's prudent to document requirements separately—perhaps restating the essential ones in ICDs.

15.4   Develop NxN and IxI Diagrams

NxN diagrams are excellent tools for making sure that we account for all interfaces, interactions, and data flows in the systems engineering process. (See Section 13.2 for a detailed discussion). NxN diagrams for interface interactions are called IxI matrices. The diagrams list the system's elements on the diagonal. The off-diagonal cells contain interface requirements and descriptions among peer subsystems along the common tier (the diagonal elements of the matrix). For such subsystems, two interface information flows occur concurrently (represented by two shades on the inset blocks). One flow involves interface requirements of subsystem A on subsystem B, plus B's interface description fed back to A. The other flow involves requirements of B on A, plus A's interface description fed back to B.

We also need lower-level diagrams, but they follow the same format. For example, the IxI matrix for propulsion and structures could contain the terms in Table 15-4, but we'd generate the same type information for each subsystem, such as avionics, thermal, and guidance, navigation, and control.

TABLE 15-4.   **Sample Entries from an IxI Matrix.** Here we show several examples of diagonal terms and their influence on other diagonal terms from an IxI matrix for typical propulsion and structures systems.

Diagonal Terms		Off-diagonal Terms
• The propulsion system to structures	Requirements	1. Thrust bearing to transfer thrust load to upper stage 2. Engine's gimballing ability for control authority (throw angle in degrees) 3. Fluid propellant input to engine (flow rate)
	Description	1. Engine's induced environments (thermal, acoustic, vibration, thrust) 2. Engine's dimensions 3. Mass characteristics 4. Flow rates
• Structural system to propulsion	Requirements	1. Mechanism attachments 2. Induced environments 3. Propellant line attachments 4. Electrical attachment
	Description	1. Propellant line flex bellows 2. Description of gimbal joint flange 3. Description of propellant line flange 4. Volumetric constraints 5. Detailed drawings of interface mating areas

I×I matrices are numerous because they're associated with each system or subsystem divided into lower-tier entities on the tree. (The lower tier is the off-diagonal elements that share interface characteristics and information with the diagonal elements.) The parts on the tree's last tiers, however, are simple enough to require no further subdivision. Figure 15-7 shows an I×I matrix for FireSAT.

N×N matrices represent information flow among the design functions and the discipline functions. The example in this case is the launch vehicle system with its associated design functions and disciplines. Notice the matrix's upper left element represents the launch vehicle system's plane, and other diagonal elements represent the remaining planes for lower design functions.

Including I×I and N×N matrices with the subsystem tree and design function stacks provides locations or placeholders for the technical information that must flow among the participants in the design process. (The subsystem tree is the decomposition of the system into its lower elements; the design function stack describes the activities associated with the tree.) It also suggests a framework for electronic information and communication to enable efficient, concurrent interactions.

These diagrams are outstanding tools to define a project's decomposition and all critical information, interactions, and requirements. They're essential to design and systems engineering and are the cradle for technical integration activities. We must define them accurately, review them for completeness, and update them periodically throughout the project's lifecycle.

Let's return to our FireSAT interface example. The propulsion system's main function is to generate thrust. Figure 15-8 shows its two sub-functions, which are at a level where we can describe detailed interfaces. We need to analyze the inputs and outputs between **control propulsion** and **manage fluids**, where fluids include helium pressurant gas and monomethyl hydrazine propellant.

An effective way to analyze the sub-functions is to create an N×N diagram just for them and study the inputs and outputs to each. Figure 15-9 shows the matrix with appropriate interactions.

Next we allocate the two sub-functions to spacecraft hardware. In this case, **control propulsion** goes to the baseplate assembly, and **manage fluids** to the propulsion module. To account for the many inputs and outputs to the two main subfunctions, we list them in an input-output and allocation table (Table 15-5). Each input and output function is allocated to one of the hardware configuration items. Figure 15-10 shows how this allocation looks for **control propulsion** and **manage fluids**. To reach the ICD-IRD inputs from the N×N matrix, we use the allocation table entries to define links between the baseplate assembly and the propulsion module.

The next step is to define interface requirements for each link. A requirements document for spacecraft-to-propulsion module interfaces captures these requirements, and controlled interface drawings (Figure 15-11) show details. Because Acme is the prime contractor, they write interface requirements for their subcontractor, PN Test Systems. "Will" statements reflect the interfaces that Acme plans to provide and "shall" statements are requirements that PN must meet. Interface requirements look like the following examples:

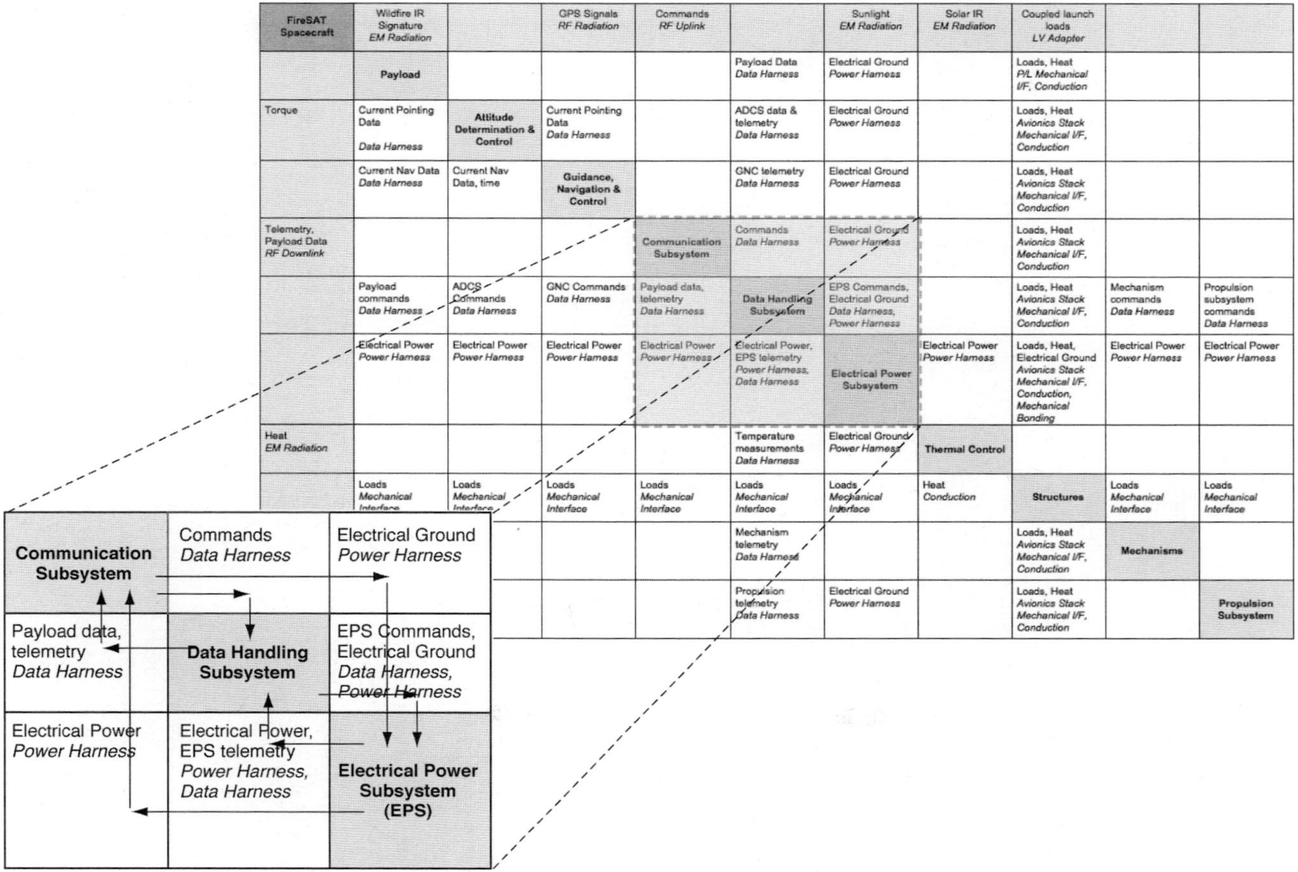

FIGURE 15-7.   FireSAT IxI Matrix. The IxI matrix is a useful tool for capturing the interfaces between system elements or between subsystems within a given system.

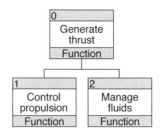

FIGURE 15-8.   FireSAT's Generate Thrust Function. Here we show how to divide the spacecraft's propulsion system function—**generate thrust**—into two sub-functions that require interfaces.

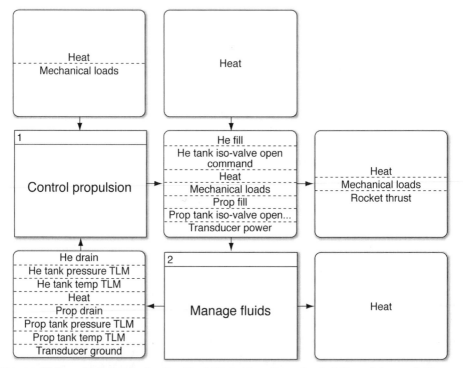

FIGURE 15-9.   N×N Matrix for the FireSAT Subfunctions Control Propulsion and Manage Fluids. Here we show inputs to the **manage fluids** subfunction, such as Helium (He) fill and He tank Iso-valve (isolation valve) open, as well as inputs to the **control propulsion** subfunction, such as He drain or He tank pressure telemetry. Also shown are their outputs (directed to the right from the subfunction boxes) and other inputs, such as heat and mechanical loads. (TLM is telemetry; He is helium.)

TABLE 15-5.   **FireSAT Input-output and Allocation Table.** This table lists inputs and outputs for **control propulsion** and **manage fluids**, allocated to the baseplate and propulsion modules. (TLM is telemetry; He is helium.)

Function	Inputs	Outputs	Done by the
Control propulsion	He drain He tank pressure TLM He tank temp TLM Heat Mechanical loads Propellant drain He drain Propellant tank pressure TLM Propellant tank temp TLM Transducer ground	He fill He tank isolation valve open command Heat Mechanical loads Propellant fill Propellant tank isolation valve open command Rocket thrust Transducer power	Baseplate assembly
Manage fluids	He fill He tank isolation valve open command Heat Mechanical loads Propellant fill Propellant tank isolation valve open command Transducer power	He drain He tank pressure TLM He tank temp TLM Heat Mechanical loads Propellant drain He drain Propellant tank pressure TLM Propellant tank temp TLM Transducer ground	Propulsion module

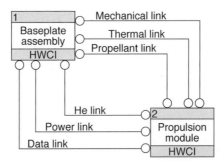

FIGURE 15-10.   **Links between FireSAT's Baseplate Assembly and Propulsion Module.** Here we show links that represent physical connections between the two hardware configuration items. Each connection represents an interface taken from the types listed in Table 15-3. Each interface must have a written requirement that describes its physical design and function.

- The propulsion module shall mechanically interface to the spacecraft using 16 #10 fasteners through the bolt-hole pattern specified in ICD XYZ

- The propulsion module contractor shall provide a 10-node thermal model for the propulsion module. This model must be compatible with the SCINDA Thermal Desk Top for coupled thermal analysis, including the surface area, thermal conductivity, and thermal capacitance for the conducted interface.

- The spacecraft will provide a 15 pin D-type connector (type ABC) for power and data interface to the propulsion module with pin definitions as per ICD XYZ

- The propulsion module shall accept the following Mil-Std 1553 [DoD, 1973] commands: 1) Helium tank isolation valve open and 2) Propellant tank isolation valve open

- The spacecraft will provide a single 28 ±1 V power supply line, current limited to 1 amp for propulsion module transducers

- The propulsion module shall provide 2 ground lines

- The propulsion module shall provide a 1/4-inch AN female fitting at the location shown in ICD XYZ for helium fill and drain

- The propulsion module shall provide a 1/4-inch AN female fitting at the location shown in ICD XYZ for propellant fill and drain

- The propulsion module shall provide a 1/4-inch AN female fitting at the location shown in ICD XYZ for propellant supply to the orbit control rocket

Interface details: Base plate module side

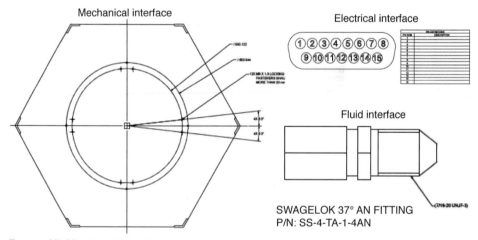

FIGURE 15-11. FireSAT's Control Drawing for Interfaces between the Baseplate Assembly and Propulsion Module. This figure shows three interface drawings with hole placement (mechanical), pin identification (electrical), and fitting design (fluid).

15.5   Develop Sub-level Work Breakdown Structure (WBS) of the Lifecycle Interface Design and Verification Tasks

In this step (Step 5 from Table 15-1) we update the WBS for the interfaces to include the information that flows from steps 1 through 4, and confirm that these are the tasks required to design, verify, and operate system interfaces. The detailed WBS relates to the tasks that are either recipients of interface data or flow directly from interface management planning. An example is system-level interfaces that we derive down to the detailed design interface at the component level.

The interface verification process incorporates the connection between the interface design, compliance with verification criteria, and verification objectives. If we verify by test, the interface's design details dictate test objectives, hardware, and procedures. The interface design and requirements are necessary parts of the compliance checklist during audits of physical and functional configurations (Section 16.6).

15.6   Develop an Interface Control Document (ICD) for Each Interface

The ICDs define and control the interface design's requirements and details (Figure 15-12). They represent three formal agreements among the participants:

- They must properly allocate requirements across, or at, the interfaces

- The designs of the interfacing hardware, software, or facilities must be consistent with the ICDs

- The interface designs must be functionally and physically compatible (Figure 15-13 depicts an example)

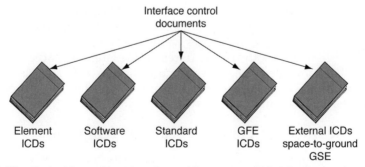

FIGURE 15-12.   Typical Sets of Interface Control Documents (ICDs) for System Interfaces. We must prepare ICDs for all interfaces required to design the system. Interfaces shown here are the main categories. (GFE is government furnished equipment; GSE is ground support equipment.)

The ICD for an interface typically consists of two parts: the first defines all interface performance requirements, and the second defines how to apply them.

For example, the ICD for a solid rocket motor (SRM) segment-to-segment interface might read:

Requirements:

1. Maintain motor pressure without leaking, verification of seal after assembly
2. Simple assembly joint at launch site
3. Structural load path between segments
4. Continuity of lateral and longitudinal stiffness of SRM case
5. Thermal insulation of joint against motor thrust

Implementation:

1. Redundant O-ring seal using a compound double clevis structural joint, grooves for O-rings. Pressure O-ring check port or ports.
2. Compound clevis joint with locking pins
3. Compound clevis joint with locking pins
4. Case thickness
5. Joint pressure sealing thermal flap

15.6.1   Software ICDs

Software ICDs are usually stand-alone documents, separate from hardware ICDs. They should contain software data word and command structures that cross the interface between two computers or devices. They may describe the operating system, code language, operating speed, frame structure, or other computer parameters. We define software interfaces with respect to the computer-to-computer, remote terminal, or other data device. Devices within an element or assembly have internal interfaces, whereas external interfaces cross interface boundaries or development boundaries within the system.

Figure 15-14 depicts software interfaces for the FireSAT spacecraft. Interfaces are part of laboratory tests of software integration as well as tests to integrate hardware and software. Derived requirements (Table 15-6) flow from original requirements, and profoundly affect system interfaces.

15.6.2   The Interface Control Document Template

After we complete the ICD for an interface, coordinating with all interested parties, we distribute it for review and then baseline it for formal control (Section 16.4). The interface control working group (ICWG) does much of this work, which we discuss in the next step, along with our approach to controlling the project's

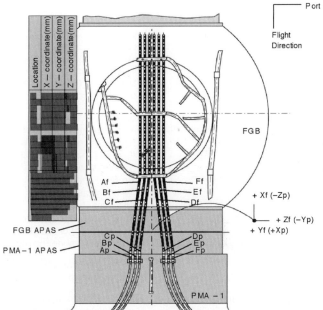

Contents
Physical interface definition
   Envelopes
   Dimensions/Tolerances
   Interface design drawings
   Fastener locations
   Connector locations

Connector/Coupling definition
   Specifications
   Part numbers
   Pin-outs
   Signal definitions

Attach mechanism designed/
      definition
   Latch mechanism designs
   Bolt patterns
   Fastener definitions

Interface surface definition
   Surface preparations
   Mating material definitions
   Bonding paths

On orbit configuration of power transfer harness assembly.

FIGURE 15-13.   Characteristics of Typical Interface Control Documents (ICDs). Here we show how detailed an ICD can be, with locations for each connector in the wire bundle that transfers power to a module on the International Space Station. (FGB is functional energy block; APAS is androgynous peripheral assembly system; PMA is pressurized mating adapter.)

configuration. An ICWG is usually more formal than the IWG mentioned in Section 15.1, though they may coincide. As we develop the ICDs, conflicts and inconsistencies always appear, so we must balance, trade, and iterate them with the system to achieve a balanced solution. The final ICD must reflect this balance.

A standardized ICD template helps to avert confusion and fosters understanding among disparate teams. It also ensures completeness of the interface definition. These documents take many forms, but the following outline and general definitions developed for the International Space Station exemplify their content. We can tailor this template to our project needs.

Interface Document Template

1.0   INTRODUCTION

   1.1   PURPOSE AND SCOPE
   State the purpose of this document and briefly identify the interface to be defined herein. For example, ("This document defines and controls the interfaces between _____ and _____").

TABLE 15-6.   **Derived Interface Requirements.** These and other derived functional requirements, industry standards, and agreements affect interface designs.

Type of Derived Requirement	Examples
Functional requirements	Utility isolation, control, monitor, and protection responsibilities Utility routing (internal or external) Acoustic noise damping
Agreements on industry, government, or agency standards and how to apply specifications	RS-170 • Video-command encryption • Video-synchronization techniques MIL-STD-1553B • 1553 bus coupling • Protocol options • Terminal fail flags • 1553 bus "box-cars" • 1553 bus extended addressing • Bus controller or remote terminal assignments • 1553 bus number and nomenclature
Bonding	Interface bonding class Requirements for surface preparation
Agreements on component standards or specifications	Fluid quick-disconnect source control and envelope drawings Specifications for fiber-optic receivers and transmitters
Agreements on physical design constraints	Reserved and keep-out areas • Utility grounding techniques • Tubing types and sizes • Alignment and indexing • Number of utility feeds and connections • Attachment and mating loads • Standard connections
Requirements and constraints for heat transport	Maximum heat flux across the interface Coolant-loop connectivity

1.2   PRECEDENCE
Define the relationship of this document to other program documents and specify which document is controlling in the event of a conflict.

1.3   RESPONSIBILITY AND CHANGE AUTHORITY
State the responsibilities of the interfacing organizations for development of this document and its contents. Define document approval authority (including change approval authority).

2.0   DOCUMENT

2.1   APPLICABLE DOCUMENTS
List binding documents that are invoked to the extent specified within this document. The latest revision or most recent version should be listed.

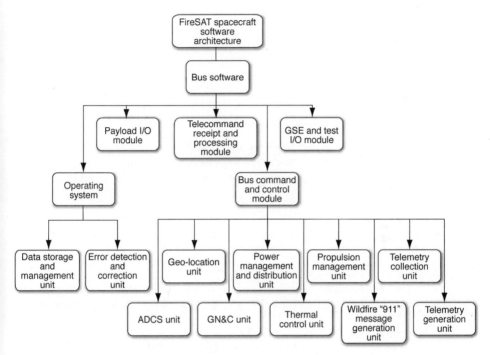

FIGURE 15-14.   FireSAT Software Architecture. This section of Figure 11-22 depicts software interfaces for the FireSAT spacecraft.

Documents and requirements imposed by higher-level documents (higher order of precedence) shall not be repeated.

2.2   REFERENCE DOCUMENTS
List any document that is referenced in the text of this section.

3.0   INTERFACES

3.1   GENERAL
In the subsections that follow, provide the detailed description, responsibilities, coordinate systems, and numerical requirements as they relate to the interface plane.

3.1.1   INTERFACE DESCRIPTION
In this section, describe the interface as defined in the System Specification. Use and refer to figures, drawings, or tables for numerical information.

3.1.2   INTERFACE RESPONSIBILITIES
Define interface hardware and interface boundary responsibilities to depict the interface plane. Use tables, figures, or drawings as appropriate.

3.1.3   COORDINATE SYSTEMS
Define the coordinate system used for each side of the interface.

3.1.4   ENGINEERING UNITS, TOLERANCES, AND CONVERSIONS
Numerical data entered in all interface documents shall be in English (inch-pound-second) units with applicable tolerances. The equivalent International System of Units (SI) value shall be added in parentheses for interface documents. For example, THA-9 lb (4.1 kg). Use standard conversions between English and SI units.

3.2   INTERFACE REQUIREMENTS or DESIGN
Define structural limiting values at the interface, such as interface loads, forcing functions, and dynamic conditions. Refer to Structural Design Loads Data Book for additional data.

3.2.1   INTERFACE PLANE
Document the interface on each side of the interface plane in the sections below.

3.2.1.1   ENVELOPE
Define the linear sizes, areas, volumes, articulating device motions, and other physical characteristics for both sides of each interface.

3.2.1.2   MASS PROPERTIES
This section of the document includes the derived interface based on the allocations contained in the applicable segment specification pertaining to that side of the interface. For example, it should cover the mass of the element in pounds or kilograms on this side of the interface.

3.2.1.3   STRUCTURAL and MECHANICAL
This section of the document includes the derived interface based on the allocations contained in the applicable segment specification pertaining to that side of the interface. For example, it should cover attachment, stiffness, latching devices, mechanisms, and other structural or mechanical interfaces.

3.2.1.4   FLUID
This section of the document includes the derived interface based on the allocations contained in the applicable segment specification pertaining to that side of the interface. For example, it should cover fluid interfaces such as thermal control, liquid O_2 and N_2 control and flow, potable and waste water uses, fuel cell water uses, atmospheric sample control, and all other fluids.

3.2.1.5   ELECTRICAL (POWER)
This section of the document includes the derived interface based on the allocations contained in the applicable segment specification pertaining to that side of the interface. For example, it should cover all

the electric currents, voltages, wattages, resistance levels, and all other electrical interfaces.

3.2.1.6   ELECTRONIC (SIGNAL)

This section of the document includes the derived interface based on the allocations contained in the applicable segment specification pertaining to that side of the interface. For example, it should cover various signal types such as audio, video, command data handling, navigation, and all other electronic signal interfaces.

3.2.1.7   SOFTWARE AND DATA

This section of the document includes the derived interface based on the allocations contained in the applicable segment specification pertaining to that side of the interface. For example, it should cover data standards, message timing, protocols, error detection and correction, functions, initialization, status, and all other software and data interfaces.

3.2.1.8   ENVIRONMENTS

This section of the document includes the derived interface based on the allocations contained in the applicable segment specification pertaining to that side of the interface. For example, it should cover the dynamic envelope measures of the element in English units or the metric equivalent on this side of the interface.

3.2.1.8.1   ELECTROMAGNETIC EFFECTS

3.2.1.8.1.1   ELECTROMAGNETIC COMPATIBILITY

The end item 1 to end item 2 interface shall meet the requirements of SSP 30243, Rev E—Space Station Requirements for Electromagnetic Compatibility [Brueggeman et al., 1998].

3.2.1.8.1.2   ELECTROMAGNETIC INTERFERENCE

The end item 1 to end item 2 interface shall meet the requirements of SSP 30237, Rev C— Space Station Electromagnetic Emission and Susceptibility Requirements [NASDA, 1996].

3.2.1.8.1.3   GROUNDING

The end item 1 to end item two 2 interface shall meet the requirements of SSP 30240, Rev D—Space Station Grounding Requirements [Jablonski et al., 2002 (1)].

3.2.1.8.1.4   BONDING

The end item 1 to end item 2 structural and mechanical interface shall meet the requirements of SSP 30245, Rev E—Space Station Electrical Bonding Requirements [Brueggeman et al., 1999].

3.2.1.8.1.5   CABLE AND WIRE DESIGN
The end item 1 to end item 2 cable and wire interface shall meet the requirements of SSP 30242, Rev F—Space Station Cable/Wire Design and Control Requirements for Electromagnetic Compatibility [Jablonski et al., 2002 (2)].

3.2.1.8.2   ACOUSTIC
The acoustic noise levels on each side of the interface shall be required to meet the program or project requirements.

3.2.1.8.3   STRUCTURAL LOADS
End item 1 to end item 2 mated loads that each end item must accommodate.

3.2.1.8.4   VIBROACOUSTICS
Account for the vibro-acoustic loads.

3.2.1.9   OTHER TYPES OF INTERFACE REQUIREMENTS
This section of the document includes other types of unique interface requirements that apply.

15.7   Manage Interfaces during Product Integration

The interface management steps aren't necessarily sequential. We do many, including this step, in parallel. Here we must ensure that project teams properly design, build, verify, and operate a space system and its interfaces. We depend on engineering for analysis and design but must plan to enhance communication and control configurations. Figure 15-15 depicts a frame of reference for the basic lifecycle flow of interface design.

This process usually starts with engineering, establishes an interface control working group (ICWG), and then makes final decisions and controls the configuration using the program requirements control board (PRCB) at each appropriate project level. We show the relationships in Figure 15-16.

The ICWG is a recommending body whose members are key people from NASA centers, involved contractors, and design specialists. NPR 7123.1a describes interface management activities in Section C.3.3.4. We don't repeat these details here but do discuss some ways to apply the management requirements. The ICWG offers a forum for assessing all interactions, risks, design maturity, trades, and so forth. It also recommends resolutions to the PRCB for issues that arise in the design process. Chairs or co-chairs coordinate the meeting and activities that engineering often performs. This group sees that the project uses best practices and identifies all major risks, as well as the best way to balance the system. Having all experts in one room ensures openness and free discussion, which lead to the best assessment. The ICWG also deals with another important, complicated set of interfaces: communication between the key disciplines and elements.

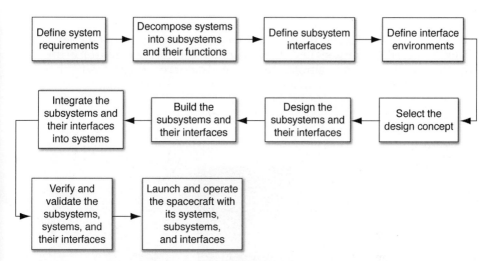

FIGURE 15-15.   **Typical Process Flow for Interface Management.** From the first definition of system requirements to spacecraft operations, we control the process using interface control documents based on standards, guidelines, criteria, philosophies, and lessons learned.

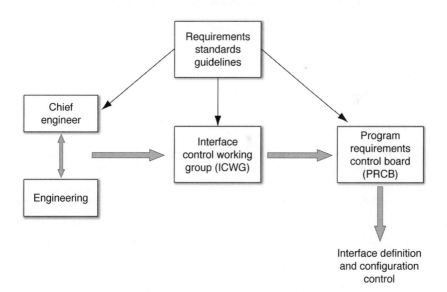

FIGURE 15-16.   **Approach to Controlling Interface Configurations.** Here we show the relationships between the designers (chief engineer and engineering) and the controllers (ICWG and PRCB). Requirements, standards, and guidelines apply to everyone in a project.

This interface design process requires members to consider not only their own discipline but also how they interact with others, so the PRCB drives project balance by developing the right project decisions at the appropriate level. Any change or decision then goes under configuration control (Chapter 16). In this way, we successfully design, control, and operate interfaces. Figure 15-17 shows an example of ICWG membership for the International Space Station.

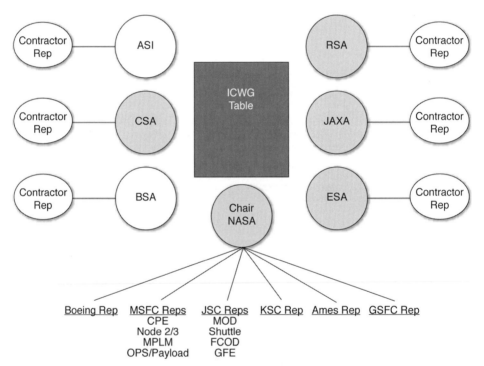

FIGURE **15-17.** **Member Organizations for the International Space Station's Interface Control Working Group (ICWG).** The ICWG includes members from all important players in the space station project. (ASI is Agenzia Spaziale Italiana; RSA is Russian Space Agency; CSA is Canadian Space Agency; JAXA is the Japan Aerospace Exploration Agency; BSA is Brazilian Space Agency; ESA is European Space Agency; MSFC is Marshall Space Flight Center; JSC is Johnson Space Center; KSC is Kennedy Space Center; GSFC is Goddard Space Flight Center; MPLM is multi-purpose logistics module; MOD is Management Operations Directorate; FCOD is Flight Crew Operations Directorate; OPS is operations; GFE is government furnished equipment; CPE is change package engineer.)

15.8   Design Interfaces, Iterate, and Trade

Interface design follows the same general approach as the design process discussed in Chapter 5. It starts by defining requirements and then selects a concept, identifies natural and induced environments, and derives requirements —including what functions must do—while iterating and trading based on risks and sensitivities. Once we've done the trades, various functions produce drawings, specifications, and the controlled design. Using these specifications, we build, verify, and operate the product. As this process continues, conflicts between other system elements and interactions require design updates, including functional and physical changes. These iterations result in updated interface requirements document (IRDs) and interface control documents (ICDs), and are part of configuration management (Chapter 16).

Every case involves form, fit, and function. Two pieces of hardware may not fit when we bring them together. Bolt-hole patterns may be different, tolerances could be off, and a host of other difficulties may crop up. Many problems have occurred when elements didn't separate during operations. We need a rigorous design process for interfaces that includes at least four main phases:

1. Define and understand what the interface must do and the resulting requirements
2. Define form and fit requirements
3. Define natural and induced environments
4. Establish and control the verification process

For example, the interfaces between two stages of a launch vehicle have several functions and multiple interface types that we must design to do them. We have interfaces for structural integrity and vehicle response; power transmission; guidance, navigation, and control; separation functions; and in some cases fluids and hydraulics. The design therefore requires that we understand the function; derive appropriate requirements; meet what form and fit dictate in terms of shape, size, and patterns; and tailor the interface to operate in all design environments. Each interface type has rules of thumb for design, but we must design everything for

- Functional simplicity
- Assembly simplicity, including processing
- Ease of inspection, check-out, and verification

Furthermore, mechanical interfaces require

- Control of mating tolerances
- Clearly defined load paths and joint definition

And human-machine interfaces require

- Simple communication
- Friendliness to human factors

The interface control working group (ICWG) coordinates these design activities and the resulting trades; government or contractor engineers typically complete them. Normally, the chief engineer contributes heavily toward ensuring that the product meets requirements and is mature.

15.9   Build Interfaces

Interface manufacturing follows the general approach discussed in Chap. 9 and must be under full process control. In other words, teams must correctly design and build interfaces according to specifications. Manufacturing specialists must be part of the design team to make the design manufacturable and to ensure product quality. Many of the interfaces have special manufacturing requirements and include separation devices, pyrotechnics, and human-machine interactions. We don't describe these processes here, but they require a lot of attention throughout the lifecycle. The design also must be inspectable and easy to check out and verify. Thus, manufacturing has to be a fundamental part of the design process, so that the interface works in all aspects of the mission as discussed in Chapter 5. To build an interface we must

1. Select an interface concept (design approach)
2. Select materials
3. Select and design a manufacturing approach
4. Establish an inspection approach
5. Determine a verification approach

Figure 15-18 shows standard materials and processing.

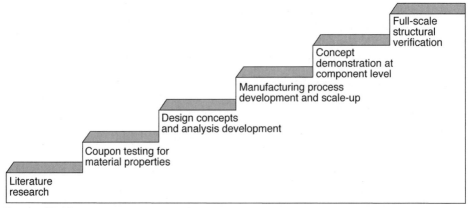

FIGURE 15-18.   Manufacturing Process. Here we show the steps for manufacturing using metals and composites, which includes selecting and characterizing the design approach, developing and verifying the manufacturing process, designing and building materials and manufacturing methods, and verifying hardware and processes. (A coupon is a small sample of the material specially shaped to fit in a testing machine.)

We must not only design the manufacturing approach but also develop the assembly plans and procedures to ensure adequate mating and so on. Simplicity of assembly is just as critical as simplicity of manufacturing, inspecting, verifying, and validating. We must document these processes well and train employees to apply them. Many problems in space hardware have occurred because workers didn't know and follow the building and inspecting procedures. For example, in the above-mentioned mission to place a satellite into GEO, the Inertial Upper Stage didn't separate its first and second stages because someone overwrapped insulation on the electrical connection between the stages, preventing them from separating. Proper training and inspecting may have saved this mission. Each of the other interface types, such as electrical, pyrotechnics, human-machine interactions, and communications, follows a similar process.

15.10   Verify Interfaces Including Integration with the System

Chapter 11 covers general verification requirements and approaches, but interfaces often require special consideration. We must verify interface functions using all combined environments. Experience has shown that if we don't have all the environments, we get the wrong answer. It's also wise to verify for abnormal and partial-failure situations.

Devising test approaches for many types of interfaces challenges many engineers' creativity and ingenuity. Only test flights can determine all the system's effects on the separation system interfaces. But other tests and simulations are vital for gaining knowledge about and confidence in the product's designed and built systems. Breadboards and simulations that mix hardware, software, and analysis are critical to verifying interfaces. Human-machine interactions require these and other simulations, as well as training for human-controlled actions, to verify the interfaces. In addition, we must use a building-block approach: complete the known component testing techniques, such as vibration and acoustic, and then verify them in system tests. Verification requires a detailed plan including testing, analysis, and inspection—as captured in Figure 15-19.

15.11   Document, Iterate, and Control the Configuration

We must document all steps and place them under configuration control (Chapter 16), then iterate them against the controlled baseline. We implement changes through the interface control process in step 7 of the process listed in Table 15-1. Figure 15-20 shows a typical change process, which applies trades and balances the system in the best way using sensitivity analysis, failure analysis (Chapter 8), and risk assessment (Chapter 8). We conduct design and manage manufacturing operations using four key tools and some others. The four tools are 1) define the uncertainties of all parameters that affect design, response etc., 2) analyze and define the system sensitivities to the parameter uncertainties, 3) develop margins to cover

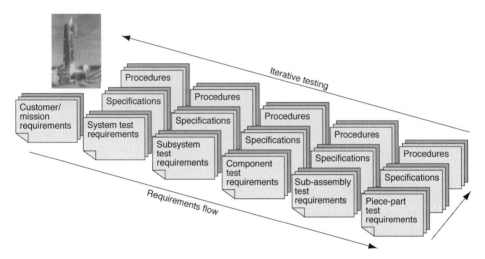

FIGURE 15-19.   Verification Flow. Verification starts at the system level and progresses through the subsystems, components, and even the piece parts. It must be consistent with specifications, procedures, and other guidance. It's iterative and is complete when the system is verified and signed off by certifications of qualification (COQs).

the unknowns we can't define in the uncertainties and sensitivities, and 4) develop risks around the above three to help manage the system. All design decisions and changes must have this kind of information. In the early design phases, design teams, working groups, engineering teams, and the project integration board help keep the system balanced.

Figure 15-20 outlines the process we use as the project matures and the design is under strict configuration control. In this case, we prepare a preliminary interface revision notice (PIRN) containing all information from the sensitivity analyses—such as risk, cost, and schedule, as well as the changes in design characteristics. We coordinate the PIRN and its effect on costs throughout the design and project organization and then send it to the interface working group for discussion and approval. In many projects, PIRNs go with members to the project control board for final approval.

15.12   Develop Operating Procedures and Training

Using the results of designing, building, and verifying the interfaces, we modify operational procedures and baseline them for mission success. Operations must be part of design, but often special mission or system constraints emerge from verification. We must add these constraints to operations, which encompasses three main areas:

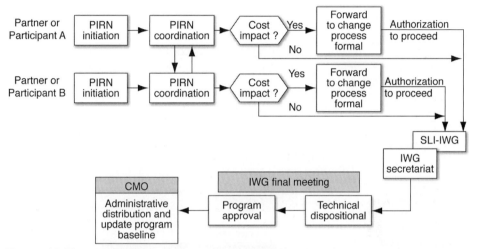

FIGURE 15-20.   Documenting and Controlling the Interface Design. Every change to an interface design requires a preliminary interface revision notice (PIRN). Assigned team members must evaluate the PIRN's effect on cost, risk, performance, and schedule. Then various working groups must decide whether the change is worthy of passing along to the Chief Mission Officer (CMO). (SLI-IWG is system-level integration-interface working group.)

- Operations functions: Handling transportation; launch preparation (analysis, hardware, and software); launch control; mission control; maintenance; logistics; and training
- Operations design: Planning and designing the mission and support structure for the above functions
- Operability: How effectively and efficiently the operations design performs

Each part and the whole are important. First, we must transport the hardware and software systems to the launch site safely and reliably (Chapter 12). In general, the transportation system's environments shouldn't drive hardware design. Thus, transportation designers must understand all system characteristics and design attachments, environmental control, and so on to keep the system within its design limits during handling, transportation, and assembly. At the launch site, teams must assemble, check out, and verify interfaces within these limits while meeting all functional requirements.

Mission control is concerned with the system's performance throughout its lifecycle. This complicated task demands detailed planning and procedures for designs and functions—including those for interfaces—to ensure mission success. Logistics is also vital to operations (Chapter 12), because it determines the best way to organize hardware, software, data, communications, and other elements so everything works efficiently.

After completing all procedures, we must develop a training plan that includes transportation, assembly, check-out, and mission operations to control interfaces and ensure mission success. The plan must contain functional characteristics, limitations, and procedures for performing each function. Training must be intensive and complete. History has shown that mission success depends on the training program's quality because human-machine interactions determine most interface design and functions, whether in building, checking out, or operating space systems. Figure 15-21 depicts the operations sequence and characteristics.

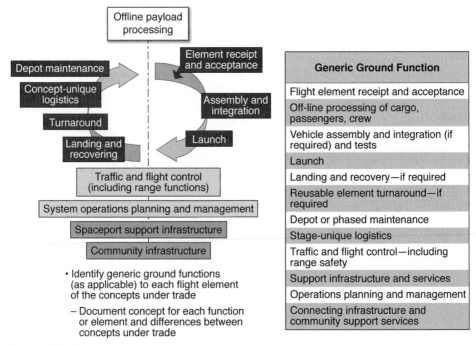

FIGURE 15-21. Functional Definition Diagram for Operations. Here we show how space systems flow after arriving at the launch site, including maintenance for reusable systems and ground support. *(Source: NASA)*

Summary

We have established the criticality of interface design and management to successful products. We developed a process for the definition of interfaces, their design, control, verification, and operations. The secret to interface design is, first and foremost, communication among the various elements of a system and with the system, followed by rigid control of the total interface using uncertainties, sensitivities, margins, and risks with a rigid quality control. The process results in quality interfaces and is essential to project success.

References

Brueggeman, James, Linda Crow, Kreg Rice, Matt McCollum, and Adam Burkey. 15 October 1999. NASDA SSP 30245, Rev E - Space Station Electrical Bonding Requirements. Fort Belvoir, VA: DAU.

Brueggeman, James, Linda Crow, Kreg Rice, Matt McCollum, and Adam Burkey. 1998. NASDA "SSP 30243, Rev E - Space Station Requirements for Electromagnetic Compatibility." Fort Belvoir, VA: DAU.

Buede, Dennis. 1992. *The Engineering Design of Systems.* New York, NJ: John Wiley and Sons, Inc.

Department of Defense (DoD). 1973. Military Standard (Mil Std) 1553B: *Structure for Bus Controllers and Remote Terminals.* Washington, DC: Government Printing Office.

Jablonski, Edward, Rebecca Chaky, Kreg Rice, Matt McCollum, and Cindy George. 31 July 2002 (1). NASDA SSP 30240, Rev D - Space Station Grounding Requirements. Fort Belvoir, VA: DAU.

Jablonski, Edward, Rebecca Chaky, Kreg Rice, Matt McCollum, and Cindy George. 31 July 2002 (2). NASDA SSP 30242, Rev F - Space Station Cable/Wire Design and Control Requirements for Electromagnetic Compatibility. Fort Belvoir, VA: DAU.

Kossiakoff, Alexander and William N. Sweet. 2002. *Systems Engineering: Principles and Practices.* New York, NY: J. Wiley.

National Space Development Agency of Japan (NASDA). 31 May 1996. SSP 30237, Rev C - Space Station Electromagnetic Emission and Susceptibility Requirements. Fort Belvoir, VA: Defense Acquisition University (DAU).

US Department of Energy (DOE). July 2006. DOE P 413.1, Program and Project Management Policy. Washington, DC: DOE.

Chapter 16

Manage Configuration

Mary Jane Cary, *Cyber Guides Business Solutions, Inc.*

Ready access to complete, accurate, and timely information enables rapid, orderly development and effective technical management. Configuration management (CM) lets us quickly and accurately communicate the ever-evolving information about our systems to our team members. It ensures that our systems achieve their intended purpose throughout their life.

16.1   The Configuration Management Process

During a project lifecycle, particularly one spanning several years, initial well-established system requirements can easily deviate toward unintended changes in scope. It may be a long time before those responsible recognize that the design intent is adrift. After all, we expect long-term projects to undergo changes. But hardware and software designs must remain traceable to their requirements, with clear, well-defined interfaces, well-coordinated changes, and a proven relationship between a system and its documentation.

Configuration management is a series of processes that include planning, identifying, managing, communicating, and verifying [OUSD(AT&L), 2001]. See Table 16-1. Like all technical planning processes, CM processes are hierarchical and iterative. We use them to communicate system requirements and definitions, including changes as they occur, to all stakeholders. Subsystem stakeholders and

639

systems engineers must review and approve subsystem-level changes to guarantee their compatibility with system requirements. We periodically review and approve requirements, designs, and specifications of major components, subsystems, and the system and release them as baselines (reference points) for subsequent project activities. This is key to successful technical management. Effective CM performance enables success, because the more we know about a system's configuration, the more effectively we can manage it.

Configuration management enables us to

- Prevent and reduce errors, bugs, and rework
- Improve customer service quality and speed
- Identify and resolve problems more quickly
- Enhance project control and logistics support

TABLE 16-1.   **Configuration Management (CM) Process Steps.** Here we list the five major configuration management processes, and the chapters offering explanations of each one.

Step	Description	Where Discussed
1	Develop CM plan	Section 16.2; Chap. 1, 13, and 15
2	Establish configuration identification	Section 16.3, Chap 19
3	Manage change	Section 10.6 and 16.4; Chaps. 13 and 18
4	Communicate configuration status	Section 16.5; Chap. 17
5	Conduct verifications and audits	Section 16.6; Chap. 11 and 18

Most importantly, understanding and executing CM allow us to manage the relationship between the evolving system and its design intent. Maintaining this balance throughout the project lifecycle results in greater mission success with reduced risk and less effort for system acquisitions, costs, and lifecycle support.

The original aim of CM was to ensure that projects defined, controlled, and verified their functional, physical, and interface technical requirements. But world-class organizations now apply its principles to manage every resource that affects quality, safety, environment, schedule, cost, and financial performance. Documenting requirements, including changes, and demanding accurate results throughout the lifecycle, let us effectively manage any resource, from systems and facilities to technical and administrative processes. Configuration management is a powerful tool for managing overall business performance.

Configuration management methods have evolved in recent years, as a result of acquisition reforms, increasing focus on continuous performance improvements, and rapid advances in information technologies. They have transitioned from post-build project inspections to in-process controls. They have also increased our ability to provide the infrastructure and information needed for collaboration among project stakeholders.

A key ingredient to mission success is sharing CM roles and responsibilities among the stakeholders. All team members must accept responsibility for delivering complete, accurate, and timely information to their colleagues. This principle is affirmed in several CM performance standards, including the International Organization for Standardization (ISO) 10007 Quality Management Guidelines for CM, the companion guide for ISO 9001 Quality Management System Standard, (both published by the International Organization for Standardization), the Software Engineering Institute's Capability Maturity Model® Integration, and the EIA649A CM National Consensus Standard, published by the Electronics Industry Association. Each of the five CM processes relies on systems engineers, project managers, quality assurance team members, and other stakeholders to provide complete, accurate, and timely information, as illustrated in Table 16-2.

TABLE 16-2.   **Configuration Management Processes and Players.** This table describes the contributors to each CM process, providing inputs and creating outputs. Typical lead contributors for each process are identified in bold. (CI is configuration item; ECP is engineering change proposal; CCB is configuration control board; FCA is functional configuration audit; PCA is physical configuration audit.)

Process	Players	Inputs	Outputs
16.2 Develop CM plan	• **CM manager** • Project manager • Systems engineers • Quality assurance engineers • Data manager • Customers • Contractors • Suppliers and subcontractors	• Stakeholder expectations • Technical requirements	• Organization CM policy and plan • Project CM plans
16.3 Establish configuration identification	• **CM manager** • **Systems engineers** • Technical team members • Project manager • Quality assurance engineers • Data manager	• CM plan • Operations concepts • Technical planning • Requirements analysis • Interface management	• CI List • Identifiers for system components and documents • CI hierarchy
16.4 Manage changes	• **CM manager** • **Systems engineers** • Quality assurance engineers • Project manager • Project team members • Interface managers • Data manager • Customers • Suppliers and subcontractors	• CM plan • Engineering change proposals and requests • Change and impact evaluations • Engineering release requests	• Baselines • Engineering change proposals • ECP Disposition • CCB Minutes • Approved, deferred, denied changes • Implementation plans • Engineering releases

TABLE 16-2. **Configuration Management Processes and Players. (Continued)** This table describes the contributors to each CM process, providing inputs and creating outputs. Typical lead contributors for each process are identified in bold. (CI is configuration item; ECP is engineering change proposal; CCB is configuration control board; FCA is functional configuration audit; PCA is physical configuration audit.)

Process	Players	Inputs	Outputs
16.5 Communicate configuration status	• **CM Manager** • Systems engineers • Project manager • Quality assurance engineers • Project team members • Data manager	• CM plan • Current baselines • CCB meeting minutes and directives	• Status reports • Outstanding change requests • Release status • Current drawing version in use
16.6 Conduct verifications and audits	• **CM manager** • **Quality assurance manager** • **Systems engineers** • Project manager • Customers • Contractors • Data manager	• CM plan • Verification and validation plans	• FCA results • PCA results • Process audit results

Configuration management processes apply during all lifecycle phases, regardless of mission size, scope, complexity, requirements, and risk. Standardized, time-proven CM processes reinforce process discipline and continuity for an organization, but we should still tailor them to unique project needs. This saves time, resources, and cost, with little added risk. In this chapter, we explore methods to standardize and tailor CM processes.

16.2   Develop the Configuration Management Plan

Core business processes must be configuration managed, with clearly defined and documented steps, established process owners for effective execution, and periodic performance verifications. Configuration management is more than an engineering discipline; it's an organization-wide tool for end-to-end process control and improvement. The first step towards this goal is an overall CM policy and plan.

Lessons Learned
Improve Organizational CM Performance

Maintain a configuration management performance team, composed of CM process owners, top managers, and other stakeholders, and charge them with responsibility for CM planning and execution. Integrating CM scope, processes, risks, and resources improves performance across the board.

16.2.1   Establish Organizational CM Policy and Plan

Configuration management planning is easiest when top managers understand the goals, concepts, benefits, and risks, and provide the resources to implement it. Common methods across the organization assure that all CM plans, whether for the organization or a project, achieve overall business objectives. The methods must address inherent differences between hardware and software. For example, interface identification and functionality typically occur at lower component levels in software, and development methods enable series and parallel development activities. All project phases require CM, but the formality, methods, and responsibilities evolve during the lifecycle.

The most effective organization configuration management plans are developed and maintained by a cross-functional team, composed of CM process owners and stakeholders with a vested interest in them. A steering committee of upper-level managers, the *CM performance team*, reviews configuration management performance and recommends changes to assure that we meet all our objectives. The resulting plan expresses the purpose of CM for the organization, identifies its scope and application through all lifecycle phases, and guides organizational discipline.

Because the organizational configuration management plan is integral to everyone's discipline, it may be embedded in another process plan. For example, in NASA, it's part of the NASA *Systems Engineering Processes and Requirements*, NPR 7123.1a. Embedded or separate, the organization's CM plan is the foundation for managing all project configurations within the enterprise, while allowing us to tailor the methods to unique project needs.

16.2.2   Tailor Project CM Plans to Meet Unique Needs

The organizational configuration management plan establishes process continuity across the enterprise, but tailoring it to individual project needs improves project and risk management. This tailoring involves documenting the activities and rationale for departing from standard practices or waiving requirements. The resulting project CM plan, including these departures and waivers, is approved by those responsible for overall technical management oversight, usually an upper management steering committee or CM performance team. This way, the plan addresses its project-unique objectives while articulating its connection to organizational objectives.

Developing the project CM plan is comparable to developing the systems engineering management plan (SEMP), as described in Section 14.3.1. All project stakeholders participate, with systems engineers (SEs), CM and program managers, and quality assurance managers providing major inputs.

16.2.3   Identify Project CM Plan Contents

Although all configuration management plans have similar content, we focus here on project-level plans to illustrate the steps for creating one.

> **Lessons Learned**
> **Improve Project CM Performance**
>
> Establish the project CM plan as a separate document, with links to the SEMP, project plan, other technical plans, and the organization's CM plan. This practice highlights project CM requirements, allows ready reference to project CM methods, and facilitates accountability.

Outline Project Purpose and Scope—Originating from customer contract and technical requirements plans, this summary defines major project objectives, and the system's configuration item hierarchy. The summary includes definitions of key terms, a list of acronyms, and all customer or contractor-specified standards, specifications, and procedures.

Set CM Roles and Responsibilities—The organizational CM plan identifies the resources required to achieve CM objectives. Because these vary with project size, scope, risks, and complexity, each such plan must describe how best to achieve its objectives. Typical roles and responsibilities are included below.

- Project team members, along with the project manager, are responsible for creating and implementing the plan. They carry out configuration identification activities, prepare interface control guidance, evaluate and approve specification changes, and participate in configuration audits.

- The *CM manager* owns the organization's CM processes and standards, guides plan implementation, manages and reports configuration management process performance, and educates and trains those involved and affected by CM activities.

- *CM librarians* or *technicians* maintain the library and database records for projects, provide status information to project members, prepare reports, and provide record archiving, access control, and security.

- *CM tool technicians* support and maintain all the CM tools, and assist productivity improvements through automation.

- A *configuration control board (CCB)* shares a vested interest in the project's outcome with the project team. This board, frequently chaired by the CM manager, includes subject matter experts and those knowledgeable of the business, the project, and its technologies. They have the authority to approve and revise configuration items and specifications; evaluate interface control issues; approve, deny, or defer proposed changes; oversee integration of hardware, software, and firmware; manage system audits; establish and coordinate efforts; and approve, deny, or defer recommendations of technical review boards.

- *Material, technical,* or *software review boards* include subject matter experts that provide unique expertise, tools, facilities, or other resources. They may meet with a project team regularly or on an as-needed basis. They focus on technical requirements, such as configuration item interrelationships,

> **Lessons Learned**
> **Using Technical Resources Effectively**
>
> - **Maintain situation list.** Identify conditions when specific technical resources are frequently required.
> - **Maintain technical resource network.** Publicize and encourage use of technical expertise, tools, models, and facilities to identify, evaluate, or resolve problems. Include alternates in key fields.
> - **Provide network with news.** Keep them informed on project progress, performance, and trend reporting.
> - **Provide rewards.** Have ways to thank people for their assistance, such as public recognition, technical collaboration opportunities, and educational conferences.

providing peer reviews of problem identification, classification, and change proposal evaluation. They recommend CM policy and documentation practices, and like the CCB, assure that any meeting minutes and project documentation they create are configuration managed.

An *interface control working group (ICWG)* is particularly valuable when multiple contractors, agencies, or organizations are involved in a system, its components, or interfaces (Section 15.7). This group comprises SEs and other technical representatives who have the necessary technical experience and authority to commit their organization's resources to technical planning and schedule solutions as defined in contractual agreements. The ICWG authors the interface control documents (ICDs); facilitates interface requirements definition, design, and control issues; and reviews and recommends resolutions to interface and integration issues.

Suppliers and sub-contractors are crucial to software and hardware configuration management performance. The technical team ensures top performance by communicating CM requirements to these stakeholders during the contract bidding process. The statement of work should include the following items: the project CM plan, contract data requirements lists, which identify all required data items, and data item descriptions, which describe each data item deliverable. But providing this information is just the first step in managing software and hardware component configurations supplied by outside organizations. We must continue to share information, including advance notices of design changes, pending obsolescence, and change notices that will affect the project.

Project size, scope, phasing, risks, and complexity significantly impact the CM organization. Some organizations delegate change management and configuration status accounting to technical project team members for small, relatively simple projects or during conceptual design and development phases. These team members perform the CM manager and technician roles, and are accountable for CM performance.

Projects beyond the design and development phases, or more complex ones, benefit from more formal, centralized, and networked CM. The configuration management function, shown in Figure 16-1, support systems engineering, subsystem project teams, and distributed CM teams with standards and workflow diagrams, common methods for numbering and revisions, data dictionaries, names of team members responsible for required data elements, and other CM and data management administration. The system CCB also provides centralized guidance, direction, and coordination for all subsystem CCBs and technical teams.

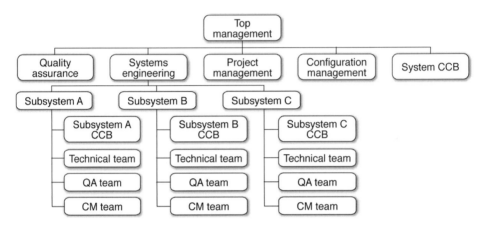

FIGURE 16-1.   **Example Organization Structure for a Complex Project.** This chart illustrates an organization that designs complex systems using networked configuration management (CM). (CCB is configuration control board; QA is quality assurance.)

Identify Lifecycle Milestones and Events—A technical review, conducted by project teams and their stakeholders at the end of each lifecycle phase, determines if the project's specifications are suitable for release and continued activity. Upon stakeholder agreement, each review concludes with the approval and release of a technical configuration baseline (see Section 13.2.2). Many Department of Defense (DoD) organizations and contractors further decompose configuration baselines into functional baselines for system-level configurations, allocated baselines for subsystems, and product or "as-built" baselines for physical design detail configurations. But they use the same phased review and release process to manage technical development.

We usually manage the business side of the project using this same phased process. Many DoD, NASA, and private industry organizations conduct project reviews in conjunction with the technical reviews. These end with the approval and release of the project baseline, which integrates the approved technical configuration baseline with the project's cost, schedule, and risk management plans and status. While SEs are involved in the review and approval of the project baseline, the project manager has lead responsibility for the entire process as well as the cost, schedule, and risk baseline elements. This shared responsibility between SEs and the project manager ensures integration of technical activities with acquisition management objectives.

The number of lifecycle phases, reviews, and baselines varies, depending on the organization. The descriptive names may differ for hardware and software CM within the same organization, or for combined CM among different organizations. Chapter 1 identifies the seven lifecycle phases, their technical reviews, and configuration baselines that NASA uses.

This project evolution, through its repetitive development, technical review, configuration baseline approval, and release cycles, is described in the organizational CM plan. Some smaller or simpler projects don't have all of the lifecycle phases shown in Figure 1-5. Inherent differences in development and production phases for hardware and software within a system further affect the timing of reviews, approvals, and releases. Project configuration management plans can waive the requirements for any unneeded lifecycle phase activities or describe any unique CM requirements.

Because project scope sometimes changes during the lifecycle, the scope and breadth of configuration management activities change as well. As a result, we must monitor CM process performance throughout the lifecycle and adjust the plan as needed to be sure it stays aligned with project objectives.

Describe CM Core Processes—This chapter describes in detail each of the processes that establish, define, manage, communicate, and verify project efforts. They include:

- Configuration identification—Describes how we identify, name, and preserve configuration items

- Change management—Explains how to initiate, evaluate, release, and implement changes

- Configuration status accounting—Describes how we record, access, report, and distribute the status of configuration items

- Configuration verifications and audits—Explain project audits, internal process audits, and supplier or subcontractor verifications and audits

Identify CM Interface Management Needs—The organizational CM plan describes the methods for identifying interface requirements, including those among hardware, software, firmware, commercial-off-the-shelf items, and government furnished equipment. It also addresses the types of interface agreements that we can establish, who is involved in the interface control working groups, and how interface items are evaluated, approved, and integrated into systems. We may even describe recommended groupings for similar functions, technologies, or designs to simplify interface management issues and improve project control. The project CM plan takes the organizational CM plan as a guide to define its unique interface management needs, while waiving all other requirements.

The increasing use of suppliers and subcontractors warrants interface management attention as well. This section of the plan describes how we incorporate items supplied by others. It also tells how supplied items are configuration managed during all lifecycle phases. Chapter 15 provides a thorough discussion of interface management.

Describe CM Data Management Processes—The methods for maintaining and preserving configuration data integrity, and providing required configuration data to customers, should be in the organizational CM plan, or in a separate organizational data management plan (see Chapter 17), which links to the organizational plan. Project-specific CM data requirements, maintenance, and preservation are in the project configuration management plan, along with any data management requirement waivers.

Define Configuration Management Metrics—We need measures to assess CM performance. The organizational CM plan describes the methods for measuring, collecting, and reporting, and describes important metrics for assessing organizational performance. The project configuration management plan defines the metrics of greatest interest to the project.

16.3   Establish Configuration Identification

Configuration identification defines the hierarchy of a project and the interrelationships among all its definition and configuration information. It establishes the foundation for definition, maintenance, and control throughout the lifecycle. Effective configuration identification assures that compatible sets of documentation are created and maintained within and among projects and the organization. Systems engineers play a critical role in configuration identification. The three step process includes:

1. Select and designate configuration items at appropriate levels

2. Assign unique identifiers, such as drawing and document identifiers and part numbers, as well as other nomenclature for hardware, software, firmware, and documentation creation and revision

3. Establish the project hierarchy and define the necessary baselines and documents

16.3.1   Select and Designate Configuration Items

This process lays the foundation for the technical management of all of a project's components and their configuration documentation. *Configuration items (CIs)* are hardware, software, or firmware items that stakeholders designate as a single entity for technical management purposes. A CI at one management level may comprise several elements that are CIs at a lower management level. These items provide a significant or unique end function for the system, and may vary in size, number of component parts, and complexity. But we document, review, manage, and approve them the same way.

The optimum number of CIs is a balance of technical, risk, and resource factors. An integrated project team of SEs works with the customer, technical project manager, logistician, and CM counterparts to identify CIs. For this they use stakeholder expectations and technical requirements definitions, logical decomposition, the work breakdown structure (WBS), interface requirements, technical planning documents, and available technical (major) components listings. Criteria to identify configuration items include:

- Item failure will significantly affect mission safety, security, environmental impact, or cost
- Item can be used in an independent, stand-alone application
- Design involves new development, higher risk, or unproven technologies
- Item will interface with an existing component, or with a component controlled by another organization
- Item is expected to be a standard, reusable, or field-replaceable component
- Item is subject to upgrade or modification during the lifecycle

For example, the FireSAT integrated project team has identified the FireSAT spacecraft as a CI, with the following six subsystem level CIs (as identified in the drawing tree): baseplate module, propulsion module; payload module, side panels, top panel, and the solar array drive assemblies (SADAs). The team has also proposed three CIs within the baseplate assembly CI: avionics stack subassembly, baseplate subassembly, and harness. These nine CIs help the team:

- Manage each CI as a single entity for CM purposes, so the team can view, review, approve, release, and manage baseline configurations for each one
- Simplify the technical management of complex systems, by identifying the significant few items within a system to manage as single entities

The FireSAT team uses the terms HWCI to designate hardware CIs and CSCI for computer software CIs, documenting each selection on the FireSAT physical makeup diagram, shown in Figure 16-2.

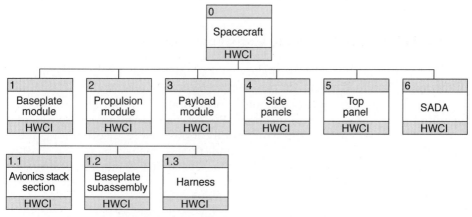

FIGURE 16-2.   FireSAT Example Physical Makeup Diagram. Here we show the FireSAT spacecraft with its six configuration items as well as the three configuration items on the baseplate module. (SADA is solar array drive assemblies.)

The FireSAT team uses this method to identify items of significance from a technical, risk, and resource perspective. The result is an approved and released CI listing, which identifies the FireSAT mission system configuration item, and the hierarchy of subordinate ones, including the satellite and auxiliary system deliverables. All CIs must be directly traceable to the work breakdown structure. The FireSAT team's proposed listing identifies a mere 300 CIs, which they will use to manage a system with over 300,000 components!

Technical documentation for each component in a CI includes such things as functional specifications, schematics, and testing requirements, all subject to configuration control through design and engineering releases. But designating an item or group of items as a CI creates the requirement for specifications, discrete identifiers, change approvals, qualification testing, and design review, approvals, and baseline releases for the item. These requirements help the team to better manage the volume and tempo of technical activity.

Lessons Learned
Tips for Establishing CIs

Allocate items to the same CI when
- Sub-assemblies share common mission applications
- Sub-assemblies have common installation and deployment requirements
- Items have functions that are highly interdependent

Separate items into unique CIs when
- They have great disparity in data input and output rates
- Added control over design requirements and changes is warranted

---

Increasing use of government furnished equipment (GFE), and commercial-off-the-shelf (COTS) items, including commercially available software, creates configuration management challenges. It's becoming common practice to designate these items as CIs to better manage and control their configurations and changes during the project lifecycle. Controls include managing software or hardware form, fit, and performance specifications, and documentation of COTS drawing and product versions. Documentation of design characteristics is required, although the type of documentation may range from a commercial product description to a detailed performance specification.

Procurement agreements should incorporate appropriate constraints, including proposed change notifications. Project teams then assess the potential impact of supplier or subcontractor modifications before incorporating them. Customers continue to challenge SEs and project managers to use existing commercial components whenever practical. Specifying fit, form, performance, and interface requirements for these components, instead of unique or detail design specifications, often saves time and money.

We identify many configuration items during the design definition phase of the project. But some CIs, such as software or maintenance manuals, and test or support equipment, are not defined and developed until later.

16.3.2   Assign Unique Identifiers

Assigning unique identifiers clearly and consistently makes it easier to manage components and their configuration documentation throughout their life. Configuration documentation includes the performance, functional, and physical attributes of a component. All other product documentation, such as test plans, assembly drawings, and maintenance manuals, originate from it. A unique identifier for every document describing a given component assures its accurate link to the configuration documentation. We identify a specific version of a document with four elements: the source, document type, document identifier, and revision identifier.

We usually identify the source of a document using a CAGE (commercial and government entity) code for all CIs, their subordinate components, and subassemblies. This code is also part of the configuration information for a component. The following is an identifier for a drawing, including its version, plus the related product serial number with any approved and released changes.

CAGE Code	PN	Serial #	Document #	Revision	Change #
81352	810810-01	SN2120	34875555	A	C3655001

Document types depend on several factors, including the type of hardware or software, and the context and environment in which the component will operate. Industry associations such as the American National Standards Institute, American Society of Mechanical Engineers and the Institute of Electrical and Electronics Engineers provide specifications of minimum acceptable performance, quality, and interface requirements for hardware and software.

The document identifier, along with its revision identifier, tells which issue of a document is currently in use, or has been released for use. Many different schemes are possible, but the customer frequently specifies the scheme, and the systems engineering team identifies the documentation types.

We also assign unique identifiers to each hardware, software, or firmware item, to assure accurate identification. This identifier normally consists of a nomenclature, CAGE code, and part or item number with a version or tab number. The method should distinguish each product configuration, link the item with its configuration documentation and other relevant data, and enable tracking of actions performed.

Project teams may assign additional identifiers, such as model, lot, or serial numbers, to CIs and their component parts to establish the change timing of each configuration and aid communication among team members. We usually create model numbers for end items, and assign a lot number or date code when we need to associate a group of units with a specific process or event. Lot numbers apply more commonly to subordinate components within a CI and we control them as we do serial numbers. When we need to distinguish individual units, we assign a serial number to each unit. The part number indicates a drawing used to build that part and a unique serial number applies to a single already built part to distinguish it from "identical" parts built from the same drawing. The part number allows us to reference drawings and design documentation, while the serial number allows us to reference quality control and verification documentation unique to a specific part.

16.3.3   Establish Project Hierarchy

Establishing the project hierarchy provides a framework for managing project requirements and data. Configuration hierarchies include the identifiers, internal structure, and interrelationships of components, plus their documentation. During the design definition phase, we decompose the system architecture to a level that enables the customer to specify and control item performance. Specifications or drawing trees and work breakdown structures (WBSs), like the one for FireSAT in Chapter 19, provide views of the project decomposition.

The WBS identifies all of the deliverable items and services, from the top system level to the lowest levels that describe hardware, software, firmware, and data products. It identifies the CIs, and links each item by its part number to each of its supporting documents by type, identifier, and revision level. A project team member has responsibility for each configuration item, and for each component within the WBS. These team members, with the assistance of CM technicians, manage the configuration documentation and see to appropriate change management and release procedures. This linked hierarchical method enables hardware and software project teams to configuration manage their CIs separately, yet lets SEs monitor and coordinate interface management and integration at every level within the system.

Lessons Learned
Tip for improving data clarity
Consider assigning co-owners for each document, an author and a user, to improve data clarity and integrity.

16.4   Manage Change

Project teams need to collaborate in a creative and flexible yet stable environment. The challenge for CM is to develop and maintain well-understood, simple, fast, and accurate processes that facilitate change in such an environment. This includes:

1. Establishing change criteria, procedures, and responsibilities
2. Receiving, recording, and evaluating change requests
3. Obtaining appropriate approvals before implementing a change
4. Incorporating approved changes in configuration items
5. Releasing revised configuration documentation for use
6. Tracking change requests to closure
7. Monitoring implementation to ascertain if changes produced the desired effects

16.4.1   Establish Change Criteria, Procedures, and Responsibilities

Establishing change criteria, procedures, and responsibilities begins with realizing that not all changes are created equal. Best practices include:

- Provide change policy statements that describe the conditions under which changes are encouraged or discouraged. For example, we may establish informal CM methods during early development, delegating methods for drawing creation and change review, approval, and release to technical project team members. This practice speeds design innovation and change during early project phases.

- Hold document authors responsible for their content

- Ensure that responsibility for coordinating and estimating change impact resides with the stakeholders most affected

- Provide simple and accurate procedures that protect configuration documentation from uncoordinated or unauthorized change

- Conduct periodic CM-led training at appropriate times for project teams

- Develop metrics to measure configuration management performance

The challenge for CM managers is to provide change management methods that mirror the project lifecycle. During early phases, project teams need methods that enable rapid change. But as the project evolves, the impacts of change broaden and implementation becomes more complex, risky, time-consuming, and expensive. Tables 19-16 and 19-26 show the FireSAT cost model that's a snapshot of cost estimates after early Phase A analysis. Managing these costs throughout the lifecycle is critical to a cost-effective mission. Configuration managers must evolve CM system formality while providing systems engineering teams with well-designed change management processes during each phase.

16.4.2   Receive, Record, and Evaluate Change Requests

During early development phases, we may receive, record, and evaluate change requests informally with technical discussions, followed by technical drawing or document revisions using date-and-time-stamped project engineer signatures. But with the approval and release of the initial configuration baseline, a more formal process begins, involving engineering change proposals, deviations, or waivers. The CM processes for managing these are essentially the same, but their effect on configuration baselines differs.

Deviations or waivers describe and enable a temporary departure from a design or performance requirement, usually for a specific number of units or a specified period of time. The main difference between a deviation and a waiver is one of timing. We use *deviations* when we discover the non-conforming issue before final assembly of the affected unit. We use *waivers* during or after final assembly or acceptance testing. Because they are temporary, neither deviations nor waivers affect an approved configuration baseline. Systems engineers play a key role in assessing whether the proposed non-conformance is a suitable alternative. Technical team members must certify that system and support operations will be unaffected, with no corrections required, before the CCB will approve any such temporary design departure.

Engineering change proposals (ECPs), or the equivalent *software problem reports (SPRs)*, propose a permanent change to the configuration baseline. The request for a permanent change to a configuration baseline often originates within the organization, but may be submitted by a customer or supplier. Reasons for an ECP or SPR include:

- Incorporate new customer requirements
- Modify performance or functionality
- Respond to a regulatory requirement
- Fix a problem, bug, or defect
- Improve producibility, operation, or maintenance

An ECP form provides space for describing the issues to be analyzed, the proposed solution, and documentation of the components and drawings affected. The originator submits the ECP to the CM technician, who records it, assigns a unique identifier for change tracking purposes, classifies the change type, and identifies the associated items, from component to system level. The technician then forwards the ECP to the appropriate project team for their evaluation. Figure 16-3 outlines this process. Software project teams submit SPRs to a software review board or software CM manager for preliminary screening before CCB review and approval.

Systems engineers and project stakeholders evaluate the merit of the ECP or SPR, including the impact of any proposed solutions on the overall project. Changes affecting approved configuration baselines, including safety, interfaces, or mission success are Class I changes, require customer program office approval, and generate a product version identification change when implemented. By contrast, Class II changes usually don't affect form, fit, or function, are approved by in-house CCBs, and do not entail product version identification changes. When evaluating changes for their impact, we must identify:

- All requirements that may conflict with or be affected by the proposed change
- The consequences to project schedule, cost, resources and risk of rejecting, deferring, or implementing the change as planned [Hass, 2003]

We categorize changes by impact, size, complexity, and required approval authority. While the project-level CCB must approve major changes, approval authority may be delegated to lower level CCBs or project teams for minor changes on internal components. Classifying changes by type also eases the task of setting implementation priorities. For example, an SPR for fixing a bug that could result in mission failure is higher priority than a change that improves interchangeability.

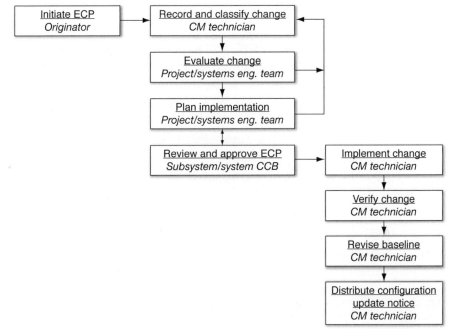

FIGURE 16-3.   Engineering Change Proposal (ECP) Process. This flow chart shows the steps and people involved in initiating, receiving, recording, evaluating, approving, and releasing ECPs. (CM is configuration management; CCB is configuration control board.)

Lessons Learned
Tips for Effective Priority Setting

Include these change classifications on the ECP and SPR forms:
- Addresses safety or reliability issue
- Requested by customer or by regulation
- Affects use, operation, or support
- Affects interchangeability
- Enhances performance
- No suitable alternate exists
- Requires field replacement or retrofit
- Implement after current inventory is consumed

This classifying enables simple and fair priority-setting for making changes. SPR is software problem report.

Addressing change classification criteria and change approval authority in the organization CM plan reinforces fairness in setting priorities and simplifies the organization's approval process. Systems engineers play a leading role in evaluating changes for potential impact across the project. They must involve all affected stakeholders in the evaluation and impact analysis, coordinating implementation planning, and reviewing the proposal for technical accuracy and completeness. They then submit the change to the configuration control board. Upon CCB approval, they work with the CM technicians to release the changes to the current configuration baseline and make them available to all stakeholders.

Lessons Learned
Tips for Effective Change Impact Analysis

Include all potential change impact issues on ECP or SPR forms, such as

- Next assembly product, interface, or process
- Software or firmware interface or function
- Tooling or fixtures
- Coding or fabrication process
- Test equipment or testing process
- Handling or support equipment
- Supplier deliverables or operations

We must assure ECP and SPR distribution to all affected stakeholders, including project and interface leads, for impact analysis. This facilitates communication, improves overall system change impact analysis, and enables coordination of effective implementation planning.

If the CCB rejects or defers an ECP or SPR, the CM technician records this disposition along with the reasons for rejection or deferral. If the change is rejected, we record the relevant evaluation documentation for historical purposes. If it's deferred, we mark documentation for re-examination with the specified date for tracking purposes. The ECP or SPR number serves as the common thread for change tracking and traceability, appearing in configuration status accounting lists and reports, and upon approval and implementation, ultimately on drawings in the revision block.

16.4.3   Obtain Approvals

While SEs have to assess the impact of changes, the CCB is ultimately responsible for determining the combined impact of changes on a system, and for approving the changes and implementation. Board membership is based on such factors as project complexity and customer requirements, with membership changing as the project evolves. The CM manager, who frequently serves as CCB chair, periodically tailors membership to system needs. For example, during the system definition phase, the CM manager delegates CCB responsibilities to systems and quality engineers. The board's formality increases during the preliminary design phase with the addition of fabrication, unit, and reliability testing managers.

The change control board has decision authority for hardware and software integration for all CIs, subsystems, and the system. It closely monitors physical and functional interfaces at all levels, and coordinates changes. Formal CM begins when the final design phase starts, with the CM manager as the chair, and the addition of the project manager and integrated logistics support stakeholders. We must also collaborate with software and other technical review boards or interface control working groups (ICWGs) to be sure that all technical performance issues and "used-on" impacts are adequately evaluated.

The CCB chair also arranges board meetings, agenda preparation, and change package review coordination, and sees that meeting minutes, including discussions, decisions, and required action items are recorded. Resulting decisions are issued as a CCB directive, which notifies all stakeholders of the decision and its implementation. Systems engineers play a major role in developing directive contents, which include:

- An implementation plan identifying all required actions by responsible stakeholders, and their sequencing, associated costs, and completion dates

- Effective date (effectivity) of each required change, which specifies when to incorporate changes to a drawing, document, software, or hardware for each CI and its contract. We express this as an incorporation date, block number, lot number, or unit number.

- All documents and drawings affected, including their change effectivity

- Any relevant directives or associated orders that must be issued

When changes affect customer-approved baselines, the customer CCB must also approve these changes before we can implement the directives.

Regularly scheduled CCB meetings can process changes efficiently when the volume of changes is consistent. The chair may call special meetings to review urgent changes. The actions of the CM manager and CCB members are a powerful indicator of an organization's configuration management performance. Board members that give top priority to meeting attendance, participation, and active involvement in guiding change evaluations to successful outcomes encourage top technical performance. Complex, cumbersome, or inefficient CM processes induce project teams to attempt work-arounds. While the need for effective change management cannot be over-stated, CM managers must see that their methods balance formality and simplicity for best results.

16.4.4   Incorporate Approved Changes

Changes to a configuration item always entail new or revised technical documentation, signified by a change in the document identifier. Documents that can change include specifications, drawings, technical manuals, parts lists, software control drawings, and test and evaluation procedures. Systems engineers frequently play a leadership role in generating the implementation plan that identifies all required actions, the responsibilities for each task, and their timing. The CM manager or technicians may help create this plan, but the project manager is responsible for seeing that it's carried out. For distribution purposes, the ECP or SPR forms provide much of the necessary information to stakeholders. They usually include only the affected design document pages, and support documents are referenced with a required date, to assure availability when needed.

16.4.5   Release Revised Configuration Documentation

We release revised configuration documentation after change approvals are issued and the documentation is modified. Configuration management technicians release and maintain each version of all component documentation, along with the reasons for superseding previous versions, and the authorizations granted by the release. Technical project team members become users of the data they developed. Releases are iterative, and authorize an activity for a component, or for a CI to transition to the next lifecycle phase. The following description of the interrelationships between lifecycle phases, design reviews, and baseline releases illustrates this point.

During the FireSAT system definition phase, SEs create technical performance specifications, drawings, and parts lists, forming the first iterations of the as-designed system configuration. In this early phase, the CM organization may delegate control of these master technical documents and their changes to the systems project team. The FireSAT CM plan specifies that the document revision method during this phase requires signatures and time-date stamps from the relevant system, hardware, software, component, and quality engineers. The technical team progresses toward the upcoming system requirements review (SRR).

The purpose of the SRR is to evaluate the progress, technical adequacy, and risk assessment for the design approach. Systems engineers, the CCB, and all stakeholders review the project configuration documentation, including all changes incorporated to date. They establish physical and functional interfaces among all FireSAT CIs and other equipment, software, hardware, facilities, and users. They also review engineering models, software builds, and test results. After SRR, an engineering release (ER) is published, notifying stakeholders that the FireSAT system baseline has been approved, and authorizing the next lifecycle phase. The system baseline serves as a verified and validated reference point.

In subsequent phases, the control of document masters and change orders transitions from the engineering project team to the CM organization. They manage software items, documents, and all changes using three software libraries. The *dynamic library* contains software items and documents that are being created or modified, and are configuration managed by the developer. The *controlled software library* contains baselines, system build files, and all of the compilers, operating systems, and other tools needed to develop and maintain software. Software CM technicians manage this library, verify and validate all new or revised item submissions as defined by the promotion policy, and assure that all changes are authorized. The *static library* is an archive of all baselines and other software items or documents that have been released for general use, as defined by the release policy.

As the FireSAT systems team nears the end of the deployment preparation phase, they have established a stable design that meets its performance requirements and is ready for the operational readiness review. Functional and physical configuration audits assure that the project fulfills all of its functional and physical specifications. Audit activities include reviews of the latest version of the as-built baseline configuration, all approved changes, completed test units and results, project schedules, and contract deliverables. After this review, the SEs publish an ER, releasing the approved as-deployed baseline for use, and authorizing deployment of FireSAT.

During the remaining lifecycle phases, all changes to components, CIs, and their documentation require formal CM approval. We may still make changes in order to correct deficiencies or enhance project performance, subject to the change order, approval, and release process in the CM plan.

16.4.6   Track Change Requests

Tracking change requests with change classes provides further insight into configuration control. Class I changes are the most critical; they usually address a safety- or mission-critical issue, or are needed for the project to meet its specification, reliability, or maintenance requirements. We usually treat Class I changes as higher priority than Class II changes. The latter simply enhance the use or operation, or reduce the lifecycle costs of the item or its repair parts. Change control for both classes is essentially the same, but the configuration control steps for Class I changes are more complex.

Class I changes to a component normally affect its form, fit, or function. When the new version of the component is no longer interchangeable with the old one, the new version must be uniquely identified, so we assign a new part number. If the component is part of a higher-level assembly, we assign new part numbers at all higher assembly levels until we can reestablish interchangeability. The part number at the interchangeable level is modified to the next revision level to signify the change in documentation. (See Figure 16-4.) Because Class II changes don't affect form, fit, or function, we usually change the part number merely by modifying it to the next revision level.

Software changes are noted with a modified revision identifier each time a file changes. We maintain historical records of all software changes in a log file or database. Software source code file versions may be archived as complete file versions or as deltas, which indicate only the changes between revisions.

Tracking approved changes on a component used in a single system is relatively simple. But on components with multiple system applications, implementation points can vary for each use. Setting change implementation effectivities allows us to manage disparate implementation, and is specified with one of two methods: either date or unit effectivity. With *date effectivity*, we select a change implementation date for incorporating the new component. This date may be based on the deployment of a given lot to control the lot configuration, or can be associated with deployment of a specific unit within a lot. This method maintains configuration control for complex assemblies in low volume production environments.

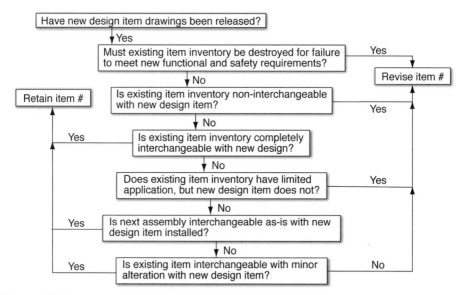

FIGURE 16-4.   Product Number Revision Process. This flow diagram describes the decisions process for retaining or revising an item number.

Unit (serial number) effectivity works on a similar principle, with the change tied to a specific end-item serial number unit. This method allows us to pre-define the configuration of each end-item serial number, and separates the configuration change from the effects of shifting schedules. The configuration manager retains the change implementation records, including effectivity information, to allow us to trace specific end-items.

16.4.7   Monitor Implementation

Monitoring implementations relies on the interrelationships of component part numbers with end-item part numbers, lot and serial numbers, and all of the configuration documentation that defines them. All changes must be approved and released for use before we permit any associated software or hardware version change. But sometimes components don't match their drawing specifications. When we discover a design non-conformance, quality assurance engineers are crucial in evaluating the root cause, and resolving the issues to prevent a recurrence.

When evaluating a non-conformance, we may conclude that temporarily waiving the design specification requirement this time is reasonable. With quality assurance (QA) and change control board approvals, we may issue a deviation or a waiver to permit this temporary relaxation in the design requirements. Section 16.4.2 gives the difference between deviations and waivers. But the conditions we use them in are the same. A deviation or waiver:

- Specifies the number of units involved, or the time period
- Dictates a return to the specified baseline design when the number of units or time period has been satisfied
- Requires QA approval to proceed
- Initiates corrective actions to uncover and resolve the root cause
- Is received, recorded, and maintained by CM as part of the project history

The government standard requiring both deviations and waivers (MIL-STD 973) has been cancelled. But some organizations continue to use both of these temporary change tools, while others have selected one as their tool of choice. Occasionally, the project team determines that the discrepancy is of no significant consequence to the design. Such a case calls for a permanent change to the design specification, permitting future acceptance of the condition.

The project team may determine that the product non-conformance is significant and the item cannot be used as is. In this case, the responsible engineer issues a discrepant material form, and physically quarantines the non-conforming product to prevent its use. Pro-active project teams review and dispose of non-conforming products as they crop up. These materials may be returned to the supplier, scrapped, or salvaged through a rework process. Customer and supplier design and quality assurance engineers frequently provide recommendations for disposition, and may be formally organized into a material review board (MRB). The MRB sees that remedial and preventive actions are taken and documented on non-conformance reports. The CM technician receives, records, and maintains the discrepant material form as part of the project history.

Over the project lifecycle, we must make sure that the configuration baseline always satisfies functional and physical performance requirements. We must also demonstrate that the delivered, as-built configuration of hardware, software, and firmware corresponds to the approved as-designed configuration. Effective configuration identification and change management methods build the foundation that assures this traceability. The desired result is the proven match between a project, design definition, and configuration documentation throughout the lifecycle.

16.5   Communicate Progress with Configuration Status Accounting

Continuous, accurate visibility of configuration documentation for every project and its CIs is the hallmark of effective configuration status accounting. Configuration management team members use *configuration status accounting (CSA)* to record, store, coordinate, and report CI status throughout the project lifecycle. The result enables traceability, shares positive and negative issues for resolution, and assures the accuracy and integrity of every CI and its configuration documentation. For reliable CSA performance we must:

- Develop and maintain a repository for storing and accessing all configuration documentation throughout the project lifecycle
- Provide the tools and processes for reliably identifying, recording, tracking, and reporting status of all CM activities for a system and its CIs

16.5.1   Develop and Maintain Configuration Management Data Repository

Developing and maintaining a CM data repository is a continual challenge, subject to changes in customer requirements and enabling technologies. Because SEs are frequent customers of CM data, they may help select repository tools, as defined in Chapter 17. The configuration status accounting (CSA) repository should be electronic-based, with all data validated upon receipt, stored in neutral or non-proprietary formats, safeguarded against unauthorized access or change, and available upon valid request. Artifact outputs include the SEMP, WBS and specification trees, system analysis documents, assembly drawings, requirements definition documents, baseline designs, test specifications and data, and status reports regarding drawing changes and releases.

Organizations developing electronic repositories usually address document control activities first, with electronic check-in and check-out capability. These activities support projects in their early phases, but we need additional capabilities in later ones. Features such as electronic document editing and markup, on-line CCB tasking, and continuous acquisition and lifecycle support data releases increase functionality and productivity. After these tools are in place, we often realize additional productivity gains through integrating design tools, business software, and enterprise resource planning systems to facilitate interactive contractor integrated technical information services.

Repository designs and data management methods as described in Chapter 17 depend on several CM realities. First, contracts may specify different CM data requirements, such as the number of lifecycle phases or tasks to be completed, along with identified roles and responsibilities. Configuration status accounting activities and required document formats may vary by contract as well. The repository design must be sufficiently flexible to support these variations.

16.5.2   Provide Processes and Tools for Sharing Configuration Management Data

Providing the tools and processes for CSA activities begins with identifying user requirements. During the CM planning process, we define the information systems and required workflow management processes. The plan also identifies the CSA documentation needed, including system deployment status information, change proposal, request applications and approvals, and lot and serial number tracking requirements. This information goes into the technical data management processes defined in Chapter 17.

Because every document created to satisfy a project deliverable requirement is subject to CM, the project team establishes an owner for each document's content, with changes subject to version control. Organizations with established document co-owners, consisting of the document author and the prime end user, enjoy improved data completeness and clarity. Team members are also delegated authority to accept, revise, and release documents according to the project CM methods. Configuration management technicians receive and record product configuration information as it evolves. Table 16-3 lists the resulting reports verifying CI status.

TABLE 16-3.   Typical Sources and Reports for Technical Processes. Here we list example sources of CM data, and common status outputs. (CI is configuration item; ECP is engineering charge proposal; SPR is software problem report.)

Information Sources	CI Configuration Status Outputs
• System performance specification • Interface control documents • CI performance specifications • Engineering drawings • Test plans and procedures • ECPs or SPRs • Deviation and waiver requests • Audit plans	• Current list of CIs for each system • Baseline (functional, allocated, or product) • Baselines as of prior date • Release and approval status of each drawing and document for a CI • Status of each ECP and SPR • Results of configuration audits

Information to create technical data packages (TDPs) originates from CSA records. These packages frequently include a historical record describing the material, physical, and performance features and characteristics of the product, as well as the assembly and acceptance test procedures done. Systems engineers, QA engineers, and other project team members consult the customer and other stakeholders during the concept phase to establish the data that will be required, the data format, and the data managers and owners.

Additional project team members, such as project management or QA team members, often tailor data collection and reporting to meet their needs. For example, the PM may need to monitor the effectivity and deployment status of configuration changes to all system configuration items at all locations. Quality assurance team members use CM records to monitor the traceability of all changes from the original baseline configuration of a CI. The configuration management and quality assurance teams also use CSA data to monitor CM process quality, and to measure overall organizational performance.

Metrics help us identify CM process strengths and areas for improvement, manage related risks, and achieve consistent CM performance. The most beneficial measurements provide positive motivation with a bias toward future performance, instead of merely a scorekeeping, historical function. Performance measurements should

- Focus time and effort on the few most important configuration management objectives
- Show a trend, with a simple and clear cause-and-effect relationship
- Support meaningful understanding of CM processes among stakeholders
- Provide insight quickly and easily, with data that's easy to gather, update, and understand

We must have buy-in from process users to secure lasting improvements. Metrics focusing on the weakest links in our CM processes allow us to isolate problems and identify their root causes. Removing non-value-added process steps, performing tasks in parallel, and improving communication accuracy and speed often result in improved performance. But we must clearly document the lessons learned, including the reasons for change, to assure continued improvement. We should record such process changes in the CM plan, thus transforming this document into a powerful working tool for translating plans into reality.

The result of CSA activities is a reliable source of configuration information to support all project activities, including project management, systems engineering, hardware and software development, fabrication, testing, and logistics. A high-performance CSA process provides project team members with the information they need to carry out any configuration management activity quickly, accurately, and efficiently.

16.6   Conduct Verifications and Audits

Configuration audits give us confidence that the organization's CM processes provide the controls necessary for project configurations. The audits validate the integrity of configuration documentation and verify its consistency with the product.

Functional and physical configuration audits are part of the operational readiness review (see Chapter 18); they verify that the product fulfills all of its functional and physical specifications. The two audits have several similarities. They both involve the first units to be deployed, take place at the contractor's facility, and establish the product baseline. We define each one with an audit plan; publish audit minutes, including any discrepancies, action items, and resolutions; and conclude it with a final certification package documenting the results. Systems engineers are responsible for the audit performance, and frequently serve as audit team leaders. The project manager, CM manager, and technical team members with their customer counterparts participate in the review and in resolving any discrepancies. Chapter 11 describes in more detail the verification and validation activities.

The *functional configuration audit (FCA)* differs in a few important ways from a physical configuration audit. An FCA verifies that the CI has met its required performance and functional capabilities. Joint reviews of the test plans, testing methods, and test data by the audit and project teams assure that all functional parameters have been tested with satisfactory results. We conduct an FCA only once for each CI, group of CIs, or system, but may perform FCA-like activities at other times. A successful FCA certification package includes these two validations:

- Verification procedures were followed, results are accurate, and the design meets all configuration item (CI) requirements
- Drawings and documentation used to procure long-lead-time products satisfy all design requirements

The *physical configuration audit (PCA)* takes place in conjunction with, or after, the FCA. It ensures that the CI's as-built configuration, including the physical location of components and their versions, conforms to its as-designed configuration, including validation of all relevant tools and documentation. Software data library compliance inspections verify and validate that all software versions are correct, that software and documentation releases have been completed according to procedure, and that all changes have been accurately identified, controlled, and tracked. The audit and project teams validate the proven match through joint reviews of the engineering specifications, released engineering documentation, and QA records.

Because project specifications evolve with the project, any differences in documentation or product between the FCA and PCA audits require explanation and validation. Some projects conduct these audits simultaneously to avoid this situation. If simultaneous audits are impractical, we can minimize the complexity by incorporating the following four activities into the PCA:

1. Compare the FCA audited configuration with the PCA configuration, noting any differences. Include a review of FCA minutes for discrepancies requiring action
2. Record in the PCA minutes any differences between documentation and product, or between FCA and PCA configurations
3. Determine the validity of the previous FCA and current PCA, and any impact on current activities
4. Accept any differences resulting from approved and tested changes that are compatible with approved specifications and recorded in engineering data. Differences caused only by test instrumentation are acceptable as well.

A successful PCA certification package normally addresses these ten validations:

- The CI product baseline specification accurately defines the CI, and its required testing, transportability, and packaging requirements
- Drawings describing equipment are complete and accurate
- The software specification listing matches the delivered software media
- Acceptance test procedures and results satisfy all specification requirements
- The software version description document completely and accurately describes all required documentation for software operation and support
- Software media satisfy contract requirements for their intended purpose

- Software manuals for loading, operating, and supporting the software configuration item accurately reflect the current software version
- The deviation request process adequately covers all shortages, unincorporated changes, and other deficiencies
- The CI contains approved parts from the applicable program parts selection list
- The engineering release and change control procedures adequately control the processing and release of engineering changes

---

Lessons Learned
Tips for Successful Audits

Write an audit plan that includes
- The audit purpose
- Who will perform it and who will be involved
- Schedule and frequency
- Scope: number of documents, processes, and method and size of sample selection
- Quality requirements: how to categorize results
- Reporting requirements: contents, method, format, responsibilities, distribution, and schedule
- Resolution requirements: expectations for root cause analysis application and results

---

We take several steps to assure successful configuration audits, starting with audit planning. Because every audit is an opportunity for learning and improvement, we use lessons learned from past successful audits to develop new audit plans. A standard set of audit forms simplifies record-keeping and aids in historical data retrieval. The configuration manager often publishes the audit plan several weeks in advance, and assists in discussion and approval of the required elements. Advance publication also gives the project team time to arrange team logistics and access to facilities, documents, and products.

We also arrange preparatory meetings to assure readiness. Daily agendas noting specific activities and their duration help further audit objectives and afford adequate time and resources for resolving action items. During the audit, project and audit team members may form smaller groups to review specific activities. Meeting minutes are recorded for sharing at project team meetings at the end of every day. These identify any issues of concern, list the next day's events, and share recommendations.

At the end of the audit, we compile a formal audit review report. It contains all audit minutes, identifies any discrepancies uncovered, and resolution actions taken. Checklists for each audit validation can help document discrepancies and assure closure of associated action items. The originator of each discrepancy identifies it by item and affected configuration documentation, records a description of the issue along with the discrepant requirement, and includes recommendations for resolution. A CM team member records each discrepancy, which is then assigned to the appropriate project team members. These members review and analyze the issue for root cause, recommend resolution plans, obtain any required approvals, and implement corrective actions. Resolution may be as simple as offering additional information that clarifies and resolves the issue, or may result in disagreement with the finding. Configuration management change control processes apply here, including approvals and oversight by the system CCB or similar audit executive panel, to assure satisfactory closure of all discrepancies.

Most FCA and PCA audits today are iterative and hierarchical. Audit teams frequently sequence these audits as "rolling" reviews, with the lowest CI levels audited first, and higher levels audited progressively as they are produced, continuing through the system level CI audits. This method of planned, sequential reviews is a more productive method for participating project teams, and less disruptive to system lifecycle management.

While we focus on configuration audits to be sure that CIs fulfill their project requirements, others carry out configuration management process audits to ensure adequate CM performance and to guide continuous improvement efforts. These audits may be done by QA team members, by customer or independent configuration management auditors as self-assessments, or some combination of these. But the purpose remains the same—to verify that our CM procedures are sufficient to meet the organization's contract and policy obligations.

Process audits focus on one of the four CM processes—request, change, build and test, or release. A written audit plan defines each audit, identifying the purpose, scope, timing, and project team members involved. Audit activities include publishing audit minutes, with identification of any discrepancies, along with their resolution. Corrective actions may include revisions to CM policies and procedures with subsequent training recommended to ensure continued CM quality performance.

References

Hass, Anne Mette Jonassen. 2003. *Configuration Management Principles and Practice*. Boston, MA: Addison-Wesley.

Office of the Undersecretary of Defense for Acquisition, Technology and Logistics (OUSD(AT&L). 7 February 2001. MIL-HDBK-61A—Configuration Management Guidance. Fort Belvoir, VA: Defense Acquisition University (DAU).

Chapter 17

Manage Technical Data

Peter C. Kent, *United Space Alliance*

Effective space systems engineering is fully electronic. Competitive hardware, software, and system designs develop in a digital environment that enables rapid product design and manufacturing cycles, as well as modeling, simulation, lean manufacturing, and other techniques that significantly reduce a system's lifecycle cost. This digital approach makes it critical to manage properly the volumes of technical data needed to describe, build, deploy, and operate systems.

This chapter reflects the perspective of a systems engineer who must establish the process and infrastructure to manage technical data for a major project. But we also consider the perspectives of project engineers who work with technical-data management (TDM) tools and infrastructure, and of contractors who operate under this infrastructure. Table 17-1 lists the TDM process as it relates to space systems engineering. At the highest level, we create, manage, and use digital information in two broad forms:

- As structured data residing in database management systems or other systems that provide standard access and retrieval

- As digital content residing in various repositories (referred to as just content from this point on)

We create, store, and manage structured data in a precisely defined way— usually in relational databases consisting of tables with logically defined rows and

TABLE 17-1. **Managing Technical Data for Applied Space Systems Engineering.** This table lists the five main steps for managing technical data related to space systems engineering and matches them to discussion sections in this chapter.

Step	Description	Where Discussed
1	Prepare a strategy for managing technical data	Section 17.1
2	Collect and store required technical-data artifacts during the engineering lifecycle	Section 17.2
3	Maintain stored technical data	Section 17.3
4	Provide technical data to authorized parties	Section 17.4
5	Collaborate by effectively using system and process artifacts	Section 17.5

columns storing instances of data. *Structured data* is part of well-defined transactions through interfaces (often user interfaces) that we develop to enter, update, and use it. Manufacturing orders and financial transactions are classic examples of generating digital information and storing it as structured data. On the other hand, *content* is digital information that takes on many forms as we create it. Examples are documents, digital images (still and video), spreadsheets, business presentations, email, web pages, and web blogs.

An *artifact* in digital engineering is data or content created during the systems engineering lifecycle that relates to the system being developed or to the lifecycle itself. Artifacts range from system requirements specifications, computer-aided design drawings and specifications, engineering bills of materials, and interactive electronic technical manuals; to design review comments and findings; to metrics on the labor hours required to complete the requirements phase of the lifecycle. Electronic artifacts are the means by which systems are defined, specified, reviewed, and in certain cases, delivered. Effective systems to create, manage, and use electronic artifacts are essential to rapid, cost-effective delivery of the space systems they support.

17.1   Prepare a Strategy for Managing Technical Data

A successful strategy for managing technical data aligns closely with the organization's goals and objectives. To succeed, systems engineers must understand engineering processes and their relationships to one another. This understanding enables them to optimize each process and align it with the organization's goals and objectives.

For example, suppose a systems engineer is in an organization that focuses on keeping a product's lifecycle short to compete better by bringing products to market quickly. This engineer creates strategies different from those in an organization that emphasizes reliability at all costs. Although both strategies cover the same space systems engineering (SSE) processes, each uses methods most appropriate for that organization. Generalizing activities and relating them in a hierarchy helps us clearly understand SSE processes and their relationships to one another. Figure 17-1 exemplifies this approach. In this manner, systems engineers define and communicate the TDM process and infrastructure that supports their projects. These processes underpin the digital-engineering environment in which engineers operate, create, and share specifications and engineering artifacts, and collaborate to produce the best designs and products.

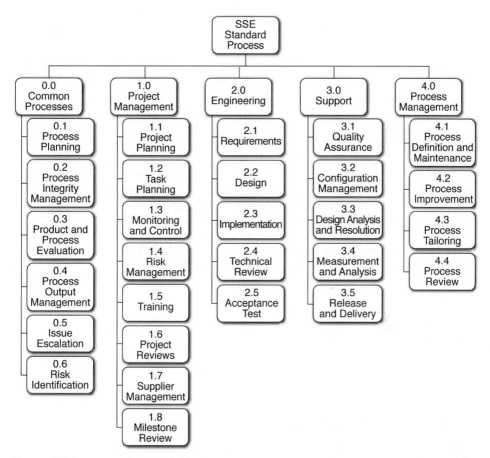

FIGURE 17-1.   Architecture for Space Systems Engineering (SSE). Here we show a sample hierarchy of SSE processes for a space operations company. It illustrates the broad array of processes surrounding mainstream engineering activities (2.0 Engineering column). Each block has artifacts that document the process results; the technical-data management system stores these artifacts and makes them accessible. (The figures and tables in this chapter are based on Engineering Policies and Procedures of United Space Alliance, LLC. Copyright © 2007. Used with permission.)

17.1.1   Identify Engineering Lifecycle Requirements for Technical Data

The SSE lifecycle determines the technical data and artifacts that systems engineers must deliver or employ. Today's costs and budgets increasingly drive SSE lifecycle requirements. Thus, objectives focus on lean processes that are flexible, agile, and responsive to the time pressure that engineering organizations face in bringing products to market or delivering services.

Effective engineering organizations use an organized, systematic approach to the SSE lifecycle. Their processes involve stakeholders from all phases of a product's lifecycle and all affected organizations to develop systems or resolve problems. Customer involvement is through detailed interaction or customer surveillance.

A tailored approach to SSE lifecycle requirements is flexible enough for systems engineers to follow appropriate procedures while ensuring all the right artifacts and coordination for a successful project. Table 17-2 presents a sample tailoring guide. It defines three levels of engineering, with coordination increasing as we move from level 1 to level 3. An activity requires more formality and coordination as it becomes larger (defined by its visibility), has more approval levels, and takes on risk (including safety, mission success or technical performance, schedule, supportability, and cost). Table 17-2 relates the project's size to the required level of systems engineering artifacts and technical data required.

TABLE 17-2.   **Tailoring Guide for Space Systems Engineering.** This matrix contains criteria for project visibility, approval, and risk that allow engineers to determine how much rigor (tailoring) they must apply to developing, reviewing, and maintaining artifacts during a project's lifecycle. We use the level of tailoring determined in Table 17-3 to establish requirements for managing technical data.

Level / Factor	1	2	3
	(Highest factor level determines the amount of SSE for a particular activity)		
Visibility	Confined within a department level or lower; including customer counterparts	Confined within a company element or command level, including customer counterparts	Crosses multiple company elements or command levels, or several customers
Approval	Director level or below	Group level	Program level
Risk identification[*]	Green level	Yellow level	Red level

[*] As defined in an appropriate scoring process to assess risk

Using Table 17-2, we can assess the amount of rigor required in the SSE effort. The highest factor level determines an activity's SSE. Table 17-3 defines requirements for managing technical data at each level of SSE outlined in the tailoring guide.

TABLE 17-3.   Requirements to Manage Technical Data for Space Systems Engineering. This matrix identifies requirements for stakeholder involvement, documentation and control, and so on, that we must satisfy during a project's engineering lifecycle. We determine the tailoring level (1, 2, or 3) from Table 17-2, based on the project's characteristics.

Characteristic	Level 1	2	3
Stakeholder involvement (It's a good idea to use a checklist similar to that in Table 17-4.)	Coordination of engineering artifacts before approval is within a department-level organization, including customer counterparts	Distribute engineering artifacts to cross-functional stakeholders for review before submitting them to board or management review. Some stakeholders may be members of development teams.	Make all key stakeholders members of development teams. They must know the engineering artifacts before the board or management acts.
Product documentation and control	Systems engineering artifacts might not be formally documented	Formally document systems-engineering artifacts only as much as required for board or management approval (if that approval is necessary). Systems engineering includes informal plans for assessing and mitigating risk.	Formally document all systems engineering artifacts, including interim results, rationale, risk assessment and mitigation plan, and evidence of applying lessons learned
Process documentation and control	Systems engineering process might not be formally documented	Document systems engineering process. Changes to engineering processes require formal approval.	Document systems engineering process and place it under program manager's control. Process changes require program manager's or designee's approval.
Progress reporting	Little review of systems engineering status	Regularly do an informal review of systems engineering status with all stakeholders	Regularly do a formal review of systems engineering status with all stakeholders

Table 17-4 is a sample checklist for determining stakeholder participation. It represents the template approach we use to detail lifecycle requirements for stakeholder reviews.

TABLE 17-4.   Sample Stakeholder Checklist. An "X" indicates required stakeholder involvement at each level of tailoring for space systems engineering.

Stakeholder	Level 1	Level 2	Level 3
Design engineer	X	X	X
Manufacturing engineer		X	X
Logistics specialist		X	X
Maintenance engineer	X		
Maintenance technician			X
Safety engineer			X
Flight operations engineer		X	X
Software engineer			X
Financial analyst			
Technical manager		X	X
Program Integration			X
Reliability engineer			X
Quality engineer			X
Systems engineer		X	X
Quality control			X
Human factors			X
Materials and processes		X	X

17.1.2   Determine Required Data Content and Form and Electronic Data Exchange Interfaces According to International Standards or Agreements

Project team members must meet the requirements for technical data defined in Section 17.1.1 as they engineer a project's systems or components and produce engineering artifacts. But for all process steps we must define each artifact's required content and form, as well as the data-exchange interfaces needed to share and transmit the artifacts.

A standard definition of each process allows us to understand the stakeholders, inputs and outputs, and process requirements. This definition provides a framework for specifying inputs and outputs, plus the flow of data and content. Figure 17-2 presents a sample standard process specification that illustrates this framework.

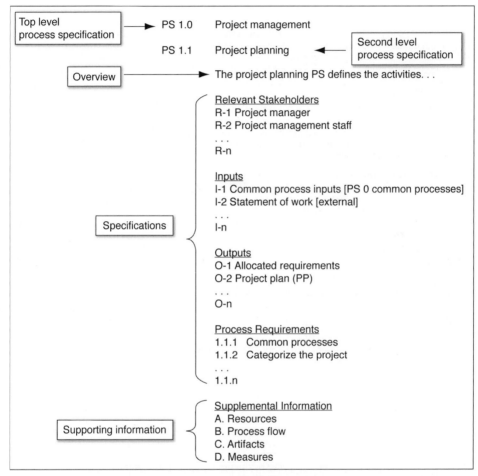

FIGURE 17-2.   Sample Specification for a Standard Process. This type of specification helps us define processes during the space systems engineering lifecycle, as well as their artifacts and flows. (PS is process specification.)

Process development teams (PDTs) normally establish specifications at the beginning of a program or project. These teams consist of systems engineers and other specialists (such as those for configuration or release management) with significant field experience and the ability to build efficient, effective processes. Once the PDT defines a systems engineering process, they shift their focus to monitoring and maintaining it. Or the project may disband the PDTs and replace them with process improvement teams chartered to optimize those areas that need it. Figure 17-3 illustrates the design process for the architecture shown in Figure 17-1, specified using the framework outlined in Figure 17-2.

PS 2.0   ENGINEERING

PS 2.2   DESIGN

The design process specification (PS) translates detailed functional requirements into a design concept (the high-level design), transforms this concept into a detailed design, and finalizes the plans for implementation

Relevant stakeholders

Definitions:

D-1.   Project team—Members of the team that conducts the project as defined in the project plan

D-2.   Engineers—Individuals assigned responsibility for designing, implementing, and sustaining the engineering work products

D-3.   Additional definitions

Responsibilities:

R-1.   Project team—Create the high-level and detailed design documents; do peer reviews, conduct the preliminary design review (PDR) and critical design review (CDR); and update the technical data package

R-2.   Engineers—Work as part of the project team to create the high-level and detailed designs

R-3.   Additional responsibilities

Inputs

I-1.   Approved project plan

I-2.   Approved system requirements specification

I-3.   Additional inputs

Outputs

O-1.   Updated work request

O-2.   Updated software design document, preliminary and detailed, as required

O-3.   Additional outputs

Process requirements

2.2.1.   **Common processes**—Perform the common process activities

2.2.2.   **Generate preliminary design (high-level design)**—The project team analyzes the requirements specification, refines the operations concept, develops detailed alternate solutions, and prototypes and develops or updates the design concept

2.2.3.   **Conduct peer reviews**—If required by the project plan, the project team does a peer review of the high-level design

2.2.4.   **Conduct a PDR**—If required by the project plan, the project team does a PDR

2.2.5.   Additional process requirements

FIGURE 17-3.   **Specification for a Sample Design Process.** This sample uses the process specification framework from Figure 17-2 for the architecture shown in Figure 17-1. (PS is process specification.)

17.1.3   Establish Framework for Technical-data Flow Within the Project's Technical Processes and to or from Contractors

We supplement the process specification framework in Figure 17-2 with information that identifies the resources, flow, artifacts, and measures in the process. Resources are standards, specifications, or other external sources of information that further define the process or help perform it. Process flows outline the steps needed to apply it. Figure 17-4 is a sample process flow for the design process specified in Figure 17-3.

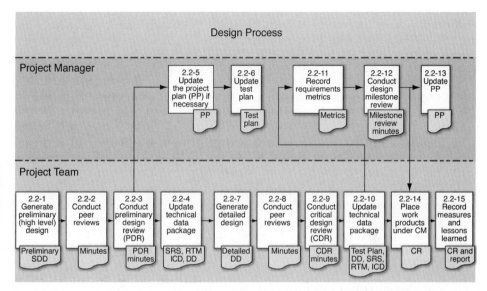

FIGURE 17-4.   Flow Diagram for a Sample Design Process. We uniquely identify each step and denote its outputs. For example, the preliminary design review in step 2.2-3 produces review minutes and triggers updating of the project plan in step 2.2-5 and the test plan in step 2.2-6. (SDD is system design document; SRS is system requirements specification; RTM is requirements traceability matrix; ICD is interface control document; DD is design document.)

To show how we apply this process flow, we consider FireSAT's critical design review (CDR, step 2.2-9 in Figure 17-4). To specify this review process, we define steps for transmitting draft artifacts to reviewers; reviewing draft content; submitting review item discrepancies (RID); reviewing, disposing of, and incorporating RIDs; and approving and releasing the final design documentation. The systems engineer developing this process can take any of several approaches, including those based on workflow, serial reviews, collaboration (with concurrent reviews), ad-hoc (relying on independent reviews and due dates), or some combination. The process decisions that the FireSAT team makes to review such items as the system requirements document or the system design document (see the project's document tree in Chapters 13 and 19) become a major driver for managing the technical data.

Artifacts are products of space systems engineering. They're the plans, specifications, drawings, models, analyses, review records and other documents, datasets, and content that define and describe the system or components we're engineering and the lifecycle model we're using. Table 17-5 shows a sample set of required artifacts for the design process in Figure 17-4. Table 17-6 defines the required artifacts for the design-review package in detail.

Depending on the organization's definition of the engineering lifecycle, processes may include manufacturing and producing the system and its sustaining engineering. We manage the output of manufacturing and production as a product—not as process artifacts. But these processes usually create artifacts: records for quality control, product discrepancies, results from the material review board, and so on. We determine how to classify and treat these types of artifacts based on the process specifications of the organization.

TABLE 17-5.   **A Sample Set of Required Artifacts for the Design Process in Figure 17-4.** We annotate each artifact with the main practitioner responsible for producing it. We retain all these artifacts for the life of the system plus two years. (SDD is system design document; PDR is preliminary design review; CDR is critical design review; SRS is system requirements specification; ICD is interface control document; RTM is requirements traceability matrix.)

Required Artifact	Practitioner
Updated work request	Project lead engineer
Updated SDD, preliminary and detailed, as required	Project lead engineer
PDR record or minutes, if applicable	Project lead engineer
CDR record or minutes, if applicable	Project lead engineer
Peer review record or minutes, if applicable	Project lead engineer
Design milestone review record or minutes, if applicable	Project manager
Updated SRS, if applicable	Project lead engineer
Updated ICD, if applicable	Project lead engineer
Updated RTM, if applicable	Project lead engineer
Prototype, if applicable	Project lead engineer
Updated test plan	Project manager
Product and process measurements	Project manager

Measures are quantifiable attributes of a process that show its performance and can result in discrete actions when we exceed defined thresholds. Table 17-7 defines a sample metric to track milestones for a design process. We set up decision criteria to warn project managers and leads whenever progress is falling behind, with adequate notice to evaluate root causes and correct problems. We usually chart and monitor measures on a project's health throughout its systems engineering lifecycle.

17.1.4   Designate Responsibilities and Authorities for Managing Technical Data

These responsibilities and authorities are typically spread among the provider of information-technology infrastructure, system operators, and users of the digital-engineering environment that feeds and uses content in the technical-data

TABLE 17-6.   A Sample Set of Artifacts Required for the Design Review Package. We annotate each artifact with the main practitioner responsible for producing it and with how it applies to the PDR and CDR. (PDR is preliminary design review; CDR is critical design review; PM is project manager; SDE is systems design engineer; DE is design engineer; A/R is as required; MAE is mission assurance engineer; LP&S is logistics planning and supportability; X means the artifact applies.)

Required Artifact	Practitioner	PDR	CDR
Project requirements document	PM	X	
Design requirements statement	SDE	X	
Schedule preparation	PM	X	X
Design drawings	DE	A/R	X
Design specifications	DE	A/R	X
Design calculations	DE	A/R	X
Design software	DE	A/R	X
Engineering cost estimate	SDE/DE	X	X
Failure effects and modes analysis (if required)	MAE	X	X
Hazard analysis (if required)	MAE	X	X
Critical item list	MAE	X	
System assurance analysis (if required)	MAE	X	
System criticality impacts (if required)	MAE	X	X
Security assessment (if required)	Security	A/R	A/R
Material requirements list (advance order list)	LP&S	A/R	
Material requirements list	LP&S		A/R
System mechanical schematic	SDE	A/R	A/R
Electro-mechanical control diagram	DE	A/R	A/R
Advanced electrical schematic	SDE/DE	A/R	A/R
Cable interconnect diagram	SDE/DE	A/R	A/R
Trade studies	SDE/DE	A/R	
Concept and sketches	SDE/DE	X	
Changes to interface control document	SDE		A/R
Environmental assessment	SDE	X	A/R
Certification requirements plan	SDE	X	

management system (TDMS). They include originating, creating, capturing, archiving, securing, maintaining privacy for, and disposing of technical-data work products.

The systems engineer's role depends on the architecture that an organization or enterprise uses to create, manage, and use technical data. Figure 17-5 shows a layered approach to data architecture that segregates functions and responsibilities into layers or partitions, which enables cost-effective assembly of the components and toolsets into a robust digital environment.

The architecture depends on a technology infrastructure encompassing the servers, workstations, data-storage networks, communication networks, and other elements that support the digital-engineering environment and TDMS. Its next layer consists of data, content, and metadata (data definitions) housed within the infrastructure. Systems, applications, and services built on the data and infrastructure form the third layer. The next layer includes data practices, methods, and toolsets

TABLE 17-7.   A **Sample Measure of Progress for the Design Process.** We track a project's major and minor design milestones to indicate progress in its design phase and to highlight any risk to completing the design on time.

		Metric: Monitor Design Progress	
Description	**Measurement Scheme**	**Measures, Thresholds, and Responses**	**Measurement Specifications**
Project managers and project leads need to see design progress weekly to address factors that may prevent achieving baselined milestones	Monitor design progress based on achieving major and minor milestones	• Variance in achieving major milestone • Variance in achieving minor milestone If any variance is more than 7 days (one reporting period), identify root causes and assign corrective actions If a major-milestone variance is more than 14 days (two reporting periods), report the variance to all stakeholders and assemble them to evaluate corrective actions and how delay will affect the project	Measurement targets: • Schedule dates and work-in-progress status of deliverables Descriptive measures: • Milestone type • Planned completion date • Actual completion date Collection techniques: • Record planned and actual dates for milestones in the project's schedule Calculations: • Compare planned to actual dates to determine achievement variance (+/- days) Results: • Variance in achieving major milestone (+/- days) • Variance in achieving minor milestone (+/- days)

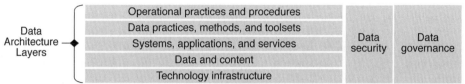

FIGURE 17-5.   A Layered Approach to Data Architecture. This approach provides flexibility to build a cost-effective digital-engineering environment and management system for technical data.

such as integrated data management, information lifecycle management, knowledge management, data warehousing, and collaboration, which help us supply data throughout the SSE lifecycle. The architecture's top layer consists of operational practices and procedures that tie its data, content, and technical aspects into the organization's engineering processes.

Spanning these layers are two crucial architectural elements: data security and data governance. *Data security* defines the standards, structures, and techniques necessary to securely access, change, and move data throughout the data infrastructure. *Data governance* defines the people, processes, and procedures that establish a total lifecycle view of data and content. These elements are essential to managing risks to engineering projects and form the foundation for all information assurance practices.

Systems engineers play a limited role in the technology infrastructure; they provide functional requirements for the systems and applications that the infrastructure supports and offer inputs on activities that create data and content. These contributions enable managers to size the infrastructure to handle the data and content that users generate. Engineers do much more in the higher layers that host data and content; systems, applications, and services; data practices and procedures; and operational practices and procedures. These layers generate, store, secure, and dispose of engineering artifacts. What systems engineers do depends on whether they are contributors, consumers, or administrators within the SSE lifecycle.

Contributors are engineers who generate or update artifacts and work products as part of SSE, such as those in Table 17-7. They follow procedures established to check artifacts into and out of the TDMS repository when creating or updating them. They also classify artifacts using their attributes, such as document number, title, filename, author name, or responsible department. Automated features of the TDMS normally key off an artifact's identity, classification, or workflow to control access, privacy, and disposal. Contributors play a major role in setting these attributes or performing workflows to see that artifacts aren't compromised or corrupted through inadvertent access.

Consumers are engineers, project leads, or other members of the project team or organization who need access to artifacts for viewing only. They usually employ portals or search sites to retrieve data and content from the digital-engineering environment and TDMS repository. Their responsibilities involve complying with access and privacy controls and protecting data and content from inappropriate disclosure while they're accessing read-only copies.

Administrators are engineers, leads, managers, or configuration managers who establish and perform data management. They are data or system owners who must define and administer the procedures and controls that ensure the integrity, availability, and quality of artifacts and work products from space systems engineering. In this role, systems engineers define the engineering processes outlined in Section 17.1.2 and illustrated in Figures 17-3 and 17-4.

Systems engineers are usually contributors and consumers during a project. They create artifacts individually or as part of a team and use artifacts that others create as inputs or references. In some instances they supply administrative support. For example, they may review requests for access and approve those justified for technical or organizational reasons.

17.1.5   Establish Rights, Obligations, and Commitments for Retaining, Transmitting, and Accessing Technical Data

These elements are part of governance for the data architecture model in Figure 17-5. Data governance focuses on establishing policies for these elements, as well as monitoring and enforcing compliance. The policies must reflect the organization's response to whatever affects retaining, transmitting, and accessing technical data—as shown in Figure 17-6.

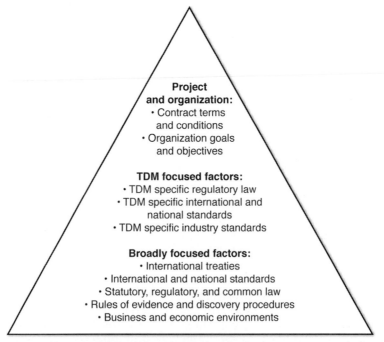

Project and organization:
· Contract terms and conditions
· Organization goals and objectives

TDM focused factors:
· TDM specific regulatory law
· TDM specific international and national standards
· TDM specific industry standards

Broadly focused factors:
· International treaties
· International and national standards
· Statutory, regulatory, and common law
· Rules of evidence and discovery procedures
· Business and economic environments

FIGURE 17-6.   **Rights, Obligations, and Commitments for Technical-data Items.** Systems engineers who define policies and procedures for data governance must consider factors ranging from very broad to very specific. (TDM is technical-data management.)

The broadest considerations draw from international treaties, national statutes, and common law that form the legal and business environment in which the organization, products, and systems operate. This legal framework establishes the context for litigation brought because of product or system failure, claims of negligence, or patent claims. Statutes on product liability, tort, and evidence discovery—such as the US federal rules of evidence—influence policies that govern how an organization creates, manages, and destroys engineering data and artifacts.

The next, more focused, factors draw from treaties, regulations, and codes for retaining and controlling data and content. These laws and standards mostly emphasize digital content, but broader regulations also affect systems engineers. The US government's regulations that control exporting and importing defense-related goods or services often affect aerospace projects. These International Traffic in Arms Regulations (ITAR) apply legislation aimed at protecting US national-security interests and advancing foreign-policy objectives.

The scope of ITAR depends on the US Munitions List, which outlines its goods and services. An organization's polices and procedures for governing engineering projects that build launch vehicle power plants must address ITAR controls, such as those precluding access to this information by foreign nationals.

The most specific factors typically arise from contractual terms and conditions, as well as the organization's goals and objectives for data or content. Contract terms and conditions often specify rights in data ownership that drive how organizations handle access to data and content. Software licenses are increasingly sources of rights in ownership for source code. The open-source movement has spawned the General Public License (GPL), which contains terms and conditions on copying, modifying, and redistributing GPL licensed software for other products or systems.

Organizational goals and objectives most specifically drive data-governance policies and procedures. They embody the organization's attitude and strategy for managing data or content. For example, some organizations manage key pieces, such as critical facets of a design, as trade secrets. They control this data very tightly. Other organizations patent designs and license them to any interested parties willing to pay acceptable compensation.

Regardless of what drives data-governance policies and procedures, we have to effectively communicate them to the project team that must adhere to them. We must also reflect them in system controls within the digital-engineering environment and TDMS. Section 17.3.3 addresses this idea in detail.

17.1.6   Establish Standards and Conventions for Storing, Transforming, Transmitting, and Presenting Data

How we treat these elements in our policy, agreements, and legislative constraints is key to applying digital engineering and its supporting TDMS. We must consider our strategy carefully because subsequent actions depend on it. Main factors in the decision are cost, flexibility, responsiveness to change, and risk.

Cost involves initial acquisition and installation as well as later maintenance and upgrade costs over the system's lifetime. Flexibility is the ability to upgrade all or parts of the system when technology, such as new tools and techniques to visualize content, evolves and reaches the marketplace. Responsiveness to change requires us to quickly modify a system to meet changing business requirements. An example is moving from business on a single contract to multiple contracts. Risk includes such factors as

- Usability, operability, supportability, scalability (and other "ilities")

- Obsolete technology or inability of technology suppliers to deliver or keep up

- Incorrect alignment of technology and business strategies

Systems engineers who set up and provision a digital-engineering environment and TDMS must work closely with software system architects, subject matter experts in institutional technology, business analysts, and others to evaluate strategies and decide on one that makes the most sense for their organization and situation.

Figure 17-7 illustrates one possible strategy: a common logical-data and content repository that's physically distributed among the organization's elements and project locations. This means two or more physical repositories in data centers (or supported by service providers) at different locations. A common logical repository means that different physical repositories communicate and synchronize data and content. Synchronizing occurs at various levels, including the

- Physical level, where content is duplicated or replicated among the physical repositories

- Application level, where applications enable users to store and access data in any of the physical repositories

- Metadata (data-definition) level, where a common metadata repository holds information about each piece of data, including where all its copies and replicas are located in the set of repositories

The strategy also entails partitioning into layers the software tools that constitute the digital-engineering environment and the TDMS. The tools interact through the data and content stored in the logical data repository, and an enterprise-wide messaging bus (not shown) allows tools to communicate with each other when necessary. This strategy loosely couples the tools and data, in that the data is independent of a particular tool.

The strategy outlined in Figure 17-7 aims to minimize lifecycle costs by favoring tools based on non-proprietary or open formats. This approach eases competition and usually keeps proprietary vendors from locking in. It also reduces upgrade costs over the system's lifespan by enabling projects to replace higher-cost tools with less expensive ones as competition drives down prices. This strategy doesn't require costly migration of data and content from one proprietary format or data structure to another. Of course, the lower-cost tool must still deliver the necessary functions, supportability, and performance.

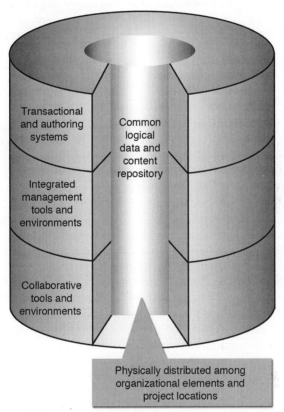

FIGURE 17-7.   **A Data-storage Strategy Based on a Common Logical Repository for Data and Content.** This physically distributed arrangement offers many benefits. Successful data-storage strategies must account for lifecycle costs, flexibility, responsiveness to change, and risk.

Flexibility means being able to add or subtract physical locations (part of the architecture's design) and tools as needs dictate. Responsiveness derives from loosely coupling tools and data, adopting open or industry standards, and using metadata scoped to the enterprise, which allows the organization to see data and content as they use it. The strategy in Figure 17-7 also limits risk in several ways:

- More diversification and redundancy among multiple physical repositories to cover system failure at a single location

- Diversification by selecting tools from multiple vendors, while avoiding the risk of poor integration and support if the number of vendors becomes unwieldy or difficult to manage

- Less obsolescence by enabling technology upgrades of components without having to replace the entire system

We can vary the strategy outlined here or devise many others, but a successful one must satisfy the organization's goals and objectives, meet the needs of project teams that depend on it, and keep abreast of a continually changing landscape.

17.1.7   Describe the Strategy for the Methods, Tools, and Metrics Used During the Technical Effort and for Managing Technical Data

We can't build a strategy for storing and accessing data, as outlined in Section 17.1.6, without considering tools for creating, managing, and using data and content. So we must also evaluate and select tools, based on the organization's style. We begin with the organization's belief in developing its own tools (make), buying commercial-off-the-shelf (COTS) tools from software vendors (buy), or buying COTS and modifying them to its own requirements and preferences (modified COTS).

Once an organization has decided on whether to make or buy tools, it must define requirements in some way, ranging from formal functional requirements, to checklists of "musts" and "wants," to loosely assembled marketing materials in which vendors feature various products. The organization's style and practices also influence these requirements. We recommend structured, formal definition processes because ad hoc, informal approaches tend to produce subjective results that depend on the biases and preferences of those involved.

After defining requirements for tools, we can use several forms of decision making to select them, ranging from a decision by one or more organizational executives; to formal decision-analysis techniques, such as Kepner-Tregoe Decision Analysis, SWOT (strengths, weaknesses, opportunities, and threats) analysis, or decision trees; to group-decision techniques such as unanimity, majority vote, range-voting, multi-voting, consensus building, or the step-ladder method. Regardless of the process, the final selection leads to applying and deploying the digital-engineering environment and technical-data management system (TDMS).

Some organizations formally specify the selection process. For example, Federal Acquisition Regulations (FARs) determine contracting processes for federal acquisition in the United States. Part 15 Subpart 3 prescribes how to select sources of products that represent the best value. In other situations, the techniques for make-versus-buy decisions apply. In many instances, the alternatives are structured so that details of the approach reside in a single process for decision making. All decisions ultimately involve evaluating alternatives against criteria or requirements, whether the process is deterministic, heuristic, or subjective. It's often tempting to skimp on or omit requirements definition, but doing so dramatically increases the risk that the digital-engineering environment and TDMS will fall far short of their intended goals. Chapter 4 discusses requirements definition in detail.

17.1.8   Prepare a Training Strategy

The training strategy for the digital-engineering environment and technical-data management must address engineering project team members, functional and IT support people, and managers from all domains in the systems engineering lifecycle. It must include

- Training needs in terms of knowledge and skills, expected levels of proficiency, and experience

- Certification methods and graduation criteria that document and validate knowledge, skills, and proficiency—examples are certification testing, convening evaluation boards, or checking qualifications

- Techniques for delivering training, such as lectures, remote video conferencing, computer-based instruction, one-on-one tutorials, or train-the-trainer sessions

- Outlines that identify training content by course or topic and enable specialists to build content that meets the organization's needs

- Training course relationships and prerequisites that establish criteria for attendance

- Training support, including subject matter expertise and consulting

- Criteria for retraining when a person's performance falls below specified thresholds

17.2   Collect and Store Required Technical-data Artifacts During the Engineering Lifecycle

The systems engineering lifecycle generates the artifacts stored in the TDMS's data and content repository. The digital-engineering environment provides the authoring and tools used to create, modify, and use these artifacts. Some examples of these tools are computer-aided design software to create drawings and do kinematical analysis; finite-element systems to create and analyze models for structure, heat loads, and fatigue; and numerical analysis packages.

17.2.1   Identify Sources of Technical-data Artifacts Designated as Outputs of Common Technical Processes

The space systems engineering lifecycle outlined in Chapters 1–3 and the organization's process specifications (Section 17.1.2) determine which data artifacts to generate. The project planning and estimating phases identify this engineering content at a high level, including updates to relevant baselined artifacts. These baselines may represent the as-designed or as-built configuration of hardware, facilities, or systems that the current project relies on, augments, or interfaces with. Or they may represent items or systems that we must decommission or demolish to make way for the current project. Sometimes these systems are very old and their engineering data isn't even in electronic form.

Project managers and leads must consider the form and format of baselines targeted for update with respect to the tools and capabilities of the organization's digital-engineering environment and TDMS. Artifact baselines, such as CAD drawings, 2-D and 3-D engineering models, or simulations, may have used earlier versions of the toolset or tools that are no longer in use. So converting to the current version may require a lot of money and effort. In fact, we may have to re-create the baselined artifacts using current tools. Project plans and cost estimates must account for this potential data translation to avoid unpleasant surprises when engineers try to use or update these artifacts.

17.2.2   Collect and Store Technical Data Following the Strategy and Procedures for Managing Technical Data

Creating or capturing content takes many forms depending on the artifact, what tools we use to author it, and its context. Tools native to the environment and integrated with the technical data management system enable systems engineers to generate new content objects and check them into the TDMS as part of the interface with tool users. Integrated tools allow us to easily check objects out of the TDMS, change them, and check them back in under a versioning scheme set up for the situation at hand.

To generate or capture content, we may use a number of support tools that aren't integrated into the digital environment. The artifacts and content that these tools generate require different ways to move them in the repository. Two examples are digital cameras used for photo documentation on site surveys and satellite imagery bought from a commercial source to support engineering analysis. In these cases, the data and content move into the TDMS through general user interfaces that it provides to navigate within its content structures, as well as to import, check out, or check in content. In some cases, general-purpose pipelines may be available to capture and move data into the TDMS from any accessible location, such as a local computer's hard drive, shared file server, or web site.

Partners, subcontractors, or affiliated organizations may also deliver technical data, based on contractual or teaming arrangements for the project. These arrangements can range from physically shipping electronic media on CDs, DVDs, tapes, and so on; to transferring files using File Transfer Protocol (FTP) or secure FTP; to formal structured interfaces for data transmission under industry standards or custom data-exchange structures built for the purpose. Several industries, such as those for electronics, manufacturing, and aerospace, have standard transfer mechanisms. Examples include

- ISO 10303—Industrial Automation Systems and Integration—Production data representation and exchange [ISO, 1994]

- EIA-836—Configuration Management Data Exchange and Interoperability [GEIA, 2002]

- IPC-2578—Sectional Requirements for Supply Chain Communication of Bill of Material and Product Design Configuration Data—Product Data eXchange (PDX) [IPC, 2001]

17.2.3   Record and Distribute Lessons Learned

To meet process-improvement goals and objectives, we must evaluate lessons learned from experience. This is step 4.2—process improvement, from Figure 17-1. To improve processes, we need a formal way to identify and capture lessons learned from completed projects, as well as from investigating and establishing root causes for anomalies, discrepancies, and failures. A database or repository is an effective way to capture and organize lessons learned. Integrating it into the digital environment is critical to ensuring that team members can access and readily apply lessons learned, rules of thumb, and experiences from previous projects.

Although internal lessons learned are vital to process improvement, we also learn from group, national, and international standards and practices for entire industries. Organizations leverage this cumulative expertise and lessons from industry and academia to improve their engineering processes. We outline below some notable programs that apply to system development.

ISO 9001 and 9004. The International Organization for Standardization (ISO) defines standards for quality-management systems within its ISO 9000 family of standards. These standards address how products are made and guide organizations in following consistent business processes that lead to quality products. ISO 9001 defines process requirements, and ISO 9004 guides process improvements.

Lean Six Sigma. Six Sigma is a business-improvement method developed at Motorola in 1986 to reduce manufacturing defects. The lean variation of Six Sigma shows how to improve service-oriented processes based on time or cycles. Although sigma denotes the measure of standard deviation (applied to defects in this case), the program depends on a define, measure, analyze, improve, and control procedure to improve processes.

Capability Maturity Model® Integration (CMMI). This is a system-development method that describes the essential elements of effective processes: in effect, the lessons learned from many systems engineers. The Software Engineering Institute, with members of industry and government, developed CMMI. It describes effective processes in 22 areas covering systems engineering, construction, and maintenance. Each area describes progressively effective practices on a 1 to 5 scale, so organizations can evaluate their own processes.

Information Technology Infrastructure Library (ITIL). ITIL contains best management practices for operating IT systems. It assumes that systems are in place and operating, so it doesn't focus on system development. But it does provide best practices for operating installed systems, managing infrastructure and systems, and supporting and delivering IT services.

17.2.4   Check Integrity of Collected Technical Data

The main integrity checks for collected or captured data must occur while we're collecting it. They give us confidence that data and content meet format standards and that metadata describing the artifact contains the information needed to store, manage, and use it. Content checks may also include

- Certifying ownership rights, such as public domain, licensed, or third-party proprietary
- Marking content as, for example, classified, restricted, proprietary, copyright, or non-disclosure
- Scanning content for restrictions on digital rights
- Watermarking content or applying other controls on digital rights to assert the organization's ownership

We should send data or content that fails integrity checks to the submitter for correction or rework, or move it to a holding area for reprocessing according to established procedures. Once we've collected and stored data and content in the TDMS, we must establish internal procedures to spot check or randomly audit them. These checks ensure that integrity processes are applied and effective.

17.2.5   Rank Order, Review, and Update Procedures for Collecting and Storing Technical Data

In most organizations, procedures for collecting and storing technical data within the systems engineering lifecycle are part of the policies and procedures that define and perform the lifecycle. So organizations should

- Periodically review and update these procedures as part of their formal processes
- Document a formal requirement that establishes the basis and criteria for these reviews
- Make sure that these criteria address the
  - Types of changes the review asks for, such as typographical, clarity of content and instruction, or actual practice experience
  - Means of review: tabletop, distributed (by email, wikis, etc.), or continuous (by electronic suggestion boxes, etc.)
  - Update approvals, tied to the nature of the changes

An example is an annual review tied to the policy or procedure's original release date and completed by the department manager who owns it. The review process might require the owner to maintain a list of subject matter experts in the topics covered in the policy or procedure and ask them through email for any material changes or enhancements it calls for. The process might allow the management owner to approve typographic changes, the next higher level of management to approve content and instruction changes, and assign approval for process changes to the owners of those processes.

In another approach, we might post a policy or procedure requiring review on a wiki-based internal web site and then assign parties to update the policy during a review window. Once the review window closes, we could open an approval window, so approvers could view the changes and sign them off online. A third approach involves electronic workflows triggered by annual review timers that route policies and procedures to identified parties for review, update, and release.

17.3   Maintain Stored Technical Data

Once we create, review, and baseline engineering artifacts, they go into the technical-data management system. Sound configuration management practices entail creating and managing baselines through documented releases and applying a revision control scheme to track and manage updates to artifacts. Any iteration of the lifecycle for space systems engineering can result in our needing to revise and re-release some or all of the baselined artifacts. Systems engineers that must establish a process and infrastructure to manage technical data for a major project have many choices of repositories for engineering artifacts and business processes to handle engineering releases and revisions. See Chapter 16 for a thorough discussion of configuration management.

17.3.1   Review and Release Systems Engineering Artifacts

In our example TDM framework, the process specifications from Section 17.1.2 cover the requirements for reviewing and releasing engineering artifacts. These specifications outline requirements that the process must satisfy, whereas process flows detail the tasks and steps. For digital engineering, we define and apply process flows as electronic workflows. To set up workflows, we use dedicated workflow engines and tools or apply workflow features of the artifact repository. We may also use the business process execution language (BPEL), which enables us to create executable process models.

BPEL is an orchestration language published by the Organization for the Advancement of Structured Information Standards. It uses industry standards such as the eXtensible Markup Language (XML) and the eXtensible Stylesheet Language for Transformations (XSLT), and is available from several vendors. BPEL has an array of constructs to execute flows such as sequence, if-then-else, if-else, while, and other flow-related commands.

Our choice of a workflow or BPEL approach often depends on the scope of the review and release flow. Generally, an organization using its own digital-design tools will document an internal flow with the tools' workflow features. Organizations use BPEL to define interactions and transitions that need to occur during a business process that they want to assemble and run as a collection of web services. A web services approach lends itself to process flows that span organizations with different digital environments. Each organization integrates its engineering environment into the review and release flow by creating web services that capture its interaction with the flow. Because BPEL is based on standards, it easily flows processes across organizations, companies, and agencies that employ different infrastructures for information technology.

Figure 17-8 illustrates a conceptual flow for reviewing and releasing engineering artifacts using workflow or BPEL. It's a sequence diagram using the Unified Modeling Language, which shows how review and release processes interact. It shows each participant with a lifeline (running vertically), the messages targeted to and from the participant (running horizontally), and in certain cases, the parameters that messages carry.

The sequence in Figure 17-8 models a simplified, success-oriented scenario. But models for a robust production process must cover all scenarios and incorporate ways to process errors, handle exceptions, iterate, and account for many other details, such as

- How to identify artifacts for revision
- Criteria to determine who the reviewers and approvers are
- Mechanics of incorporating notices and comments
- Whether artifacts will carry effectivity—and if so, what type (date, lot, serial number, and so on)

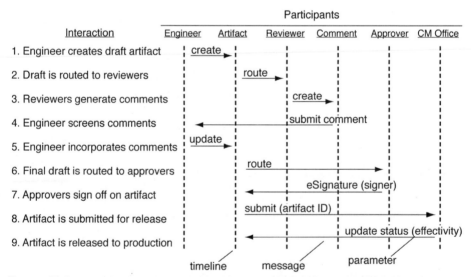

FIGURE 17-8.   A Sample Process for Reviewing and Releasing Engineering Data, Expressed as a Sequence Diagram in the Unified Modeling Language. The diagram lists each participant across the top and specifies interactions among participants as messages to and from them. (CM is configuration management.)

17.3.2   Update and Revise Systems Engineering Artifacts

Many processes in managing technical data have common underpinnings. Updating and revising systems engineering artifacts embody nearly every process step for reviewing and releasing them. Of course, they focus on updating versus creating, which can mean huge differences in the reviewer's attention, engagement, and rigor applied to each step. But from the perspective of TDMS, processing the artifacts is nearly identical.

In such cases, a service-oriented architecture has advantages over alternatives in its ability to easily construct processes and flows from libraries of standardized services. It enables us to abstract or generalize these services, and easily create derivative process flows at low cost and effort. Figure 17-9 illustrates a conceptual flow for updating and revising engineering artifacts. It parallels the flow in Figure 17-8, with only minor changes showing that we're updating, not creating. We could easily apply this flow using BPEL and the services in hand for review and release.

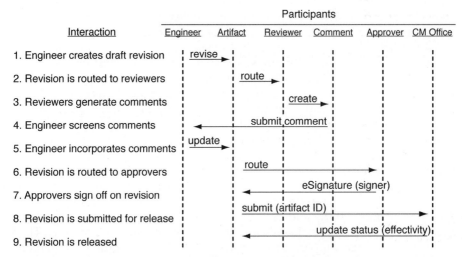

FIGURE 17-9.   A Sample Process for Updating and Releasing Engineering Artifacts. This sequence diagram in the universal modeling language is nearly identical to the process flow for creating artifacts in Figure 17-8. We could easily apply it using a services-oriented architecture. (CM is configuration management.)

17.3.3   Manage Databases for Quality, Integrity, Security, and Availability to Authorized Users

Managing the TDMS and its databases requires focus on the users authorized to access the system and on its data and content. We begin managing the system when it goes live, but that requires a lot of preparation and planning. Going live involves starting many system structures, including five key processes:

- Setting up user accounts

- Setting up groups, roles, and permissions
- Establishing storage schemes and structures
- Creating access controls
- Loading initial content and data

To go live, user registration must be in place as a system feature or a supporting process. For a technical data management system with many initial users, it's a good idea to preload the first user accounts to make sure registration doesn't become a bottleneck. To do so, the project team must assemble user-account data before going live, and the development team must build and run a one-shot job to populate user-account structures within the system. When the number of initial users goes beyond several hundred, the benefits of having accounts in place outweighs its cost in effort and coordination.

Groups, roles, and permissions are essential features for any TDMS whose mission goes beyond narrow, specialized data capture and management. They are general capabilities that system administrators must use to create specific instances that will control the system at go-live. Setting up groups and roles and loading user accounts into groups can require a lot of time if the number of groups and users is large. Mapping permissions to roles can also be time-consuming if the scheme is complex. In these situations, we must preload user accounts because the system won't start operating without them. Some form of automation to support administrators is essential for systems that must handle hundreds or thousands of users. That automation could be one-shot system updates or an administrative interface that streamlines the configuration change process.

The storage scheme that a TDMS uses depends on its design, but any robust TDMS is flexible and extensible. The storage schemes and structures we want in place for go-live must adjust to the beginning content and data load, as well as the projected expansion for operations. For example, if content resides in virtual cabinets and folders we must establish in advance the cabinets and folders that can hold go-live content. If the folders' names depend on content, we must set up the scheme to extend immediately as more content flows into the TDMS. Administrators must prepare these structures and may require automation to complete extensive tasks on time.

We can use any of a half dozen or so models to control access for a TDMS. A popular one employs access-control lists that associate users or groups with instances or collections of content, such as all content in a folder. Many access-control schemes include inheritance, so higher-level lists can flow down to lower levels. As with groups and roles, access controls must be in place when we go live.

17.3.4   Maintain Technical Data

We should operate the digital-engineering environment and TDMS under a maintenance and operations plan created for this purpose. The plan defines organizational, procedural, and contingency structures that ensure the system's health and reliability.

Operational roles and responsibilities. These define who does what to operate the system. Some examples of groups typically called out are

- Network communications—Set up access to the TDMS through the wide-area network (WAN) or local-area network (LAN), and monitor the network traffic to and from the TDMS

- Infrastructure services—Operate the servers, data-storage farms, and application services that make up the TDMS; monitor health and performance of the system and TDMS

- Information Technology (IT) security—Responds to system threats, such as intrusion by hackers or denial-of-service attacks, and monitor the system's security controls

- IT service desk—Provides a single point of contact to users for all service requests and problems; works with users to resolve incidents caused by known errors with known solutions; escalates other incidents through procedures in the TDMS

Operational processes and procedures. These are essential to consistent, reliable TDMS services. They define the steps for doing tasks and responding to foreseeable situations. Documented procedures eliminate ambiguity and enable rapid response to situations that make the TDMS unavailable. Examples are procedures to start, shut down, and back up the system.

Operational measures and metrics. These are the pulse of a TDMS. We must design metrics to support the needs of each role or responsibility, and today, many commercial-off-the-shelf solutions provide real time monitoring, alerts, and the measures and indices that form the metrics. System uptime or availability, network bandwidth use, network latency, average and peak response times, and response time to resolve problems are just a few examples of the metrics required to monitor the health of a TDMS.

Service-level Agreements (SLAs). Providers and consumers of services must agree on levels and types of service. The agreements focus on either internal or external activities. External SLAs for TDMS usually spotlight users' concerns. Examples are

- Performance, such as average and peak response times for standard interactions

- Availability, such as being available more than 99.9% of standard operating hours

- Incident resolution, such as replying to problems received by email in less than six hours

Internal SLAs center on infrastructure and system concerns. Examples are

- Recovery, such as system recovery from backup tapes in less than 12 hours

- Troubleshooting, such as owning and investigating high-priority problem tickets within one hour

- Maintenance, such as keeping maintenance outages to less than three hours, between midnight and 6:00 a.m., and not more than once per month

Problem Management, Escalation, and Tracking. Incidents reported by users to the service desk are often resolved by tracking known errors and solutions. For example, users report they can't log in and are typing their password correctly. The service desk agent asks them to check the caps lock on their keyboard because passwords are case sensitive. Users report caps lock is on, turn it off, and are now able to log in. Any incident the service desk can't resolve gets documented as a problem and escalated within the TDMS support organization. Procedures for tracking, escalating, and managing problems are a crucial part of the maintenance operation plan (MOP). As a rule, organizations that operate many systems have dedicated documents that define procedures for resolving problems. If so, the MOP should reference the standard procedure and outline any unique characteristics of the TDMS that may affect it. For example, the plan might instruct service-desk agents to always include the name, type, and format of content files associated with TDMS problems.

The MOP outlines the procedures for operating and maintaining the technical data management system, but the content also requires maintenance. The content's nature and volume determine the maintenance-support structure and procedures. Maintenance usually focuses on the content files and the data attributes describing each piece of content. We do maintenance at a physical level and at a logical or administrative level. The physical level encompasses the servers, disk drives, file systems, and database-management systems that support the TDMS. The logical or administrative level comprises the TDMS structures and the constructs that it provides to grant and control access and help people use the system. Logical or administrative maintenance always takes place through the TDMS itself or the system tools.

Most TDMS designs store content in computer files. Maintaining this content thus falls within the practices and procedures for maintaining file systems or file sharing, which include

- Monitoring and resolving file fragmentation
- Monitoring and expanding free space as needed to accept growing content
- Monitoring rights to access files, so those granted match those specified in the TDMS
- Monitoring system event logs and file system metadata to secure and control access to the underlying content files

Most robust TDMS designs store content attributes in a relational database management system (RDBMS), for which we apply practices and procedures of database maintenance. Key maintenance activities for a RDBMS are

- Managing support resources, such as disk space, system memory, and system bus configurations

- Regularly collecting or updating the statistics needed to optimize query and access paths
- Tuning its performance to help meet the service-level agreements for system performance, responsiveness, and availability

Content administration (maintenance at a logical level) aims to keep people from inappropriately using or accessing content. In certain settings, such as government classification or controlled export, access of content by unauthorized people has significant consequences. If users have low turnover and static access needs, administering content is relatively simple. Once set up, access controls (groups, roles, permissions, and so on) require few changes, so administration basically becomes monitoring. A dynamic setting, in which users come and go often and access needs change continually, is more challenging. In these cases, the computer user registration form (CURF) and similar systems automate creating, destroying, or deactivating user accounts with little effort. They also allow processing of requests for access changes using defined workflows that streamline administration and produce standard audit reports.

17.3.5   Keep People from Inappropriately Using or Accessing Stored Data

We have to configure and apply access controls properly to keep people from inappropriately using or accessing data and content. That means configuring the system's features to make them open, restrictive, and maintainable. The TDMS design determines which access controls it provides the systems engineer, so it's often a deciding factor in an engineering organization's choice of tools. Whether the access-control model draws on a mandatory, discretionary, role, or capability approach, it must support the organization's data-security policies and architecture.

Figure 17-10 shows typical controls that systems engineers use to manage access to engineering artifacts. It's based on implementing discretionary access control—allowing those who "own" a piece of data or content to alter its access controls at their discretion or transfer controls from one piece to another through inheritance. The figure outlines the access-control features available to system administrators and data owners. An essential entity is the user: an authenticated subject interacting with the system. We organize users into groups to make administering access controls easier. Each group has many users, and a user can be in many groups (a many-to-many association). Central to this example is the access-control list (ACL). Each ACL calls out one or more groups or individual users, along with one or more rights or privileges (such as read access, write access, the ability to create new versions, or the ability to delete an object).

In this example, ACLs provide control when they are assigned to instances of data, content, tools, or services (technical-data resources) that the system manages or provides. Each resource matches an ACL that controls access to and operations on it, such as creating new versions. A default ACL could match any object that is otherwise unassigned. Conversely, we can assign one ACL to many technical-data resources, making each one reusable and easing administration of access controls.

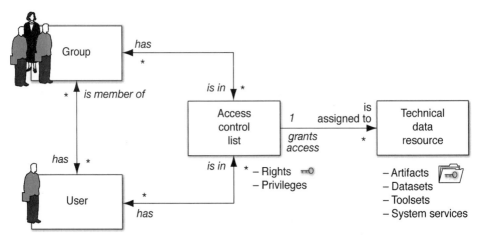

FIGURE 17-10. **Access Controls Can Use Any of Several Control Models.** In this example, access-control lists associate users—individually or by groups—to technical-data resources, such as artifacts, datasets, or tools. These lists also detail the rights and privileges that each user has to the resource. The key symbol represents lockable access, requiring approval. The number 1 means access is granted to only one. The asterisk symbol represents a multiplication by 0 or more between the boxes.

Figure 17-10 has four associations; three are bi-directional, one is uni-directional. The asterisks mean "zero or more." The "1" means "one only." The arrow at the end of an association shows its direction. Bi-directional associations have arrows at both ends. For example, the bi-directional association between group and user at left signifies that from the group perspective "a group has zero or more users," and from the user perspective "a user is a member of zero or more groups."

The example illustrated in Figure 17-10 is one of many access-control configurations possible with today's digital-engineering environments and TDMSs, but we must use a robust scheme to manage technical data successfully. So we should treat access control as a separate systems engineering project.

17.3.6   Maintain Stored Technical Data to Protect It Against Foreseeable Hazards

A TDMS often represents an organizational investment of millions of dollars and is the backbone of the digital-engineering environment that the organization depends on daily. Success can quickly turn to failure in the face of such events as fires, floods, hurricanes, earthquakes, riots, or malicious acts. We must prepare for these events.

A disaster recovery and continuity of operations plan (DRCOP) prepares an organization to deal with potential disasters, as well as such common mistakes as an engineer's accidentally deleting a working file or a configuration management administrator's deleting a folder structure holding released drawings. Mistakes occur; we need to be ready for them.

A DRCOP is very simple in principle. First, it assumes a destructive event has occurred (whether foreseen or unforeseen). Second, it outlines processes and procedures to continue business operations and to bring the system back to operation. But these processes and procedures must be in place before a destructive event occurs! Once a disaster happens, it's too late to begin thinking about backing up the system or working on alternate or failover operating sites.

Systems engineers who establish the TDM process for an engineering organization are rarely solely responsible for a DRCOP. Planning for disaster recovery is a collaborative effort spanning the provider of the information technology infrastructure, the system operator, and possibly other organizations. Participants depend on the acquisition strategy that the organization employs to host the TDMS. Systems engineers play a key role. They specify what TDMS service levels are necessary to support the organization's engineering processes, as well as how the system's unavailability under various scenarios will affect costs. We use these cost impacts to determine the most effective backup and recovery strategy.

A sound DRCOP must reflect the TDMS's platform and system architecture. It must also balance the costs and benefits of mitigating risk. Otherwise, it will falter under the burdensome cost of implementation. For a DRCOP to be effective, we must periodically test and validate it. An elegant, detailed, but untested DRCOP gives us the same false sense of security as a smoke detector with dead batteries. A robust DRCOP has at least six elements:

Data and content backups. Backups offer the fallback position when all else fails to resolve a problem or recover the system. We can also use backups to recover files or folders that have been accidentally deleted or corrupted.

Backup frequency and schedule. Operational constraints dictate

- Backup frequencies and the schedule they run on (when the system must be available to support users, when it needs maintenance, and so on)
- The backup system's performance and configuration (backup speed, whether it includes verifying backup media, and the like)
- The system's size at any time (because size determines how long the backups will run, all other factors being equal)

Backup media storage and retrieval. Storing backup media is a crucial decision in the DRCOP. Storing it in the same location as the system puts it at risk under certain disaster scenarios, such as hurricanes, earthquakes, fires, or acts of terrorism. Storing it at an offsite remote location is prudent. The distance between the storage location and the system, as well as the way we access backup media, helps determine how quickly we can restore the system from backups.

Hot standby and failover. If the TDMS is critical to the enterprise's mission, say for revenue or safety reasons, it may not be able to accept unavailability to recover from backups. In these cases, we can install hot-standby hardware or establish failover sites that replicate the system's infrastructure and enable it to continue operations on very short notice. The timeliness of the switch to hot-standby hardware or failover sites depends on the TDMS's architecture and the design of alternate support elements. Generally, hot standby and failover greatly increase the cost of the TDMS. Analyzing costs versus benefits and return on investment helps determine how aggressively to employ these solutions.

Content and data recovery. Having backups in place and hot standby or failover sites ready to step in is meaningless if we don't know how to use them. A sound DRCOP details recovery procedures step by step to show how to failover a system or recover part or all of it. Best are procedures that segregate the steps by role and offer an integration process that coordinates actions among the roles.

Test scripts for content and data recovery. Using a recovery procedure for the first time when we need it is very risky. The unexpected often trips up best-laid plans. The solution is to develop test scripts that exercise recovery procedures before we need them. This approach also trains system operators and support staff on recovering the system, builds their proficiency in applying the procedures, decreases the time needed to recover, and gives system users some confidence in it.

17.3.7   Back Up and Archive Artifacts for Recovery and Future Uses

Archives are records created during the systems engineering lifecycle and set aside for long-term preservation. Backing up data and content for archives requires processes and procedures distinct from backups for system recovery in three main areas:

- The format of an archived record, which generally has a long life with respect to the technology used to create and access it
- The archived record's location, usually an independent repository dedicated to archiving and set up for long-term reliability of data-storage media
- Procedures used to access archived records, which normally involve a librarian department, organization, or function that creates and maintains the archive

We must write procedures for archiving and make sure that the digital-engineering environment or TDMS supports it. Archiving requirements depend on several perspectives, including the organization's goals and objectives, the contract's terms and conditions, regulations from government or industry, and the organization's policies on retaining records to reduce litigation and risk. One such regulation is the Code of Federal Regulations (CFR) Title 36 Part 1234.30 — *Standards for the Creation, Use, Preservation, and Disposition of Electronic Records*.

We must balance the future usefulness of archiving a project's history and artifacts against the cost to create and maintain the archive. Because knowledge changes rapidly in space systems engineering, we're unlikely to apply much of an archive to future projects. But lessons learned, data acquired, and historical preservation still justify it.

17.4   Provide Technical Data to Authorized Parties

Our goal in managing technical data is to capture, manage, and give it to authorized people quickly and effectively. We need to maintain a library of engineering artifacts, content, and reference data relevant to users' needs. So strong user interfaces to quickly and conveniently find, retrieve, and use the information are critical to its success.

17.4.1   Maintain Information Library or Reference Index to Provide Data Available and Access Instructions

A large organization's technical data management system and digital-engineering environment house enormous volumes of data and content, commonly amounting to tens of millions of data objects taking up many terabytes of disk space. Even a mid-sized organization's repository holds thousands of drawings, documents, and models. Fortunately, we have many tools and techniques to plug into the TDM infrastructure that host user interfaces for data delivery based on popular web search engines such as Google, or to build portals tailored to systems engineering. Systems for business intelligence, data mining, data warehousing, decision support, and knowledge management augment basic search and retrieval capabilities in the TDMS.

Tools are no good if they're awkward to use, assume unlikely user knowledge or behavior, or provide inadequate or inaccurate results. So our data-delivery strategy must focus on usability. Figure 17-11 illustrates a systems engineering portal that indexes engineering process documents, based on the hierarchy presented in Figure 17-1. It color codes each process area and divides the process into sub-processes. Next, it lists process specifications as internal operating procedures (IOP), as well as the templates and forms that the IOPs call out to run engineering tasks.

Each IOP, template, and form callout is a hyperlink that users can click to bring up the item from the TDMS. Links to other portals, document trees, artifact trees, and templates offer convenient access to engineering content that SEs need when working an assignment. Other links include tools for computer-based training and digital engineering, as well as to other websites and web-based tools that help engineers work more effectively.

17.4.2   Receive and Evaluate Requests for Technical Data and Delivery Instructions

Receiving and evaluating requests is easier with automated user registration to administer user accounts and access levels across many tools and data sets. A registration system offers

- To users a single focal point for requesting and updating access

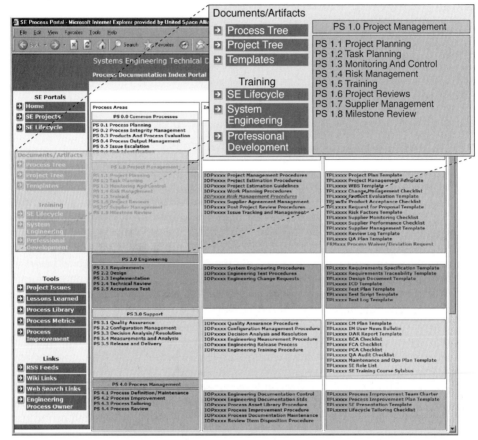

FIGURE 17-11. **A Web-based Portal that Gives Systems Engineers a Reference Index to Engineering Processes, Internal Operating Procedures, Templates, and Forms.** Each procedure, template, and form callout is a hyperlink that brings up the item. Hyperlinks in the leftmost column offer easy access to other systems engineering portals, documents, artifacts, online training, tools, and related websites.

- To organizations
  - A consolidated view of who has access to what
  - A framework for control procedures such as account-approval workflows, periodic revalidation of system access, and license management in cases where access requires licensed commercial-off-the-shelf (COTS) software
  - A way to establish access rights, limits, obligations, and commitments with users through agreements they enter into when accessing or registering for accounts—essential in certain government settings and useful in others

Today, the SE or IT organization preparing automated registration has many COTS tools to choose from for identity management. This covers user accounts, passwords, access based on policies and roles, and directory services, among others. Creating user accounts when needed and making them inactive when appropriate is referred to as provisioning and de-provisioning. Tools for identity management usually provide workflows to provision user accounts and access, as well as workflows or triggers to de-provision (inactivate) accounts and access as required.

Figures 17-12 through 17-14 show a hypothetical user interface for requesting access to managed systems that constitute a typical digital-engineering environment and TDMS. The process begins by identifying a person who is requesting an account or access. People without an account (new users) are prompted to enter information that uniquely identifies them. The organization's business policies and practices determine the information required. For people with an account (current users requesting access to another service), the system retrieves information and displays it for the user to confirm.

The system then prompts the user for the application or facility to which access is requested. It may also ask for justification of the request, what groups the user needs to be in, or the required profile. The request submittal ends with the user validating the data entered and acknowledging any responsibilities under the organization's policies and procedures. Pressing the "submit" button means electronically accepting these responsibilities.

Submitting the request could kick off a workflow that routes it for approval and implementation. Typical steps may include approval by a sponsor or agent, approval by a data owner who oversees the data and content requested, and processing by a system administrator or automated routine to set up the user's account, permissions, and rights. The request process must inform the user and others involved in the approval workflow as the request proceeds. It must also securely send account and access credentials to users when the request is approved (or the reason for rejection if not approved).

17.4.3   Confirm that Required and Requested Technical Data Is Appropriately Distributed

A confirmation shows that the system has satisfied the user's needs and met established procedures, directives, and agreements. Sound account management includes periodically revalidating user accounts and access rights—typically once or twice a year. It involves routing approved requests to sponsors and data owners, so they can review and recertify each active account. Organizations should design an automated, efficient process for revalidating accounts that includes at least six key features:

- Lists online the active accounts due for revalidation, filtered by approval role
- Filters accounts due for revalidation by specified criteria
- Displays account information, history, permissions, and rights

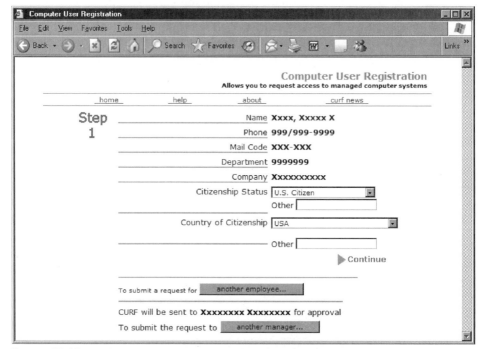

FIGURE 17-12.   **Identifying a User for Automated User Access Registration.** The identity management system captures or displays attribute data that uniquely identifies a person. For existing users, the system simply fills in this information.

- Re-approves or disapproves an individual account, a selected list of accounts, or a block of accounts
- Automatically notifies sponsors and users of account disapproval
- Automatically deactivates disapproved accounts by direct disapproval or expiration (not recertifying an account within a specified grace period from the recertification date)

Figure 17-15 shows a web-based revalidation page that allows system managers to review active accounts due for revalidation and to approve or disapprove their continued use. This page displays the access requests approved by revalidating managers and allows them to easily update each account with their disposition.

In certain situations we may need to remove access immediately. Examples are an employee fired for cause, an accident investigation, or a student being expelled. Procedures for managing user accounts should provide the criteria, contacts, and processes for immediate removal.

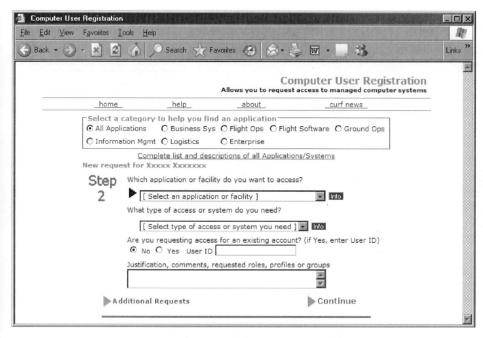

FIGURE 17-13.   Continuing Registration. The process continues as users select an application or facility and provide information on their roles, profiles, groups, and justification.

17.4.4   Confirm Conformance to Rules for Electronic Access

Designing and configuring access controls (Section 17.3.5) and procedures for requesting and approving access (Sections 17.4.2 and 17.4.3) must work together to be sure that access to engineering data meets the organization's data-governance policies. In particular, the system must confirm that all rules are followed before allowing access to the database and before releasing or transferring data electronically to the requester. Granting access involves adding approved users to the appropriate groups, roles, or other control structures set out in the TDMS's access model. Once we establish user accounts, we must maintain them and the groups, roles, and access rules as people come and go or circumstances change (such as promotions, role changes, or absences).

17.4.5   Prove Correctness, Reliability, and Security of Technical Data Sent to Recipients

Audits determine if people are following established procedures, directives, and agreements. Systems engineering audits evaluate the systems and processes of the digital-engineering environment and TDMS, and sample content and data from projects or products managed by the TDMS.

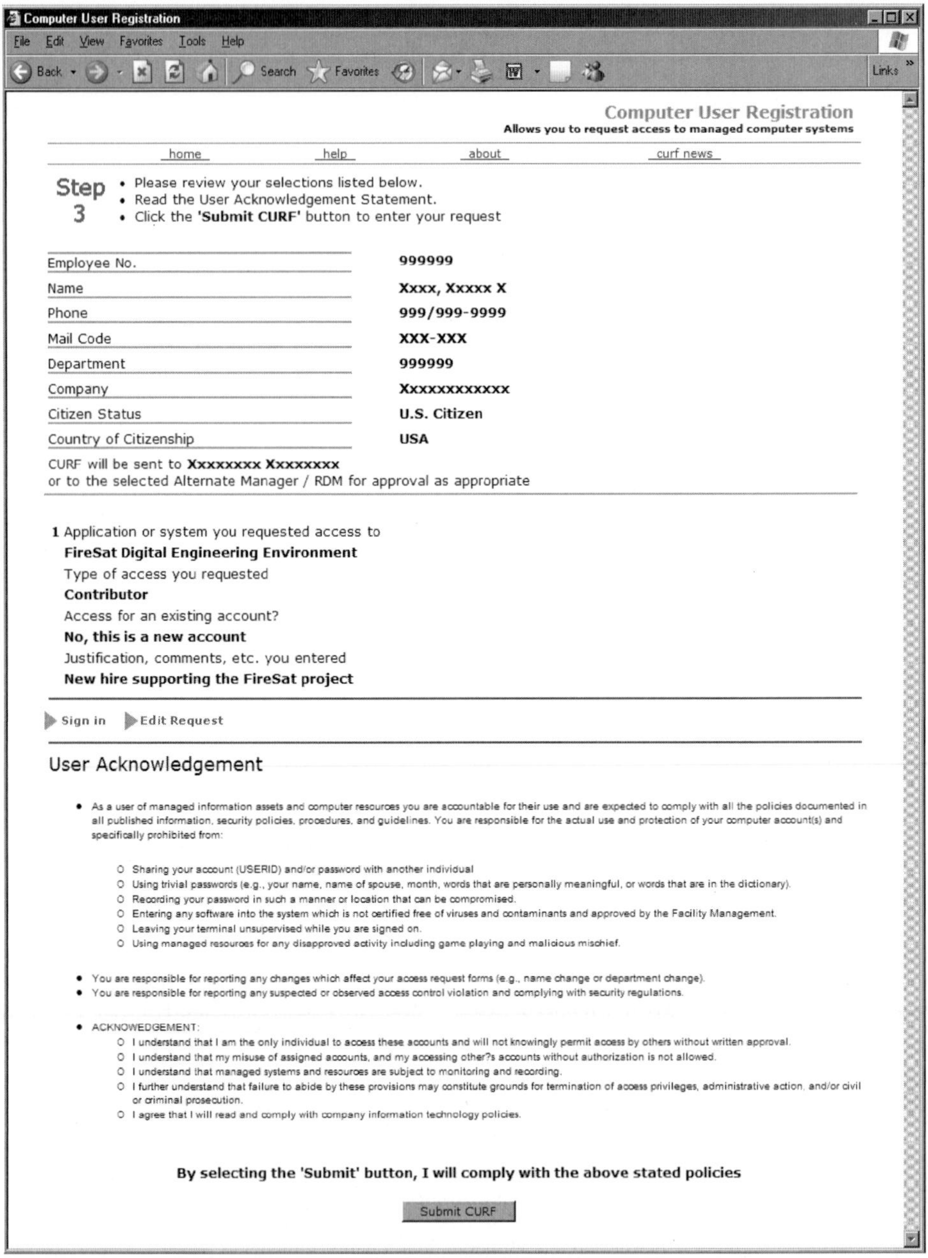

FIGURE 17-14. Submitting the Request for Processing. By clicking the "submit" button, the user electronically acknowledges an understanding of the policies and procedures that must be followed.

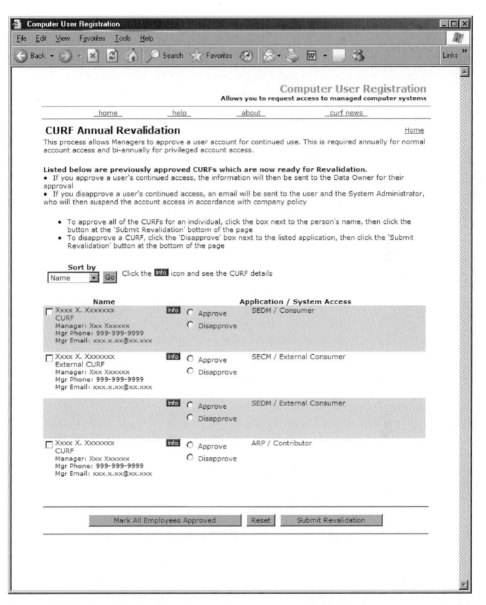

FIGURE 17-15.   Revalidating User Accounts and Access Grants. Having an automated user interface simplifies this process and is essential when we must periodically revalidate thousands of accounts and access grants.

In most large organizations, an internal audit group does the audits. They focus on business processes to ensure compliance with all standards and procedures, as well as generally accepted practices. They also evaluate supporting systems and tools to be sure the artifacts produced are correct and reliable. Organizations that don't have an internal audit group can contract with external auditors.

Audits of the systems engineering lifecycle and TDMS evaluate SE practices, procedures, and systems to confirm that required technical data is appropriately captured, managed, and distributed to satisfy the needs of project teams and their customers. Audits also verify that lifecycles follow established procedures, directives and agreements. They typically carry out these eight actions:

- Review content and data submissions against technical data management system quality controls and standards, which evaluates the accuracy and integrity of content and data, and gauges the effectiveness of input edit checks and validation features
- Review exception-handling processes to gauge how effectively the system detects and resolves exceptions
- Review data backups and the DRCOP's recovery procedures; review the results of periodic testing under the recovery plan
- Review users' accounts and evaluate how effectively the system manages them; randomly sample approvals and responses to computer users' requests to gauge how effectively the system provisions and de-provisions accounts
- Review access controls and mechanisms; take random samples of access controls to gauge the effectiveness of these processes
- Review all privileged-user accounts for rationale and need
- Document all audit results and findings
- Track findings through resolution and retesting

An audit assesses the correctness, reliability, and security of data and content stored and managed by the TDMS. It provides a level of confidence to stakeholders of the TDMS's services and functions.

17.5   Collaborate by Effectively Using System and Process Artifacts

Systems engineering is rarely an individual effort. Integrated product teams and other forms of teaming are normal. These project structures bring together skills and expertise from multiple disciplines to engineer the complex systems of our digitally driven world. Modern digital-engineering environments and data architectures (Figure 17-5) are the infrastructure that enables success. But electronic collaboration and systems for integrated data management—whether they follow physical, functional, or other lines—often augment this infrastructure to support engineering project teams.

Collaboration tools and environments range from widely available separate tools such as email and distributed file sharing to COTS packages designed and marketed for collaboration. High-end packages usually offer integrated tools using familiar paradigms, such as electronic rooms, shared cabinets and folders, or virtual spaces. They're operational at installation. Low-end collaboration environments more often come from separate components or tools; we have to integrate these individual tools into the environment.

17.5.1   Use Collaboration Tools and Techniques

Effective collaboration requires both synchronous and asynchronous communication. Examples of the former are phone calls, Internet chats, or instant messages; examples of the latter are voice mail, email, or fax. We also need ways to interact in groups and subgroups. Examples of group interaction are audio-video conferencing, virtual whiteboards, or teleconferences; examples of subgroup interaction are wikis, blogs, or Internet forums. Coordinating all these efforts requires such management tools and techniques as scheduling systems; shared calendars; tools to manage tasks, actions, and issues; workflows; and content lifecycles.

As the web evolves, its second-generation tools, techniques, and underpinnings can also help us collaborate. For example, social-networking sites develop online communities linked through interests or themes defined by the sites or their users. Their methods of interacting, forming groups, and sharing information evolve rapidly and offer chances to apply them in systems engineering.

17.5.2   Use Search and Retrieval Tools and Techniques

To capture and manage digital artifacts for systems engineering, and to provide this data and content to authorized people, we must have a strong user interface for searching, retrieving, and using space systems information. Search and retrieval tools are central to the user's experience, so they determine the system's effectiveness. Figure 17-16 illustrates a web portal for search and retrieval tailored to systems engineering policies and procedures in NASA's Space Shuttle program.

In many cases people from several disciplines make up the project team developing a product or system. Each discipline brings different skills, vocabulary, and perspective on the systems engineering lifecycle and its artifacts. This diversity makes search and retrieval challenging when people search by attribute or keyword. Different people often refer differently to the same item or concept, so users must understand these differences and reflect them in their searches.

Figures 17-17 through 17-19 illustrate a solution to this challenge: setting up search and retrieval so users can search by drilling down into illustrations of hardware or topical baselines. Using a visual "dashboard," they can set search filters and then retrieve results by clicking on displayed results hot spots. This way, they don't have to understand names and identifiers; the system creates fully qualified queries in the background, based on their visual contexts and filters.

FIGURE 17-16.   A Website for Integrated Data Management Based on Visual Search and Retrieval Techniques. This portal supports searching different data and content domains by establishing a visual context and setting filters to narrow the search results.

Processes, systems, and techniques for managing technical data result from an evolution that began with the computer age. It planted the seeds for an infrastructure that has substituted integrated digital design environments for drafting boards and computer-based modeling systems for slide rules. These modeling systems can accurately analyze virtually every physical, economic, and managerial aspect of a space system's lifecycle. Engineering has steadily shifted from paper artifacts to digital ones. Functionally aligned organizations whose members interact face-to-face have given way to virtual project teams spread across continents, collaborating through web-based tools. These changes have radically changed how space systems engineering operates and how we create, manage, and use its products and deliverables.

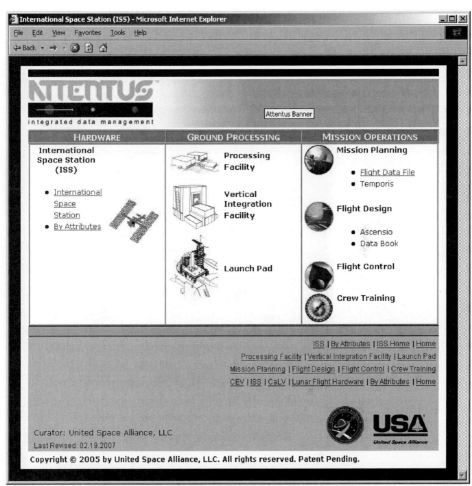

FIGURE 17-17.   Web Portal for Integrated Data Management (IDM). This IDM portal searches collections of hardware baselines, ground processing facilities, and mission-operations topics for the ISS. Users can drill into each baseline through hyperlinks or hotspots on the web page.

This radical shift in engineering processes and products has changed technical-data management from mere administrative support to a critical, pervasive function that translates into competitive advantage when properly designed, constructed, and applied. Those who understand its potential use it to deliver the seemingly impossible; those who ignore it risk failure in even the most mundane activities.

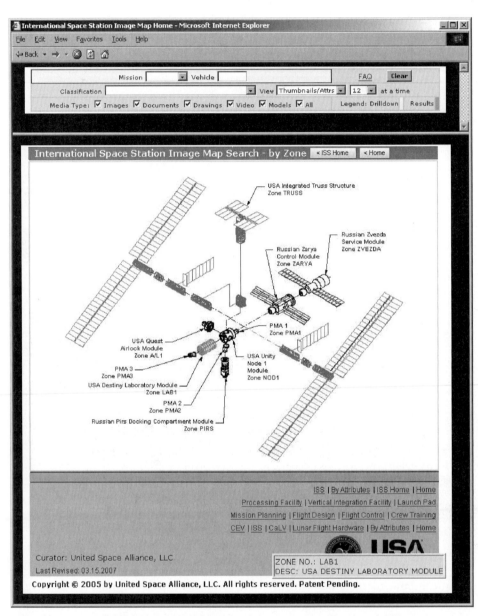

FIGURE 17-18.   **Baseline for Hardware on the International Space Station (ISS).** The baseline for ISS hardware enables users to drill down into any modules or assemblies on orbit or being processed for launch.

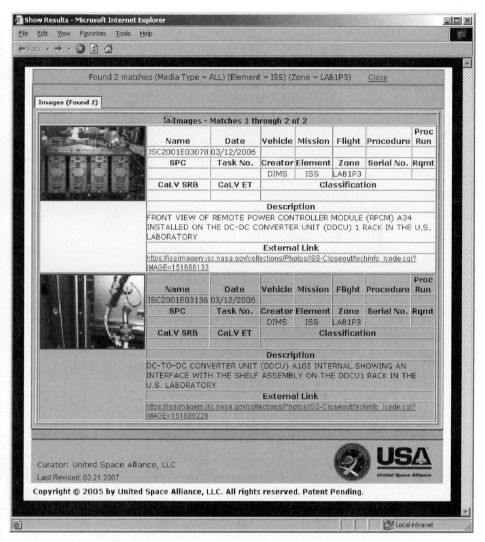

FIGURE 17-19.   Data Search Results for the International Space Station (ISS). Search results are in a grid with data attributes stored in the system that describe each item. The item's thumbnail enables users to launch the viewer associated with this content.

References

Government Electronics and Information Technology Association (GEIA). 15 June 2002. EIA-836—*Configuration Management Data Exchange and Interoperability.* GEIA website: www.geia.org

Institute for Printed Circuits (IPC). November 2001. IPC-2578—*Sectional Requirements for Supply Chain Communication of Bill of Material and Product Design Configuration Data — Product Data eXchange (PDX).* Bannockburn, IL: IPC website: www.ipc.org

International Organization for Standardization (ISO). 1994. ISO 10303—*Industrial Automation Systems and Integration—Production Data Representation and Exchange,* also known as *Standard for the Exchange of Product Model Data (STEP).* ISO website: www.iso.org

Chapter 18

Technical Assessment and Reviews

Michael C. Pennotti, Ph.D., *Stevens Institute of Technology*
Mark P. Saunders, *NASA Langley Research Center*

System development is highly iterative, involving lots of looping back and revisiting of earlier work as new information becomes available. Still, nearly all of its processes divide into periods of activity, or stages, punctuated by formal reviews, also referred to as gates. Formal reviews at key project milestones answer three basic questions [Cooper, 2001]:

- Have we done the activities well in the preceding stage?
- Does continuing with the project make sense?
- Are the proposed actions, resources, and funding enough to complete the next stage?

Technical reviews address these questions for the evolving system design, whereas program, project, and other reviews address other aspects of a project. But all reviews have the same intended outcome: the entire team's shared understanding of the project's status and a clear go/no-go decision about its next stage. Most projects emphasize formal reviews, but the same process is useful for informal ones, if the technical team thinks an independent assessment of their progress would be valuable. Informal reviews offer insight, perspective, and alternate approaches that those closest to the work may miss. They often clarify technical issues and suggest alternative approaches that the team may not have considered. Table 18-1 shows a process for effective technical reviews.

TABLE 18-1.   Technical Assessment Process. Here we list the process steps for assessing a project's progress during its lifecycle.

Step	Description	Where Discussed
1	Define the subject and scope of the review	Section 18.1
2	Establish the entrance criteria	Section 18.2
3	Identify and invite the review team	Section 18.3
4	Conduct the review	Section 18.4
5	Ensure that the system meets success criteria	Section 18.5
6	Specify and document the decisions	Section 18.6
7	Clarify and document the action items	Section 18.7
8	Baseline the design	Section 18.8
9	Improve the process for technical reviews	Section 18.9

18.1   Define the Subject and Scope of the Review

In planning a technical review, we first decide on its purpose. Formal reviews require certain decisions, as discussed in Section 18.8. Informal reviews often have broader, less defined objectives. For the review to be effective, its intended outcome should be clear to all participants from the beginning.

Every technical review examines some system, and every system has a context. A design team controls the system, but the context is everything else that acts on it and within which it must fulfill its purpose. In planning the review, we must differentiate between the two—to establish the system boundary. For example, the Constellation Program identifies five major elements to support crewed space flight to the moon, Mars, and beyond: the Orion crew exploration vehicle, the Ares 1 launch vehicle, ground operations, mission operations, and the extravehicular activity system. Reviewers and presenters must recognize that an Ares 1 review focuses on the launch vehicle. The other program elements are part of the context— fixed elements that constrain the launch vehicle.

Another consideration is the system's position in the lifecycle, which affects how reviewers see the technical design's maturity. Uncertainty early in a project reflects an open design process, one not overly influenced by preconceived solutions or existing systems. The same amount of uncertainty late in a program indicates inadequate progress or unwarranted risk.

For example, at the system definition review for the Mars Science Laboratory, the independent review team was relatively comfortable. They knew the teams hadn't fully assessed entry, descent, and landing risks. And they may not have verified and validated their designs adequately. But the schedule allowed plenty of time to address these issues before the preliminary design review. If the same deficiencies remained at critical design review, it would have called for a more drastic set of actions.

Different systems engineering processes define technical reviews that all projects must pass through. Processes differ, and some organizations use different words to mean the same thing, but their intent is the same: to ensure the orderly spending of resources to produce what users want within cost and schedule constraints. This chapter focuses on the key reviews in NPR 7123.1a [NASA, 2007 (1)] concerning the mission concept, system requirements, system definition, preliminary design, and critical design reviews (Figure 18-1). We mention other terms where appropriate.

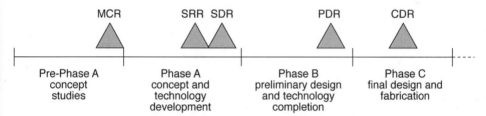

FIGURE 18-1.   Major Technical Reviews. NASA's lifecycle model (from NPR 7123.1a) specifies technical reviews at major development points. (MCR is mission concept review; SRR is system requirements review; SDR is system definition review; PDR is preliminary design review; CDR is critical design review.)

In Section 18.8 we discuss other formal technical reviews, including those for test readiness, system acceptance, flight readiness, operational readiness, and decommissioning. We also describe informal technical reviews that project teams may hold at any time they consider appropriate. But first we describe the rest of the review process.

18.2   Establish the Entrance Criteria

Entrance criteria specify the inputs or products we must include in a technical review. For formal reviews, they're the deliverables that must come from the development's preceding stage [Cooper, 2001]. For informal reviews, they may be less formal, but should be no less specific. If the development team hasn't specified and met these entrance criteria in advance, a technical review can turn into a design meeting. Reviews are often unsuccessful if the calendar, not the availability of required work products, drives them.

For example, the Orion Project held its system definition review as scheduled in August 2007, even though the spacecraft architecture couldn't meet its mass allocation. So the review had to remain open for five more months until they met that milestone. This extension increased cost, delayed the decision to proceed to the next phase, and left uncertain how design changes would affect other projects. Project teams must produce required artifacts early enough so reviewers have plenty of time to digest them before the review meeting.

Too often, those who define and manage development processes focus more on activities than on outputs. They would be far more effective if they concentrated on deliverables for each stage and on thoroughly reviewing these deliverables at the next gate. Otherwise, the teams' activities may be unfocused, and they may show up for the subsequent review with a long story about how they spent their time instead of the products they were supposed to deliver. The purpose of a technical review is not to listen to such stories, but to assess the design's maturity and its fitness for the next development stage.

TABLE 18-2.   Sample Review Criteria. The entrance and success criteria for a mission concept review are specific and clearly defined. (MOE is measure of effectiveness; MOP is measure of performance.)

Mission Concept Review	
Entrance Criteria	**Success Criteria**
• Mission goals and objectives • Analysis of alternative concepts to show that at least one is feasible • Concept of operations • Preliminary mission de-scope options • Preliminary risk assessment, including technologies and associated strategies and options for managing or reducing risk • Conceptual strategy test and evaluation • Preliminary technical plans to achieve next phase • Defined MOEs and MOPs • Conceptual lifecycle support strategies (logistics, manufacturing, and operation)	• We have clear, internally consistent mission objectives • Our preliminary requirements provide a system that will meet the mission objectives • The project has a feasible mission and technical solution at an acceptable cost (rough estimate) • We stated and rank-ordered criteria for evaluating candidate systems • We clearly identified the need for the mission • The cost and schedule estimates are credible • We did an updated technical search to find assets or products that could satisfy the mission or parts of it • Technical planning is sufficient to proceed to the next phase • We identified acceptable risk and mitigation strategies, based on technical risk assessments

18.3   Identify and Invite the Review Team

The people who participate in a review are even more important than its structure. Key members of the development team must be present to describe the deliverables and answer the reviewers' questions. They must also explain the reviewers' decisions and guidance to other members of the team. An effective technical review requires representatives from other organizations, as well—particularly ones that the design will affect, such as manufacturing, test, logistics, support, or training. Other experts not directly involved in the project should offer independent perspectives [Starr and Zimmerman, 2002]. Examples are:

- Someone with general design expertise
- Someone familiar with the technology
- A program manager from another, unrelated program
- An expert on the application domain
- Someone with general knowledge of architectural principles, whom we count on to ask questions that are a little "off-the-wall"

For example, the review board for the Mars Science Laboratory includes experts in the technical disciplines of guidance, navigation and control; propulsion; nuclear power; telecommunications; robotics; science instrumentation; systems engineering; and other technical disciplines, as well as experts in mission operations; project management; cost, schedule and resource analysis; and mission assurance.

In some cases, sharp reviewers from outside the project team join a review and recognize technical problems. For example, during the system definition review for the Space Interferometry Mission, deficiencies emerged in the two-dimensional metrology approach to establish the mirrors' relative positions. But it wasn't an optics expert who pointed them out—a reviewer only partly familiar with the technology found them. So, inviting outside reviewers often brings valuable talent to the review process.

Formal reviews must also include those who can commit the resources required for the next development stage [Cooper, 2001]. These are usually senior managers or administrators with demanding schedules who are often difficult to engage. But if they don't participate, decisions made during the review won't be binding, and their absence suggests to other reviewers and the development team that the review isn't important. On the other hand, their participation greatly strengthens a formal review's effect and its contribution to project success.

18.4   Conduct the Review

Technical reviews aren't a place for problem solving; they should verify the solutions [NAVAIR, 2006]. Reviews often aren't as effective as they might be because reviewers and presenters become adversaries. Presenters try to get through their material with minimum questions and no action items. Reviewers, assuming this intent, try to ferret out what they believe the presenters may be hiding. This adversarial relationship interferes with the intended outcome. Reviewers and presenters must remember that they're on the same side. They presumably want the best product, system, or solution for the customer and other stakeholders, and each needs the other's support to bring about that outcome.

The Phoenix Mars Mission exemplifies how this collaborative relationship breaks down. (Phoenix was to study the arctic region of Mars's surface.) At system definition review, the chair of the independent review board asserted: "This project will not go forward until I say so." Despite prompt action to temper his decree, this remark harmed the relationship between the project team and review board for months. It became much more difficult for the review board to get accurate, timely information from the project team, and the project team resisted the board's suggestions and recommendations. Both wasted valuable time and energy worrying about each other when they should have focused on designing the mission. Although the situation gradually improved, it would have been far better if the relationship hadn't been damaged in the first place.

Reviewers can help prevent adversarial relationships. For example, mentioning all the flaws in a presentation seldom builds trust and a collaborative spirit. Most teams produce some ideas worthy of praise, so reviewers should first say what they like. Acknowledging the positive elements reinforces the team's accomplishments, helps keep these contributions in subsequent work, and makes the presenters more receptive to constructive criticism. Couching that criticism as "what else we'd like you to do" instead of "what's wrong with what you've done" also maintains a positive atmosphere. Teams can always do more, no matter how well they've done. A comment from this perspective emphasizes what needs to happen in the future, which the team can influence, not what's wrong with the past, which the team can't change.

Though important for an effective review, including the senior managers creates a special challenge because hierarchical relationships between reviewers and presenters may stifle debate and minority views [Surowiecki, 2004]. This happens when a leader, intentionally or unintentionally, expresses opinions that others fear to challenge. Effective leaders encourage debate, minority opinions, objective evidence, and all possible interpretations before deciding. They clearly value others' opinions, even if those opinions contradict their own or if they decide to go in a different direction. All types of disasters have occurred when reviews didn't foster such debate. In fact, that's the reason NASA's Space Flight Program and Project Management Requirements [NASA, 2007 (2)] now require a way to document dissenting opinions and carry them forward.

Nearly all reviewers should address material presented to them, but it's often far more difficult to critique what isn't presented—and that may be the most important contribution they make. An effective review bores under the surface to:

- Question the critical assumptions that underlie a design
- Make sure that we've considered a broad range of options before selecting the preferred one
- Examine every assertion to ascertain the facts and data behind it
- Explore relationships among the system's elements and between different presentations
- Identify blind spots that people immersed in the details may have missed

Example: Mars Pathfinder, the first successful Mars landing since Viking in the 1970s, relied on a parachute to control the spacecraft's final descent to the surface. At the critical design review, one of the reviewers questioned whether the Kalman filters in the radar altimeter properly accounted for fluctuations from the spacecraft's swinging beneath the chute as it approached the uneven terrain below. The question produced a strong response—an extensive test program to be sure that the altimeter worked properly.

18.5   Ensure that the System Meets Success Criteria

Success criteria are standards for evaluating the deliverables. Just as with entrance criteria, we must clearly specify them in advance and see that the development team and reviewers understand them. Success criteria must be specific, not vague or ambiguous, so we must clearly compare them to the deliverables. Specific criteria eliminate surprises—deliverables either conform to the standards or they don't. Vague criteria lead to subjective, qualitative assessments, which don't result in a strong review. For example, JPL's design principles for robotic missions are well-posed criteria because they specify requirements to meet certain margins depending on the design's technical maturity. At system definition review, for instance, they require a mass margin of 30 percent.

18.6   Specify and Document the Decisions

The most important output from a formal review is the decision about whether to continue to the next development stage. This decision should be unambiguous. According to Cooper, only four possible decisions exist [Cooper, 2001]:

- **Go** means we approve the project and have committed the resources required for the next development stage
- **Kill** means we should end the project and spend no more money on it
- **Hold** means the deliverables met the technical success criteria, but we should suspend the project for a fixed period, or indefinitely. The resources required to proceed may not be available, or a higher-priority use for them may have emerged, or the customer's requirements may have changed. We should put the project on the shelf until priorities change or the required resources become available.
- **Recycle** means the deliverables didn't meet our success criteria or weren't good enough to support a clear decision. So we must repeat the previous stage and do another review before allowing the project to proceed.

Some would add a fifth option: **Proceed with conditions**. This means letting the project proceed even though deliverables didn't fully meet the success criteria, so we have to redo some deliverables and resubmit them for approval. Particularly under aggressive schedules, reviewers may have a strong incentive to choose this option, especially if "recycle" will result in a slip in a major project milestone or a missed market opportunity. But it's a dangerous choice. Too often, despite best intentions, no one revises the deficient deliverables. Having passed through the gate, and under intense schedule pressure, the development team may simply move forward without satisfactorily completing some design element that reviewers considered critical. Unless a disciplined process holds the development team strictly accountable, "proceed with conditions" can thwart the review's purpose.

For example, at the preliminary design review (PDR) for Mars Science Laboratory, the independent review team said the authorized funds weren't enough to complete the project as planned, but the development team decided to proceed with only a modest increase. At the critical design review, three major systems weren't at the required maturity level and authorized funds were still inadequate. As a result, the project received authorization to proceed but with a smaller scope. If that decision had been made at PDR, it would have left more time to respond.

18.7   Clarify and Document the Action Items

Besides a clear go/no-go decision, every review generates action items that we must capture at the time and then reassess at the end of the review. People sometimes suggest action items that become unimportant or inappropriate later in the review. Or action items seem clear when first captured, but later reviews find them ambiguous. So reviewing all the action items at the end of a review gives us confidence that everyone agrees they are pertinent. We must clearly document each action item, including the name of the person responsible for its completion and a target date. No action item is completely specified until someone commits to doing it by a specific date. Sample action items taken from the Constellation Program's system requirements review were to:

- Flow down the need, goals, and objectives for improving the efficiency and reducing the difficulty of handling hardware between missions
- Add more systems engineers to the program team
- Add an independent agent to double-check Marshall Space Flight Center's analysis of flight performance to orbit

18.8   Baseline the Design

Besides the clear go/no-go decision, another important output of a formal technical review is a *technical baseline*. This is an agreed-on definition of some system aspect that we then place under configuration control and formal change-control procedures. The formal technical review affirms that the design has matured enough to justify this baseline and that all team members know what the new baseline is.

The timing of technical baselines—and therefore of formal reviews—is critical to effective system development. Declaring baselines too early can needlessly constrain innovation and add unwarranted bureaucracy when frequent changes become necessary. Waiting too long to declare a baseline results in excessive churn and rework. It may lead to design errors that drive up costs and, in the worst case, scuttle the project.

As with technical reviews, different systems engineering processes define distinct baselines, and some define the same baseline using different terms. Again, we focus on baselines produced by reviews in NPR 7123.1a and we reference alternate terms where appropriate. A baseline is usually one of these types: mission or operational; requirements or system; functional; allocated or component; and design. We describe each one below with the technical review that produces it.

18.8.1   Reviewing the Mission Baseline Concept

The mission baseline defines the need that a proposed system intends to address and the operational concept for meeting that need. It focuses on the system's role in its context and identifies the external systems it will interact with. This first of the technical baselines tells us more about the problem to be solved than about the nature of the solution. In it we specify the system's objectives before starting the design. We may refer to the mission baseline as a business or customer baseline because it defines the key stakeholders for the system and the requirements of each stakeholder. It represents agreed-on conditions that the system must satisfy for success.

The mission baseline comes from the mission concept review. Entrance criteria include stating the mission's goals and objectives, determining measures of effectiveness and performance, analyzing concepts for meeting those objectives, and establishing a concept of operations for the selected approach. The development team should also assess risks and appropriate mitigation plans; a conceptual strategy for test, evaluation, and lifecycle support; and preliminary technical plans for completing the next development phase. Success criteria for assessing the mission concept review's deliverables fall into four categories:

- Have we clearly identified the need for this mission? Have we explored other ways to meet the need and rejected them in favor of the preferred approach?

- Have we clearly defined and stated the mission objectives, and kept them internally consistent? Are they measurable or observable, so we'll know when we've met them?

- Is the mission feasible? Is the proposed concept technically feasible at technology and system levels? Can we reasonably provide the solution with resources budgeted or likely to be made available?

- Have we done enough technical planning to move to the next development phase? Have we identified the risks and developed acceptable mitigation plans?

If we meet these success criteria, the mission baseline is defined and the system moves to concept development. For a complete list of entrance and success criteria for the mission concept review, see Table 18-2.

18.8.2   Reviewing the System Baseline and Requirements

Sometimes called the requirements baseline, the system baseline specifies the functional and performance requirements a developed system must meet. These requirements translate the stakeholders' requirements, contained in the mission baseline, into technical specifications for the system. They move from the stakeholders' language to the language of the engineers who will complete the design, so they must trace directly to stakeholder requirements and characterize the system at its highest level. Although a requirements baseline sometimes includes lower-level specifications, we recommend that it describe the system as a black box, without subsystems or components. It must identify the system's inputs, outputs, and other performance characteristics.

The system baseline is the product of a system requirements review (SRR). The entrance criteria include a system requirements document and a complete specification of what the system must do. It commonly produces a first-level functional architecture that allocates system requirements to the top-level functions, but we should be sure that the system requirements are correct before beginning to derive and flow down the lower-level requirements. Other entrance criteria for an SRR are:

- Incorporate any changes made to the mission requirements and concept since mission concept review (MCR)
- Have a preliminary maintenance concept for the system
- Complete project-management and systems engineering plans for the next development phase

Success criteria for assessing the SRR's deliverables emphasize five main areas:

- Have we satisfactorily completed all action items from the MCR?
- Are the system requirements complete, clearly written, and traceable to stakeholder requirements in the mission baseline? Is each system requirement verifiable and do we have a preliminary plan for its verification?
- Is a sound process in place for allocating and managing requirements at every level of the design?
- Are the system and subsystem design approaches reasonable and are they likely to result in a system that satisfies the mission needs? If we've allocated system requirements to subsystems, is the allocation traceable and are the resulting subsystem requirements well written?
- Have we done enough technical planning to go to the next development phase? Have we identified risks and figured out how to handle them?

If we meet these conditions, the system baseline is complete and preliminary design can begin. Table 19-18 lists the entrance and success criteria for the system requirements review. The following is an example of planning for a system requirements review.

Example: NASA's Constellation Program, headquartered at the Johnson Space Center, manages, develops, and integrates the flight and ground infrastructure, systems, and supporting elements that enable continued human access to space. It covers crewed missions to the Moon, Mars, and beyond once the Space Shuttle retires. Some of its deliverables are the crew exploration vehicle, exploration launch projects, mission operations, ground operations, and the extravehicular activity system.

In early 2007 NASA applied new policies and established independent review teams (IRTs) for its programs and projects. With the mission concept already complete as part of the Exploration Systems Architecture Study, these teams began their independent work at the system requirements review (SRR). New policies specified the SRR's top-level scope. But each IRT had terms of reference, including the entrance and success criteria, so team members and managers knew what to expect for the review. Everyone agreed to these terms, including the agency stakeholders. As they developed the terms, they also chose team members by relating their technical and program disciplines to each project.

Because the projects within the program are coupled, the results of the SRR for each project roll up to synchronize at the program level and ensure integration of the overall system. The IRTs match this structure, and the chairs of the project IRTs are the principal members of the program's independent review team. This approach helps us be sure that the strengths and weaknesses noted during each project's review are considered at the program level.

18.8.3   Reviewing the Functional Baseline and System Definition

The functional baseline consists of a fully specified functional architecture, with all requirements flowed down to the lowest functions. It states in detail what the system must do to complete its mission, as well as the necessary relationships among its tasks. The baseline develops from a system definition review. Entrance criteria for this review include a functional architecture, plus supporting trade analyses and data. To assess its deliverables, we consider four main types of success criteria:

- Have we satisfactorily completed all action items from the system requirements review?

- Is the functional architecture complete? Can we trace the required system-level scenarios through every level of the functional architecture? Have we flowed performance requirements down to the lowest functions?

- Can the proposed functional architecture accomplish what the system must do? Is the technical approach reasonable? Do the timing and sequencing of lower tasks match higher functional requirements?
- Have we done enough technical planning to go to the next development phase? Have we identified risks and planned how to handle them?

If we've met these criteria, the functional baseline is complete and preliminary design continues. Table 19-27 gives a full list of entrance and success criteria for the system definition review.

18.8.4   Reviewing the Allocated Baseline and Preliminary Design

The allocated baseline is a complete architectural model of the system, with all functions in the functional baseline allocated to components and all nonfunctional system requirements flowed down to the lowest functions and components. (See Chapters 4 and 5 for a detailed discussion of functional and nonfunctional requirements.) Sometimes referred to as the component baseline, the allocated baseline is an output of the preliminary design review (PDR).

Entrance criteria for a PDR include subsystem design specifications for each configuration item (Chapter 16). These specifications must be as complete for components as the system specifications were for the system during the system requirements review. Design engineers use them to complete their detailed designs, and test engineers use them to make sure that each delivered item meets its requirements. This point is crucial in development, because a PDR is the gateway to detailed design and is essential to wisely spending a much larger amount of money and other resources.

A PDR also requires complete interface specifications, which must show how the configuration items interconnect and interact so the resulting system meets its top-level requirements. Interface specifications are particularly critical for the final system. Experience shows that most system problems occur at interfaces (see Chapter 15).

An example of effectively handling interface specifications comes from the Magnetospheric Multiscale Mission, designed to study magnetic reconnection in Earth's magnetosphere. The project team identified as critical the interface between four stacked spacecraft and the Evolved Expendable Launch Vehicle because the decision about whether to base the launch vehicle on a Delta IV or an Atlas V rocket would come one year after the critical design review (CDR). They decided the Agency should seriously consider moving the launch vehicle decision to nine months before CDR, so the project wouldn't have to develop interface designs for both options.

An example of ineffective handling comes from the Mars Polar Lander project. On this spacecraft, a sensor was supposed to determine when the spacecraft's legs had touched down on the Martian surface and signal the engines to shut down. Unfortunately, project teams had tested all the components separately, but no one did an end-to-end test. During the landing, the sensor apparently responded to a transient touch-down signal generated when the spacecraft's legs deployed, leading to premature engine shutdown and loss of the mission.

A third vital input to PDR is a preliminary specification for the software design, including a complete software architecture and a preliminary data design. This specification drives the detailed software design. To assess deliverables from a PDR, we consider four main types of success criteria:

- Have we satisfactorily completed all action items from the system definition review?

- Has everyone agreed to the system-level requirements and does the preliminary design meet those requirements?

- Have we flowed down requirements completely and appropriately, and are all configuration item requirements traceable to system requirements and well written?

- Are all development risks well understood, acceptable, and managed in an effective mitigation plan?

If we meet these success criteria, the allocated baseline is complete and detailed design begins. For a complete list of entrance and success criteria for the preliminary design review, see Table 19-37. The following is an example of a preliminary design review and a critical design review.

Example: The Jet Propulsion Lab (JPL) is designing the Mars Science Laboratory Mission to send the next-generation rover to Mars in 2009. They did a preliminary mission and systems review (PMSR)—a different name for the system definition review—in December 2006. The independent review team (IRT) consisted of system experts from JPL, Goddard Space Flight Center, and other external organizations. The project team presented their design and development plans to the IRT, and then responded to the IRT's questions and requests for action.

As part of this review, the IRT found several technical areas to improve through further project analysis and work. For example, they believed the project team hadn't finished assessing the risks of entry, descent, and landing (EDL) at Mars and had taken a potentially faulty approach to verification and validation. The IRT also thought the design appeared to carry more single-string systems than was warranted for a $1.5 billion mission and recommended that the project reexamine both areas.

At the next milestone—the preliminary design review (PDR)—the IRT reassessed the project for entrance and success criteria mentioned above. Having taken the PMSR's issues and actions seriously, the project offered analyses and system changes as part of their PDR presentations. The project showed the IRT the results of their EDL analyses and convinced the IRT that their validation and verification approach was adequate to demonstrate the EDL performance. The project had concurred with the IRT's assessment of single-string systems and had added avionics hardware to increase the spacecraft's redundancy.

The IRT was satisfied with these changes and found the technical design to be in good shape, but they had noted in the previous review that the schedule looked extremely tight. During this later review, they strongly recommended the project increase their schedule margin by pulling forward their critical design review (CDR), moving their software deliveries to their test beds earlier, and making their actuator deliveries as early as possible, along with several other suggestions. The project leader concurred.

At the CDR the independent review team again assessed the project's progress and their plans for development and operations, as specified in paragraph 18.8.5. The IRT said the end-to-end system design held together with adequate (though tight) technical margins, but several critical systems weren't yet at the CDR level. Just before CDR, the project discovered that the actuator design wouldn't meet the design specifications so they had started developing a new design. Coupled with late software development, design issues with the sample handling equipment chain, and a few other late designs, this actuator problem had added schedule pressure to the planetary-constrained launch window. The IRT suggested several unique ways to resolve this schedule pressure, which the project agreed to investigate as part of their continuing planning for workarounds.

Finally, the independent review team recognized that the project understood the risks they faced and had plans in place to address them, but they didn't have enough money to complete the spacecraft development. The review team communicated this important fact to management in their report, so the latter could correct the problem.

18.8.5   Completing the Design Baseline and Critical Design

The design baseline is the complete system design, including all configuration items, software, and interfaces. It results from a critical design review (CDR) and is the input to fabrication, assembly, integration, and test. Entrance criteria for CDR include build-to specifications for each hardware and software configuration item, software design documents, and complete plans for fabrication, assembly, integration, test, and verification. To assess the CDR deliverables, we emphasize four types of success criteria:

- The design meets requirements with adequate margins in all technical performance measures
- Interface control documents are sufficient to begin fabrication, assembly, integration, and test
- Requirements and plans for verification and validation are complete
- Everyone understands mission risks, and plans and required resources are in place to manage them

If we meet these conditions, the design baseline is complete—and fabrication, assembly, integration, and test begin. Table 19-44 lists all the entrance and success criteria for the critical design review.

18.8.6   Other Technical Reviews

The reviews we've discussed so far produce the key technical baselines for development. As indicated above, they're common to almost all development processes, although the mission concept review and system definition review in particular may appear under several different names. But NPR 7123.1a defines several other technical reviews that often occur in developing space systems.

A *test readiness review* occurs before verification testing. It confirms that an article is ready for testing, test plans are in place, and facilities and people are ready. It can be particularly important when the test costs a lot or uses other scarce resources. For example, pre-flight vibration tests damaged the High Energy Solar Spectroscopic Imager spacecraft because the vibration test system malfunctioned. The root cause was having no procedure requiring the test team to look for shaker performance problems in the pre-test data.

A *systems acceptance review* verifies that an end item meets its stakeholder requirements either before shipping or before the stakeholders accept delivery. It usually occurs after testing is complete and often includes reviewing the results of a system acceptance test that stakeholders have performed or witnessed.

A *flight readiness review* occurs before a spacecraft launches to prove that it's ready for a safe, successful flight. Besides the spacecraft and its launch vehicle, it addresses associated ground support equipment and people.

An *operational readiness review* is the final review before a system starts operating. It examines the system and verifies that final documentation matches the system as built and delivered.

A *decommissioning review* confirms the decision to decommission a system, with particular attention to this operation's safety. For example, during Hubble Telescope Servicing Mission 4 (scheduled as of this writing for launch in May 2009), the Shuttle crew will attach to the spacecraft a soft-capture mechanism to help control its reentry into the Earth's atmosphere in 2013.

18.8.7   Detailed Technical Reviews

The technical reviews discussed above are all examples of formal reviews matched to particular purposes and development milestones. But technical reviews can take place at any time during a project—whenever the project manager, chief engineer, or even the development team thinks the project would benefit if people from several disciplines assessed some aspect of the emerging design. Although such informal reviews don't have published specifications, we can easily create a specification using the same steps as those for formal reviews:

1. Define the review's purpose and select the reviewers

2. Develop success criteria and specify the outcomes we expect reviewers to produce

3. Determine the entrance criteria, which specify what the reviewers must deliver

4. Distribute the list of deliverables to reviewers before the first review meeting

5. Follow the formal-review process to do the informal review: present exhibits, ask questions, decide on actions, and capture action items for appropriate follow-up. Keep discussions open, so all can express their views.

The process to specify and complete a technical review is straightforward, but carrying it out in a disciplined way is difficult. As a positive example, the team who developed the Tracking Data Relay Satellite decided to do informal reviews of the system definition and system requirements to address obsolete components. They were disciplined enough to add these reviews, even though the first review required for these "build-to-print" satellites was the preliminary design review.

18.9   Improve the Process for Technical Reviews

As for all key business processes, we must continually manage and improve the process for technical reviews. Entrance criteria specify the inputs to a technical review; they're the deliverables a development team produces and presents at the review. The outputs of a review are a go/no-go decision, assessments of the deliverables against the success criteria, and action items produced from those assessments. We can improve each of these elements.

The most important aspect of effective continuous improvement is well-defined process metrics that we continually monitor and use to identify problems and measure improvement. For inputs to a technical review, metrics might include the number of entrance criteria satisfied, how soon deliverables get to the reviewers, and the number of deficiencies in the deliverables. Teams should collect these metrics in every technical review. If they find problems, they should analyze root causes and start improvements to address the most important ones.

The review process offers more potential metrics, including especially the number of reviewers participating and the percent of time each was present. If reviewers aren't present throughout the review, they lose the context and devalue their contributions. In addition, discussion needs to be open during the review. If participants are reluctant to express their opinions, asking them directly may produce unreliable data. Anonymous surveys or asking an independent third party to gather opinions may be necessary to acquire valid information. Whatever the mechanism, open dialog and credible information are critical to effective reviews.

Finally, for outputs, a key metric might be the number of go/no-go decisions people adhere to in the weeks and months following the review. Because of their importance, not applying these decisions undermines the review's credibility and effectiveness. Other output metrics might include the number of action items completed on time and even the number completed at all.

If we collect these metrics regularly, analyze them objectively, and use them consistently for improvement, the review process is certain to be effective. Also, our attention to continuously improving the process sends a powerful message to all concerned about the importance of technical reviews.

References

Cooper, R. G. 2001. *Winning at New Products: Accelerating the Process from Idea to Launch.* Cambridge, MA: Perseus Publishing.

NASA. March 26, 2007 (1). NPR 7123.1a - Systems Engineering Processes and Requirements. Washington, DC: NASA.

NASA. March 6, 2007 (2). NPR 7120.5d - Space Flight Program and Project Management Requirements. Washington, DC: NASA.

Naval Air Systems Command (NAVAIR). April 10, 2006. NAVAIRINST 4355.19C - (AIR-4.0/5.0/6.0) Systems Engineering Technical Review Process. Patuxent River, MD: Department of the Navy.

Starr, Daniel, and Gus Zimmerman. July/August, 2002. "A Blueprint for Success: Implementing an Architectural Review System," STQE Magazine.

Surowiecki, J. 2004. *The Wisdom of Crowds: Why the Many Are Smarter Than the Few and How Collective Wisdom Shapes Business, Economies, Societies and Nations.* New York, NY: Doubleday.

Chapter 19

FireSAT End-to-End Case Study
A Parable of Space Systems Engineering in Seven Parts

Jerry Jon Sellers, *Teaching Science & Technology, Inc.*
Jody Fluhr, *Fluhr Engineering, LLC*
Peter Van Wirt, *Teaching Science & Technology, Inc.*

FireSAT is arguably the most famous space mission that never flew. The idea for this fictitious mission was first developed in *Space Mission Analysis and Design* (SMAD) [Larson and Wertz, 1999] as a running example throughout each chapter to highlight how theory applies to a concrete problem. Later, *Understanding Space: An Introduction to Astronautics* (US) [Sellers, 2004] adapted the FireSAT idea to illustrate concepts of space systems engineering and subsystem design. FireSAT has since become a workhorse in other books and in classrooms the world around. The incarnation of FireSAT in this chapter will look familiar to anyone who has encountered the mission in either SMAD or US, but the authors took academic license with the mission and system definition for the sake of consistency and to illustrate different technical points.

This chapter uses FireSAT to highlight significant ideas and issues in both the science and art of space systems engineering. The science issues are underscored through numerous exhibits of systems engineering products such as requirements documents and verification plans. For each technical baseline, we present issues and anecdotes to give a flavor for the art of systems engineering. These examples come from real-world experiences but "the names have been changed to protect the innocent." The reader may ask why certain products or examples are discussed at one point in the lifecycle but not others. Of course, one of the biggest challenges of space systems engineering (and any book on the subject) is that in the real world, many things are happening at once. Life happens in parallel but books can only be written one page at a time. So the authors chose to highlight various issues at different points in the lifecycle to even out the discussion.

It's impossible to fully plumb the depths of a real-life mission in a single book, let alone a single chapter. This forced the authors to pick and choose the products and issues they presented. Our aim is to provide enough depth to give a feel for the real complexity of the problem (often not appreciated until one really gets into this business) without overwhelming the reader with detail. Table 19-1 lists the activities described in this chapter along with cross-references to other chapters. Unlike the rest of the book, which takes a process view of space systems engineering, this chapter takes us longitudinally through a project lifecycle to show how to apply those processes at each technical baseline. The objectives of this chapter are to:

- Tie together all the 17 systems engineering processes by illustrating their application throughout a project lifecycle
- Provide a stand-alone systems engineering case study

TABLE 19-1.   Chapter Activities. This is a list of activities addressed in this chapter along with cross-references to other relevant chapters.

Activity	Where Discussed
Certifying the as-deployed baseline	Section 19.1
Defining the mission baseline	Section 19.2, Chaps. 2, 3, and 13
Create the system baseline	Section 19.3, Chap. 4
Establish the functional baseline	Section 19.4, Chap. 5
Reach the design-to baseline	Section 19.5
Set the build-to baseline	Section 19.6, Chaps. 6 and 9
Achieve the as-built baseline	Section 19.7, Chaps. 10, 11, and 12

Here we try to get a grasp of the daunting challenge of taking a space mission from a blank piece of paper through launch and early orbit operations. Of course, such a story, even for a relatively simple system such as FireSAT, would span years and the systems engineering products would fill volumes. Our approach is to give the entire project a broad brush to cover the cradle-to-grave activities and products in an inclusive way, zooming in on selected items, exhibits, and anecdotes to give the reader enough concrete examples to get through the tough spots of project development.

We've organized this chapter around the seven technical baselines that all systems engineering processes support. As Chapter 13 describes, these baselines serve as official "lines in the sand" that fully describe the current state of the system. The baselines follow the mission lifecycle, and so each successive one reflects greater design maturity. Table 19-2 shows all these baselines.

TABLE 19-2.   **Major Project Technical Baselines.** These are the seven major technical baselines in the project lifecycle and the major reviews at which they're defined.

Technical Baseline	Established at
Mission baseline	Mission concept review (MCR)
System baseline	System requirements review (SRR)
Functional baseline	System definition review (SDR)
Design-to baseline	Preliminary design review (PDR)
Build-to baseline	Critical design review (CDR)
As-built baseline	System acceptance review (SAR)
As-deployed baseline	Post-launch assessment review (PLAR)

We start our story at the end of FireSAT design and development, which is the beginning of transition to normal operations. We then travel back in time to the outset of the project to trace its evolution from initial customer need to delivery for launch.

19.1   Certify the As-deployed Baseline

Our discussion begins at the final technical baseline. The as-deployed baseline represents the final snapshot of what the operators have to work with to perform the mission on orbit. This is the ultimate "end game" of the FireSAT project and a fitting place to focus on through every phase of the system development.

19.1.1   What Got Us Here?

To be sitting at the post launch assessment review (PLAR), ready to give the thumbs up to begin normal operations, a lot of things first have to happen exactly right. To have successfully made it this far is an achievement combining strong project leadership, skilled project management, effective systems engineering, and some amount of luck. To be at this point, the satellites that the FireSAT team has been slaving away at for the past five years have to be:

- Designed from initial concept through nuts and bolts
- Built or procured

- Assembled and integrated
- Verified and validated
- Transitioned to the launch site and integrated on the launch vehicle
- Successfully launched into their parking orbit by the launch vehicle
- Successfully turned on, sent telemetry, received commands, deployed solar arrays, produced power, determined and controlled attitude, and maneuvered to their mission orbit at an altitude of 700 km
- Calibrated and operated to sufficiently verify payload performance requirements in the space environment, i.e., detecting real fires on the ground

Flight controllers can now operate the spacecraft in its mission configuration, as shown in Figure 19-1, happily going about its day-to-day mission of keeping watch over the US, hunting for wildfires.

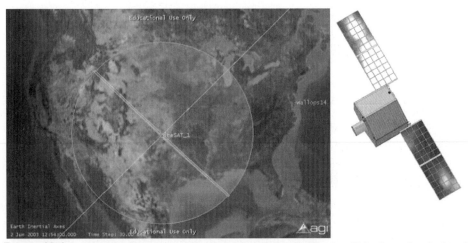

FIGURE 19-1.   **FireSAT in Deployed Configuration Doing its Mission.** This shows the deployed configuration of the FireSAT spacecraft along with a ground track illustrating the zone of coverage for the satellite as it searches for wildfires.

With all that behind us, we're ready to "turn over the keys" of the system to day-to-day operators. Now is a good time to reflect on the effort that brought us to this point and look at some of the issues that we may still face.

19.1.2   The End Game—PLAR Products

Table 19-3 describes entrance and exit criteria for the PLAR. To complete the PLAR we must satisfy all the success criteria. This means completing the relevant documents, also listed in the table. The next subsection looks at the systems engineering level of effort to expect at this point in the project.

TABLE 19-3.   Entrance and Success Criteria for the Post-launch Assessment Review (PLAR). Included are the supporting documents needed to complete each success criterion. Adapted from NASA [2007].

Post-launch Assessment Review	
Entrance Criteria	**Success Criteria**
• The launch and early operations performance, including (when appropriate) the early propulsive maneuver results, are available • The observed spacecraft and science instrument performance, including instrument calibration plans and status, are available • The launch vehicle performance assessment and mission implications, including launch sequence assessment, and launch operations experience with lessons learned, are completed • The mission operations and ground data system experience, including tracking and data acquisition support as well as spacecraft telemetry data analysis, is available • The mission operations organization, including status of staffing, facilities, tools, and mission software (e.g., spacecraft analysis and sequencing), is available • In-flight anomalies and the responsive actions taken, including any autonomous fault protection actions taken by the spacecraft or any unexplained spacecraft telemetry, including alarms, are documented • The need for any significant changes to procedures, interface agreements, software, and staffing has been documented • Documentation is updated, including any updates originating from the early operations experience • Future development and test plans have been made	• The observed spacecraft and science payload performance agrees with prediction, or if not, is adequately understood so that future behavior can be predicted with confidence • All anomalies have been adequately documented, and their impact on operations assessed. Further, anomalies impacting spacecraft health and safety or critical flight operations have been properly disposed. • The mission operations capabilities, including staffing and plans, are adequate to accommodate the flight performance • Liens, if any, on operations, identified as part of the operational readiness review, have been satisfactorily disposed

Table 19-4 summarizes each of the documents that support the success criteria and defines their top-level contents.

19.1.3   Process Level of Effort

Following the PLAR, most of the major systems engineering (SE) processes throttle down to a low idle. Sustaining SE is mainly focused on spacecraft and ground system software. Depending on the life of the mission and the nature of the project, we may also need ground system hardware upgrades, requiring all SE processes to move back into high gear. Table 19-5 shows these relative levels of effort for each of the 17 processes. Here and in similar tables in this chapter, we depict the effort expended on each process as analogous to a cell phone signal—zero bars represents no or very little effort, and five bars represents maximum effort for a given process. Because the magnitude of effort differs between processes, one should not compare the number of bars for one process to those for another; rather, the "signal strength" depicts the relative effort for that process.

TABLE 19-4.   Top-Level Contents for Key Post-launch Assessment Review (PLAR) Documents. This table outlines the top-level contents for each of the documents needed to address PLAR success criteria.

Product	Contents
System activation results report	• Launch and early operations performance • Results of orbit maneuvers • Current spacecraft health and status • In-flight anomalies and resolutions • Activated autonomous fault protections and resolutions • Activated spacecraft alarms • Ground anomalies and resolutions
Payload calibration results report	• Instrument calibration results and status • Actual versus predicted performance
Integration and test anomaly resolution report	• Anomalies encountered during integration and test phase • Equipment change-outs as part of anomaly resolutions • Equipment exercised outside limits
Mission support training and simulation results report	• Mission operations capabilities • Baselined operations procedures • Any outstanding operations procedures • Staff certifications • Shift support plan • Anomaly tracking system and boards • Simulation events, anomalies introduced, and team's performance in resolving simulated anomalies

19.1.4   Representative Issues, Items, and Anecdotes

In an ideal world, all anomalies and issues would be resolved before the PLAR and the decision to begin routine operations would be free of controversy and drama. Unfortunately, in the real world things are seldom that easy. Let's look at one small representative anomaly that vexed the board at the FireSAT Post Launch Assessment Review to get an idea of how the various elements come together to uncover issues.

During integrated testing, the magnetometer, used for attitude determination, was activated. Using a "test like you fly" approach, this was done during electromagnetic compatibility and interference tests as part of routine verification of all subsystems of the attitude determination and control system (ADCS), including the magnetorquers. Unfortunately, the success criteria focused on the actual sensor output data, NOT on how the ADCS software would use that data. Because of this narrow view, the integration and test anomaly resolution report didn't identify any magnetometer anomalies. (Figure 19-2 gives an internal view of the spacecraft showing the locations of the magnetorquers and magnetometer.)

However, during the PLAR an issue came up regarding the interaction between the magnetometer and magnetorquers in the operational environment. During spacecraft commissioning, the data from the magnetometer could be

TABLE 19-5.   Post-launch Assessment Review (PLAR) Systems Engineering Process Relative Level of Effort. This table illustrates the relative level of effort for each of the 17 systems engineering processes after the PLAR. Analogous to a cell phone signal, zero bars represents no or very low resource requirements and five bars represents maximum resource use. Resource requirements apply for a particular process only; a given number of bars for one process does not imply the same level of resources as the same number of bars for a different process.

Process	Effort	Comments
1. Stakeholder expectation definition	■□□□□	Ideally, stakeholders won't need to change their expectations based on actual in-orbit performance
2. Technical requirements definition	■□□□□	Some effort to capture and implement software requirements continues throughout the life of the program
3. Logical decomposition	□□□□□	Only in support of ongoing software refinement and development
4. Physical solution	□□□□□	Only in support of ongoing software refinement and development
5. Product implementation	■□□□□	All buy, build, and reuse decisions have already been made. Operational contracts, including software maintenance, should already be in place. We may have to revisit these as ground station upgrades and other sustaining engineering requirements evolve throughout the mission operations phase.
6. Product integration	■□□□□	Only in support of ongoing software refinement and development
7. Product verification	■□□□□	Only in support of ongoing software refinement and development
8. Product validation	■□□□□	Validation in support of ongoing software refinement and development continues. True and final product validation comes about throughout the life of the program as the system demonstrates its usefulness to the stakeholders.
9. Product transition	■□□□□	Only in support of ongoing software refinement and development
10. Technical planning	■□□□□	Replaced by operational planning
11. Requirements management	■□□□□	Only in support of ongoing software refinement and development
12. Interface management	■□□□□	Only in support of ongoing software refinement and development
13. Technical risk management	■■□□□	By this point, technical risks and mitigations should largely be embodied in mission or flight rules for the operators; additional risk management is on a contingency basis
14. Configuration management	■□□□□	Only in support of ongoing software refinement and development
15. Technical data management	■□□□□	Ongoing throughout the life of the program, especially in archiving and managing engineering and payload data
16. Technical assessment	■■□□□	Formal technical assessment mostly ends with PLAR, but periodic project reviews continue at some level through final disposal
17. Decision analysis	■□□□□	By this point, key decisions are embodied in flight rules and other procedures for the operators; additional decision analysis is on a contingency basis

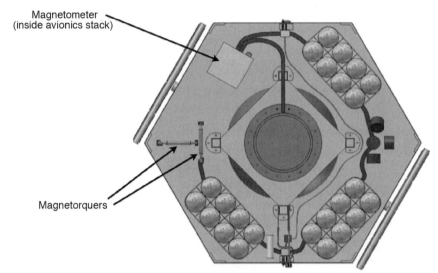

Magnetometer
(inside avionics stack)

Magnetorquers

FIGURE 19-2.   FireSAT Magnetorquers and Magnetometers. Anomalous interactions among these components necessitated both short- and long-term solutions.

compared to expected data based on standard models of the Earth's magnetic field. As luck would have it, commissioning procedures collected this data when the torque rods were inactive. Once again, the magnetometer data was assessed to be accurate and the system activation results report identified no anomalies.

During payload calibration activities, however, something funny happened. Most of the time, when the spacecraft was commanded to point at and track a known fire source on the ground, pointing stability was within expected parameters. But occasionally, sudden large pointing oscillations of up to 1° occurred. The payload calibration results report noted this erratic behavior, which could seriously hamper normal operations.

A tiger team was formed to investigate the issue. At the PLAR, they presented their preliminary findings—existing software prevented the magnetometer from taking a measurement while a torque rod was firing (during which the electromagnet was energized, producing a strong magnetic field). However, the software interrupt **did** allow measurements immediately after a firing. But there was a short recovery time before the residual magnetic field fully dissipated. If the magnetometer took readings too soon after the torque rod fired, it produced erroneous data. To make matters worse, since magnetometer data may not have been included in the navigation state filter for some time due to the torque rod operation, the filter would give this erroneous data too much weight (or "gain" in Kalman filter parlance). This would lead to large attitude excursions.

To work around the problem, the tiger team recommended some short-term changes to operational procedures that were documented in the mission support training and simulation results report. In addition, the project team submitted an engineering change proposal for a long-term fix to the software, requiring a short recovery time between torque rod firing and magnetometer data collection.

19.1.5   Conclusions

The above is only a small example of the kind of issues that project managers and engineers must grapple with both during the early stages of on-orbit operations and throughout the mission life. We have to recognize how much the decisions made during operations depend on the vast knowledge gained about the system over many years of development, integration, and testing. At each formal baseline, the knowledge about the system grows.

At the PLAR, the decision authorities had to review all the information included in the project exhibits, giving special attention to the above issue. They then had to decide if it was prudent to accept the short term operational work-around and put FireSAT on line for the US Forest Service. Once in full service, FireSAT could finally fulfill its intended purpose, answering the need for a more effective means to detect and monitor wildfires throughout the US.

But how did we get here? How did we start with this need and turn it into this complex system, ready to go online to search for wildfires? Smokey Bear meets Buck Rogers? In the next sections, we turn back the clock on the project to see how it progressed from concept to launch.

19.2   Define the Mission Baseline

This chapter began the FireSAT story near the end of the lifecycle, partly to set the stage for how we get there from here, but also because it is with this final system in mind that systems engineers should approach the first baby steps of a new space project. Here we go back to square one to see how to apply from the beginning the space systems engineering processes described in this book. We start by looking at the hurdles we must first jump to enter Phase A and start a real project. The technical results from Pre-phase A form the basis for the first project baseline, known as the mission baseline, established at the mission concept review (MCR).

Pre-phase A is the proving ground for new projects. It's a Darwinian experiment to explore new ideas and concepts, to see which ones can survive the demanding environments of cost, schedule, performance, and risk (and politics). Depending on the organization, dozens of potential projects may successfully complete Pre-phase A, but only a few actually go on to Phase A. What separates the "fit" projects from the ones that get relegated to a file drawer typically depends on the need (e.g., national security versus scientific curiosity), availability of funding, state of dependent technology, political will, project leadership, timing, and luck. Here we explore ways to better ensure our chances for survival by seeing how FireSAT braved these environs.

Many mission concept studies end up as internal reports, technical journal articles, or presentations at conferences. Studies that we want to go beyond the thinking-about-it phase and get real money to do serious design must pass some critical decision gate as described in Chapter 13. The gate keepers vary by organization and by the scope of the potential project. NASA, the European Space Agency, and DoD use a mission concept review (MCR) to pass from Pre-phase A to a full-fledged Phase A project.

19.2.1   What Got Us Here?

Organizations initiate Pre-phase A studies for any number of reasons. The government may request them from industry (either funded or unfunded) via industrial requests for information, or through formal study contracts such as analyses of alternatives. They may be done internally, as NASA does with announcements of opportunity, or within industry as part of internal R&D efforts exploring new business opportunities.

Pre-phase A studies represent a low-risk way to flesh out some new idea or concept to see if it has merit. Because no formal project (with associated project offices and funding) has yet been established, there's usually little political fallout from canceling a Pre-phase A study part way through or accepting the results and doing nothing with them. This isn't normally the case once a project enters Phase A, after considerable political and financial capital has gone into it.

We point out that early on, perhaps well before this effort was even defined as a project, the decision to go with a space-based approach for addressing the USFS need may not have been obvious. The USFS, working internally and with other agencies, may have conducted numerous trade studies to arrive at the conclusion that a "Fire Imaging REconnaissance Satellite" a.k.a. FireSAT, was the best approach. Here we take that conclusion as given.

For our purposes, we imagine that NASA did the FireSAT Pre-phase A study at the request of the US Forest Service and NOAA. Such a request (and requisite funding) begins with intergovernmental agreements such as memorandums of agreement (MOAs) and memorandums of understanding (MOUs). So now this task lands on our desk: *Develop the mission concept for an on-orbit fire detection system*. Where do we start? The first step is to define the problem by laying out the deliverable products needed to satisfy the success criteria for Pre-phase A. From there, we can lay out a plan to develop and deliver these products using our systems engineering tools.

19.2.2   The End Game—MCR Products

The FireSAT Pre-phase A study manager coming to grips with the task begins by looking at the organizational guidelines for entering, and surviving, a mission concept review. Table 19-6 lists the entrance and success criteria for an MCR. Table 19-7 maps these success criteria into the necessary supporting documents to address each one. These documents must be in hand before the review starts. Table 19-8 lists the top-level contents for each document.

TABLE 19-6. Mission Concept Review (MCR) Entrance and Success Criteria. These considerations are in addition to the programmatic matters faced by a newly-approved project. Adapted from NASA [2007].

Mission Concept Review	
Entrance Criteria	**Success Criteria**
1. Mission goals and objectives. 2. Analysis of alternative concepts to show at least one is feasible. 3. Concept of operations. 4. Preliminary mission descope options. 5. Preliminary risk assessment, including technologies and associated risk management/ mitigation strategies and options. 6. Conceptual test and evaluation strategy. 7. Preliminary technical plans to achieve next phase. 8. Defined MOEs and MOPs. 9. Conceptual lifecycle support strategies (logistics, manufacturing, and operation).	1. Mission objectives are clearly defined and stated and are unambiguous and internally consistent. 2. The preliminary set of requirements satisfactorily provides a system that will meet the mission objectives. 3. The mission is feasible. A solution has been identified that is technically feasible. A rough cost estimate is within an acceptable cost range. 4. The concept evaluation criteria to be used in candidate systems evaluation have been identified and prioritized. 5. The need for the mission has been clearly identified. 6. The cost and schedule estimates are credible. 7. An updated technical search was done to identify existing assets or products that could satisfy the mission or parts of the mission. 8. Technical planning is sufficient to proceed to the next phase. 9. Risk and mitigation strategies have been identified and are acceptable based on technical risk assessments.

19.2.3 Process Level of Effort

Once we know what we need, we step back to assess the relative level of effort necessary for each SE process. Table 19-9 summarizes the 17 systems engineering processes that this book describes along with the relative level of effort to be expended on them during Pre-phase A.

19.2.4 Results

The key results of Pre-phase A closely follow the major documents list described in Table 19-8. We start with the one document that underlies the entire definition of the mission: the scope document. We then look at the results captured in the FireSAT mission design report, the draft operations concepts document, the draft risk management plan, the SEMP focused on the project document tree, and finally the plan to achieve SRR. The remaining documents, while important to a good MCR, are more important in later phases; for now we hold off on describing their detailed contents. The following subsections detail the main results.

FireSAT Scope Document

Chapter 4 describes the need for a clearly defined project scope document. Engineers often have a "ready, fire, aim" tendency to want to hurry up and get to the meat of the technical requirements so they can get busy building stuff. But if we jump into requirements before getting complete buy-in on the project scope by

TABLE 19-7.   **Mission Concept Review (MCR) Success Criteria Versus Supporting Documents.**
From this brief analysis, we can put together the list of paper products that we'll need to develop throughout the course of the study with their top-level contents identified, including the items specifically called for in the entrance criteria.

MCR Success Criteria	Supporting Document
1. Mission objectives are clearly defined and stated and are unambiguous and internally consistent	FireSAT scope document
2. The preliminary set of requirements satisfactorily provides a system which will meet the mission objectives.	FireSAT mission design study report including analysis of alternatives, cost estimates
3. The mission is feasible. A solution has been identified which is technically feasible. A rough cost estimate is within an acceptable cost range.	FireSAT mission design study report, draft mission operations concept document
4. The concept evaluation criteria to be used in candidate systems evaluation have been identified and prioritized.	FireSAT mission design study report
5. The need for the mission has been clearly identified.	FireSAT scope document
6. The cost and schedule estimates are credible.	FireSAT mission design study report (credibility to be judged by the review board)
7. A technical search was done to identify existing assets or products that could satisfy the mission or parts of the mission.	FireSAT mission design study report
8. Technical planning is sufficient to proceed to the next phase.	FireSAT DRAFT SEMP (sufficiency to be judged by the review board)
9. Risk and mitigation strategies have been identified and are acceptable.	FireSAT risk management plan

the customer and other stakeholders, the requirements engineering effort will simply chase its tail, never able to converge on a clear picture of what everyone really wants.

To define the project scope we follow the processes described in Chapters 2 and 3. We start by capturing the customer expectations as the need we're trying to address along with associated goals and objectives (collectively called need, goals, and objectives—NGOs). After extensive discussions with the FireSAT project customer—the USFS—we identify the need statement, goals, and objectives, the basis for the project scope document as summarized in Table 19-10. These goals and objectives form the basis for the mission-level measures of effectiveness (MOEs).

In developing the NGOs, we also have to identify the relevant stakeholders. Chapter 2 describes this fully, and Table 19-11 identifies the FireSAT stakeholders.

With our understanding of the customer's expectations in the form of NGOs, and appreciation for the project stakeholders in mind, we turn our attention to the concept of operations. First we must get our arms around the pieces of the problem we want to solve, then look at how the pieces fit together. As Chapter 3 describes,

TABLE 19-8.   **Major Mission Concepts Review (MCR) Products and Their Contents.** These documents are associated with the review's success criteria. (MOE is measure of effectiveness.)

Product	Contents	Product	Contents
Scope document	• Mission stakeholders • Mission need statement • Goals and objectives (rank ordered for descope) • Constraints and driving assumptions • Concept of operations (e.g., OV-1) • Applicable documents list	Draft risk management plan	• Preliminary risk assessment • Technologies and associated risk management plus mitigation strategies and options

Mission design study report	• Preliminary mission-level requirements traceable to the scope document
	• Key performance parameters (KPPs)
	• Analysis of alternatives trade-tree, including
	  – Alternative mission concepts
	  – Identification of system drivers for each concept and architecture
	  – Results of characterization for each alternative
	  – Critical requirements identified or derived
	  – Lifecycle cost model report
	  – Concept or architecture utility assessment, including evaluation criteria, such as cost, technical maturity, MOEs
	• Proposed baseline mission concept
	• Preliminary mission-level systems requirements document (SRD) with allocated system requirements
	• Conclusions and recommendations
		Draft systems engineering management plan (SEMP)	• Plan for managing the SE effort
			• Documents driving the SEMP
			• Technical summary
			• Integration technical effort (including conceptual lifecycle support strategies logistics, manufacturing, operations, etc.)
			• The 17 systems engineering processes
			• Technology insertion
			• SE activities
			• Project plan integration
			• Waivers
			• Appendices

		Draft master verification plan	• Integration and verification (I&V) • Organization and staffing • Verification team operational relationships • Project model philosophy • Launch site verification • On-orbit verification • Post-mission and disposal verification • Verification documentation • Verification methodology • Support equipment requirements • Facility requirements
Draft operations concept document	• Design reference mission (including operational scenarios) • Timelines • Data flow diagram • Organizational and team responsibilities • Cost and complexity drivers for a given set of inputs • Requirements and derived requirements • Technology development plan	Plan to achieve system requirements review readiness	• Rank ordered list of studies and trades-offs • Forward work mapped to program schedule • Technology gap closure plan mapped to program schedule • Requirements update and management plan • To-be-determined and to-be-resolved (TBD/TBR) burn-down plan • MCR issue resolution plan

context diagrams define the boundaries of the system of interest and help us focus on the inputs (both the ones we can control and the ones we can't) and the outputs that the customer is paying for. Figure 19-3 provides a simple view of the FireSAT system context.

TABLE 19-9. **Systems Engineering Process Level of Effort During Pre-phase A.** Analogous to a cell phone signal, zero bars represents no or very low resource requirements and five bars represents maximum resource use. Resource requirements apply for a particular process only; a given number of bars for one process does not imply the same level of resources as the same number of bars for a different process.

Process	Effort	Comments
1. Stakeholder expectation definition	■■■■■	In Pre-phase A it's crucial to define the stakeholder's expectations. If the project doesn't capture these from the start, then whatever comes after won't be what they want. Key to this is defining the ultimate success criteria for the project.
2. Technical requirements definition	■■□□□	For any project going on to Phase A, the preliminary requirements are one of the most important hand-overs. Pre-phase A is especially important in identifying the critical requirements, the ones that have the biggest impact on the mission definition. The reason this effort isn't 100% is that until a project becomes "real," it's not prudent to invest too much in detailed requirements definition.
3. Logical decomposition	■■□□□	Minor effort during this phase aimed at capturing top-level functions
4. Physical solution	■□□□□	The main objective of Pre-phase A is not to come up with the final physical solution. It's to define a possible physical solution that can be shown to be feasible. Time and resources are usually not sufficient, nor does it make sense in this phase, to invest too much effort in detailed design.
5. Product implementation	■□□□□	It is important in Pre-phase A to at least examine buy, build, or reuse options. This is part of establishing the maturity of enabling technologies. If it can't be reused or bought somewhere, then it has to be built or invented from scratch.
6. Product integration	□□□□□	Only minimum effort expended on this process, mainly to identify enabling products for integration that may affect lifecycle cost
7. Product verification	□□□□□	Only minimum effort expended on this process, mainly to identify enabling products for verification that may affect lifecycle cost
8. Product validation	■□□□□	This is a critical but often overlooked process that needs to be addressed to some level during Pre-phase A. While we have the attention of the stakeholders to define need, goals, and objectives, we must address ultimate project success criteria. These form the basis for eventual system validation.
9. Product transition	■□□□□	Only minimum effort expended on this process, mainly to identify enabling products for transition that may impact lifecycle cost

TABLE 19-9. **Systems Engineering Process Level of Effort During Pre-phase A. (Continued)** Analogous to a cell phone signal, zero bars represents no or very low resource requirements and five bars represents maximum resource use. Resource requirements apply for a particular process only; a given number of bars for one process does not imply the same level of resources as the same number of bars for a different process.

Process	Effort	Comments
10. Technical planning	■■□□□	For any project going on to Phase A, the preliminary technical planning products are another important hand-over. They form the basis for defining the nature of the project to be formally established in Phase A. Pre-phase A is especially important in defining the initial integrated master plan (IMP), systems engineering management plan (SEMP) (also known as a systems engineering plan (SEP)), and master verification plan (MVP) (also known as the test and evaluation master plan (TEMP)). The only reason this effort isn't 100% is that until a project becomes "real," it's not prudent to invest too much in detailed planning.
11. Requirements management	■□□□□	Little or no effort is made to formally manage requirements until a project officially kicks off in Phase A—only enough to keep study teams in synch
12. Interface management	□□□□□	Little or no effort is made to formally manage interface requirements until a project officially kicks off in Phase A—only enough to keep study teams in synch
13. Technical risk management	■□□□□	During Pre-phase A, top-level programmatic and technical risks should be fully identified to avoid kicking off a project that has unforeseen and unmitigated risks. But little formal effort is needed here to actively mitigate these risks.
14. Configuration management	■□□□□	Little or no effort is made to formally manage configuration until a project officially kicks off in Phase A—only enough to keep study teams in synch
15. Technical data management	■□□□□	Little or no effort is made to formally manage technical data until a project officially kicks off in Phase A—only enough to keep study teams in synch
16. Technical assessment	■□□□□	Pre-phase A technical assessment culminates in the mission concept review. So we need to prepare for it.
17. Decision analysis	■■□□□	Rigorous technical decision analysis is key to developing credible Pre-phase A results. It's impossible to do all the myriad trade studies that may be identified, so good decision analysis selects the most critical ones.

This context diagram shows clearly that the FireSAT System will take inputs from wildfires and deliver information to the USFS. This is a good start, and as obvious as it may seem, it's extremely useful to get all stakeholders to agree even to this much. Of course, this simple view barely scratches the surface of the problem. To dig deeper, we look at a top-level mission architecture description. As discussed in *Space Mission Analysis and Design* (SMAD) [Larson and Wertz, 1999] and illustrated in Figure 19-4, space missions share a common set of architectural elements. Expanding on this approach, we lay out a notional FireSAT mission architecture block diagram, as shown in Figure 19-5.

TABLE **19-10.**   **FireSAT Need, Goals, and Objectives.** This is the beginning of the process of understanding stakeholder expectations.

Mission Need: *The US Forest Service needs a more effective means to detect and monitor potentially dangerous wildfires*	
Goals	**Objectives**
1. Provide timely detection and notification of potentially dangerous wildfires	1.1. Detect a potentially dangerous wildfire in less than 1 day (threshold), 12 hours (objective)
	1.2. Provide notification to USFS within 1 hour of detection (threshold), 30 minutes (objective)
2. Provide continuous monitoring of dangerous and potentially dangerous wildfires	2.1. Provide 24/7 monitoring of high priority dangerous and potentially dangerous wildfires
3. Reduce the economic impact of wildfires	3.1. Reduce the average annual cost of fighting wildfires by 20% from 2006 average baseline
	3.2. Reduce the annual property losses due to wildfires by 25% over 2006 baseline
4. Reduce the risk to firefighting personal	4.1. Reduce the average size of fire at first contact by firefighters by 20% from 2006 average baseline
	4.2. Develop a wildfire notification system with greater than 90% user satisfaction rating
5. Collect statistical data on the outbreak, spread, speed, and duration of wildfires	
6. Detect and monitor wildfires in other countries	
7. Collect other forest management data	
8. Demonstrate to the public that positive action is underway to contain wildfires	

This exercise helps us better understand the potential players (physical, personal, and contextual) that have an active role in the mission. However, this is somewhat of a "chicken and egg" process. The goal is to develop a cost-effective architecture, with a detailed design for each part of every element, to deliver the needed capability the customer wants. But to define the scope of that problem we have to assume the skeleton of the solution in the form of the notional architecture based on experience and judgment. So this isn't necessarily the final solution.

As we go further into the design process and analyze the numerous trade-offs, we may have to add, subtract, or combine elements to construct a more effective architecture. With these basic building blocks and caveats in mind, we now consider how all the pieces of our notional architecture interact to carry out the mission. Figure 19-6 delves deeper into the system of interest to show the relationships between some of the composite elements.

TABLE 19-11.   **Stakeholders and Their Roles for the FireSAT Mission.** Sponsors provide funding and may be either active or passive.

Stakeholder	Type
Congress	Sponsor, Passive
Forest Service	Active
NOAA	Active
NASA	Passive
Prime contractor	Passive
Taxpayer	Sponsor, Passive
People living near forests	Active
State government	Passive
Wildlife organizations	Active

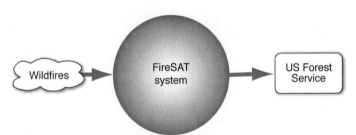

FIGURE 19-3.   **A Simple Context Diagram for the FireSAT System.** A context diagram reflects the boundary of the system of interest and the active stakeholders.

At this point, we envision a system composed of space elements, launch elements, mission operations, and ground elements that interact in specific ways. In our next step toward developing a concept of operations for the mission, we look at alternative mission concepts. SMAD uses the term *mission concept* to describe the decisions made about data delivery; communications; tasking; scheduling and control; and mission timeline. Different choices for any of these can lead to very different views on how to conduct the mission. In approaching these decisions, we must be clear about where we are. Currently, to detect wildfires, the USFS employs fire towers staffed by keen-eyed observers, who constantly scan for traces of fires, along with reports from the general populace, and even unpiloted aerial vehicles during very high threat periods. We summarize this current concept of operations in what we sometimes refer to as an operational view 1 (OV-1) as described in Chapter 3. The OV-1 provides a simple way of presenting the complex interactions between the elements of the concept of operations. Figure 19-7 shows the current fire detection concept of operations, including the major elements and how they interact.

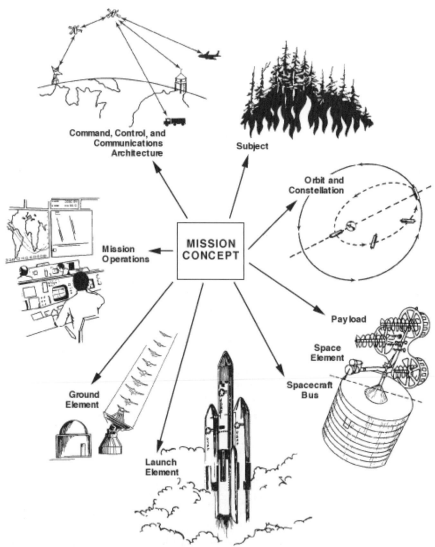

FIGURE 19-4.   Space Mission Architecture. All space missions include these basic elements. See text for definitions. Requirements for the system flow from the operator, end user, and developer and are allocated to the mission elements.

Besides understanding what we have, we must also be clear about any rules of engagement that constrain what we're allowed to do. Brainstorming and imaginative out-of-the-box thinking may come up with any number of innovative ways to improve on the existing capability. But harsh realities—technical, political,

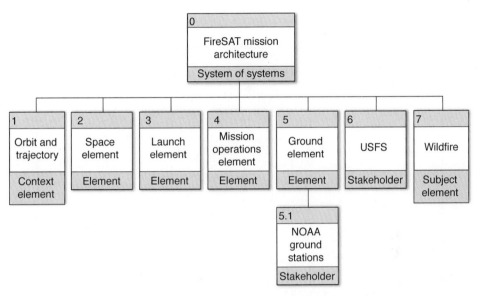

FIGURE 19-5.   FireSAT System of Systems Architecture Diagram. The FireSAT mission architecture, or "system of systems," comprises a number of interrelated elements that must all work together to achieve the mission. (USFS is US Forest Service.)

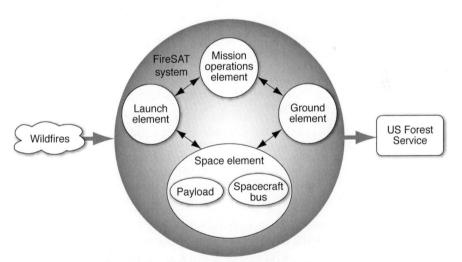

FIGURE 19-6.   Context Diagram for the FireSAT System, Including Likely Reference System Elements. If we have a reference architecture for the system of interest, the context diagram can reflect it, as shown here.

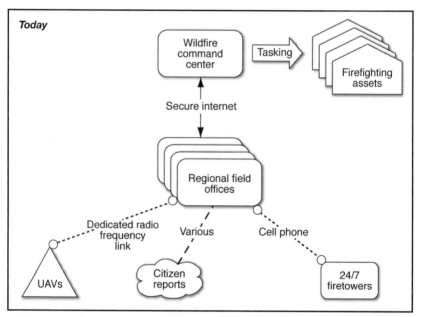

FIGURE 19-7.   Current Fire Detection Concept of Operations. This figure gives us a view of the current USFS fire detection concept of operations. Knowing where we are helps define where we want to go with the FireSAT system.

and budgetary—constrain the possible to the practical. For FireSAT, the customer and other stakeholders set the following constraints:

- C1: The FireSAT System shall achieve initial operating capability (IOC) within 5 years of authority to proceed (ATP), and full operating capability (FOC) within 6 years of ATP

- C2: The FireSAT system total lifecycle cost, including 5 years of on-orbit operations, shall not exceed $200M (in FY 2008 dollars)

- C3: The FireSAT System shall use existing NOAA ground stations at Wallops Island, Virginia and Fairbanks, Alaska for all mission command and control. Detailed technical interface is defined in NOAA GS-ISD-XYX.

With this background, and our understanding of the notional architecture, we capture the alternatives for the new mission concept as summarized in Table 19-12.

This analysis helps us focus on a finite set of realistic options, which then goes into a new concept of operations picture. The new FireSAT OV-1 is shown in Figure 19-8.

Establishing the basic concept of operations prepares us to move on to a more formal description of the interfaces between the different mission functions, as shown in Figure 19-9, and different physical elements, as shown in Figure 19-8. In

TABLE 19-12.   **FireSAT Mission Concept Analysis.** (Adapted from SMAD Table 2-1 [Larson and Wertz, 1999].) The mission concept is defined by major decisions in each of these four areas. This shows the options for FireSAT. (IOC is initial operating capability; RFI is request for information)

Mission Concept Element	Definition	FireSAT Issues	Alternatives
Data delivery	How mission and housekeeping data are generated or collected, distributed, and used	1. How are wildfires detected? 2. How are the results transmitted to the firefighter in the field?	1.1 Identified by satellite, OR 1.2 Identified by ground from raw satellite data 2.1 Fire notifications sent via NOAA ground station, OR 2.2 Fire notifications sent directly to USFS field offices
Communications architecture	How the various components of the system talk to each other	1. What communications network transmits forest fire data to the users in the field? 2. When are spacecraft in view of ground stations?	1. Use of NOAA ground stations dictated by stakeholders 2. Spacecraft access to ground stations determined by orbit and constellation design
Tasking, scheduling, and control	How the system decides what to do in the long term and short term. What sensors are active and when is data being transmitted and processed?	1. Which forested areas are receiving attention this month?	1.1 Continuous monitoring of entire US, OR 1.2 Tailored observation campaign concentrating on potential hot-spots
Mission timeline	The overall schedule for planning, building, deployment, operations, replacement, and end-of-life	1. When will the first FireSAT become operational? 2. What is the schedule for satellite replenishment?	1. IOC dictated by customer 2. Outside of project scope

many ways, systems engineering is "all about the interfaces," so defining how the elements of the system are linked is one of the first steps in understanding the system. Focusing just on the space element for illustration, Figure 19-9 shows how information and energy move in and out, from where and to where. Chapter 5 describes how these inputs and outputs form the basis for more detailed functional analysis of the space element as a system in its own right. With this firm foundation, the project turns its attention to the more detailed technical definition of the mission as detailed in the mission design study report.

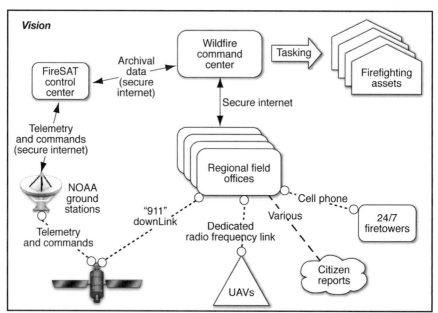

FIGURE 19-8.   FireSAT Mission Concept of Operations Presented as an Operational View 1 (OV-1). The FireSAT capability will allow the USFS to integrate existing ground and air fire detection assets with space-based assets to detect and track potentially dangerous wildfires in a more cost-effective way that promises to save lives and property.

FireSAT Mission Design Study Report

The FireSAT mission design study report captures and summarizes the myriad technical assumptions, decisions and trade-offs that go into the proposed mission concept. Those unfamiliar with the technical basis for conceptual space mission design may refer to SMAD. Chapter 4 of that book expands on the process of requirements engineering, which turns stakeholder goals and objectives into actionable technical requirements.

Given the project scope, the next task of the FireSAT Pre-phase A study is to apply technical decision processes to analyze alternative mission concepts. To better guide the concept selection, designers first need to codify some preliminary requirements, as developed in Chapter 2 and shown again in Table 19-13. Some of these requirements apply to the entire FireSAT mission architecture (i.e., both the space element and the ground element), while others are already logically allocated to the space element.

At this point in the system development, we don't have an exhaustive or final list of requirements, only the ones most relevant to shaping the overall architecture and determining mission feasibility. Trying to capture too many requirements too soon is counterproductive, as requirements management can quickly overcome

FireSAT space element	Telemetry *(packetized data, RF link)*		"911" Messages *(packetized data, RF link)*		
Commands *(packetized data, RF link)*	**NOAA ground stations**	Telemetry *(packetized data, secure internet link)*			
	Commands *(packetized data, secure internet link)*	**FireSAT command and control**		Archival data *(storage media, e.g., DVD)*	
			Regional field offices	Recommend- ations and requests *(email)*	
		Archival data requests *(email)*		**Wildfire command center**	Taskings *(email)*
					Firefighting assets

FIGURE 19-9.   **An IxI Diagram Showing the Connectivity Among Envisioned System Elements.** This diagram further clarifies the interfaces among the system elements. IxI matrices are explained in Chapter 14. Items in regular font describe "what" is interfaced between elements. Items in italics tell "how." (RF is radio frequency.)

TABLE 19-13.   **Preliminary FireSAT Mission Requirements and Associated Rationale.** Requirements in Pre-phase-A should be just detailed enough to allow us to understand the system. Requirements in bold are considered Key Performance Parameters (KPPs).

Requirement	Description	Rationale
1. **Detection**	The FireSAT system shall detect potentially dangerous wildfires (defined to be greater than 150 m in any linear dimension) with a confidence interval of 95%	The US Forest Service (USFS) has determined that a 95% confidence interval is sufficient for the scope of FireSAT
2. **Coverage**	The FireSAT system shall cover the entire United States, including Alaska and Hawaii	For a US Government funded program, coverage of all 50 states is a political necessity
3. **Persistence**	The FireSAT system shall monitor the coverage area for potentially dangerous wildfires at least once per 12-hour period	The USFS set a mission objective to detect fires within 24 hours (threshold) and within 12 hours (objective). By requiring a the coverage area to be monitored at least once per 12 hour period, the maximum time between observations will be 24 hours.

TABLE 19-13. **Preliminary FireSAT Mission Requirements and Associated Rationale. (Continued)** Requirements in Pre-phase-A should be just detailed enough to allow us to understand the system. Requirements in bold are considered Key Performance Parameters (KPPs).

Requirement	Description	Rationale
4. **Timeliness**	The FireSAT system shall send fire notifications to users within 30 minutes of fire detection (objective), 1 hour (threshold)	The USFS has determined that a 1-hour to 30-minute notification time is sufficient to meet mission objectives for the available budget
5. **Geo-location**	The FireSAT system shall geo-locate potentially dangerous wildfires to within 500 m (objective), 5 km (threshold)	The USFS has determined that a 500-m to 5-km geo-location accuracy on detected wildfires will support the goal of reducing firefighting costs
6. Reliability	FireSAT space elements shall be single-fault tolerant ("fail-ops") for critical mission functions consistent with a Class B NASA mission as per NASA NPR 8705.4	FireSAT is considered class B based on its priority, national significance, cost, complexity, and lifetime
7. Design Life	The FireSAT system shall have an operational on-orbit lifetime of 5 years. The system should have an operational on-orbit lifetime of 7 years.	The USFS has determined that a minimum 5-year design life is technically feasible. Seven years is a design objective.
8. Initial/Full Operational Capability	The FireSAT system initial operational capability (IOC) shall be within 5 years of authority to proceed (ATP) with full operational capability within 6 years of ATP	The on-going cost of fighting wildfires demands a capability as soon as possible. A 5-year IOC was deemed to be reasonable given the scope of the FireSAT system compared to other spacecraft of similar complexity.
9. **End of Life Disposal**	FireSAT space elements shall have sufficient end of life delta-V margin to de-orbit to a mean altitude of <200 km (for low-Earth orbit missions) or >450 km above geostationary belt (for GEO missions)	End of life disposal of satellites is required by NASA policy.
10. Ground System Interface	The FireSAT system shall use existing NOAA ground stations at Wallops Island, Virginia and Fairbanks, Alaska for all mission command and control. Detailed technical interface is defined in NOAA GS-ISD-XYX.	The NOAA ground stations represent a considerable investment in infrastructure. By using these existing assets the FireSAT project save time, money, and effort.
11. Budget	The FireSAT system total mission lifecycle cost, including 5 years of on-orbit operations, shall not exceed $200M (in FY 2008 dollars)	This is the budget constraint levied on the project based on projected funding availability

requirements engineering and system design. The challenge in Pre-phase A is to grasp enough of the most relevant requirements to understand the nature of the system but not so many as to be distracting.

From these preliminary requirements, the systems engineering team works with the customer and other stakeholders to define the mission key performance parameters (KPPs). As described in Chapter 2, KPPs represent the "5 plus or minus 2" capabilities or characteristics the system **must** have. They're the sacrosanct mission requirements. If the system, for what ever reason, can't achieve one or more of these KPPs, we should consider scrapping the entire project. For FireSAT, after much negotiation and gnashing of teeth, stakeholders agree that the set of KPPs will be requirements 1 through 5 and 9 from Table 19-13.

The preliminary requirements, refined by the KPPs, form the basis for **what** the system should do and be. The design engineers then turn their attention to **how**. The customer and the scope document concept of operations already define much of the mission concept. Building on this, the design team begins to brainstorm. Several space-based concepts are put forth, but in the end the ideas converge to two competing mission concepts: (1) the *GEO Fire Tower* and (2) the *LEO Fire Scout*. These two options differ fundamentally in their communications architecture as determined by the orbit and constellation approach within the mission architecture. The GEO Fire Tower option is illustrated in Figure 19-10 and the LEO Fire Scout concept in Figure 19-11. Table 19-14 summarizes their similarities and differences.

FIGURE 19-10.   FireSAT GEO Fire Tower Concept. This concept features a single satellite in geostationary orbit to provide round-the-clock coverage of the United States.

Armed with the preliminary requirements, and two alternative mission concepts, mission designers take a systematic approach to analyze the different mission architectures. To tackle this most effectively, engineers apply some of the technical decision techniques described in Chapter 6. They begin by constructing an architecture trade tree. In developing the tree, we start with an architectural element, then look at variations between it and each of the other elements subject to trade. One could arguably start with any of the elements, but in the case of FireSAT the orbit and constellation decision is the most fundamental, logically driving many other decisions. So we start the trade tree with two major branches representing the two mission concepts, and the remaining architecture elements

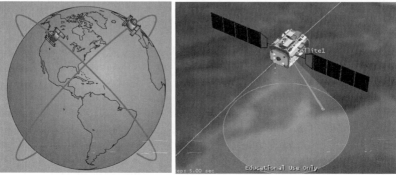

FIGURE 19-11.   **FireSAT LEO Fire Scout Concept.** This concept features two satellites in an inclined orbit, providing periodic revisit of the United States.

TABLE 19-14.   **Summary of Competing FireSAT Mission Concepts.** This table summarize the primary differences between the GEO Fire Tower and LEO Fire Scout mission concepts.

Concept	Data Delivery	Tasking, Scheduling, and Controlling	Communications Architecture
GEO Fire Tower	Spacecraft downlinks directly to NOAA stations	Shared operations teams at NOAA ground station, managed by USFS	Use a single large satellite in geostationary orbit to monitor and downlink directly to NOAA ground stations. Use existing communications infrastructure (e.g., internet) to distribute data from NOAA ground stations to offices and users.
LEO Fire Scout	Satellites broadcast fire warning "911" messages directly to USFS regional field offices	Same	Use multiple small satellites in low-Earth orbit to monitor and detect fires. Immediately broadcast fire notification "911" message to USFS regional field offices. Less timely, more detailed fire information will be stored onboard, along with health and status telemetry, for downlink to NOAA ground stations during the next pass.

flowing from there. Figure 19-12 illustrates the highest level of trade-off, followed by the more detailed architectural trade for the GEO Fire Tower. Figure 19-13 illustrates the detailed trade-offs for the LEO Fire Scout.

For both options, some elements of the trade tree (e.g., operations concepts) have no significant trade space, as they flow directly from the choice of mission concept. Other elements, such as launch vehicle selection, have several discrete options. Orbit altitude, on the other hand, in the case of the LEO Fire Scout, has an infinite number of options between the lowest practical altitude (~300 km) to the edge of the lower Van Allen Radiation Belt (~1200 km). For this trade, we examine several options within this range to identify trends or performance plateaus. As for the wildfires, characterizing them involves a separate trade study to look at smoke, atmospheric

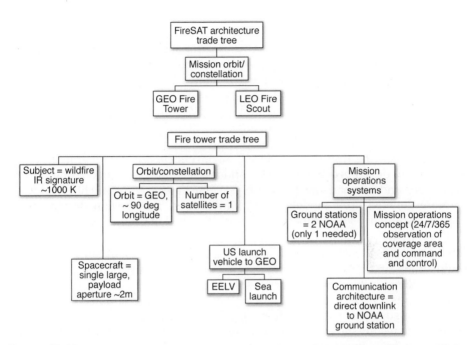

FIGURE 19-12.   **FireSAT Trade Tree Part 1.** This part of the FireSAT mission architecture trade tree shows the top-level trade-offs between the two competing mission concepts (top). At bottom is the detailed trade tree for the GEO Fire Tower option.

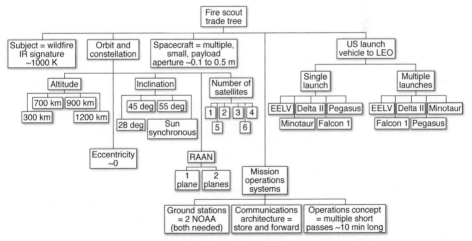

FIGURE 19-13.   **FireSAT Trade Tree Part 2.** This part of the FireSAT mission architecture trade tree shows the detailed trade-offs for the LEO Fire Scout option. (RAAN is right ascension of the ascending node.)

disturbances, infrared signature, atmospheric attenuation, and other possible observation characteristics. This study concludes that a sensor tuned to 4.2 μm would be best for wildfire detection; this forms the basis for payload design for both the Fire Tower and Fire Scout options. Finally, we derive the fundamental characteristics of the space element, case by case, based on the choices made for the other options.

With the trade space defined, the team starts the first design and analysis cycle (DAC) of the project. Project-wide planning and scheduling for DACs is described in Chapter 13. The DAC depends on a set of tools and models ranging in complexity from back-of-the-envelope calculations to detailed integrated simulations. For the first FireSAT DAC, engineers choose an integrated system design model developed in an Excel™ spreadsheet (and validated using off-line calculations and comparisons to existing systems) along with Satellite Tool Kit™ for more detailed orbit modeling and simulations. This system design model includes parametric cost estimating tools as described in Chapter 8.

Central to any effective mission design report is a concise summary of alternatives considered and their relative merits. Table 19-15 summarizes the results of the first DAC, comparing the two competing mission concepts against how well they meet the preliminary system requirements. For the LEO Fire Scout option, results are presented for the overall best performing option based on orbit coverage versus mission cost with an altitude of 700 km and an inclination of 55 degrees. The results show that from a performance standpoint, both options could theoretically meet the mission requirements. The GEO Fire Tower has a decided persistence advantage, since it can stare continuously at the US. The two options diverge most in price and practicality.

TABLE 19-15.   **Summary of Comparisons of Mission Concept Options Versus Preliminary Requirements.** This summary compares the merits of the two FireSAT mission concepts.

Requirement	GEO Fire Tower	LEO Fire Scout	Requirement	GEO Fire Tower	LEO Fire Scout
1. Detection	✔	✔	6. Design life	✔	✔
2. Coverage	✔	✔	7. Initial and full operational capability	X	✔
3. Persistence	✔	✔	8. End of life disposal	✔	✔
4. Timeliness	✔	✔	9. Ground system interface	✔	✔
5. Reliability	✔	✔	10. Budget	X	✔

Spacecraft size (and hence cost) is driven by payload size (see Chapter 7). The payload sizing model depends on the necessary aperture, which in turn is derived from the basic laws of optics (the bigger the lens, the better you can see). For sufficient spatial resolution to detect 150 m sized wildfires, the payload aperture at geosynchronous altitude would be over 2.4 m. That's the size of the Hubble Space

Telescope! Simple analogous cost modeling has priced the Fire Tower system at several billion dollars, far beyond the available budget. In addition, such a large-scale project probably wouldn't meet the IOC requirement. Thus, the Fire Tower concept is eliminated from further consideration.

After the down-select to the Fire Scout concept, engineers continue to iterate through the branches of the trade-tree to characterize each option. They focus primarily on trade-offs between orbit altitude and the number and size of satellites needed to do the mission. But the constrained time and resources of Pre-phase A also force them to make numerous assumptions, including some implementation details. All these assumptions must be analyzed and second-guessed once the project begins for real in Phase A. The next section describes in more detail the nature of these trade-offs, and details of the decisions, as part of the requirements engineering process.

To estimate the cost of the Fire Scout option, project managers rely on parametric cost estimating relationships (CERs) appropriate for a satellite of this size. Table 19-16 shows these CERs. Applying them to the design data developed during the Pre-phase A study results in the lifecycle cost estimates shown in Table 19-17. These figures are in FY2000 dollars (the year the CERs were validated for). With an inflation factor of 1.148 (SMAD), the estimated lifecycle cost is $157.9M in FY2007 dollars. So the Fire Scout option gives the project a nearly 25% cost margin.

TABLE 19-16. Cost Estimating Relationships (CERs) for Earth-orbiting Small Satellites Including Research, Development, Test, and Evaluation (RDT&E) and Theoretical First Unit. Total subsystem cost in FY00$M is a function of the independent variable X [Larson and Wertz, 1999]. (BOL is beginning of life; EOL is end of life.)

Cost Component	Parameter, X (Units)	Input Data Range	Subsystem Cost CER (FY00$K)	Standard Error SE (FY00$K)
1. Payload	Spacecraft total cost (FY00$K)	1.922 – 50.661	$0.4X$	$0.4 \times SE_{bus}$
2. Spacecraft	Satellite bus dry mass (kg)	20 – 400	$781 – 26.1X^{1.261}$	3,696
2.1 Structure	Structure mass (kg)	5 – 100	$299 – 14.2X\ln(X)$	1,097
2.2 Thermal	Thermal control mass (kg)	5 – 12	$246 + 4.2X^2$	119
	Average power (W)	5 – 410	$-183 + 181X^{0.22}$	127
2.3 Electrical power system	Power system mass (kg)	7 – 70	$-926 + 396X^{0.72}$	910
	Solar array area (m²)	0.3 – 11	$-210,631 + 213,527X^{0.0066}$	1,647
	Battery capacity (A·hr)	5 – 32	$375 – 494X^{0.764}$	1,554

TABLE 19-16. Cost Estimating Relationships (CERs) for Earth-orbiting Small Satellites Including Research, Development, Test, and Evaluation (RDT&E) and Theoretical First Unit. **(Continued)** Total subsystem cost in FY00$M is a function of the independent variable X [Larson and Wertz, 1999]. (BOL is beginning of life; EOL is end of life.)

Cost Component	Parameter, X (Units)	Input Data Range	Subsystem Cost CER (FY00$K)	Standard Error SE (FY00$K)
	BOL power (W)	20 – 480	$-5,850 + 4,629X^{0.15}$	1,585
	EOL power (W)	5 – 440	$131 - 401X^{0.452}$	1,633
2.4a Telemetry tracking and command (TT&C)	TT&C + DH mass (kg)	3 – 30	$367 + 40.6X^{1.35}$	629
	Downlink data rate (Kbps)	1 – 1,000	$3.638 - 3.057X^{-0.23}$	1,246
2.4b Command and data handling (C&DH)	TT&C + DH mass (kg)	3 – 30	$484 + 55X^{1.35}$	854
	Data storage capacity (MB)	0.02 – 100	$-27,235 + 29,388X^{0.0079}$	1,606
2.5 Attitude determination and control system (ADCS)	ADCS dry mass (kg)	1 – 25	$1.358 + 8.58X^2$	1.113
	Pointing accuracy (deg)	0.25 – 2	$341 + 2,651X^{-0.5}$	1,505
	Pointing knowledge (deg)	0.1 – 3	$2,643 - 1,364\ln(X)$	1,795
2.6 Propulsion	Satellite bus dry mass (kg)	20 – 400	$65.5 + 2.19X^{1.261}$	310
	Satellite volume (m^3)	0.03 – 1.3	$1,539 - 434 \ln X$	398
	Number of thrusters	1 – 8	$4,303 - 3.903X^{-0.5}$	834
3. Integration, assembly, and test	Spacecraft total cost (FY00$K)	1.922 – 50.661	3.139X	$0.139 \times SE_{bus}$
4. Program level	Spacecraft total cost (FY00$K)	1.922 – 50.661	3.229X	$0.229 \times SE_{bus}$
5. Ground support equipment	Spacecraft total cost (FY00$K)	1.922 – 50.661	0.066X	$0.066 \times SE_{bus}$
6. Launch and orbital operations support	Spacecraft total cost (FY00$K)	1.922 – 50.661	0.061X	$0.061 \times SE_{bus}$

TABLE 19-17. **Estimated FireSAT Lifecycle Cost.** The table shows estimated FireSAT mission lifecycle costs based on the Fire Scout concept. Estimates use Aerospace Corp. Small Satellite Cost Model (SSCM 8.0). [Larson and Wertz, 1999]. (RDT&E is research, development, test, and evaluation; ADCS is attitude determination and control system; C&DH is command and data handling; TT&C is telemetry, tracking, and control.)

	RDT&E Costs (FY00 $M)	Standard Error	First Unit Cost (FY00 $M)	Standard Error	Additional Units Cost (FY00 $M)	Total Cost (FY00 $M)
Payload	$7.07	2.97%	$4.71	1.98%	$4.24	$16.02
Spacecraft bus	$17.69	7.43%	$11.80	4.96%	$10.61	$40.07
ADCS	$2.01	0.84%	$3.42	1.44%	$3.07	$8.50
C&DH	$3.51	1.47%	$1.50	0.63%	$1.35	$6.36
Power	$4.26	1.79%	$2.61	1.10%	$2.35	$9.22
Propulsion	$1.24	0.52%	$1.24	0.52%	$1.11	$3.59
Structure	$3.77	1.59%	$1.62	0.68%	$1.46	$6.85
Thermal	$0.30	0.12%	$0.30	0.12%	$0.27	$0.87
TT&C	$2.60	1.09%	$1.11	0.47%	$1.00	$4.71
Integration, assembly, and test	$0	0%	$4.10	1.72%	$3.69	$7.79
Project management	$3.37	1.42%	$3.37	1.42%	$3.04	$9.78
Ground support equipment	$1.94	0.82%	$0	0%	$0	$1.94
Launch integration and early orbit operations support	$0	0%	$1.80	0.76%	$1.62	$3.42
Total space segment cost	$30.07		$25.78		$23.20	$79.05
Launch costs						$28
Operations costs (5 years)						$30.50
Total mission lifecycle cost						$157.90

The recommended baseline mission architecture resulting from the Pre-phase A study is summarized in Figure 19-14. The project now has a definitive baseline mission concept for which the design is closed, meaning that all key elements have been modeled and analyzed from end to end. The necessary performance for each is at or below known values or within the reach of available technology. Concurrent with the design effort, the engineering team, in consultation with veteran operators from NOAA, NASA, USFS, and industry, begin to expand on the basic concept of operations proposed at the beginning of the study. This operational insight proves invaluable to the design team and helps them to better derive and allocate requirements for all mission elements. We document these results in the *operations concept document*.

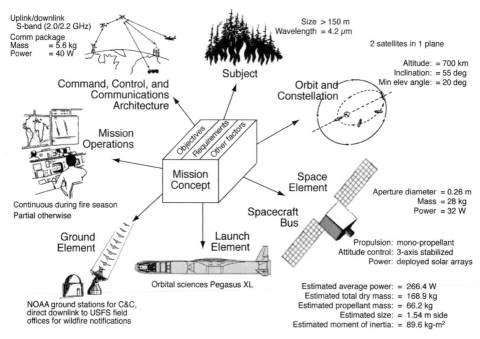

Mission: Detect and monitor potentially dangerous wildfires

FIGURE 19-14.   **Summary of Recommended FireSAT Mission Architecture Based on Pre-phase A Trade Studies.** Iterative trade studies yield mission characteristics that are quantified and more detailed. (C&C is command and control; USFS is US Forest Service.)

Draft Operations Concept Document

Chapter 3 gives a discussion of the operations concept development. For a detailed description of developing an operations plan, see *Cost Effective Space Mission Operations* (CESMO) [Squibb, 2006]. Development of the draft operations concept document begins with establishing a design reference mission (DRM) sequence as illustrated in Figure 19-15. The DRM for FireSAT is similar to other scientific earth observations missions. Normal operations are preceded by a series of spacecraft and payload commissioning steps and followed by disposal at the end of the mission, years in the future.

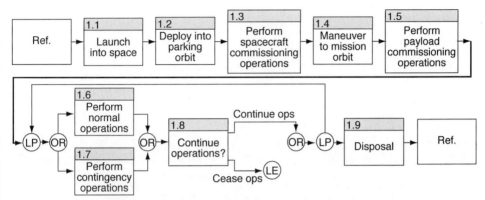

FIGURE 19-15.   FireSAT Design Reference Mission (DRM). The DRM embodies the operations concept. The one for FireSAT is fairly standard.

The DRM definition bounds the operational activities of the mission from cradle (launch) to grave (disposal). However, with time and resources being short in Pre-phase A, most of the operational planning focuses on what everyone hopes will be the bulk of the mission: normal operations. Drilling down on normal operations, the focus turns to the mission's main function—to detect wildfires. The integrated design and operations team starts by sketching a storyboard of what the detection scenario might look like, as shown in Figure 19-16. From there, they define a more formal sequence of functions, shown in Figure 19-17, along with the hand-offs between space and ground elements as illustrated on the context diagram in Figure 19-18. Finally, they build upon this operational insight to expand on the basic scenario timeline in Figure 19-16, to develop the much more detailed timeline shown in Figure 19-19. This detailed timeline allocates the 30-minute notification time to the major elements of the architecture. These derived response times will later be driving requirements for various mission elements.

Based on these and other analyses, the draft operations concept document derives additional requirements for ground systems, including command and control infrastructure at the operations center, necessary staff levels, and training plans.

Draft Systems Engineering Management Plan

The development and use of the *systems engineering management plan* (SEMP) is described in detail in Chapter 14. We explore more details of this critical project document later in the chapter. Here we address one piece of the technical planning puzzle in the SEMP: the project document tree. In this and the previous section, we've zeroed in on the handful of crucial products that pace the system development. The *project document tree* is the big picture look at all the major documents that we must generate to move the project forward on sound technical and managerial grounds. Table 19-18 shows the FireSAT document tree and identifies the delivery scheduling and pedigree for each document from baseline to baseline.

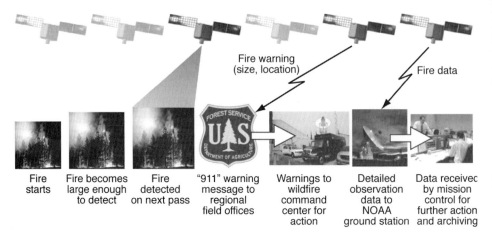

FIGURE 19-16. **Simple Storyboard of Timeline for "Detect Wildfire" Capability.** Pictures quickly capture a simple scenario that shows how a fire starts, becomes larger, and is detected by the FireSAT system, which then alerts NOAA and the firefighters.

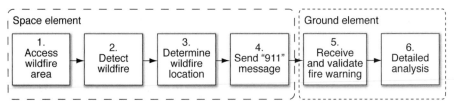

FIGURE 19-17. **Block Diagram for an Operational Scenario That Responds to the "Detect Wildfire" Capability.** We use many informal and formal graphical methods to describe operational scenarios.

Draft Risk Management Plan

Another essential task during Pre-phase A is to establish the risk management philosophy as defined in the *risk management plan*. Chapter 8 describes risk management in detail. One of the main activities in this early assessment of project risk is to brainstorm and rank order the top risks, along with proposed mitigation approaches. The "stop light chart" in Figure 19-20 summarizes the top five. Other parts of the plan summarize the approach for capturing, reporting, tracking, and dispositioning risks during the project.

Plan to Achieve Readiness for the System Requirements Review

The final document in developing the mission baseline is a catch-all set of plans that map out how the project will reach the next technical baseline, the system baseline at the system requirements review (SRR) during Phase A. From a

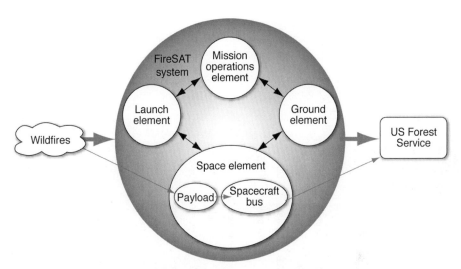

FIGURE 19-18.   Tracing an Operational Scenario on a Context Diagram. This type of diagram brings users and systems engineers together in understanding required capabilities and the system that will realize them.

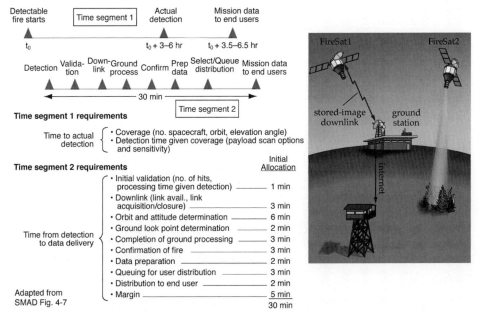

FIGURE 19-19.   Fire Detection and Notification Timeline. This is a more detailed timeline analysis for the operational scenario. This input is important for assessing and applying implementation concepts for the system elements. Time segment 2 is the second part of Time segment 1 (Adapted from SMAD [Larson and Wertz, 1999]).

TABLE 19-18. FireSAT Project Document Tree. Here we see the principal project documents and how they mature through major reviews. (MCR is mission concepts review; SRR is system requirements review; SDR is system definition review; PDR is preliminary design review; CDR is critical design review; SAR is system acceptance review; PLAR is post-launch assessment review.)

Document Name	Number	MCR	SRR	SDR	PDR	CDR	SAR	PLAR
Project scope document	FS-10012	Baseline						
Mission design study report	FS-10001	Baseline						
Spacecraft system design study report	FS-10005	Draft	Draft	Baseline				
Concept of operations	FS-80010	Draft	Baseline	Update	Update	Update	Update	Final
Systems requirements document	FS-80020	Draft	Baseline	Update	Update	Final		
Systems engineering management plan	FS-70010	Draft	Baseline	Update	Update	Final		
Software development plan	FS-70050		Draft	Baseline	Update	Update	Update	Update
Flight hardware, software, and ground support equipment specifications	FS-8x000		Draft	Baseline	Baseline	Final		
Ground system specifications	FS-84000		Draft	Draft	Baseline	Update	Final	
Mission operations specifications	FS-81000		Draft	Draft	Baseline	Update	Final	
Safety and mission assurance plan	FS-70070	Draft	Baseline	Update	Update	Update	Update	Update
External interface requirement and control documents	FS-8x010	Draft	Baseline	Update	Update	Final		
Spacecraft to payload interface requirement and control documents	FS-8x020			Draft	Baseline	Final		
Spacecraft to launch vehicle interface requirement and control document	FS-8x030			Draft	Baseline	Final		
Spacecraft to ground system interface requirement and control document	FS-8x040			Draft	Baseline	Update	Final	
Launch vehicle to ground system interface requirement and control document	FS-8x050			Draft	Baseline	Final		

TABLE 19-18.   FireSAT Project Document Tree. (Continued) Here we see the principal project documents and how they mature through major reviews. (MCR is mission concepts review; SRR is system requirements review; SDR is system definition review; PDR is preliminary design review; CDR is critical design review; SAR is system acceptance review; PLAR is post-launch assessment review.)

Document Name	Number	MCR	SRR	SDR	PDR	CDR	SAR	PLAR
Ground system to mission operations interface requirement and control document	FS-8x060			Draft	Baseline	Final		
Natural environment definition document	FS-8x070	Draft	Baseline	Update	Update	Final		
Configuration management plan	FS-70060	Draft	Baseline	Update	Update	Update	Final	
Data management plan	FS-70070	Draft	Baseline	Update	Update	Update	Final	
Electromagnetic compatibility and interference control plan	FS-70080		Draft	Baseline	Update	Final		
Mass properties control plan	FS-70090		Draft	Baseline	Update	Final		
Mass properties reports	FS-70091			Update	Update	Update	Update	Final
Manufacturing and assembly plan	FS-70075		Draft	Baseline	Update	Final		
Spacecraft systems analysis plan	FS-70076	Draft	Baseline	Update	Update	Final		
Spacecraft systems analysis reports	FS-70077		Submit	Update	Update	Update	Update	Update
System configuration document	FS-89000		Draft	Baseline	Update	Update	Final	
Engineering drawings	FS-89xxxx				Draft	Baseline	Update	
Risk management plan	FS-70780	Draft	Baseline	Update	Update	Update	Final	
Risk management reports	FS-70781		Submit	Update	Update	Update	Update	Update
Reliability, maintainability, and supportability plan	FS-70073	Draft	Baseline	Update	Update	Update	Update	Update
Instrumentation and command list	FS-98000				Draft	Baseline	Final	
Master verification plan	FS-70002	Draft	Baseline	Update	Update	Final		
Verification compliance reports	FS-90001						Submit	
Acceptance data package	FS-95000						Submit	

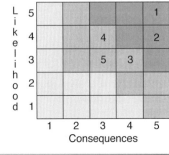

Risk	ID	Trend	Approach	Title
1	FS-023	⇨	R	Payload focal plane technology
2	FS-003	⇧	M	Fire detection software development
3	FS-014	⇩	M	Fire "911" notification method
4	FS-120	☐	W	USFS, NOAA, NASA MOA
5	FS-115	⇧	A	NOAA ground station interface protocols

FIGURE 19-20.   Preliminary FireSAT Risk Assessment. We easily see the top five risks, as well as their likelihood, consequences, and status. (USFS is US Forest Service; NOAA is National Oceanic and Atmospheric Administration; MOA is memorandum of agreement.)

technical standpoint, it's not necessary to have all the system answers by the end of Pre-phase A. The real achievement is identifying the major questions. A rank-ordered list of trade-offs to do in Phase A guides the planning and execution of the subsequent design and analysis cycles (DACs). If we plan another DAC to support development of products for SRR, this list drives requirements for DAC resources (e.g., models, personnel, and development hardware and software).

Since the system requirements document is one of the key documents baselined at SRR, many analyses and trade studies focus on assessing assumptions from the Pre-phase A study and on deriving additional system and subsystem level requirements. One aim is to identify and assess technology readiness levels for enabling system technologies to craft early developmental testing or other risk reduction strategies.

Finally, because much of the emphasis during the early part of Phase A is on requirements development, we must have plans to manage these requirements, and all the other technical data that will be created at a faster pace beginning at the start of the phase. Chapter 4 discusses requirements management. Details of the bigger picture of technical data management are in Chapter 17.

19.2.5   Representative Issues, Items, and Anecdotes

As in any Pre-phase A study, limited time, budget, and resources severely constrain the number and type of detailed trade studies that can be done. While the engineers involved rightly feel proud of the proposed mission architecture and the sound basis for its recommendation, several unresolved issues come up that could significantly affect how best to achieve customer expectations. The first involves some potential conflicts among four of the preliminary mission requirements, depending on their interpretation:

- The FireSAT spacecraft shall cover the entire United States, including Alaska and Hawaii
- The FireSAT spacecraft shall monitor the coverage area on a daily basis
- The FireSAT spacecraft shall provide fire notifications within one hour (threshold), 30 minutes (objective) of fire detection
- The FireSAT project shall use existing NOAA ground stations at Wallops Island, Virginia and Fairbanks, Alaska for all mission command and control

Basic orbit analysis shows that the study team's proposed mission architecture can easily achieve daily coverage of the all US territories up to and including 70 degrees latitude (there are no wildfires above this latitude since it's just tundra). Further, "daily" is somewhat vague and therefore hard to verify. In addition, since a constellation of two satellites is now envisioned, daily coverage by a single satellite isn't required as long as the constellation provides sufficient coverage as a whole. So, in consultation with the stakeholders, the project agrees on the following modified preliminary requirements:

- The FireSAT spacecraft shall cover the entire United States, including Alaska and Hawaii, up to a latitude of 70 degrees
- The FireSAT constellation revisit time to the coverage area shall be less than 24 hours

When considering the requirement for fire notification limits, engineers initially consider it redundant in the face of the revised revisit requirement. After all, if we're revisiting every 24 hours or less, why impose an additional 30- to 60-minute notification requirement? But further consultation with the stakeholders, notably experienced firefighters, produces a counterargument. Strong evidence, based on past history, suggests that just because a system **knows** about a fire, without a requirement to **release** this information in a timely manner, it might hold onto that information for further study or other reasons. More practically, the notification requirement compels the entire team to work together to develop detailed timelines for fire detection, information processing, and distribution as described in the operations concept document. With these arguments in mind, the notification requirements remain as written.

The constraint to use NOAA ground stations also raises issues. It offers the potential for considerable cost savings by eliminating the need to separately procure such facilities. But preliminary orbit analysis indicates that with only two locations in Alaska and Virginia, this constraint will limit the system's ability to meet the 30- to 60-minute notification requirement. For example, a FireSAT spacecraft on a descending node pass over Hawaii won't be in contact with either NOAA site. Thus, if a fire is detected, it will be more than one orbit later before the data can be sent to one of the two stations, violating the 60-minute requirement.

Furthermore, initial analyses of the data formats reveals that it might not be necessary to send complete images to the ground. Rather, the spacecraft could generate simple "911" fire warnings with time and location information for each fire detected. This would allow the USFS to immediately cue local fire detection assets for rapid confirmation of each fire warning. Such a simple, low data rate message could easily be received by regional field offices scattered all over the US with relatively inexpensive equipment. Thus, any time FireSAT was in view of the US, it would also be in view of a one or more USFS field offices. The requirement to force this simple data package to first pass through one of only two NOAA sites appears to impose unnecessary overhead and delay on a time-critical message. All stakeholders agree. So this interface requirement is defined as follows:

- The FireSAT spacecraft shall broadcast fire notification warnings so they can be received by USFS field offices. Detailed technical interface is defined in USFS GS-ISD-XYX.

The requirement does force all USFS field offices to be outfitted with a simple FireSAT warning reception station. Fortunately, the cost for these stations is estimated to be well within the project budget.

19.2.6   Conclusions

With great relief, the FireSAT study team walks out of the mission concept review. They have convinced the milestone decision authorities that the FireSAT system is feasible. It has a high probability of achieving its KPPs within the allowable budget. The joint project board, comprising senior NASA, NOAA, and USFS personnel, has stamped the project APPROVED FOR NEW START. But this is only the first small step on the long road toward the PLAR described at the beginning of the chapter. The next critical steps take place in Phase A.

19.3   Create the System Baseline

With Pre-phase A behind us, and the thrill of being named as an official "new start" project still buzzing through the office, the real work begins in earnest at the start of Phase A. During this phase, two important technical baselines are defined. The first is the *system baseline*, established after successful completion of the system requirements review (SRR). The second one, the *functional baseline*, is addressed in the next section.

As the project enters Phase A, it's committed to performing key steps of the systems engineering process to successfully complete the SRR. In this phase, we identify hardware and software system elements, each defined by products needed for the review. These products will be analyzed by the SRR teams with inputs collected via review item dispositions. In this phase:

- We further refine the operations concept to derive operational requirements
- Operational requirements drive creation of new requirements and refinement of existing system-level ones
- We develop an operational model to capture high-level flows that identify operations performed by various operational nodes
- We develop a functional model for each system and tie them together at the system-of-systems (SoS) level functional model
- We capture functional and physical interfaces
- We generate draft interface requirements documents for each system-to-system pairing
- We generate draft system requirements documents
- We put together draft SoS-level and system-level verification plans, containing verification requirements and configurations, support equipment and facilities, and verification schedules
- We draft risk management plans for each element
- We produce technology plans for each element—which identify technology maturity, and plans to mature key technologies as needed

19.3.1   What Got Us Here?

All this is a tall order. To tackle this problem we'll break it up into bite-sized pieces starting out by seeing how we got here, then defining the scope of the deliverables needed. To reach the start of Phase A, the project first has to:

- Complete a credible Pre-phase A study that survives the mission concept review
- Secure funding, and associated political commitment, to proceed with a new-start Phase A program

The organizations involved in the FireSAT project (NASA, NOAA, and the US Forest Service) must also complete an inter-agency memorandum of agreement defining their individual roles and responsibilities, and commitments for funding or contributions in-kind (e.g., NOAA ground station facilities and personnel, NASA spacecraft test facilities). In addition, the project secures one or more contractors to support systems engineering and integration (SE&I) and other tasks.

Clearly, details of the planned acquisition have a huge impact on the entire lifecycle. The contract award process (e.g., open competition or indefinite-deliverable, indefinite quantity), the nature of the contract (e.g., fixed price or cost-plus), and other factors all shape the implementation of Phase A and how government and contractor roles and responsibilities are defined. Our focus here is on practical implementation of the systems engineering processes to achieve a successful space mission. So we have the luxury of side-stepping these critical and extremely difficult contracting issues. For our discussion, we assume that a single prime contractor, Acme Astronautics Corp., is put on contact early in Phase A after a brief (and uncontested) bidding and award process. Throughout our discussion, we stick largely to engineering-oriented issues, avoiding the inevitable contracting issues that also arise. However, the reader should not conclude that the contracting challenges of FireSAT, or any space project, are trivial—quite the opposite. Contracting decisions largely shape the systems engineering task.

19.3.2   The End Game—SRR Projects

Phase A contains two important gates, the system requirements review and the system definition review. So the new FireSAT lead systems engineer or project manager has to start by looking at the guidelines for entering, and surviving, the first of these—the SRR—as shown in Table 19-19.

TABLE 19-19.   **System Requirements Review (SRR) Entrance and Success Criteria.** These considerations are in addition to the programmatic matters faced by a newly-approved project. Adapted from NASA [2007].

System Requirements Review	
Entrance Criteria	**Success Criteria**

• The project has successfully completed the mission concepts review (MCR) and responded to all MCR requests for action and review item discrepancies
• A preliminary SRR agenda, success criteria, and charge to the board have been agreed to by the technical team, project manager, and review chair before the SRR
• The following technical products for hardware and software system elements are available to the review participants in advance:
  a. System requirements document
  b. System software functionality description
  c. Updated concept of operations
  d. Updated mission requirements, if applicable
  e. Baselined systems engineering management plan
  f. Risk management plan
  g. Preliminary system requirements allocation to the next lower level system
  h. Updated cost estimate
  i. Technology development maturity assessment plan
  j. Updated risk assessment and mitigations (including probabilistic risk assessment as applicable)
  k. Logistics documentation (e.g., preliminary maintenance plan)
  l. Preliminary human rating plan, if applicable
  m. Software development plan
  n. System safety and mission assurance plan
  o. Configuration management plan
  p. Initial document tree
  q. Verification and validation approach
  r. Preliminary system safety analysis
  s. Other specialty disciplines, as required
	• The project uses a sound process for allocating and controlling requirements throughout all levels, and has a plan to complete the definition activity on schedule
	• Requirements definition is complete with respect to top-level mission and science requirements, and interfaces with external entities and between major internal elements have been defined
	• Requirements allocation and flow-down of key driving requirements have been defined down to subsystems
	• Preliminary approaches have been determined for verifying and validating requirements down to the subsystem level
	• Major risks have been identified and technically assessed, and viable mitigation strategies have been defined

As we did for the MCR, we begin by laying out success criteria versus products to answer the mail on each criterion, as shown in Table 19-20. From this analysis, we put together the list of paper products that need to be developed throughout the first part of Phase A along with their top-level contents. These include the items identified above as well as specific items called for in the entrance criteria, as shown in Table 19-21.

TABLE 19-20.   System Requirements Review (SRR) Versus Success Criteria. Each criterion calls for one or more documents as evidence that the project meets the criterion.

System Requirements Review Success Criterion	Supporting Documents
1. The project uses a sound process for allocating and controlling requirements through all levels, and has a plan to complete the definition activity on schedule	Systems engineering management plan (SEMP), software development plan
2. Requirements definition is complete with respect to top-level mission and science requirements, and interfaces with external entities and between major internal elements have been defined	Mission design study report, system design study report, Mission-level system requirements document (SRD), and interface requirements documents (IRDs), operations concept document
3. Requirements allocation and flow-down of key driving requirements have been defined down to subsystems	System-level SRDs, draft hardware, software, and ground support equipment SRDs
4. Preliminary approaches have been determined for verifying and validating requirements down to the subsystem level	Mission-level SRD, system-level SRD, master verification plan
5. Major risks have been identified, and viable mitigation strategies have been defined	Risk management plan

19.3.3   Process Level of Effort

Table 19-22 illustrates the relative level of effort for each of the 17 systems engineering processes needed to produce these deliverables during this part of Phase A.

19.3.4   Results

In some cases, Phase A starts with the momentum already gained during Pre-phase A. This momentum can be maintained if the mission scope stays the same and scope creep doesn't fundamentally change the validity of the Pre-phase A study results. If it does, Phase A starts from behind, trying to play catch-up by reanalyzing results from the concept study to determine which results are still valid and which mission concepts need to be reexamined. An insidious type of scope or requirements creep that can easily sneak in during Phase A (or almost any time for that matter) is what the authors call "justs." These take the form of "We like all of your Pre-phase A ideas; *just* use XYZ existing spacecraft bus instead." Or "The current operations concept is great; *just* add the ability to downlink anywhere in the world." These seemingly innocent additions can become major constraints on the system and may necessitate a complete redesign.

TABLE 19-21. **Top-level Products for System Requirements Review (SRR) Major Contents.** This table details the contents of the documents listed in Table 19-20. (TBD/TBR is to be determined/to be resolved.)

Product	Contents	Product	Contents
FireSAT scope document	• Merged into system requirements document; see discussion below	**Risk management plan**	• Preliminary risk assessment • Technologies and associated risk management and mitigation strategies and options

Mission design study report (UPDATE)	• Analysis of alternatives trade-tree including:
	  – Alternate mission concepts
	  – Identification of system drivers for each concept or architecture
	  – Results of characterization for each alternative
	  – Critical requirements identified or derived
	  – Concept or architecture utility assessment, including evaluation criteria such as cost, technical maturity, and measures of effectiveness (MOEs)
	• Proposed baseline mission concept
	• Conclusions and recommendations
		Systems engineering management plan (SEMP) (BASELINE)	• Plan for managing the SE effort
			• Documents driving SEMP
			• Technical summary
			• Integration technical effort (including conceptual lifecycle support strategies, logistics, manufacturing, operation, etc.)
			• The 17 engineering processes
			• Technology insertion
			• SE activities
			• Project plan integration
			• Waivers
			• Appendices

System design study report (for each system element within the mission architecture as appropriate) (DRAFT)	• Analysis of alternatives trade-tree including—for each system in the architecture—spacecraft, ground system, etc. • Alternative system concepts • Identification of system drivers for each concept or architecture • Results of characterization for each alternative • Critical requirements identified or derived • Concept or architecture utility assessment, including evaluation criteria for cost, technical maturity, and MOEs • Detailed functional analysis • Proposed baseline system concept with functional allocation	**Operations concept document (UPDATE)**	• Operational scenarios • Timelines • Data flow diagram • Organization and team responsibilities • Cost and complexity drivers for a given set of inputs • Requirements and derived requirements • Technology development plan

TABLE 19-21. **Top-level Products for System Requirements Review (SRR) Major Contents. (Continued)** This table details the contents of the documents listed in Table 19-20. (TBD/TBR is to be determined/to be resolved.)

Product	Contents	Product	Contents

Software development plan (DRAFT)	See discussion in functional baseline section	FireSAT Mission-level and systems-level system requirements documents (BASELINE)	• Scope
			• Applicable documents
			• Requirements
			  – Prime item definition
			  – System capabilities
			  – System characteristics
			  – Design and construction
			  – Logistics
			  – Personnel and training
			  – Subsystem requirements
			  – Precedence
			  – Qualification
			  – Standard sample
			• Verification requirements
			• Notes
			• Appendices

Interface requirements documents	See discussion in functional baseline section	Master verification plan	See discussion in as-built baseline section
Plan to achieve system definition review (SDR) readiness	• Rank ordered list of studies and trades-offs • Forward work mapped to program schedule • Technology gap closure plan mapped to program schedule • Requirements update and management plan • TBD/TBR burn-down plan • SRR issue resolution plan		

TABLE 19-22. **Systems Engineering Process Level of Effort Leading to System Requirements Review (SRR).** Analogous to a cell phone signal, zero bars represents no or very low resource use and five bars represents maximum resource use. Resource requirements apply to a particular process only; a given number of bars for one process does not imply the same level of resources as the same number of bars for a different process.

Process	Level of Effort	Comments
1. Stakeholder expectation definition	■■■■□	Stakeholder expectations are constantly revisited as technical requirements are developed. This serves as a requirements validation.
2. Technical requirements definition	■■■■■	This is a main focus of preparation for SRR
3. Logical decomposition	■■■■■	This supports requirements development

TABLE 19-22. **Systems Engineering Process Level of Effort Leading to System Requirements Review (SRR). (Continued)** Analogous to a cell phone signal, zero bars represents no or very low resource use and five bars represents maximum resource use. Resource requirements apply to a particular process only; a given number of bars for one process does not imply the same level of resources as the same number of bars for a different process.

Process	Level of Effort	Comments
4. Physical solution	■■□□□	As the logical decomposition matures, allocation of functionality drives the definition of the physical solution. Many trades begin and continue past SRR. Special emphasis goes to preventing a premature definition of the physical architecture. The logical decomposition leads the physical solution design activities.
5. Product implementation	■□□□□	Identifying existing assets and assessing the cost-benefit and risk-reduction aspects of existing systems are standard trades during this phase. The project must also carefully determine the need for long-lead procurement items.
6. Product integration	■□□□□	As complement to the logical decomposition process, this takes into account all lifecycle phases for each system element. We identify and coordinate with stakeholders to ensure that we ascertain all requirements and interfaces.
7. Product verification	■□□□□	Verification planning is crucial to assure that requirements defined during this phase are testable and that cost and schedule estimates account for this critical phase. Since the physical solution is not completely fixed in this phase, we can't fully define the verification program.
8. Product validation	■■□□□	Validation planning is matured in this phase
9. Product transition	□□□□□	Only minimum effort expended on this process, mainly to identify enabling products for transition that may impact lifecycle cost
10. Technical planning	■■■■■	These processes must be thoroughly established for a successful SRR, as they determine how the program will be executed in the next phase
11. Requirements management	■■■□□	The requirements management process must be mature and in place to control the emerging set of requirements
12. Interface management	■□□□□	Interface management is addressed in the requirements management processes
13. Technical risk management	■■□□□	Risk planning and management remain at a vigorous level
14. Configuration management	■□□□□	End item control and requirements management controls must be in place
15. Technical data management	■■□□□	Tools and repositories must be fully deployed along with the processes to guide engineering teams in concurrently defining systems
16. Technical assessment	■■□□□	Initial planning to support the upcoming SRR and subsequent reviews begins
17. Decision analysis	■■■■■	Rigorous technical decision analysis is key to developing credible Phase A results. It's impossible to complete all the myriad trade studies that may be identified, so good decision analysis identifies the most critical ones.

Assuming FireSAT survives (or avoids) such major scope or requirements creep, we return our attention to the first major review of the official new project—the system requirements review. As described above, the SRR is more than a review of the system requirements. But because capturing and documenting system requirements is one of most high profile activities during this part of the lifecycle, we devote most of our attention in this section to the FireSAT mission-level and system-level system requirements documents.

System Requirements Document

Even for a relatively small project like FireSAT, the number of mission-level, system-level (e.g., spacecraft) and allocated subsystem-level requirements could easily reach many hundreds. We can't practically review them all here. Instead, we pick out a few requirement threads that illustrate the top-down process of requirements engineering, from mission to system to subsystem-level, with associated issues and assumptions that crop up along the way. We then look at how to organize these requirements to better manage them.

Chapter 4 describes the complete process of requirements engineering. We apply this process here to see how some of the FireSAT spacecraft requirements are developed. For purposes of discussion, our system of interest is the FireSAT spacecraft, but we'll also see how to allocate mission-level requirements to this level of the system. The two broad types of requirements described in Chapter 4 are functional and nonfunctional. To define the functional requirements, we need to go back to the system context. We develop a number of system context diagrams during Pre-phase A, and present them here in a slightly different form to better guide our analysis. Figure 19-21 shows the context diagram for the space element. We begin with the most important functional requirement of the system, that of sending out fire notification "911" messages to the users. But we still need to know "what," "when," "how," and "who" for this requirement. For guidance on these details, we return to several of the stakeholders' mission objectives that serve as measures of effectiveness (MOEs) listed earlier:

- Cost decrease: Average annual cost of fighting wildfires shall decease by 20% in 2008 dollars

- Property loss: Average annual property losses due to wildfires shall decrease by 25% in 2008 dollars

- Size at first contact: Average size of fire at first contact by firefighters shall decrease by 20%

All of these customer expectations relate to defining the size, location, and timeliness of notification of wildfires. The smaller the fire detected, the more accurately it is located. And the sooner the USFS knows about it, the more likely they can contain it. Thus, as part of the requirements engineering process, we can think of these three MOEs as generating an "issue." That issue is a need for a detailed trade study, called the *FireSAT Wildfire Definition Study Report*, to

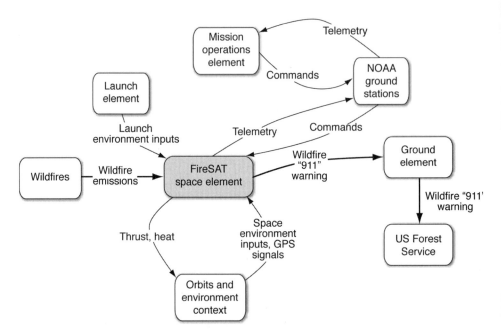

FIGURE 19-21.   **Context Diagram for FireSAT System with Focus on the Space Element.** This expanded context diagram allows us to focus on the space element and understand the major interfaces between it and other elements of the mission. (GPS is Global Positioning System.)

determine 1) minimum wildfire size to be detected— essentially the definition of a "potentially dangerous wildfire", 2) the geolocation accuracy needed, and 3) how quickly firefighters need to receive notifications. The results of this study generate three corresponding derived requirements, as Figure 19-22 illustrates. These form the basis for three of the key performance parameters (KPPs) defined earlier. (The numbers correspond to the requirements listed in Table 19-13):

- KPP 1: Detection—System shall detect potentially dangerous wildfires (defined to be greater than 150 m in any linear dimension) with a confidence of 95%

- KPP 3: Timeliness—The FireSAT system shall send fire notifications to users within 30 minutes of fire detection (objective), 1 hour (threshold)

- KPP 4: Geo-location—The FireSAT system shall geo-locate potentially dangerous wildfires to within 500 m (objective), 5 km (threshold)

The "system" referred to here is the system of systems physical architecture that includes the space element along with ground and launch elements. This analysis helps us define "what" and "when." But we still need to decide "who" and "how." The space segment alone can't meet any of these requirements. Even the first

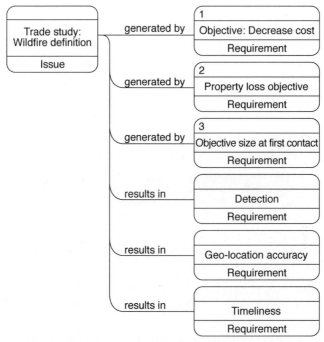

FIGURE 19-22.   Relationship between the Wildfire Definition Trade Study and Requirements.
The last three requirements are derived from the first three.

requires some coordination and confirmation by the ground to reduce the number of false-positive warnings. Thus, we must further refine these mission-level requirements and allocate them down to the system level. Let's pick out two threads to examine further—detection and geo-location accuracy.

We start with the detection KPP, where wildfire size (150 m) and confidence level (95%) were determined by the trade study. This KPP is a good starting point, but engineers need to further decompose it to a design-to requirement for the spacecraft payload sensor. This trade study determines that to meet the KPP we need to derive two separate payload requirements, one for spatial resolution (the smallest thing the payload can see), and one for spectral resolution (the signature energy wavelength of a wildfire). This relationship is illustrated in Figure 19-23.

Now we turn to the second thread that derives from the requirement to accurately geo-locate the wildfires. As described in SMAD, satellite remote sensing geolocation accuracy is a function of:

- Satellite position knowledge (in 3 axes)
- Satellite pointing knowledge (in 2 axes)
- Satellite timing accuracy (how good the onboard clock is)

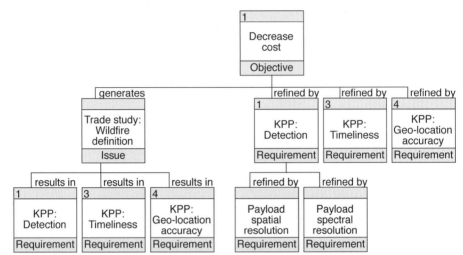

FIGURE 19-23. Relationship Between Top-Level and Lower-Level Requirements. A top-level mission desire, in this case the measure of effectiveness to decrease the cost of fighting fires, generates trade studies and spawns more detailed requirements in the form of key performance parameters (KPPs). These in turn are further defined by component-level requirements such as the payload spatial and spectral resolution.

- Inherent altitude error for ground targets (how good our 3-dimensional map of the Earth's surface is)

The last one depends on the accuracy of the map database. This requirement can't even be allocated within the FireSAT system of systems, as the USFS depends on maps provided by the US Geological Survey. Fortunately, most of these maps are highly accurate and so constitute only a small source of error, but it's one that's beyond the control of the FireSAT project.

Now let's look at the first element listed above, satellite position knowledge in 3 axes. Even if the mission plans on using onboard GPS-derived navigation data, a fairly common practice, we would still need mission control to do some oversight and backup using ground tracking. Therefore, part of the accuracy requirement must be allocated to both the spacecraft and ground systems.

The middle two error sources, pointing knowledge and timing, arguably depend solely on the spacecraft. But let's drill down deeper on pointing knowledge to make a point. Onboard pointing knowledge is determined by a software estimator such as a Kalman Filter, which in turn is driven by inputs from a suite of attitude sensors (some collection of star sensors, Sun sensors, Earth horizon sensors, gyroscopes, and magnetometers). So whatever requirement is derived for onboard pointing knowledge must ultimately be further allocated down to the software and hardware elements that make up the attitude determination and control subsystem. This requirement allocation from mission-level to system-level, to subsystem-level, and finally to detailed specifications at the component level is illustrated in Figure 19-24. (For our discussion a requirement defines a "what" while a specification defines a "how.")

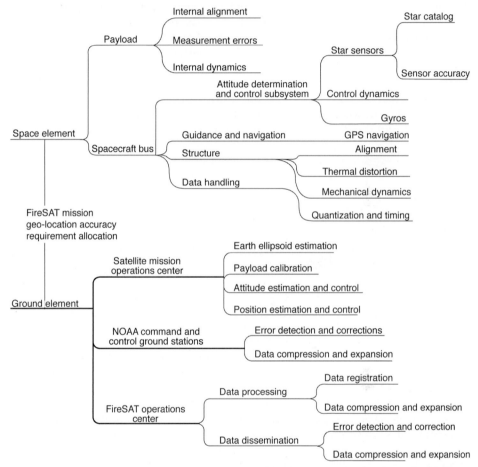

FIGURE 19-24. Example Requirements Allocation Tree for Geo-position Error Requirement. This illustrates how a single requirement, geo-location accuracy, can be allocated first to space or ground elements, then down to hardware or software components. (Adapted from *Space Mission Analysis and Design* [Larson and Wertz, 1999].)

This example underscores the critical importance of requirements management, especially with respect to traceability. For example, if the allocated star sensor accuracy specification is not achieved for some reason during system development, we need to understand the impact on the geolocation source requirement. This way systems engineers can determine how best to reallocate the accuracy budget among the other parts of the system.

These examples illustrate but two threads in the dozens that made up the complete FireSAT spacecraft system and subsystem functional requirements. Similarly, we need to develop additional threads for the nonfunctional requirements such as launch loads, thermal conductivity, memory capacity, mass, cost, and redundancy architecture. Building on the results from Pre-phase A, and applying the techniques described in Chapters 2 and 4, systems engineers seek to capture all of the requirements for every element of the mission architecture. By developing the detailed mission operations concept, we derive additional functional and nonfunctional requirements to address the usability of the system. Table 19-23 summarizes some of the trade-offs, derived requirements, and FireSAT design decisions. Each technical requirement comes with a number of critical attributes that must be managed along with it. These include:

- Rationale
- Source
- Trace
- Verification method
- Verification phase
- Priority
- Increment
- Requirement owner

With all these requirements and their attributes defined, derived, and modified early in Phase A, it's critical to organize, manage, and communicate them efficiently. The defining requirements document is usually called the system requirements document (SRD) or the System/Segment Specification (SSS). Different organizations have their own standards or practices for writing these documents. For example, Mil-Std-961C is widely used throughout the aerospace industry mainly because of the strong military roots in even civil space programs. FireSAT, like many projects, has tailored these guidelines to develop the following outline for the FireSAT space segment SRD:

1.0   Scope
2.0   Applicable documents
3.0   Requirements
3.1   Prime item definition
3.2   System capabilities
3.3   System characteristics
3.4   Design and construction
3.5   Logistics
3.6   Personnel and training
3.7   Subsystem requirements
3.8   Precedence

TABLE 19-23.   Summary of FireSAT Design Decisions Based on Preliminary and Derived Requirements. These trade-offs help flesh out and specify the system design. (EIRP is effective isotropic radiated power.)

Mission Requirements	Description	Derived System Requirements	Issue or Trade	FireSAT Design Decision
1. Detection	The FireSAT spacecraft shall detect potentially dangerous wildfires with a 95% confidence	• Spectral resolution • Aperture size	Spectral resolution depends on wildfire temperature, atmospheric windows, and aperture diffraction limitation	• $\lambda = 4.2\ \mu m$ • Aperture = 0.26 m for assumed altitude of 700 km
2. Coverage	The FireSAT spacecraft shall cover the entire United States, including Alaska and Hawaii, to a latitude of 70°	Orbit inclination	Higher inclination reduces launch vehicle capacity	• Altitude = 700 km • Orbit inclination = 55°
3. Persistence	The FireSAT spacecraft shall monitor the coverage area on a daily basis	• Orbit altitude • Number of satellites	• Higher altitude gives greater persistence but drives up sensor size • More satellites drives up cost, complexity, and number of launches	• h = 700 km • 2 satellites in one plane separated 180° in true anomaly
4. Timeliness	The FireSAT spacecraft shall provide fire notifications within 60 minutes (threshold), 30 minutes (objective) of fire detection	Ground station locations	• Using only two NOAA ground stations limits access for downlink • Direct downlink to USFS field offices decreases latency	• Require FireSAT spacecraft to broadcast "911" messages to USFS field offices • All field offices to be equipped with necessary receiver
5. Geo-location	The system shall provide geolocation information on fires to within 5000 meters (threshold), 500 meters (objective), 3 sigma	• Attitude determination • Orbit determination • Timing • Target altitude error	Geolocation is a function of: • Orbit determination accuracy • Attitude determination accuracy • Timing error • Target altitude error	Derived requirements for: • Orbit determination < 0.2 km along track, 0.2 km across track, 0.1 km radial • Attitude determination < 0.06° azimuth, 0.03° nadir • Time error < 0.5 s Target altitude error < 1 km

TABLE 19-23. Summary of FireSAT Design Decisions Based on Preliminary and Derived Requirements. (Continued) These trade-offs help flesh out and specify the system design. (EIRP is effective isotropic radiated power.)

Mission Requirements	Description	Derived System Requirements	Issue or Trade	FireSAT Design Decision
6. Minimum elevation angle	The spacecraft shall support payload operations out to 20° elevation angle	• Sensor field of regard (FOR) • Field of view (FOV)	• Greater FOR gives better coverage, persistence • Higher FOV increases sensor data rate, focal plane complexity	Sensor field of regard = 115°
7. Reliability	The system shall have an overall design reliability of 95% (threshold), 98% (objective) consistent with Class B NASA mission as per NASA NPR 8705.4	Flow down to system	Higher reliability drives up cost, complexity	• Flow requirement down to subsystem level • Perform rigorous FMEA
8. Design life	The system shall have an operational on-orbit lifetime of 5 years (threshold), 7 years (objective)	Flow down to system	Longer lifetime increases consumables	delta-V budget, power budget includes 5-year life with margin
9. End of life disposal	System shall have sufficient end of life delta-V margin to de-orbit to a mean altitude of <200 km	delta-V	Increased delta-V drives up loaded mass, propulsion system complexity	Total delta-V budget = 513 m/s
10. Ground station (GS) interface	FireSAT shall use existing NOAA ground stations at Wallops Island, Virginia and Fairbanks, Alaska for all mission command and control. Detailed technical interface is defined in NOAA GS-ISD-XYX	Uplink and downlink frequencies, data rates, modulation schemes, GS EIRP, system noise	Constrains communication subsystem technology options	Use compatible uplink/downlink hardware, flow down to spacecraft communication subsystem requirements

3.9   Qualification

3.10   Standard sample

4.0   Verification requirements

5.0   Notes

Appendices

The scope described in Section 1.0 of this document reiterates the project need, goals, objectives, and concept of operations, as well as the scope of the document itself. Thus, it replaces the FireSAT scope document baselined in Pre-phase A. This provides important project guidance and one-stop-shopping for managing it along with the requirements. As discussed earlier, nothing derails requirements faster than scope creep.

Table 19-24 shows an example requirements matrix with samples from each of the sections of the SRD except 3.7: Subsystem requirements and 4.0: Verification requirements. The subsystem requirements are addressed in more detail in the functional baseline discussion. Verification requirements are one of the subjects of the build-to baseline discussion later in this chapter. The matrix does not show requirements traceability. However, by using requirements management tools, such as DOORS, or a systems engineering tool, such as CORE or CRADLE, we can readily track the detailed level-by-level traceability, as shown in the example back in Figure 19-23.

Each requirement in the SRD must also be validated before being baselined. A major part of the SRR is critically evaluating each requirement against the VALID criteria. This assessment determines if each requirement, as written, is:

- **V** erifiable
- **A** chievable
- **L** ogical
- **I** ntegral
- **D** efinitive

An example of this requirements validation exercise for FireSAT is described in Chapter 11. Most of this effort, of course, has to be completed before the formal SRR, so the results are presented along with open items still to be resolved. As described in Chapter 4, it's not uncommon to have numerous TBRs and TBDs included within formal requirements statements at this point. But a realistic strategy to resolve these open issues in a timely manner must be part of the technical planning effort presented at the review.

The SRD and the other documents summarized above, as well as the mission and system study reports, the SEMP, the operations concept, and the risk analysis report, are all featured at the SRR. There may be a temptation to simply "recycle" some of the documents, especially the SEMP, at each major review. We should avoid this and make a clean scrub of all these documents to bring them in line with the evolving state of the design and other inevitable project changes. Another document called out at the SRR is the software development plan. Like the others,

TABLE 19-24. **FireSAT Requirements Matrix.** (SV is FireSAT space vehicle; TRL is technology readiness level; PDR is preliminary design review; EEE is electronic, electrical, and electromechanical; OPR is office of primary responsibility; KSC is Kennedy Space Center.)

SRD Section	Requirement	Rationale	Source	Priority	OPR
3.1.3 Prime item definition	The space vehicle will consist of 1) the wildfire detection payload and 2) the spacecraft bus	This is an accepted architecture definition for a spacecraft	FS-10005	A	SE
3.2.1 Spatial resolution	The space vehicle shall have a spatial resolution of less than or equal to 2.14×10^{-4} radians (0.0123°)	This requirement comes from the FireSAT wildfire definition trade study and corresponds to a 150 m resolution at 700 km	FS-TS-004A	A	USFS
3.2.2 Spectral resolution	The space vehicle shall detect wildfire emissions in the wavelength range of 4.2 µm +/– 0.2 µm	This requirement comes from the FireSAT wildfire definition trade study and corresponds to a wildfire temperature of 690K. Optimum wavelength for a fire at 1000 K would be 2.9 µm; however, there's no atmospheric window at that wavelength so 4.2 µm represents a good compromise.	FS-TS-004A	A	USFS
3.2.7 Availability	The space vehicle shall have an operational availability of 98%, excluding outages due to weather, with a maximum continuous outage of no more than 72 hours	The USFS has determined that an average availability value of 98% suffices to meet project goals. Some planned outage is necessary to provide for software updates and station keeping maneuvers. 72 hours is short enough to minimize impact on the mission, especially if the time corresponds to seasons of relatively low wildfire risk.	USFS-TR-086	B	USFS
3.3.1 Space vehicle mass	The space vehicle loaded mass shall not exceed 250 kg	This is a maximum mass budget based on launching two satellites on a single Pegasus launch into a parking orbit of 150 km at 55° inclination.	GSFC-TIM-043	B	SE
3.3.2 Launch environment	The space vehicle shall meet its requirements after exposure to the induced launch environment as defined in the Pegasus User's Handbook.	The induced environment during launch is typically the most severe from a mechanical loads standpoint. The SV must be able to function after being subjected to this environment.	GSFC-TIM-021	A	SE

TABLE 19-24.   FireSAT Requirements Matrix. (Continued) (SV is FireSAT space vehicle; TRL is technology readiness level; PDR is preliminary design review; EEE is electronic, electrical, and electromechanical; OPR is office of primary responsibility; KSC is Kennedy Space Center.)

SRD Section	Requirement	Rationale	Source	Priority	OPR
3.3.2.1 Space vehicle natural frequency	The space vehicle first-mode natural frequency shall be greater than 20 Hz	This constraint is based on the launch vehicle induced environment. The SV natural frequency must be above the induced environment frequency to avoid resonance problems.	FS-8×030	A	SE
3.3.4 On-orbit design lifetime	The space vehicle shall be designed for an operational on-orbit lifetime of 5 years (threshold), 7 years (objective)	The USFS has determined that a minimum 5-year design life is technically feasible. Seven years is a design objective.	USFS-TR-054	B	SE
3.3.8 Space environment	The SV shall meet its requirements after exposure to the natural space environment at 700 km mean altitude, 55° inclination as defined in the NASA Space Environment Handbook	Spacecraft must be designed and constructed to operate in the anticipated space environment to successfully fulfill its mission	GSFC-TIM-021	A	SE
3.4.1 Space environment effects on material selection	Thorough evaluation of the environmental effects of the trajectory paths and orbits shall be assessed for the impact on space vehicle materials selection and design	Understanding the trajectory and orbital environmental effects (e.g., electrostatic discharge, radiation, atomic oxygen) on the spacecraft will eliminate costly redesign and fixes, and minimize on-orbit failures due to environmental interaction with spacecraft materials	GSFC–STD –1000	A	GSFC
3.4.6 Maturity of new technologies	All space vehicle technologies shall achieve a TRL 6 by PDR. This doesn't apply to technology demonstration opportunities.	The use of new and unproven technologies requires a thorough qualification program to reduce risk to an acceptable level	GSFC–STD –1000	A	GSFC
3.5.1 Supply	Project shall define a plan for required spare units (including spare EEE parts) that is compatible with available resources and acceptable risk	An inadequate spare parts program leads to part shortages during the development phase and has a direct impact on potential workarounds or retrofit plans	GSFC–STD –1000	B	GSFC

TABLE 19-24. FireSAT Requirements Matrix. (Continued) (SV is FireSAT space vehicle; TRL is technology readiness level; PDR is preliminary design review; EEE is electronic, electrical, and electromechanical; OPR is office of primary responsibility; KSC is Kennedy Space Center.)

SRD Section	Requirement	Rationale	Source	Priority	OPR
3.6.1 Maintenance personnel	The space vehicle shall be maintainable by a Class 2 - engineering technician as defined in KSC WP-3556, Technician Qualifications Manual	This is the usual level of training for KSC personnel	LV-TIM-003	C	LV
3.7 Subsystem requirements	Addressed in next section				
3.8.1 Precedence	When in conflict, FireSAT project-level requirements shall take precedence over system-level requirements	All project-level requirements are overriding	Project manager	B	PM
3.9.1 Qualification of heritage flight hardware	All space vehicle heritage flight hardware shall be fully qualified and verified for use in its new application. This qualification shall take into consideration necessary design modifications, changes to expected environments, and differences in operational use.	All hardware, whether heritage or not, needs to be qualified for its expected environment and operational uses	GSFC–STD –1000	B	GSFC
3.9.3 Structural qualification	Structural tests that demonstrate that SV flight hardware is compatible with expected mission environments shall be conducted in compliance with the NASA/GSFC General Environmental Verification Specifications (GEVS-SE Rev A 1996)	Demonstration of structural requirements is a key risk reduction activity during mission development	GSFC–STD –1000	A	GSFC
3.10.1 Standard sample	Two flight model space vehicles shall be delivered	Architecture analysis indicates a two- satellite constellation is needed to perform the mission	Project management	A	PM

it may have been drafted during Pre-phase A. It's another useful tool for laying out the system software effort. Because software issues come to a head during build up to the SDR, we defer their discussion until the next section.

19.3.5   Representative Issues, Items, and Anecdotes

In building up to the FireSAT system readiness review, the team is feeling pretty good about most of the nuts and bolts technical aspects of the project. But they still face some nagging issues related to technical planning and project cost modeling. These issues focus principally on several sections of the systems engineering management plan. The SEMP difficulties arise from misunderstandings between the various organizations involved in the project. These organizations and their primary responsibilities are summarized in the Section 3.3 of the SEMP—Product Integration—as shown in Table 19-25.

TABLE 19-25.   Initial Allocation of Organization Roles and Responsibilities. Culled from the original list of FireSAT stakeholders, these four have the most direct roles in the project.

Organization	Location	Roles and Responsibilities
NASA/GSFC	Greenbelt, Maryland	Lead government agency for spacecraft and payload procurement
NOAA	Washington, D.C.	Lead government agency for operations
US Forest Service Fire Fighting Office	Colorado Springs, Colorado	Customer and user
Acme Astronautics Corp.	Harlan, Iowa	Prime contractor for spacecraft and payload

In the project WBS, the project organization was straightforward, as shown in Figure 19-25. Figure 19-26 shows the payload work breakdown structure. The SEMP issue that surfaced at the SRR centered around the sensor development work package, defined as follows:

- 1.05.4.1 Sensor Development: This element includes the equipment, data, services, and facilities required to design, develop, produce, and test through certification and acceptance, a sensor that meets all subsystem level and external interface requirements. Payload sensor development includes subsystem requirements generation and allocation to the component level; design and development of breadboards, engineering, qualification, and acceptance test equipment and facilities; wiring; flight hardware; spares; flight and ground software; and subsystem test and integration.

This work package has been assigned to the prime contractor, Acme Astronautics Corp., with NASA/GSFC as the lead agency. The NASA chief systems

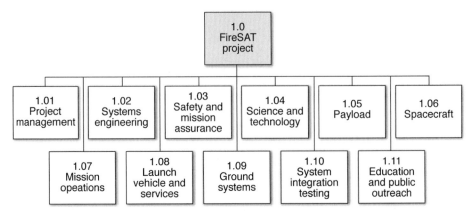

FIGURE 19-25. Project-Level Work Breakdown Structure. The project-level WBS includes both programmatic and technical elements.

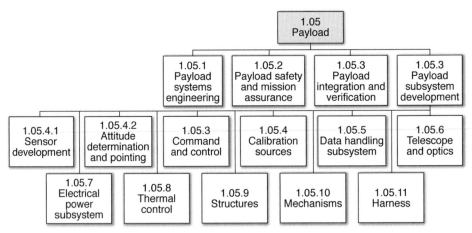

FIGURE 19-26. Payload Work Breakdown Structure. This figure depicts the second-level payload element, all the third-level elements associated with it, and the fourth-level elements under "Payload subsystem development."

engineer for the project, working with the FireSAT team cadre from both NASA and Acme, explores some existing sensor technologies that have a high degree of flight heritage. They want hardware with the highest possible technology readiness level (TRL) to reduce risk. The issue begins when the other project stakeholder, NOAA, formally asked to lead the sensor effort instead. They already have an ongoing development effort for a sensor with nearly identical characteristics for a weather instrument on an NPOESS follow-on mission. Inserting this technology into FireSAT promises to increase its sensitivity and scan agility while removing a large part of the payload R&D cost from the FireSAT budget. NOAA argues this as a win-win situation where the FireSAT project can improve performance and reduce cost, while NPOESS benefits from an early risk-reduction flight of their new instrument.

The FireSAT lead systems engineer at NASA knows that when something sounds too good to be true, it probably is. The downside of this arrangement is to accept significant downstream cost, schedule, and performance risks. The project will incur these risks if the promised hardware isn't delivered on time (a greater risk with lower TRL technology) or if the promised higher performance isn't achieved. The motto "Better is the enemy of good," crosses the minds of the systems engineering team. But with the giddiness and optimism usually present early in a new project (and a fair amount of political influence from the headquarters of both agencies and interested Congressional delegations), they decide to restructure the project to baseline the NOAA-provided instrument. The caveat is that the instrument must be at TRL 6 or above by the preliminary design review. So they redefine the SEMP roles and responsibilities, as shown in Table 19-26.

TABLE 19-26.   Revised Allocation of Organization Roles and Responsibilities Based on NOAA Offer to Lead Payload Development. At this point in the project lifecycle, the systems engineering team judges the benefits of going with the NOAA instrument to outweigh the risks.

Organization	Location	Roles and Responsibilities
NASA/GSFC	Greenbelt, Maryland	Lead government agency for spacecraft and payload procurement
NOAA	Washington, D.C.	Lead government agency for payload development and mission operations
US Forest Service Fire Fighting Office	Colorado Springs, Colorado	Customer and user
Acme Astronautics Corp.	Harlan, Iowa	Prime contractor for spacecraft, integrating contractor for payload

Cognizant of the potential risks of this decision (and smarting from being relegated to the role of integrator rather than lead designer for the payload), Acme Astronautics musters a small amount of internal funding to continue sensor development activities in parallel, albeit at a lower level than previously planned. They make this decision with the tacit support of the NASA engineering team. The NOAA team, while they consider this parallel effort a wasteful distraction, aren't in a position to officially object.

As with any project, especially a relatively small one like FireSAT, cost is always in the forefront. A combination of analogous systems cost models and parametric cost models based on historical data guides project budget estimates throughout Pre-phase A and well into Phase A. The decision to assign project responsibility for payload development to NOAA has a significant impact on the project bottom line. The Small Satellite Cost Model (SSCM) developed by the Aerospace Corp. [Larson and Wertz, 1999], which was used to develop the initial project's cost estimate, indicates fractional lifecycle costs, as shown in Table 19-27.

TABLE 19-27.   **Fractional Spacecraft Costs Based on Small Satellite Cost Model [Larson and Wertz, 1999].** Costs for a space project include far more than what gets launched into space. (TT&C is telemetry, tracking, and control; C&DH is command and data handling; ADCS is attitude determination and control system.)

Subsystem or Activity	Percentage of Spacecraft Bus Cost	Non-Recurring Percentage	Recurring Percentage
Payload	40%	60.0%	40.0%
Spacecraft bus total	100%	60.0%	40.0%
Structure	*18.3%*	*70.0%*	*30.0%*
Thermal	*2.0%*	*50.0%*	*50.0%*
Power	*23.3%*	*62.0%*	*38.0%*
TT&C	*12.6%*	*71.0%*	*29.0%*
C&DH	*17.1%*	*71.0%*	*29.0%*
ADCS	*18.4%*	*37.0%*	*63.0%*
Propulsion	*8.4%*	*50.0%*	*50.0%*
Integration, assembly, and test	13.9%	0.0%	100.0%
Program level	22.9%	50.0%	50.0%
Ground support equipment	6.6%	100.0%	0.0%
Launch and orbital operations support	6.1%	0.0%	100.0%
Total	**189.5%**	**92%**	**97.5%**

As the table shows, the payload represents 40% of the total spacecraft cost. So with a 60/40 split between non-recurring and recurring engineering on the payload, the FireSAT project can potentially save around 24% of the cost of the first spacecraft (60% of 40%) by assigning the payload development responsibility to NOAA. Headquarters budget watchers want to immediately take this much out of the project budget. But the FireSAT project manager argues that since insertion of the NOAA payload is contingent on it being at TRL 6 by the preliminary design review, the budget shouldn't be reduced until the project meets that milestone. Unfortunately, he loses that argument and the funding is removed, with an "every effort" will be made to add back funding later if there is a problem.

19.3.6   Conclusions

Leaving the system requirements review, the project management and systems engineering teams are only slightly battered and bruised. Fortunately, the ground work laid down in Pre-phase A has helped make the transition from project concept to project reality less painful. This is only possible because no stakeholders attempt to significantly redirect the project scope. Building on the Pre-phase A concept, the mission and system-level requirements can be baselined at the SRR, giving the project a sound basis for moving forward to the next major milestone in Phase A: the functional baseline.

19.4   Establish the Functional Baseline

While the SRR represents a significant milestone in the project development, it's only the first lap in a long race to the finish line. The second technical baseline during Phase A, the *functional baseline*, further defines system design details. The goal is to develop the necessary systems engineering and project management evidence that the system is ready to transition to preliminary design in Phase B. In this phase, we continue to decompose the top-level understanding of the system down to the complete specification of the system at the build-to baseline.

At this point it's useful to review what we know about the spacecraft up to and including the system definition review to better understand how we employ the SE processes during this part of the lifecycle. Ideally, the results at SDR represent the logical maturation of design decisions made during Pre-phase A and the consequences of the requirements work baselined at SRR.

During Pre-phase A, we took a comprehensive look at the overall mission feasibility, including the conceptual design of a spacecraft that could fulfill the basic mission objectives. This was only a **conceptual** design, not a detailed or even preliminary one. At that level, we had to make many assumptions about subsystem performance, mass and power allocations, interfaces, and a host of other parameters to find a solution that was in the ballpark. Some of these assumptions could lead to requirements to be baselined at SRR, which in turn necessitate the use of specific technologies or kick off full-blown technology development programs. Pre-phase A studies, by their nature, take a big-picture view of the project that includes some aspect of nearly every systems engineering process. The goal is to uncover any hidden show-stoppers that may cause the project to exceed acceptable cost, schedule, performance, or risk boundaries.

As we moved into Phase A, we directed most of our efforts to requirements engineering. Traditional systems design efforts didn't stop during this part of Phase A (in fact, as we'll see, an entire mini-design and analysis cycle was completed before SRR). But the focus of these design and analysis efforts was in support of requirements engineering.

Following SRR, during the latter part of Phase A, attention shifts to using the baseline requirements as a starting point for more detailed functional analysis, and physical architecture definition to lay the groundwork for preliminary design in Phase B. One way to think about systems engineering is to divide it into four domains: 1) the requirements domain, 2) the functional domain, 3) the physical domain, and 4) the verification and validation domain, as illustrated in Figure 19-27.

Of course, all these domains are highly coupled, but their emphasis does change throughout the project lifecycle. Leading up to SRR, our focus was on the system requirements domain. But in the run up to the system definition review (SDR), we concentrate on the functional domain. We want to drive out uncertainty at the system level and start the design flow-down effort. During this period, we undertake major trade studies as part of a second major design and analysis cycle (DAC) to further define system functional behavior as allocated to specific subsystems. During this part of Phase A, we:

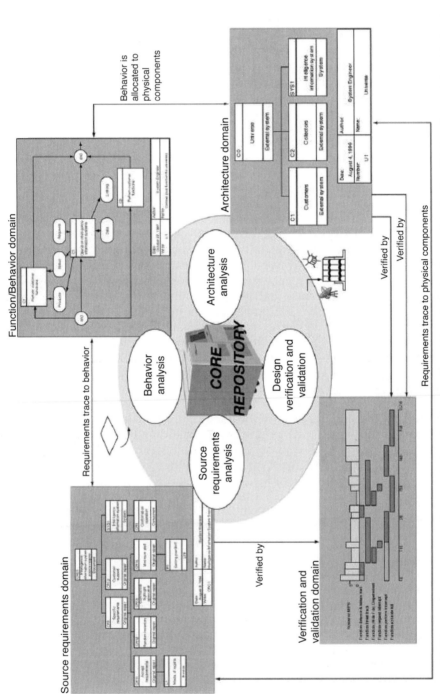

FIGURE 19-27.   Domain Views of Model-based Systems Engineering. Although depicted separately, the four SE domains are interrelated and iterative. Adapted from an image provided by Vitech Corp. Used with permission.

- Develop a functional description of the system down to the subsystem level
- Define system software functional requirements
- Complete major trade studies to determine final configurations and functional distribution
- Update the SEMP with emphasis on technology development, maturity, and assessment planning
- Develop interface control documents to include software
- Update risk assessments and mitigation strategies based on trade study results and project policy
- Update safety and mission assurance plan using improved safety analysis information

19.4.1   What Got Us Here?

To begin work on the functional baseline, the project first has to complete a successful system requirements review, which

- Defines and validates high level systems requirements
- Specifies external system interfaces
- Establishes plans for controlling requirements and risks
- Identifies system implementation approaches

The SRR has established the initial definition of FireSAT system requirements. In preparation for SDR, we must subsequently refine our understanding of those requirements through functional analysis and physical decomposition.

19.4.2   The End Game—SDR Products

By the end of the system definition review, the systems engineering tools necessary for design flow-down are established, the system functions are identified and distributed to the subsystem level, and our knowledge of the risks, trades, and technology developments are matured. Table 19-28 gives the entrance and success criteria for a system definition review. Some projects also complete a mission design review before or instead of the SDR; we can tailor either or both reviews to the project. We consider only the SDR here because there's a lot of overlap between the two.

19.4.3   SDR Products

We now translate these criteria into tangible deliverable products, as shown in Table 19-29. The new products identified in this table show that many of the exit criteria are satisfied by the further development of the system architecture and the expansion of the functional baseline to the subsystem level. Table 19-30 lists the contents of these documents. As we develop these products, the new information triggers updates in the SEMP; the technology development, maturity, and assessment plan; risk assessment and mitigation planning, etc.

Table **19-28.**   **System Definition Review Entrance and Success Criteria.** The functional domain receives special emphasis during this part of the lifecycle. Adapted from NASA [2007].

System Definition Review	
Entrance Criteria	**Success Criteria**

1. Complete the SRR and respond to all requests for action and review item discrepancies
2. The technical team, project manager, and review chair agree to a preliminary SDR agenda, success criteria, and charge to the board
3. Make SDR technical products listed below available to the relevant participants in advance:
  a. System architecture
  b. Preferred system solution definition, including major tradeoffs and options
  c. Updated baselined documentation, as required
  d. Preliminary functional baseline (with supporting trade-off analyses and data)
  e. Preliminary system software functional requirements
  f. Changes to the systems engineering management plan, if any
  g. Updated risk management plan
  h. Updated risk assessment and mitigations (including probabilistic risk analyses, as applicable)
  i. Updated technology development, maturity, and assessment plan
  j. Updated cost and schedule data
  k. Updated logistics documentation
  l. Updated human rating plan, if applicable
  m. Software test plan
  n. Software requirements documents
  o. Interface requirements documents (including software)
  p. Technical resource use estimates and margins
  q. Updated safety and mission assurance plan
  r. Updated preliminary safety analysis
	1. Systems requirements, including mission success criteria and any sponsor-imposed constraints, are defined and form the basis for the proposed conceptual design
	2. All technical requirements are allocated and the flow down to subsystems is adequate. The requirements, design approaches, and conceptual design will fulfill the mission needs consistent with the available resources (cost, schedule, mass, and power).
	3. The requirements process is sound and can reasonably be expected to continue to identify and flow detailed requirements in a manner timely for development
	4. The technical approach is credible and responsive to the identified requirements
	5. Technical plans have been updated, as necessary
	6. The tradeoffs are complete, and those planned for Phase B adequately address the option space
	7. Significant development, mission, and safety risks are identified and technically assessed, and we have a process and the resources to manage the risks
	8. We've planned adequately for the development of any enabling new technology
	9. The operations concept is consistent with proposed design concepts and aligns with the mission requirements

19.4.4   Process Level of Effort

Table 19-31 illustrates the relative level of effort for each of the 17 SE processes needed to produce the key deliverables during this latter part of Phase A. The main difference between this and pre-SRR is greater focus on logical decomposition and physical solution activities.

19.4.5   Results

For this discussion, the system of interest at the system definition review is the FireSAT space element. Here we focus on a few of the principal documents that define it as the project nears the end of Phase A: 1) the spacecraft system design study report, 2) the software development plan, and 3) key sections of the systems engineering management plan (SEMP).

TABLE 19-29.   **Supporting Documents for System Definition Review Success Criteria.** This table expands on the second column of Table 19-28 above. These documents are essential for fulfilling their associated criteria.

SDR Success Criterion	Supporting Documents
1. System requirements, including mission success criteria and any sponsor-imposed constraints, are defined and form the basis for the proposed conceptual design	• System requirements document • System design study reports
2. All technical requirements are allocated and the flow down to subsystems is adequate. The requirements, design approaches, and conceptual design will fulfill the mission needs consistent with available resources (cost, schedule, mass, and power).	• System design study reports • Mission and system-level requirements documents
3. The requirements process is sound and can reasonably be expected to continue to identify and flow detailed requirements in a manner timely for development	• Updated SEMP • Software management plan
4. The technical approach is credible and responsive to the identified requirements	• System design reports • Updated SEMP • Software management plan
5. Technical plans have been updated, as necessary	• Updated SEMP
6. The tradeoffs are complete and those planned for Phase B adequately address the option space	• System design study reports • Updated SEMP (containing DAC plans)
7. Significant development, mission, and safety risks are identified and technically assessed, and a process and resources exist to manage the risks	• Technical risk reports
8. We've planned adequately for the development of any enabling new technology	• Updated SEMP (containing technology development plan)
9. The operations concept is consistent with proposed design concepts and aligns with the mission requirements	• System design reports • Operations concept document

FireSAT Spacecraft System Design Study Report

The FireSAT spacecraft system design study report is the document that captures the numerous design decisions and rationale that encompass the baseline system definition. (Other projects or programs split out some of these results into other configuration-controlled documents). We presented many of the conceptual design results for the spacecraft as part of the Pre-phase A discussion. Here we concentrate on three important results to highlight at the SDR: 1) the spacecraft functional analysis, 2) the spacecraft physical architecture (with functional allocations), and 3) the results of system margin analysis.

Chapter 5 presents various techniques for approaching system functional analysis. One way to begin this analysis is to examine system inputs and outputs.

TABLE 19-30.   Contents for Primary System Definition Review (SDR) Supporting Documents. Here we show the most important documents listed above, further specifying them for the FireSAT example. (TBD is to be determined; TBR is to be resolved; SEMP is systems engineering management plan; PDR is preliminary design review.)

Product	Contents	Product	Contents
FireSAT system design study report (for each system element within the mission architecture as appropriate) (BASELINE)	• Analysis of alternatives trade-tree, including (for each system in the architecture) spacecraft, ground system, etc. • Alternative system concepts • Identification of system drivers for each concept or architecture • Results of characterization for each alternative • Critical requirements identified or derived • Concept or architecture utility assessment, including evaluation criteria such as cost, technical maturity, and measures of effectiveness • Detailed functional analysis • Proposed baseline system configuration with functional allocation	**SEMP (UPDATE)**	• Plan for managing the SE effort • Documents driving SEMP • Technical summary • Integration technical effort (including conceptual lifecycle support strategies, logistics, manufacturing, operation, etc.) • The 17 systems engineering processes • Technology insertion • SE activities • Project plan integration • Waivers • Appendices

Risk management plan (UPDATE)	• Preliminary risk assessment
	 • Technologies and associated risk management and mitigation strategies and options
		FireSAT mission-level and systems-level system requirements documents (SRD) (UPDATE)	• Scope
			 • Applicable documents
			 • Requirements
			   – Prime item definition
			   – System capabilities
			   – System characteristics
			   – Design and construction
			   – Logistics
			   – Personnel and training
			   – Subsystem requirements
			   – Precedence
			   – Qualification
			   – Standard sample
			 • Verification requirements
			 • Notes
			 • Appendices

TABLE 19-30.   Contents for Primary System Definition Review (SDR) Supporting Documents. (Continued) Here we show the most important documents listed above, further specifying them for the FireSAT example. (TBD is to be determined; TBR is to be resolved; SEMP is systems engineering management plan; PDR is preliminary design review.)

Product	Contents	Product	Contents
Software development plan (BASELINE)	See discussion below	**Operations concept document (UPDATE)**	• Operational scenarios • Timelines • Data flow diagram • Organization and team responsibilities • Cost and complexity drivers for a given set of inputs • Requirements and derived requirements • Technology development plan
Plan to achieve PDR readiness	• Rank ordered list of studies and trades-offs • Forward work mapped to program schedule • Technology gap closure plan mapped to program schedule • Requirements update and management plan • TBD and TBR burn-down plan • SRR issue resolution plan		

TABLE 19-31. **Systems Engineering Process Level of Effort Leading up to System Definition Review (SDR).** Analogous to a cell phone signal, zero bars represents no or very low resource requirements and five bars represent maximum resource utilization. Resource requirements are for a particular process only; a given number of bars for one process does not imply the same level of resources as that number of bars for a different process. (SRR is system requirements review.)

Process	Level of Effort	Comments
1. Stakeholder expectation definition	■■■□□	We continually revisit stakeholder expectations as we develop technical requirements. This serves as a requirements validation.
2. Technical requirements definition	■■■■■	We refine requirements to lower levels as a result of functional analysis
3. Logical decomposition	■■■■■	This is a main focus for preparation for SDR
4. Physical solution	■■■□□	As the logical decomposition matures, allocation of functionality drives the definition of the physical solution. We complete additional trade studies leading up to SDR. The logical decomposition leads the physical solution design activities.
5. Product implementation	■□□□□	Identifying existing assets and the assessment of the cost-benefit and risk-reduction aspects of existing systems are standard trades during this phase. We need to carefully consider requirements for long-lead procurement items.
6. Product integration	■□□□□	During the logical decomposition process, we consider all lifecycle phases for each system element. We identify stakeholders and coordinate with them to be sure that we recognize all requirements and interfaces (such as to existing and planned facilities).
7. Product verification	■■□□□	Verification planning is key to assuring that requirements defined during this phase are testable and that cost and schedule estimates account for this critical phase. Since the physical solution isn't fully defined in this phase, we can't completely fix the verification program. But it's essential to begin to identify long-lead requirements such as test facilities and ground support equipment.
8. Product validation	■■□□□	Validation planning is matured in this phase
9. Product transition	□□□□□	Only minimum effort expended on this process, mainly to identify enabling products for transition that may affect lifecycle cost
10. Technical planning	■■■■■	These processes must be fully defined to complete SRR successfully, as they determine how we execute the program in the next phase
11. Requirements management	■■■□□	The requirements management process must be mature and in place to control the emerging set of requirements
12. Interface management	■□□□□	Interface management is addressed in the requirements management processes

TABLE 19-31. **Systems Engineering Process Level of Effort Leading up to System Definition Review (SDR). (Continued)** Analogous to a cell phone signal, zero bars represents no or very low resource requirements and five bars represent maximum resource utilization. Resource requirements are for a particular process only; a given number of bars for one process does not imply the same level of resources as that number of bars for a different process. (SRR is system requirements review.)

Process	Level of Effort	Comments
13. Technical risk management	■■☐☐☐	Risk planning and management remain at a vigorous level during this phase
14. Configuration management	■☐☐☐☐	End item control and requirements management controls must be in place
15. Technical data management	■■☐☐☐	Tools and repositories must be fully deployed along with the processes to guide engineering teams in concurrently defining systems
16. Technical assessment	■■☐☐☐	Initial planning to support the upcoming SRR and subsequent reviews begins
17. Decision analysis	■■■■■	Rigorous technical decision analysis is crucial to developing credible Phase A results. It's impossible to complete all the myriad possible trade studies, so good decision analysis identifies the most critical ones.

For that, we construct a simple spacecraft-level context diagram, as shown in Figure 19-28.

FIGURE 19-28. **FireSAT Space Element Context Diagram.** This simple system context diagram for the FireSAT space element is a starting point for functional analysis based on system inputs and outputs.

Building on this simple context, we develop a functional architecture based on inputs and outputs, as shown in Figure 19-29. The hierarchical view is a good starting point, but to truly understand functional relationships, we must follow the various "threads" that weave their way through the system. For example, to better understand how the all-important wildfire "911" notifications are generated, we follow that thread from beginning to end through each of the sub-functions, as shown in Figure 19-30.

Following all the major threads first identified by our system context diagram, and discerning all the external and internal "triggers" for each function, we then develop an enhanced functional flow block diagram (EFFBD), as shown in Figure 19-31. This provides a variety of insights into the FireSAT spacecraft functions. To begin with, it illustrates the sequential relationships among the functions

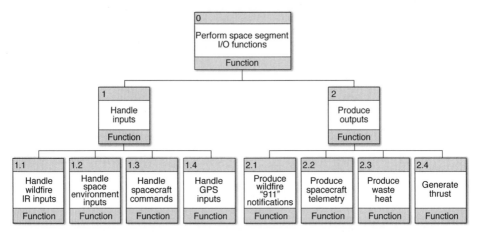

FIGURE 19-29.   Space Element Functional Decomposition Using Inputs and Outputs. This view of the FireSAT spacecraft functional architecture focuses on system inputs and outputs. (I/O is input/output; IR is infrared; GPS is Global Positioning System.)

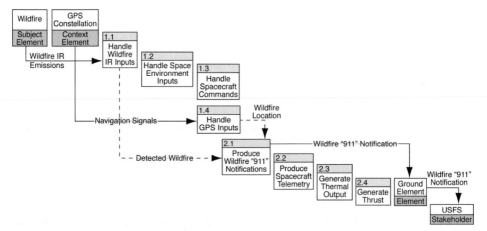

FIGURE 19-30.   Functional Flow Block Diagram of the Wildfire "911" Thread. This functional flow diagram illustrates the roles played by the elements that constitute the FireSAT spacecraft functional architecture in executing the wildfire "911" notifications. We can develop similar diagrams for all of the functional threads performed by the system. Here links external to the spacecraft are shown as solid lines. Internal links are shown as dashed lines. (GPS is Global Positioning System; IR is infrared; USFS is US Forest Service.)

identified by the I/O functional hierarchy. The input functions occur in parallel with one another, as do the output functions. But input and output functions occur in series as part of an ongoing loop of space operations. And the triggers that go into and out of each sub-function provide insight as to how the inputs are handled and how important outputs are generated.

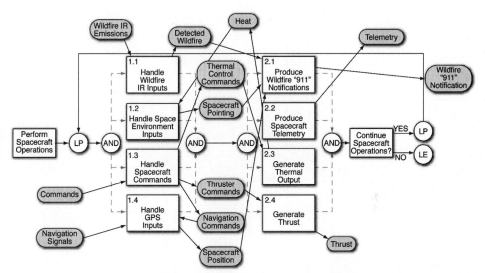

FIGURE 19-31.   FireSAT Enhanced Functional Flow Block Diagram (EFFBD). The EFFBD provides a variety of insights into the FireSAT spacecraft functions. (LP is loop; LE is loop end; IR is infrared; GPS is Global Positioning System.)

Let's take a more detailed look at the critical function of producing the "911" messages. We consider what such a message must contain: detection of a fire **and** its location and time. So the "911" message has two components, the detection portion and the geolocation/timing function. Notionally, the first could be allocated to the payload while the second could be allocated to the bus. Refining these two functions still further, as shown in Figure 19-32, we see the additional complexity needed to perform each one.

This understanding is important when we look further at how best to allocate these functions to the correct physical piece of the system (many functions are performed in software). Turning from the functional architecture to the physical, we can start with a very simple, top-level decomposition of the spacecraft into bus and payload, as illustrated in Figure 19-33.

Now, our goal is to allocate functions to the spacecraft physical architecture. Looking at the collection of spacecraft functions, some logical groupings emerge. Handling wildfire IR inputs is logically the function of the spacecraft payload. Handling environmental inputs, commands, and GPS signals traditionally goes to the spacecraft bus. On the output side, producing the "911" notifications is handled jointly by the payload and bus. This decomposition and allocation task is aided by analyzing the necessary functional behavior of the system in increasing detail, and then assigning these sub-functions to various parts of the system. By

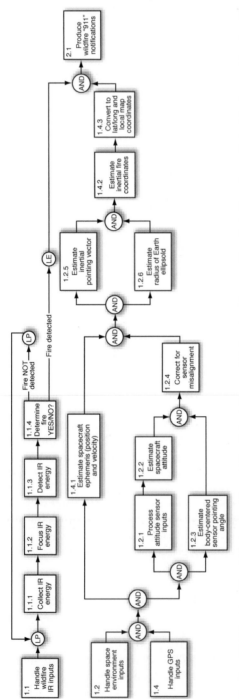

FIGURE 19-32. **Functional Decomposition of the Detect Fire and Determine Geo-position Data Functions.** Both detection and location (in space and time) are critical to notification. (IR is infrared; GPS is Global Positioning System.)

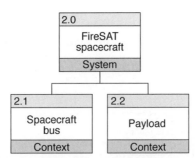

FIGURE 19-33.   FireSAT Spacecraft Top-level Physical Architecture. We traditionally divide spacecraft into two parts, bus and payload. The payload deals with all purely mission functions (e.g., generate mission data), while the bus takes care of all the "overhead" (power, thermal control, etc.).

grouping like functions into subsystems, we capture a rationale for each subsystem, and identify its scope as well. Engineers tend to know the subsystems of a space system *a priori,* and to discount the functional decomposition. This lack of functional insight is particularly dangerous for the software development, as functions that are necessarily handled by software may not be fully identified until much later in the system development, adding cost and schedule risk. Figure 19-34 depicts the more detailed FireSAT physical architecture.

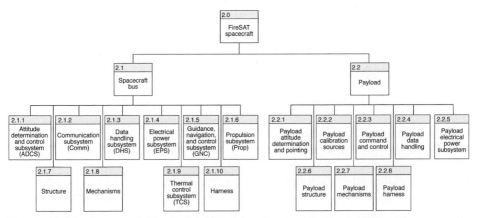

FIGURE 19-34.   Spacecraft Physical Architecture, Bus, and Payload. Here we decompose the FireSAT spacecraft into the subsystems that constitute the bus and payload.

We can see the allocation of the functional architecture to the physical system by substituting our physical system for the functions described earlier and looking again at the inputs and outputs—but this time from a subsystem perspective, as illustrated in Figure 19-35. This helps us identify the links between subsystems that we will rigorously define as the system-to-system and subsystem-to-subsystem interfaces. With the detailed analysis complete, we examine the relationships between requirements, functions, and physical allocations as shown in Figure 19-36, using the example of the geo-location accuracy requirement.

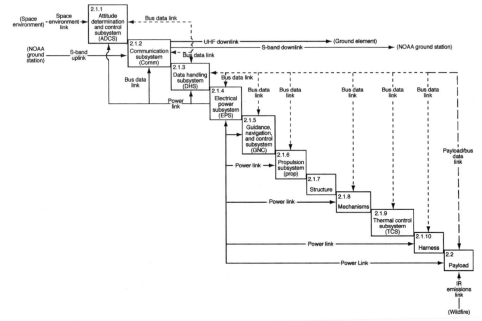

FIGURE 19-35.   Allocated Functions. This diagram illustrates how the subsystems that constitute the spacecraft bus and payload handle the inputs and outputs that the functional analysis identifies. This way, we can see the allocation of functional to physical. The links between the system and external elements, as well as the internal links, form the basis for more rigorous interface descriptions. Here links external to the spacecraft are shown as solid lines. Internal links are shown as dashed lines.

The final important set of results in the system design report presented at the SDR is the margins associated with key performance parameters (KPPs) and technical performance measures (TPMs). Let's look at an example of each to see how to compute and track them, starting with the KPP for geolocation accuracy. As stated above, satellite remote sensing geolocation accuracy is a function of:

- Satellite position knowledge (in 3 axes)
- Satellite pointing knowledge (in 2 axes)
- Satellite timing accuracy (how good the onboard clock is)
- Inherent altitude error for ground targets (how good our 3-dimensional map of the Earth's surface is)

To meet the 5 km requirement, we've made several assumptions, some of which have become derived requirements, for all of these parameters. There's a

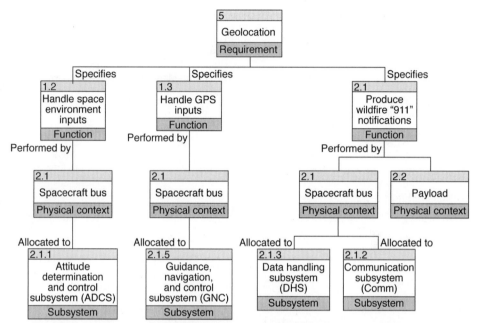

FIGURE 19-36. Relationship Among Requirements, Functions, and Physical Allocations. Once we determine the physical context for a function, we further allocate it to a specific subsystem within that context.

straightforward set of analytic equations that relate errors in each of these parameters to total geo-location error. Table 19-32 summarizes these results as part of the analysis in the system design report. It shows that the best estimate indicates a little over 1 km margin in this KPP, assuming that each of the contributing error sources can be kept at or below its assumed value.

The system design report captures similar results for technical performance measures (TPMs) such as mass, power, and link margin. During conceptual design, the payload size was estimated using parametric scaling rules for optical instruments [Larson and Wertz, 1999]. The spacecraft mass was then computed based on similar scaling rules for spacecraft of comparable size. This also provided estimates for allocated mass and power for each subsystem. All calculations included 25% mass, 30% power, and 15% propellant margin. Table 19-33 summarizes these conceptual design results.

Of course, these results beg the question of how much margin is enough at this point in the design cycle. While there are several schools of thought in this area, the GSFC "Gold Standards" [GSFC, 2005] provides guidelines. Per this directive, the resource margins are found to be adequate for the end of Phase A, as shown in Table 19-34.

All these results further refine the system and the derived subsystem requirements. So another important document update presented at the system

TABLE 19-32. Results of Analytic Calculations Indicating Estimated Geo-position Error. These results are based on analytic relationships between error sources and overall geo-position error. As of the SDR, the estimated geo-position error is not quite 4 km. This leaves a bit more than 1 km margin, assuming that we can keep the errors in each of the contributing parameters at or below these values.

Error Sources	Error Budget	Resulting Mapping Errors (km)	Resulting Pointing Errors (degrees)
Spacecraft Position Errors			
Along-track	0.2 km	0.18	0.007
Cross-track	0.2 km	0.18	0.007
Radial	0.1 km	0.25	0.003
Orientation Errors			
Azimuth	0.06 deg	1.405	0.051
Nadir angle	0.03 deg	2.425	0.03
Other Errors			
Target altitude	1 km	2.747	N/A
Spacecraft clock	0.5 sec	0.232	0.008
Total Error (RSS)		**3.947**	**0.061**

definition review is the system requirements document (SRD). Some projects include the subsystem requirements in a separate SRD. For a relatively small system such as FireSAT, it makes sense to minimize controlled documents and include these allocated requirements in the spacecraft-level SRD. Table 19-35 shows some of these additional requirements.

In parallel with capturing the system-level requirements, the team is also keenly aware of the project's software challenges. The next section discusses how they tackle these.

FireSAT Software Development Plan

The software development plan, the software test plan, and the software requirements document all mature during Phase A. At this point in the project, we have just started to identify the software requirements. Software risks, development costs, and development timelines are just becoming available. This is one reason that software development tends to lag system development.

This happens for FireSAT as well. While most of the results at the SRR are very well received, the team scores low marks for software planning. The project decision authority warns them to "get their software act together" by SDR or face serious consequences ("heads will roll"). The team starts to organize the effort by writing a software development plan (SDP), using the SDP template shown in Table 19-36.

TABLE 19-33.   **FireSAT Conceptual Design Results.** These indicate mass, power, and propellant margins and initial allocated mass and power for each subsystem. (IR is infrared; GPS is Global Positioning System.)

Design Inputs and Assumptions		Component and Results		Mass (kg)	Average Power (W)
Payload mass	28.1 kg	Payload		28.1	31.6
Percentage of spacecraft mass taken up by payload	20%	Spacecraft subsystems		112.5	73.8
Payload power	31.6 W		ADCS	12.5	14.0
Percentage of spacecraft power used by payload	30%		C&DH	6.4	5.8
			Power	43.4	27.3
Spacecraft mass margin	25%		Propulsion	5.8	4.6
Spacecraft power margin	30%		Structure	33.8	0
			Thermal	5.3	4.6
Orbit delta-V budget	513 m/s		TT&C (communications)	5.3	17.5
Attitude control percentage	5%	Margin		35.2	31.6
Propellant margin	15%				
Propellant residual	2%	Spacecraft dry mass		175.8	
Propulsion subsystem specific impulse	210 s	Propellant		61.3	
			Orbital maneuvers	49.8	
			Attitude control	2.5	
			Margin	7.8	
			Residual	1.2	
		Total spacecraft loaded mass		237.1	
		Spacecraft average power			137.0

To illustrate the importance of software planning, let's focus on a single critical software function. From our earlier functional analysis, we decomposed the system-level function "Detect Fires" into several sub-functions, as shown again in Figure 19-37. One critical sub-function is "Determine Fire Yes/No?".

To meet the stakeholders' need for rapid detection and notification, engineers decide to allocate this sub-function to software. But which software? Should the payload software make this call or should the raw data be sent to the bus to be

TABLE 19-34.   Technical Resource Margins. All values are given at the end of the phase [GSFC, 2005].

Resource	Pre-phase A	Phase A	Phase B	Phase C	Phase D
Mass	≥30%	≥25%	≥20%	≥15%	0
Power (with respect to end-of-life capability)	≥30%	≥25%	≥15%	≥15%	≥10%*
Propellant		3σ†			3σ
Telemetry and command hardware channels**	≥25%	≥20%	≥15%	≥10%	0
RF link	3dB	3dB	3dB	3dB	3dB

Margin (in percent) = available resource − estimated value of resource/(estimated resource) × 100

At launch, there shall be 10% predicted power margin for mission-critical, cruise, and safing modes, and to accommodate in-flight operational uncertainties
† The three-sigma variation is due to the following: 1) worst-case spacecraft mass properties; 2) 3σ low performance of the launch vehicle; 3) 3σ low performance of the propulsion subsystem (thruster performance or alignment, propellant residuals); 4) 3σ flight dynamics errors and constraints; 5) thruster failure (applies only to single-fault-tolerant systems)
** Telemetry and command hardware channels read data from such hardware as thermistors, heaters, switches, motors, etc.

analyzed and assessed for the critical "Yes/No" fire detection decision? The choice is further complicated by the earlier programmatic decision to have NOAA provide the payload separate from NASA and the prime contractor, Acme. But any potential political squabbling over this decision is trumped by the simple technical limitation of transferring the necessary data over the spacecraft data harness at a rate sufficient to meet the requirements. So from a technical standpoint, it's expedient to perform this sub-function within the payload image processing software. We document this rationale in the SDP section 4.2.4—Handling Critical Requirements.

Section 5.2.2 of the SDP establishes requirements for a dedicated software test bed that uses simulated infrared imagery to verify the necessary digital signal processing algorithms to detect fires with high confidence. Sections 5.3 through 5.5 deconstruct the problem from the payload focal plane level through the data capture and analysis software design. At FireSAT's orbital velocity, using detector technology to achieve infrared sensitivity in the 4.2-micrometer range at a resolution of 30–180 m, the sensor data rate could reach 100 megabits per second. This is an enormous volume of data to sift through, leading more than one daunted engineer to speak of needles and US-sized hay stacks.

Fortunately, rather than having to completely reinvent the wheel to solve this problem, the systems engineers and software designers are able to adapt unclassified digital signal processing algorithms that were originally developed for the USAF Defense Support Program to detect enemy missile launches from space. With these pieces of the software plan in place, the systems engineers can move confidently forward to the SDR.

TABLE 19-35.   Example Subsystem Requirements. These go into the updated spacecraft system requirements document at the system definition review.

System Requirements Document (SRD) Section	Requirement	Rationale	Source	Priority	Office of Primary Responsibility (OPR)
3.1.3.1 Payload Definition	The wildfire detection payload will consist of: 1) attitude determination and pointing, 2) command and control, 3) calibration sources, 4) data handling subsystem, 5) telescope and optics, 6) electrical power subsystem, 7) thermal control, 8) structures and mechanisms, and 9) harness	This payload architecture is derived from the functional-to-physical allocation in the conceptual design	FS-10005	B	SE
3.1.3.2 Spacecraft Bus Definition	Spacecraft bus will consist of: 1) attitude determination and control subsystem, 2) guidance, navigation, and control subsystem, 3) communication subsystem, 4) data handling subsystem, 5) flight software, 6) electrical power subsystem, 7) thermal control, 8) structures and mechanisms, 9) propulsion, and 10) harness	This is an accepted subsystem architecture definition for a spacecraft and is consistent with the project's WBS	FS-10005	B	PM
3.7.1.2 Attitude Determination and Control Subsystem	Attitude determination and control subsystem (ADCS) shall determine attitude to within +/– 0.06 degrees in the azimuth axis and within 0.03 degrees in the nadir axis (3 sigma)	This requirement was derived from an analysis of the mission-level geolocation accuracy requirement. It represents an allocated error budget to the ADCS	FS-3045	B	ADCS Eng.

TABLE 19-36.   **Software Development Plan Template.** The FireSAT team uses this template to begin organizing the project software tasks. (CSCI is computer software configuration item; HWCI is hardware configuration item.) Adapted from NWSC [2005].

Section 1. Scope	5.5 Software Requirements Analysis
Section 2. Referenced Documents	5.6 Software Design
Section 3. Overview of Required Work	5.7 Software Implementation and Unit Testing
Section 4. Plans for Performing General Software Development Activities	5.8 Unit Integration and Testing
4.1 Software Development Process	5.9 CSCI Qualification Testing
4.2 General Plans for Software Development	5.10 CSCI/HWCI Integration and Testing
4.2.1 Software Development Methods	5.11 System Qualification Testing
4.2.2 Standards for Software Products	5.12 Preparing for Software Use
4.2.3 Reusable Software Products	5.13 Preparing for Software Transition
4.2.4 Handling of Critical Requirements	5.14 Software Configuration Management
4.2.5 Computer Hardware Resource Use	5.15 Software Product Evaluation
4.2.6 Recording of Rationale	5.16 Software Quality Assurance
4.2.7 Access for Acquirer Review	5.17 Corrective Action
Section 5. Plans for Performing Detailed Software Development Activities	5.18 Joint Technical and Management Reviews
5.1 Project Planning and Oversight	5.19 Other Software Development Activities
5.2 Establishing a Software Development Environment	5.19.1 Risk Management
5.2.1 Software Engineering Environment	5.19.2 Software Management Indicators
5.2.2 Software Test Environment	5.19.3 Security and Privacy
5.2.3 Software Development Library	5.19.4 Subcontractor Management
5.2.4 Software Development Files	5.19.5 Interface With Software Independent Verification and Validation (IV&V) Agents
5.2.5 Non-Deliverable Software	5.19.6 Coordination With Associate Developers
5.3 System Requirements Analysis	5.19.7 Improvement of Project Processes
5.4 System Design	Section 6. Schedules and Activity Network
	Section 7. Project Organization and Resources

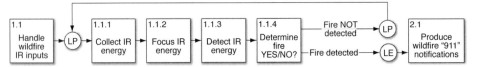

FIGURE 19-37.   **Logical Decomposition of the System-Level "Detect Fires" Function.** Here we see the "Detect Fires" function decomposed into a series of sub-functions. A critical one allocated to software is the "Determine Fire Yes/No?" function. (IR is infrared.)

The Systems Engineering Management Plan

Chapters 13 and 14 summarize key project guidance contained in the SEMP. Here we focus on the technical planning aspects of this document: how to organize and manage the design and analysis cycles (DACs). A DAC logically organizes the technical work to be done on the system design. It synchronizes engineering teams working in geographically separate locations on different but interdependent parts of the problem. Well-planned design and analysis cycles help to better coordinate the tools, assumptions, and results these teams use for major design decisions and future analysis. For example, mechanical engineers doing stress analysis need the latest coupled loads calculations and system configuration information. Analysis done with the incorrect assumptions or erroneous data wastes time and contributes to cost, schedule, and performance risk. The DAC plan for FireSAT, as contained in the project SEMP, is shown in Table 19-37. Along with the simple DAC schedule is a summary of the focus and types of tools for each DAC. This information is essential for resource planning to ensure that the necessary tools are available when needed.

TABLE 19-37. FireSAT Design and Analysis Cycle (DAC) Planning. As the project progresses through its lifecycle, the DACs become increasingly comprehensive. (COTS is commercial-off-the-shelf; STK is Satellite Tool Kit.)

Pre-phase A	M C R	Phase A	S R R	Phase A	S D R	Phase B	P D R	Phase C
DAC 0								
		DAC 1.1						
				DAC 1.2				
						DAC 2.0		
								DAC 3.0

DAC	Focus	Tools
0	• Mission concept and architecture trade studies • Requirements sensitivity analysis • Mission effectiveness	• COTS software tools (e.g., STK, MS Office) • Simple custom analytic tools (e.g., spreadsheets, Matlab)
1.1	• Detailed system trade studies • Requirements analysis	All of the above plus • Custom simulations and CAD based on COTS tools (e.g., Simulink, Pro-E)
1.2	• Functional analysis • System design trade studies • Margin and sensitivity analysis	All of the above plus • COTS SE and PM tools (e.g., Core, MS project)
2	• System and subsystem design • System and subsystem analysis • Stress analysis, interactive behavior analysis	All of the above plus • COTS design and analysis tools (e.g., AutoCAD, Thermal Desk Top, NASTRAN)
3	• Component and part-level design • Manufacturing analysis • Integrated system behavior modeling	All of the above plus • Custom simulations • Bread boards, component emulators • Software test beds

19.4.6   Representative Issues, Items, and Anecdotes

During the system definition review, the development team senses that the project is well-scoped, with practical and achievable requirements, given the project's budget and schedule. The trade studies have identified some risk items to put on the watch list, and actions are assigned to mitigate those risks. Among them is the interface to the NOAA ground stations. One of the FireSAT system requirements is that the satellite be able to communicate with the NOAA ground stations.

In allocating the reliability requirement to the telemetry, tracking, and commanding subsystem, and with an eye toward reducing lifecycle operations costs, the lead systems engineer decides to make this a largely autonomous function within the spacecraft bus software. Unfortunately, the NOAA ground stations require a physical initiation of request by the operator before the spacecraft-to-ground station downlink operation. Thus, the FireSAT requirements are incompatible with the ground station interface requirement. FireSAT can either operate autonomously, or it can meet the requirement to use the NOAA ground stations as is.

Since the matter primarily affects NOAA, a NOAA-led working group is assigned to resolve this conflict. Neither choice impacts the system hardware development. However, because the external interface would change, this further delays software development. The NOAA team also finds a significant disconnect in the payload development. Because they are treating the FireSAT payload as a risk reduction flight for a future NOAA project, they've been using their own work breakdown structure, logical decomposition, and software development plan. As a result, their technical products don't align with the technical requirements for the rest of the FireSAT project. The payload development doesn't break out the payload functions to the level where software, firmware, and hardware allocations can be tracked and managed. This leads to some uncertainty in capturing the functions that FireSAT has to allocate to the payload.

The concern is that some "glueware" will be needed to patch the payload to the FireSAT interface once integration starts. For this reason, the FireSAT project manager flags this interface as a high risk item to be tracked. As part of the risk mitigation, the configuration control effort on the payload-to-bus interface is elevated and weekly technical tag-ups between the bus and payload teams are mandated. Neither team welcomes additional meetings and reports, as they're already laboring to stay on schedule. But they recognize the reasons behind the decision, and so redouble their efforts.

19.4.7   Conclusions

The adrenaline rush leading up to the SDR fades as the team celebrates another successful review, and systems engineers and project managers reflect on what they've accomplished and what work remains. Enormous efforts during Pre-phase A have served the project well in Phase A by completing a rigorous analysis of alternatives to identify a solid technical solution. Early in Phase A, this enabled engineers to focus less on redesign and more on analysis to confirm or modify basic assumptions to derive a valid set of requirements baselined at SRR. And the rigorous functional analysis performed during the DAC following the SRR helped define the system architecture and allocate critical functions to hardware and software. As the team prepares to enter Phase B, all these results, along with the collective skills of the entire team, will be needed to complete the system preliminary design leading to the preliminary design review (PDR).

19.5   Reach the Design-to Baseline

By the time a project starts Phase B, all the important players are in place and the tempo is fast. From a nuts and bolts engineering perspective, this phase is where all the main mission and system trade-offs are analyzed and resolved. Major changes to system design decisions after this phase get incredibly painful and expensive. The end of Phase B, months or years from the start, is a preliminary design review (PDR) that establishes the project "design-to" baseline, giving design engineers license to start doing production drawings and other detail work. (However, some long-lead items may need to reach critical design well in advance of PDR to ensure they can be produced in time to meet the program schedule.)

The differences between preliminary design and detailed design are sometimes in the eye of the beholder. To make an informed preliminary design decision, detailed analyses or even complete baselining of a single design element, down to the part level, is often necessary. For example, the size of the FireSAT solar arrays affects the overall system configuration, and is largely fixed by the efficiency of the solar cells making up the arrays. So during preliminary design, the detailed specification of the cell efficiency (possibly the exact cell and part number, given the limited number of space-qualified suppliers) needs to be baselined to allow more top-level system configuration issues to proceed.

19.5.1   What Got Us Here?

In Pre-phase A we established the major characteristics of the mission. After completing both the SRR and SDR in Phase A, the project now has both a functional baseline for how the system is to behave and a system baseline that describes the architecture and the interfaces between architecture elements. In Phase B, we further define the characteristics of each element of the system—and at least one level down, sufficient to allow designers to start working internal interfaces and associated analyses.

19.5.2   The End Game—PDR Products

The PDR inputs and success criteria are shown in Table 19-38. We now translate these criteria into tangible deliverable products, as shown in Table 19-39. The new products that this table identifies show that many of the exit criteria are satisfied by the further development of the system architecture and the expansion of the functional baseline to the subsystem level. As these products are developed, this new information triggers updates in the SEMP; the technology development, maturity, and assessment plan; risk assessment and mitigation planning, etc. More detailed contents of some of these documents are shown in Table 19-40.

TABLE 19-38.   Preliminary Design Review (PDR) Entrance and Success Criteria. (SRR is system requirements review; SDR is system definition review; MDR is mission definition review; RID is review item discrepancy; TPM is technical performance measure; RFA is request for action; EEE is electronic, electrical, and electromechanical; EMI is electromagnetic interference; PRA is probability risk analysis; EMC is electromagnetic compatibility; S&MA is safety and mission assurance.) Adapted from NASA [2007].

Preliminary Design Review

Entrance Criteria

1. Project has successfully completed the SDR and SRR or MDR and responded to all RFAs and RIDs, or has a timely closure plan for those remaining open
2. A preliminary PDR agenda, success criteria, and charge to the board have been agreed to by the technical team, project manager, and review chair before the PDR
3. PDR technical products listed below for both hardware and software system elements have been made available to the cognizant participants before the review:
   a. Updated baselined documentation, as required
   b. Preliminary subsystem design specifications for each configuration item (hardware and software), with supporting trade-off analyses and data, as required. The preliminary software design specification should include a completed definition of the software architecture and a preliminary database design description, as applicable.
   c. Updated technology development maturity assessment plan
   d. Updated risk assessment and mitigation
   e. Updated cost and schedule data
   f. Updated logistics documentation, as required
   g. Applicable technical plans (e.g., technical performance measurement plan, contamination control plan, parts management plan, environments control plan, EMI/EMC control plan, payload-to-carrier integration plan, producibility/manufacturability program plan, reliability program plan, quality assurance plan)
   h. Applicable standards
   i. Safety analyses and plans
   j. Engineering drawing tree
   k. Interface control documents
   l. Verification and validation plan
   m. Plans to respond to regulatory requirements (e.g., environmental impact statement), as required
   n. Disposal plan
   o. Technical resource use estimates and margins
   p. System-level safety analysis
   q. Preliminary limited life items list (LLIL)

Success Criteria

1. The top-level requirements—including mission success criteria, TPMs, and any sponsor-imposed constraints—are agreed upon, finalized, stated clearly, and consistent with the preliminary design
2. The flow down of verifiable requirements is complete and proper or if not, we have an adequate plan for timely resolution of open items. Requirements are traceable to mission goals and objectives.
3. The preliminary design is expected to meet the requirements at an acceptable level of risk
4. Definition of the technical interfaces is consistent with the overall technical maturity and provides an acceptable level of risk
5. We have adequate technical margins with respect to TPMs
6. Any required new technology has been developed to an adequate state of readiness, or we have viable back-up options
7. The project risks are understood and have been credibly assessed, and we have plans, a process, and resources to effectively manage them
8. Safety and mission assurance (e.g., safety, reliability, maintainability, quality, and EEE parts) have been adequately addressed in preliminary designs and any applicable S&MA products (e.g., PRA, system safety analysis, and failure modes and effects analysis) have been approved
9. The operational concept is technically sound, includes (where appropriate) human factors, and includes the flow-down of requirements for its execution

TABLE 19-39.   Supporting Documents for Preliminary Design Review Success Criteria. (TPM is technical performance measure; GSE is ground support equipment; SEMP is systems engineering management plan; EEE is electronic, electrical, and electromechanical; PRA is probability risk analysis; S&MA is safety and mission assurance.)

Preliminary Design Review (PDR) Success Criterion	Supporting Document
1. The top-level requirements—including mission success criteria, TPMs, and any sponsor-imposed constraints—are agreed upon, finalized, stated clearly, and are consistent with the preliminary design	• System design study reports • System requirements documents • Technical performance measurement plan and reports
2. The flow down of verifiable requirements is complete and proper or if not, we have an adequate plan for timely resolution of open items. Requirements are traceable to mission goals and objectives.	• System requirements documents (including hardware, software, and GSE)
3. The preliminary design is expected to meet the requirements at an acceptable level of risk	• System design reports • Software development plan • Drawing tree • Top-level release drawings
4. Definition of the technical interfaces is consistent with the overall technical maturity and provides an acceptable level of risk	• Interface requirements documents • Draft interface control documents
5. The TPMs have adequate technical margins	• System design reports
6. Any required new technology has been developed to adequate readiness, or we have back-up options with enough support to make them a viable alternative	• Updated SEMP (containing technology development plan)
7. The project risks are understood and have been credibly assessed, and we have plans, a process, and resources to effectively manage them	• System design reports • Project risk plan and risk reports
8. Safety and mission assurance (e.g., safety, reliability, maintainability, quality, and EEE parts) have been adequately addressed in preliminary designs and any applicable S&MA products (e.g., PRA, system safety analyses, and failure modes and effects analyses) have been approved	• Reliability, maintainability, and supportability plan • Safety and mission assurance plan
9. The operational concept is technically sound, includes (where appropriate) human factors, and encompasses the flow down of requirements for its execution	• Operations concept document

19.5.3   Process Level of Effort

Table 19-41 illustrates the relative level of effort for each of the 17 SE processes needed to produce these deliverables during Phase B. The main difference between this and pre-SDR is expanded focus on requirements management and similar activities.

TABLE 19-40. Contents for Primary Preliminary Design Review (PDR) Supporting Documents. (MOE is measure of effectiveness; GSE is ground support equipment; CDR is critical design review; TBD/TBR is to be determined and to be resolved.)

Product	Contents	Product	Contents
FireSAT system design study report (for each system element within the mission architecture, as appropriate) (UPDATE)	• Analysis of alternatives trade-tree for each system in the architecture (e.g., spacecraft, ground system, etc.) • Alternate system concepts • Identification of system drivers for each concept or architecture • Results of characterization for each alternative • Critical requirements identified or derived • Concept or architecture utility assessment, including evaluation criteria such as cost, technical maturity, MOEs • Detailed functional analysis • Proposed baseline system configuration with functional allocation	**Systems engineering management plan (SEMP) (UPDATE)**	• Plan for managing the SE effort • Documents driving SEMP • Technical summary • Integration technical effort (Including conceptual lifecycle support strategies, logistics, manufacturing, operation, etc.) • 17 systems engineering processes • Technology insertion • SE activities • Project plan integration • Waivers • Appendices

Spacecraft-to-launch vehicle, spacecraft-to-ground systems, spacecraft-to-GSE interface control documents (BASELINE)	• See discussion below	**FireSAT mission-level and system-level system requirements documents (UPDATE)**	• Scope
			• Applicable documents
			• Requirements
			  – Prime item definition
			  – System capabilities
			  – System characteristics
			  – Design and construction
			  – Logistics
			  – Personnel and training
			  – Subsystem requirements
			  – Precedence
			  – Qualification
			  – Standard sample
			• Verification requirements
			• Notes and appendices

TABLE 19-40. **Contents for Primary Preliminary Design Review (PDR) Supporting Documents. (Continued)** (MOE is measure of effectiveness; GSE is ground support equipment; CDR is critical design review; TBD/TBR is to be determined and to be resolved.)

Product	Contents	Product	Contents
Software development plan (UPDATE)	See discussion in previous section	**Engineering drawing tree and release drawings**	• See discussion below
Plan to achieve CDR readiness	• Rank-ordered list of studies and trades-offs • Forward work mapped to program schedule • Technology gap closure plan mapped to program schedule • Requirements update and management plan • TBD/TBR burn-down plan • PDR issue resolution plan		

TABLE 19-41. **Systems Engineering Process Level of Effort Leading up to PDR.** Analogous to a cell phone signal, zero bars represents no or very low resource requirements and five bars represents maximum resource use. Resource requirements apply for a particular process only; a given number of bars for one process does not imply the same level of resources as the same number of bars for a different process.

Process	Level of Effort	Comments
1. Stakeholder expectation definition	■■■□□	Stakeholder expectations should now be adequately captured in technical requirements. More negotiations may be needed as lower-level trade-offs turn up.
2. Technical requirements definition	■■■■■	We refine requirements to lower levels as a result of preliminary design and analysis
3. Logical decomposition	■■■■■	Continues to be important in support of preliminary design and analysis
4. Physical solution	■■■□□	During preliminary design, the final form of the physical solution begins to emerge
5. Product implementation	■□□□□	Identifying existing assets and assessing the cost-benefit and risk-reduction aspects of existing systems are standard during this phase. We also plan for necessary long-lead procurement items.
6. Product integration	■■□□□	As part of the logical decomposition process, we consider all lifecycle phases for each system element. We identify and coordinate with stakeholders to ensure that all requirements and interfaces (such as to existing or planned facilities) are captured.
7. Product verification	■□□□□	Verification planning ramps up to slightly higher levels as a trade-off consideration during design

TABLE 19-41. **Systems Engineering Process Level of Effort Leading up to PDR. (Continued)** Analogous to a cell phone signal, zero bars represents no or very low resource requirements and five bars represents maximum resource use. Resource requirements apply for a particular process only; a given number of bars for one process does not imply the same level of resources as the same number of bars for a different process.

Process	Level of Effort	Comments
8. Product validation	■■□□□	Validation planning continues at a low level as a trade-off during design
9. Product transition	■□□□□	Only minimum effort on this process, mainly to identify enabling products for transition and how they may impact lifecycle cost
10. Technical planning	■■■□□	These processes must be fully defined for a successful completion of PDR as they determine how the program will be executed in the next phase
11. Requirements management	■■■■■	The requirements management process must be fixed and in place to control the evolving and expanding set of requirements
12. Interface management	■■■□□	We address interface management in the requirements management processes
13. Technical risk management	■■■□□	Risk planning and management remain at a vigorous level
14. Configuration management	■■■□□	End item control and requirements management controls must be in place
15. Technical data management	■■■□□	Tools and repositories must be fully deployed along with the processes to guide engineering teams in concurrently defining systems
16. Technical assessment	■■■□□	Planning to support the upcoming PDR and subsequent reviews must begin
17. Decision analysis	■■■□□	Rigorous technical decision analysis is key to developing credible Phase B results. It's impossible to complete all the trade studies that we identify, so good decision analysis identifies the most critical ones.

19.5.4   Results

During Phase B, FireSAT project engineers complete the first major design and analysis cycle (DAC), focused on subsystem design and interactive system behavior. The Pre-phase A and Phase A DACs were more conceptual in nature. That level of analysis often features gross assumptions, and first-order analytic models tend to be the rule. In Phase B, we start questioning some of these assumptions in earnest as second-order analysis tools and techniques begin to uncover unexpected results and unintended consequences of initial design choices. To understand this further, in this section we look at exhibits from 1) the system design report, 2) the configuration management plan, 3) engineering drawings, and 4) the spacecraft-to-launch vehicle interface requirements document (IRD).

FireSAT Spacecraft System Design Report

For the FireSAT spacecraft, the project continues to rely on the system design report for many of the design decisions asked for at the PDR. Let's look at two results that mature during Phase B—the system configuration (which can also be split out into a separate system configuration document) and the internal interfaces. The spacecraft system architecture was defined at the SDR, as shown in Figure 19-38.

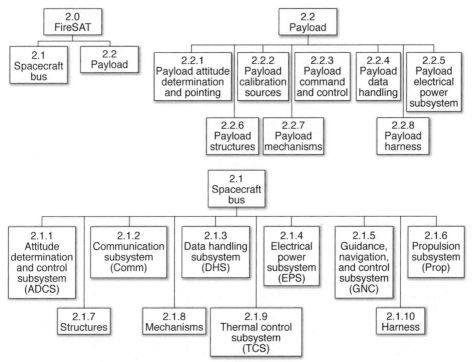

FIGURE 19-38.   FireSAT Spacecraft Architecture. The system definition review yielded this architecture, which undergoes further refinement and definition in Phase B.

Block diagrams and tables, however, are an unsatisfying way of visualizing a system. So, during Phase B (or more likely, back in Pre-phase A to help grab everyone's attention and focus their thinking), we expand on this basic view to develop both external and internal configuration concepts. The external spacecraft configuration, stowed and deployed, is shown in Figure 19-39.

To develop the internal configuration, we must first put together an equipment list of the major components that make up the spacecraft. During

Deployed
configuration

Stowed
configuration

FIGURE 19-39.   FireSAT Spacecraft External Configuration, Deployed and Stowed. Conceptual
drawings complement diagrams and tables. They help everyone involved in the
project to get an idea of the final system.

preliminary design, these components are identified, designed (or designated as
off-the-shelf) and characterized (e.g., mass, power, dimensions). The resulting
FireSAT spacecraft equipment list, along with component masses, is shown in
Table 19-42.

TABLE 19-42.   FireSAT Spacecraft Equipment List. We roll up the component masses to get the
subsystem masses, and these to arrive at the system mass. (MON is mixed oxides of
nitrogen; ISO is isolation; OD is outside diameter; ADCS is attitude determination and
control system; GPS is Global Positioning System; GNC is guidance, navigation, and
control; EPS is electrical power system; HPA is high-power antenna; Rx is receive;
Tx is transmit; He is helium.)

Item	Quantity	Mass (kg)	Item	Quantity	Mass (kg)
Payload	1	28.1	**Propulsion**		44.14
Bus			Propulsion electronics	1	1.00
ADCS and GNC		9.74	Propellant tank	1	20.00
ADCS control electronics	1	2.00	Pressurant tank (including gas)	1	10.00
Reaction wheels	4	0.79	Orbit control engine	1	1.00
Magnetorquers	3	1.50	Reaction control engines	6	3.00
Sun sensor	1	0.60	He fill and drain valve	1	0.25
Earth sensor	1	2.75	He regulator	1	1.00
Magnetometer	1	0.85	He tank ISO valve	2	1.00
GPS receiver	1	1.00	MON fill and drain valve	1	0.50
GPS antenna	1	0.25	MON tank ISO valve	2	1.00
EPS		41.68	Lines and fittings (6.35 mm OD line)	1	5.39
Battery (24 cells)	1	9.31	**Structures and mechanisms**		33.37

TABLE 19-42.   **FireSAT Spacecraft Equipment List. (Continued)** We roll up the component masses to get the subsystem masses, and these to arrive at the system mass. (MON is mixed oxides of nitrogen; ISO is isolation; OD is outside diameter; ADCS is attitude determination and control system; GPS is Global Positioning System; GNC is guidance, navigation, and control; EPS is electrical power system; HPA is high-power antenna; Rx is receive; Tx is transmit; He is helium.)

Item	Quantity	Mass (kg)	Item	Quantity	Mass (kg)
Power control unit	1	5.96	Primary structure	1	18.87
Regulators and converters	1	7.45	Secondary structure	Various	3.50
EPS harness	1	7.03	Launch vehicle adapter	1	5.00
Solar arrays	2	11.93	Separation system	1	2.00
Communications		3.26	Solar array drive assemblies	2	4.00
Receiver	1	1.00	**Thermal control**		3.90
Rx antenna	1	0.08	Coatings and blankets	Various	3.90
Transmitter (including HPA)	1	2.10			
Tx antenna	1	0.08	**Dry mass with margin**		168.89
Command and data handling		4.70	**Propellant mass**		66.21
Primary flight computer	1	2.35			
Data harness	1	2.35	**Total mass**		235.10

Once we know what's inside, we have to lay out the internal configuration, what goes where. This is both a creative and an analytic activity, since the quantifiable requirements (component A connects to component B) and the "-ility" requirements (such as maintainability) all need to be met. The "-ilities" are difficult to quantify but easy to spot when there is a problem. The book *Spacecraft Structures and Mechanisms* [Sarafin and Larson, 1995] provides guidelines for achieving an optimal spacecraft configuration. For FireSAT, the internal configuration is laid out with integration in mind (more on this later) and can be seen in Figure 19-40.

With the internal configuration understood, the preliminary design effort focuses on how all of the pieces interact. We begin with the instantiated physical block diagram as described in Chapter 5 and depicted in Figure 19-41. This diagram helps us focus on the physical interfaces between the subsystems and between the system and the outside world.

Another useful tool for capturing and defining these interactions is the I×I matrix as described in Chapter 15. Figure 19-42 shows the I×I matrix for the FireSAT spacecraft. The matrix is read clockwise; that is, items come out of the subsystems clockwise and into the associated subsystem on the diagonals.

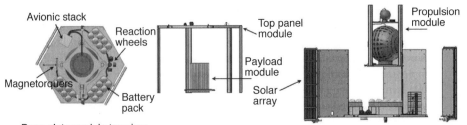

Base plate module top view

FIGURE 19-40.   FireSAT Spacecraft Internal Configuration. The inside of a spacecraft is crowded! We have to keep interfaces in mind when arranging the subsystems and components.

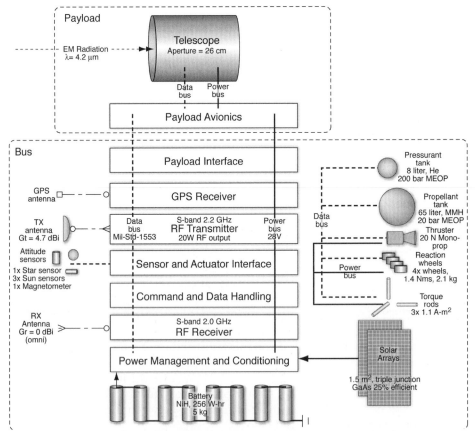

FIGURE 19-41.   Instantiated Physical Architecture. An instantiated physical architecture such as this provides a variety of useful insights into the system structure, performance, and interfaces. (MEOP is maximum expected operating pressure; MMH is monomethyl hydrazine; GPS is Global Positioning System; TX is transmit; RX is receive; Gr is receiver gain; RF is radio frequency; NiH is nickel hydride; GaAs is gallium arsenide.)

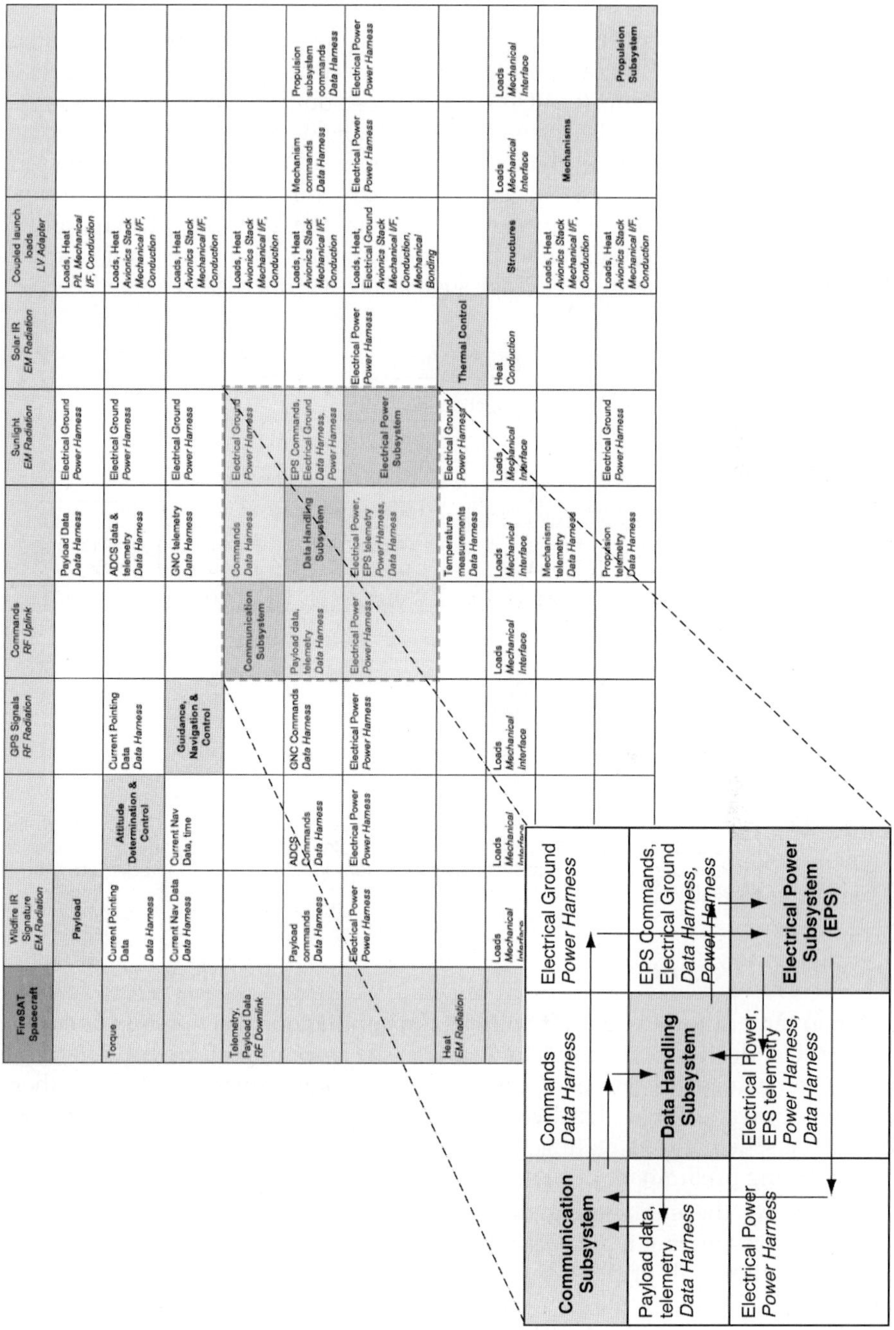

FIGURE 19-42.   FireSAT Spacecraft I×I Matrix. A matrix like this presents a lot of interaction information in a compact form.

Configuration Management Plan

Knowing the system configuration allows us to generate the system configuration management plan. We logically divide the system into manageable hardware configuration items (HWCIs). Figure 19-43 outlines how the spacecraft is grouped into six major modules (assemblies), and how we decompose the baseplate assembly into three additional HWCIs. As different HWCIs come from different vendors, and all come together at the prime contractor, this makes configuration and interface management easy. The configuration definition, including the equipment list, together with the configuration management plan, also helps track critical technical performance measures (TPMs), such as mass properties.

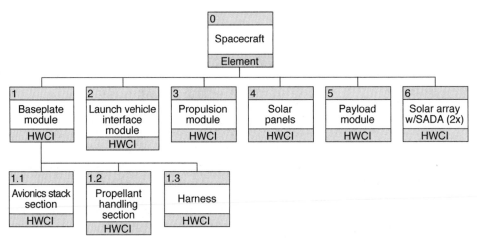

FIGURE 19-43.   FireSAT Spacecraft Hardware Configuration Items (HWCIs). Decomposing the HWCIs in the configuration management plan assists the prime contractor in coordinating with the many configuration item vendors. (SADA is solar array drive assembly)

Engineering Drawings

"If you can't draw it, you can't build it" is a familiar quip on the shop floor. The goal of critical design is to completely specify the system to sufficient detail that someone can actually obtain all the pieces by building, reusing, or buying them from some source. All these hardware end-item specifications appear as drawings or electronic schematics. Creating these drawings is a huge effort that begins to ramp up during preliminary design.

In planning the engineering drawing tasks, the first step is to bound the problem by constructing a *drawing tree*. A drawing tree lays out the hierarchical relationship among all of the drawings for the system of interest and defines a drawing number methodology. One useful way to delineate the drawing tree structure is to start with the hardware configuration items defined in the configuration control process (the next section discusses how this is also useful for integration). Figure 19-44 depicts the FireSAT spacecraft drawing tree.

The completion status of planned drawings serves as an important metric of overall design completeness as the percentage of drawings signed and released represents an earned value for the overall level of effort needed to finish the design. Preliminary, top-level drawings are made and released during Phase B for planning and design purposes. Figure 19-45 shows the stowed configuration drawing. At this point, we may also begin drawings, especially interface control drawings, on long-lead items such as solar arrays. As a general rule of thumb, projects aim for about 40% of the drawings, mostly top-level and interface drawings, to be completed by the PDR.

Spacecraft-to-Launch Vehicle Interface Requirements Document (IRD)

The previous discussions on preliminary design results have considered mostly the spacecraft configuration and internal interfaces, but that's certainly not the only focus during Phase B. At the PDR, or preferably before, we also baseline key interfaces between the spacecraft and the "outside world." These interfaces are defined by interface requirements documents (IRDs). A good example of these is the spacecraft-to-launch vehicle IRD. Table 19-43 shows an example table of contents from this IRD. Apparent from this lengthy example is the amount of detail about the interface, both static and operational, that must be baselined by the PDR. This highlights an often annoying truth about the space systems engineering process: the order and pace at which we pursue design decisions may be forced upon us by the demands of external interfaces. Or design decisions we thought were ours to make may in fact be fixed by the selection of a given external interface, such as the choice of launch vehicle.

Having to confront many of these design decisions earlier than we like is somewhat disruptive to schedule. But in the long run it's better to establish these interfaces, and their inevitable design consequences, as early as possible. Of course, the requirements that need to be defined in the IRD should not come as a surprise. A draft IRD should have been put together during Phase A with initial allocated requirements addressed as part of the SRR. Knowing what type of data is needed for the IRD helps guide planning for the Phase-B DAC. For example, mass properties and finite element models are likely deliverable data packages as part of the IRD.

These analyses, and necessary tools, need to be factored into the DAC planning as early as possible. Other interface issues, such as connector definitions and battery charging voltage characteristics, may force specific detailed design decisions during Phase B or sooner.

19.5.5   Representative Issues, Items, and Anecdotes

We recall that back in Pre-phase A, project managers decided to transfer the development of the FireSAT payload to NOAA. This move saved considerable funds from the project budget by leveraging an infrared instrument being

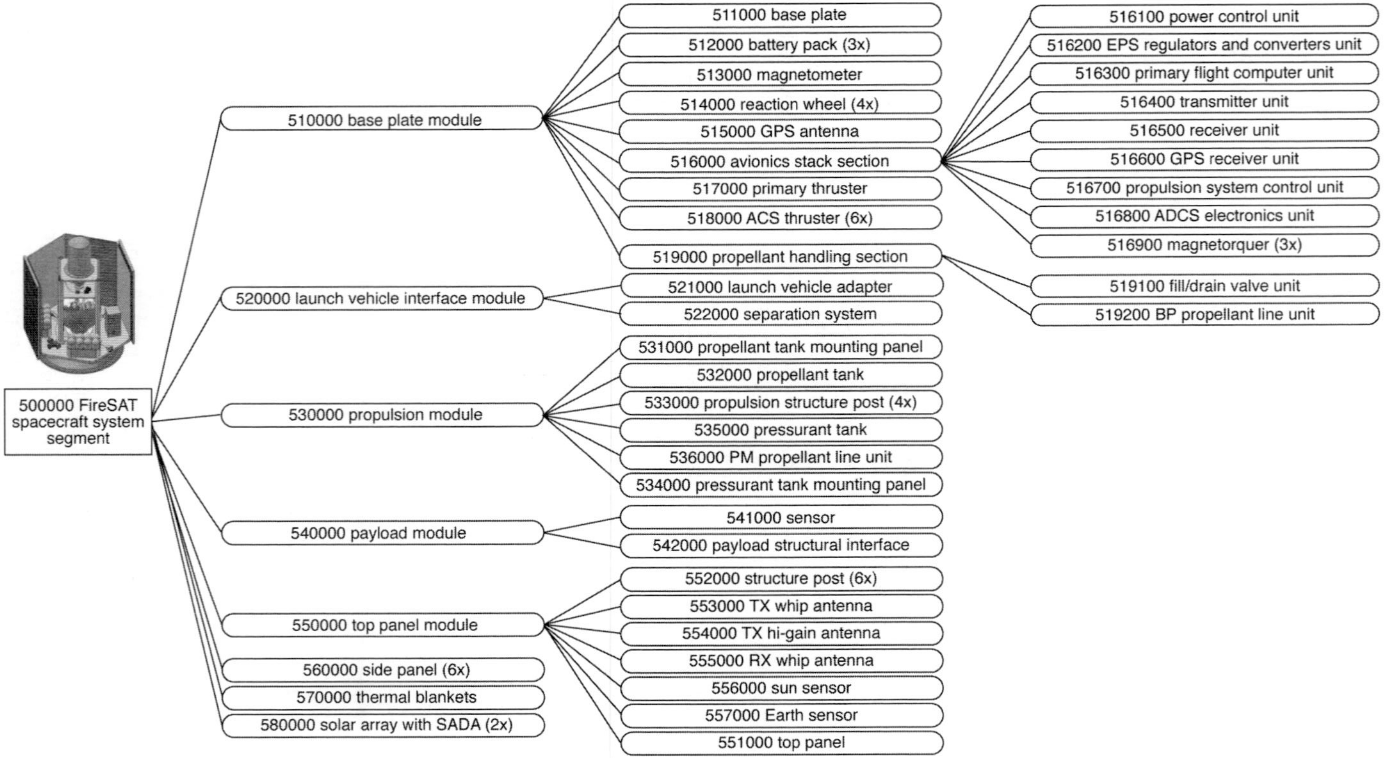

FIGURE 19-44.   FireSAT Spacecraft Drawing Tree. In FireSAT, we group the planned drawings according to the diagram of the hardware configuration items shown in Figure 19-43 and define an overall drawing numbering strategy for the entire system. (SADA is solar array drive assembly; GPS is Global Positioning System; ACS is attitude control subsystem; TX is transmit; RX is receive; ADCS is attitude determination and control subsystem.)

Drawing tree items:

500000 FireSAT spacecraft system segment

510000 base plate module
- 511000 base plate
- 512000 battery pack (3x)
- 513000 magnetometer
- 514000 reaction wheel (4x)
- 515000 GPS antenna
- 516000 avionics stack section
  - 516100 power control unit
  - 516200 EPS regulators and converters unit
  - 516300 primary flight computer unit
  - 516400 transmitter unit
  - 516500 receiver unit
  - 516600 GPS receiver unit
  - 516700 propulsion system control unit
  - 516800 ADCS electronics unit
  - 516900 magnetorquer (3x)
- 517000 primary thruster
- 518000 ACS thruster (6x)

519000 propellant handling section
- 519100 fill/drain valve unit
- 519200 BP propellant line unit

520000 launch vehicle interface module
- 521000 launch vehicle adapter
- 522000 separation system

530000 propulsion module
- 531000 propellant tank mounting panel
- 532000 propellant tank
- 533000 propulsion structure post (4x)
- 535000 pressurant tank
- 536000 PM propellant line unit
- 534000 pressurant tank mounting panel

540000 payload module
- 541000 sensor
- 542000 payload structural interface

550000 top panel module
- 551000 top panel
- 552000 structure post (6x)
- 553000 TX whip antenna
- 554000 TX hi-gain antenna
- 555000 RX whip antenna
- 556000 sun sensor
- 557000 Earth sensor

560000 side panel (6x)

570000 thermal blankets

580000 solar array with SADA (2x)

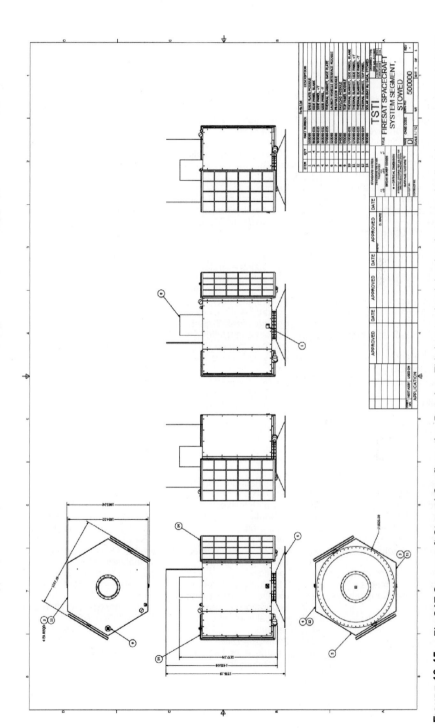

FIGURE 19-45.   FireSAT Spacecraft Stowed Configuration Drawing. This is a top-level drawing, but more detailed than a conceptual picture. For example, it includes several spacecraft dimensions.

TABLE 19-43. FireSAT-to-Launch Vehicle Interface Requirements Document Contents. (Adapted from FalconSAT-3 Evolved Expendable Launch Vehicle Secondary Payload Adapter IRD, courtesy USAF Academy Space Systems Research Center.) This is only one of many IRDs for the FireSAT project.

Acronyms and Abbreviations	3.2.5.1 Special Trajectory Requirements	3.3.3.2 FireSAT Harness Characteristics
1.0 SCOPE	3.2.5.1.1 Telemetry Maneuvers	3.3.3.3 Bonding Requirements
1.1 Identification	3.2.5.1.2 Free Molecular Heating Restraints	3.3.3.4 Voltage of the Spacecraft Battery and Polarity of the Battery Ground
1.2 Documentation Overview		
1.3 Interface Verification		
2.0 REFERENCED DOCUMENTS	3.2.5.1.3 Spacecraft Separation Requirements	3.3.3.5 Spacecraft Separation Characteristics
3.0 REQUIREMENTS	3.2.5.1.4 Position	3.3.3.6 External Signal Requirements
3.1 Spacecraft Configuration	3.2.5.1.5 Attitude	
3.1.1 Spacecraft Description/Mission Objectives	3.2.5.1.6 Sequence and Timing	3.3.3.7 Spacecraft Separation Sensed by Launch Vehicle
	3.2.5.1.7 Contamination and Collision Avoidance Maneuver (CCAM)	3.3.3.8 Connector Pin Assignments for the Electrical Ground Support Equipment
3.1.2 Standard Coordinate Systems		
3.1.3 Spacecraft Mass Properties	3.2.6 Launch and Flight Operation Requirements	3.4 Spacecraft Handling and Processing
3.1.4 Spacecraft Hazardous Systems	3.2.6.1 Operations – Prelaunch	3.4.4 Security
3.1.4.1 Propulsion System	3.2.6.1.1 Spacecraft Ground Station Interface Requirements	3.4.4.1 Payload Processing Facility Security
3.1.4.2 Radio Frequency (RF) Systems	3.2.6.1.2 Launch Countdown	3.4.4.2 Transportation Security
	3.2.6.2 Operations – Launch Through Spacecraft Separation	3.4.4.3 Pad Security
3.1.4.3 Spacecraft Batteries		3.4.5 Special Handling Requirements
3.1.4.4 Electro-Explosive Devices (EED)	3.2.6.2.1 Spacecraft Uplink and Downlink Requirements	
3.1.4.5 Deployable Systems	3.2.6.2.2 Launch Vehicle Tracking Stations	3.4.6 Supplied Equipment and Facilities
3.1.4.6 Spacecraft Systems Activated Prior to Spacecraft Separation	3.2.6.3 Operations—Post Spacecraft Separation	3.4.6.1 Payload Processing Facility
	3.2.6.3.1 Spacecraft Tracking Station	3.4.6.1.1 Facility Access Training
3.1.4.7 Other Hazardous Systems	3.2.6.3.2 Spacecraft Acquisition Assistance Requirements	3.4.6.1.2 Facility Requirements
3.1.5 Spacecraft Volume	3.4.1 Environmental Requirements	3.4.6.1.3 Crane Requirements
3.1.5.1 Spacecraft Venting	3.4.2 Spacecraft Airflow and Purges	3.4.6.1.4 Electrical Requirements
3.1.5.2 Non-ventable Volume		
3.2 Mission Parameters		3.4.6.1.5 Integrated Activities
3.2.1 Mission Description	3.4.3 Contamination/Cleanliness Requirements	3.4.6.2 Launch Pad
3.2.2 Orbit Characteristics		3.4.6.2.1 Launch Pad Access Training
3.2.2.1 Other Orbit Requirements	3.3 Spacecraft Interface Requirements	
3.2.3 Launch Date and Time	3.3.1 Responsibility	3.4.6.2.2 Launch Pad Facility Requirements
3.2.3.1 Launch Windows	3.3.2 Mechanical Interfaces	
3.2.3.2 Launch Exclusion Dates	3.3.2.1 Spacecraft Protuberances	3.4.6.2.3 Pad Electrical Requirements
	3.3.2.1.1 Fairing Envelope Protuberances	
3.2.4 Spacecraft Constraints on Mission Parameters		3.4.6.2.4 Pad Activities
3.2.4.1 Sun-Angle Constraints	3.3.2.1.2 LV-to-Spacecraft Interface Plane Protuberances	3.4.7 Range Safety Console Interface
3.2.4.2 Eclipse	3.3.2.2 Separation System	3.4.8 Other Spacecraft Testing
3.2.4.3 Ascending Node	3.3.3 Electrical Interface Requirements	4.0 VERIFICATION
3.2.4.4 Telemetry Constraint		4.1 Verification Method
3.2.4.5 Thermal Attitude Constraints	3.3.3.1 Connector Pin Assignments in the Launch Vehicle Wiring Harness	4.1.1 Test
		4.1.2 Analysis
3.2.4.6 Other Spacecraft Constraints on Mission Parameters		4.1.3 Inspection
		4.2 Verification Matrix
3.2.5 Trajectory and Spacecraft Separation Requirement		

developed for a different NOAA program. At the time, it was stipulated that the NOAA infrared payload must meet Technology Readiness Level (TRL) 6 before the FireSAT PDR. One month before the PDR, a non-advocacy board for technology readiness conducts a formal review of the state of the payload development. Unfortunately, it doesn't rate the development at TRL 6. To complete the additional tests mandated by the board, NOAA estimates that their payload team will need an additional three months.

Rather than slip the PDR, the project manager, in consultation with the lead systems engineer, decides to hold the PDR as scheduled and deal with the consequences of the payload schedule slip separately. At the PDR, two competing get-well arguments are put forth.

When the decision was made to give NOAA responsibility for the payload, Acme committed internal research and development funds to continue their own instrument development in-house. NASA strongly supported this effort as a risk reduction effort. Acme now argues that since the NOAA development has failed to meet the required milestone, the project office should reassign payload project responsibility back to Acme as originally planned. According to their estimates, a modest change in contract scope, with funding in line with their original proposal, could accelerate their internally-developed payload to reach PDR-level within two months. Their instrument design is based on existing, proven technology with a notional TRL of 7 or better. Bringing the design to the required level of detail is only a matter of applying sufficient engineering manpower.

As expected, the primary team, led by NOAA, disagrees strongly with Acme's argument and puts forward one of their own. They feel that the new payload they've developed will be a significant improvement in quality over the existing technology favored by Acme. It would offer better temperature resolution, and require only 50% of the power necessary to support the Acme design, which is based on older detector technology. In addition, NOAA (and some members of Congress) are applying political pressure not to abandon the NOAA design, as its use on FireSAT would be an important risk reduction for a much higher profile program.

The FireSAT project manager and the systems engineer face a tough decision. Continuing to bet on the NOAA instrument will lead to at least a three-month schedule slip, assuming their estimates are correct. If they fail to meet their goal, or more likely, get very close and then request even more time, the project could face uncontrolled delays. But switching to the Acme design means having to add back funding that, while available at the start of the project, has been reallocated after the decision to use the NOAA payload. And even that option will take at least an additional two months, assuming the funds can even be found and added to the contract immediately (unlikely in the world of government contracting).

In the end, the managers choose to leave both options open as long as possible. They schedule a Delta-PDR for three months after the first PDR and also delay the official transition to Phase C. Because NOAA doesn't want to be responsible for delaying a project to combat wildfires (and Congressmen from wildfire-prone western states are starting to ask tough questions), they agree to find stop-gap funding for Acme to continue their effort on the back-up payload. Meanwhile, a tiger team of industry and government "grey beards" are brought together to find ways to accelerate the NOAA payload development.

While this compromise decision allows for continued development of both payload options, the project still faces a delay of at least three months, and incurs extra schedule, cost, and performance risk. While this decision doesn't satisfy the Acme team, the project manager feels that the potential payoff of better resolution and lower power is worth the risk. The argument that these benefits aren't necessary to meet the current requirements is true, but the added resolution could find smaller wildfires, ultimately saving more lives and property. Furthermore, the long payload duty cycle means that the potential power savings of the NOAA payload could be significant.

19.5.6   Conclusions

Preliminary design review epilogue: the risk pays off. The combination of the tiger team expertise and the competition from Acme spurred on the NOAA payload team, which manages to resolve all open issues from the technology readiness review board. At the Delta-PDR the preliminary designs for both payload options are presented. Both options can do the job, but the NOAA option with increased performance at lower power (and outside the FireSAT project budget) wins the day. Although disappointed, the Acme team manager assures his superiors that their investment of IR&D is more than returned in the form of good will among all members of the project team, including NOAA (and positions them to bid for a commercial mission that could leverage the same sensor technology). While hypothetically in competition, Acme engineers actively worked with the NOAA team to help them resolve technical issues, and some of the innovative solutions Acme had developed for their own payload have found their way into the NOAA design. Furthermore, because of their efforts at payload design, the Acme team is intimately familiar with both sides of the interface, making the eventual integration of bus and payload that much easier. With this major drama behind them, and the *design-to baseline* firmly established, the FireSAT team is ready to move into the critical design phase.

19.6   Set the Build-to Baseline

Following the Delta-PDR, the FireSAT project is back on track. The system is now ready to be completely specified down to its lowest level. Every resistor, capacitor, nut, and bolt has to be defined, along with software unit-level functionality. The project is about to reach the bottom of the systems engineering "V." The critical design review will demonstrate that the FireSAT design is complete to the level necessary to begin full-scale fabrication, assembly, integration, and test. But how do we get there from our delayed preliminary design review?

As any project enters Phase C, focus shifts from defining what to design to actually designing it. During Phase C, any "hand waving" or other unfounded assumptions should be uncovered as the team digs into the literal nuts and bolts of the system. The PDR establishes a *design-to baseline* defined by a strong set of requirements and other supporting project documents; the CDR has to achieve a *build-to baseline* with sufficient technical detail for contracting officers to start buying components and parts and for machinists to start bending metal. During this phase, the project:

- Develops a product build-to specification for each hardware and software system element, with supporting trades and data
- Completes all technical data packages, including IRDs, integrated schematics, spares provisioning list, specifications, and drawings
- Updates baselined documents, such as the SEMP
- Completes software design documents
- Finalizes the master verification plan and supporting documents
- Develops a launch site operations plan
- Finalizes an on-orbit check-out and activation plan
- Updates risk assessments, cost and schedule data, and logistics documentation

19.6.1   What Got Us Here?

At the entrance to this phase, the system design satisfies the exit criteria of the preliminary design review. This includes:

- All top-level requirements are agreed upon, finalized, and validated and are consistent with the preliminary design
- Technical interface definitions are captured and are consistent with the design maturity
- Technical margins are defined and deemed adequate for the design maturity with acceptable risk
- Any required new technology has been developed to an adequate technology readiness level (TRL), typically 6 by the PDR
- The mission operations concept has evolved with the design and system requirements as necessary and remains technically sound

This also means that any open items left over from the PDR are either closed or a timely closure plan has been defined.

19.6.2   The End Game—CDR Products

As we did in previous sections, we start Phase C planning by looking at what we need to have done when it's all over. Table 19-44 shows the CDR inputs and success criteria.

Mapping these criteria into products, we get the results shown in Table 19-45. The new products in this table show that many of the exit criteria are satisfied by the further development of the system architecture and the expansion of the

TABLE 19-44. Critical Design Review (CDR) Inputs and Success Criteria. Phase C takes the project to CDR, after which the project is ready for actual building. (RFA is request for action; RID is review item disposition; EEE is electronic, electrical, and electromechanical; PRA is probability risk analysis.) Adapted from NASA [2007].

Critical Design Review

Entrance Criteria	Success Criteria

1. The project has successfully completed the PDR and responded to all PDR RFAs and RIDs, or has a timely closure plan for those remaining open
 	
2. A preliminary CDR agenda, success criteria, and charge to the board have been agreed to by the technical team, project manager, and review chair before the CDR
 	
3. The CDR technical work products listed below for both hardware and software system elements have been made available to the cognizant participants before the review:
  a. Updated baselined documents, as required
  b. Product build-to specifications for each hardware and software configuration item, along with supporting trade-off analyses and data
  c. Fabrication, assembly, integration, and test plans and procedures
  d. Technical data package (e.g., integrated schematics, spares provisioning list, interface control documents, engineering analyses, and specifications)
  e. Operational limits and constraints
  f. Technical resource use estimates and margins
  g. Acceptance criteria
  h. Command and telemetry list
  i. Verification plan (including requirements and specification)
  j. Validation plan
  k. Launch site operations plan
  l. Check-out and activation plan
  m. Disposal plan (including decommissioning or termination)
  n. Updated technology development maturity assessment plan
  o. Updated risk assessment and mitigation plan
  p. Updated reliability analyses and assessments
  q. Updated cost and schedule data
  r. Updated logistics documentation
  s. Software design documents (including interface design documents)
  t. Updated limited life items list
  u. Subsystem-level and preliminary operations safety analyses
  v. System and subsystem certification plans and requirements (as needed)
  w. System safety analysis with associated verifications
	1. The detailed design is expected to meet the requirements with adequate margins at an acceptable level of risk
	
	2. Interface control documents are sufficiently mature to proceed with fabrication, assembly, integration, and test, and plans are in place to manage any open items
	
	3. We have high confidence in the product baseline, and we have (or will soon have) adequate documentation to allow proceeding with fabrication, assembly, integration, and test
	
	4. The product verification and product validation requirements and plans are complete
	
	5. The testing approach is comprehensive, and the planning for system assembly, integration, test, and launch site and mission operations is sufficient to progress into the next phase
	
	6. There are adequate technical and programmatic margins and resources to complete the development within budget, schedule, and risk constraints
	
	7. Risks to mission success are understood and credibly assessed, and we have the plans and resources to effectively manage them
	
	8. Safety and mission assurance (S&MA) (e.g., safety, reliability, maintainability, quality, and EEE parts) have been adequately addressed in system and operational designs, and any applicable S&MA products (e.g., PRA, system safety analysis, and failure modes and effects analysis) have been approved

design-to baseline to the component and part level. As we develop these products, this new information triggers updates in the SEMP; the technology development, maturity, and assessment plan; risk assessment and mitigation planning, etc. Table 19-46 shows more detailed contents of some of these documents.

TABLE 19-45.   Products to Satisfy the Critical Design Review (CDR) Success Criteria. Here we list the documentation supporting the success criteria enumerated in Table 19-44. (GSE is ground support equipment; EEE is electronic, electrical, and electromechanical; PRA is probability risk analysis.)

CDR Success Criterion	Supporting Documents
1. The detailed design is expected to meet the requirements with adequate margins at an acceptable level of risk	• Technical measurement plan and reports • Requirements validation matrices
2. Interface control documents are sufficiently mature to proceed with fabrication, assembly, integration, and test, and plans are in place to manage any open items	• Interface requirements and control documentation • Baseline configuration document
3. We have high confidence in the product baseline, and have (or will soon have) adequate documentation to proceed with fabrication, assembly, integration, and test	• System design reports • Software development plan • Released engineering drawings • Flight hardware, software, and ground support equipment specifications
4. Product verification and validation requirements and plans are complete	• System requirements documents (hardware, software, and GSE) • Master verification plan
5. The testing approach is comprehensive, and the planning for system assembly, integration, test, and launch site and mission operations is sufficient to progress into the next phase	• Master verification plan • Manufacturing and assembly plan
6. We have adequate technical and programmatic margins and resources to complete the development within budget, schedule, and risk constraints	• System design reports • Integrated master plan • Integrated master schedule
7. Risks to mission success are understood and credibly assessed, and we have plans and resources to effectively manage them	• Technical risk reports
8. Safety and mission assurance (S&MA) (e.g., safety, reliability, maintainability, quality, and EEE parts) have been adequately addressed in system and operational designs, and any applicable S&MA products (e.g., PRA, system safety analysis, and failure modes and effects analysis) have been approved	• Safety and mission assurance plan • Failure modes and effects analysis

TABLE 19-46.   Contents of Primary CDR Supporting Documents. This table details some of the more important documents mentioned in Table 19-45. (MOE is measure of effectiveness; TBD/TBR is to be determined/to be resolved; SAR is system acceptance review.)

Product	Contents	Product	Contents
FireSAT system design study report (for each system element within the mission architecture as appropriate) (UPDATE)	• Analysis of alternatives trade-tree including (for each system in the architecture, e.g., spacecraft, ground system, etc.) • Alternate system concepts • Identification of system drivers for each concept or architecture • Results of characterization for each alternative • Critical requirements identified or derived • Concept and architecture utility assessment including evaluation criteria, e.g., cost, technical maturity, MOEs • Detailed functional analysis • Proposed baseline system configuration with functional allocation	Master verification plan, test, and verification requirements	• See discussion below

Spacecraft-to-launch vehicle, spacecraft-to-ground systems, spacecraft-to-user interface control documents (ICDs) (BASELINE)	• See discussion in Section 19.5	FireSAT mission-level and systems-level system requirements documents (SRD) (UPDATE)	• Scope
			• Applicable documents
			• Requirements
			  – Prime item definition
			  – System capabilities
			  – System characteristics
			  – Design and construction
			  – Logistics
			  – Personnel and training
			  – Subsystem requirements
			  – Precedence
			  – Qualification
			  – Standard sample
			• Verification requirements
			• Notes and appendices

Software development plan (SDP) (UPDATE)	• See discussion in Section 19.5	Detailed engineering release drawings	• See discussion below
Plan to achieve SAR readiness	• Rank-ordered list of studies and trade-offs • Forward work mapped to program schedule • Technology gap closure plan mapped to program schedule • Requirements update and management plan • TBD/TBR burn-down plan • CDR issue resolution plan	Manufacture and assembly plan	• See discussion in SAR section

19.6.3   Process Level of Effort

Table 19-47 illustrates the relative level of effort for each of the 17 SE processes during Phase C needed to produce these deliverables. The main difference between this and Phase B is expanded focus on the physical solution and interface control.

TABLE 19-47.   Systems Engineering Process Level of Effort Leading up to CDR. Analogous to a cell phone signal, zero bars represents no or very low resource requirements and five bars represent maximum resource use. Resource requirements are for a particular process only; a given number of bars for one process does not imply the same level of resources as the same number of bars for a different process.

Process	Level of Effort	Comments
1. Stakeholder expectation definition	■□□□□	Stakeholder expectations should now be adequately captured in technical requirements. More negotiations may be needed as we uncover lower-level trade-offs.
2. Technical requirements definition	■■■□□	Major system-level requirements should now be well-defined. Some refinement may be needed as we make new trade-offs during detailed design.
3. Logical decomposition	■□□□□	Largely completed during earlier phases but still useful in refining lower-level specifications, especially for software
4. Physical solution	■■■■■	During critical design, the final form of the physical solution is defined
5. Product implementation	■■■□□	Identifying existing assets and the assessment of the cost-benefit and risk-reduction aspects of existing systems are standard trades during this phase
6. Product integration	■■□□□	Product integration planning and scheduling trade-offs are important considerations during detailed design
7. Product verification	■■□□□	Verification planning moves to high gear during this phase as plans are finalized
8. Product validation	■■□□□	Validation planning continues apace with verification planning
9. Product transition	■□□□□	Only minimum effort on this process, mainly to identify enabling products for transition and their impact on lifecycle cost
10. Technical planning	■■□□□	Technical planning continues, largely as replanning when unforeseen issues arise
11. Requirements management	■■■■■	The requirements management process must be fully defined and in place to control the evolving and expanding set of requirements and detail specifications
12. Interface management	■■■■■	Interface management efforts move into high gear as part of detailed design
13. Technical risk management	■■■■□	Risk planning and management remain at a vigorous level

Table 19-47. **Systems Engineering Process Level of Effort Leading up to CDR. (Continued)**
Analogous to a cell phone signal, zero bars represents no or very low resource requirements and five bars represent maximum resource use. Resource requirements are for a particular process only; a given number of bars for one process does not imply the same level of resources as the same number of bars for a different process.

Process	Level of Effort	Comments
14. Configuration management	■■■■■	End item configuration control processes are critical throughout detailed design
15. Technical data management	■■■■□	Tools and repositories must be fully deployed along with the processes to guide engineering teams in concurrently defining systems
16. Technical assessment	■■■■□	Planning to support the upcoming CDR and subsequent reviews must begin
17. Decision analysis	■■□□□	Decision analysis processes continue but at a lower level as (we hope) most major project issues have been resolved by this point

19.6.4   Results

As the name implies, the build-to (and code-to) baselines define all the people, processes, and design definition products needed to "turn the crank" to produce the necessary hardware and software items that make up the system. But equally important is knowing whether, once the crank has been turned, we've actually built the system we wanted. For FireSAT Phase C results, we focus on these two aspects of the system lifecycle—design and verification. The top-level view of the design, including trade-offs, architecture, and configuration are in the system design report described in earlier sections. In this section, we concentrate first on how the engineering drawings capture the design details. We then turn our attention to verification planning, looking at three key documents, the *system requirements document* (SRD Section 4), the *master verification plan*, and the *test and verification requirements*.

Engineering Models and Drawings

One of the most important tools for capturing the hardware design is models and drawings. These include:

- Computer-aided design (CAD) models and configuration drawings
- Manufacturing, schematics, or detail drawings (showing how to fabricate a specific part). Includes individual details on finishes, treatments, etc. (for example, alodining of aluminum parts)
- Assembly drawings (showing how individual parts go together) with parts list

- Instructions on how to create the assembly from its component parts along with detailed procedure steps, required tooling, torque values, etc.

We organize the drawings in a drawing tree as described in the last section. Manufacturing drawings must have enough information for a machinist to fabricate a part from raw materials, or include detailed process descriptions for adhesive bonding or other techniques. Electrical schematics must provide sufficient detail that the metal traces can be laid on the printed circuit board and surface mount components added (either by hand or by machine). Figure 19-46 shows the manufacturing drawing for the FireSAT Sun sensor bracket.

Of course, these days it's not likely that parts will actually be hand-machined by a skilled machinist. Rather, the drawing details are programmed into a computer-aided mill that machines the part precisely with less touch-labor. But a skilled machinist must still translate the CAD drawing, or three-dimensional CAD model, into sensible computer-aided manufacturing routines. The same is true of printed circuit board (PCB) traces. This process is now almost completely automated, as many small companies specialize in turning out PCBs based on standard e-CAD files. For low production-rate surface mount boards, all the soldering of miniscule parts is done under a microscope by skilled technicians (who typically must hold NASA, European Space Agency, or similar certifications). One manufacturing requirement that NASA has levied on the spacecraft is that all electronic flight hardware be assembled by NASA-certified technicians.

For FireSAT, like most spacecraft, a few parts are custom-made. These include primary and secondary structural parts and some spacecraft-specific PCBs (such as the payload avionics). The vast majority of piece parts, including fasteners, batteries, and connectors, come from vendors. The same is true as we work our way up the assembly chain. The prime contractor procures many major components (such as transmitters) or even whole assemblies (such as solar arrays) as single items from subcontractors. These subcontractors, in turn, procure lower-level components or parts from other vendors. But at some level of the supply chain, someone actually has to make something. So manufacturing drawings or similar detailed specifications must exist for every single part. These specifications are mostly not controlled at the project level. Instead, the prime contractor relies on military or industry standards (such as ISO-9002) or other methods to ensure high quality from the bottom of the supply chain to the top.

Once we have the parts, they must be assembled into components and other higher-level assemblies. Figure 19-46 shows the assembly drawing for the Sun sensor. After all the components are assembled, the spacecraft is ready for final integration. Figures 19-47, 19-48, and 19-49 are all part of the single assembly drawing for the spacecraft. Next to Figure 19-50 is a blow-up of the parts list. Final spacecraft assembly can only take place when all parts are ready for integration.

Ideally, all the detail drawings, schematics, and software unit design would be completed by the CDR. But this rarely happens. The usual rule of thumb is to try to have the drawings about 80–90% complete. Which 80-90% (and the potential unknown problems lurking in the other 10–20%) is one of the risks to assess during

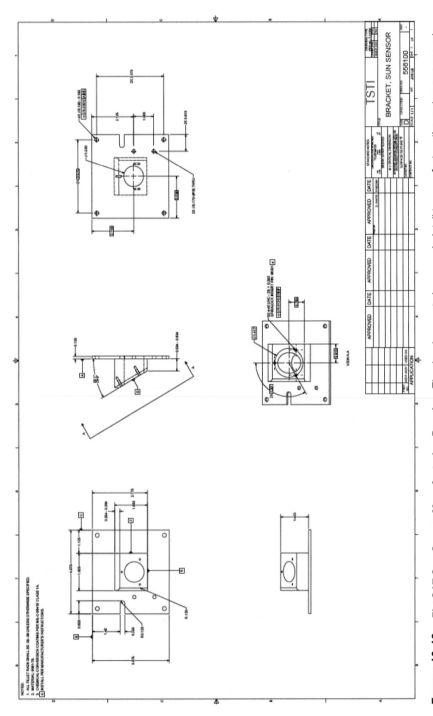

FIGURE 19-46. **FireSAT Sun Sensor Manufacturing Drawing.** The drawing must give enough detail to manufacture the part or component.

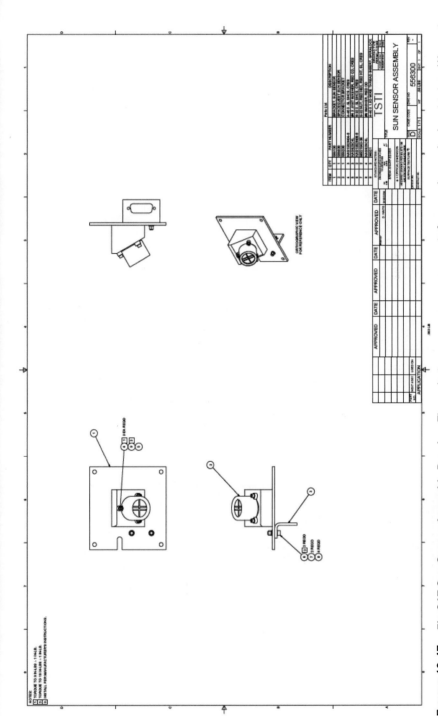

FIGURE 19-47.   FireSAT Sun Sensor Assembly Drawing. The project procures many components from vendors and subcontractors. We must have drawings for all of these.

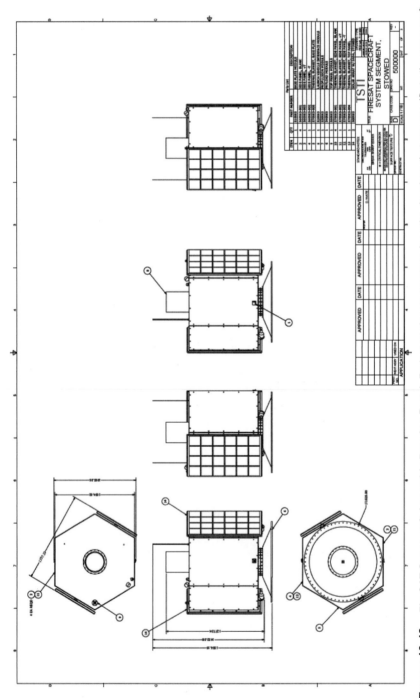

FIGURE 19-48.   FireSAT Spacecraft Assembly Drawing Part 1. Project engineers must produce the drawings that are specific to the spacecraft we're building.

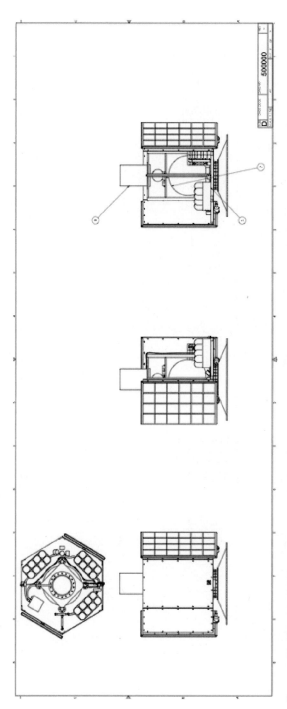

FIGURE 19-49.   FireSAT Spacecraft Assembly Drawing Part 2. This figure complements Figure 19-48.

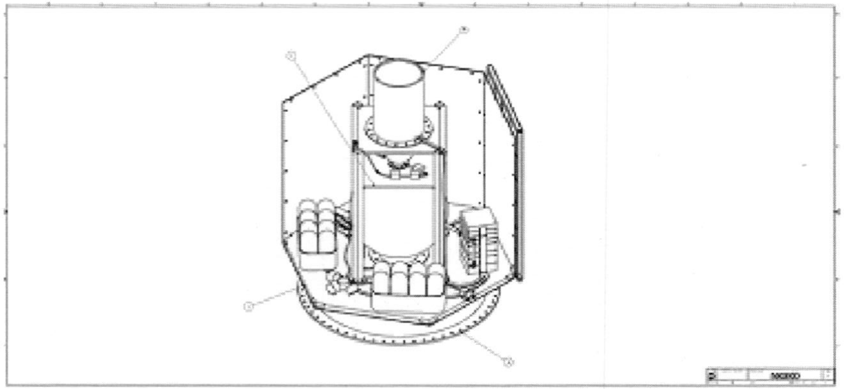

Parts List

ITEM	QTY	PART NUMBER	DESCRIPTION
1	1	510000	BASE PLATE MODULE
2	4	560000-001	SIDE PANEL, BLANK
3	1	560000-003	SIDE PANEL, +Y
4	1	560000-002	SIDE PANEL, -Y
5	1	570000-005	THERMAL BLANKET, BASE PLATE
6	1	520000	LAUNCH VEHICLE INTERFACE MODULE
7	1	530000	PROPULSION MODULE
8	1	540000	PAYLOAD MODULE
9	1	550000	TOP PANEL MODULE
10	4	570000-001	THERMAL BLANKET, SIDE PANEL BLANK
11	1	570000-002	THERMAL BLANKET, SIDE PANEL, +Y
12	1	570000-003	THERMAL BLANKET, SIDE PANEL, -Y
13	1	570000-004	THERMAL BLANKET, TOP PANEL
14	2	580000	SOLAR ARRAY W/ SADA, STOWED

FIGURE 19-50.   FireSAT Assembly Drawing Part 3 and the Parts List. Every view drawn provides essential information for spacecraft integration.

the CDR. For mechanical or electrical parts, gauging design completeness is straightforward—when the drawing is done, the design is done. But with software, we have no absolute metric for percent completion. Metrics such as lines of code (LOC), earned value, or other tools are at best only indirect indicators.

Verification Documents

Fully specifying the FireSAT spacecraft and support system designs through drawings, schematics, and software state models is only half the battle. The other half is deciding how to know that we actually get what we asked for. That's the purpose of verification. Chapter 11 describes how to go from an initial design requirement through final close-out of that requirement via a verification closure notice (VCN) or similar document. So how do we know the system has been built right? To illustrate these processes, we take the example of a single system-level requirement and trace how the verification planning is captured in three key documents: 1) the system requirements document, 2) the master verification plan, and 3) the test and verification requirements document. Each document features a different level of detail, and we need all of them to fully define the big picture. All these documents were drafted or baselined at earlier reviews, but we've delayed focusing on them until the discussion on the CDR, when they must be finalized before being implemented in the next phase.

System Requirements Document (SRD) (Section 4.0)

Product verification begins with requirements. In Section 19.3, we describe the purpose and organization of the system requirements document (SRD). Section 3.0 of this document captures the system requirements, and we present several examples. Now we discuss Section 4.0 of the SRD. Traditionally, Section 4 requirements detail the verification approach for the requirements in Section 3. A single verification requirement contains this approach. The three main elements of a verification requirement (what we're verifying, how we're verifying it, and success criteria) are described in Chapter 11. Ideally, there should be a one-to-one relationship between each Section 3 requirement and each Section 4 verification requirement. As an example, we'll trace the requirement for the space vehicle natural frequency (see Table 19-48) through the verification process.

TABLE 19-48.   **Excerpt from the Requirement Matrix.** This matrix is in Section 3 of the system requirements document.

Requirement	Description	Rationale	Source	Priority	POC
3.3.2 Space Vehicle Natural Frequency	Space vehicle first-mode natural frequency shall be greater than 20 Hz	This constraint is based on the launch vehicle induced environment. The natural frequency must be above the induced environment frequency to avoid resonance issues.	FS-8x030	A	SE

We start by developing the verification requirement for this design requirement and documenting it in Section 4 of the SRD, either in straight text or in a verification requirements matrix, as shown in Table 19-49. We recall that the SRD, including Section 4, is baselined at the SRR. This means that *way* back in early Phase A, engineers had to start thinking about verification. This can be a painful and time-consuming process, especially when heaped on top of the already daunting task of gathering and validating all the Section 3 system requirements. The task is often further hampered by a lack of verification expertise among the largely design-oriented engineers involved with developing design requirements. Even so, the best way to be sure a requirement is verifiable is to force engineers to think through its verification requirement. This exercise is also essential in identifying the special test equipment and facilities that must be factored into the budget and technical planning schedules. All this information goes into the baseline master verification plan (MVP) (also called a test and evaluation master plan (TEMP) by some organizations).

TABLE 19-49.   Excerpt from the Verification Requirements Matrix. This matrix is in Section 4 of the system requirements document.

Verification Requirement	Source Requirement	Description	Verification Rationale	Units
4.3.2 Space Vehicle Natural Frequency Verification	3.3.2 Space vehicle natural frequency: Space vehicle first-mode natural frequency shall be greater than 20 Hz	The space vehicle first-mode natural frequency shall be verified by analysis and test. The analysis shall develop a multi-node finite element model to estimate natural modes. The test shall conduct a modal survey (sine sweep) of the vehicle using a vibration table. The analysis and test shall be considered successful if the estimated and measured first mode is greater than 20 Hz.	Finite element models are a proven analysis technique for predicting the natural frequency of complex structures. The design cycle will depend on the results of this analysis through Phase C. Modal survey is a industry-standard test method for measuring natural frequency as per GEVS-XX-Y.	Qualification model, flight model

Master Verification Plan

The SRD Section 4 verification requirements capture "what," "how," and "how well" for each design requirement, and are baselined at the SRR. At the SDR, we also baseline the master verification plan. The development of this document is also described in Chapter 11. Table 19-50 shows the MVP contents.

The MVP is a strategic plan because it looks at the end-to-end big picture for all verification activities. It schedules, coordinates, and derives project

TABLE 19-50.   Master Verification Plan Contents. Planning for verification must begin in Phase A.

1.0 Introduction	5.0 Systems qualification verification
1.1 Scope	5.1 Tests
1.2 Applicable documents	5.2 Analyses
1.3 Document maintenance and control	5.3 Inspections
2.0 Program or project description	5.4 Demonstrations
2.1 Program or project overview and	6.0 Systems acceptance verification
verification master schedule	6.1 Tests
2.2 Systems descriptions	6.2 Analyses
2.3 Subsystems descriptions	6.3 Inspections
3.0 Integration and verification (I&V)	6.4 Demonstrations
organization and staffing	7.0 Launch site verification
3.1 Program or project management	8.0 On-orbit verification
offices	9.0 Post-mission and disposal verification
3.2 Base or field center I&V organizations	10.0 Verification documentation
3.3 International partner I&V organizations	11.0 Verification methodology
3.4 Prime contractor I&V organization	12.0 Support equipment
3.5 Subcontractor I&V organizations	12.1 Ground support equipment
4.0 Verification team operational relationships	12.2 Flight support equipment
4.1 Verification team scheduling and	12.3 Transportation, handling, and other
review meetings	logistics support
4.2 Verification and design reviews	12.4 Tracking station and operations
4.3 Data discrepancy reporting and	center support
resolution procedures	13.0 Facilities

requirements for facilities and equipment. Sections 1 through 4 of the plan lay out the responsibilities—"who"—across the project for managing and conducting these activities. Sections 5 through 13 define "what" is to be done, "when" and "using what" in terms of facilities and equipment. One important aspect of the MVP is the description of the project verification methodology, sometimes called model philosophy. This discussion defines the fundamental approach to risk management and verification that the project will take. Figure 19-51 illustrates the approach used for the FireSAT spacecraft.

As captured by the verification requirement, the design requirement for natural frequency will be verified by both test and analysis. The test will use a modal survey. Let's trace the verification plan for the natural frequency requirement (allocated to the spacecraft structure) in Figure 19-51 to see where and with what model the modal survey tests occur. We first see a modal survey for the structural test model under Engineering Development Units. This is a planned risk reduction effort and doesn't formally close out the requirement. A second modal survey is planned for the qualification model and a third with the flight model. The verification closure notice for this requirement is signed off after the qualification model test and analyses are complete.

While it emphasizes the test flow for the flight model, the MVP also lays out how to sequence the tests in Figure 19-51. This sequence is shown in Figure 19-52. Here we see the baseline modal survey, the test that verifies the requirement along with the analysis results, in sequence 2 after installation on the vibration test stand. In standard practice, modal surveys are repeated after each static load or random vibration

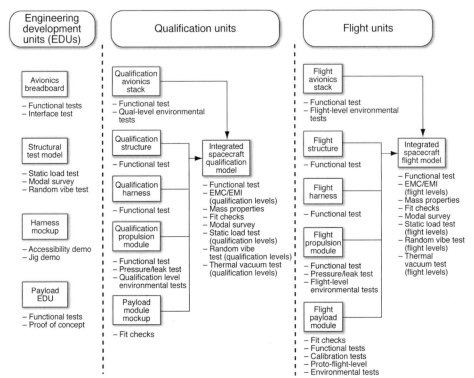

FIGURE 19-51. **Example Development and Test Campaign Planning Diagram for FireSAT.** This figure illustrates the number and type of engineering development, qualification and flight or proto-flight units to be built, their evolution, and the type of verification activities they'll be subjected to. To verify the natural frequency requirement, for example, we conduct modal surveys on all three types of units shown at the top of the figure. (EMC/EMI is electromagnetic compatibility/electromagnetic interference.)

sequence to determine if any change has occurred in the structure (a change in the frequency of the modes indicates something has been broken by overtest).

Test and Verification Requirements

The verification requirements in Section 4 of the SRD tell what to verify and generally how and how well. The MVP puts this in the context of the larger plan for all verification activities. Together these define our strategic plan. The test and verification requirements (TVRs) go a step further to represent the tactical plan by defining the details of "when," "who," "with what," and "by doing this..." The TVRs get down to the practical level by laying out in detail the test plans, procedures, and test configurations and equipment needed to verify all requirements.

Let's look at the TVR associated with the natural frequency requirement example we've been following. The plan is to verify it during the flight model baseline modal survey. From Figure 19-52 we know that the baseline modal survey is one part of a series of tests using a vibration test stand. The vibration test stand simulates the mechanical environment (static load and random vibration) that the spacecraft will experience on the launch vehicle. It also provides inputs of known frequency to measure vehicle response (indicating natural frequency). So it makes sense to group together into one TVR all of the requirements associated with the induced launch environment that will be verified using the vibration test stand. This way, the TVR provides a logical grouping of verification requirements into single large campaigns made up of similar events (or at least events using similar equipment). This streamlines our planning, helps make the most efficient use of expensive test equipment, and minimizes the number of transitions needed on flight hardware. Table 19-51 shows the organization for a sample TVR.

TABLE 19-51.   **Test and Verification Requirement Example for the Induced Environment Requirements.** Judicious grouping of tests helps maximize test and verification efficiency.

TVR ID	Source Requirements	Hardware Assembly Level	TVR Office of Primary Responsibility
1.0	3.3.2 Space vehicle (SV) shall meet its requirements after exposure to the induced launch environment as defined in the Pegasus User's Handbook 3.3.2.1 SV natural frequency: SV first-mode natural frequency shall be greater than 20 Hz 3.3.2.2 Static load... 3.3.2.3 Random vibration...	FireSAT SV Flight Model (PN 2001-1), Flight Software Version 1.1.2	FireSAT Project Office

TVR Execution	Input Products	Output Products	Success Criteria
Acme Astro Corp.	• FireSAT finite element analysis report • FireSAT qualification model vibration campaign test plan (as executed) • FireSAT qualification model vibration test campaign report • Vehicle modal survey configuration • Vibration test stand • Vibration test stand instrumentation • SV- to-vibration test stand adapter	• FireSAT flight model vibration campaign test plan (including detailed test procedures) • FireSAT flight model vibration test campaign report	The analysis and test shall be considered successful if the estimated and measured first mode is greater than 20 Hz.

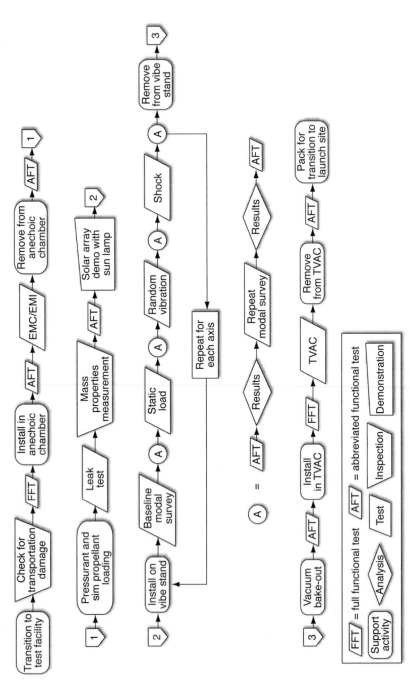

FIGURE 19-52.   FireSAT Environmental Test Campaign Sequence. This shows where the modal survey (sequence 2) fits into the complete test sequence. (TVAC is thermal vacuum.)

We've discussed the development of the verification requirements, the MVP, and the TVR one after the other. In practice, though, we develop all these documents in parallel, with multiple iterations. Strategic planning needs tactical inputs and vise versa. Throughout the development of these FireSAT verification documents, detailed planning is aided by use of model-based systems engineering (MBSE) tools, requirements management tools, data bases, or other methods for capturing and managing technical data as described in Chapter 17.

Figure 19-53 illustrates how an MBSE tool (such as CORE™, developed by Vitech Corp.) organizes the relationships among requirements, verification requirements, verification events, and their associated configurations and procedures. Ultimately, the TVRs capture these relationships in formal reports. One might argue that the TVRs as separate documents are unnecessary if there is project-wide access to the SE model. A great advantage of using sophisticated SE modeling tools, especially for complex systems, is automated cross-checking that can uncover orphaned requirements or other loose ends. These may be difficult to discover by tedious manual review and may only show themselves late in the project, leading to delays or costly last-minute fixes.

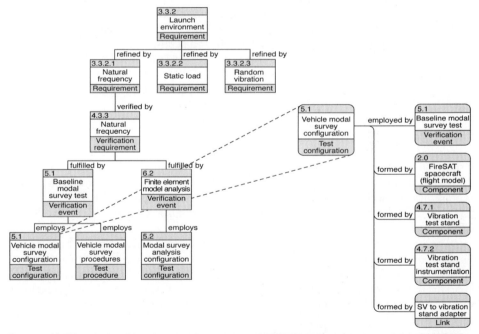

FIGURE 19-53.   Model-based Systems Engineering (MBSE) Representation of the Verification Traceability. An MBSE tool provides top-to-bottom traceability from source requirement to test configuration and procedures. Here the launch environment source requirement is traced to its verification requirement. Fulfilling the verification requirement entails both a test and analysis. The test is part of a dedicated verification event that requires a specific test configuration along with detailed procedures.

Our discussion has focused on only one requirement, natural frequency. This is only one of hundreds of requirements for the FireSAT spacecraft. For each one, the project defines (and validates) a verification requirement. Together these form part of the development of the MVP and TVRs. All this meticulous planning pays off both during the initial planning phases (Phase A, B, and C) and when it comes time to implement during Phase D. During the planning phases, the focus on verification helps derive requirements for test facilities and specialized handling equipment with long lead times. More important, it fosters detailed discussions between the system designers and the system verifiers on the practical aspects of how to conduct specific tests. These discussions lead to many changes to the design itself to add in testing "hooks," both physical and in software, to enable many tests to be feasible. During Phase D, when the team starts implementing the plan, the planning effort also proves invaluable, as we describe in the next section.

19.6.5   Representative Issues, Items, and Anecdotes

Just when the FireSAT systems engineering team thinks things are going well, the project office threw a spanner in the works. Two months before the CDR, the project incurs a significant out-year budget cut. To mitigate the budget shortfall, the project management team, in consultation with the major stakeholders, decides on a relatively risky, yet powerful, budget savings measure. Instead of launching each FireSAT on a dedicated Pegasus launch as baselined in the original mission architecture in Pre-phase A, the project will instead launch both satellites at the same time on a single Falcon-1 launch vehicle.

The first impact of this decision is on the project technical planning. The project manager and systems engineer have to decide whether to delay the CDR or hold the review as planned with a huge set of open issues. In the end, they decide that rather than disturb the momentum of the entire team, the CDR will be held as scheduled based on the original requirements. In parallel, a tiger team of mainly structural engineers and launch vehicle interface experts is formed to initiate a mini-DAC to analyze the launch vehicle and deployment changes and to identify the significant impacts to the current design baseline. Following the regularly scheduled CDR, they present their preliminary results, with a full-blown Delta-CDR to be held six months later.

The tiger team reports that the launch capability of the Falcon-1 vehicle to the FireSAT orbit will allow both spacecraft to fit onto one vehicle, but with a far more complex launch adapter deployment sequence. The shroud volume can fit the two spacecraft, but the omni-antenna on top of the spacecraft will have to be modified to be deployable to stay within the dynamic envelope. Besides these operational and configuration changes, the team also reports on the changes to the induced launch environment and its impact on the current design. The effect of changing this single requirement is:

3.3.2 Launch Environment: The space vehicle (SV) shall meet its requirements after exposure to the induced launch environment as defined in the *Falcon-1 User's Handbook*.

Both the axial and lateral accelerations are lower on the Falcon-1 than on the Pegasus XL, but the natural frequency requirement for the spacecraft has increased from 20 Hz to 25 Hz, entailing a change in this requirement:

3.3.2.1 SV Natural Frequency: SV first-mode natural frequency shall be greater than *25 Hz*.

The change demands more detailed analysis to show that the current design will meet this increase in stiffness, and a change to the verification events is necessary to ensure compliance with this requirement. Fortunately, based on initial results from the structural test modal survey completed only a week before the CDR, the margin in the current design turns out to be sufficient to easily meet the 25 Hz requirement for each vehicle. It still remains to be shown that the vertically stacked, dual satellite configuration is sufficiently stiff. To better understand this problem, a detailed coupled loads analysis is required that will add six months to the schedule.

In addition, the mass of the two spacecraft plus payload adapter must be under 540 kg. The best estimate at this time is within this requirement, but with a 40 kg payload adapter, mass growth from this point forward will have to be carefully monitored and controlled. This adds complexity to the design change process and to the risk of a work package cost overrun during the next phase.

There is one silver lining in the cloud. While incorporating these changes into the project plan will cause a six month slip in the planned initial operating capability (IOC), it will actually have little effect on the final operating capability (FOC) and might actually move it ahead. Because the new launch campaign plan involves only one launch instead of two, the additional time between launches, predicted to be at least six months, is eliminated from the schedule, along with the considerable expense of a second launch campaign.

19.6.6   Conclusions

Somewhat the worse for wear, the FireSAT systems engineering team walks away from the Delta-CDR confident that they've laid to rest all of the open issues precipitated by the last-minute launch vehicle change. They're thankful for the efforts they made back in Phase A to put in place rigorous requirements processes and technical data management systems. These tools make it easier, and less risky, to trace out the impact of the top-level induced environment requirement change on all the child, grandchild, and great-grandchild requirements. Risk-reduction testing during Phase C with various engineering models also gives the team confidence in their test procedures and verification teams. With the final go-ahead from the milestone authority, they're ready to start bending metal!

19.7   Achieve the As-built Baseline

Finally, the day arrives when all the stakeholders can peer through the cleanroom glass at the gleaming spacecraft hardware! All the years of late nights, long meetings, and lost weekends have come together as the spacecraft, ground support equipment, and ground station hardware and software are assembled and tested, piece by piece. At the system acceptance review, the hardware and software are officially delivered to the customer and one of the final major milestones before launch and on-orbit operations is achieved.

In phases A, B, and C we planned, designed, and prepared for the true creation of the system in Phase D. The previous phases focused primarily on the left side of the systems engineering "V," taking us from stakeholder expectations to a complete definition of the system down to the smallest detail. In Phase D, we work our way up the right side of the "V" to finalize all parts and components, integrate them, verify and validate them, and finally transition them from the builder to the customer for integration onto the launch vehicle and the beginning of real mission operations. During Phase D:

- Any remaining detailed design work is completed, including the final 10–20% of drawings and other detailed specifications not completed by the CDR
- All parts are fabricated or procured
- All components are built, bought, or reused
- All flight and ground system software is finished and delivered (or at least initial versions sufficient to support testing other elements)
- All facilities and equipment needed to support assembly, integration, and testing are available
- The project manufacturing and assembly plan is fully implemented to produce the end-item hardware and software
- The project master verification plan is fully implemented, along with supporting documents defined by the test and verification requirements. These documents produce objective evidence that all the hardware and software end items meet their requirements and will actually deliver the needed capabilities.

19.7.1   What Got Us Here?

At the entrance to Phase D, the system design satisfies the exit criteria of the critical design review, and any needed Delta-CDRs. This included:

- Definition of a detailed design with sufficient margin at acceptable risk
- Interface definitions and detail drawings and models of sufficient detail to start fabrication or procurement
- Detailed verification and validation planning with facilities, procedures and personnel identified

This means that any open items left over from the CDR are either closed or a timely closure plan defined.

19.7.2   The End Game—SAR Products

We start Phase D planning by looking at the entrance and success criteria in Table 19-52 for the major review waiting for us at the end—the system acceptance review (SAR).

TABLE 19-52.   System Acceptance Review (SAR) Entrance and Success Criteria. This review closes out Phase D; successful completion permits the project to continue to the operational phase.

System Acceptance Review	
Entrance Criteria	**Success Criteria**

1. A preliminary agenda has been coordinated (nominally) before the SAR
2. The following SAR technical products have been made available to the cognizant participants before the review:
   a. results of the SARs conducted at the major suppliers
   b. transition to production and manufacturing plan
   c. product verification results
   d. product validation results
   e. documentation that the delivered system complies with the established acceptance criteria
   f. documentation that the system will perform properly in the expected operational environment
   g. technical data package updated to include all test results
   h. certification package
   i. updated risk assessment and mitigation
   j. successfully completed previous milestone reviews
   k. remaining liens or unclosed actions and plans for closure
	1. Required tests and analyses are complete and indicate that the system will perform properly in the expected operational environment
	2. Risks are known and manageable
	3. System meets the established acceptance criteria
	4. Required safe shipping, handling, check-out, and operational plans and procedures are complete and ready for use
	5. Technical data package is complete and reflects the delivered system
	6. All applicable lessons learned for organizational improvement and system operations are captured

Mapping these criteria into products yields the results shown in Table 19-53. In Phase D, we have a major shift from planning and design oriented "this is what we think..." type documents to factual reports focused on "this is what we know..." Now we exercise all those arduously developed plans and report on the results. More detailed contents of some of these key documents are shown in Table 19-54.

19.7.3   Process Level of Effort

Table 19-55 illustrates the relative level of effort for each of the 17 SE processes during Phase D needed to produce these key deliverables. The main difference

TABLE 19-53. Products to Satisfy the System Acceptance Review (SAR) Success Criteria. The documents listed give evidence of meeting the associated success criteria.

SAR Success Criterion	Supporting Documents
1. Required tests and analyses are complete and indicate that the system will perform properly in the expected operational environment	• Test reports • Verification compliance reports
2. Risks are known and manageable	• Risk mitigation report • Open work report
3. System meets the established acceptance criteria	• Validation compliance reports
4. Required shipping, handling, check-out, and operational plans and procedures are complete and ready for use	• Operations procedures • Activation and in-flight check-out plans • User manuals • Operational limits and constraints • DD250—Property transfer authorization • System transition plan
5. Technical data package is complete and reflects the delivered system	• Acceptance data package
6. All applicable lessons learned for organizational improvement and system operations are captured	• Project lessons learned report

between this and Phase C is expanded focus on implementation, verification, validation, and transition as the team starts working with real hardware and software.

19.7.4   Results

In Phase D we build and test the end items of the project. First we focus on the manufacturing and assembly plan. We then look at documented results from the assembly, integration, and test (AIT) campaign.

Manufacturing and Assembly Plan

As we saw during our Phase B discussion, the engineers were keenly aware of potential manufacturing and assembly issues during preliminary design. Thinking ahead to assembly and integration, the designers strategically divided the spacecraft into several major modules that were defined as major configuration items (see Figure 19-54). This approach facilitates configuration control by logically dividing the system into stand-alone pieces, each with a single responsible contractor or organization. More importantly, it keeps each module as independent as possible for as long as possible during manufacture and AIT to avoid having a glitch in one of them impact the schedule for all of them. Following this approach to the spacecraft configuration, we define the top-level assembly and integration sequence as follows:

1.   Assemble base plate module

TABLE **19-54.**   **Contents for Primary System Acceptance Review (SAR) Supporting Documents.** Details of some of the documents listed in Table 19-52 are shown here. (TBD is to be determined; TBR is to be resolved; FRR is flight readiness review.)

Product	Contents	Product	Contents
Test plans	• See discussion below	**Test reports**	• Activity description • Test article configuration • Support equipment used • Test results and performance data • Summary of deviations, problems, failures, and re-tests • Copy of as-run procedures • Summary and authentication
Verification compliance matrix	• See discussion below	**Verification closure notices**	• See discussion below
System transition plan	• System description • Support item descriptions (packing materials, support equipment) • Handling and environmental control requirements • Detailed procedures and timelines • Deliverable lists	**Detailed engineering release drawing (as built)**	• As described in previous section but annotated with any variances during actual construction
Plan to achieve FRR readiness	• Rank ordered list of studies and trades-offs • Forward work mapped to program schedule • Technology gap closure plan mapped to program schedule • Requirements update and management plan • TBD/TBR burn-down plan • SAR issue resolution plan	**Manufacture and assembly plan**	• See discussion below

2.   Assemble propulsion module

3.   Assemble top-panel module

4.   Assemble payload module

5.   Integrate modules with side panels, solar arrays, and thermal blankets

During detailed AIT planning, the team develops integrated verification fishbone (IVF) diagrams (as described in Chapter 11) to logically lay out the sequence of integration and verification activities. Figure 19-55 shows the IVF for integrating the propulsion module to the baseplate. From the IVF diagrams, the team develops a detailed AIT schedule, as shown in Table 19-56.

Table 19-55. Systems Engineering Process Level of Effort Leading to System Acceptance Review (SAR). Analogous to a cell phone signal, zero bars represents no or very low resource requirements and five bars represent maximum resource use. Resource requirements are for a particular process only; a given number of bars for one process does not imply the same level of resources as the same number of bars for a different process.

Process	Level of Effort	Comments
1. Stakeholder expectation definition	■□□□□	By now, the technical requirements should fully capture stakeholder expectations. Any changes would be too late to fix.
2. Technical requirements definition	■□□□□	Major system-level requirements should now be well-defined. Requirements changes, except to fix flaws in the system uncovered during verification and validation should not be accepted.
3. Logical decomposition	□□□□□	Largely completed during earlier phases but still useful in refining lower-level specifications, especially for software
4. Physical solution	■■□□□	As the system completes final assembly, we define the final physical form of the system
5. Product implementation	■■■■■	This phase is the final realization of all implementation planning
6. Product integration	■■■■■	Product integration activities peak during this phase
7. Product verification	■■■■■	Verification activities peak during this phase as plans are executed
8. Product validation	■■■■■	Validation activities peak during this phase, once a sufficient system assembly maturity is achieved
9. Product transition	■■■■■	Transition activities peak during this phase as plans are executed
10. Technical planning	■□□□□	Technical planning continues, largely as replanning when unforeseen issues arise
11. Requirements management	■■■■□	The requirements management process is used to manage requirements changes required to fix system flaws
12. Interface management	■□□□□	Interface management activities peak during this phase as all system interfaces are finally put together
13. Technical risk management	■■□□□	Risk planning and management remain at a vigorous level during this phase in response to contingencies
14. Configuration management	■■■■■	End item configuration control processes are critical throughout final assembly, integration, and verification
15. Technical data management	■■■■■	Ready access to the most current data is critical to smooth assembly, integration, and verification
16. Technical assessment	■■■■■	Planning to support upcoming SAR
17. Decision analysis	■□□□□	Decision analysis processes continue but at a lower level as (we hope) most major project issues have been resolved by this point

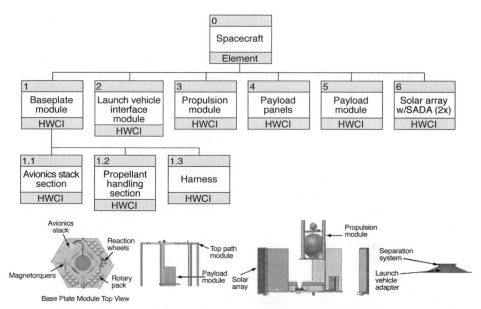

FIGURE 19-54.   FireSAT Spacecraft Hardware Configuration Item (HWCI) Definition. This includes a CAD model of the major items. (SADA is solar array drive assembly.)

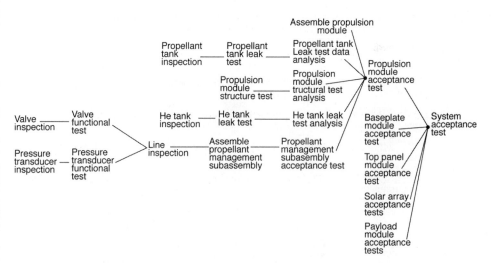

FIGURE 19-55.   Integrated Verification Fishbone Logic Diagram for FireSAT Propulsion Module. The other four modules have similar integration sequences.

TABLE **19-56.   FireSAT Spacecraft Integration Plan.** The fishbone diagrams show the AIT sequence; the integration plan gives the schedule. (ACS is attitude control subsystem; GPS is Global Positioning System; TX is transmit; RX is receive.)

Item	1 Quarter			2 Quarter			3 Quarter			4 Quarter		
	J	F	M	A	M	J	J	A	S	O	N	D
Spacecraft assembly, integration, and testing (AIT)	▓	▓	▓	▓	▓	▓	▓	▓	▓	▓	▓	
Bus integration	▓	▓	▓	▓	▓	▓						
Base plate subassembly	▓	▓	▓	▓	▓							
Base plate fabrication	▓											
Avionics stack	▓											
Fill/drain valves		▓										
Propellant lines and fittings AIT			▓									
Primary thruster			▓									
Magnetometer			▓									
Launch vehicle interface		▓										
ACS thrusters (6x)		▓										
Battery 8-packs (3x)		▓										
Reaction wheel unit				▓								
Solar array units (2x)				▓	▓							
GPS antenna			▓									
Top panel subassembly	▓	▓	▓	▓	▓	▓						
Antennas	▓											
TX whip antenna		▓										
TX hi-gain antenna	▓											
RX whip antenna	▓											
Top panel fabrication	▓											
Attitude sensors			▓	▓	▓	▓						
Side panels (6x) fabrication		▓	▓	▓								
Propulsion assembly	▓	▓	▓	▓	▓							
Truss structure fabrication		▓	▓									
Propellant tank		▓										
Pressurant tank	▓											
Propulsion structural interface		▓	▓	▓								
Valves, regulators, lines, connectors		▓	▓	▓								
Payload assembly	▓	▓	▓	▓	▓	▓	▓	▓	▓			
Sensor	▓	▓	▓	▓	▓	▓	▓	▓	▓			
Payload structural interface	▓	▓										
Integrated testing							▓	▓	▓	▓	▓	

Example Test Plan

The FireSAT verification effort entails numerous test plans, starting with engineering development units and carrying through the flight model testing. A representative example is the vibration test campaign plan. The table of contents for this plan is shown below:

Let's look at an excerpt from a few sections to see the level required by such a plan.

6.2 Test Configuration, Equipment, and Fixtures

The qualification model (QM) will attach with a flight-like band clamp, preloaded to 2800 lb (1.4 times the flight preload of 2000 lb), to a test fixture machined to provide an appropriate interface for the clamp. This fixture will bolt to a second fixture, which will be a 1.5"-thick plate with a hole pattern matching that of the shaker table. The second fixture will bolt to the table. Figure 19-56 shows the configuration for vibration testing. During these tests, the QM will be turned off (no power) and all umbilical connections shall be disconnected from the top panel.

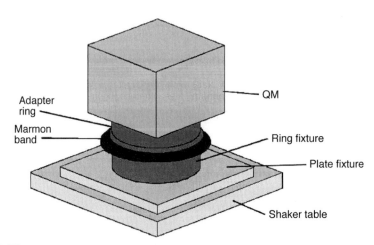

FIGURE 19-56. **Vibration test stand configuration.** Specifications for the test configuration must be clear and unambiguous.

6.3 Instrumentation and Data Acquisition

Accelerometers shall be mounted to the QM in the locations defined in Figure 19-57. Actual locations of accelerometers shall be documented with sketches, with measured dimensions, by Acme Astronautics Corp. personnel and included in the test report. Table 19-57 specifies when the channels are to be monitored. In addition, accelerometers needed to control the input will be used, as determined by test-lab personnel.

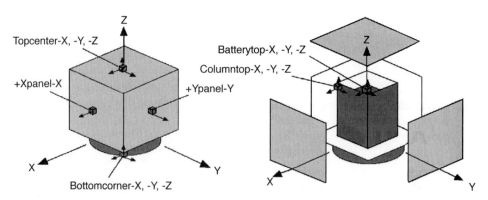

FIGURE 19-57. **Accelerometer locations and labels for vibration testing.** The arrows indicate the directions in which acceleration is to be measured.

TABLE 19-57.   **Minimum Data to be Monitored during Vibration Testing.** This table corresponds to Figure 19-57 above. (M is monitored.)

Channel Label	Z-Axis Tests	X-Axis Tests	Y-Axis Tests
Topcenter-X	M	M	
Topcenter-Y	M		M
Topcenter-Z	M		
+Xpanel-X	M	M	
+Ypanel-Y	M		M
Bottomcorner-X		M	M
Bottomcorner-Y		M	M
Bottomcorner-Z	M	M	M
Columntop-X	M	M	
Columntop-Y	M		M
Columntop-Z	M		
Batterytop-X		M	
Batterytop-Y			M
Batterytop-Z	M		

Data shall be collected over a frequency range of 20 Hz to 2000 Hz. For sine sweep and sine burst testing, data shall be plotted on a log-log plot of acceleration in g versus frequency. For random vibration testing, data shall be plotted on a log-log plot of acceleration power spectral density (PSD) in g^2/Hz versus frequency.

6.4 Sine Sweep/Baseline Modal Survey

The objective of this test is to determine natural frequencies and provide information for understanding the modes of vibration. The test shall be done for each axis before the other vibration tests in order to establish a baseline and shall be repeated after high-level tests, as specified in Section 11.7, to identify any changes in structural behavior.

The sine-sweep test shall be done by introducing sinusoidal excitation to the base of the QM over the frequency range of 20 to 2000 Hz at a rate of approximately 3 octaves/minute. Peak input acceleration across this range will be 0.25 +/– 0.05 g. Such low input acceleration should ensure safe response levels and negligible fatigue damage to the test article.

Example Summary Test Report

A test without a report of its results effectively never happened. As described in Chapter 11, a detailed test report includes:

- Activity description
- Test article configuration
- Support equipment used
- Test results and performance data
- Summary of deviations, problems, failures, and retests
- Copy of as-run procedures
- Summary and authentication

In addition to detailed test reports for each test, the FireSAT project finds it useful to create a "quick-look" summary verification report for each detailed report. These provide an executive summary of the test along with top-level results. Figure 19-58 shows an example for the flight model baseline modal survey test. Results for this test are not conclusive and so pass/fail is deferred. The details raised by this issue are discussed in the next section.

Example Verification Compliance Matrix

We can "roll up" all verification activities in a verification compliance matrix as shown in Table 19-58 for a few representative requirements. The compliance matrix provides a quick summary of all verification close-outs and an easy means to track test reports and other close-out data.

19.7.5   Representative Issues, Items, and Anecdotes

The first issue encountered during this phase is one that impacts any project. The natural order of how design proceeds, from the top down, is the opposite of how manufacturing and assembly take place. Tom Kelly, in his book *Moon Lander* [Kelly, 2001] put it best: "Unfortunately, the natural order in which design drawings are produced is opposite from Manufacturing's needs. A designer visualized the whole unit or assembly he was designing, producing sketches and an assembly drawing showing how the device fits together. From this the individual detailed parts making up the assembly were pulled out and drawn up, complete with the dimensions, material specifications, and instructions that allowed the shop to make them. Manufacturing naturally wants the detailed parts drawings first, so they can obtain the raw materials and purchase components and then fabricate the parts." Figure 19-59 illustrates this dichotomy.

Another issue involves change management. The customer makes a late request for some additional operational autonomy in the command and control (C&C) CSCI software. This change generates both programmatic and technical risks. The spacecraft test schedule is already packed with critical test activities (loads, thermal vacuum testing, and EMI). Most of the management reserve in the schedule has already been consumed to deal with several technical issues driven mostly by late hardware deliveries. Specifically, one avionics box has been sold as a simple reuse of a previous design. But after the project commits to this hardware, it turns out that

SUMMARY VERIFICATION REPORT

PROJECT NAME: ___FireSAT___   TEST ITEM DESCRIPTION: ___Spacecraft___
MANUFACTURER: ___Acme Astro Corp___ SERIAL NUMBER: ___50000-1A___

LEVEL OF ASSEMBLY	HARDWARE	TEST
☐ PART	☐ ENGINEERING MODEL	☑ INITIAL TEST
☐ UNIT/COMPONENT	☐ QUALIFICATION MODEL	STARTING DATE _3/21/08_
☐ SECTION	☐ PROTO-FLIGHT	☐ RE-TEST
☐ SUBSYSTEM/INSTRUMENT	☐ FLIGHT-SPARE	☐ PARTIAL
☐ MODULE	☑ FLIGHT MODEL	☐ FULL
☑ SPACECRAFT/SYSTEM		STARTING DATE _____

STRUCTURAL - MECHANICAL	ELECTROMAGNETIC COMPATIBILITY/ RADIO FREQUENCY	THERMAL/PRESSURE
☐ STRUCTURAL LOADS		☐ THERMAL-VACUUM
☐ STATIC ☐ ACCEL.		(Number of Cycles ____)
☐ SINE BURST	☐ CONDUCTED EMISSIONS	☐ THERMAL CYCLING
☐ VIBRATION	☐ RADIATED EMISSIONS	(Number of Cycles ____)
☐ RANDOM ☐ SINE	☐ CONDUCTED SUSCEPTIBILITY	☐ THERMAL BALANCE
☐ ACOUSTIC	☐ RADIATED SUSCEPTIBILITY	☐ TEMPERATURE-HUMIDITY
☐ MECHANICAL SHOCK	☐ MAGNETIC PROPERTIES	☐ LEAKAGE
☐ ACTUATION ☐ SIMULATED	☐ GAIN/ANTENNA PATTERN	☐ OTHER (describe): _____
☐ MECHANICAL FUNCTION		
☑ MODAL SURVEY	**ELECTRICAL PERFORMANCE**	
☐ PRESSURE PROFILE		
☐ MASS PROPERTIES	☐ FULL FUNCTIONAL TEST	**OPTICAL/INSTRUMENT UNIQUE**
☐ FIT CHECK	☐ ABBREVIATED FUNCT. TEST	☐ DESCRIBE _____
☐ OTHER (describe): _____	☐ STRESS TEST	
	☐ END-TO-END	
	☐ COMPATIBILITY TEST	
	☐ MISSION SIMULATION	

VERIFICATION PROCEDURE NO: _FS-VP-037_ REV: _B_ DATE: _2/28/08_
APPLICABLE VERIFICATION PLAN: _FireSAT Vibration Test Campaign Plan_
FACILITY DESCRIPTION: _Vibration Table_ LOCATION: _NASA/GSFC_
TEST LOG/DATA REFERENCE: _fi/Firesat/Test Data/FM/Vibration_

RESULTS: ☐ PASS ☐ FAIL ☑ DEFER
COMMENTS: ___Margin Error on measurement exceeds Success___
___criteria___

COGNIZANT ENGINEER: _Joe I did it_   Q/A REP: _Mary I saw it_
DATE: _3/22/08_   DATE: _3/22/08_

PAGE _1_ OF _1_

FIGURE 19-58. Summary Verification Report from FireSAT Spacecraft Modal Survey Test. FireSAT, like many projects, develops some custom report forms tailored to the project.

while it's the "same" box, it's now being manufactured by a different supplier. The systems engineer orders additional inspections and tests to ensure that the materials and processes are unchanged. This delay causes the spacecraft to be unavailable to load the new C&C software with the customer's additional autonomy. Rather than delay the SAR, the project manager and systems engineer elect to hold the review as scheduled, and add this issue to the traveling open-work list. The contractor will

TABLE 19-58. **Example FireSAT Verification Compliance Matrix.** (A is analysis; T is test; I is inspection; D is demonstration.)

Requirement	Verification Methods	Compliance Data	Non-Conformance	Status
3.2.1 Spatial resolution: Space vehicle shall have a spatial resolution of less than or equal to 2.14×10^{-4} radians (0.0115°)	A	FS/TR-0445 Payload Resolution Analysis Report		Closed
3.2.2 Induced environment: Space vehicle shall meet its requirements after exposure to the induced launch environment as defined in the Falcon-1 User's Handbook.	A, T	FS/TR-0123 FireSAT Induced Environment Verification Report		Open
3.3.2.1 Natural frequency: Space vehicle first-mode natural frequency shall be greater than 25 Hz	A, T	FS/TR-0115 FireSAT Vibration Test Campaign Report		Open
3.3.1 Space vehicle mass: Space vehicle loaded mass shall not exceed 250 kg	A, T	FS/TR-0108 FireSAT Flight Model Mass Properties Report		Closed
3.4.6 TRL: All space vehicle technologies shall achieve a TRL 6 by the PDR. Not applicable to technology demonstration opportunities.	I	FS-6004 FireSAT Preliminary Design Review Action Close Out Report		Closed
3.6.1 Maintenance personnel: Space vehicle shall be maintainable by a Class 2 - Engineering Technician as defined in KSC WP-3556, Technician Qualifications Manual.	D	FS/TR-00089 FireSAT Qualification Model Maintenance Demonstration Report		Closed

then be required to perform a Delta-SAR to report the results of regressions testing with the new software after physical delivery to the launch site.

The most significant issue at the SAR is the result of changing the launch vehicle to the Falcon-1. Now that two FireSATs will be launched simultaneously, the stacked configuration of two spacecraft must exhibit the required physical characteristics (center of gravity and moment of inertia (CG/MOI), natural frequency, etc.). Since a vibration test series on two flight models is deemed too risky, the vibration testing is done with the engineering model and the qualification model attached to the new payload adapter/separation mechanism.

When this test is run, the results show acceptable center of gravity and moment of inertia and acceptable accelerations on the top spacecraft. But the natural frequency test is inconclusive. The stack does exhibit a natural frequency of 26 Hz, which meets the launch provider's requirements. But the uncertainty in the test due to test equipment, plus the fact that non-flight hardware was being tested, indicates

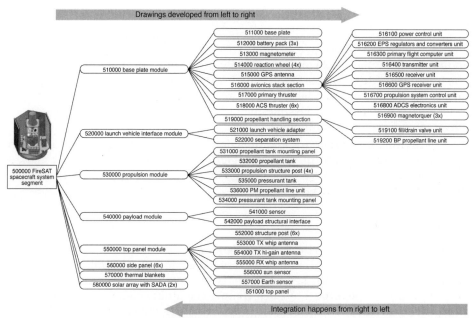

FIGURE 19-59. Drawing Tree and Integration. The nature of the design process produces drawings from the top down. However, manufacturing and assembly occurs from the bottom up. This creates a tension between the design team working in one direction and the manufacturing and assembly team, who are waiting to start from the other direction. (SADA is solar array drive assembly; GPS is Global Positioning System; ACS is attitude control subsystem; TX is transmit; RX is receive; ADCS is attitude determination and control subsystem.)

that the natural frequency was within 2 Hz of the actual flight configuration. So the requirement for natural frequency to be greater than 25 Hz is not conclusively confirmed. The launch provider has to be consulted and then convinced that this potential 24 Hz spacecraft stack is within the acceptable margin for the Falcon-1. After several meetings, the launch provider accepts the new limits.

The final issue involves the transition plan. Early in the project, the US Department of Defense (DoD), which works cooperatively with NOAA on the NPOESS program, agreed to provide C-17 aircraft transportation for the spacecraft to the launch site for integration. But world events have increased demand for these assets to support the US Amy, and Air Force officials have threatened to back away on commitments to deliver the spacecraft. Several options are available but all would impact schedule and slip the delivery date by up to two weeks, putting both the customer and contractor in a bad light with the congressional subcommittee following the program. The program is also dangerously close to the automatic initiation of congressional subcommittee review due to schedule growth in accordance with congressionally mandated project trip wires.

Fortunately, this being the first-ever "big" satellite contract for Acme Astronautics Corp., and one of the first-ever for the great state of Iowa, the governor of the state steps in and offers Air National Guard C-130 transport aircraft as part of a "training exercise." This offer makes great headlines. But the landing loads on the C-130 are significantly different from those on the C-17 for which the spacecraft transportation container has been designed. Engineers must quickly design, build, and qualify a shock absorbing vibration damper to fit on the underside of the container. Despite the last minute scramble, the spacecraft are delivered on time for integration.

The culmination of the SAR and final delivery is the signing and presentation of the DD-250 receiving form, shown in Figure 19-60. Not only does this mark a major milestone in the project, it means that the prime contractor can receive near-final payment for their hard work!

19.7.6   Conclusions

As the C-130 rolls down the runway, the champagne corks start to fly. The long road from mission concept to system delivery has passed another milestone. There's still much to do. Many of the key project personnel only sip at their glasses; as they have to get across the airport to catch commercial airline connections to meet the spacecraft on the other end to start final preparations for launch. Once unpacked and checked out at the launch site, flight batteries must be charged and propellant loaded before the spacecraft can be integrated onto the launch vehicle. While much risk still looms ahead, the team has good reason to celebrate their accomplishments. From those heady, optimistic days of Pre-phase A, through the cold realities of Phase D, the team has steadfastly adhered to rigorous systems engineering discipline. A skillful mix of the science and art of systems engineering has been crucial to finally deliver a verified system that meets the customer's expectations while maintaining the project on schedule and cost with manageable risk.

Epilogue

Exactly 9 years, 4 months, 3 days, and 7 hours after launch of the first FireSAT spacecraft, some of the original designers are on hand in the mission control center when the final command is sent to the spacecraft. Days before, controllers ordered it to fire its thrusters one last time to deorbit down to a perigee altitude of 100 km. Now, it will only be a matter of hours before the hardworking little satellite burns up somewhere over the Pacific Ocean. It has already operated more than two years past its design life and, along with its sibling launched at the same time (and still going strong), has far exceeded the customer's expectations. FireSAT has helped to bring firefighting into the Space Age. But while these pioneers have blazed the trail, the FireSAT block II satellites, launched three years earlier with vastly improved sensors, are setting a new standard in fire detection and tracking. The hard-learned lessons from the first FireSAT systems engineering processes helped the team deliver the block II versions on time and well within budget. With the

MATERIAL INSPECTION AND RECEIVING REPORT						Form Approved OMB No. 0704-0248

The public reporting burden for this collection of information is estimated to average 30 minutes per response, including the time for reviewing instructions, searching existing data sources, gathering and maintaining the data needed, and completing and reviewing the collection of information. Send comments regarding this burden estimate or any other aspect of this collection of information, including suggestions for reducing the burden, to Department of Defense, Washington Headquarters Services, Directorate for Information Operations and Reports (0704-0248), 1215 Jefferson Davis Highway, Suite 1204, Arlington, VA 22202-4302. Respondents should be aware that notwithstanding any other provision of law, no person shall be subject to any penalty for failing to comply with a collection of information if it does not display a currently valid OMB control number.

PLEASE DO NOT RETURN YOUR COMPLETED FORM TO EITHER OF THESE ADDRESSES.
SEND THIS FORM IN ACCORDANCE WITH THE INSTRUCTIONS CONTAINED IN THE DFARS, APPENDIX F-401.

1. PROCUREMENT INSTRUMENT IDENTIFICATION (CONTRACT) NO.		ORDER NO.		6. INVOICE NO./DATE	7. PAGE	OF	8. ACCEPTANCE POINT
GS-99858858		85646464		20100401	1	1	85656
2. SHIPMENT NO.	3. DATE SHIPPED	4. B/L		5. DISCOUNT TERMS			
002	20100328	TCN		N/A			
9. PRIME CONTRACTOR CODE				10. ADMINISTERED BY CODE			
Acme Astro Corp. KW9999				NASA/GSFC 8765			
11. SHIPPED FROM (If other than 9) CODE		FOB:		12. PAYMENT WILL BE MADE BY CODE			
Harlan, IA				Direct Deposit 9886			
13. SHIPPED TO CODE				14. MARKED FOR CODE			
NASA/KSC 8875				Acceptance & Processing			

15. ITEM NO.	16. STOCK/PART NO. DESCRIPTION (Indicate number of shipping containers - type of container - container number.)	17. QUANTITY SHIP/REC'D*	18. UNIT	19. UNIT PRICE	20. AMOUNT
002 A	FireSAT Flight Model Spacecraft	1	02	20,035,158	20,035,158

21. CONTRACT QUALITY ASSURANCE

a. ORIGIN	b. DESTINATION	22. RECEIVER'S USE
☐ CQA ☒ ACCEPTANCE of listed items has been made by me or under my supervision and they conform to contract, except as noted herein or on supporting documents.	☐ CQA ☒ ACCEPTANCE of listed items has been made by me or under my supervision and they conform to contract, except as noted herein or on supporting documents.	Quantities shown in column 17 were received in apparent good condition except as noted.
03/21/10	03/28/10	DATE RECEIVED 03/04/10 SIGNATURE OF AUTHORIZED GOVERNMENT REPRESENTATIVE
DATE SIGNATURE OF AUTHORIZED GOVERNMENT REPRESENTATIVE	DATE SIGNATURE OF AUTHORIZED GOVERNMENT REPRESENTATIVE	TYPED NAME: I. Received all
TYPED NAME: I. Sawi tall	TYPED NAME: I Shipped tall	TITLE: Payload Processor
TITLE: QA	TITLE: Acceptor	MAILING ADDRESS: NASA/KSC FLA
MAILING ADDRESS: 3457 Ave A DCMA CO	MAILING ADDRESS: NASA/GSFC Greenbelt MD	COMMERCIAL TELEPHONE 123-456-7899 NUMBER:
COMMERCIAL TELEPHONE 303-123-4567 NUMBER:	COMMERCIAL TELEPHONE 211-312-3127 NUMBER:	* If quantity received by the Government is the same as quantity shipped, indicate by (X) mark; if different, enter actual quantity received below quantity shipped and encircle.
23. CONTRACTOR USE ONLY		

DD FORM 250, AUG 2000   PREVIOUS EDITION IS OBSOLETE.

FIGURE 19-60.   FireSAT DD-250. This form marks the transition of the spacecraft from the builder to the user.

team still somewhat exhausted from the block II effort, there's word on the street that the USFS is ready to put out a request for information for a new and improved version. A Pre-Phase A study is scheduled to kick-off next month. Time to dig out those systems engineering notes again.

References

Boden, Daryl G., Gael Squibb, and Wiley J. Larson. 2006. *Cost-effective Space Mission Operations*. 2nd Ed. Boston, MA: McGraw-Hill.

GSFC. 2005. Rules for Design, Development, Verification and Operation of Flight Systems, GSFC-STD-1000, May 30, 2005.

Kelly, Thomas J. 2001. *Moon Lander.* New York, NY: Harper Collins.

Larson, Wiley J. and James R. Wertz. 1999. *Space Mission Analysis and Design.* 3rd Ed. Dordrecht, Netherlands: Kluwer Academic Publishers.

NASA. 2007. NPR 7123.1a — NASA Systems Engineering Processes and Requirements. Washington, DC: NASA.

NWSC, 2005. Software Development Plan Template. TM-SPP-02 v2.0, Space and Naval Warfare Systems Center, April 5, 2005.

Sarafin, Thomas and Wiley J. Larson. 1995. *Spacecraft Structures and Mechanisms.* New York, NY: McGraw Hill.

Sellers, Jerry Jon. 2004. *Understanding Space.* 3rd Ed. New York, NY: McGraw Hill.

Index

Symbols

Numerics

A